Behavioral Neuroscience

GEORGE SPILICH
Washington College

WILEY

VICE PRESIDENT	Amanda Miller
EDITORIAL DIRECTOR	Justin Jeffryes
EXECUTIVE EDITOR	Glenn A. Wilson
SR. MARKETING MANAGER	Carolyn Wells
SR. MANAGER, COURSE CONTENT & PRODUCTION OPERATIONS	Dorothy Sinclair
SR. COURSE PRODUCTION OPERATION SPECIALIST	Valerie A. Vargas
SR. CREATIVE PRODUCT DESIGNER	Wendy Lai

Cover Photo: © PIXOLOGICSTUDIO/Getty Images

This book was typeset in 9.5/12 STIX Two at Lumina Datamatics, Inc.

Wiley is a global leader in research and education, unlocking human potential by enabling discovery, powering education, and shaping workforces. For over 200 years, Wiley has fueled the world's knowledge ecosystem. Today, our high-impact content, platforms, and services help researchers, learners, institutions, and corporations achieve their goals in an ever-changing world. Visit us at Wiley.com.

EPUB ISBN: 978-1-11950566-2

The inside back cover will contain printing identification and country of origin if omitted from this page. In addition, if the ISBN on the cover differs from the ISBN on this page, the one on the cover is correct.

Library of Congress Cataloging-in-Publication Data

Names: Spilich, George, author.
Title: Behavioral neuroscience / George Spilich, Washington College.
Description: First edition. | [Hoboken, NJ] : Wiley, [2023] | Includes index.
Identifiers: LCCN 2022053829 (print) | LCCN 2022053830 (ebook) | ISBN 9781118547380 (print) | ISBN 9781119889625 (adobe pdf) | ISBN 9781119505662 (epub)
Subjects: LCSH: Psychobiology—Textbooks. | Neuropsychology—Textbooks.
Classification: LCC QP360 .S695 2023 (print) | LCC QP360 (ebook) | DDC 612.8—dc23/eng/20221215
LC record available at https://lccn.loc.gov/2022053829
LC ebook record available at https://lccn.loc.gov/2022053830

Printed in the United States of America.

SKY10044487_031523

About the Author

GEORGE SPILICH is a behavioral neuroscientist who earned degrees from the University of Wisconsin, the University of Texas at El Paso, and the University of Pittsburgh, where he was at the Learning Research and Development Center. He is currently an emeritus faculty at Washington College, the first college founded after the American Revolution, where he was the department chair for 20 years, was the first John Toll Professor, and won the Lindback Foundation Teaching Award. In academic year 1988–1989, he replaced two-time Nobel laureate Linus Pauling as the Fulbright Researcher for Yugoslavia and spent the year as visiting professor in the departments of neurology and nuclear medicine at university hospitals. He was the inaugural curator of the Cromwell Center for Teaching and Learning at Washington College. His research interests explore the processes that underlie encoding, storage, and retrieval of information in an effort to understand how a biological system creates and stores expertise and then how disease and damage disrupt that mechanism.

The human brain has 100 billion neurons, each neuron is connected to 10 thousand other neurons. Sitting on your shoulders is the most complicated object in the known Universe.

—MICHIO KAKU

It is essential to understand our brains in some detail if we are to assess correctly our place in this vast and complicated universe we see all around us.

—FRANCIS CRICK, NOBEL LAUREATE

I'm just looking for clues at the scene of the crime.

—JOSEPH FIDLER WALSH

Preface

Understanding human behavior from a biological perspective is the goal of behavioral neuroscience. The field incorporates biology, genetics, developmental biology, gerontology, psychopharmacology, cognitive psychology, and even sociology to create a multidisciplinary approach to understanding behavior.

Behavioral neuroscientists consider questions such as, How does our evolutionary past influence our behavior today? To what extent is our behavior controlled by our genes? How do our senses convert raw input into what we call reality? How do we regulate our internal processes to stay alive? Why are some substances so addictive? What does it mean to have a sex or gender identity? Why do I sleep (or not!), and why do I dream? Why can't I remember what I learned last year? Why do some people become depressed, anxious, or even have a total break from reality? Is beauty only a construction of my mind? These and thousands of other questions are the purview of behavioral neuroscience, one of the most exciting of human endeavors.

Features Unique to *Behavioral Neuroscience*

Meets Students Where They Are: Some students will take this course in a traditional face-to-face lecture/discussion format; some will enroll in an online class, and others will take a hybrid course. Students choosing to enroll in behavioral neuroscience often come from two groups: those preparing for a career in medicine or science, and those students who are fulfilling a general education science requirement and choose behavioral neuroscience to learn more about themselves. *Behavioral Neuroscience* is written to reach all those audiences.

Making Science Approachable and Inviting: Students following a course of study leading to a career as a physician, medical/physician's assistant, nurse, or substance abuse/addictions counselor; students preparing for graduate work in behavioral neuroscience; and those preparing to take the MCAT will benefit from the focus on science. However, all students, including those following a different career path, will benefit from an accessible writing style and frequent examples of how behavioral neuroscience speaks to the issues of everyday life.

Encourages Active Learning through Effective Use of Media: *Behavioral Neuroscience* engages students with a wealth of multimedia created or curated by the author and puts that media to use often at the service of conveying traditionally difficult-to-master concepts. Every chapter contains multiple activities that engage students' imagination while conveying content.

Videos, Dynamic Visualizations, and Online Demonstrations: The *Check It Out*, *Try It Out*, and *Think About It* activities include video interviews, dynamic visualizations, interactive images, and links to online demonstrations that can be watched as many times as the student feels is necessary.

Check It Out activities direct students to watch a short video that conveys concepts and then the student answers questions which reinforce their understanding of the material.

Check It Out | What Is It Like to Be Face blind?

wavebreakmedia/Adobe Stock

It can be hard to imagine the everyday life of an individual with prosopagnosia given that recognizing the faces of others seems to be at the heart of our social experience. How might face blindness change your behavior? Watch the video **What is it like to be face blind?** and then answer these questions.

1. Was Kristen a child or a young adult when she learned she had prosopagnosia? What does that tell you about the adaptability of humans?
2. Kristen spoke about how her face blindness was an impediment and how it might be an asset. How might you have accommodated being face blind in a similar situation?
3. In behavioral neuroscience we often see that behaviors or functions that are essential for our survival are hardwired into their own dedicated structure. What does the existence of the FFA suggest about the importance of the ability to recognize who is in our clan and who is not in the evolution of humans?

Try It Out activities have the student perform some activity related to the chapter content that deepens their understanding of the content and makes it memorable.

Try It Out | The Landolt C Task

You can test your own vision using a Landolt C task by downloading the Freiburg vision and contrast test (FrACT), which is available from your instructor. Watch the brief video Freiburg Visual Acuity Task, collect your own measure of visual acuity, interpret it in light of the worksheet for this activity, and then answer these questions.

1. What advantages might there be to the Landolt C task over typical paper tests such as the Snellen chart? What are the disadvantages?
2. Visual acuity is very important to your success. Search the internet using the phrase "How to protect my vision" and identify two risks to your vision and what you can do to protect your vision from them.

Think About It activities provide the student with a topical issue in behavioral neuroscience and then guides them to consider the implications of what they are learning in their life.

Think About It | Has Civilization Changed How We Sleep?

Gorodenkoff / Adobe Stock

Humans are a remarkably diverse species, but if there is one trait that binds us all together, it is sleep. Everyone sleeps, regardless of sex, culture, or personal beliefs. Given the universality of sleep, you might think that how we sleep has not changed over time, but there is considerable evidence to suggest that our current sleep habits are a recent development in our history as a species.

In modern industrialized nations, people often wake after dawn, go to work or school, come home and stay awake several hours after sunset, and then sleep through the night in one continuous stretch. This sleep pattern is known as monophasic sleep because it consists of a single continuous event. But what if monophasic sleep is not our natural sleep pattern? Before the invention of electric lights, candles and lanterns were an expensive and not very effective source of illumination, and before they were invented, our only source of illumination after dark was a fire.

Historians have recognized in reading diaries and other personal accounts of life in the preindustrial age that many people engaged in biphasic sleep. They went to bed as night fell, slept 2–4 hours, awoke around midnight and looked after their children or perhaps tended the fire, and then went back to sleep for 4–5 more hours before awakening for their day. The Industrial Age and its appetite for workers who could be present for a long and continuous shift plus the easy availability of electric lights may have dramatically changed a basic human behavior.

Questions

1. **Search the internet using the terms "biphasic sleep or second sleep in history." Do you think the evidence for second sleep in the preindustrial age is convincing? Why or why not?**
2. **As a general rule, predators sleep more than prey. Humans today sleep like predators, yet some argue that our natural sleep pattern is biphasic sleep. Would biphasic sleep be more adaptive for our distant ancestors whose existence in the African savanna was more tenuous?**
3. **It is assumed that the Industrial Age created an environment in which it was advantageous to replace biphasic sleep with monophasic sleep. In today's world, people may be part of a team that is distributed across several time zones. Would biphasic sleep be an advantage for a person whose career requires them to interact part of the day with people on the other side of the globe?**

Vibrant Illustrations and Interactivities: *Behavioral Neuroscience* sets a new standard for quality of images used in the text to depict concepts and neuroanatomy.

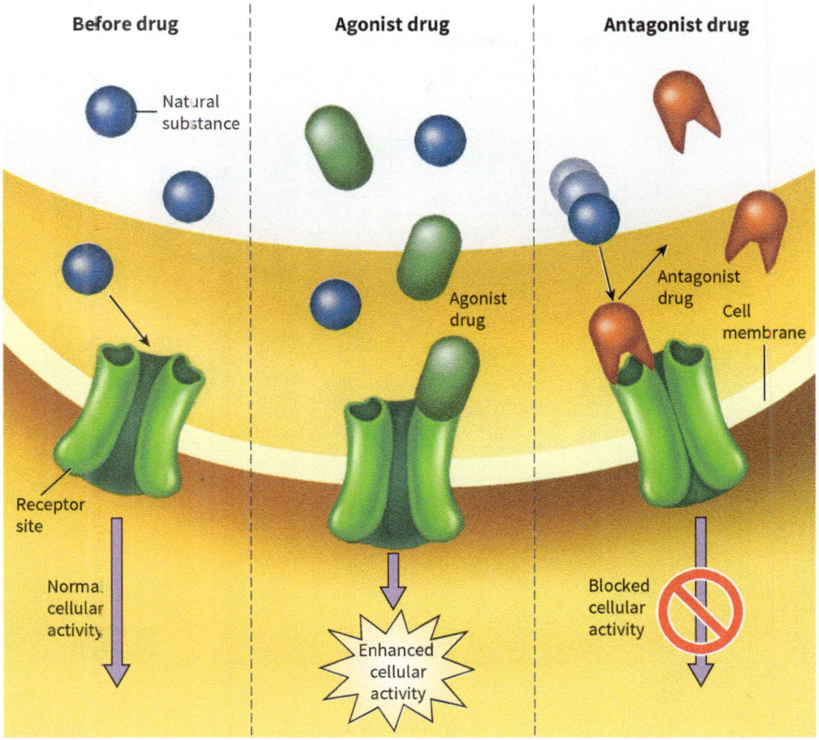

FIGURE 3.13 **Agonists and antagonists at the receptor** An agonist docks to a receptor and activates it, whereas an antagonist blocks the receptor from activating.

Process images help students grasp the mechanisms that underlie complex processes such as olfaction.

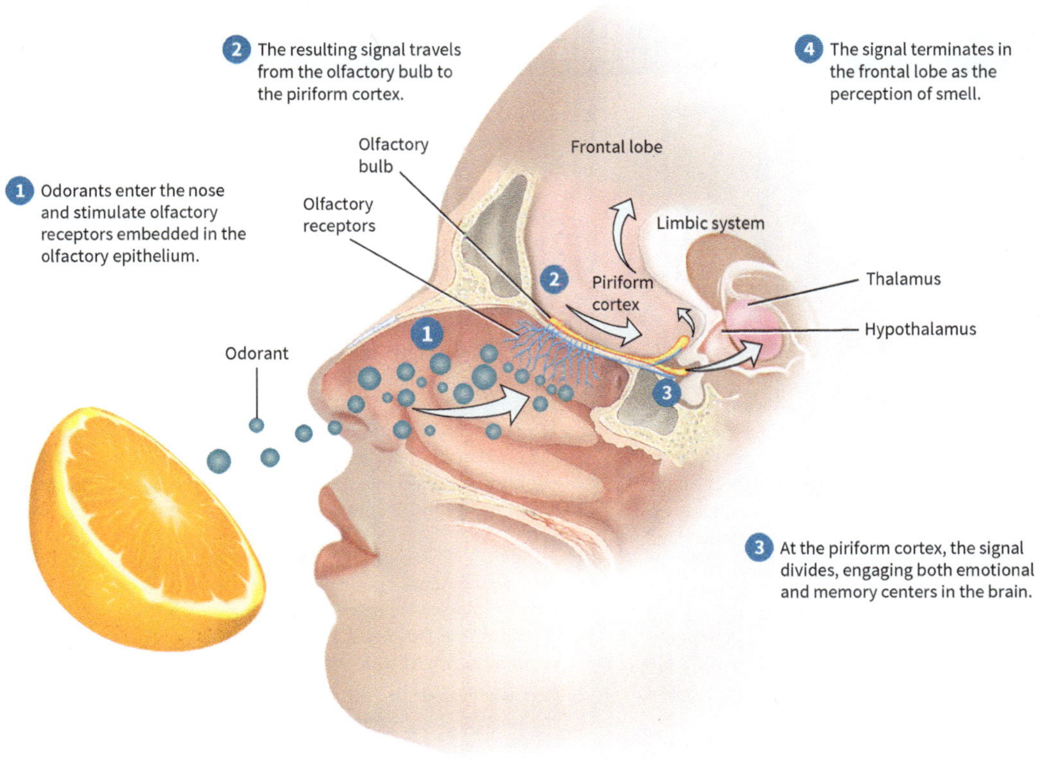

FIGURE 6.16 **The olfactory circuit** The olfactory circuit begins in the nasal cavity and ends in the brain.

In addition, the enhanced e-text includes 35 interactivities distributed throughout the chapters that bring illustrations to life, reinforcing the active learning opportunities this text offers.

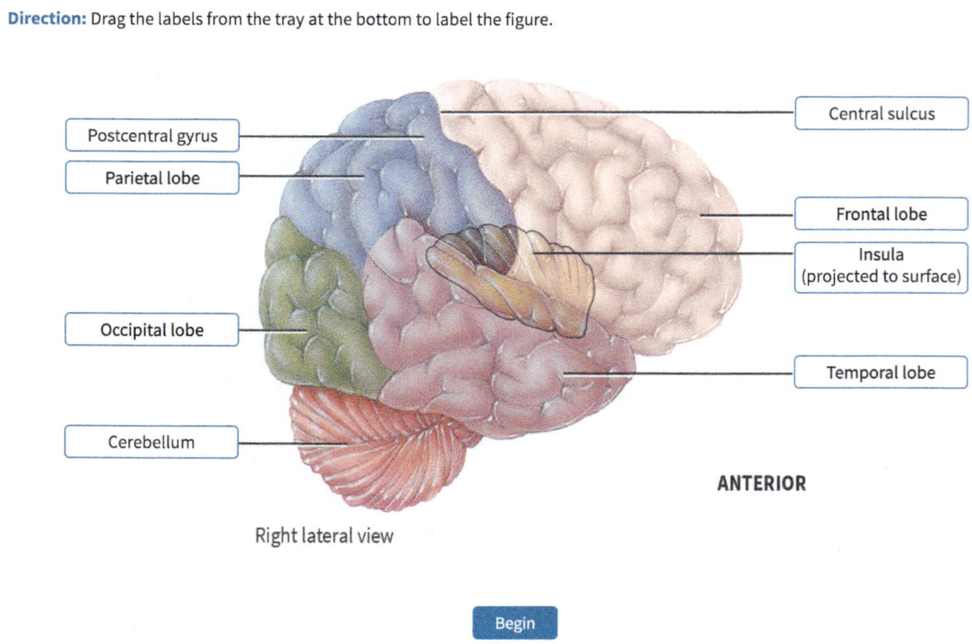

Advances in Behavioral Neuroscience: These profiles in each chapter highlight an individual whose work has significantly enhanced our understanding of the field. Students are directed to the researcher's lab web page and answer brief questions about their work that are designed to encourage the student to see the connection between science in the lab and their own lives.

Advances in Behavioral Neuroscience | Candace Pert and Opioid Receptors

Everett Collection Inc.

Sometimes in life, you have to listen to your own inner voice and do what you think is right, even when others disagree. Candace Pert was an internationally renowned neuroscientist whose curiosity and intuition about nature led her to rewrite our understanding of the link between our brain and our health.

Dr. Pert was born in 1946 in New York City; she graduated from Bryn Mawr College in 1970 and then entered the Ph.D. program at Johns Hopkins, where she worked in the laboratory of Solomon Snyder, one of the world's premier pharmacologists. Pert earned her doctorate in pharmacology from the Johns Hopkins School of Medicine in 1974.

In 1972, while still a graduate student in Snyder's lab, Pert began to think seriously about a vexing question: what is the neural mechanism that allows us to feel pleasure? Pert believed there must be an endogenous or innate opioid receptor in the human brain. Although she was told that the search for this receptor was a dead end and she should go back to her other work, she came into the lab on a weekend and conducted the experiment that made her world famous. Other discoveries would soon follow.

In a career that spanned four decades, Pert published over 250 research articles and changed the way we think about how the brain affects our overall health. In her later work, she was a pioneer of mind–body medicine and a proponent of the concept that emotions have a powerful effect on physical health. She was affectionately known as the Mother of Psychoneuroimmunology and the Goddess of Neuroscience and encouraged her students and collaborators to think outside the box of accepted knowledge. Although Dr. Pert died in 2013, her discoveries and insights continue to influence the field today.

Questions

1. **One way to end opioid addiction would be to permanently blockade opioid receptors. Assuming for a moment that this is even possible, what unpleasant consequences might this cure create?**
2. **Pert was one of the founders of the field of psychoneuroendocrinology. Search the internet for the term, and (a) summarize in one sentence what the field is about, and (b) note one disorder that the field may provide a treatment for.**

Access to Actual Data: *Behavioral Neuroscience* provides instructors and their students with access to data from actual experiments. Examples of data sets include:

- The effects of mild concussions upon attention
- Differences in visual acuity that distinguish varsity athletes from nonathletes
- The effects of low levels of alcohol upon visual search

Depending upon the course prerequisites regarding statistics, the student could be asked to interpret a graph; interpret the results of a statistical test; or determine the independent and dependent variables in a data set, select the appropriate statistical test, perform the analysis themselves, and interpret the results. Instructors are provided with a completed and annotated solution to all three levels of analysis so that students at every level of statistical preparation can think critically about the analysis and interpretation of data.

Ethics: Where appropriate, students are encouraged to consider the ethical implications of scientific and clinical issues.

Emphasis on Major Themes: Major themes that are threaded throughout each chapter include inhibitory and excitatory mechanisms that operate at multiple levels to regulate behavior; the influence of our evolutionary past upon human behavior today; and clinical and real-world implications.

Behavioral Neuroscience in the Real World: Your brain is fascinating in and of itself but seeing how behavioral neuroscience speaks to everyday issues is an important part of its appeal. Why are sugars and fats so appealing when consuming too much of them is unhealthy? Why are some people unable to carry a tune? How is it that some people can perceive the seams of a baseball being thrown at them at 90 miles an hour from 60 feet away in less than two-tenths of a second? Why do some people have to engage in elaborate behavioral rituals to leave their house? Why are some people suspicious of people that they have never met? *Behavioral Neuroscience* considers these and many other meaningful questions.

MCAT Behavioral neuroscience is a popular major as well as a popular course for students preparing for a career in medicine, and for that reason, topics that are tested on the MCAT are clearly identified throughout the text. The Instructor's Companion Website contains a complete list of topics covered on the MCAT at the time of writing, and those topics which *Behavioral Neuroscience* covers are clearly indicated.

To the Student

For most of my adult life, I have been fascinated by the human brain and how it creates behavior, particularly higher-order cognitive behaviors such as memory and skilled visual perception, and how changes in one's underlying neurology lead to cognitive disorders. I have spent literally thousands of hours thinking about how I can best convey the subject of behavioral neuroscience to students, and I have also spent more than 15,000 hours in front of students trying my best to explain complex concepts in an interesting and accessible way. My goal in writing this book is to make behavioral neuroscience as interesting and exciting to you as it is to me, and to do so in a way that you will carry that understanding forward in your life. Conversations with my students have taught me much and so I hope to engage in conversation with you also. Feel free to contact me directly at BNbyGS@gmail.com.

To the Instructor

In addition to the unique features listed previously, you will find that *Behavioral Neuroscience* has the pedagogical learning aids that you have come to rely on.

Chapters are organized into manageable sections with learning objectives clearly spelled out and addressed in each section. Every section ends with a section summary and comprehension checks in the form of multiple-choice questions that verify the student has acquired the main themes. The art sets a new standard for quality, and interactive images that students can use to check their comprehension are included in each chapter. There are dozens of videos associated with the text that illustrate concepts; these can be folded into class presentations or made available via your school's content management system. The Instructor's Companion Website includes many other videos that, while not directly referenced in the textbook, will be of use to instructors. There is a computerized test bank for the easy construction of multiple-choice exams, PowerPoint presentations for every chapter, and an Instructor's Manual with a variety of tools and tips to benefit both seasoned faculty and those who may be teaching behavioral neuroscience for the first time. Conversations with departmental colleagues about new discoveries that we want to share with our students were always an important part of my professional life and I hope to engage with you in such dialogues going forward. Feel free to contact me directly at BNbyGS@gmail.com.

Acknowledgements

Writing a textbook is an enormous undertaking that involves many people. I'd like to thank my professors, whose wisdom and knowledge greatly influenced me: Tony Caggiula, Jim Devine, Judith Goggin, Harry Harlow, Dick Keesey, Alan Lesgold, Dominic Massaro, Don Moss, Fred Mote, Chuck Perfetti, Ed Striker, Jim Voss, and Randy Whitworth. During the writing process, I benefited from illuminating email conversations with hundreds of colleagues, ranging from Nobel laureates to first-year faculty at colleges and universities across the globe. My colleagues in my own department and across academic divisions at Washington College provided me with a model of how to be a scholar and a professional educator. Every professor knows that students teach us as much as we teach them, and I am grateful for all that my students have taught me. They continue to do so.

I want to thank the colleagues who reviewed the original proposal and early samples of a book envisioned several years ago and whose feedback helped shape the development of what is now *Behavioral Neuroscience*; among them:

J. Timothy Cannon, *University of Scranton*
Jeffrey Carlton, *California State University, Sacramento*
Rolf Greenwald, *Central Washington University*
Jeffrey Grimm, *Western Washington University*
Michael Hoane, *Augusta University*
Andreas Keil, *University of Florida*
Mark Kristal, *SUNY Buffalo*
Charles Kutscher, *Syracuse University*
Donald Leitner, *Saint Joseph's University*
Daniel McConnell, *University of Central Florida*
Maura Mitrushina, *California State University, Northridge*
Michael Owren, *Emory University*
Patricia Sharp, *Bowling Green State*

I was fortunate to fall into the hands of a talented team at Wiley. My development editor, Rebecca Heider, was invaluable in creating this text and taught me the nuts and bolts of academic publishing. Lauren Olesky, Dorothy Sinclair, Valerie Vargas, and Carolyn Wells have provided expert guidance and advice at every point in the process. Rajeev Kumar and Sukhwinder Singh oversaw the creation of media that is essential to the text. This project started with a phone call from Glenn Wilson, and his wisdom and energy have propelled this project across the finish line. Thank you, Glenn. You are the very best.

Finally, I am indebted to Becky, my wife and partner in all things, who tolerated my physical and mental absences throughout the writing. I could not have done this without you.

Brief Contents

1 Evolution and Genetics 1

2 The Neuron and Action Potential 27

3 Synapses and Neurotransmitters 51

4 The Nervous System 78

5 Vision 110

6 Audition, Chemosenses, and Cutaneous Senses 144

7 Homeostatic Regulation of the Internal Environment 174

8 The Endocrine System 200

9 Reward, Addiction, and Psychopharmacology 227

10 Sexual Reproduction and Behavior 261

11 Neurodevelopment Across the Life Span 289

12 Learning and Conditioning 314

13 Sleep, Dreaming, and Circadian Rhythms 338

14 Language, Music, and the Brain 365

15 Cognitive Neuroscience 391

16 Psychiatric and Neurological Disorders 423

17 Social Neuroscience 456

INDEX 485

Contents

1 Evolution and Genetics 1

Stories of the Brain 2

1.1 Human Origins 2
The History of the Earth 2
The Evolution of Mammals 3
The Evolution of Primates 4
The Evolution of Humans 4
Check It Out: Are You Part Neanderthal? 5
1.2 The Genetic Basis of Evolution 7
Mendel and the Theory of Heritability 7
Try It Out: Mendelian Traits 9
Modern Genetics 9
Think About It: Genetic Engineering 11
Genetic Variation 12
Advances in Behavioral Neuroscience: Doudna, Charpentier, and CRISPR-Cas9 14
1.3 Evolution and the Environment 15
Natural Selection 15
The Evolution of Color Vision 16
Adaptations to Extreme Environments 17
Try It Out: Are You Ready to Climb Everest? 18
Evolution and the Human Diet 19
1.4 The Evolution of Social Behavior 21
Evidence for the Heritability of Behavior 21
The Survival Benefits of Social Behavior 22
The Biological Basis for Modern Behavior 23
Check It Out: My Genetic History 24
Conclusion 25
Critical Thinking Questions 25 / Dig Deeper into Behavioral Neuroscience 25 / Chapter 1 Before You Go On Answers 26 / Chapter 1 Glossary 26

2 The Neuron and Action Potential 27

Stories of the Brain 28

2.1 An Introduction to the Neuron 28
An Overview of the Nervous System 28
The Discovery of the Neuron 29
Check It Out: Staining a Neuron 30
2.2 The Structure of a Neuron 31
Eukaryotic Cell Structure 31
Neuron Components 32
Advances in Behavioral Neuroscience: Mark Harnett and the Role of Dendrites 33
Types of Neurons 34
Check It Out: Guillain-Barré Syndrome 37
2.3 Neural Transmission of a Signal 38
Energy Produced through Imbalances 38
Ion Exchange 39
The Sodium-Potassium Pump 40
Think About It: Disrupting the Sodium-Potassium Pump 42
The Action Potential 42
The All-or-None Law 44
2.4 White Matter, Gray Matter, and Glia 45
White and Gray Matter 46
Glia 46
Think About It: White or Gray Matter? 47
Conclusion 48
Critical Thinking Questions 49 / Dig Deeper into Behavioral Neuroscience 49 / Chapter 2 Before You Go On Answers 49 / Chapter 2 Glossary 50

3 Synapses and Neurotransmitters 51

Stories of the Brain 52

3.1 The Synapse 52
The Presynaptic Mechanism 53
The Postsynaptic Mechanism 53
A Closer Look at Interneuron Communication 55
Chemical versus Electrical Synapses 56
Check It Out: Electrical Synapses and Atrial Fibrillation 58
3.2 Receptors and Postsynaptic Effects 59
Ionotropic and Metabotropic Receptors 59
Downregulation and Upregulation 61
Think About It: How Upregulation Explains Nicotine Cravings 62
Autoreceptors and Heteroreceptors 62
3.3 An Overview of Major Neurotransmitters 64
Four Key Neurotransmitter Systems 64
Check It Out: The Sleepy Sickness Epidemic 66
Other Neurotransmitters of Note 68
3.4 Disorders and Treatments Related to Neurotransmission 71
Synaptopathies 71
Advances in Behavioral Neuroscience: Dimitri Kullmann and the Silent Synapse 72
Alcohol, Drugs, and Addiction 73
Drug Treatments Targeting Neurotransmission 73
Conclusion 75
Critical Thinking Questions 75 / Dig Deeper into Behavioral Neuroscience 76 / Chapter 3 Before You Go On Answers 76 / Chapter 3 Glossary 77

4 The Nervous System 78

Stories of the Brain 79

4.1 An Overview of the Nervous System 79
Divisions of the Nervous System 80
Gross Anatomy of the Brain 82
Check It Out: Gross Anatomy of the Brain 83
The Brainstem 84
The Spinal Cord 85
Advances in Behavioral Neuroscience: Nicholas Hatsopoulos and Spinal Cord Injury Repair 87

4.2 Major Structures and Support Systems of the CNS 88
The Cerebral Cortex 88
The Cerebellum 91
Essential Support Systems of the CNS 92
Check It Out: Comparing Sheep and Human Brains 93
Check It Out: The Ventricular and Glymphatic Systems 95

4.3 Major Subcortical Processing Circuits 95
The Corpus Callosum: Information Transfer 96
The Thalamus and Hypothalamus: Integration and Regulation 96
The Limbic System: Emotion and Memory 96
The Basal Ganglia: Initiation and Control of Movement 98
Check It Out: Delgado's Bull 99

4.4 The Peripheral Nervous System 100
Divisions of the Peripheral Nervous System 101
The Neuromuscular Junction 102
The Cranial Nerves 103
Think About It: Myasthenia Gravis and the Neuromuscular Junction 104
Conclusion 106
Critical Thinking Questions 107 / Dig Deeper into Behavioral Neuroscience 107 / Chapter 4 Before You Go On Answers 108 / Chapter 4 Glossary 108

5 Vision 110

Stories of the Brain 111

5.1 An Introduction to Vision 112
The Physics of Light 112
The Architecture of the Human Eye 113
How the Focal Plane Affects Our Vision 114
Measuring Visual Acuity and Contrast Sensitivity 115
Try It Out: The Landolt C Task 115
The Visual Processing Stream 116

5.2 The Retina 117
Photoreceptors 117
Retinal Processing Beyond Rod and Cones 119
Check It Out: Peeking into the Retina Through OCT 119

5.3 The Subcortical Centers 122
The Geniculostriate Pathway 122
The Tectopulvinar Pathway 124
Check It Out: The Eye of the Batter 124
Damage to the Visual Stream 125

5.4 The Visual Cortex 126
The Primary Visual Cortex 127
Processing in the Visual Cortex 128
Advances in Behavioral Neuroscience: Margaret Wong-Riley and the PVC 130
The Extrastriate Cortex 131
Check It Out: Motion Aftereffects 133
The Ventral and Dorsal Streams 133

5.5 Color Vision 134
The Trichromatic Theory 135
Color Vision Abnormalities 136
Color Sensitivity Across the Visual Spectrum 138
Think About It: Safety and Color Vision 139
The Opponent Process Theory 139
Try It Out: The Checkerboard Illusion 140
Conclusion 141
Critical Thinking Questions 141 / Dig Deeper into Behavioral Neuroscience 142 / Chapter 5 Before You Go On Answers 142 / Chapter 5 Glossary 143

6 Audition, Chemosenses, and Cutaneous Senses 144

Stories of the Brain 145

6.1 Audition: Hearing and the Ear 145
The Physics of Sound 145
The Anatomy of the Ear 146
Try It Out: Test the Limits of Your Hearing 146
Transmission of Sound Waves Through the Ear 149
Addressing Hearing Deficits 151
Check It Out: Experience a Hearing Deficit and a Cochlear Implant 152

6.2 Auditory Processing and the Vestibular System 153
Auditory Processing in the Brain 153
Amplitude and Frequency Coding 154
Think About It: Auditory Sensitivity and Age 156
Auditory Localization 156
The Vestibular System 157
Try It Out: Experience Auditory Localization 157

6.3 The Chemosenses: Olfaction 159
The Anatomy and Physiology of Olfaction 159
Olfactory Receptors 161
Individual Differences in Olfaction 161
Check It Out: Conversation with an Anosmic 162

6.4 The Chemosenses: Gustation 163
The Anatomy and Physiology of Taste 163
Taste Receptors 165
Individual Differences in Taste 165

Try It Out: Are You a Picky Eater? 165
Advances in Behavioral Neuroscience: Linda Bartoshuk and Supertasters 166
6.5 **The Cutaneous Senses** 167
Pressure and Vibration 168
Pain 169
Check It Out: Global Kinesthetic Deficit 169
Conclusion 171
Critical Thinking Questions 171 / Dig Deeper into Behavioral Neuroscience 171 / Chapter 6 Before You Go On Answers 172 / Chapter 6 Glossary 173

7 Homeostatic Regulation of the Internal Environment 174

Stories of the Brain 175
7.1 **An Overview of Homeostasis** 175
Balance and Change in the Body 175
Feedback Systems 177
Try It Out: Negative Feedback Systems in Nature and Technology 177
7.2 **Homeostatic Regulation of Temperature** 178
Ectothermic versus Endothermic Thermoregulation 178
The Neural Basis of Thermoregulation 178
Hyperthermia and Hypothermia 180
Think About It: Avoiding Exercise-Induced Hyperthermia 180
7.3 **Homeostatic Regulation of Fluids Through Thirst** 183
Volumetric Thirst and the Renin–Angiotensin System 183
Osmotic Thirst and Cellular Sodium Concentration 184
Disorders Related to Fluid Imbalance 185
Try It Out: Recognizing the Dangers of Fluid Imbalance 185
7.4 **Homeostatic Regulation of Hunger and Satiety** 186
The Gastrointestinal Tract 186
The Neural Basis of Hunger and Satiety 188
Advances in Behavioral Neuroscience: Philip Teitelbaum and the Motivation to Eat 188
Hormonal Regulation of Hunger 189
Hormonal Regulation of Satiety 190
7.5 **Disorders of Hunger and Satiety** 191
Try It Out: Measure Your BMI 192
Obesity 192
Prader-Willi Syndrome 194
Anorexia Nervosa 195
The Challenges of Treating Disordered Eating 196
Conclusion 197
Critical Thinking Questions 197 / Dig Deeper into Behavioral Neuroscience 198 / Chapter 7 Before You Go On Answers 199 / Chapter 7 Glossary 199

8 The Endocrine System 200

Stories of the Brain 201
8.1 **An Introduction to the Endocrine System** 201
The Discovery of Hormones as Messengers 201
Comparing the Endocrine and Nervous Systems 202
Try It Out: Fast- and Slow-Response Systems 203
Glands and Hormones 203
Negative Feedback and Endocrine Function 205
8.2 **Endocrine Glands in the CNS** 206
The Hypothalamus, Pituitary Gland, and Pineal Gland 206
Pituitary Hormones 208
Endocrine Regulation of Growth 209
Think About It: Synthetic Growth Hormone Use and Abuse 210
8.3 **Endocrine Glands in the Periphery** 212
The Thymus and Thyroid 212
The Pancreas and Glucoregulation 213
Check It Out: Glucoregulation 214
Think About It: Preventing Type 2 Diabetes 215
The Adrenal Glands and Stress 216
The Gonads and Sexual Development 217
8.4 **The Stress Response** 218
The HPA Axis 218
The General Adaptational Syndrome 221
Post-Traumatic Stress Disorder 222
Advances in Behavioral Neuroscience: Carrie Hill Kennedy and PTSD 223
Conclusion 224
Critical Thinking Questions 224 / Dig Deeper into Behavioral Neuroscience 225 / Chapter 8 Before You Go On Answers 225 / Chapter 8 Glossary 226

9 Reward, Addiction, and Psychopharmacology 227

Stories of the Brain 228
9.1 **The Biological Basis for Reward** 228
Discovering the Biological Basis for Reward 229
Reward Centers in the Brain 229
Think About It: The Ethics of Deep Brain Stimulation 230
The Mesolimbic and Mesocortical Pathways 231
The Reward Circuit and Human Behavior 232
9.2 **Addiction as Dysfunction of the Reward Circuit** 234
The Genetic Basis for Addiction 234
Substance Abuse Disorders 235
Risk Factors for Addiction 236
Try It Out: Diagnosing Addictive Behavior 236
9.3 **The Essentials of Psychopharmacology** 238
How Drugs Enter the Body 238
Pharmacodynamics 240

Think About It: Behavioral Tolerance and Drug Overdoses 242
Pharmacokinetics 242
9.4 **Intoxicants** 244
Alcohol 244
Check It Out: The Effects of Alcohol on Visual Search 246
Opioids 247
Advances in Behavioral Neuroscience: Candace Pert and Opioid Receptors 248
Marijuana 250
9.5 **Stimulants and Other Commonly Abused Drugs** 252
Nicotine 252
Caffeine 253
Think About It: Do We Abuse Caffeine? 254
Other Commonly Abused Drugs 254
Conclusion 257
Critical Thinking Questions 258 / Dig Deeper into Behavioral Neuroscience 258 / Chapter 9 Before You Go On Answers 259 / Chapter 9 Glossary 259

10 Sexual Reproduction and Behavior 261

Stories of the Brain 262
10.1 **An Overview of Sex and Reproduction** 262
Asexual and Sexual Reproduction 263
Male and Female Reproductive Anatomy 264
Sex versus Gender 265
Try It Out: Sex Roles and Gender Roles 267
10.2 **Prenatal Sexual Organization** 268
Genetic Influences on Biological Sex 268
Sexual Differentiation *in Utero* 269
Hormonal Influences on Prenatal Sexual Development 270
Try It Out: The D2:D4 Ratio 271
10.3 **The HPG Axis and Life-Span Sexual Development** 273
Puberty 273
Think About It: Environmental Effects on Puberty Onset 274
Hormonal Control of the Menstrual Cycle 276
Hormonal Activation of Parental Behavior 277
Sexuality and Aging 278
10.4 **Sexual and Reproductive Behaviors** 279
The Human Sexual Response 279
Sexual Orientation 282
Advances in Behavioral Neuroscience: Simon LeVay and the Neural Basis of Sexual Orientation 283
The Biological Basis of Attraction 284
Check It Out: Are Average Faces More Attractive? 285
Conclusion 286
Critical Thinking Questions 286 / Dig Deeper into Behavioral Neuroscience 286 / Chapter 10 Before You Go On Answers 287 / Chapter 10 Glossary 288

11 Neurodevelopment Across the Life Span 289

Stories of the Brain 290
11.1 **Prenatal Development of Brain Structures** 290
Embryonic Neurodevelopment 291
Think About It: Developing Brains at Risk 292
Fetal Neurodevelopment 293
11.2 **Prenatal Neurodevelopment at the Cellular Level** 295
Neurodevelopment as an Organizational Process 295
Check It Out: Visualizing Neuromigration 297
Guidance for Neuromigration 300
11.3 **Postnatal Neurodevelopment** 301
Environmental Influences on Postnatal Neurodevelopment 301
Advances in Behavioral Neuroscience: Marian Diamond and Neuroplasticity 302
The Adolescent Brain 304
Try It Out: Are You a Risk Taker? 305
The Aging Brain 305
11.4 **Neurodevelopmental Disorders** 307
Defects in Neurulation 307
Check It Out: Spina Bifida 308
Fetal Alcohol Syndrome 308
Attention-Deficit/Hyperactivity Disorder 309
Williams Syndrome 310
Schizophrenia 310
Conclusion 311
Critical Thinking Questions 311 / Dig Deeper into Behavioral Neuroscience 312 / Chapter 11 Before You Go On Answers 312 / Chapter 11 Glossary 313

12 Learning and Conditioning 314

Stories of the Brain 315
12.1 **Classical Conditioning** 315
Innate Versus Learned Behavior 315
Pavlov and the Discovery of Classical Conditioning 316
Check It Out: Experience Conditioning 317
Try It Out: Classical Conditioning in Everyday Life 319
Watson and the Little Albert Experiment 319
Think About It: Ethics in Psychology Research 319
Conditioned Taste Aversion 320
12.2 **Operant Conditioning** 322
The Discovery of Operant Conditioning 322

Reinforcement and Punishment 322
Conditioning Phenomena 323
Try It Out: Turning Vices into Virtues 324
Reinforcement Schedules 325
Try It Out: Operant Conditioning in Everyday Life 326
12.3 Learning and the Brain 327
Where Memories Are Stored 327
The Neuronal Basis of Learning 329
Long-Term Potentiation 330
Check It Out: Long-Term Potentiation 331
NMDA and AMPA Receptors 332
Advances in Behavioral Neuroscience: Karl Deisseroth and Optogenetics 334
Conclusion 335
Critical Thinking Questions 335 / Dig Deeper into Behavioral Neuroscience 335 / Chapter 12 Before You Go On Answers 336 / Chapter 12 Glossary 336

13 Sleep, Dreaming, and Circadian Rhythms 338

Stories of the Brain 339
13.1 The Mystery of Sleep 339
Evolutionary Patterns of Sleep 339
Think About It: Has Civilization Changed How We Sleep? 341
The Purpose of Sleep 342
13.2 Sleep Stages and Dreaming 344
The Study of Sleep 344
The Stages of Sleep 345
Try It Out: What's Your Sleep IQ? 347
Dreaming 347
Try It Out: Are You a Lucid Dreamer? 348
13.3 The Biological Basis of Sleep 349
Sensory Stimulation and Sleep 349
The Reticular Activating System and Sleep 350
The Brainstem, Hypothalamus, and Sleep 351
Neurotransmitters and Sleep 351
Advances in Behavioral Neuroscience: Gina Poe and the Role of Sleep in Memory 352
13.4 Sleep Disruptions 353
Drugs and Sleep 353
Sleep Disorders 355
Think About It: Sleep Insufficiency 356
13.5 Circadian Rhythms 357
Melatonin and the Sleep–Wake Cycle 358
The Suprachiasmatic Nucleus 359
When Diurnal and Circadian Rhythms Are Mismatched 359
Try It Out: The Dangers of Microsleep 361
Conclusion 361
Critical Thinking Questions 362 / Dig Deeper into Behavioral Neuroscience 362 / Chapter 13 Before You Go On Answers 363 / Chapter 13 Glossary 363

14 Language, Music, and the Brain 365

Stories of the Brain 366
14.1 The Evolution of Human Language 366
Communication in the Animal Kingdom 366
Check It Out: The McGurk Effect 367
The *FOXP2* Gene 368
Evidence for Speech in Our Ancestry 368
14.2 The Neural Basis of Language 370
The Discovery of Broca's and Wernicke's Areas 370
The Geschwind–Wernicke Model of Language 372
The Two-Circuit Model of Language Comprehension and Production 373
Contributions of the Right and Left Hemispheres 374
Try it Out: The Dichotic Listening Task 375
14.3 The Neural Basis of Reading and Writing 377
The Invention of Reading and Writing 377
Reading Processes 378
Eye Movements and Reading 379
Reading and Writing Disorders 380
Check It Out: Reading Through the Eyes of a Person with Dyslexia 381
14.4 The Neural Basis of Music 383
The Evolutionary Basis of Music 383
Characteristics of Music 384
Think About It: Williams Syndrome 384
Music Processing in the Brain 385
Advances in Behavioral Neuroscience: Isabelle Peretz and Neural Processing of Music 385
Try It Out: Test Your Own Tone Discrimination 386
Conclusion 387
Critical Thinking Questions 388 / Dig Deeper into Behavioral Neuroscience 388 / Chapter 14 Before You Go On Answers 389 / Chapter 14 Glossary 389

15 Cognitive Neuroscience 391

Stories of the Brain 392
15.1 Memory 392
The Three-Store Model of Memory 392
Sensory Memory 393
Check It Out: Experience Echoic Memory 394
Working Memory 395
Try It Out: Measure Your Working Memory 396
Long-Term Memory 396
Amnesia 397
Think About It: A Life Lived Moment by Moment 398
15.2 Attention 399
Aspects of Attention 399
The Neural Basis of Attention 400
Check It Out: Change Blindness 401
Hemineglect 401

Think About It: Why Do Actors Enter from Stage Left? 402
15.3 **Object Recognition** 404
Visual Object Recognition 404
Challenges to Object Recognition 405
Visual Agnosia 406
Prosopagnosia 407
Check It Out: What Is It Like to Be Face blind? 408
15.4 **Executive Function** 409
Aspects of Executive Function 409
The Neural Basis of Executive Function 410
Advances in Behavioral Neuroscience: Marie Banich and Executive Function 412
Evaluating Executive Function 412
Psychosurgery as a Treatment for Executive Dysfunction 413
Try It Out: The Stroop Task 413
Think About It: Moniz, Psychosurgery, and the Lobotomy 414
15.5 **Hemispheric Specialization** 415
Anatomical Differences Between the Hemispheres 415
Sperry and the Split-Brain Procedure 416
Check It Out: Do We Have Two Brains? 418
The Mystery of Left-Handedness 418
Try It Out: Are You Right- or Left-Handed? 419
Conclusion 420
Critical Thinking Questions 420 / Dig Deeper into Behavioral Neuroscience 420 / Chapter 15 Before You Go On Answers 421 / Chapter 15 Glossary 421

16 Psychiatric and Neurological Disorders 423

Stories of the Brain 424
16.1 **Mood Disorders** 424
An Overview of Depression 425
The Neural Basis of Depression 426
Advances in Behavioral Neuroscience: Avshalom Caspi and the Genetic Basis of Depression 427
Treatment of Depression 428
Bipolar Disorder 430
16.2 **Anxiety Disorders** 432
General Anxiety Disorder 433
Panic Disorder 434
Obsessive–Compulsive Disorder 435
Think About It: OCD, Compulsions, and Rituals 436
16.3 **Consciousness Disorders** 437
Schizophrenia 438
Think About It: The Cost of Schizophrenia 441
Delusions of Identity 442
16.4 **Neurological Disorders** 443
Bacterial and Viral Infections 443
Cerebrovascular Accidents 443
Check It Out: The Facts About Strokes 445
Traumatic Brain Injury 445
Try It Out: Concussions and Cognition 447
Dementia 447
Movement Disorders 449
Check It Out: Understanding Tourette's Syndrome 451
Conclusion 452
Critical Thinking Questions 452 / Dig Deeper into Behavioral Neuroscience 453 / Chapter 16 Before You Go On Answers 454 / Chapter 16 Glossary 454

17 Social Neuroscience 456

Stories of the Brain 457
17.1 **Emotions** 457
Defining Emotion 457
The Evolutionary Basis of Emotions 458
Theories of Emotions 459
Advances in Behavioral Neuroscience: Leda Cosmides and Evolutionary Psychology 459
The Neural Basis of Emotions 460
17.2 **Communicating Internal States** 463
Facial Expressions 463
Social and Duchenne Smiles 464
Try It Out: Cultural Differences in Emotional Expressions 464
Deception as an Adaptive Strategy 465
Check It Out: What's in a Smile? 465
Theory of Mind 466
Check It Out: Can You Tell the Truth from a Lie? 466
17.3 **Beauty and Attraction** 469
External Tells for Health and Reproductive Fitness 469
Ovulation and Attractiveness 471
Hormonal Influences on Pair Bonding 472
17.4 **Human Social Networks** 473
Evolutionary Advantages of Group Membership 473
Signaling Group Membership 474
Ingroups and Outgroups 475
Prejudice and Stereotypes 475
Think About It: The Implicit Association Test 477
17.5 **Autism Spectrum Disorder and Emotional Blindness** 478
Autism Spectrum Disorder 479
Difficulties in Processing Emotional Cues 480
The Neurology of Asperger's Syndrome 480
Check It Out: Lindsay's Social World 481
Conclusion 482
Critical Thinking Questions 482 / Dig Deeper into Behavioral Neuroscience 482 / Chapter 17 Before You Go On Answers 483 / Chapter 17 Glossary 484

INDEX 485

CHAPTER 1

Evolution and Genetics

Our modern environment is dramatically different from the environment where our earliest ancestors evolved.

CHAPTER OUTLINE

1.1 Human Origins

- The History of the Earth
- The Evolution of Mammals
- The Evolution of Primates
- The Evolution of Humans

  Check It Out: Are You Part Neanderthal?

1.2 The Genetic Basis of Evolution

- Mendel and the Theory of Heritability

  Try It Out: Mendelian Traits

- Modern Genetics

  Think About It: Genetic Engineering

- Genetic Variation

  Advances in Behavioral Neuroscience: Doudna, Charpentier, and CRISPR-Cas9

1.3 Evolution and the Environment

- Natural Selection
- The Evolution of Color Vision
- Adaptations to Extreme Environments

  Try It Out: Are You Ready to Climb Everest?

- Evolution and the Human Diet

1.4 The Evolution of Social Behavior

- Evidence for the Heritability of Behavior
- The Survival Benefits of Social Behavior
- The Biological Basis of Modern Behavior

  Check It Out: What's Your Ancestral Journey?

LEARNING OBJECTIVES

LO 1.1 **Describe** the history of the Earth and the evolutionary path that led to humans.

LO 1.2 **Outline** the genetic principles that lead to evolution.

LO 1.3 **Explain** how evolution is affected by the environment.

LO 1.4 **Describe** the genetic basis of social behavior.

Stories of the Brain

In 1968, Nancy Wexler was walking across the street one morning with her mother when a policeman, seeing her mother stumble and stagger a bit, scolded her mom "for being drunk" so early in the day. In fact, Nancy's mom was not drunk but suffered from a neurological disorder that ran in her family. Nancy had just graduated from college with a degree in psychology; she went on to earn a doctorate in clinical psychology but spent her professional career investigating Huntington's disease (HD), the devastating neurological disorder that afflicted her family and others.

HD is caused by a mutation in a specific gene that then creates a mutated protein, which gradually damages brain cells. Individuals with HD exhibit uncontrolled jerky movements and a loss of muscle coordination that progresses until the individual can no longer swallow or breathe. Cognitive functions such as planning, abstract thinking, and inhibition of inappropriate actions are also affected in HD. A sad aspect of HD is that symptoms typically do not appear until the middle 30s, at which time most individuals with HD who are unaware of their situation have already had children and potentially passed on the defective gene to their progeny. Dr. Wexler devoted her life to understanding HD and was instrumental in the research that led to identifying the gene that causes HD. In 1993, Dr. Wexler won the Lasker Public Service Award for research policy and advocacy toward understanding HD.

HD shows us that genes can explain some neurological and behavioral conditions. Many factors influence the structural and behavioral aspects of the brain, and you may have come to the erroneous conclusion that genes explain everything about human behavior. The truth is more interesting: our genes interact with the environment, and this is illustrated by the difficulties some of us experience digesting milk and also by the unusual physiology of some ethnic groups. For example, you may know that the Dutch are the tallest nationality, and Sri Lankans are the shortest.

We begin this chapter with two questions: how did humans get here, and how does our evolutionary past influence us today? To address these questions, we will first trace what we know of the human story and then discuss the biological mechanism of heritability. Understanding this material will be essential when we turn our attention to the biological basis for human behaviors including eating disorders, psychiatric illnesses, and sexuality. There are two key concepts that this chapter will introduce: first, you and I are the result of millions of years of adaptation to our environment, and some adaptations from our past are no longer helpful in our modern world; second, genes guide not only the expression of physical traits but also behavior. Our human ancestors evolved in the open savannas of sub-Saharan Africa. How does this genetic heritage affect us in the modern world?

1.1 Human Origins

LO 1.1 **Describe** the history of the Earth and the evolutionary path that led to humans.

The history of life on Earth leading up to humans is billions of years long. This history has been marked by environmental changes that were both sudden and severe. At times, almost the entire **biome**, or community of animal and plant species, of the Earth disappeared only to be replaced later by very different organisms. As the climate changed in response to these often-catastrophic changes, life adapted. The history of life on Earth is one of challenge and response, extinction or survival.

The History of the Earth

As near as we can determine, our planet was formed about 4.5 billion years ago from interstellar dust and debris left over from the Big Bang in which our universe was born. At that time our planet was significantly warmer than it is today. The Moon was much closer than it is now, so tides were more dramatic, and the land was bleak and desolate beyond belief. The atmosphere at this time was likely composed mostly of hydrogen sulfide and methane; it would be deadly poisonous to us. Frequent volcanoes released enormous amounts of steam and carbon dioxide into the atmosphere. No plants existed to break up the rock into fertile soil, so what land existed was either barren rock or sand. Eons upon eons passed. The Earth slowly cooled, and the oceans deepened on this inhospitable planet.

Then, about 3.8 billion years ago, something truly remarkable happened: *life began*. The fragmentary evidence supports the idea that for the first 2 billion years, life on Earth consisted of simple microbes that formed into gooey colonies or mats that floated about aimlessly. As more eons passed, life diversified into more complex forms such as plants, animals, fungi, and single-celled organisms known as protists. Eventually, more complex organisms such as *Dickinsonia* appeared (**Figure 1.1**). These very primitive animals ranged in size from a fraction of an inch (2–3 mm) to around four and a half feet in length (1.5 m).

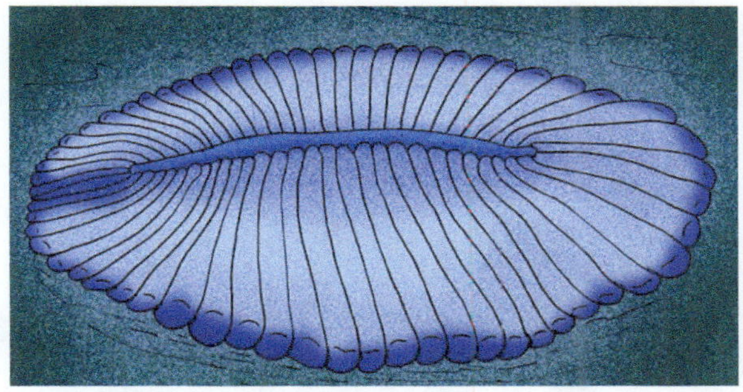

FIGURE 1.1 **Dickinsonia costata** The fossil record suggests that the first animals were small, flat, and soft bodied like *Dickinsonia costata*, shown here. *Dickinsonia* is thought to have lived in the oceans and seas of our ancient world 580 million years ago. If you are unsure which end is the head and which is the tail of this ancient animal, you are in good company because paleontologists are not sure either.

It would be easy to imagine that life evolved and diversified gradually and continually over the eons, but that would be a mistake. The history of life on Earth has been marked by a series of extinction events in which a sudden and severe change in the Earth's climate led to a dramatic and rapid change in the quantity and diversity of species.

One of the largest extinction events occurred about 252 million years ago. The **Permian–Triassic (P–T) extinction**, or the Great Dying, was the greatest extinction in the history of the world. Up to 80% of all marine species disappeared, and about 70% of all land species were lost. The causes of this catastrophe are not clear, but possibilities include a collision with a mile-wide meteor, the release of methane sequestered in the sea floor, a sudden increase in volcanic activity, or some combination of these events. Nature abhors a vacuum, so when most life had been erased from the planet, something had to take its place. That something was primarily dinosaurs. For about 150 million years, dinosaurs ruled the Earth. The early mammals that are our ancestors emerged toward the end of this epoch; they were small and lived their lives on the edge of survival.

Just as dinosaurs benefited from an extinction that wiped out the competition, mammals benefited from another catastrophic event. Around 66 million years ago, in the **Cretaceous–Tertiary (K–T) extinction**, 75% of all plant and animal species disappeared. Because the K-T extinction was relatively recent (at least to a geologist) when compared to the P-T extinction, we know more about the circumstances of this die-off. Again, the evidence suggests a massive asteroid impact, a sudden increase in volcanic activity, or both. Whatever its cause, the K-T extinction opened the door for mammals to thrive, and their numbers and diversity exploded. From those small lemur-like animals huddling in the underbrush and hiding from dinosaurs in dank burrows evolved the rich panoply of mammals who are our distant ancestors.

The Evolution of Mammals

When the P-T extinction cleared away most of the land-dwelling vertebrates, whole ecosystems collapsed. Those animals that managed to survive began to compete anew for resources in a dramatically changed landscape. Among the survivors were the ancestors of mammals who successfully adapted to the new environment. **Mammals** are defined by several key adaptations including lactation, hair or fur, a larger brain, and social behaviors such as cooperation.

First among these adaptations is **lactation**, when females have mammary glands to provide milk for their young. The offspring of mammals are also born live as opposed to in an egg. The evolution of live births and lactation tied offspring to the mother after birth for a period of dependency and nurturing. This extended period of dependency by the newborn favored the development of a larger and more complex brain.

A second mammalian adaptation is hair or fur to assist with internal temperature regulation. Furthermore, mammals are warm blooded, which means they have both the ability to generate heat internally and the ability to maintain a constant internal temperature. Fur and warm-bloodedness allowed mammals to extend their habitats into regions previously closed to their ancestors.

FIGURE 1.2 **An early mammal** *Morganucodon watsoni* is one of our earliest mammalian ancestors. A tiny, shrew-like animal, it lived in burrows and mostly ate insects. Although generally classified as mammalian, they maintained some of the characteristics of earlier forms such as birthing their young in eggs.

Perhaps the most important mammalian adaption was the appearance of the neocortex in the brain, which supports complex sensory processes and problem-solving cognition. The emergence of a larger brain placed increased caloric demands on early mammals, so a spiral developed where larger brains enabled these animals to find more food, which then led to a bigger brain, which led to more success at finding food.

Finally, the increase in caloric needs coupled with the extended period of dependency of the young fostered the development of social behaviors that involved cooperation and reciprocity. These powerful adaptations are a part of the long line that led to you and me. Figure 1.2 depicts one of the first in the mammalian line, *Morganucodon*, which appeared about 200 million years ago and was about the size of a modern mouse.

The Evolution of Primates

The continual evolution of life on this planet did not culminate in humans in the same way that a stairway will eventually lead you from the noisy bustle of the street up to the serene penthouse. Rather, many lines are continuously evolving from a common source just as the many branches of a tree shoot out from the main trunk. Some of these evolutionary branches lead toward us, some go off in other directions and end in organisms we see today, and others are now extinct. Our goal is to understand the evolution of our brain and how our past as a species affects us today, so we will continue to follow the evolutionary stream that leads from animals through mammals to the **primates**.

It is believed that primates branched off from the mammalian line somewhere between 80 and 65 million years ago (Figure 1.3). Because our ancestors lived in an arboreal habitat, we see a number of adaptations among primates that would be useful to an organism that spends time both in the trees and on the ground. For example, primates are typically **bipedal**; that is, they walk on their hind legs, which frees up the front limbs for other activities. Primates benefit from shoulder joints and opposable thumbs, which provide a wide range of motion and the ability to grasp and manipulate objects. Primate eyes evolved to be at the front of the face instead of the sides, which provides superior depth perception, and primates have some version of color vision. Another significant extension of the mammalian line is an expanded period of dependency in infancy, which permits considerable brain development to occur outside the mother's body. Finally, primates benefitted from development of a large brain with an emphasis on expanding the cortex, the region of the brain where complex thinking and problem solving occurs.

Let us end our discussion of the defining characteristics of primates by keeping the idea in mind that some traits we think of as human emerged because they conferred a survival advantage to our ancestors.

The Evolution of Humans

Humans (also known as *homo sapiens*, or wise person) demonstrate their mammalian heritage though many characteristics such as breastfeeding their young and being warm blooded. We humans also honor our primate heritage though highly developed bipedalism, excellent visual acuity coupled with color vision, and dexterous arms and hands with opposable thumbs and a sensitive grip capable of fine movements. However, there are important additions to

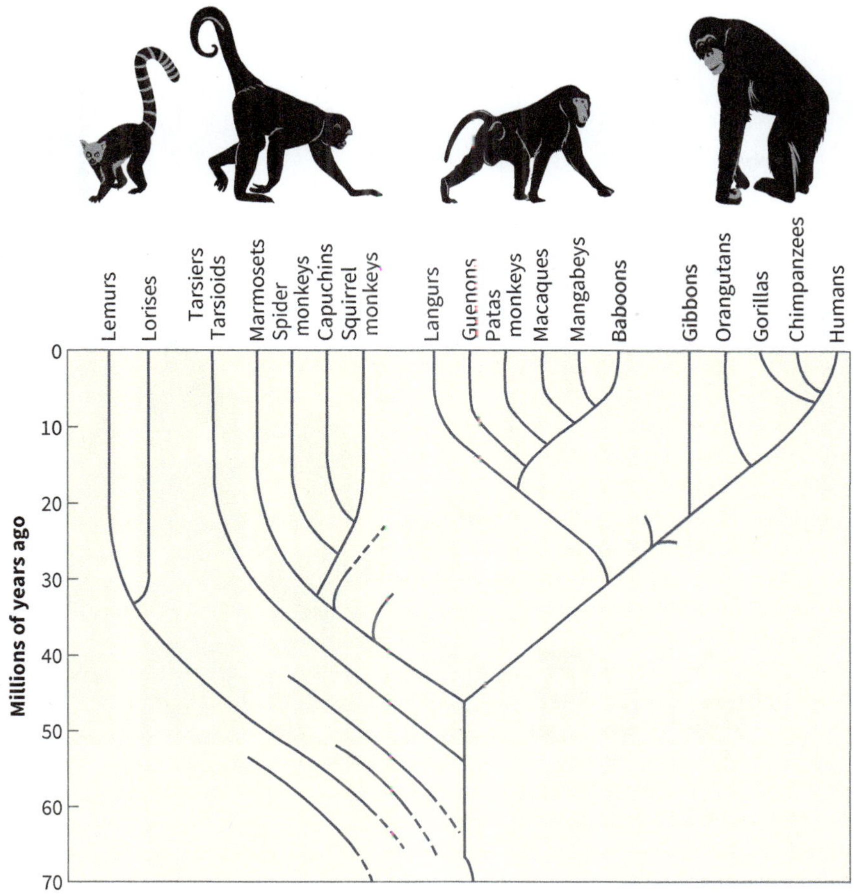

FIGURE 1.3 The primate family tree The evolutionary line that led to you began with an ancestor born more than 60 million years ago when primates first appeared. Primates include species with a wide range of appearances such as lemurs, monkeys, apes, and humans.

this model that humans enjoy, and perhaps the most important are a very large brain in comparison to our body and a further lengthening of the period of dependency after birth. See **Check It Out: Are You Part Neanderthal?** to explore your own connections to our distant ancestors.

Check It Out | Are You Part Neanderthal?

Aaron Rutten/Shutterstock

Do we carry genetic information from distant ancestors who are now extinct, such as Neanderthals? Watch the video **Are You Part Neanderthal?** to learn about these ancestors, why you might still carry their DNA in your genome, and what that might mean for you and me today. Then answer these questions.

1. List two current medical or behavioral disorders that are linked to Neanderthal DNA.
2. Why might hypercoagulability have been a useful trait in our distant ancestors' time?
3. Search online for "John Anthony Capra" and mention one fact about him or his research that you found interesting.

Image courtesy of bioninja.com.au

FIGURE 1.4 A comparison of mammalian brains Humans have a large brain in comparison to their body size, but the most important distinction is the relatively large number of neurons in their cortex.

Humans do not have the largest brain in the animal kingdom; for example, an elephant's brain is three to four times larger than ours. What distinguishes humans is not just a large brain but also the substantial folding necessary to fit the large cortex into the human head. Humans enjoy the greatest cortex-to-brain mass ratio of any animal, and this is the key to their success as a species (**Figure 1.4**).

Now that you have an overview of how the human species evolved from the earliest life forms over millennia of adaptations, we will turn to the genetic basis for this evolution.

1.1 Before You Go On

Section Summary

Life has existed on this planet for about 4 billion years, increasing in complexity, diversifying in form, and shaped by cataclysmic events. After the extinction of the dinosaurs, mammals flourished, characterized by lactation, fur or hair, a larger brain, and a longer period of dependency after birth. Later, primates adapted to life among the trees with bipedalism, depth perception, and opposable thumbs. Finally, our human ancestors evolved a large brain capable of problem solving.

Comprehension Check

1. Which is the best description of the history of life on Earth?
   - **a.** rapid evolution from simple to complex
   - **b.** slow change interrupted by occasional dramatic change
   - **c.** a straight line with a sudden drop at the end
   - **d.** a series of repeating waves
2. Which of these are defining traits of all mammals?
   - **a.** two eyes, bipedalism, and opposable thumbs
   - **b.** bipedalism, lactation, and a neocortex
   - **c.** warm-bloodedness, lactation, and live births
   - **d.** color vision, increased metabolism, and live births
3. Select the statement that is NOT correct.
   - **a.** Humans and apes have a common ancestor.
   - **b.** Apes and monkeys share a common ancestor.
   - **c.** Humans evolved from monkeys.
   - **d.** Humans are more closely related to a lemur than to an elephant.
4. Compared to other mammals, the human _____.
   - **a.** offspring has a shorter period of dependency after birth
   - **b.** hand has a stronger grip
   - **c.** offspring has a longer gestational period
   - **d.** cerebral cortex has more neurons

1.2 The Genetic Basis of Evolution

LO 1.2 Outline the genetic principles that lead to evolution.

Why do we consider evolution and genetics as we learn about the brain and behavior? The answer is that by considering the evolutionary basis of human behavior, we can better understand human behaviors such as why we eat more than we should, why some people seem more attractive than others, and why people can become addicted to harmful drugs and behaviors. To understand the mechanism behind evolution, you must have at least a passing acquaintance with genetics.

Understanding how this process works is essential to understanding how the brain works and how the brain controls basic behavioral processes. In subsequent chapters, you will be asked to consider to what extent our genes influence behaviors such as attention span, food preferences, propensity for violence, and even cognitive abilities or social behaviors such as empathy.

Mendel and the Theory of Heritability MCAT

You have surely noticed that people tend to physically resemble their biological relatives. Tall parents often produce tall offspring, and the opposite is also generally true. Our ancestors were aware that physical characteristics of parents were likely to appear in their progeny, yet they also recognized that offspring were likely to differ from their parents. Most learned individuals thought that the characteristics of the two parents were somehow equally blended in their offspring. One theory, proposed by the French naturalist Lamarck (1744–1829), was that an organism can acquire characteristics in its lifetime and then pass them on to offspring. This concept was used to explain the long neck of the giraffe; as a giraffe repeatedly extended its neck to reach distant leaves, the neck stretched, and that characteristic was passed on to its progeny. Despite the best efforts of great minds to understand heritability, it was the work of a humble cleric that showed science the way forward.

Gregor Mendel (1822–1884) was both a scientist and a member of a religious order in what is now the Czech Republic. His fellow monks raised their own food, and Mendel's university studies in natural history led him to apply what he learned to the materials at hand, namely the pea plants in the garden.

Mendel noticed that pea seeds varied in their color (yellow or green) and texture (wrinkly or round). Peas reproduce sexually, with genetic material contributed by each of two parents. By hand-pollinating his plants over multiple generations, Mendel eventually arrived at two strains of pea seeds whose traits (color and texture) never varied. Today we would call those strains purebred.

Over the course of several decades, Mendel conducted multiple experiments to investigate how physical characteristics were passed from one generation of pea plants to the next. It is likely that others had observed similar results before, but part of Mendel's genius was his careful record keeping and thoroughness. By crossing and recrossing these strains of pea plants and noting the characteristics of each generation, Mendel deduced three basic laws of genetics: first, information about traits comes in specific units that are passed from one generation to the next; second, some traits are more likely to be expressed than others; and third, a plant can contain heritable information that it does not express.

Mendel's first insight was that there was some entity inside his pea plants that contained heritability information that was passed from parent to offspring. Today we call that entity a **gene**, and a few other terms from modern genetics will help us discuss Mendel's observation. Genes that control traits come in different varieties, known as **alleles**, for example, one allele determining yellow color in pea plants and another green. When each parent contributes the same allele to their offspring (for example, yellow/yellow or green/green), the resulting offspring is **homozygous**. However, when each parent contributes a different allele (e.g., yellow/green or round/wrinkled), that offspring is **heterozygous** for that trait.

Mendel's second insight was that all alleles are not equal. In a heterozygous offspring, the **dominant trait** will be expressed in preference to the **recessive trait**. Recessive traits are only expressed in homozygous offspring. In the pea example, yellow color is a dominant trait, whereas green is recessive. When two purebred parents are crossed, one homozygous for the dominant trait of yellow and the other homozygous for the recessive trait of green, the first-generation offspring are all yellow (**Figure 1.5a**).

Mendel's third insight was that an organism can contain genetic information that it does not overtly express. He came to this conclusion when he took the offspring of the first generation and crossed them again. What he observed in the second-generation offspring was totally unexpected. Crossing the first-generation yellow pea plants created a second generation with a predictable 3:1 ratio of offspring that was a clue to the underlying genetic mechanism (**Figure 1.5b**). Notice that in this crossing, some offspring are yellow but carry the gene for green, which is not expressed but is passed on to the successive generations. We call the visible expression of genetic traits (the yellow color) the **phenotype** and the actual genetic information (the yellow/yellow or yellow/green alleles) the **genotype**.

As you consider the distinction between dominant and recessive traits, you may wonder about such traits in humans. In this **Try It Out: Mendelian Traits**, you can identify phenotypical traits in yourself that are dominant or recessive. Keep in mind that dominant traits are not necessarily desirable, nor are recessive traits necessarily undesirable.

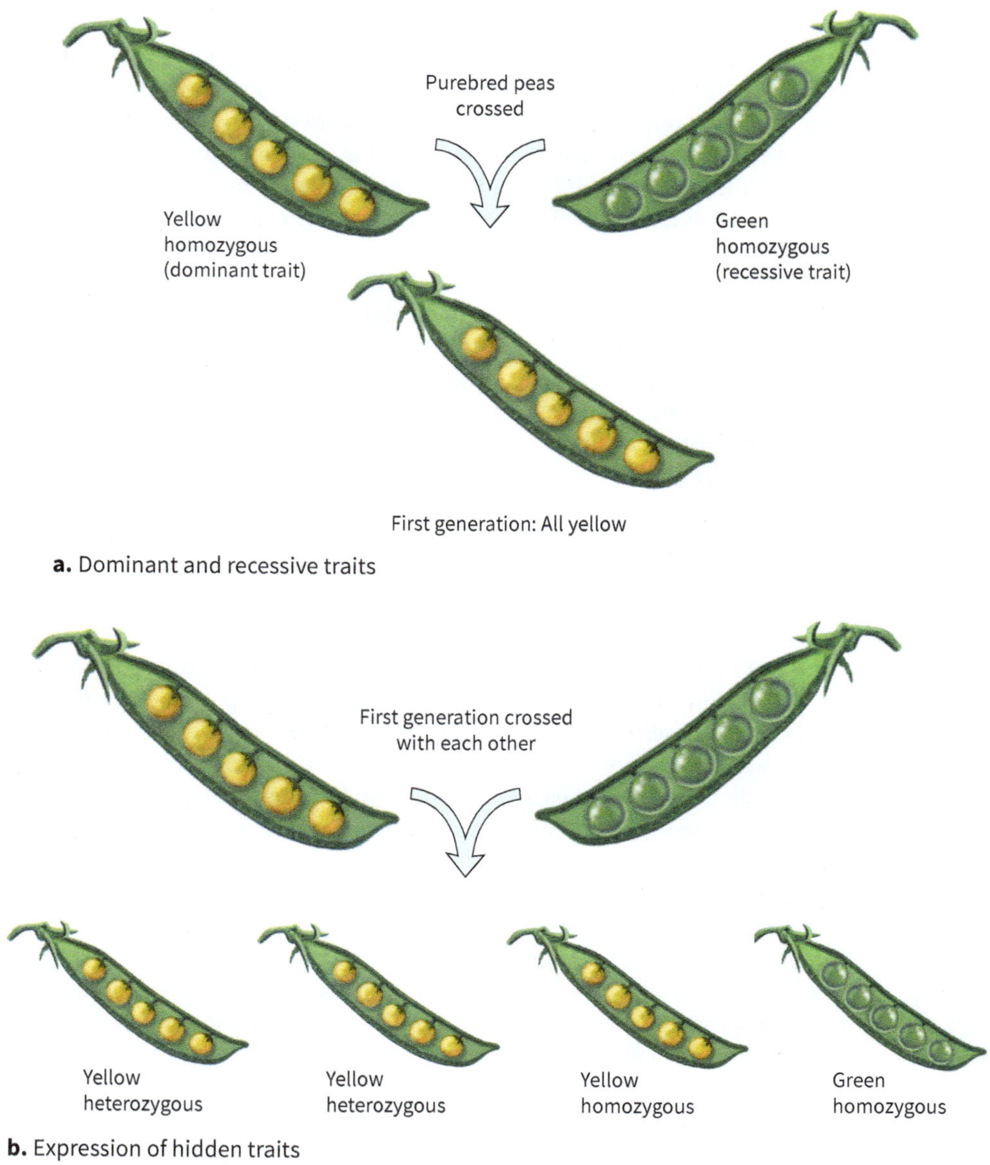

FIGURE 1.5 Mendelian heritability a. Mendel crossed two homozygous strains of pea plant, one of which only produced yellow peas and the other of which only produced green peas. In the first generation, the offspring expressed the traits of only the parent expressing the dominant trait of yellow peas. **b.** When Mendel then crossed the first-generation plants, the offspring expressed the characteristics of the original parents in a predictable pattern, with three offspring expressing the dominant trait for every one that expressed the recessive trait.

Try It Out | Mendelian Traits

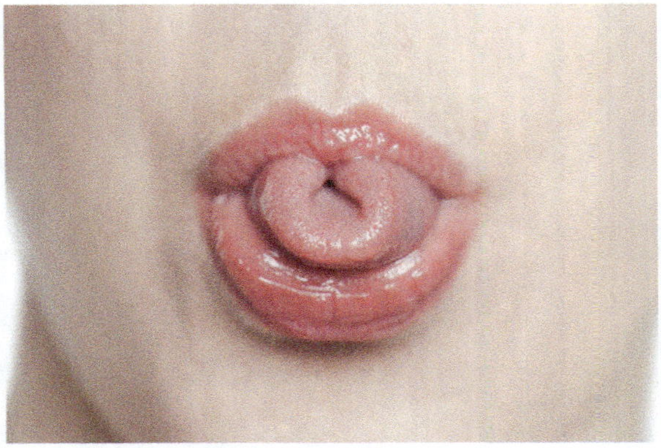
Dan Kosmayer/Shutterstock

Geneticists have identified several phenotypical traits that are controlled by a single gene or at most a small number of genes. These traits follow the laws of Mendelian heritability, so you are likely to see evidence of the heritability of such traits in yourself and family members. Download and complete the activity **Mendelian Traits**.

Mendel established the broad principles of heritability, which are that we carry in our genes information (our genotype) that we may not directly express (our phenotype) and that we pass that information on to our offspring. Today we understand the molecular basis of these principles, which is the topic of the next section.

Modern Genetics MCAT

The cells in your body contain a complete set of directions for making you. The nucleus of each cell contains long strands of genetic material known as **chromosomes**, which are comprised of individual units known as genes. Genes consist of **DNA (deoxyribonucleic acid)**, which contains the basic directions for the creation of **proteins**. Proteins are complex molecules comprised of amino acids that create the tissue, bone, and blood that make up you and me.

Chromosomes are threadlike structures inside the nucleus of every animal or plant cell that contain the library of genetic information necessary to create that organism. When examined under a microscope, they appear to be comprised of a series of colored bands; hence the name chromosome, which comes from the Greek *chromo* for color and *soma* for body. Each chromosome is divided into two arms, the p (for petite) short arm and the longer arm, q. Geneticists identify the physical location of a gene by the number of the chromosome and the arm on which it is located; a gene identified as 7q31.2 would be located on the long arm of chromosome 7 at band 31.2 (**Figure 1.6**).

Species differ in the number of chromosomes within the nucleus. For instance, the common pigeon has 80 pairs, yeast has 32, the Adders-tongue fern sports 1,440 pairs, and a mosquito manages to torment us with a life based on only six pairs of chromosomes. Take a moment to view a typical human **karyogram** (**Figure 1.7**). Karyograms are used to identify and evaluate the size, shape, and number of chromosomes in a representative cell. Extra or missing chromosomes or any other structural abnormality will show up in the karyogram.

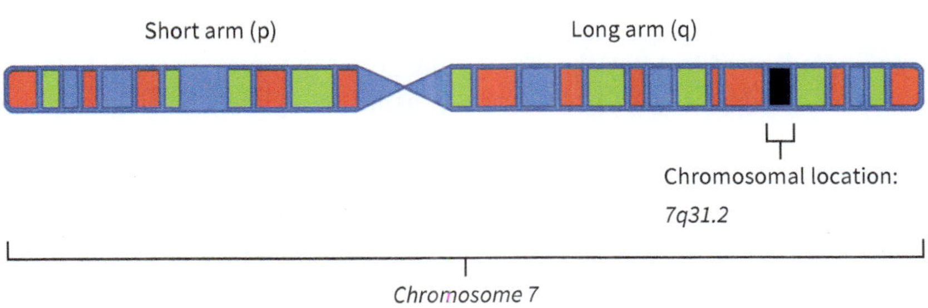

FIGURE 1.6 A human chromosome The location of genetic information on a chromosome is specified by the number of the chromosome, the arm on which it resides, and finally on the specific band where it can be found. The gene 7q31.2 resides on the long arm of chromosome 7, at band 31.2.

FIGURE 1.7 Human karyograms **a.** In the karyogram for a human female, there are two X chromosomes in position 23. **b.** In the karyogram for a human male, the corresponding allosomes consist of an X and a Y.

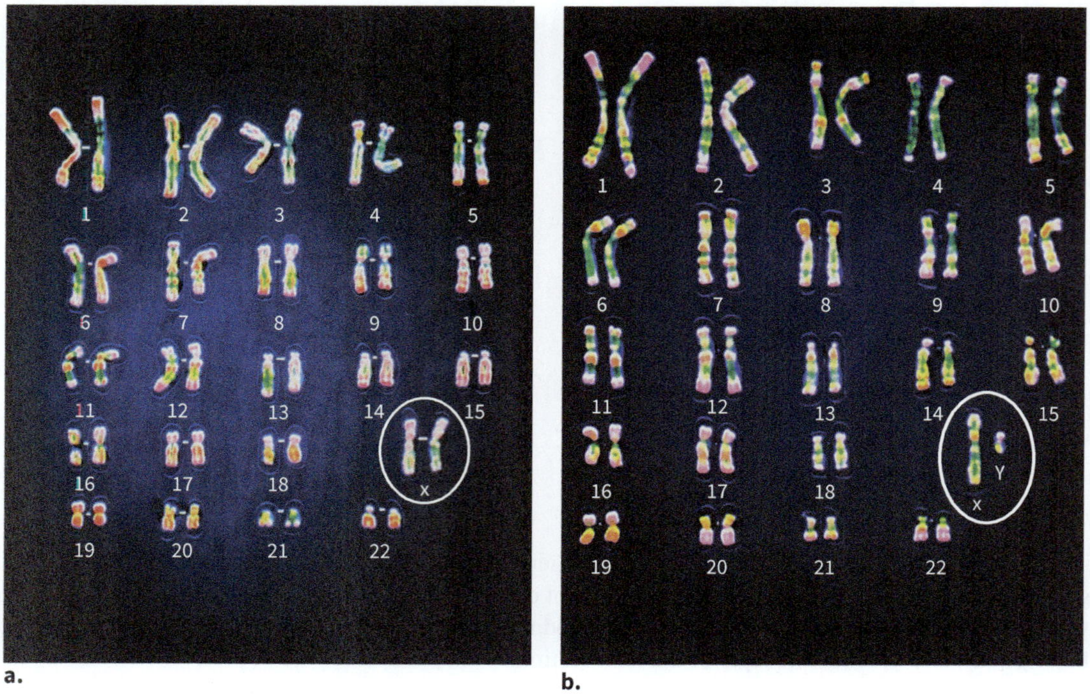

In a karyogram, the 23 chromosome pairs are arranged from largest to smallest. Of these, 22 pairs are referred to as **autosomes** and the 23rd pair consists of **allosomes**, which are sex linked. Allosomes carry the genetic information necessary for sex selection as well as the rest of the information in our DNA. Autosome pairs are labeled 1–22, and the sex-linked 23rd pair is also labeled with X and/or Y. Figure 1.7a shows the typical karyogram of a human female with an XX pair, and Figure 1.7b shows the corresponding karyogram of a male with an XY pair.

It has been known for some time that the mechanism of heredity is contained within the chromosome itself, but how that occurred was not clear until the middle of the last century. Advances in our understanding of the genetic basis of physical traits have allowed scientists to manipulate genes to produce desired results (see **Think About It: Genetic Engineering**).

The discovery of the physical configuration of DNA was a major milestone in molecular biology. James Watson and Francis Crick won the Nobel Prize in Medicine for their discovery of the structure of DNA (**Figure 1.8**). One way to think of DNA is to imagine a spiral staircase with steps of two different colors. The helical supports that spiral upward are the sugar and phosphate backbone, and the steps that connect the supports are comprised of bases. The steps are always composed of either an adenine–thymine (AT) pair or a guanine–cytosine (GC) base pair. The unit of one step and its support together (a sugar–phosphate backbone with a single base) is known as a **nucleotide**.

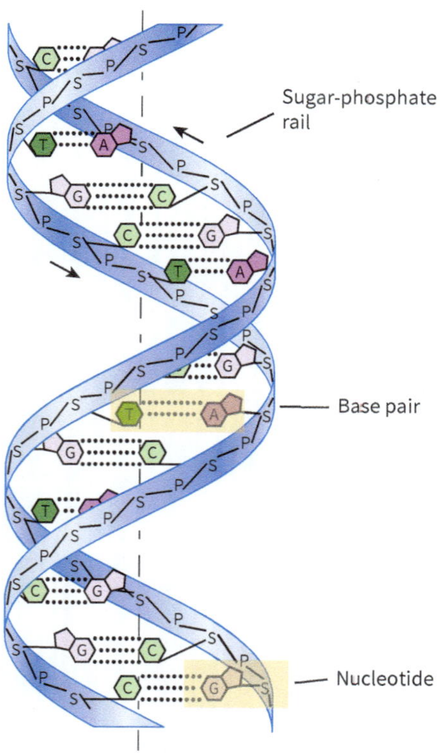

FIGURE 1.8 The helical structure of DNA Chromosomal DNA normally consists of two intertwined helical strands comprised of sugars and phosphates linked by a combination of the four bases A, C, G, and T. A link between a single base and its sugar-phosphate rail is a nucleotide.

It is worth a moment to stop and consider what you have just learned. Humans are amazingly complicated organisms. Your brain has close to 100 billion neurons, and they are arranged and connected in a very specific fashion according to a complex set of instructions. That entire set of instructions for creating a person is contained in the DNA of every cell, using an alphabet of only four letters that represent base pairs. If you took all the DNA out of one cell and stretched it out, it would be a strand about 6 feet (1.8 m) long but only 50 trillionths of an inch wide. If you were to unwrap all the DNA in your body and create a single strand, that strand would stretch to the Moon and back 6,000 times.

Now that you now understand how much information is contained in your DNA, we can discuss how DNA transfers information from one cell to another in

Think About It | Genetic Engineering

Kazyavka/Adobe Stock

Genetic engineering techniques have led to the development of genetically modified organisms (GMOs) to serve human needs. Between 20% and 40% of crops worldwide are lost to insect pests and diseases each year, and crops that have been genetically engineered to resist such threats reduce that loss, providing an economic boon and helping to feed a hungry planet. Some GMOs are designed to improve an existing organism. For example, natural tomatoes often lose their flavor in the interval between being picked and being delivered to the grocer's shelf, so scientists have tried to genetically modify tomatoes to ripen more slowly and provide the consumer with a better product.

Genetically modified bacteria have been created to produce insulin, the hormone essential to life for diabetics, and gene editing is used to create animal models for research into human diseases such as obesity, heart disease, diabetes, addiction, and anxiety. It is thought that someday genetic engineering could be used to eliminate cancers by turning off the genes that are responsible for uncontrolled cell division.

It is clear that there are potential benefits to genetic engineering, but there are also dangers, especially in genetically engineering humans. Humans are genetically complex organisms. Traits such as intelligence and athletic ability that we might want to enhance are controlled by many genes, so influencing development in one gene would likely have unwanted effects elsewhere.

Questions

1. **Do you think it would be a good idea or a bad idea for parents to genetically modify their children to express specific traits? Why or why not?**
2. **If genetic engineering were widely available to parents, do you think the end result would be to increase diversity or reduce diversity? That is, would succeeding generations of humans become more similar or more varied, and would the result be an improvement? Explain your answer.**

a process called replication. Let us return for a moment to our analogy of a spiral staircase. When the cell is dividing, one rail of the staircase peels away, taking one half of the step pair (a nucleotide) with it. An identical nucleotide is then constructed out of raw material (**Figure 1.9**). The end result can also be conceptualized as a zipper that unzips into two halves, and then creates a second matching zipper half for each of the original segments before zipping them both back up. It has been estimated that there will be 10^{16} DNA replications in your body during the course of your life.

It is surprising to realize that the complexity of living organisms can be explained through a set of directions that, although almost unimaginably long, consist of only the letters A, C, G, and T. Yet consider that the document you read on a computer screen, the movie you watch on your television, and the voice you hear on your cell-phone are all represented in your computer as a complex series of only zeros and ones. Given that insight, it may not seem so far-fetched that in your DNA there are instructions for creating your brain that are merely a long string of four characters.

How are the instructions in DNA carried out to build our specific cells, tissues, and bones? A process known as transcription takes a section of the DNA and creates a specific strand known as **messenger RNA (mRNA)**. Messenger RNA then leaves the nucleus and enters the cell body or cytoplasm, where something truly amazing occurs. A structure inside each cell known as a **ribosome** is literally a machine that reads the instructions on the mRNA. By

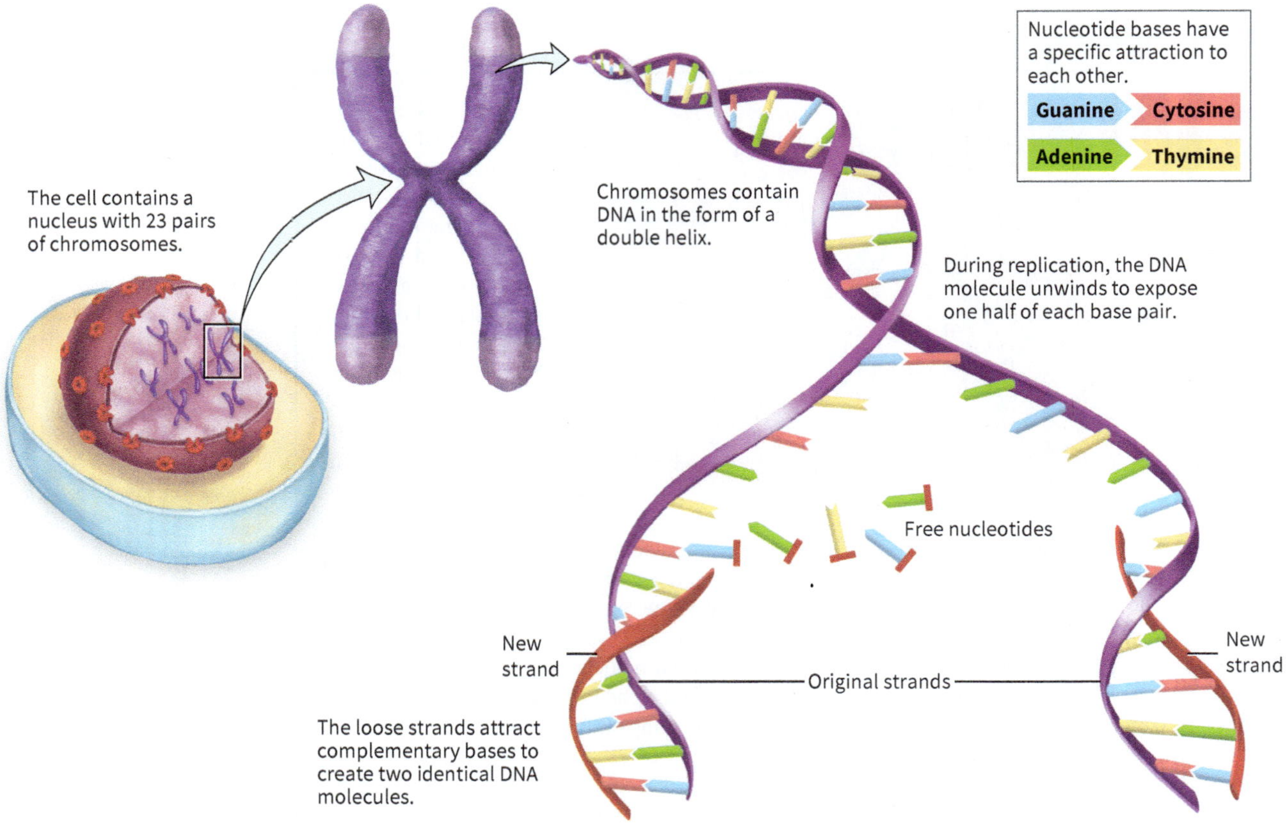

FIGURE 1.9 **DNA replication** Chromosomal DNA normally consists of two strands intertwined in a helix. During the replication process, the helix unwinds into two forks. Each fork then attracts the appropriate base to complete a nucleotide, and in doing so, a new copy of the original DNA is created.

collecting amino acids carried by another form of RNA known as transfer RNA (tRNA), the ribosome builds proteins. To summarize, DNA creates RNA, and RNA creates proteins. Living things are made of cells, and cells are made up of proteins, and so DNA creates us.

Genetic Variation MCAT

How then does our genetic mechanism create diversity and explain genetic disorders? Now we can flesh out our discussion of genetics and behavior by revealing how variation occurs during the replication procedure. In theory, the genetic material in your chromosomes replicates itself exactly, but the reality is sometimes different. Errors occur very rarely, but sometimes certain bases are left out, added, or replaced. Figure 1.10 illustrates three common mechanisms responsible for genetic mutations.

Although the DNA replication process creates millions of copies without a single error, errors do rarely occur, and these errors form a basis for variation within a species. Random mutations continuously introduce changes into the genetic code. The genetic code has redundancies to prevent serious problems, and the replication process has safeguards, so most

FIGURE 1.10 **How transcription errors create genetic diversity**

	Insertion	Deletion	Substitution
Original Sequence	GGTCAC	GAC~~TT~~G	TGGGAC
Mutated Sequence	GGTGACAC	GACG	TGGAAC

Insertion errors involve adding bases where they do not belong. In a deletion error, an expected replication does not occur. Substitution errors involve replacing one base with another.

FIGURE 1.11 **Karyogram of trisomy 21** **a.** A transcription error at chromosome 21 leads to a trisomy, or three-body chromosome, instead of the typical two-body configuration. **b.** This difference results in the constellation of physical and cognitive symptoms associated with Down syndrome.

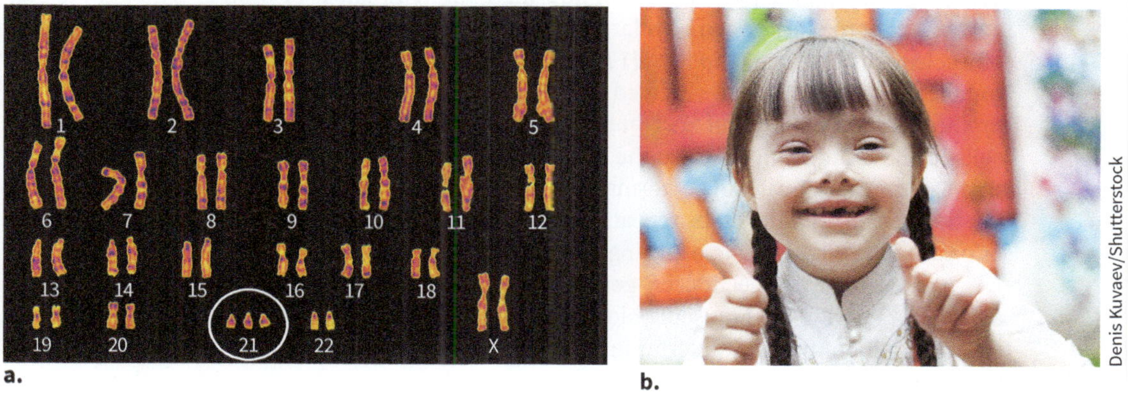

mutations are neutral with no impact on the organism. Sometimes mutations are beneficial and create adaptive traits that help a species survive. Maladaptive mutations can lead to disorders such as trisomy 21 (also known as Down syndrome). In such cases, a transcription error at chromosome 21 leads to three chromosomes, and the result is a collection of physical and intellectual traits that vary from mild to severe. **Figure 1.11** presents a karyogram and an image of an individual with trisomy 21.

In the overview of the mechanism of genetics, you saw that a complex process replicates the instructions for creating the proteins that make up you and me. You may also have come to visualize the initial division as an unzipping of a zipper that then reassembles itself by selecting unattached zipper pieces and fitting them into their appropriate place. Now we come to the rub. We all have had the experience of getting a zipper out of alignment and then struggling to zip it up. A similar displacement during the transcription process can have devastating results.

Huntington disease (HD), mentioned in the *Stories of the Brain* that opened this chapter, illustrates the dangers involved in an unfortunate displacement.

Individuals afflicted with Huntington's initially show symptoms of clumsiness and uncontrolled motor ticks. HD is a progressive disease, so over time, the severity of these symptoms increases as a result of damage to structures in the basal ganglia. In more advanced cases, atrophy of the cortex appears, accompanied by cognitive decline as well as psychiatric symptoms such as anxiety and depression. **Figure 1.12** depicts two coronal MRIs: one from a neurotypical individual and one from an individual with advanced HD.

FIGURE 1.12 **A neurotypical brain and a brain with HD** **a.** This MRI shows the brain of a neurotypical adult. **b.** In comparison, the brain of an individual with HD shows enlargement of ventricles at the expense of the basal ganglia in addition to widespread cortical atrophy. These brain changes explain the symptoms of motor and cognitive disorder associated with HD.

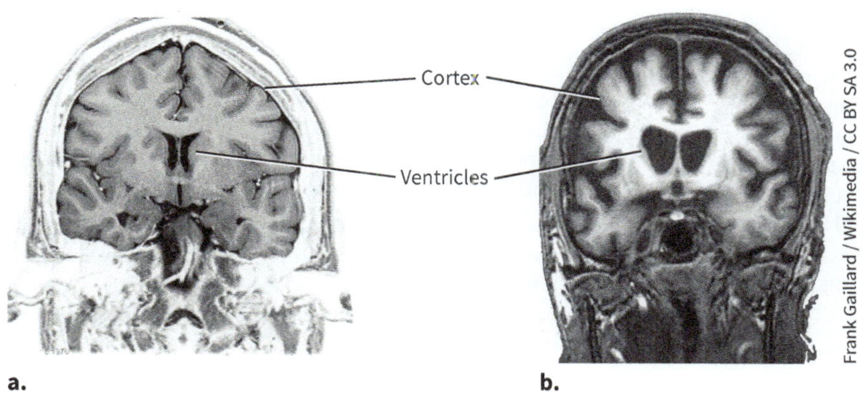

Let's take a moment to see how a transcription error can lead to HD. You know now that genes create proteins, and these proteins create us and regulate how our bodies work. Every human has two copies of a gene known as *HTT*, or the *huntingtin* gene at location 4p16.3, and this gene codes for the protein HTT, or huntingtin protein. The HTT protein is thought to play an important role in normal cellular processes, especially in the human brain. The sequence of nucleotides that make up the *HTT* gene varies in length among individuals with different numbers of repeats. If the repeated section exceeds a certain threshold, the protein that is produced is not the normal HTT but a mutant form known as mHTT. This protein damages cells in the brain, which causes the symptoms of HD. If the nucleotide repeat count is less than 26, the individual will be normal. Between 27 and 39 repeats, the individual may or may not be affected, but at 40 plus repeats, the individual will certainly develop HD, and the odds of passing the disease on to offspring are 50%.

As you read this section on genetics, you were likely amazed that a mechanism that arranges a handful of minuscule biologic compounds in the specific order we call DNA can result in something as wonderful and complex as you and me. As you learned more about genes, you realized that errors in the process can lead to serious disorders and even death when the order and arrangement of that DNA structure becomes disrupted, just as a computer file can become corrupted. But there are tools that can go into a corrupted file and recover the original data. Why can't science create a tool that allows us to repair DNA or even make improvements?

The answer is that scientists have been thinking along the same lines for some time, and a tool to modify genes has been created and is now being used in many beneficial ways (see **Advances in Behavioral Neuroscience: Doudna, Charpentier, and CRISPR-Cas9**).

In this section we have discussed the genetic mechanisms that convey inheritance while simultaneously enabling variation. In the next section we turn to the question of how random genetic variations lead to evolution.

Advances in Behavioral Neuroscience | Doudna, Charpentier, and CRISPR-Cas9

dpa picture alliance/Alamy Images

Scientists Jennifer Doudna of the University of California–Berkley and Emmanuelle Charpentier of the Max Planck Institute in Berlin, Germany, wondered if they could use nature to edit genes selectively. Imagine that a section of DNA contains a transcription error that causes a physiological or behavioral problem. In order to correct the error, we would need a tool that could locate the target string, cut that string out, and then splice a new string into its place.

Doudna and Charpentier saw that they could use a naturally occurring bacterial defense mechanism in a purposeful way to edit genes. When a virus invades a bacterium, its immune system captures snippets of DNA from the invading virus and uses them to create what are known as CRISPR arrays. The next time the bacterium encounters the same virus, it recognizes these CRISPR arrays and defends itself by creating an enzyme to cut the invader's DNA apart. Doudna and Charpentier recognized that they could use CRISPRs to identify a specific target in an organism's DNA. They engineered a snippet of RNA known as the Cas9 that could recognize CRISPRs and snip the sequence at the correct place. The cell's own repair mechanisms then add, delete, or change the target DNA by replacing the original DNA segment with a customized sequence. This tool is known as CRISPR-Cas9, or simply Cas9.

Cas9 has already been successfully used to genetically modify crops to increase their yield and make them more resistant to pests. The use of Cas9 in humans has been tightly regulated because there are significant risks involved in genetic manipulation. One use being actively explored is using Cas9 to treat sickle cell disease, a potentially fatal genetic disorder that affects millions of individuals worldwide, resulting in over 100,000 deaths annually. Experimental trials using Cas9 to treat eye disorders have also met with some success.

Drs. Doudna and Charpentier were awarded the Nobel Prize in Chemistry in 2020 for developing CRISPR-Cas9, a tool that has revolutionized genetic engineering and opened the door to treatments for many genetic disorders.

Questions

1. **Here is a partial list of genetic disorders that have serious if not fatal consequences: cystic fibrosis, Down syndrome, hemophilia, Huntington disease, Tourette syndrome, and Duchenne muscular dystrophy. Pick one to read about, and then answer these questions about the disease you chose: what are the most common symptoms, what is the approximate frequency of occurrence, and is there currently a reasonable treatment?**
2. **Genetic engineering is a two-edged sword in that it can have negative consequences. Think of a way that genetic engineering might be used that you would find unappealing or even illegal or unethical. Briefly explain why you came to this conclusion.**

1.2 Before You Go On

Section Summary

Mendel demonstrated that each parent contributes genes to their offspring, that some traits are dominant over other traits, and that we have information in our genotype that we do not express in our phenotype. Modern genetics shows that DNA is carried in chromosomes, and humans have 23 pairs of chromosomes, including one pair of sex-linked allosomes. The DNA transcription process introduces random variation into the genetic code of offspring, which might be adaptive, maladaptive, or neutral in their impact on survivability.

Comprehension Check

1. If I am homozygous for brown eyes, which statement is true?

- **a.** I will certainly have brown eyes.
- **b.** I may have brown eyes, but it's not certain.
- **c.** If my children's other parent has blue eyes, at least half of them will have blue eyes.
- **d.** I may carry the gene for blue eyes but not express it.

2. Place the following structures in order from largest to smallest.

- **a.** DNA, nucleotide, chromosome, gene
- **b.** gene, chromosome, DNA, nucleotide
- **c.** chromosome, nucleotide, gene, DNA
- **d.** DNA, gene, chromosome, nucleotide

3. Humans have _____ chromosome pairs, of which _____ decide(s) the sex of that individual.

- **a.** 12, 1
- **b.** 22, 2
- **c.** 23, 1
- **d.** 46, 2

4. What consequences do transcription errors in DNA replication have for an organism?

- **a.** a positive effect, a negative effect, or no effect
- **b.** a positive effect that leads to a new species
- **c.** a negative effect that leads to disease or death
- **d.** no meaningful effect

1.3 Evolution and the Environment

LO 1.3 Explain how evolution is affected by the environment.

The success of humans in their environment is due in great measure to the evolution of a complex and powerful brain. This brain enabled both skilled problem solving and complex social mechanisms. You and I might assume that logical and social functions are voluntary, when in fact they have a genetic basis. To make that point, we will discuss how our environment from 100,000 years ago shaped us to succeed in that world. At the end of this discussion, you may ask yourself if the behaviors that were adaptive 100,000 years ago on the African savanna are still adaptive today in today's world.

Natural Selection MCAT

We are surrounded by a panoply of species in the natural world. Plants and animals appear in an almost limitless variety of sizes, colors, and shapes. The question of how that all came about was considered by philosophers and scientists without a clear consensus until the naturalist Charles Darwin embarked in 1831 on a voyage around the world. Darwin observed that animals that were closely related had developed structural adaptations that uniquely served their environment and their particular needs. For example, on the Galapagos Islands, Darwin noted that finches on different islands had evolved specialized beaks that allowed them to feed in the unique environment each island provided (Figure 1.13).

FIGURE 1.13 Darwin's finches Although all four finches are closely related genetically, each evolved a distinctive beak in response to their food preferences as well as local environmental conditions such as drought or the impact of humans on that island. A long, pointed beak made some finches more fit for picking seeds out of the cactus flower, whereas shorter beaks were more efficient for eating seeds and insects found on the ground.

The basic principles of evolution apply to you and me just as they apply to finches in the Galapagos Islands. These principles are variation, inheritance, selection, and time. Variation in the form of genetic mutations naturally occurs in the DNA replication process, and that variation can be passed from generation to generation. Organisms with mutations that are adaptive tend to thrive, whereas those with mutations that are neutral or maladaptive tend to disappear in a relatively short period of time. In other words, random variations in DNA bring about physical changes in one's descendants, and the utility of these random variations is tested by the subsequent survivability of those descendants. This process is known as **natural selection**.

When coupled with spontaneous mutations, environmental pressures lead to genetic diversity. For instance, assume a random permutation in the genes of an organism that confers a 10% greater lung capacity than is common, and furthermore, this variation is passed on to offspring. It is likely that this variation will be adaptive because more oxygen in the bloodstream translates into greater vitality and energy, which means the descendants of that individual will have a greater probability of surviving to reproduce, thereby passing the same genetic mutation on to a subsequent generation. In fact, if the variation is highly adaptive, then the descendants of this one individual organism (all other factors being equal) will replace their competitors over time. Eventually, what once was an unusual adaptation will become the norm. An example of an evolutionary adaptation that conferred a benefit on humans is color vision.

The Evolution of Color Vision

The evolution of color vision in humans in a useful example of natural selection at work. In addition to depth perception and a keen ability to visually separate an object from its background, a third visual skill useful in the struggle to survive is color vision. Many animals have some form of color vision. Tropical fish, birds, and also insects can distinguish colors. In fact, some insects have better color vision than humans.

Many organisms get along just fine without it, so what evolutionary advantage does chromatic sensitivity confer? Our early ancestors may have developed color vision to better distinguish tasty fruit from the surrounding leaves. Another obvious advantage that color vision provides is a greater ability to distinguish food that is spoiled or unripe from food that is edible (**Figure 1.14**). Other organisms use their sense of smell to make this judgment, but humans rely on vision, so we need a visual cue.

A second hypothetical advantage to color vision is subtler. For those organisms that come into sexual receptivity on a cyclical basis, it is advantageous for the female to signal when she is fertile and receptive. It is equally advantageous for the male to know when his romantic interest is likely to be met with success or failure. Many female mammals and especially primates experience a change in color somewhere on their body that signals ovulation. One hypothesis is that the evolution of color vision is at least partially due to our ancestors using color as a signal for sexual receptivity. You might well respond to this hypothesis by noting that humans do not

FIGURE 1.14 **Advantages of color vision** Color vision provides an organism with additional information about ripeness because food that is unsafe to eat is often a different color from food that is ripe. Compare the view on the left in black and white with the same view in color on the right and decide which view provides the most information.

appear to change color when they are in the mood to mate. In a subsequent chapter, we will discuss the evidence that humans may respond to subtle perceptual cues that signal ovulation.

Among animals who perceive color, the range and sensitivity of color vision varies significantly. Species evolved different types of color vision in response to their particular environments and needs. You will discover in the chapter on vision that we construct our visual world based on the output of a set of highly specialized receptor cells and structures, some of which are designed to create lines and edges out of raw sensory data. This, in turn, helps us distinguish objects from their surroundings. Structures in the eye known as cones are responsible for color vision. Each type of cone is sensitive to a particular frequency (color) of light such as red, green, and blue. Having fewer cones means less variation among the colors you can perceive, whereas having more cones allows finer discrimination. More cones sometimes extend the range of sensitivity into new regions of the electromagnetic spectrum.

Paleobiologists believe that our vertebrate ancestors had four different cones. At some time around the emergence of the dinosaurs, however, our mammalian ancestors lost one of those cones. It is believed that this ancestor hid during the day and therefore had little need for complex color vision. We can imagine that a random genetic mutation caused the loss of that cone in an individual. Because the mutation was not maladaptive in that environment, it gradually spread through the population. Living descendants of the dinosaurs such as birds and reptiles (which share a common ancestor with dinosaurs) still have four-cone systems, whereas today's mammals in general only have three cones. The extremely sensitive color vision of birds and insects, some of whom have five cones, help a bee find a pollen source or a hawk sight a mouse in the leaves. **Table 1.1** provides an idea of the wide range of color perception across species.

Color vision is just one example of how humans have evolved for greater success in our environment. It may surprise you to learn that groups of humans can experience relatively rapid genetic adaptations to their local environmental conditions, the topic of our next section.

Adaptations to Extreme Environments

In this chapter, we have sped through literally billions of years of the history of the world and of the evolution of life on Earth. A reasonable person would assume that the time course of evolution is slow; certainly in the thousands of years if not longer. Surprisingly, changes in response to environmental pressure can be relatively rapid. We can look at human populations in extreme environments for evidence of rapid evolutionary adaptation. Two examples of this principle are the Inuit of the Arctic and the Sherpa of Nepal.

All humans need vitamin D to survive, and most of us get our daily dose from being outside in the sun. In fact, it is believed that skin pigment evolved as a protective mechanism to control the amount of vitamin D absorbed by the skin; humans living near the equator who enjoy long days with bright sun evolved darker skin, whereas humans living in more northern regions with shorter days in winter and less intense sunlight evolved lighter skin in order to absorb more vitamin D.

The Inuit live in the northern polar region, where the sun never sets for half the year and never rises for the rest. Vitamin D is essential for life, but it does not remain in the body for long. So how do they survive in this harsh climate given that there is almost no sunlight for six months to replenish their vitamin D supply?

TABLE 1.1 Color Vision across Species

Number of cones	Colors perceived (approx.)	Species
1	100	Marine mammals, some primates
2	10,000	Most terrestrial nonprimates
3	10 million	Most primates including apes and humans
4	100 million	Most reptiles, birds, and insects
5	10 billion	Some butterflies and some birds such as pigeons

FIGURE 1.15 Sherpa crossing a chasm Because their respiratory and cardiovascular systems have adapted to living at high altitudes, the Sherpa are able to climb with heavy packs despite low levels of oxygen.

Part of their adaptation has to do with their diet rich in fish and blubber, foods that have high levels of vitamin D. In fact, the levels of vitamin D in the Inuit diet are so high that they would be toxic to you or me. Now comes the interesting evolutionary twist: the Inuit have evolved a very inefficient process of metabolizing and storing vitamin D as a result of their naturally high dietary intake. If they were to metabolize vitamin D as efficiently as you and I do, they would become very ill from the high levels of vitamin D in their blood. Along the same lines, if they adopted the typical diet of a North American or European, they would become seriously vitamin D deficient. Over a relatively short period of time, the Inuit have evolved so that the combination of available sunlight and food sources gives them the right amount of vitamin D to survive and thrive.

A similar example of adaptation to extreme environments occurs at high altitudes. The Himalayas contain the world's highest peak, Everest (also known as Sagarmatha) along with more than 110 peaks that reach over 23,000 feet (7,100 m) into the sky. At such high altitudes, the air is thinner, and consequently, it is difficult to get enough oxygen through your lungs and into your bloodstream. This oxygen deficit causes low energy, shortness of breath, headaches, and even high-altitude pulmonary edema (HAPE), a condition that can be fatal if untreated. You may know that mountain climbers who ascend to altitudes about 16,000 feet (~5,000 m) may experience considerable difficulties, and above 20,000 feet (~6,000 m) every person experiences some mental confusion. Few climbers can reach the summit of Everest without supplemental oxygen.

Expeditions to the top of Everest engage an indigenous ethnic group known as the Sherpa to serve as porters and climbing guides. Some Sherpa come from tiny hamlets situated between 14,000 and 16,000 feet (4,267–4,877 m). The Sherpa can climb to about 20,000 feet without oxygen while carrying loads of 50 pounds or more on their back, whereas highly athletic professional non-Sherpa climbers with oxygen tanks struggle and sometimes perish while attempting to lift just their own body weight to the top. Only a few non-Sherpa have reached the summit of Everest more than once, but some Sherpa (**Figure 1.15**) have summited Everest many times, and sometimes without supplemental oxygen.

Archaeological data suggest that the Sherpa moved to the Nepalese region within the last 3,000 years. They then colonized the highest areas of Nepal in the last 300–400 years. This means the physical adaptations that allow the Sherpa to live at high altitudes have evolved within the past 3,000 years.

The remarkable capability of the Sherpa to climb at high altitudes without supplemental oxygen has to do with physical adaptations in their pulmonary and cardiovascular systems. Compared with other ethnic groups, the Sherpa have a larger lung capacity and important advantages in their circulatory system allow them to climb at altitudes where the low air pressure and oxygen levels would render many people unconscious. For example, their muscle cells' mitochondria (which create energy to power the cells) are more efficient, and their ability to forcefully push air out of their lungs (known as peak flow) is exceptional (see **Try It Out: Are You Ready to Climb Everest?**).

Try It Out | Are You Ready to Climb Everest?

You cannot easily measure your mitochondria or pulmonary peak flow, but you can compare your lung volume to that of the Sherpa as one measure of high-altitude fitness. Download the activity **Are You Ready to Climb Everest?** to compare your lung volume to that of the world's best high-altitude athletes.

Evolution and the Human Diet

Another example of how environment has shaped our evolutionary history is related to the availability of different food sources.

Fat and sugar If you are like most adults, you struggle somewhat to maintain a healthy weight. Why do unhealthy foods taste so good? Why doesn't the average person find a rice cake, tofu, or a low-fat wheat biscuit as pleasing as a pint of chocolate ice cream or a double patty bacon cheeseburger with a side of fries? By now you should realize that the real question is not why unhealthy foods taste good, but instead what in our evolutionary past has caused humans with a genetic profile that included a preference for fats and sugars to be more likely to survive and reproduce than those without such a preference.

Humans have a large brain that demands a substantial supply of calories as fuel. It makes sense that humans who have a preference for energy-rich foods are more likely to survive. Two foods that are compact sources of energy are sugars and fats, both of which provide a great deal of calories in a small package. In the African savanna that was home to our distant ancestors, where would you find fats and sugars? Fats would most likely be a part of an animal kill. If the image of meat that comes to your mind is a nice steak marbled with fat lying in its plastic tray at your local market awaiting your pleasure, think again. Every animal was free range back in the day. Early humans spent their time running away from predators and running towards prey. Even though our ancient ancestors would find little fat in their meat, any individual who had a preference for the taste of fat would have access to a compact source of energy unavailable to those who avoided it.

Obtaining sugars would be somewhat less problematic than finding fats, but it would still require effort. Fruit and honey were typical sources of sugar in our past. Fruits contain the sugar fructose. Ancient fruits would be considerably smaller and have less food value. That being said, those ancestors who favored the taste of calorie-rich foods such as fruits and honey and were motivated to seek them out would have an advantage over those who did not. Wild fruit was often available to our ancestors for the taking, but obtaining honey was quite a chore before beekeeping was invented. If you were a person with a sweet tooth that longed to be satisfied, you might have carefully weighed the dangers involved in robbing a beehive in order to satisfy that desire. Evolution works for bees just as it does for humans, and hives that defended their honey aggressively survived to reproduce and pass on their defensive genes.

We will return in a later chapter to the question of food preferences and our brain, but for now you can simply understand that human food preferences today have a genetic basis in the drive of early humans to survive and reproduce. With this in mind, you can see that simply saying to a client who is struggling with obesity that they should choose to eat less fattening food is akin to urging the tide to roll back out and stay there. There are ways we can use our understanding of how our brain interprets tastes and signals satiety to our advantage, but we must learn to work with our body and brain and not against it.

Lactose intolerance Another example of how evolution and the human diet interact is in the tolerance of lactose, the sugar found in milk. Individuals with lactose intolerance experience significant gastrointestinal distress, vomiting, or cramps when they drink milk and, to a lesser extent, eat milk products like butter and cheese. Lactose intolerance is caused by a deficiency of the enzyme lactase, which enables the absorption of lactose through the intestinal wall.

Most humans can digest milk during infancy, but as the infant matures and is no longer nursing, the genes that control lactase production are turned off. Without lactase, lactose passes undigested into the colon, where it becomes a fertile breeding ground for bacteria, leading to illness. The gene that controls lactase production, *2q21* or the *LCT* gene, is located on the long arm (q) of chromosome 2, at location 21. When the gene is turned on, the ability to create lactase persists past infancy to allow digestion of dairy products into adulthood.

Lactase persistence had little utility for most of our history as nomadic hunter-gatherers, but about 10,000 years ago some human groups began to domesticate cattle. Once milk was easily available, adults who had the genetic makeup that allowed them to digest lactose had an additional food source. Those lucky few were more likely to survive, reproduce, and pass this mutation on to their offspring. Although lactose intolerance is the genetic default across much of the world, lactase persistence is more frequently found in ethnic groups that developed dairy farming early. For this reason, lactase persistence is found among roughly 80% of Europeans and Americans of European ancestry, whereas it is relatively rare among the Chinese and individuals whose origin is sub-Saharan Africa (**Figure 1.16**).

Evolution occurs when adaptive genetic variations in a species persist over generations. This is true not only for Galapagos finches and pea plants but also for humans as the vitamin D metabolism of the Inuit, the unusual physiological ability of the Sherpa, and the emergence of lactose tolerance demonstrate. It seems fair to say that all species, including humans, will continue to evolve over generations in response to environmental pressures.

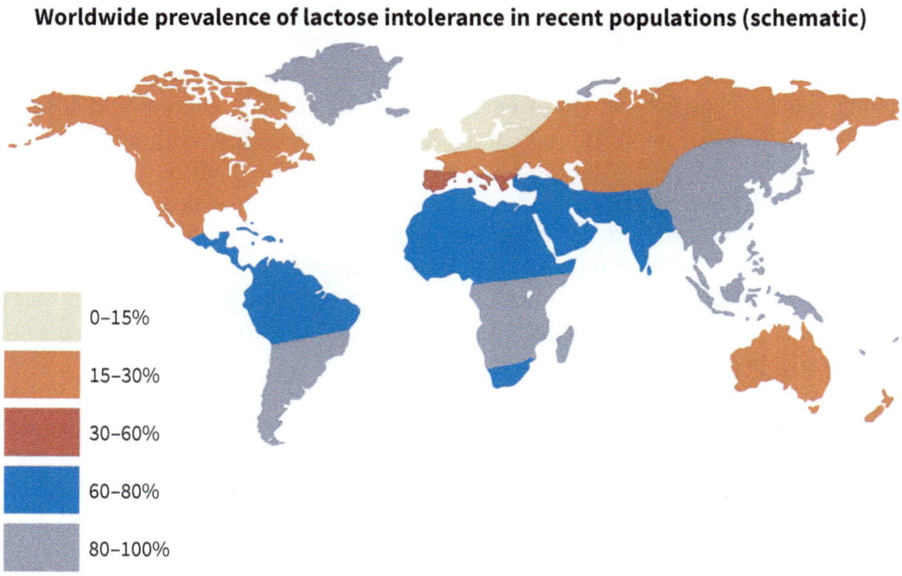

Food Intolerance Network, 2013 / Food Intolerance Network.

FIGURE 1.16 Global prevalence of lactose intolerance Around 8,000 years ago, humans in some regions began to domesticate cows, sheep, and goats. In those ethnic groups, the ability to digest lactose provided an advantage. Ethnic groups that did not raise animals for their milk make up the rest of the world's population, and they did not experience pressure to evolve this adaptation.

1.3 Before You Go On

Section Summary

Natural selection is the evolutionary process through which random DNA mutations that are adaptive are passed on and those that are maladaptive die out. Over time, humans have evolved adaptations to help them survive in their environment, including color vision, which makes it easier to find food. Some human populations have evolved relatively recently in response to extreme conditions such as living in low sunlight or at high altitude. In just a few hundred years, physical adaptations such as lactose tolerance enhanced survival and began to take hold in specific groups, and these adaptations persist today.

Comprehension Check

1. According to Darwin's theory of natural selection, organisms that survive to reproduce are the ones best adapted to their environment because they have _____.

- **a.** more strength and ferocity
- **b.** little in the way of competition from other species
- **c.** adaptations that maximize their fitness for that environment
- **d.** attributes and traits developed through experience

2. One advantage that color vision confers on humans is greater ability to _____.

- **a.** detect a predator in the bush
- **b.** distinguish ripe fruit from its leafy background
- **c.** distinguish clan members from strangers
- **d.** perceive depth while moving about

3. The Inuit and Sherpa demonstrate unusual physiological abilities as a result of what?

- **a.** medical advances
- **b.** environmental pressures
- **c.** many years of training
- **d.** culture that values certain behaviors

4. What is the evolutionary behaviorist understanding of the appeal of foods like donuts, candy bars, and cookies?

- **a.** Fats and sugars taste good, so we evolved the means to make these foods.
- **b.** Advertising induces us to eat fats and sugars.
- **c.** Those ancestors who preferred fats and sugars were more likely to survive.
- **d.** Those ancestors who learned to cook fats and sugars had an advantage.

1.4 The Evolution of Social Behavior

LO 1.4 Describe the genetic basis of social behavior.

Physical attributes such as height, skin, eye color, and the ability to digest lactose are all attributes of individuals that fit nicely within the common view of genetic inheritance. Now comes an idea that may be quite startling to you: behavior can also be heritable. This does not mean that if you favor peppermint ice cream, Belle Époque art, or Norwegian thrash metal, your children will share the same passions. What it does mean is that some aspects of social behavior have a genetic component to them. Genetic transmission of behavioral propensities to one's offspring can affect the survivability of descendants as surely as quail having coloring that helps them blend into the brush helps them survive.

FIGURE 1.17 Dimitri Belyayev with foxes Wild silver foxes are aggressive and do not stand being handled by humans. By selectively breeding those foxes in each litter that were most approachable, over some years, Dimitri Belyayev bred silver foxes who enjoyed interacting with humans.

Evidence for the Heritability of Behavior

Early evidence of the heritability of behavior came in the 1950s from the Russian geneticist Dmitri Belyayev, who was given the task of breeding silver foxes for their pelts. Silver foxes are wild animals that bite when handled. Normally, a pelt farmer would keep all the offspring from each litter, but Belyayev kept only the animals from each litter that were most willing to tolerate handling by humans. Belyayev repeated this process of selecting for docile foxes with each successive litter. By ten generations, the majority of the pups born from his breeding stock enjoyed being handled and even licked their keepers (**Figure 1.17**).

The behavioral changes in the foxes were impressive, and the physical changes that accompanied the increase in docility was just as startling. The foxes began to exhibit the physical characteristics of dogs. Their faces changed, their ears were floppier, and the foxes came to resemble domestic dogs more than their feral fox ancestors.

The evolution of domesticated dogs is an example of natural selection of behavioral traits that occurred over millennia. Research that compares the DNA of modern dogs to wolves suggests that dogs diverged from wolves at least 40,000 years ago. The story might go that at some point in the distant past, a band of wild dogs approached a human campfire, and one dog that demonstrated more curiosity than aggression was rewarded with food and attention. Perhaps this one dog came back night after night for scraps, and an adventurous hunter decided to see just how close they could bring the dog to the fire. The dog then gave birth, and the clan decided not to eat the friendliest pups but instead to raise them. Any pup that was aggressive went into the stew, but the more docile pups were kept because our human ancestors noticed the dog pups provided a warning system by reacting to the presence of unseen predators out in the darkness. Over generations of canine pups, the innate traits of aggression and wariness toward humans were discarded, and traits such as docility and attachment to humans were selected.

Over the past few thousand years, humans have selectively bred dogs for both physical and behavioral traits. As a result, dog fanciers can choose from mastiffs, which may weigh 150–200 lbs (68–90 kg), and Chihuahuas, which weigh as little as 7 lbs (3.2 kg) as well as dogs with strong behavioral traits such as sheep dogs and other herders that are always looking to keep their entire family in sight (**Figure 1.18**).

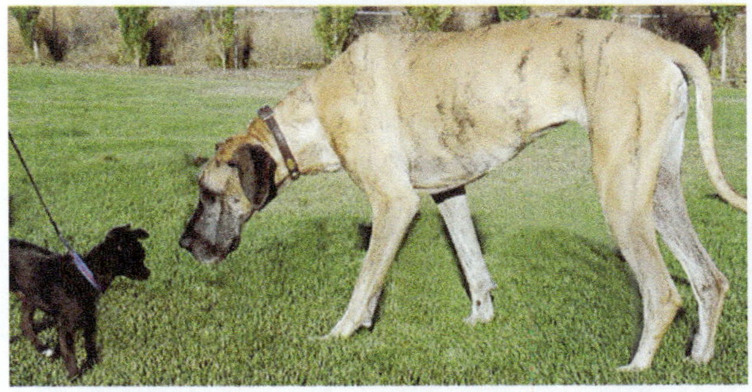

FIGURE 1.18 Divergence due to selective breeding All modern dogs as well as modern wolves share a common ancestor. The significant differences in appearance and behavior of this Great Dane and this Chihuahua are the result of selective breeding for certain traits and behaviors.

Archaic canines were wild animals that were hostile to humans, but due to selective breeding on the part of humans, they changed both physically and behaviorally. This fact demonstrates that not only physical traits such as height, hair, and skin color but also behaviors have a genetic basis and may evolve over time in response to environmental pressures.

The Survival Benefits of Social Behavior

In our overview of the evolution of humans and how that process shapes our behavior today, let us return to the savanna of 100,000 years ago, this time paying special attention to potential prey and predators. Take a moment to examine the landscape of the Rift Valley in Africa, the place where the fossil evidence suggests that humans evolved (**Figure 1.19**). There are few places to hide from predators, and prey can see you coming from a long way off.

It doesn't take long to realize that we humans were in serious hot water. Compared to our competitors in the wild, humans are neither strong nor fleet of foot. We don't have claws that can rip and tear, nor do we have fangs that can deter a predator or bring down a tasty wildebeest. Even a very fit runner can attest to the fact that most dogs can catch a human in a race of less than a few hundred yards. When you get right down to it, we can neither outrun the competition nor outfight it in single unarmed combat. How then did we ever survive?

There are several answers to this question. One is that humans have a large brain in relation to the size of our bodies, which supports complex cognitive skills such as abstract thought and problem solving. Another answer to the puzzle is that humans are social beings at our very core. Our large brain allows us to create new ways to adapt and survive by being social and not solitary, and we use that larger brain to coordinate our actions with others. A single human standing alone and empty handed in the African savanna has a very short life expectancy. However, humans who can create tools and also organize their behavior with other clan members are quite the match for any predator.

Humans share many characteristics with other mammals and primates, but we often express them in an exceptional fashion that makes us uniquely suited to complex social interactions. So, what cognitive skills support social behaviors in humans? Entire textbooks have been written on this topic, but for the purposes of our discussion, we can identify four general skills that distinguish humans from other animals and form the foundation of human social behaviors: memory, problem solving, language, and theory of mind.

First, many animals recognize litter or clan members, but humans' visual and memory capabilities allow us to recognize thousands of other people, and that ability enables reciprocal obligations. For example, humans can be aware of which clan members have cooperated with us in the past, which individuals have been uncooperative, and which ones request our help but do not reciprocate. Memory also helps build a database of possible food sources.

Joseph Sohm-Visions of America/Getty Images

FIGURE 1.19 The Rift Valley in Africa Our earliest human ancestors evolved in a landscape like this. The climate was hot, there was not a great deal of vegetation, and both predator and prey had to be resourceful to survive.

Second, memory creates the data for skilled problem solving. Problem-solving behavior in humans relies on the ability to think abstractly, plan complex behaviors, seek out new solutions to problems, and manipulate information gained from prior experience.

Third, the ability to communicate permits coordination of group action at a distance. Many organisms can communicate intentions through either gestures or utterances, but humans have developed language, which permits the development of culture, enabling us to pass on experience and wisdom from one generation to the next.

All of these skills are important, but the most important might be what psychologists call **theory of mind (TOM)**, a phrase that refers to the ability to put oneself in another's shoes and gauge intentions or goals that may or may not be readily apparent. If you are a good bluffer in poker or recognize that a person who is smiling at you may not have your best intentions at heart, you have a grasp of the importance of TOM.

Now we have established why social behavior is adaptive and which cognitive skills support that behavior. Next we will turn to how our evolutionary past informs our behavior today.

The Biological Basis for Modern Behavior

In subsequent chapters, we will touch upon some interesting questions that consider how evolutionary pressures on our brain to adapt might play a role in guiding current human behavior in ways that will likely surprise you. For now, we will just touch briefly on a few examples of how evolutionary principles affect our social behavior today.

An obvious example of social behavior that is influenced by evolution is how humans choose a reproductive partner. What does it mean for a person to be physically attractive? We all can pick attractive individuals out of a crowd, but what physical features are we evaluating when we make that decision? From an evolutionary perspective, attractiveness in itself is irrelevant; what matters is selecting a healthy and fertile partner. It would be advantageous for any physical or behavioral traits that signal health and reproductive fitness to be translated into what you and I call attractiveness.

The ability to read the intentions of another and to present a false presentation of one's own intentions are examples of complex human skills that would have been favored by evolution. From an evolutionary perspective, it makes sense that some individuals would learn to project sincerity while plotting something in their own self-interest. The modern practices of catfishing and ghosting can be seen as examples of the continuing struggle of each organism to gain an advantage over competitors, however slight. In the same vein, however, humans have developed lie-detection abilities for their own protection.

It's important to realize that evolution has no purpose or goal; it is a process wherein random variations in the genetic code bring about changes in physical structures or a propensity toward a certain behavior. Some changes are deleterious and lead to the death of the individual or even the disappearance of an entire species, and other changes have no apparent effect. Some changes are adaptive and confer an advantage on that individual; if the trait is passed on to offspring, the odds of their survival are enhanced.

Next, we want to keep in mind that from the perspective of each species, life is a competitive struggle. One organism develops a claw or fang, and in response, its prey develops an adaptation that allows it to run faster than before to avoid that claw or fang. One organism discovers a new environment filled with prey, and in response, those prey develop a taste that is sickening to the predator. For every snake that can strike like lightning, there is a mongoose slightly faster. Acacia trees evolved long thorns to protect their leaves, and in response, giraffes evolved tough but labile lips to remove those succulent leaves without injury.

Finally, we need to remember that evolution is a continuous process that is still happening today. Traits that were adaptive 50,000 or 100,000 years ago may no longer be beneficial in today's world. Our genetic makeup was forged in the crucible of an open savanna, but for the vast majority of humankind, that environment is no longer our home. See **Check It Out: My Genetic History** to see a part of the genetic history of one individual, your author.

Paleobiologists believe that our ancestors from 20,000 or more years ago likely never encountered more than 500 people in their entire lives. You might be in the presence of more people than that every day, and that change in social environment has implications for human behavior. A powerful need for affiliation to clan members over strangers may have been useful in our evolutionary past, but in today's multicultural world, important members of our so-called tribe may not look the same as you and I and yet be quite reliable and trustworthy. A distinct preference for foods that are high in energy was adaptive in our past when such foods were scarce or difficult to obtain; today, unlimited access to snack cakes and fried pork rinds works against our health. One way to capture the issue is as follows: you and I struggle the problems of a nuclear world, but we do so with a Stone Age brain.

Check It Out | My Genetic History

There are several companies that provide an ancestry analysis based on a DNA sample to better understand the history of the human family. Watch the brief video **My Genetic History**, and answer these questions:

1. If you had the opportunity to have your own genetic history revealed, would you choose to do so? Explain your answer.
2. In your own view, does your genetics or your environment play a greater role in guiding your behavior? There is no right or wrong answer, but briefly state your opinion.

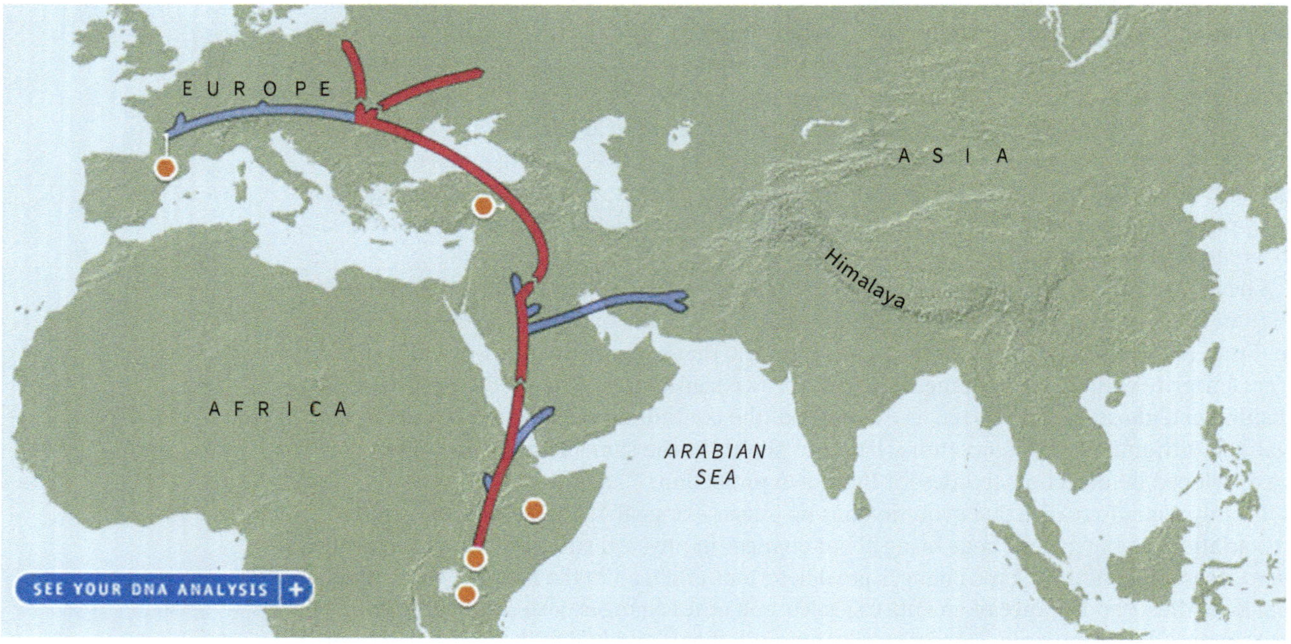

1.4 Before You Go On

Section Summary

Behavior has a genetic basis just as physical traits like skin and hair color do, as is demonstrated by the ability to breed friendly foxes or dogs. Social behavior provides survival advantages as human memory, language, and theory of mind allow us to cooperate with other humans for mutual benefit and defense. Modern behaviors such as who we find attractive or our propensity to lie are rooted in our evolutionary drive to compete and survive.

Comprehension Check

1. Which of these is an important insight from evolutionary theory?
   - **a.** Humans are no longer evolving because we are the dominant species.
   - **b.** Evolutionary change occurs through the females of the species.
   - **c.** Evolutionary change can occur in as little as one generation.
   - **d.** Behavior can also be a part of evolutionary change.
2. Which of these responses demonstrates theory of mind (TOM)?
   - **a.** I am hungry, so I reach for an apple.
   - **b.** A friend says they are hungry, so I hand them an apple.
   - **c.** A friend just lost their expensive phone. They are smiling, but I try to cheer them up.
   - **d.** A friend is smiling as they watch a comedy, and I assume they enjoy the show.
3. Select the statement that is NOT true of the world of our ancestors 20,000 years ago.
   - **a.** They found it difficult to obtain energy-rich foods.
   - **b.** They enjoyed close social bonds with thousands of clan members.
   - **c.** They likely led a more physically strenuous life than we do.
   - **d.** They were likely to encounter fierce animal predators.

Conclusion

We began this chapter with the story of Nancy Wexler, a scientist with a personal interest in Huntington disease, a progressive neurological disorder with a genetic root. Our ancestors evolved in a world that no longer exists, but their genetic heritage still influences us today. In this chapter, we considered two important and intertwined themes: the person we are today has been shaped by the environment of our ancestors through evolution, and the mechanism that creates that change is still active today. Genetics interact with the environment to create who we are, physically and behaviorally. Some behavioral tendencies that were necessary for our ancestors' survival are no longer adaptive in the modern world.

In the past, environmental pressures affected the human genome over hundreds or thousands of years, but technological advances such as Cas9, which have the promise of curing diseases such as Huntington disease, create the possibility of rapid genomic change in a generation or two. This type of genetic change would not be shaped by environmental pressures but by personal preferences.

As you continue your study of behavioral neuroscience, keep two questions from this chapter in mind: why might evolution have favored a behavior, and is this behavior still adaptive in the modern world?

Critical Thinking Questions

1. Crocodiles and mosquitoes have hardly changed over the last tens of millions of years, yet you and I look nothing like our distant ancestors. Why do you think that might be?
2. If you had to choose two characteristics of primates that you think are most important for humans, which two would you choose and why?
3. Search online for the phrase "Belyayev foxes" and read more about his experiment. Do you agree with his conclusion that behavior and physical appearance may be directed by the same genes? Why or why not?
4. The DNA replication process is designed to create a copy that may have small alterations from the original. Why are those alterations an essential part of the evolutionary process? If the replication process always made an exact copy of the parent genetic code, would that make humans more or less adaptable?
5. What would you do if you discovered that Huntington disease ran in your family? Would you want to know if you carry the disorder, or would you prefer not to know? Would you choose to not have biological children, or would you take the risk? Briefly explain your answers.

Dig Deeper into Behavioral Neuroscience

1.1 Human Origins

The following sources explore the origins of humans. The Bronowski text, although a bit dated, is well written, and a BBC series accompanies it. The two Darwin texts are classics, and the Harari text is a very modern perspective.

Bronowski, J. (1973). *The Ascent of Man*. Boston: Little Brown.
Darwin, C. (1859). *On the Origin of Species*. London: John Murray.
Darwin, C. (1871). *The Descent of Man*. London: John Murray.
Harari, Y. (2018). *Sapiens: A Brief History of Humankind*.

1.2 The Genetic Basis of Evolution

A general introduction to genetics:

Genetics primer. (2012). *Alcohol Research: Current Reviews, 34*(3), 270–271. https://www.ncbi.nlm.nih.gov/pmc/articles/PMC3860415/

To learn more about the field of behavioral genetics:

Knopik, V. S., Neiderhiser, J. M., DeFries, J. C., & Plomin, R. (2016). *Behavioral Genetics* (7th ed.). New York: Worth.

Understanding DNA replication can be confusing, so the video **DNA Replication** (available with this book) illustrating the process may be helpful.

1.3 Evolution and the Environment

The genetic basis for high-altitude adaptation:

Havryk, A. P., Gilbert, M., & Burgess, K. R. (2002, August 30). Spirometry values in Himalayan high altitude residents (Sherpas). *Respiratory Physiology & Neurobiology, 132*(2), 223–232.
Greksa, L. P. (1996, February). Evidence for a genetic basis to the enhanced total lung capacity of Andean highlanders. *Human Biology, 68*(1), 119–129.

To read more about lactose intolerance, its causes, and its treatment:

Malik, T. F., & Panuganti, K. K. (2020). Lactose intolerance. *Statpearls.* https://pubmed.ncbi.nlm.nih.gov/30335318

1.4 The Evolution of Social Behavior

This article explains how when wolves entered into a partnership with humans, their offspring eventually became household pets.

Savolainen, P., Zhang, Y. P., Luo, J., Lundeberg, J., & Leitner, T. (2002, November 22). Genetic evidence for an East Asian origin of domestic dogs. *Science, 298*(5598), 1610–1613. https://doi.org/10.1126/science.1073906

A key concept in behavioral neuroscience is that biology and behavior are intertwined. When a behavioral trait such as docility is selected for, physical traits also change.

Wade, N. (2006, July 25). Nice rats, Nasty rats: Maybe it's all in the genes. *New York Times*.

Chapter 1 Before You Go On Answers

1.1 Before You Go On Answers

1. b. slow change interrupted by occasional dramatic change

2. c. warm-bloodedness, lactation, and live births

3. c. Humans evolved from monkeys.

4. d. cerebral cortex has more neurons

1.2 Before You Go On Answers

1. a. I will certainly have brown eyes.

2. b. gene, chromosome, DNA, nucleotide

3. c. 23, 1

4. a. a positive effect, a negative effect, or no effect

1.3 Before You Go On Answers

1. c. adaptations that maximize their fitness for that environment

2. b. distinguish ripe fruit from its leafy background

3. b. environmental pressures

4. c. Those ancestors who preferred fats and sugars were more likely to survive.

1.4 Before You Go On Answers

1. d. Behavior can also be a part of evolutionary change.

2. c. A friend just lost their expensive phone. They are smiling, but I try to cheer them up.

3. b. They enjoyed close social bonds with thousands of clan members.

Chapter 1 Glossary

allele A variant form of a specific gene, for example, one that codes for male-pattern baldness or not.

allosome The chromosome that determines the sex of an individual; also known as the sex chromosome.

autosome A chromosome that is not sex linked.

biome A naturally occurring community of flora (plants) and fauna (animals) that inhabit a major habitat such as a forest, tundra, or shoreline.

bipedal Travelling by two legs or rear limbs.

chromosome A physical structure comprised of DNA and other material that contains the instructions for creating proteins.

Cretaceous–Tertiary (K–T) event The sudden mass extinction event that extinguished three-quarters of the aquatic and land-based species on Earth approximately 66 million years ago.

DNA (deoxyribonucleic acid) The molecule that carries the basic genetic instructions for making an organism.

dominant trait A variant or allele of a gene that overrides or masks the effect of a different variant.

gene The physical region on a chromosome that operates as one functional unit of heritability.

genotype The genetic makeup of a cell that determines its observable characteristics, or phenotype.

heterozygous Having two different alleles of a gene.

homozygous Containing two identical alleles of a gene.

Huntington disease (HD) A progressive neurological disorder that causes involuntary jerky motor movements and cognitive disturbances.

karyogram An image of the full set of chromosomes for an organism.

lactation The secretion of milk from mammillary glands that is common among mammals.

mammal A class of vertebrates who have the following characteristics: the presence of mammary glands in females (from which they get their name), a neocortex, fur or hair, and warm-bloodedness.

messenger RNA (mRNA) A type of RNA molecule that collects genetic information from DNA and conveys it to a ribosome where proteins are constructed from amino acids.

natural selection The process by which organisms that are better adapted to their environment are more likely to survive and produce offspring.

nucleotide A unit of DNA consisting of a base, a sugar, and a phosphate group.

Permian–Triassic extinction (P–T) event The mass extinction that occurred about 250 million years ago, causing the disappearance of an estimated 96% of marine life and 70% of land vertebrates.

phenotype The observable characteristics of an organism that are the expression of its genotype.

primate One of a diverse group of mammals that are bipedal with an opposable thumb, a large brain, and stereoscopic vision.

protein A large molecule comprised of amino acids that perform essential functions such as catalyzing metabolic reactions, replication, and creation of structures.

recessive trait The weak or unexpressed trait in a dichotomous pair of alleles that has no effect on the phenotype of the organism.

ribosome A molecular machine found in all living cells that is the site of protein synthesis.

theory of mind (TOM) The ability to recognize that others have emotions, desires, intents, and wishes different from one's own, which is a sign of complex cognitive development.

CHAPTER 2

The Neuron and Action Potential

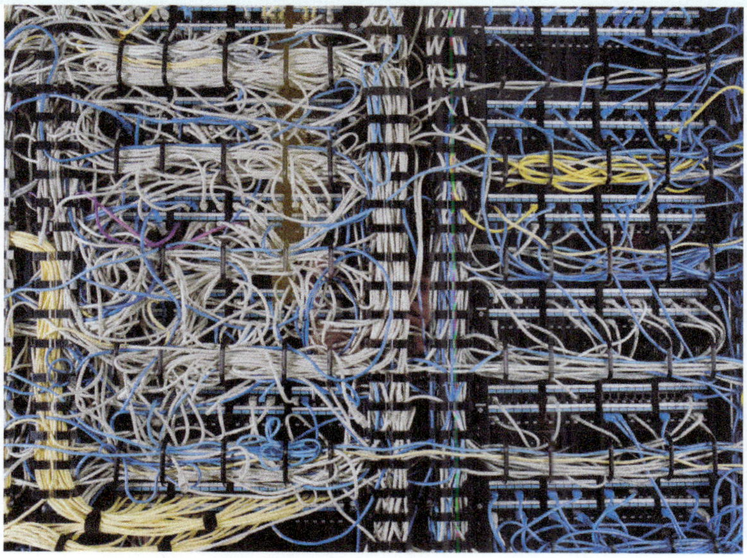

Leland Bobbe/DigitalVision/Getty Images

Complex information processing in the human nervous system requires elaborate connections just like a computer system.

CHAPTER OUTLINE

2.1 An Introduction to the Neuron

- An Overview of the Nervous System
- The Discovery of the Neuron

Check It Out: Staining a Neuron

2.2 The Structure of a Neuron

- Eukaryotic Cell Structure
- Neuron Components

Advances in Behavioral Neuroscience: Mark Harnett and the Role of Dendrites

- Types of Neurons

Check It Out: Guillain-Barré Syndrome

2.3 Neural Transmission of a Signal

- Energy Produced through Imbalances
- Ion Exchange
- The Sodium-Potassium Pump

Think About It: Disrupting the Sodium-Potassium Pump

- The Action Potential
- The All-or-None Law

2.4 White Matter, Gray Matter, and Glia

- White and Gray Matter

Think About It: White or Gray Matter?

- Glia

LEARNING OBJECTIVES

LO 2.1 **Understand** the importance of the neuron and why it was a mystery for so long.

LO 2.2 **Describe** the characteristics of different types of neurons.

LO 2.3 **Describe** how the sodium-potassium pump and the action potential propagate a neural signal.

LO 2.4 **Identify** the individual roles of white matter, gray matter, and glia.

Stories of the Brain

Maria was a mother of two young children who worked part-time in a bank. Her life was like that of any young parent, divided between work and family responsibilities. That all changed one fateful day.

Maria had always been an energetic individual, able to juggle the demands of work and home with ease and never too tired to take in a movie or see friends after the workday was done. But now she often felt weak and listless, too tired to go to her job or play with her children. This alone was worrisome, but there were other disturbing signs. Some days her skin felt prickly as if it were covered with ants, and at other times, a leg or an arm might feel numb. Although she had always been nimble, when she was tired, she could lose her balance and stumble.

Maria assumed her problems were due to a flu that was going around, but her symptoms never completely disappeared. A few days or even weeks would go by without symptoms, and then her problems would return. The day she realized that something was seriously wrong was the day she recognized that her vision had changed. She had been grumbling that her house was dark, so she replaced the light bulbs in her fixtures with higher wattage bulbs. Eventually, she noticed that the problem wasn't her light bulbs—something was wrong with her left eye. Everything slightly above the middle of her field of vision was somehow darker and less detailed than the rest. When she understood that her difficulties in reading the paper and seeing the television were due to her eyes, she made an appointment to see her ophthalmologist, who asked her about her symptoms and her family history. Her ophthalmologist sent Maria to a neurologist, who ordered a series of MRI images of Maria's brain. The diagnosis was multiple sclerosis (MS), a disease that disrupts the communication between neurons and can lead to vision loss, fatigue, loss of muscle coordination, and balance problems. MS is a chronic disorder and cannot be cured, but many MS patients live well into their 70s. Maria has responded well to drug treatment and lives a full life as a teacher and a parent.

What do Maria's fatigue, loss of balance, and vision problems have to do with the neuron? In this chapter, you will see that proper functioning of neurons is the basis for all cognitive and motor functions. When neural communication breaks down, as it does in MS, the results can be devastating. As we move ahead in our discussion of behavior, having a solid grasp of the neuron as the basic unit of communication will help you understand how the constant information sharing among our bodily systems keeps us alive and allows us to thrive.

2.1 An Introduction to the Neuron

LO 2.1 Understand the importance of the neuron and why it was a mystery for so long.

In neuroscience, the basic unit of the biology of your mind is the neuron. In trying to understand how the brain creates behavior, we must first grasp the fundamentals of the neuron as a unit. If you understand how a neuron works, that knowledge will allow you to understand the biological basis of how our brain regulates normal bodily functions such as blood pressure, hunger and thirst, and also higher order behaviors such as cognition, emotions, and neurological and psychiatric disorders. You may be surprised to discover that most of what we know about the neuron was discovered within the last 200 years, and important discoveries are still being made today.

An Overview of the Nervous System

We know from our study of evolution that behavior can effectively increase the odds of surviving and reproducing, and for behavior to occur, a **nervous system** is necessary to take in sensory information, make decisions, and coordinate actions in response to sensory input and decision processes. All vertebrates, that is, animals with a spine or spinal cord, share a two-part template

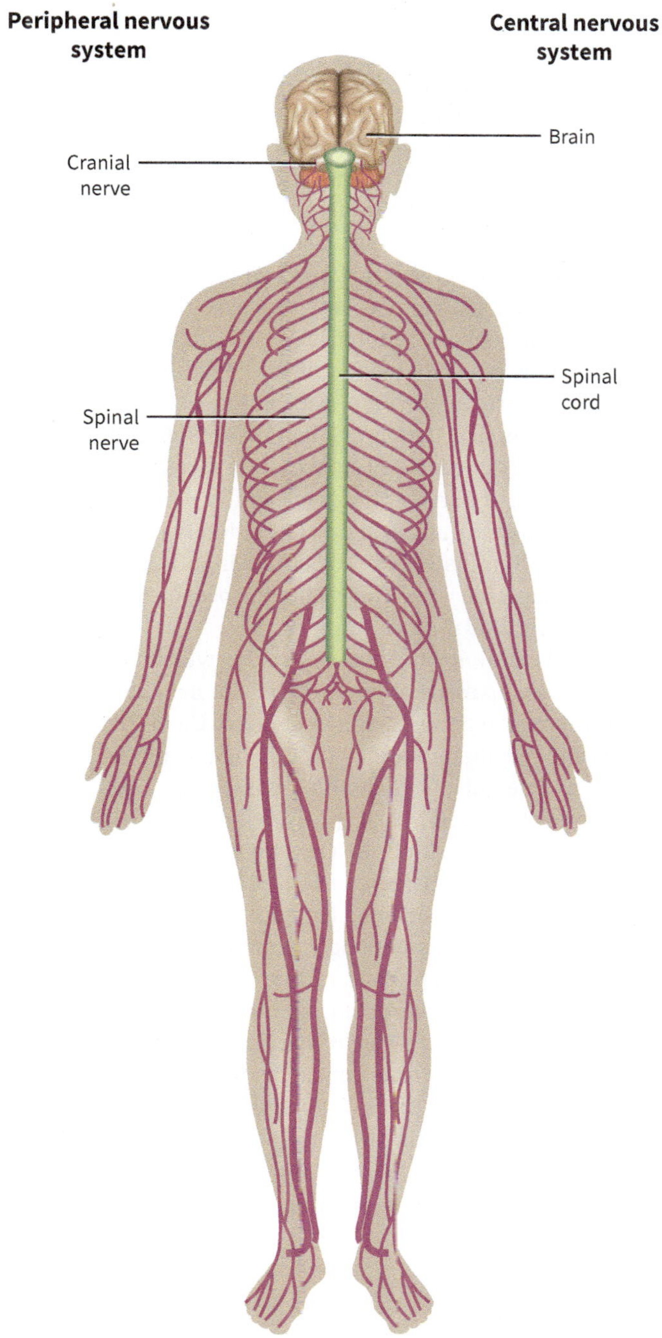

FIGURE 2.1 The human nervous system The CNS is comprised of the brain and spinal cord. The PNS is made up of an extensive network of cranial and spinal nerves originating in the spine that form the essential link between the CNS and the body.

for a nervous system: a central nervous system (CNS) and a peripheral nervous system (PNS). The CNS consists of a brain and spinal cord, and the PNS is comprised of a complex network of nerves that connect the CNS to the rest of the body. The human nervous system is depicted in Figure 2.1. We will discuss the CNS and PNS in detail in other chapters. Our current interest is in the basic building block of the nervous system—the neuron. It may surprise you to discover that it took researchers quite a while to even begin to understand how neurons work.

The Discovery of the Neuron

In 1685, Robert Hooke discovered that living organisms are made up of cells, but it was not until the advent of powerful microscopes in the early 19th century that scientists were able to peer inside a cell and explore its architecture. When anatomists turned their attention to the nervous system, they were stymied at first by its complex arrangement. In order to see the microstructure of the brain, an anatomist must extract a slice thin enough that light can shine

FIGURE 2.2 A model of neural complexity This single red strand of pasta demonstrates how difficult it would be to deduce the organization of a neuron embedded in a complex tangle of cells if all you had available to you was a series of thin, two-dimensional slices.

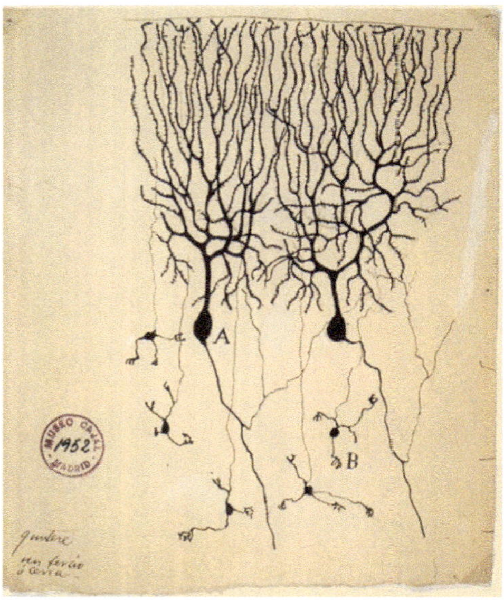

Cajal / Wikimedia / Public Domain

FIGURE 2.3 A hand-drawn neuron by Cajal Before the advent of sophisticated microscopes with photographic capabilities, scientists had to hand draw what they saw. This drawing of a neuron by Cajal in 1899 rivals images created with modern technology.

through. These slices, called sections, were difficult to interpret because the two-dimensional view afforded by the section consisted of cells and parts of cells all jumbled together.

Imagine for a moment that your task is to visualize the complete structure of one individual item in the tangled mass of pasta in **Figure 2.2**. It is not so hard to do when your three-dimensional view allows you to easily distinguish one noodle from another. Now imagine that the noodles are microscopic in size and almost as transparent as glass and that there are hundreds of thousands of them, all tangled together and crammed into a very small space. Finally, imagine you can only see a single, very thin slice that cuts through each and every strand at a different and random angle. Would understanding the overall structure from such images be a difficult task? Figure 2.2 helps you to appreciate the obstacles the early neuroanatomists faced when they began to study the brain.

Anatomists like Camillo Golgi (1843–1926) struggled with this problem and discovered that some substances were taken up by cells in a way that made individual cells or parts of those cells visible. Over 100 years ago, Golgi discovered that a silver nitrate stain would randomly illuminate only a few neurons, but it would illuminate those cells in their entirety, thereby reducing the jumble to a manageable view. This technique became known as Golgi staining (see **Check It Out: Staining a Neuron**).

As cell-staining procedures became widely available, neuroanatomists began to tease apart the complexity of the nervous system. Foremost among these early explorers was Santiago Ramón y Cajal (1852–1934). Cajal was a precocious and rebellious youth. At age 11 he was briefly jailed after he destroyed the gate to a neighbor's yard with a homemade cannon. Despite this inauspicious start to his career, Cajal earned both a medical degree and a doctorate in anatomy and held positions at a several universities in Spain. Cajal's drawings of neurons are still considered to be among the best images of neurons, even in our era of powerful imaging techniques (**Figure 2.3**).

Cajal proposed the seemingly paradoxical idea that neurons are not physically connected to each other; rather, there is a small gap, now known as the **synapse**, that separates one neuron from another. Cajal also proposed that one part of a neuron extends part of itself to physically seek out another neuron to communicate with; this extension is now known as a growth cone, and we will learn more about these structures when we discuss neurodevelopment and also the neural basis of learning. Cajal and his colleague Camillo Golgi shared the 1906 Nobel Prize in Physiology or Medicine for their discovery of fundamental concepts in neuroscience.

Check It Out | Staining a Neuron

Josef Spacek

The Golgi stain illuminates a single neuron within a three-dimensional matrix of neurons. Watch the short video **The Golgi Neural Stain**, and then answer these questions:

1. The Golgi stain randomly illuminates only a few neurons but does so in their entirety. How does the stain help neuroanatomists understand a neuron?
2. X-rays and MRI provide a view of large structures, whereas Golgi and other stains illuminate features at a microscopic level. Which view do you think is most useful and why?
3. Golgi and other stains are used at autopsy to diagnose diseases such as Alzheimer's. Do you think it is a good or bad idea to collect a sample of brain tissue and perform an appropriate test when it is believed that a person may have died of a neurodegenerative disease? Explain your answer.

Now that you have a sense of how neurons were discovered and visualized, we will begin our discussion of the neuron by describing its physical structure and then continue to an exploration of how it functions.

2.1 Before You Go On

Section Summary

The nervous systems of all vertebrates consist of central and peripheral systems. In both systems, the neuron is the basic unit of function. Although cells were identified by 1685, significant understanding of the neuron did not occur until almost 200 years later when neural staining techniques were developed, beginning with the Golgi stain.

Comprehension Check

1. What are the two divisions of the vertebrate nervous system?

- **a.** the excitatory nervous system and the inhibitory nervous system
- **b.** the peripheral nervous system and the central nervous system
- **c.** the inner nervous system and the outer nervous system
- **d.** the brain and the spinal cord

2. An important reason why our understanding of the neuron occurred relatively late is that _____.

- **a.** people believed that investigating the basis of life was unethical
- **b.** microscopes were only invented in the last century
- **c.** neurons posed unique challenges in visualizing their structure
- **d.** it was felt that understanding behavior was less important than other goals

3. A unique and unexplained characteristic of the Golgi stain is that it illuminates _____.

- **a.** only the cell bodies of neurons
- **b.** every neuron in great detail
- **c.** a three-dimensional view of neurons
- **d.** a small and random number of neurons in their entirety

2.2 The Structure of a Neuron

LO 2.2 Describe the characteristics of different types of neurons.

Before we begin, take a moment to consider what we are about to discuss and why it is important to understand this material. Your brain is a complicated information-processing device, more complex than anything else we know of in nature. That complex entity is comprised of individual units known as neurons. At some level, every behavior depends on the proper functioning of neurons. We cannot understand how your heart continues to beat, how you perceive a rainbow, appreciate a melody, choose what and when to eat, or understand the root causes of schizophrenia, concussions, and addictions without understanding the neuron. To understand the neuron, we first need to know a bit about cells in general.

Eukaryotic Cell Structure MCAT

The cell is the basic functional unit of life, and all living organisms are made up of cells. Humans and other animals are made up of **eukaryotic cells**, which are distinguished from bacterial cells, for example, because they have distinct membrane-bound structures called **organelles** (Figure 2.4). The most prominent among these is the **nucleus**, which contains genetic information. Everything within the cell besides the nucleus is termed the **cytoplasm**.

The cells that make up all living organisms are complicated structures, but for our purposes, only a few of these structures are relevant. Table 2.1 defines the most basic cell structures that are relevant to the current discussion.

As you learn about the nervous system and how it creates and controls behavior, you will encounter concepts that are built on understanding the workings of a eukaryotic cell. For example, Golgi body dysfunction is implicated in both Huntington and Alzheimer's diseases,

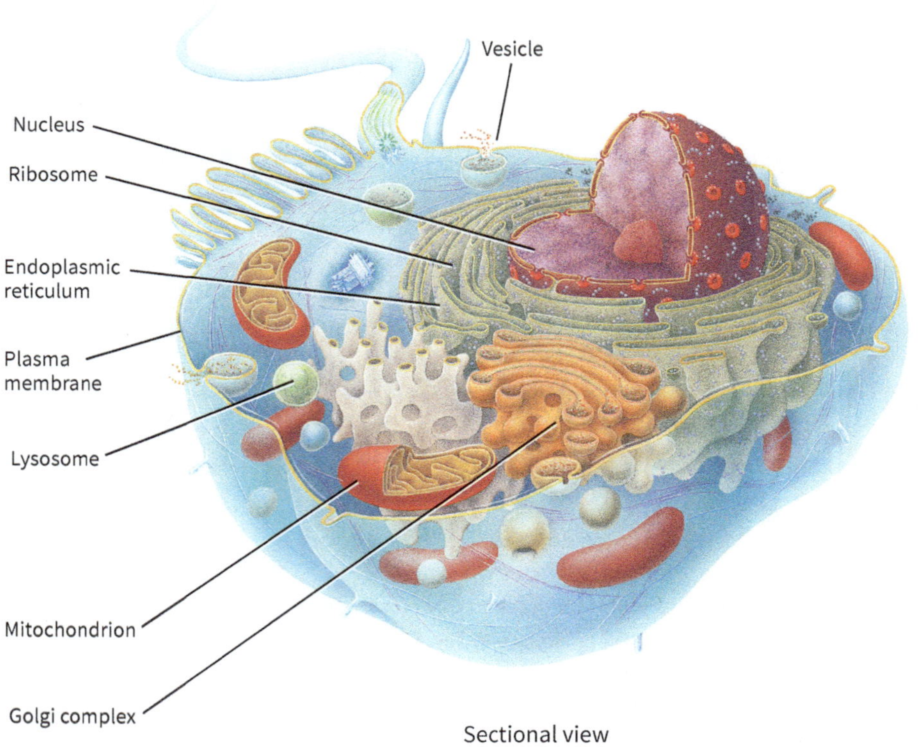

FIGURE 2.4 A simplified eukaryotic cell All eukaryotic cells, whether they are brain, bone, muscle, or any other type of cell, share the common elements shown here.

TABLE 2.1 Important Cell Structures and Their Functions

Structure	Function
Endoplasmic reticulum (ER)	Transporting proteins within the cell
Golgi body or Golgi complex	Creating and packaging proteins into vesicles
Lysosome	Removing metabolic waste from the cell
Mitochondrion (plural mitochondria)	Generating ATP, the energy source within the cell
Nucleus	Containing DNA, which holds the cell's genetic information
Plasma membrane	The membrane that separates the inside of the cell from the outer environment
Ribosome	Creating proteins
Vesicle	Containing proteins destined for use outside the cell

and mitochondrial dysfunction is associated with Huntington, Alzheimer's, epilepsy, schizophrenia, and multiple sclerosis. With a basic idea of what constitutes a cell in hand, we now turn our focus to the neuron, which is a very specialized type of cell.

Neuron Components MCAT

A neuron is the basic information-processing unit of the nervous system. Neurons have a complex mechanism to signal back and forth to each other in a sort of conversation in which each party has something to communicate that affects the other party. Each of us has over 85 billion neurons of various types, but by and large they share a common structure (Figure 2.5).

A typical neuron consists of a cell body, or **soma**, that contains the mechanisms that maintain the cell. If you examine a typical neuron such as the one illustrated in Figure 2.5,

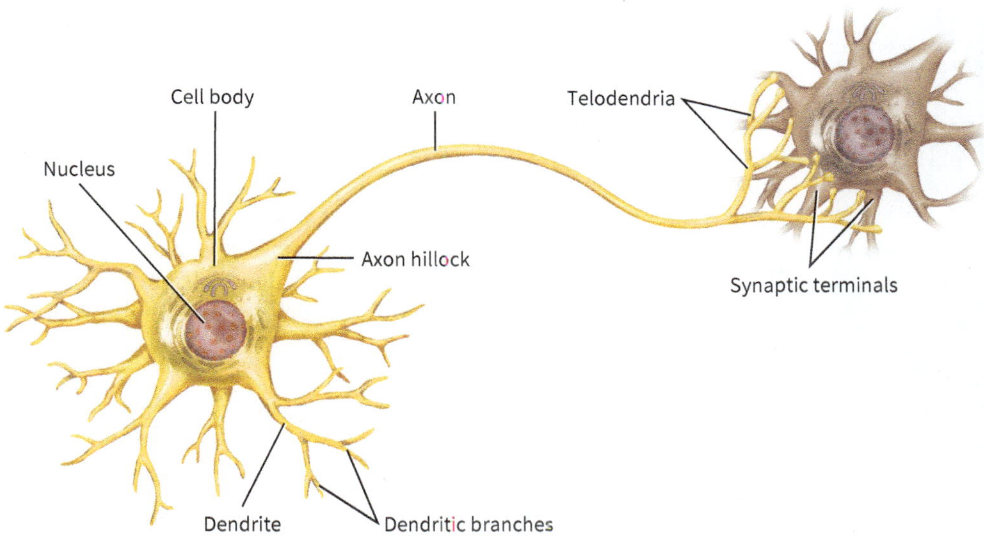

FIGURE 2.5 A typical neuron A typical neuron consists of a soma, with dendrites and an axon that project outward from the soma. The axon terminates in telodendria, which have terminal buttons or synaptic terminals in communication with another neuron.

you will see that the perimeter of the cell body is surrounded by extensions called **dendrites** (from the Greek *déndron* for tree), which are treelike branches that project from the soma. The **axon**, which emerges from the soma at a small bump known as the axon hillock, is a long nerve fiber. At the end of the axon are structures known as **telodendria**, which have at their endpoint synaptic terminals or terminal buttons. These terminal structures are the point at which information traveling down the axon is transmitted to the next neuron. There are additional structures that complete a neuron, but these are the most basic. Now we will investigate several of these structures in greater detail.

Dendrites

Dendrites collect information from the environment or from another neuron. If you think of neurons transmitting data from one place to another in the nervous system like traffic on a highway, then dendrites are the on-ramps where the signal begins. Some neurons do not initiate a signal but instead collect information from other neurons and transmit that information elsewhere in the same way that a cellphone tower picks up the signal from your phone as you drive and transfers that signal to another tower. In both cases, the dendrite is the site on the neuron where the transmission of information begins its journey. Recent research suggests that the role of dendrites is more complex than originally thought (see **Advances in Behavioral Neuroscience: Mark Harnett and the Role of Dendrites**).

Advances in Behavioral Neuroscience | Mark Harnett and the Role of Dendrites

Reproduced with permission of Dr. Mark Harnett

Dendrites are specialized projections of a neuron that receive input from sensory receptors or other neurons and carry that information into the soma for subsequent processing. Mark Harnett and his colleagues have demonstrated that dendrites are more active participants in signal processing than originally thought. Harnett is a neuroscientist whose investigations into the function of human dendrites have led to a much deeper understanding of their function.

Dendrites may be thought of like transistors in a computer; they perform simple logical operations on incoming signals, and the results of those elementary operations are one of the first steps in the complex chain of information processing that keeps us alive and creates consciousness.

Much of what we know about dendrites comes from research using rodent neural tissue, but Harnett questioned whether rodent dendrites work in the same way as human dendrites. The Harnett lab discovered that human dendrites may function in a fundamentally different fashion from rodent dendrites in that the signal from a distant part of the dendrite may be weaker than a signal from a part of the dendrites close to the cell body. Human neurons may use this information to create a more complex signal in the soma. You can learn more about Harnett's work by watching the video **Are Human Dendrites Special?**

Question

1. **Think of one research question that involves dendrites. It might involve comparing two or more groups of individuals or looking at a disorder. State your research question and briefly mention why you chose that topic to investigate.**

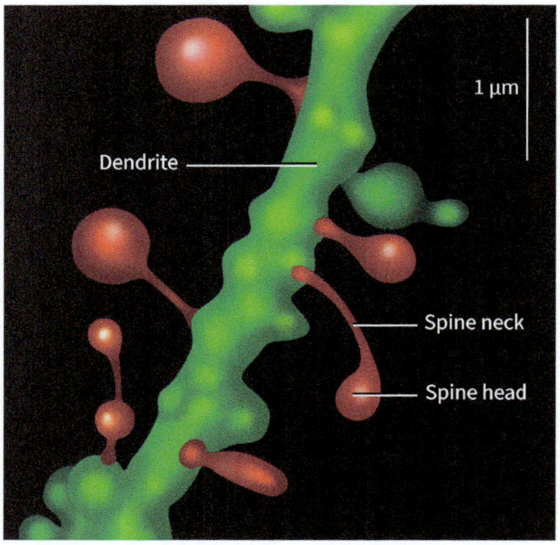

FIGURE 2.6 Dendritic spines Dendritic spines are small spikes or protuberances that are created when a link between two neurons is stable. It is believed that the creation and maintenance of long-term memories is in some way related to dendritic spine formation.

Dendritic spines In the early days of optical microscopy, anatomists could see that neurons had dendrites. As technology improved, they were able to see small structures that looked like spines with a thin head and neck that were attached to the dendrites (Figure 2.6). A **dendritic spine** is miniscule; yet, as you are about to see, sometimes good things come in exceedingly small packages. In general, each dendritic spine connects with only one other neuron. You have around 85 billion neurons, and each neuron has between 1 and 24 dendrites, and each dendrite has several branches, and each dendritic branch can have thousands of spines. The potential number of neural connections in your brain is in the trillions.

Axons The axon is a long nerve fiber that conducts electrical impulses from the dendrites and soma to the telodendria at the end of the axon. It was originally thought that the axon hillock was the place on the neuron where the signal originates, but we now believe the site of origin is between the hillock and the first section of the axon.

Some axons are microscopically short, and others are quite long; in fact, the longest axon in the human body is the sciatic nerve, which originates at the base of the spinal cord and extends all the way to the big toe. In a subsequent section, we will discuss how the axon sends a signal down its length and into the terminus of the neuron and then into the synapse. That process is key to understanding the importance of the neuron.

Now that you know the basic components of all types of neurons, we turn to a discussion of the different types of neurons encountered in the nervous system.

Types of Neurons MCAT

Neurons can be classified in various ways depending on what characteristic we focus on, including structure, function, and whether or not they have a myelin sheath.

Structural classification When we focus on the structure of neurons, there are four basic types: unipolar, bipolar, multipolar, and pseudounipolar (Figure 2.7). Polarity when referring to cells refers to their structural symmetry and not their electrical characteristics. Unipolar neurons have only one axon projecting from the cell body. Bipolar neurons have both an axon and a dendrite projecting from the cell body. Multipolar neurons have a single axon, but many dendrites and pseudounipolar neurons have axons that are split into two branches.

Functional classification Neurons are also classified by their function in the nervous system. Neurons can be sensory (or afferent) neurons, motor (or efferent) neurons, or interneurons (Figure 2.8). **Sensory neurons** (also called **afferent neurons**), transmit

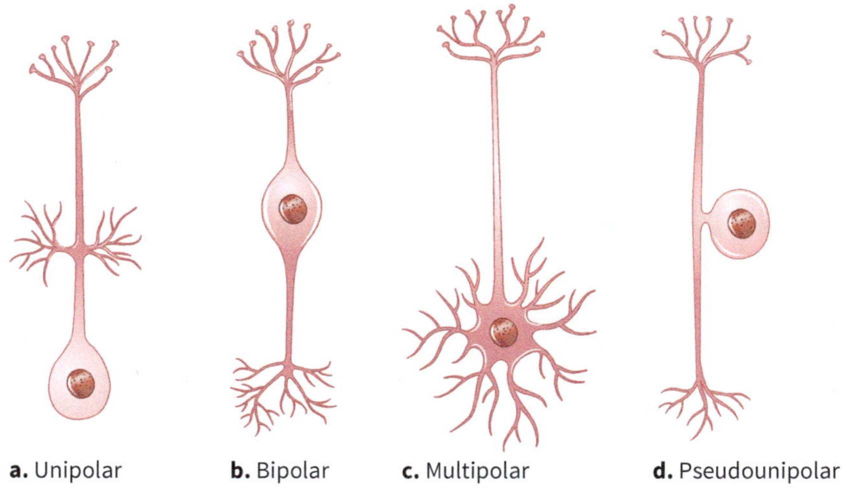

FIGURE 2.7 Neurons categorized by structure **a.** A unipolar neuron has one process, or structure, that extends from the soma. **b.** A bipolar neuron has one dendrite and one axon. **c.** A multipolar neuron has multiple dendrites and axons extending from the soma. **d.** A pseudounipolar neuron has an axon that is split into two.

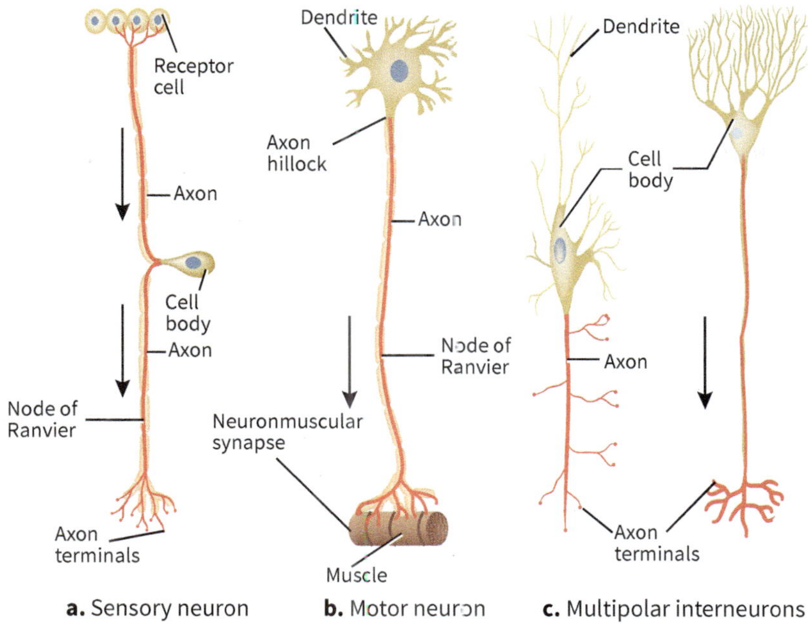

FIGURE 2.8 Neurons categorized by function
a. Sensory neurons take in signals from receptors for analysis.
b. Motor neurons stimulate muscles to act. **c.** Multipolar interneurons connect different brain regions.

information from bodily receptor cells into the brain. For example, neurons that send data from your eyes, ears, skin, and gut are all sensory neurons. Just as data must flow from the body into the brain, so must data flow from the brain into the body. **Motor neurons** (also called **efferent neurons**) transmit data from the brain and toward muscles and glands. One way to remember the difference between afferent and efferent neurons is to use the words arrive and exit. Afferent neurons send signals that *arrive* in the brain, and efferent neurons send signals that *exit* the brain and travel out into the body. Finally, **interneurons** create a connection between neurons that is neither afferent nor efferent. Some interneurons connect sensory neurons with motor neurons, whereas other interneurons form circuits with nearby neurons, creating local analyzing structures. Interneurons are typically multipolar.

Sensory neurons, motor neurons, and interneurons work together to take in a signal, transmit it to other neurons, and then create behavior. For example, a mosquito bites your arm, which stimulates a skin receptor. That sensory neuron transmits its information to an interneuron, which then stimulates a motor neuron, causing you to move your arm to smack the mosquito. With this overall picture in hand, we can explore the role of the myelin sheath.

The myelin sheath In addition to structural and functional variations in neurons, their axons come in two varieties: myelinated and unmyelinated. **Myelin** is a lipid, or fatty substance, that insulates some neurons. Myelinated axons in the CNS are sheathed in myelin produced by specialized cells known as glia, which we will discuss later in the chapter. The function of myelin is to insulate the axon electrically from surrounding cells, much as the colored plastic covering on wires insulates one wire from another to prevent short circuits. The myelin surrounding the axon resembles a string of beads on a necklace, with gaps in the myelin that are known as **nodes of Ranvier**.

These gaps speed the conduction of an electrical impulse down the axon through two mechanisms. The first is that electrons have the property of jumping from one node to the next, thereby speeding the impulse (**Figure 2.9a**). This mode of conduction is known as **saltatory conduction** (from the Latin *saltare*, meaning to jump or hop). The second is that the propagation of the impulse relies on the presence of sodium and potassium ions (more on that later), and the nodes serve to concentrate those essential ions where they will have the greatest effect on nerve transmission speed. Notice that in our stylized depiction, the signal requires only a few hops (as illustrated by the six blue arrows) to travel down the axon.

In contrast, unmyelinated neurons do not have the myelin sheath that their myelinated counterparts enjoy. As a result, the propagation of information down the unmyelinated axon is slower than it is for myelinated axons (**Figure 2.9b**) because in this case, the signal must be passed not from node to node, but from individual point to point, as indicated by the many plus and negative potential signs. If you consider myelinated neurons to be an express train

FIGURE 2.9 **Conduction in myelinated and unmyelinated axons** **a.** In a myelinated axon, the signal moves down the axon by jumping over the beads of myelin sheath. **b.** In an unmyelinated axon, the signal travels more slowly.

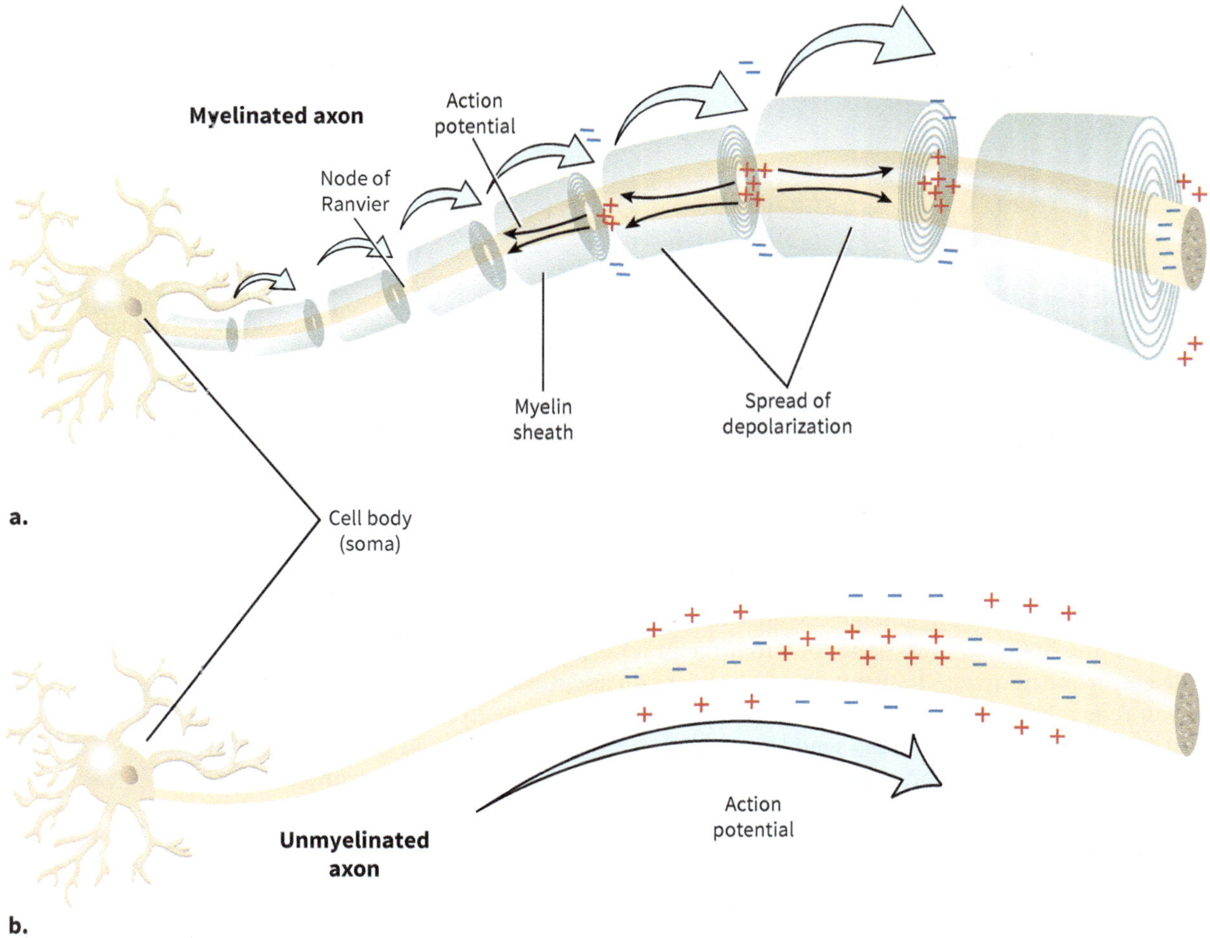

and unmyelinated neurons to be the local line, you have a good grasp of the concept; both will get you there, but one takes quite a bit longer to reach the destination. Nerve conduction speed differs with age, size of nerve, and many other variables. The slowest transmission rate is about 0.5 m/sec, and the fastest is over 100 m/sec. The majority of nerve fibers in the CNS are myelinated, but many nerve fibers in the PNS are unmyelinated.

Some disorders are specific to the myelin itself, for example, multiple sclerosis, which was discussed in this chapter's *Stories of the Brain*. If an axon is damaged but the myelin sheath is intact, the sheath can sometimes serve as a guide for regeneration. Unfortunately, when an axon is damaged, the myelin sheath is often also damaged, and nerves do not always regenerate after injury. **Guillain-Barré syndrome (GBS)** is a rare disorder in which a person's autoimmune system attacks the myelin sheath in the peripheral nerves leading from the brain to the muscles and other organs in the body (see **Check It Out: Guillain-Barré Syndrome**). These nerves, known as peripheral nerves because they are outside the brain, support communication between the brain and your muscles and organs. When that communication is disrupted, the results can be severe, and individuals with GBS typically experience problems with muscle coordination and balance.

Individuals suffering from GBS experience muscle weakness that can spread throughout the body and in the most extreme cases can lead to respiratory failure and even death. The cause of GBS is clearly understood at this time although the exact mechanisms are not. GBS symptoms usually appear soon after a digestive or respiratory infection, so it is thought that somehow the infection triggers an immune system attack on peripheral nerve cells. GBS is quite rare, occurring in fewer than two individuals in 100,000.

Check It Out | Guillain-Barré Syndrome

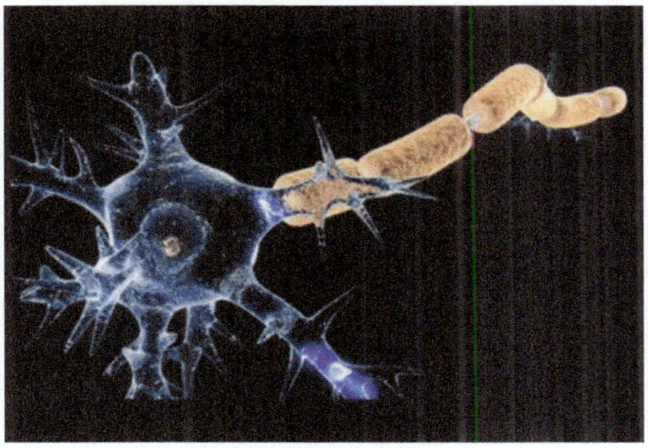

Watch the brief video **Guillain-Barré Syndrome**, which illustrates how myelin is damaged by GBS, and then answer these questions.

1. GBS affects only the myelinated nerves in the body, whereas multiple sclerosis (MS) affects myelinated nerves in the brain. Think of one symptom that individuals with MS might experience that you would not expect with GBS.
2. Having a disorder such as GBS might prevent you from pursuing some careers but would leave many others open. Think of a career that a person with GBS would have no difficulties pursuing.

At present, there is no process for regenerating myelin, but this problem has attracted the attention of researchers. One day, you may be the person who solves the problem of regenerating the myelin sheathing of nerve fibers to cure GBS and other diseases.

We now turn to a discussion of how a signal is transmitted down the axon of a neuron to provide a mechanism for creating behavior and consciousness. That involves two important concepts: the sodium-potassium pump and the action potential.

2.2 Before You Go On

Section Summary

Humans are comprised of eukaryotic cells, which have a distinct internal structure, including a nucleus. Neurons share these characteristics but have additional specialized structures to enable them to perform their function of information transmission: dendrites, a soma, an axon, and telodendria. Neurons can be classified in several ways based on their function and physical characteristics, including the presence or absence of a myelin sheath.

Comprehension Check

1. Which of these is an important feature of eukaryotic cells?
   - **a.** Their cytoplasm holds their genetic material.
   - **b.** Their nucleus has a double-walled membrane.
   - **c.** They do not have organelles.
   - **d.** They have a well-defined nucleus.
2. What is the role of the dendrite in a neuron?
   - **a.** carrying genetic information forward to the next generation
   - **b.** synthesizing proteins, which are essential for life
   - **c.** amplifying the transmission of information
   - **d.** taking in a signal from outside the neuron
3. What are the four structural varieties of neurons?
   - **a.** unipolar, bipolar, multipolar, and pseudounipolar
   - **b.** excitatory, inhibitory, resting, and interneurons
   - **c.** excitatory, inhibitory, interneurons, and myelinated
   - **d.** excitatory, inhibitory, myelinated, and unmyelinated
4. What are the three functional varieties of neurons?
   - **a.** sensory, receptive, and interneurons
   - **b.** sensory, motor, and interneurons
   - **c.** sensory, activity, and interneurons
   - **d.** sensory, white matter, and gray matter
5. Which of these statements is true of a myelinated neuron?
   - **a.** It is considered to be gray matter.
   - **b.** It transmits a stronger signal.
   - **c.** It is fast acting.
   - **d.** It only exists in the brain.

2.3 Neural Transmission of a Signal

LO 2.3 Describe how the sodium-potassium pump and the action potential propagate a neural signal.

In the middle of the 18th century, scientists were trying to understand the basics of what we now call neurophysiology, and a prime question was how the brain conveys signals throughout the body. A watershed moment occurred in the laboratory of Luigi Galvani (1737–1798) when he was at the University of Bologna. Galvani was interested in the newly discovered phenomenon of electricity. While investigating the effects of a static electric charge on the skin of a frog, he accidently touched a scalpel that had collected a static charge to the dissected leg of a frog. As a spark flowed from the scalpel to the frog's leg muscle, the leg twitched, and Galvani realized that electricity was somehow involved in nerve conduction.

In the late 1950s, John Eccles (1903–1997) won the Nobel Prize in Physiology or Medicine for devising a way to record synaptic transmission from a single neuron. At about the same time, Hubel and Wiesel's pioneering studies of single-cell recordings from a cat's visual cortex showed others what was possible and also led to the Nobel Prize. What these luminaries and their successors developed was a microelectrode small enough to pierce but not destroy a single neural cell and the electronic capability to amplify the cell's signal. When those pieces of the puzzle fell into place, the process by which neurons transmit information began to come into focus. We now know that neurons use an electrochemical process to transmit a signal, a process described in the next section.

Energy Produced through Imbalances MCAT

It is useful to think of a neuron as a tiny machine that performs an enormously important task and, like any machine, a neuron requires energy to accomplish its goal. That energy is the result of imbalances. To understand how imbalances provide energy, imagine how a dam produces electrical power (**Figure 2.10**). Just as water stored behind a dam's wall represents stored, or potential, energy available for release as kinetic energy, imbalances between the intracellular and extracellular environments of neurons are a source of energy. The two types of imbalances that power neurons are an imbalance in electrical charge and an imbalance in ion concentration.

Electrical potential One imbalance is due to the difference in the electrical charge of ions between the inside and outside of a neuron. Ions are atoms or molecules with either a positive or negative electrical charge known as **electrical potential**. In this case, the ions in question are sodium and potassium. The intracellular environment of a neuron has a slightly

GYRO PHOTOGRAPHY/amanaimagesRF/Getty Images

a.

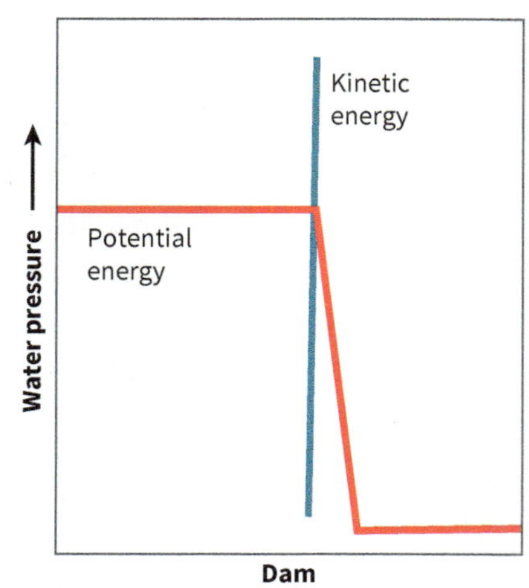

b.

FIGURE 2.10 An example of stored energy through imbalance a. The water pressure behind a dam wall can be in the order of tens of thousands of pounds per square foot and converting that potential energy into kinetic energy that spins turbines that create electricity can light a whole city. **b.** In a similar fashion, imbalances in electrical charge and ion concentration on either side of a neuronal membrane can power our nervous system.

negative electrical potential when compared with the extracellular environment, which has a slightly positive electrical potential. Opposite charges attract, so there is a force acting on the axonal membrane to exchange sodium and potassium ions and thereby eliminate the difference in electrical potential between the inside and outside cellular environments.

The magnitude of the difference between the electrical potentials inside and outside the cell is known as the **membrane potential**. The membrane potential is high before the neuron fires and low afterward, just as the stored energy is high at the top of the dam and much lower at the bottom.

Concentration gradient In addition to the electrical potential, there is an imbalance in chemical concentrations between the internal and external environments of a neuron. The internal cellular environment has a greater concentration of potassium, whereas the external environment has a greater concentration of sodium. This difference is known as a **concentration gradient**. Solutions with different chemical concentrations on either side of a semipermeable barrier will tend to equalize through diffusion. If you have ever placed a drop of food coloring in a glass of water and seen all the water gradually turn color, you understand the principle that fluids and gases diffuse from regions of higher concentrations to regions of lower concentrations (Figure 2.11). Similarly, there is a force acting to equalize the concentrations of sodium and potassium on either side of the axonal membrane by exchanging sodium and potassium ions.

Resting potential Think back to the example of the hydroelectric dam, where the water behind the dam was a source of potential energy. Similarly, the electrical potential and concentration gradient are sources of potential energy (approximately –65 to –70 mV) ready to be released. When the inside and outside of the cell are at different electrical potentials and concentration gradients, this causes a state of **polarization**. An easy way to remember this is to think of the two very separate poles of the Earth. This polarized state of readiness is known as the **resting potential**, and a neuron's natural state is to be in this state of imbalance unless stimulated. How this energy is released in response to a stimulus is the topic of our next sections.

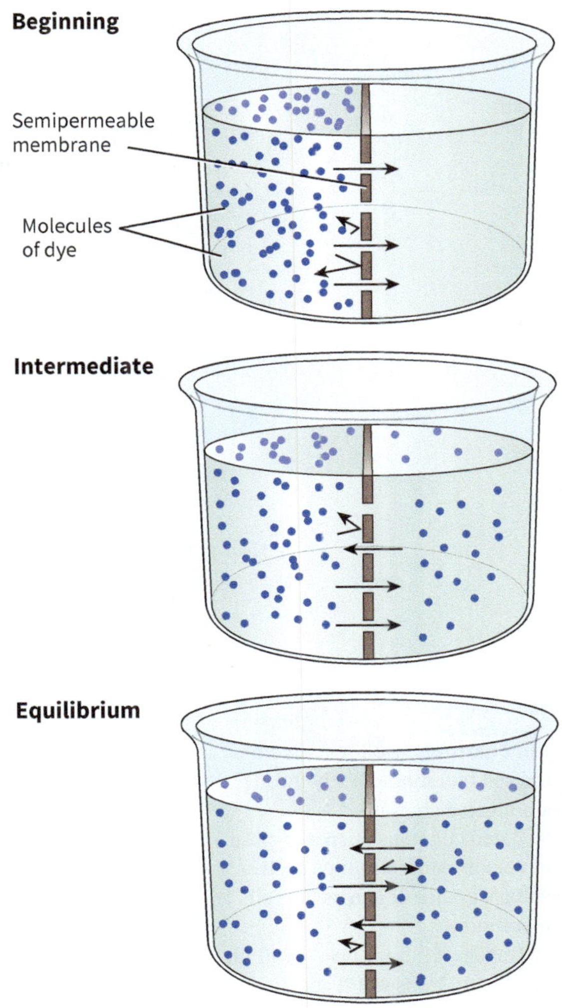

FIGURE 2.11 **Diffusion** Diffusion results from the movement of molecules from higher to lower concentrations to achieve equilibrium. When molecules of dye are placed in a beaker that is separated into two components by a semipermeable membrane, those molecules will diffuse across the membrane until the concentrations on both sides of the membrane are equal.

Ion Exchange MCAT

In the last section you saw how differences in charge and concentration cause sodium and potassium ions to move in and out of the neuron. You also saw that the difference in ion concentration and/or electrical charge serves as stored (potential) energy, much as a compressed spring stores energy. In this section we explore *how* these ions move across the membrane. Ion exchange across the neuronal membrane is the mechanism that converts stored energy into an impulse that powers signal transmission through our nervous system. In principle, it is identical to the simple example you just read describing diffusion of dye molecules across the membrane in a beaker.

Ion channels are specialized proteins in the cell membrane that allow ions to pass through in response to differences in the extracellular and intracellular electrochemical gradient. In general, such channels are selective, which means they only transport one type of ion, for example sodium, potassium, or calcium. There are three different mechanisms involved in ion exchange: nongated (leaky) channels, gated channels, and transport pumps.

Nongated channels are open pores through the cell membrane that allow the passage of ions (Figure 2.12a). Their openings are so small that ions seem to leak through one at a time. Gated channels open and close either spontaneously or in response to a stimulus (Figure 2.12b). For example, voltage-gated channels open or close in response to changes in electrical charge on either side of the membrane. When there is an excess of a molecule such as sodium ions in the extracellular environment, this causes electrochemical and concentration imbalances, which open a voltage-gated ion channel that is specific for sodium to allow sodium ions to pass through the membrane and restore balance.

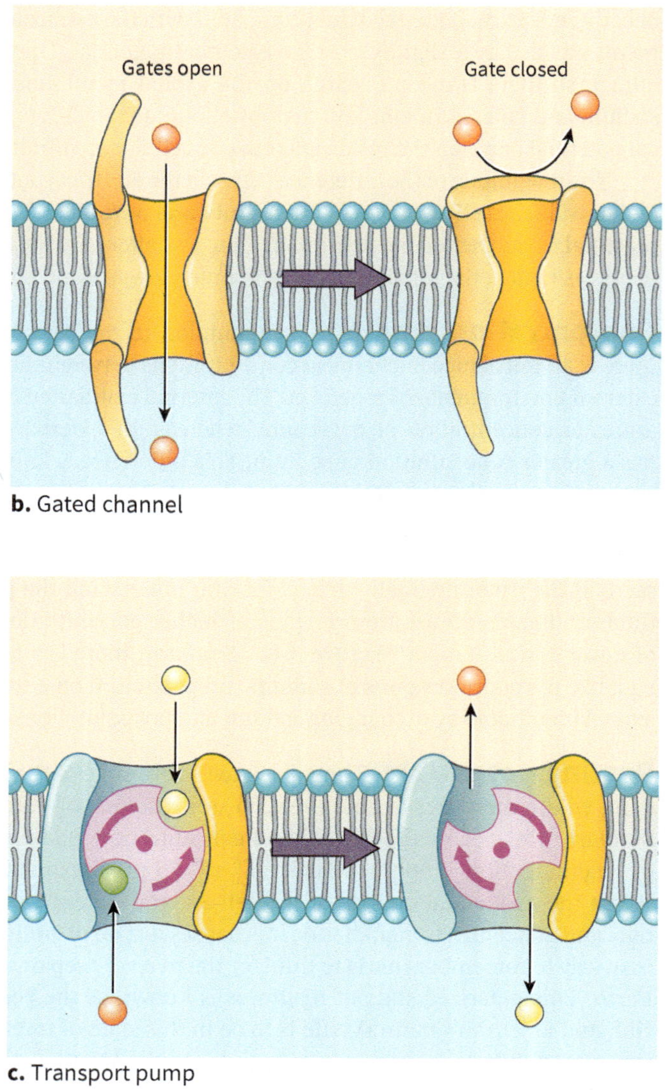

FIGURE 2.12 Three types of ion channels **a.** Ions can cross the cell membrane freely through a nongated channel.
b. Gated channels open and close in response to a stimulus such as voltage gradients across the cell membrane.
c. Transport pumps open to admit an ion and transport it across the cell membrane against the concentration or electrochemical gradient.

Both nongated and gated ion channels serve to reduce the chemical or electrical gradient across the two sides of the cell membrane. These ion exchanges are powered, so to speak, by electrochemical and/or concentration imbalances. You probably have already realized that these two ion channels continuously function to eliminate the difference between the extracellular and intracellular environments. In keeping with a principle of biology that you will see again and again, there must be a contrasting or opposing mechanism that moves ions back against the gradient. That mechanism is the transport pump (also known as a transport protein).

In contrast to nongated and gated ion channels, which are always working to reduce the differences between the extracellular and intracellular areas, transport pumps actively move ions against that gradient (Figure 2.12c). Because they move molecules from areas of low concentration to areas of high concentration, they are known as **active transport** channels.

There are several types of ion pumps in the body, but we have a particular interest in the sodium-potassium pump because its operation is essential to neural function. We now turn to a discussion of this elegant mechanism.

The Sodium-Potassium Pump MCAT

In contrast to the ion channels that continually work toward electrochemical and concentration balance across the neuronal membrane, the **sodium-potassium pump**, also known

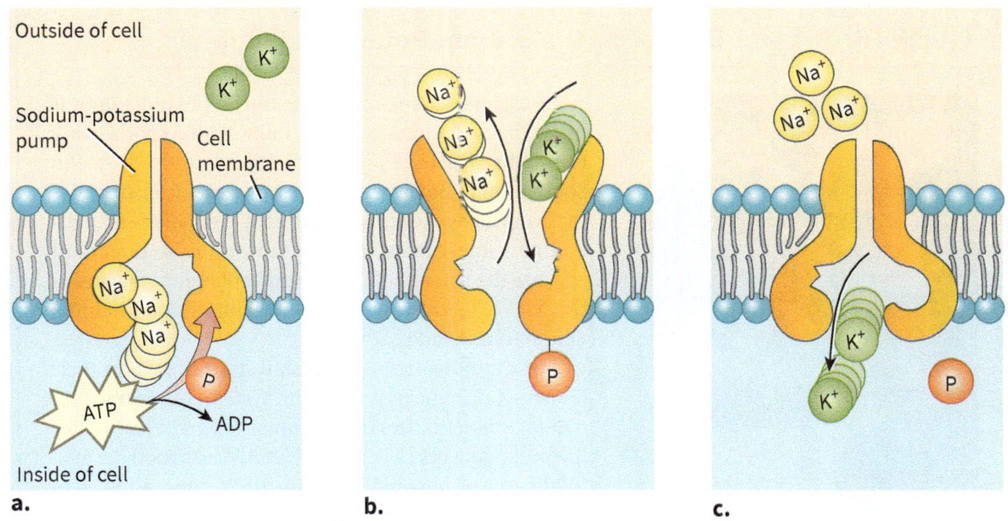

FIGURE 2.13 One cycle of the sodium-potassium pump
a. When the pump opens, three sodium ions inside the cell are attracted to the pump and energy is released by the conversion of ATP to ADP.
b. As the pump opens to release those sodium ions, two potassium ions are attracted to the pump.
c. The two potassium ions are then transported inside the cell.

as the Na^+/K^+ pump, engages in active transport of ions across the cell membrane. It works against the gradient, moving sodium ions into the cell environment while moving potassium ions out.

At the start of a cycle of the sodium-potassium pump, there is a concentration difference between the extracellular and intracellular environments with more sodium ions (Na^+) inside and more potassium ions (K^+) outside the cellular membrane. The cycle begins when the ion channel opens to the intracellular environment. Three sodium ions respond to the electrical potential and concentration gradients by moving into the channel and attaching to binding sites specific to sodium. When a molecule of ATP binds to its receptor site on the channel, ATP releases a phosphate (P), thereby becoming ADP but in the process also releases energy that is now available to the cell (**Figure 2.13a**). This causes the protein channel to change its configuration by closing at the internal side and opening at the external side, releasing the three sodium ions (Na^+) into the extracellular environment. The channel is now open to the extracellular environment, and two potassium ions (K^+) are attracted by electrical potential and concentration gradient to fill the two potassium-binding sites in the channel (**Figure 2.13b**). This docking of potassium ions causes the channel to change configuration again, resulting in the release of the two potassium ions into the cell (**Figure 2.13c**). The pump is now back at its initial state, ready to perform the cycle again. One entire cycle of the sodium-potassium pump from beginning to end takes about one half of a millisecond.

To summarize, for every three sodium ions moved from inside the cell to the outside, two potassium ions are exchanged from outside to inside. This process maintains an imbalance in the membrane potential that is essential for neurons to function properly. One of the lessons you are learning in your study of behavioral neuroscience is that the brain and body have specific requirements and that too much or too little of a good thing can both be bad. The sodium-potassium pump depends on a fairly specific level of sodium in the extracellular environment in order to function, and when the proper exchange of these and other ions are disrupted, the consequences can be severe or even fatal (**Think About It: Disrupting the Sodium-Potassium Pump**).

Before we move on, take a moment to consider what you have just learned. At this very moment, you may be sitting quietly and not moving a muscle as you contemplate the sodium-potassium pump; yet at every moment, roughly 86 trillion ion channels are busily keeping you alive and conscious. You now can see that at the molecular level, a complex mechanism is creating and then maintaining electrical and chemical imbalances at the axonal membrane. How does what happens at the level of a single ion channel create the transmission of a signal down the axon? To answer that question, we now turn to the second half of this process, the action potential.

Think About It | Disrupting the Sodium-Potassium Pump

btrenkel/Getty Images

After exercise or spending time out in the hot sun, it is likely that you quenched your thirst with a sports drink that touted itself as providing electrolyte replacement. Sodium and potassium are examples of electrolytes that your body requires to maintain basic life functions. You now know that these electrolytes are so important because the balance of sodium and potassium that keeps the Na^+/K^+ pump running is essential to maintaining neural functioning.

If an individual consumes water faster than their kidneys can remove it from the system, the extracellular environment loses the sodium gradient necessary for the pump to work, a condition known as hyponatremia. Mild cases of hyponatremia are what cause athletes to develop muscle cramps on the athletic field due to low electrolyte levels. These can be ameliorated with stretching coupled with an electrolyte replacement drink. In the most severe cases, hyponatremia can lead to a breakdown of the neurons that control heartbeat, which can be fatal. This explains why hazing rituals that require individuals to consume large quantities of water in a short period of time have sometimes been fatal.

If too little sodium in the extracellular environment causes problems, it stands to reason that too much can also be dangerous, a condition known as hypernatremia. Hypernatremia typically results from diseases that cause uncontrolled vomiting or diarrhea or from being denied fluids for long periods of time as one might experience in a desert. As the extracellular sodium concentration exceeds the ability of the pump to manage the conditions, neural functions cease, and death can result.

Conditions like anorexia, which can involve malnutrition and weight-regulating behaviors such as forced vomiting and consuming diuretics, can cause a potassium electrolyte imbalance, often leading to severe effects.

Questions

1. **Search the internet for "causes of hyponatremia" and "causes of hypernatremia" and list at least one cause for each condition.**
2. **Search using the terms "Anorexia and electrolyte imbalances" and explain the different ways that anorexia can affect the Na^+/K^+ pump.**
3. **"Water, water everywhere and not a drop to drink" is a line from "The Rime of the Ancient Mariner," a famous poem by Samuel Coleridge. Sea water has a high concentration of sodium. Mariners are warned that should they be adrift at sea without fresh water, they should not drink sea water. Explain.**

The Action Potential MCAT

At this point in our discussion, you can see that the sodium-potassium pump begins its cycle with a resting potential of around –65 mV, and then it cycles through an exchange of ions. How does that process result in a neuron communicating with another neuron? The answer is the action potential.

At resting potential, your neurons are polarized with imbalances in their electrical potential and concentration gradient. A stimulation of sufficient size will trigger an ion exchange that leads to **depolarization** of the cellular membrane. This depolarization begins at the axon's initial segment and moves down the axon to the terminal buttons, which represent one side of the connection with another neuron. As one segment after another of the axon depolarizes, there is a steep increase in membrane potential known as the **action potential** (Figure 2.14). When the action potential reaches its peak of about +40 mV, **repolarization**, or a return to the initial resting potential, begins. Before that return to the initial state can occur, however, the membrane polarity overshoots the initial value of −65 mV and is in a state of **hyperpolarization**. When a membrane is hyperpolarized, there is a period during which it is very difficult for the neuron to fire again known as the **refractory period** before the region of the axonal membrane returns to the resting potential and the impulse travels down the axon.

Now that you understand that the action potential is generated by a process of depolarization, let's take a closer look at what happens at the axon membrane as ion channels and the sodium-potassium pump open and close, moving ions back and forth across the plasma membrane (**Figure 2.15**). At rest, the resting potential is lower than the threshold, so the two ion channels and the pump are all closed (**step 1**). There is an initial chemical and electrical charge imbalance with more sodium ions outside the cell and more potassium ions inside, giving the extracellular environment a positive charge while the inside has a negative charge. This imbalance represents stored energy that is available to the cell, like the water behind a dam.

When a neuron is stimulated sufficiently, the axon membrane depolarizes, and the charge imbalance attracts sodium ions in the extracellular fluid to enter the sodium-ion channel at the right and be transported into the cell (**step 2**). Notice that this ion exchange reverses the potential of the two sides of the membrane so that the outside environment is now negative while the inside is positive. This charge imbalance causes potassium ions inside the cell to be attracted to the potassium-ion channel at far left where they are transported out of the cell (**step 3**).

So far you have seen how the ion channels move sodium and potassium across the axon membrane in an attempt to achieve an

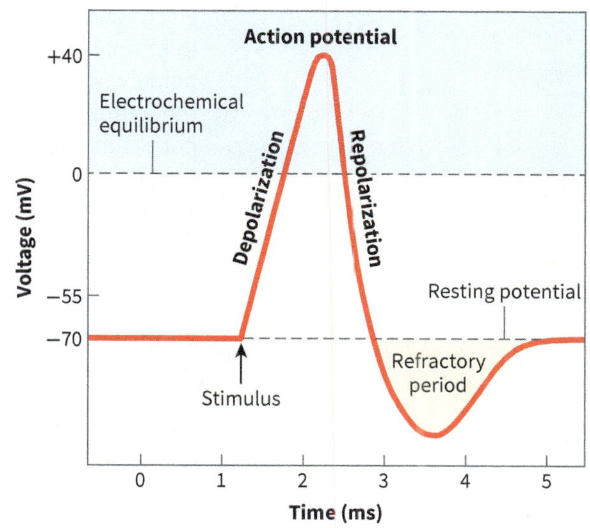

FIGURE 2.14 The action potential A stimulus of sufficient strength triggers an electrochemical impulse in the neuron that travels down the axon as one section of the axon after another depolarizes. After the potential has reached its peak voltage, segments of the axon gradually repolarize. There is a period of hyperpolarization before returning to the polarized resting potential.

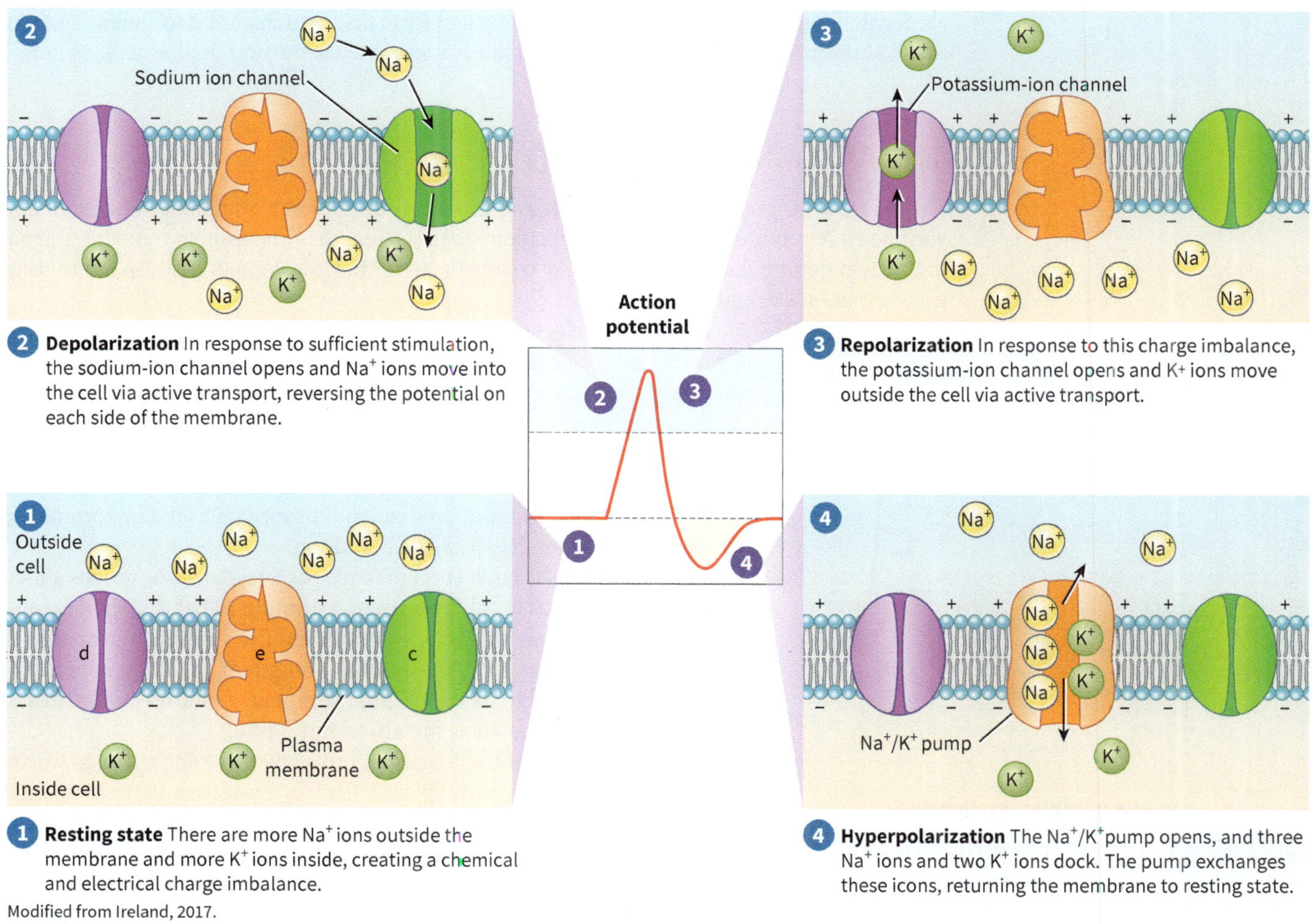

FIGURE 2.15 The action potential at a molecular level The changes in electrical charge that occur in the action potential (illustrated by the central graph) are caused by the physical opening and closing of ion channels and the sodium-potassium pump.

electrochemical and concentration balance between the environments on either side. Keep in mind, however, that for a neuron to fire, it must be in a state of electrochemical imbalance, namely polarization. Returning the neuron to this polarized state requires active transport by the sodium-potassium pump. Three sodium ions and two potassium ions dock onto a sodium-potassium pump, which then releases the sodium into the extracellular fluid while returning the potassium ions to the cytosol (**step 4**). The system is now returned to the resting state (step 1 again), where it is polarized and ready to fire again. We have walked through one cycle of depolarization and repolarization, but keep in mind that this depolarization cascades down adjacent segments of an axon and then on to the neighboring neurons.

To fully understand the relationship between ion channels and the sodium-potassium pump, you just need to remember that systems out of equilibrium tend toward returning to balance, and that it takes energy to restore that imbalance. In short, the resting state of the axon membrane is one of concentration and electrical imbalance; the ion channels permit the passage of ions to eliminate that imbalance while the pump restores it. These two processes, one attempting to eliminate an imbalance and the other to restore it, are at the heart of a neuron's function.

To connect the action potential to real-world sensation, we can take a moment to follow the course of a single neural impulse. Imagine you are sitting outside under a tree, contemplating the mysteries of behavioral neuroscience and especially the sodium-potassium pump. As you sit there, a leaf slowly falls and lands ever so lightly on your arm. That leaf stimulates sensory neurons in your skin that then fire, sending an action potential down the cells' axons that eventually create the awareness of pressure on your skin. Signals travel down a nerve at a rate of about 55–60 m/sec, so from the time that leaf touched your arm to the moment the sensation reached your brain was perhaps a tenth of a second.

You now know that when a neuron is sufficiently stimulated, an electrochemical signal travels down the axon and this information-carrying signal can then be transmitted to another neuron, much as the momentum of one domino in a line is transmitted to the next domino as each falls.

The All-or-None Law MCAT

The purpose of a neuron is to transmit and process information throughout your brain and your body. A stimulus triggers depolarization along the length of the axon to transmit a signal to the next neuron, but stimuli can vary widely in strength. How do neurons respond to stimuli of different strengths?

In order to trigger a response, the intensity of the stimulus must be above that neuron's **threshold potential**, which is typically around –55 mV. When the initial stimulation is not sufficient to trigger an action potential, the membrane remains at its initial resting state (Figure 2.16). Imagine for a moment what it would be like if the threshold membrane potential were much lower or even nonexistent. Our nervous system would be constantly flooded with stimulation that would overwhelm the important signals that we need to safely navigate our world.

Once a stimulus is large enough to trigger a response, how does a neuron convey differences in the strength of a stimulus? That is, does a neuron fire weakly in response to a weak stimulus and fire strongly in response to a strong stimulus? The answer is that the strength of an action potential is unrelated to the strength of the original stimulus; a neuron either fires or does not fire. This is known as the **all-or-none law**.

There is a model for understanding this law involving a device we are all familiar with: the toilet. A typical toilet has a reservoir that holds water and a handle that triggers flushing. Depressing the handle will initiate the flush, but only if the handle is sufficiently depressed. Lesser movements of the handle have no effect. Once the handle is depressed, the flush procedure begins and must continue until the water reservoir is completely empty. When that happens, the process cannot be repeated until the tank has completely refilled. One can flush repeatedly, but the rate of recycling depends on the time it takes to refill the tank.

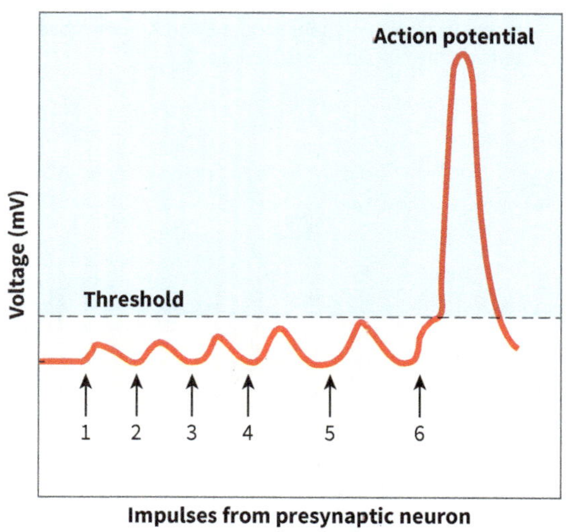

FIGURE 2.16 Threshold potential and the all-or-none law An action potential is only generated when a stimulus exceeds the threshold membrane potential of the neuron. In this example, five excitatory potentials that were subthreshold did not create a response before the sixth exceeded the threshold potential, causing the cell to fire.

With this reminder of how the common toilet works, let us return to the action potential. Given that a neuron either fires or does not fire, you might wonder how our nervous system can code for differences in the strength of a stimulus. There must be a mechanism for communicating the strength of a stimulus because you can certainly tell the difference between the roar of a jet airplane at takeoff and the gentle whisper of wind through the trees of a forest. There are two parts to the answer. The first is that neurons can fire more frequently (within limits), so frequency of discharge is a signal for strength. In addition, your nervous system interprets the number of neurons firing in response to a stimulus as a measure of intensity. To understand how mass firings can be interpreted as a measure of intensity, imagine a single person singing softly. Now imagine a choir of 100 people, all singing softly at the same time. You would perceive the choir as being much louder even though each person is singing softly. It is remarkable to consider that although neurons are quite simple devices capable only of either firing or not firing, when acting in concert with a multitude of other neurons, they are capable of conveying information as complex as a Shakespearean sonnet, the perception of a fine wine, or the meaning conveyed by a single raised eyebrow.

2.3 Before You Go On

Section Summary

An essential property of the neuron is its ability to transmit a signal to another neuron. The sodium-potassium pump is a mechanism that is powered by a difference in electrical potential and a concentration gradient. This pump creates an action potential that travels down the neuron to transmit a signal to neighboring neurons. Stimuli beneath the threshold do not result in an action potential, and after the neuron fires, there is a refractory period in which is it difficult for the neuron to fire again.

Comprehension Check

1. The sodium-potassium pump relies on which two principles?
   - **a.** the attraction of opposites and nature abhorring a vacuum
   - **b.** Brownian motion and dispersion in fluids
   - **c.** electrical imbalances and sodium ions
   - **d.** electrical potential and concentration gradient
2. The resting membrane potential is _____.
   - **a.** always in balance with its environment
   - **b.** the difference between sodium and calcium ion concentrations
   - **c.** the electrical potential inside a neuron that is resting after firing
   - **d.** the difference between the intra- and extracellular environments
3. During the refractory period of the action potential, the neuron is _____.
   - **a.** unlikely to fire
   - **b.** preparing to fire
   - **c.** in the process of firing
   - **d.** building a growth cone
4. When a neuron fires, it signals the strength of the stimulus by _____.
   - **a.** firing weakly for weak stimuli and strongly for intense stimuli
   - **b.** inhibiting other neurons from firing
   - **c.** exciting nearby neurons to fire
   - **d.** firing more rapidly

2.4 White Matter, Gray Matter, and Glia

LO 2.4 Identify the individual roles of white matter, gray matter, and glia.

By this point in our discussion, you have learned quite a bit about how neurons work, but we now need to change the level of our focus from the individual neuron to a more global view. Individual neurons combine into tissues, so we now turn to a discussion of white and gray matter and finally to a type of support cell known as glia.

White and Gray Matter MCAT

Earlier in our discussion you learned that some axons are myelinated and that this myelin serves to significantly increase the speed of signal conduction down the axon. Myelinated fibers are white in appearance, and unmyelinated fibers are gray (**Figure 2.17**).

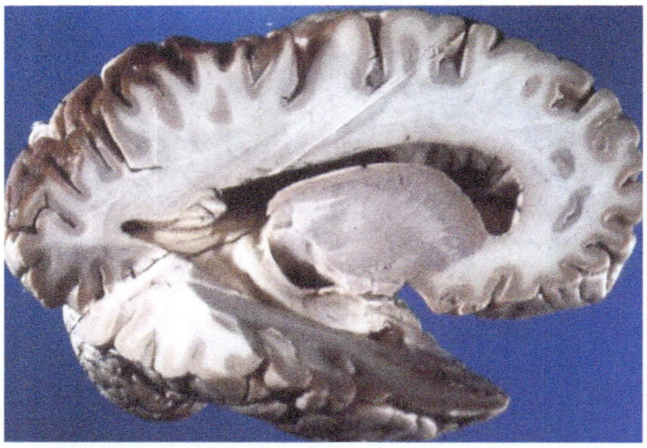

John Beal

FIGURE 2.17 Lateral view of human brain showing white and gray matter White matter is white in appearance due to the lipid content of myelinated axons. Gray matter is darker without the whitening effect of myelin.

White matter consists mainly of myelinated axons that speed neural processes. Think of white matter as transmission lines carrying signals from one part of the brain to another. To give you an idea of the extent of white matter in your brain, consider this: if you laid each myelinated neuron in your brain end to end, the line would extend about 100 miles (160 km).

One of the most important advances in neuroimaging in the last quarter of a century was the development of diffusion tensor imaging (DTI), which provides a complex image of the white matter tracts in the brain. These images allow scientists and clinicians to trace the myelinated tracts through the brain in a three-dimensional fashion. DTI has provided us with a view of the brain's organization that was difficult to extract from two-dimensional views, and the perspective this provides will inform our discussions in other chapters (**Figure 2.18**). As you compare Figure 2.17 with Figure 2.18, you see that the key to understanding the function of white matter lies in its role as an information highway. We will discuss this function in greater detail in our coverage of the structures of the central nervous system.

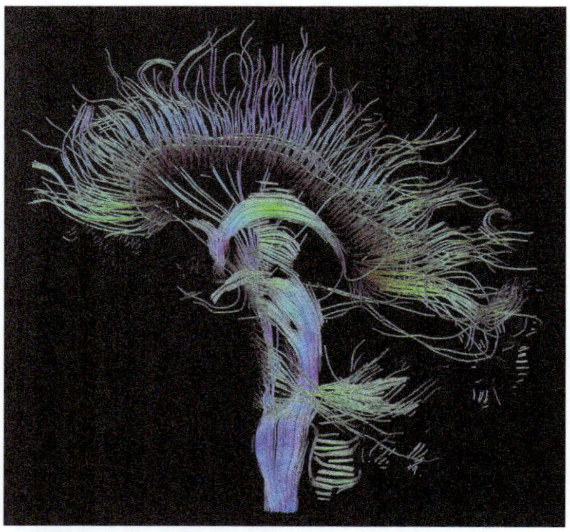

Thomas Schultz / Wikimedia / CC BY SA 3.0

FIGURE 2.18 White matter structure in the brain In contrast with the image of white fibers portrayed in Figure 2.17, this image created by DTI more accurately illustrates the complex distribution of white matter in the brain.

Gray matter, on the other hand, consists mainly of the cell bodies and dendrites associated with individual neurons and glia. Gray matter is involved in sensory perception, memory, emotions, speech production and comprehension, decision making, and muscle control. You might think of gray matter as the place where all your thinking occurs. An oversimplified but useful sentence to summarize these two tissues is this: gray matter is responsible for computation and white matter for transmitting the output of those computations elsewhere in the brain or the body (**Think About It: White or Gray Matter?**).

Now that you know the physiological differences between white and gray matter and the different role each plays in the brain, you should be able to make some predictions about the character of disorders in each. You could also pose some interesting research questions about what role white and gray matter might play in individual behavior differences. Before you read on, stop and think about these questions for just a moment. You have enough data to be able to think about this issue.

There are a number of fascinating research questions that our understanding of white and gray matter might help us understand. For example, there is a slight but steady decline with age in short-term memory and semantic memory (memory of words) that is correlated with the volume of gray matter in the cortex. Other research has suggested that there is a link between the thickness of gray matter and performance on intellectual tests of memory, attention, and language. Some researchers have found that adolescents who experience a psychotic break have less gray matter than healthy control subjects.

Contrast those research reports with the consequences of white matter disorders. Demyelinization is associated with multiple sclerosis, as well as infectious inflammatory diseases such as encephalitis, vitamin B_{12} and folate or iron deficiencies, and other deficiencies caused by malnutrition. Alzheimer's disease, autism, and bipolar disorder have all been linked to changes in white matter structure. At this point, we have discussed the structure of an individual neuron, the several types of neurons, and how a signal propagates down the axon of a neuron. But there is more to neuroanatomy at the cellular level, as you are about to see.

Glia MCAT

Although our focus has been primarily on neurons, they are not the whole story. **Glia**, also known as neuroglia, are cells in the nervous system that were discovered as recently as 1856.

Think About It | White or Gray Matter?

White and gray matter have different functions, and comparing your brain to a computer can provide a way to think about that difference. Gray matter is like the processing components in a computer that perform the actual computations, whereas white matter serves the purpose of the cables that connect the many different processing units. Both are important; computations need to occur, but the results also need to be transmitted to other processors.

Are there interesting individual differences in white and gray matter that would help us better understand the biology behind behavior? One study measured the amount of gray matter in the human cortex at autopsy and correlated it with each individual's IQ when alive; the data indicated that more gray matter was associated with higher IQ (Haier et al., 2004).

White matter, on the other hand, plays the part of information transfer. Although speed of computation certainly is an important aspect of intelligence, the ability to rapidly move information from one processing center in the brain to another is also a part of intelligence. One study that used sophisticated imaging techniques to trace the network of white matter tracts through the human brain reported that highly organized networks were associated with higher scholastic achievement (Bathelt et al., 2019). Clearly, speed of processing and efficient movement of information are both a part of intelligence.

Questions

1. **Which do you think is more important to intelligence, speed of computation or speed of moving those computations around in the brain? Provide evidence for your answer.**
2. **If you could increase just one of these two dimensions of intelligence in yourself, which would you choose and why?**

Because they were thought to simply keep neurons glued in place, they were named after the Greek for glue. For most of the last century, glia were thought to merely provide a scaffold for neurons. We now know that glia are intimately involved in the day-to-day operation of the brain. Glia are considered to be a part of gray matter.

Glia do provide structural support for neurons, but they should be considered co-stars with neurons rather than simply supporting actors. Glia provide nutrients to neurons, participate in the regulation of the internal environment of the brain, and also play a role in the brain's response to damage. Recent evidence suggests that glia may in fact play an important function in modulating or adjusting neurotransmission by neurons. In fact, glia release chemical messengers known as gliotransmitters to control neurotransmission. Glia in the human digestive system, known as enteric glia, regulate digestion, and evidence suggests they may play a role in obesity and Parkinson's disease. Undoubtedly, glia will be an important research focus in the years ahead.

There are several types of glia, and their most important characteristics are summarized in Table 2.2. Figure 2.19 illustrates the different types of glia found in the CNS.

TABLE 2.2 Characteristics of the Major Glia

Glia type	Location	Function
Microglia	CNS	Cleaning away debris from damaged cells in the CNS and spinal cord
Astrocytes	CNS	Recycling neurotransmitters, forming the blood–brain barrier, and regulating blood flow in the brain
Oligodendrocytes	CNS	Creating the myelin sheath for neurons in the CNS
Ependymal cells	CNS	Creating cerebrospinal fluid that flows through ventricles in the brain
Radial cells	CNS	Providing a scaffolding for growth in developing brain; still active in adult cerebellum and retina
Schwann cells	PNS	Creating myelin sheath for neurons in the PNS
Satellite cells	PNS	Regulating the chemical environment in some neurons
Enteric glia cells	PNS	Helping regulate homeostasis in the digestive system

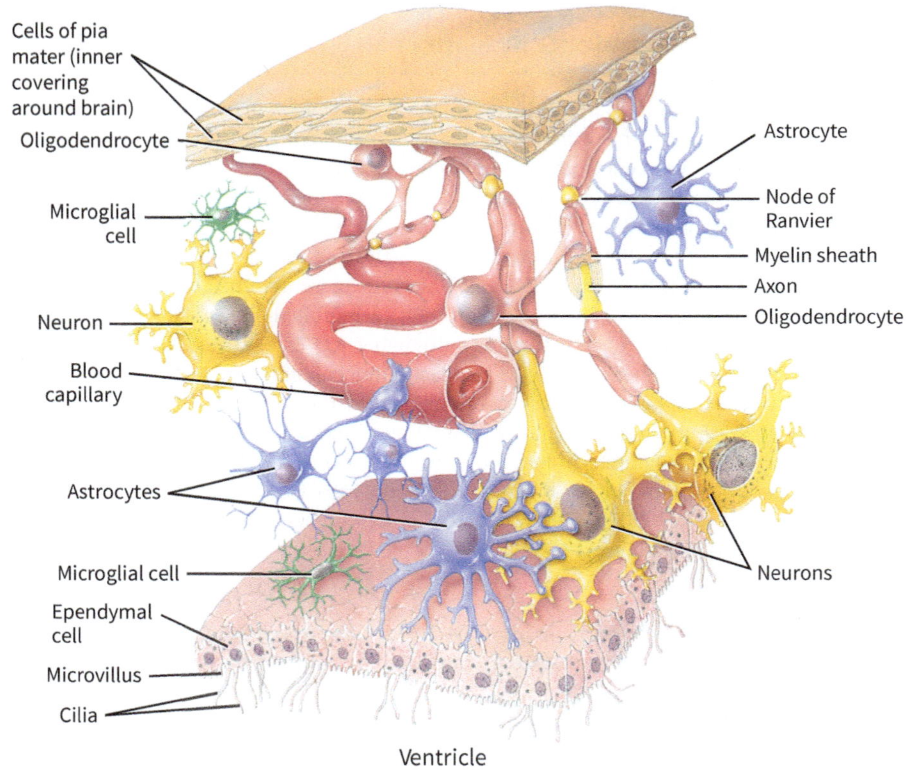

FIGURE 2.19 Types of glia Glia are cells that provide support functions to neurons such as maintaining homeostasis, forming myelin, and providing physical support and protection. Once considered relatively unimportant, glia are now recognized to be essential to the proper functioning of the nervous system.

2.4 Before You Go On

Section Summary

White matter refers to tracts of myelinated neurons that transmit information throughout the CNS and PNS. Gray matter consists of neurons that are involved in computation and also glia and blood vessels. Glia are support cells that maintain the neural environment by providing a physical scaffolding, delivering nutrients, and removing metabolic waste. Glia also generate myelin in both the CNS and PNS.

Comprehension Check

1. A task likely to draw on white matter is _____, and a task likely to draw on gray matter is _____.

- **a.** answering a phone call; catching a baseball
- **b.** reacting to a bug bite; blinking in response to a bright light
- **c.** deciding what to eat; catching a baseball
- **d.** reacting to a bug bite; playing chess

2. Which of these could be a symptom of a disease that damages white matter?

- **a.** problems with balance and coordination
- **b.** an increased appetite
- **c.** a decrease in the desire to sleep
- **d.** a sudden change from right-handedness to left-handedness

3. What is the role of glia in the nervous system?

- **a.** providing essential support functions to neurons
- **b.** increasing the speed of neural transmission
- **c.** facilitating the sodium-potassium pump
- **d.** attenuating the resting potential

Conclusion

We began this chapter with the story of Maria, who was diagnosed with multiple sclerosis, a neurodegenerative disorder that strips away the myelin sheath from neurons in the brain. Her symptoms included visual difficulties, balance problems, fatigue, and tingling and numbness in her extremities. In this chapter, you learned about the physical structure of neurons and the internal mechanism that make them work to create behavior. For many of us, a goal of behavioral neuroscience is to understand everyday behavior as well as the cause of neurological and psychiatric disorders. As complex as these topics are, at some level, all behavior rests upon the proper functioning of neurons.

Critical Thinking Questions

1. You learned in this chapter that there are sensory neurons, motor neurons, and neurons that connect to other neurons. It is likely that we all differ slightly in the relative amounts of each type. If you could increase the number of one of these neuron types you had, which would it be and why?
2. In this chapter, you learned that dendrites convey information from one neuron to another, and that the human nervous system enjoys a luxury of dendrites compared with other mammals. What activities might you engage in throughout your life to build more dendrites?
3. The all-or-none law states that a neuron will not fire unless its threshold potential has been exceeded; small stimulations are suppressed. How might this work to create a nervous system in which signals are clearly distinguished from the background noise of cell activity?
4. Humans have quite a bit more white matter than other mammals. Why might that contribute to the human ability to think at an abstract level?

Dig Deeper into Behavioral Neuroscience

2.1 An Introduction to the Neuron

There are several excellent online sources to read more about discovery of the neuron:

https://cerebromente.org.br/n17/history/neurons2_i.htm

This book describes the life and work of Cajal as well as the importance of his discoveries.

Erhlich, B. (2016). *The Dreams of Santiago Ramon y Cajal.* Oxford Press. ISBN-10: 9780190619619

2.2 The Structure of a Neuron

This text is a comprehensive overview of the neuron with emphasis on the biological mechanisms that define it.

Levitan, I. B., & Kaczmarek, L. K. (2015). *The Neuron: Cell and Molecular Biology* (4th ed.). Oxford Press. ISBN-10: 9780199773893

2.3 Neural Transmission of a Signal

To read more about the sodium pump and the biological and chemical processes that create it:

Clausen, M. V., Hilbers, F., & Poulsen, H. (2017, June). The structure and function of the Na,K-ATPase isoforms in health and disease. *Frontiers in Physiology, 8*, 371. https://doi.org/10.3389/fphys.2017.00371. PMC: 5459889. PMID: 28634454.

Here is an inexpensive Kindle-based tutorial on the action potential written by a Ph.D. in Physiology:

Tam, N. (2014). *The action potential: A tutorial study.* Kindle edition. ASIN: B00PBRBSSI

2.4 White Matter, Gray Matter, and Glia

Do individual differences in white or gray matter make a difference? Here are sources that can help you to explore that issue.

Gennatas, E. D., Avants, B. B., Wolf, D. H., Satterthwaite, T. D., Ruparel, K., Ciric, R., Hakonarson, H., Gur, R. E., & Gur, R. C. (May 2017). Age-related effects and sex differences in gray matter density, volume, mass, and cortical thickness from childhood to young adulthood. *The Journal of Neuroscience, 37*(20), 5065–5073. https://doi.10.1523/JNEUROSCI.3550-16.2017. PMC: 5444192. PMID: 28432144.

Fields, R. D. (2008). White Matter Matters. *Scientific American, 298*(3), 54–61. Bibcode:2008SciAm.298c..54D. doi:10.1038/scientificamerican0308-54.

Haier, R. J., Jung, R. E., Yeo, R. A., Head, K., & Alkire, M. T. (2004). Structural brain variation and general intelligence. *NeuroImage, 23*(1), 425–433. https://doi.org/10.1016/j.neuroimage.2004.04.025

Bathelt, J., Scerif, G., Nobre, G. A., & Astle, D. E. (2019). Whole-brain white matter organization, intelligence, and educational attainment. *Trends in Neuroscience and Education, 15*, 38–47.

To learn more about glia in an engaging but informative fashion.

Koob, A. (2009). *The root of thought* (Kindle ed.). Person FT Press. ASIN: B002BVLQHM.

Chapter 2 Before You Go On Answers

2.1 Before You Go On Answers

1. b. the peripheral nervous system and the central nervous system
2. c. neurons posed unique challenges in visualizing their structure
3. d. a small and random number of neurons in their entirety

2.2 Before You Go On Answers

1. d. They have a well-defined nucleus.
2. d. taking in a signal from outside the neuron
3. a. unipolar, bipolar, multipolar, and pseudounipolar

4. b. sensory, motor, and interneurons
5. c. It is fast acting.

2.3 Before You Go On Answers

1. d. electrical potential and concentration gradient
2. c. the electrical potential inside a neuron that is resting after firing
3. a. unlikely to fire
4. d. firing more rapidly

2.4 Before You Go On Answers

1. d. reacting to a bug bite; playing chess
2. c. a decrease in the desire to sleep
3. a. providing essential support functions to neurons

Chapter 2 Glossary

action potential A rapid rise and then fall in the electrical potential of a neuron's membrane that results in information transmission.

active transport The movement of ions across a cell membrane from a region of lower ion concentration into a region of higher ion concentration.

afferent neuron A neuron that detects a change in the external or internal environment and sends a signal to the central nervous system; also known as a sensory neuron.

all-or-none law A principle in neurophysiology stating that the response of a nerve cell is not dependent on the intensity of the original stimulus.

axon A long, slender fiber in the vertebrate neuron that conducts impulses known as action potentials.

concentration gradient The force generated when the concentration of a particle is higher in one region of a solution than in another region.

cytoplasm The material inside a cell, with the exception of the nucleus.

dendrite A short, branched extension of a nerve cell that receives input from another cell.

dendritic spine A small bump or spike at the end of a dendrite that attaches to one other neuron.

depolarization A change in a membrane's potential that makes it more positive.

efferent neuron A neuron responsible for carrying information away from the brain to effect behavior; also known as a motor neuron.

electrical potential The difference in electrical charge on either side of a membrane that drives an ion in a specific direction.

eukaryotic cell An animal cell containing membrane-bound structures such as a nucleus.

glia Non-neural cells in the CNS and PNS that provide support for neurons; also known as neuroglia.

Guillain-Barré syndrome (GBS) A rare disease in which the immune system attacks the myelin on neurons.

hyperpolarization A change in a membrane's potential that makes it more negative than resting potential.

interneuron A neuron located in the central nervous system that is neither efferent nor afferent.

membrane potential The difference in electrical potential between the inside and the outside of a cell.

motor neuron A neuron located in the central nervous system that controls a muscle or a gland in the body; also known as an afferent neuron.

myelin An insulating sheath around some neurons that speeds signal transmission.

nervous system A complex network of nerves and cells that transfer signals between the brain and the body through the spinal cord.

node of Ranvier One of a series of gaps in the myelin sheath of myelinated neurons that speeds signal transmission.

nucleus An organelle in a eukaryotic cell that contains most of the cell's genetic material.

organelle An organized structure in a cell that is contained within a membrane.

polarization The state when the electrical potential outside a neuron is positive and the inside is negative.

refractory period The period immediately following an action potential in which the cell is incapable of generating another action potential.

repolarization A change in the electrical potential of a membrane that returns the potential from a positive to a negative state.

resting potential The state of a neuron in which a –65 mV difference exists between the inner and outer environments.

saltatory conduction The propagation of an impulse along a myelinated neuron.

sensory neuron A neuron that detects a change in the external or internal environment and sends a signal to the central nervous system; also known as an afferent neuron.

sodium-potassium pump A pump that generates an electrochemical potential that is responsible for the action potential; also known as a Na^+/K^+ pump.

soma The cell body of a neuron.

synapse The functional junction between two neurons where information transmission occurs.

telodendria The branches at the end of an axon.

threshold potential The membrane potential that must be exceeded to depolarize a membrane and create an action potential.

CHAPTER 3

Synapses and Neuro-transmitters

PeopleImages/iStock/Getty Images

Information transfer from one neuron to the next occurs across the synapse, like the gap between these two hands.

CHAPTER OUTLINE

3.1 The Synapse
- The Presynaptic Mechanism
- The Postsynaptic Mechanism
- A Closer Look at Interneuron Communication
- Chemical versus Electrical Synapses

Check It Out: Electrical Synapses and Atrial Fibrillation

3.2 Receptors and Postsynaptic Effects
- Ionotropic and Metabotropic Receptors
- Downregulation and Upregulation

Think About It: How Upregulation Explains Nicotine Cravings

- Autoreceptors and Heteroreceptors

3.3 An Overview of Major Neurotransmitters
- Four Key Neurotransmitter Systems

Check It Out: The Sleepy Sickness Epidemic

- Other Neurotransmitters of Note

3.4 Disorders and Treatments Related to Neurotransmission
- Synaptopathies

Advances in Behavioral Neuroscience: Dimitri Kullmann and the Silent Synapse

- Alcohol, Drugs, and Addiction
- Drug Treatments Targeting Neurotransmission

LEARNING OBJECTIVES

LO 3.1 **Describe** how information is transmitted between neurons through neurotransmitters, synapses, and receptor sites.

LO 3.2 **Outline** the different regulatory processes involved in neurotransmission.

LO 3.3 **Identify** the most common neurotransmitters, explaining the functions of each.

LO 3.4 **Integrate** your understanding of synaptic processes to understand related disorders and their treatments.

Stories of the Brain

Kurt was a professor of mathematics at a small college. He loved to teach his students about the many ways mathematics can solve real-world problems. After a long career, he retired to enjoy his hobbies of gardening and playing the violin in the town's adult orchestra. One morning while he was eating breakfast, his coffee cup slipped out of his hand and fell to the floor, shattering into pieces. "How careless of me," he thought and gave it no mind. But then he began to notice that he was often dropping knives, forks, cups, and glasses. He also noticed that he was struggling to play pieces on his violin that in the past he had easily managed; his playing was slower, and he sometimes lost his grip on the bow because his hands were shaking. One day on their daily walk, his wife noticed that he could not keep up even a modest pace although he had always been a vigorous walker. He walked with an awkward shuffle, and when they stopped at a crosswalk, he hesitated to start walking again. It was as if he could not initiate voluntary movements without a struggle.

Kurt had always been an active person, quick to enter into conversations and eager to be out and about, but now he was quiet and avoided social contact. After he fell several times for no good reason, his wife took him to see a neurologist. The diagnosis was Parkinson's disease.

In a different example, Jean Marie was a woman in her mid-50s who lived with her brother in a modest row home in Pittsburgh. Jean Marie had been diagnosed with mild schizophrenia, and her psychiatrist had prescribed antipsychotic medications to help her live a normal life. However, after many years on these medications, Jean Marie begin to experience some disturbing motor behaviors. She would grimace and smack her lips over and over, and she continuously blinked her eyes. Her greatest pleasure in life was to sit on a rocking chair on her front porch, smoke a cigarette, and watch the world go by. But as she rocked for as much as 10 hours a day, she eventually broke her wooden rocker. Her brother then bought her a cast iron rocker so that she could rock to her heart's content, which she did all day, winter or summer, until her unrelenting rocking wore a groove in the concrete porch. Her brother took her to see a neurologist who diagnosed her with tardive dyskinesia.

In contrast with Kurt, who could not initiate voluntary movements, Jean Marie was consumed by involuntary movements. Both of these examples illustrate the malfunction of communication between neurons. To fully understand how our brain creates consciousness, coordinates our movements, and allows us to live a life of purpose, we must understand how information is transferred from one neuron to another.

3.1 The Synapse

LO 3.1 Describe how information is transmitted between neurons through neurotransmitters, synapses, and receptor sites.

In the previous chapter, you learned about the structure of a typical neuron and that when a signal of sufficient strength stimulates the dendrites of that neuron, an action potential travels down the axon until it reaches the terminal buttons. If we hope to understand how information is transmitted throughout the nervous system, we must understand what happens next; that is, how information is transferred from one neuron to the next. To do so, we will focus on the **synapse**, which is the connection between two neurons. From here on, we will call the neuron transmitting the signal the presynaptic (before the synapse) neuron and the neuron receiving the signal the postsynaptic (after the synapse) neuron. As a signal travels from one neuron to the next, of course, each neuron serves in both roles.

What you are about to learn is one of the more complicated processes in neuroscience, so a simple metaphor may help you grasp the material more quickly. Imagine for a moment that you have an erasable whiteboard on which you have written a note. You hand that whiteboard to a friend, who reads the note, then erases the board, and finally hands it back to you so that it can be used again. This simple mental image of two people passing an object containing a message back and forth will help you understand how one neuron communicates with another.

The Presynaptic Mechanism MCAT

When a neuron has been sufficiently stimulated, a signal travels down the axon and into the terminal button. That signal causes small vessels known as **synaptic vesicles** to move toward the end of the terminal button and merge with the presynaptic membrane (**Figure 3.1**). Synaptic vesicles are created by the Golgi bodies in the neuron's soma, or cell body. These vesicles are about 40 nm in diameter and are found close to the terminal membrane of the axon button, near the edge of the presynaptic neuron. In response to a signal of sufficient strength, synaptic vesicles are transported through microtubules along the axon until they reach a synapse in the telodendria at the end of the axon.

Synaptic vesicles contain proteins with two different mechanisms—transport and trafficking. The **transport proteins**, or simply transporters, regulate the absorption, or uptake, into the vesicle of chemical messengers known as **neurotransmitters**. The role of the trafficking proteins is not as well understood, but it appears that they perform essential housekeeping chores that maintain the integrity of the vesicle. When a vesicle reaches the membrane of the terminal button, it releases its neurotransmitter into a gap between the two neurons called the **synaptic cleft**. The neurotransmitter then crosses the synaptic cleft and may dock at a receptor in the postsynaptic neuron. This is the point at which we shift our attention from the presynaptic neuron (the friend who is handing over the whiteboard with its message) to the postsynaptic neuron (the friend who is accepting it).

The Postsynaptic Mechanism MCAT

Once released into the synapse by the presynaptic neuron, a neurotransmitter may reach a neurotransmitter receptor located on the surface of the postsynaptic neuron or glia. Interactions between the neurotransmitter and proteins in the postsynaptic neuron cause the receptor channels in the postsynaptic neuron to open, allowing the neurotransmitter to dock at that receptor.

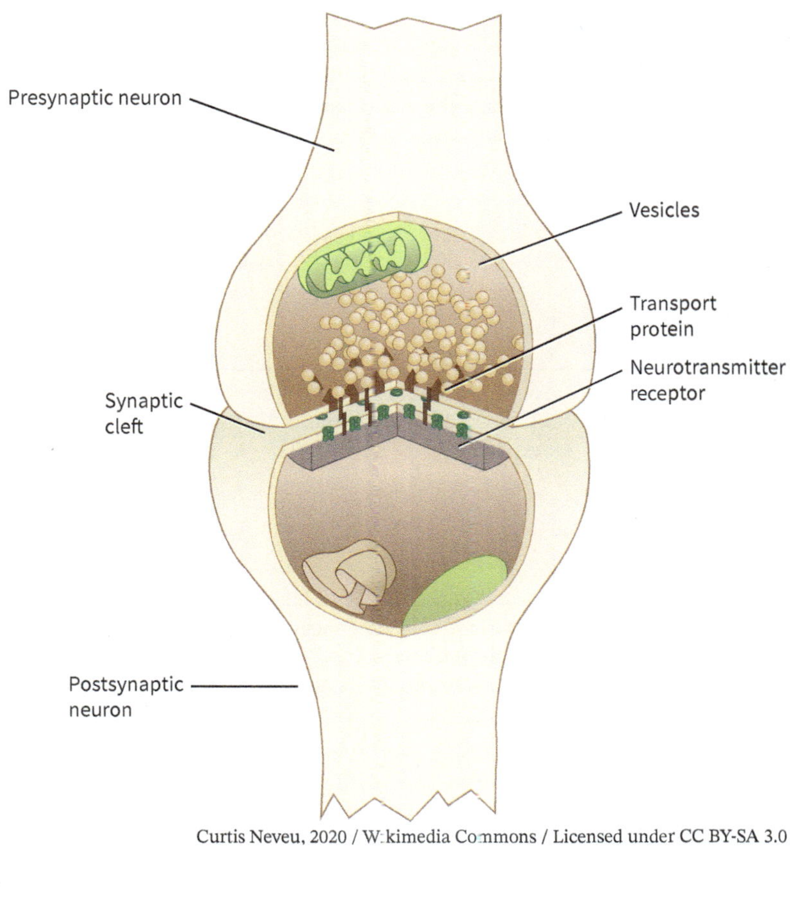

FIGURE 3.1 The synapse Neurotransmitters created in the presynaptic neuron are transported to the synapse and released into the synaptic cleft, where they dock at the postsynaptic receptor.

What you already know about voltage-gated ion channels from our discussion of the sodium-potassium pump in Chapter 2 can serve as a model for understanding what happens at a typical receptor site. These postsynaptic receptors are protein channels that have an affinity for specific neurotransmitters. When that neurotransmitter comes into contact with the receptor, the contact causes a conformational change in the receptor that brings about an exchange of ions between the extracellular and intracellular environments. This leads to a change in the postsynaptic membrane potential.

If enough receptors in the postsynaptic neuron are stimulated, there is a change in the membrane potential of the postsynaptic neuron, and this triggers an action potential. More frequent firing by a receptor or more receptors firing are coded as a stronger signal. At this point, information has now been transmitted from one neuron to the next. This process of transmitting information from one neuron to the next is the basis for everything our nervous system does to maintain life and create consciousness.

You might ask what happens after the neurotransmitter docks in the receptor site, and that is a worthwhile question. The answer is that the receptor site releases the neurotransmitter back into the synaptic cleft, where it is deactivated by enzymes. It may then be taken back up into the presynaptic neuron by a transport protein. Transporters return the neurotransmitter to the presynaptic axon terminal where it can be reabsorbed into a synaptic vesicle. Then the cycle can begin again.

Table 3.1 returns to our metaphor of using a whiteboard to transfer information from you to a friend and summarizes what you have just learned in a way that makes it easier to understand and recall.

TABLE 3.1 Steps in the Transfer of a Neurotransmitter across the Synapse

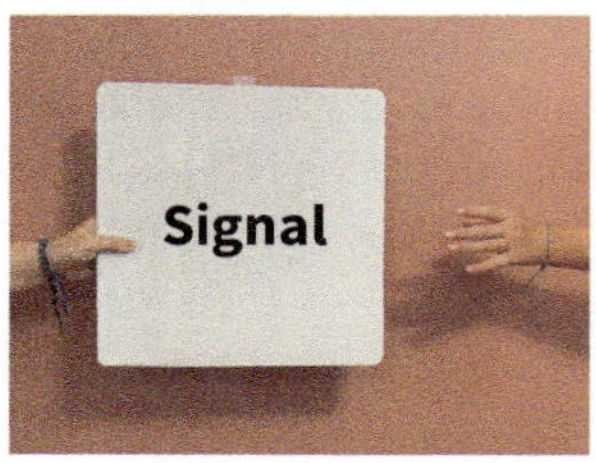	You compose a message on the whiteboard.	A neurotransmitter is synthesized and transported into a vesicle, which moves toward the synapse.
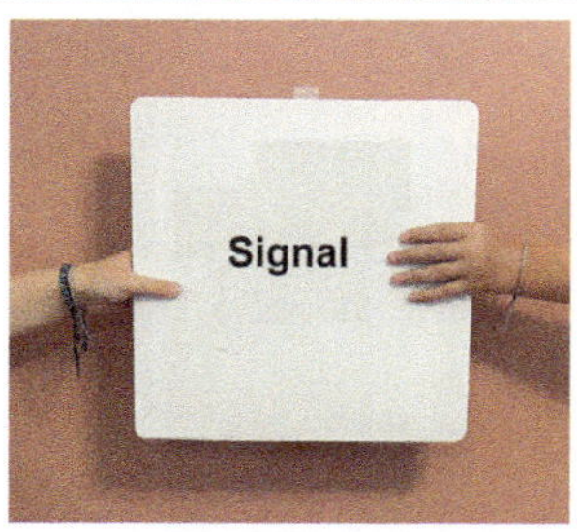	You hand the whiteboard to a friend.	The vesicle releases its contents into the synaptic cleft.
	Your friend takes the whiteboard and reads the message.	The neurotransmitter reaches and activates the postsynaptic receptor, thereby transferring information.
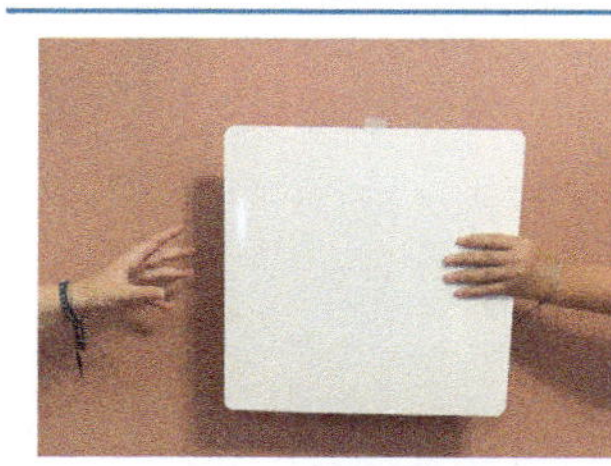	Your friend clears the message from the whiteboard and returns it to you for its next use.	The neurotransmitter is released from the receptor, deactivated by enzymes, and then taken up again by the presynaptic neuron.

A Closer Look at Interneuron Communication MCAT

We continue our exploration of synaptic mechanism with a more detailed look into this complex process. As we do so, it may help you to refer to Figure 3.2 before you read the next section.

Neurotransmitters are synthesized in the cytoplasm from components such as amino acids. Synaptic vesicles containing neurotransmitters migrate down the axon until they reach the axon terminal in the presynaptic neuron. When an action potential reaches the axon terminal (**step 1**), a change in the axon's membrane potential begins a cascade of actions. You already understand the concept of voltage-gated ion channels that can open to create a passage for transfer of sodium and potassium ions between the inner and outer cellular environments that have a concentration gradient. A similar situation exists at the axon terminal, where a concentration gradient exists for calcium ions (Ca^{2+}) available in the inner and outer cellular environments. At the axon terminal, there are voltage-gated channels that open to allow Ca^{2+} ions to travel from the extracellular environment into the axon terminal, which increases the calcium concentration inside the axon (**step 2**). In response to this change in the inner environment, calcium-sensitive proteins on the surface of the vesicles are attracted to the synapse membrane where they dock with the membrane and release the neurotransmitter into the synaptic cleft (**step 3**).

Neurotransmitters then enter the synaptic cleft, where some simply diffuse away, but others bind to a receptor on the dendritic spine of the postsynaptic neuron (**step 4**). When a neurotransmitter binds to a receptor on the surface of the postsynaptic neuron, that causes ion channels to open, allowing sodium ions to pass through and creating a change in the membrane potential of that neuron (**step 5**). Sufficient change in the membrane potential will

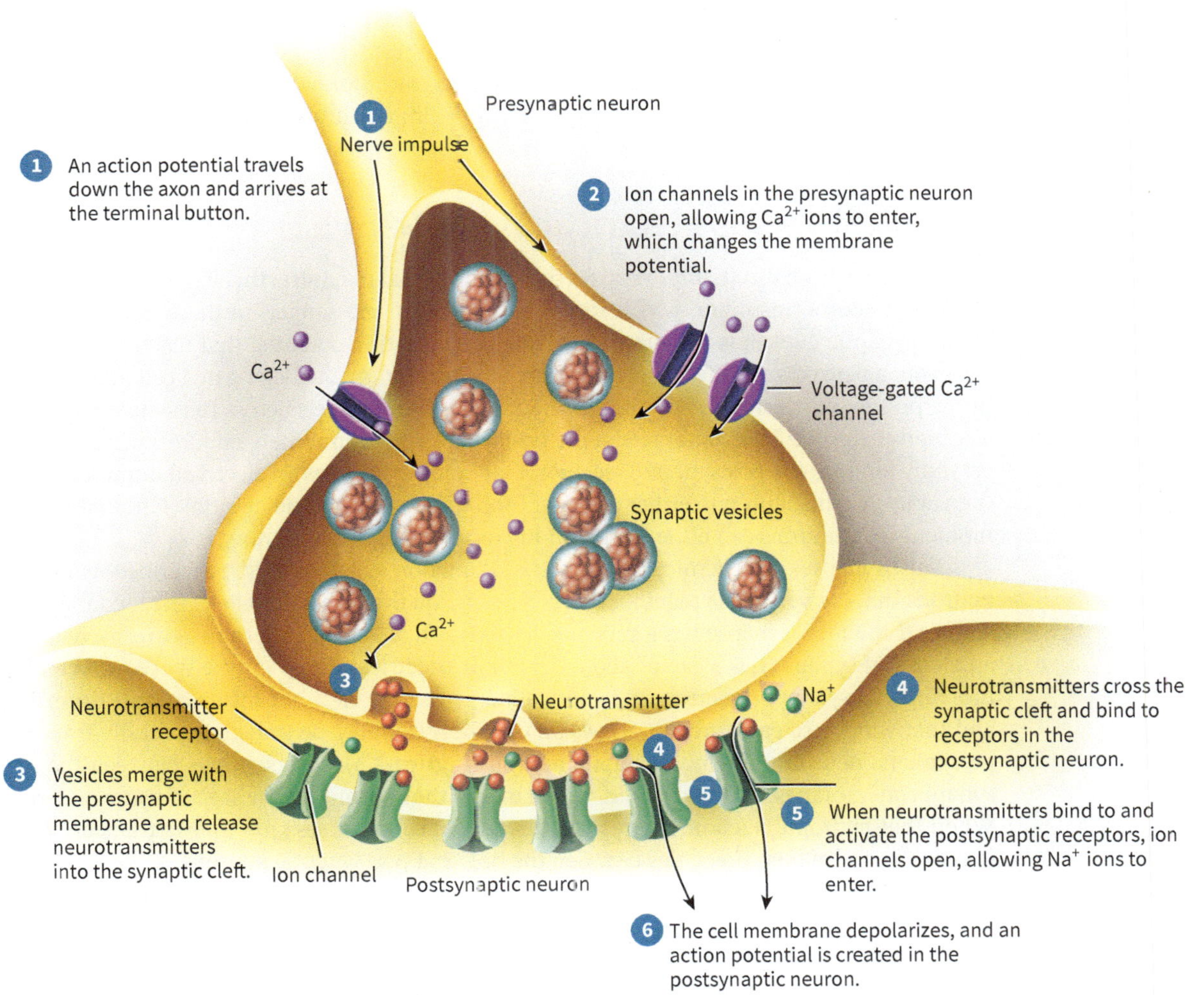

FIGURE 3.2 **Neurotransmitter transfer at the synapse** A nerve impulse is transferred from one neuron to the next as ion channels open and membrane potentials change.

result in an action potential (**step 6**). If it does, information transmission has now occurred. Before we move on, it is worth noting that receptors can exist not only on the surface of the postsynaptic neuron, but also on the soma or axon.

Once a neurotransmitter from the presynaptic side of the synapse has docked with a receptor on the postsynaptic side and information has been transferred, that neurotransmitter must be removed. You can see that there must be a mechanism that removes the neurotransmitter key, so to speak, from the receptor lock. Otherwise, the receptor would never fire again. In fact, enzymes in the postsynaptic environment break apart the neurotransmitter, thereby releasing it from the receptor. Diffusion can also remove the neurotransmitter from the receptor. The entire process of the neurotransmitter binding with the receptor and then releasing it takes just a fraction of a millisecond.

Deactivated neurotransmitters are then reabsorbed back into the presynaptic axon terminal by the action of **reuptake pumps**, which are protein pores that attract and absorb neurotransmitters from the synaptic cleft. Finally, note that there is a collection of fibers in the dendritic spine of the postsynaptic neuron known as postsynaptic density. These fibers are not well understood, but they appear to organize and maintain the postsynaptic receptors.

You now have a global view of how one neuron communicates with another neuron. A signal originates in a neuron and travels down the axon until it reaches the axon terminal where a vesicle filled with neurotransmitters releases its contents into the synaptic cleft. If that neurotransmitter reaches a receptor, it docks and activates the receptor. In doing so, it creates a change in the membrane potential of the second neuron that may lead to information transmission. Once that happens, the neurotransmitter is released from the receptor and taken up again by the presynaptic neuron.

As complicated as this might seem, if you visualize the process, you will quickly see that the basics are simple and even obvious. A signal from the presynaptic neuron causes a vessel filled with a chemical messenger to dock on the edge of the presynaptic synapse and then release its contents into the synapse. If the neurotransmitter encounters the appropriate receptor, that neurotransmitter docks, activates the receptor, and then disengages, thereby leaving that receptor site available for future, if not almost immediate, reuse.

Chemical versus Electrical Synapses MCAT

So far we have been discussing synapses that are chemical in nature; that is, they transmit data through a neurotransmitter, which serves as a chemical messenger. In these neurons, when the presynaptic neuron is sufficiently activated by an action potential that has traveled down the axon, a neurotransmitter is released via the activation of a voltage-gated calcium channel in the presynaptic terminal button. This neurotransmitter travels across the synapse and ideally binds to a specific receptor site in the postsynaptic neuron. When a neurotransmitter successfully binds to the postsynaptic receptor, that binding initiates an electrical impulse, which then activates the neuron to either excite or inhibit the postsynaptic neuron. (Remember for a moment that neurons can either be gas or brake, excitatory or inhibitory.)

Most synapses are chemical in nature and function as we have just described. However, neurons that are very close together can communicate through a direct electrical connection. A typical chemical synapse has a gap, or synaptic cleft, of 20 to 40 micrometers (μm), whereas an electrical synapse has a gap of about 3.5 μm. In electrical synapses, the space between the pre- and postsynaptic membranes is not called a synaptic cleft but instead a **gap junction**, and that very close gap junction allows an electrical signal to be transmitted easily from the presynaptic neuron to the postsynaptic neuron (**Figure 3.3**). The channels between the two synaptic membranes are hydrophilic, which means they permit the passage of water and ions. When that hydrophilic channel opens, a direct connection of 1–2 μm allows transfer of the ions and even smaller molecules between the pre- and postsynaptic neurons.

There are several key differences between chemical and electrical synapses. Chemical synapses take about a millisecond for the neurotransmitter to cross the synaptic cleft, whereas signal transmission for an electrical synapse is almost instantaneous. Chemical synapses transmit a signal in one direction only, whereas electrical synapses can signal in either direction. Chemical synapses convey only ions from one side to the other, whereas electrical synapses can also allow molecules such as ATP to migrate across the synapse. Chemical synapses

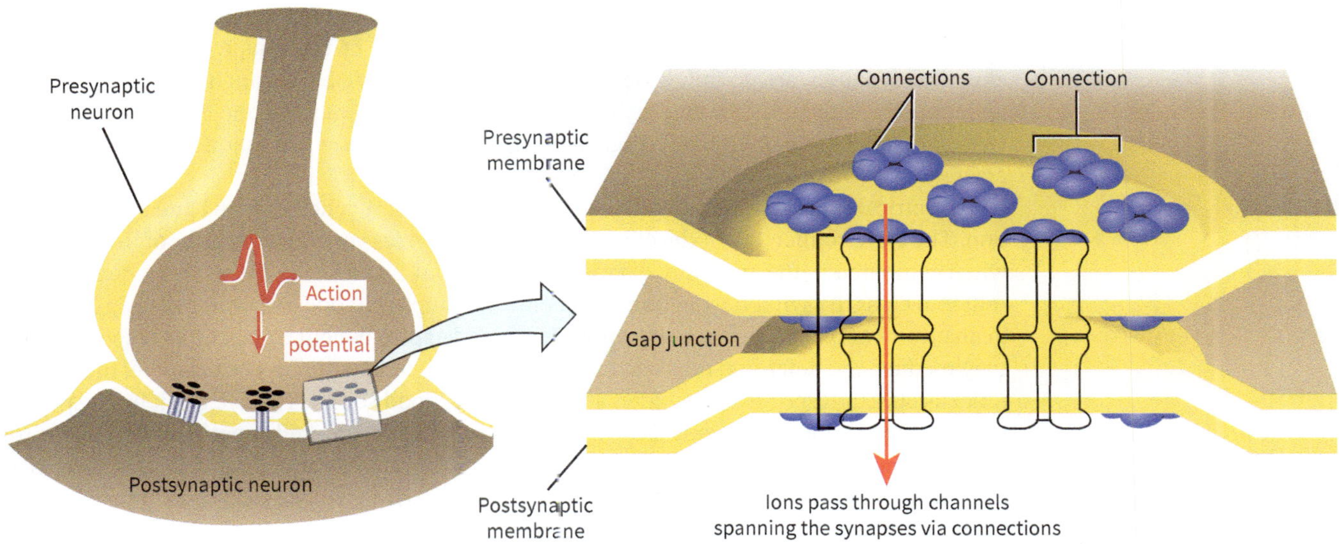

FIGURE 3.3 The gap junction of an electrical synapse The pre- and postsynaptic membranes of the electrical synapse almost touch at the gap junction, where they are pierced by hydrophilic channels. When the neuron fires, these channels open to permit the passage of ions in either direction.

are much more common than electrical synapses, but as you will see in this chapter, their mechanism is much more complicated, and they are more varied in how they function. The essential differences between chemical and electrical synapses are summarized in Table 3.2.

You should immediately ask yourself why we have two different types of synapse. The chemical synapse has a large gap and requires the transmission of information through a chemical messenger, whereas the electrical synapse is a direct physical connection. What would be the advantage of one over the other? Why not have just one type of synapse? The answer is that each type of connection has its own purpose.

In contrast with electrical synapses, chemical synapses can code for strength of signal through the number of receptor sites that are activated at any one time. This principle is easy to understand through this metaphor. Imagine you are sitting among a group of friends, and one person very quietly says that it is time for lunch. One voice does not create a consensus for lunch, and no one makes a move to leave. A while later, most of the group quietly begins to murmur about being hungry. Now you should be convinced that it is time for lunch, and the whole group gets up. In the first case, you have one small signal from one source, whereas in the second, you have many small signals from many sources. The more signals sent between two neurons, the stronger the connection grows. Because chemical synapses modify themselves in response to continued stimulation, they are the biological basis for learning.

TABLE 3.2 Comparing Chemical and Electrical Synapses

	Chemical	Electrical
Speed	~1 millisecond	Nearly instantaneous
Cleft or gap width	20–40 μm	1–2 μm
Directionality	Pre- to postsynaptic neuron only	Either direction
Functionality	Only has effect at receptor site	Affects many cell processes
Frequency	Most common synapse	Not as common
Advantages	Finer modulation	Fast and coordinated
Disadvantages	Slow	Unmodulated
Location	Throughout the body, including for learning and perception.	Defensive reflexes, cardiac muscles, and retina

If electrical synapses are faster, why do we even have chemical synapses? The answer to that question is a major theme of this chapter. By the end of this chapter, you will see that chemical synapses are complex mechanisms that are modifiable through pre- and postsynaptic mechanisms that may amplify or attenuate the signal sent from one neuron to the next. In fact, chemical synapses can change in response to experience and modify their function to create what we call learning.

Electrical synapses are thought to confer two important benefits: speed and synchronization. The close connection of the electrical synapse provides very rapid transfer of information when compared to the chemical synapse. It takes about 2 milliseconds (ms) for a neurotransmitter to cross the synaptic junction in a chemical synapse but only 0.2 ms for the gap junction in an electrical synapse to open. Because electrical synapses benefit from a rapid response rate, they mediate the defensive reflexes that protect the life of an animal. For example, the startle reflex occurs in response to a sudden and unexpected loud sound. The eye blink is a response to some object looming at your face, and the nociceptive flexion reflex occurs when you rapidly and unconsciously withdraw your arm or leg in response to a painful stimulus such as a burn or cut. Because of their rapid response rate, electrical synapses are also more likely to fire synchronously (at the same time), and this property makes them essential in tasks such as coordinating cardiac muscles. The signals to coordinate the action of the muscles that move blood through the heart originate in a small bundle of tissue at the top of the right atrium known as the sinoatrial (SA) node. The SA node as well as much of the cardiac muscle is populated with electrical synapses, which have the characteristic of coordinating their action with adjacent neurons (see **Check It Out: The Importance of Electrical Synapses**). Electrical synapses are also found in the retina, where rapid integration of the thousands of neurons provides us with a rich perceptual experience.

Check It Out | Electrical Synapses and Atrial Fibrillation

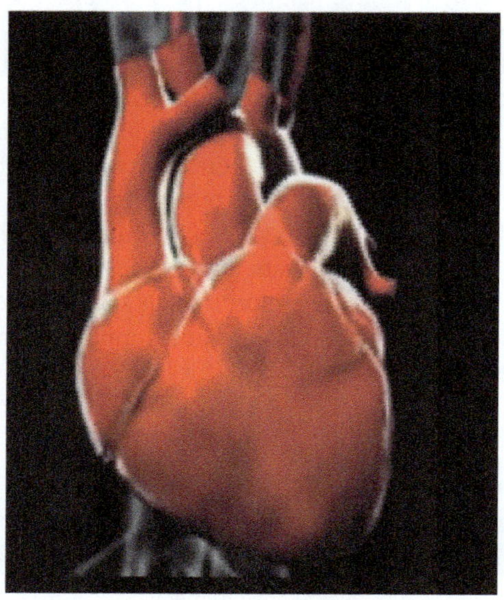

The heart muscles must contract in the proper order to move blood through the heart and into the body keep us alive. Without this coordination of action, a serious condition known as atrial fibrillation (A-fib) may occur. Watch the video **Atrial Fibrillation**, and then answer these questions.

1. How common is A-fib? How many fatalities per year result from A-fib? List two risk factors for A-fib.
2. The video mentioned one consequence of A-fib that directly affects the CNS. What is that consequence, and why is it dangerous?
3. Electrical synapses use calcium ions, whereas chemical synapses use sodium or potassium ions. What is an implication of a lack of calcium on cardiac function?

3.1 Before You Go On

Section Summary

When a signal is received, synaptic vesicles move down to the axon terminal of the presynaptic neuron, where they release neurotransmitters into the synaptic cleft. Neurotransmitters may dock with a receptor on the postsynaptic neuron, thereby transferring the signal before being deactivated and reabsorbed into the presynaptic neuron so the process can begin again. Chemical synapses provide a more complex signal, whereas electrical synapses are much quicker, lending themselves to situations where speed and coordination are paramount.

Comprehension Check

1. Neurotransmitters are synthesized in the _____ and stored in structures known as _____.

- **a.** cytoplasm; synaptic vesicles
- **b.** Golgi bodies; mitochondria
- **c.** synaptic vesicles; Golgi bodies
- **d.** microtubules; neuromodulators

2. What is the name for the small gap between the pre- and postsynaptic neurons?

- **a.** neural gap
- **b.** synaptic cleft
- **c.** synaptic vesicle
- **d.** ion channel

3. The neural process known as reuptake is important because it _____.

- **a.** amplifies the effect of the neurotransmitter at the synapse
- **b.** prevents a neurotransmitter from docking at the wrong receptor
- **c.** guides the neurotransmitter to the receptor where it can be taken up
- **d.** returns the neurotransmitter to the presynaptic neuron for reuse

4. A major difference between chemical and electrical synapses is that electrical synapses _____.

- **a.** are much larger in size
- **b.** are disrupted by magnetic fields
- **c.** operate much faster
- **d.** are found more frequently in the brain

3.2 Receptors and Postsynaptic Effects

LO 3.2 Outline the different regulatory processes involved in neurotransmission.

Up to this point, we have focused on the release of a chemical messenger known as a neurotransmitter that crosses the synapse to transfer information from the presynaptic neuron to the postsynaptic neuron. We now turn our attention to what happens when that neurotransmitter makes contact with the receptor on the postsynaptic neuron. Understanding this mechanism is essential for your subsequent comprehension of the treatment of many psychiatric illnesses, so we now turn to a discussion of receptor types and how these structures work to create complex behavior.

Ionotropic and Metabotropic Receptors MCAT

Just as there are different types of neurons, there are also different types of receptor sites. **Ionotropic receptors**, also known as ligand-gated ion channels, are receptors that are characterized by a chemical gate that opens in response to the presence of the neurotransmitter. Recall from Chapter 2 that gated ion channels open and close in response to a stimulus. The voltage-gated ion channels that are part of the presynaptic mechanism open or close in response to differences in electrical charge, whereas ligand-gated channels open or close in response to the binding of a ligand. A ligand is a molecule that binds to another molecule, and neurotransmitters are one example of ligands. When the neurotransmitter binds to the receptor site, the channel opens, thereby allowing ions to flow from the extracellular environment into the neural environment. This causes a change in the membrane potential, which can then lead to an action potential in the postsynaptic neuron.

Metabotropic receptors, on the other hand, work in a very different fashion. Unlike the direct action of an ionotropic receptor, when a metabotropic receptor is activated by a neurotransmitter it acts indirectly by releasing a **G protein** that is affixed to the inside of the postsynaptic neuron. When released, the G protein initiates activity within the postsynaptic neuron. This G protein serves as a second messenger (we will discuss these in a moment), and metabotropic receptors are also known as G protein-coupled receptors.

Ionotropic receptors are characterized by rapid response rates on the order of a few milliseconds, whereas metabotropic receptors often remain open for seconds, minutes, and possibly even days. Together, the two receptor types provide both a rapid response to a stimulus and a slow, gradual response. Figure 3.4 illustrates the physical differences between the two

FIGURE 3.4 Ionotropic and metabotropic receptors **a.** An ionotropic receptor affects the postsynaptic neuron directly, opening a channel that allows ions in the extracellular environment to enter the neuron, which changes the membrane potential. **b.** A metabotropic receptor changes cellular activity indirectly, releasing a G protein that then activates other messengers that open the receptor pore to allow the exchange of ions.

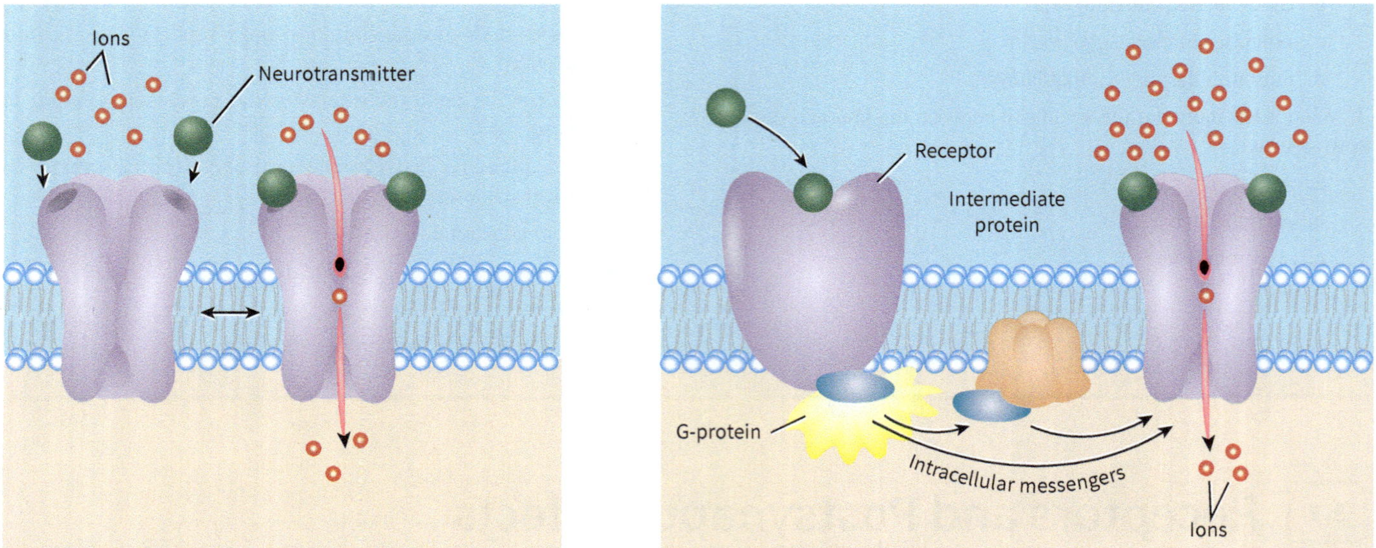

a. Ionotropic receptors **b.** Metabotropic receptors

receptor types. When an ionotropic channel opens in response to a neurotransmitter docking at the receptor, that opening allows ions in the extracellular environment to enter the neuron, thereby creating a change in that neuron's membrane potential. A metabotropic receptor works differently. When a neurotransmitter docks at a G protein-coupled receptor, that receptor releases a G protein. The G protein then activates other messengers within the cell that open an ion receptor pore. This allows the exchange of ions that creates a change in membrane potential, which in turn creates the action potential essential for life.

Here is a simplistic but perhaps effective visual metaphor for remembering the difference between the two receptor types. An ionotropic receptor is like a hostess in a restaurant: you enter the synaptic establishment, and she instantly takes you across the membrane of the lobby and into the table area. In contrast, a metabotropic receptor is like a hotel manager: when they see you entering, they signal a variety of workers who then spring into action across the cell body of the hotel; the front desk to check you in, a bell boy to carry your bags to your room, and the concierge to provide information about local attractions. If you think of the many hotel employees with their distinct functions as being activated by the G proteins released by the metabotropic channel, you have the basics down.

Neurotransmitters are considered first messengers because they bring a message from outside the cell. In contrast, metabotropic receptors are considered **second messengers** because they relay a message from the neurotransmitter within the cell to create effects throughout the postsynaptic neuron. When a metabotropic receptor fires and the G proteins are released into the cell body, they affect a range of intracellular processes.

Second messengers may affect the manufacture and release of neurotransmitters within the postsynaptic neuron, act on the genetic material in the cell nucleus, and even open ion channels in the postsynaptic neuron. Unlike ionotropic receptors, which operate within about 5 milliseconds, metabotropic receptors create second-messenger effects that can continue for seconds and even longer. It is believed that keeping first and second messenger systems in balance is important. For example, disruptions of that balance play a role in affective disorders such as bipolar disorder, mania, and depression.

If you are confused about the difference between a first and second messenger, here is a simple analogy. A friend from your childhood messages you that they will soon be in town. In response, you first tell your friends about the visit and second, clean your home and shop for treats. When a signal came in from outside (a signal from another neuron) it triggered a communication to others (the signal passed to other neurons) as well as an internal response that facilitates the event (a change in internal cellular functioning). Calling your friends in

FIGURE 3.5 **Upregulation and downregulation of receptors** **a.** Upregulation of receptors increases sensitivity by adding receptors. **b.** Downregulation decreases sensitivity by removing receptors.

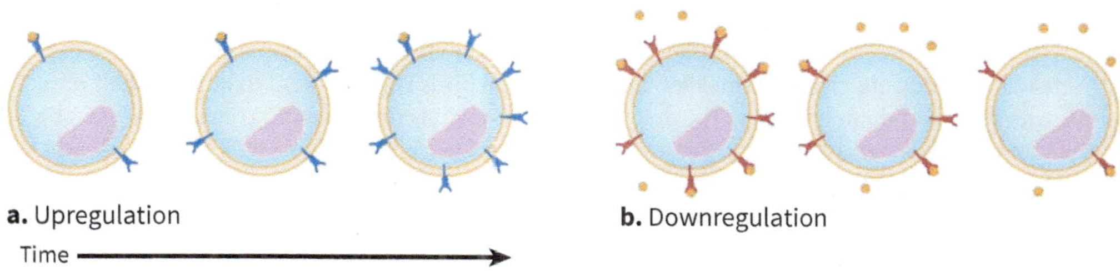

response to the initial signal is a first messenger response, and straightening up your home and shopping for snacks is a second messenger response. The common thread in the metaphor of the friend's visit and second messenger mechanisms is that both cases, the secondary effect is to enhance the function of the primary response.

Downregulation and Upregulation MCAT

In the room where I write, an overhead fan is gently moving the air about me. Because that sound is constant, my brain has chosen to reduce my conscious perception of the sound. At the same time, the weather forecast contains the threat of strong thunderstorms, so I am more likely to be aware of the wind gathering outside my window. This tuning of my consciousness to reduce my response to some stimuli while enhancing my response to others is a wonderful feature of human consciousness. A similar mechanism of regulation occurs at the receptor level.

When a neurotransmitter is chronically absent or blocked, the receptor may become supersensitive, and this is known as **upregulation** (**Figure 3.5a**). Upregulation occurs when, over time and in response to the decrease or absence of a neurotransmitter, the postsynaptic neuron responds by increasing the number of available receptors. It is common in physiology that mechanisms come in opposing pairs, so there is also a process of desensitization known as **downregulation** (**Figure 3.5b**). Downregulation occurs when a receptor has been exposed over a long period of time to an excessive amount of a particular neurotransmitter, and the receptor becomes desensitized by the loss of receptors in the postsynaptic neuron.

Examples of downregulation and upregulation as a result of drug effects may help you understand these important neural mechanisms. When excessive amounts of a neurotransmitter are available at the synapse, the neuron responds by reducing the number of receptors in an effort to maintain a balance, or homeostasis. An individual who takes an opioid continuously for pain relief floods their opioid receptors, so these neurons downregulate by reducing the number of opioid receptors. The result is that the individual gradually develops a tolerance to the drug so that increasing amounts of the drug are required to achieve the same level of pain reduction.

In our second example, an individual being treated for a psychiatric disorder such as schizophrenia is often prescribed an antipsychotic drug that blocks certain dopamine (DA) receptors that are implicated in schizophrenic behavior. Over time, this continued blockade of DA receptors causes these neurons to upregulate by increasing the number of DA receptors available. As a result, these neurons are supersensitive to dopamine. Should the antipsychotic be discontinued for any reason, a rebound effect may occur, leading to a different disorder. This was the case with Jean Marie in this chapter's *Stories of the Brain*, who developed a repetitive motion disorder called tardive dyskinesia as a result of mismanagement of drugs she took for schizophrenia. Nicotine use results in an unusual form of upregulation, which you can explore further in **Think About It: How Upregulation Explains Nicotine Cravings**.

Beyond understanding down- and upregulation, there are two important insights to be learned here. The first is that our brain and body are always striving to be in homeostasis, or balance. As a result, neurons respond to over- or undersupply of a neurotransmitter by adjusting their function, which helps maintain homeostasis but can sometimes have an unintended negative outcome. The second insight comes from the first; treatment of psychiatric and neurological disorders and addictions is difficult because our neural functions respond to treatment by changing their function. This dynamic response typically results in less therapeutic effect. We will discuss these matters in greater detail when we consider the questions of addiction and psychiatric disorders.

Think About It | How Upregulation Explains Nicotine Cravings

According to the Centers for Disease Control, of the millions of American adults who are estimated to be current cigarette smokers, more than two-thirds say they want to quit and half have made a serious attempt to quit in the past year. Despite this, cigarette smoking remains the leading cause of preventable disease and death in the United States, killing almost half a million Americans per year. In addition, smoking-related illnesses continue to have a huge societal cost in terms of medical expenses and lost productivity.

We can look to neurotransmitters to explain in part why smoking is so addictive and why it is so hard to quit. When a drug that mimics the effects of a neurotransmitter becomes available in great quantities over an extended period of time, receptors for that drug will respond typically by downregulating. Nicotine, a stimulant found in tobacco, is unusual in that long-term exposure to nicotine results in upregulation of receptors. Nicotine receptors that are denied nicotine create the perception of craving, which is one reason that it is so difficult to quit smoking.

In one study, researchers imaged active nicotine receptors in nonsmokers and in smokers as they went through a 12-week smoking cessation program. Active nicotine receptors are shown in green in this image. Notice how the number of nicotine receptors decreased in smokers as they abstained from smoking for twelve weeks.

Understanding how and why nicotine is so addictive helps explain why governments use regulations and public health campaigns to try to prevent young people from taking up smoking.

Questions

1. **It has been reported that for smokers trying to quit, the first week of abstinence is the hardest and that if a smoker can abstain from smoking for 3 months, they are much more likely to break their habit. Do the images of nicotine receptors from this study support or refute those observations? Explain.**
2. **Imagine that a friend who has been smoking for some years has decided to quit and asks you for advice. Based on what you saw in this imaging study, what advice can you give about what this person might expect as they go through a 12-week smoking-cessation program?**
3. **Research the regulations on smoking in your area, including age restrictions and e-cigarettes. Given what you now know about nicotine, do you think these regulations are fair and effective? What changes would you make? Explain your answer.**

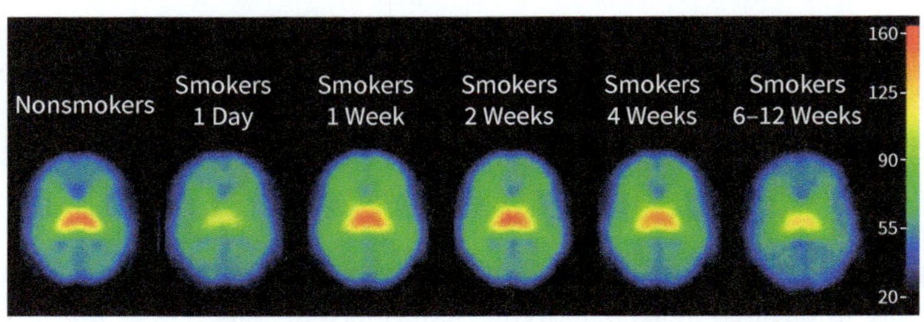

Autoreceptors and Heteroreceptors MCAT

Let us take a moment to move our focus to the receptor site, because it would be a mistake to consider it a mere lock that sits passively awaiting a key. You now know that the release of a neurotransmitter into the synapse initiates a complex process including docking of the neurotransmitter at the receptor site, the neurotransmitter initiating a response in the postsynaptic neuron, the deactivation and release of the neurotransmitter from the receptor, and finally the reabsorption of the neurotransmitter into the presynaptic neuron. As complicated as all this is, there is one more mechanism to consider, which is that the presynaptic neuron acts as a governor on the release of neurotransmitter into the synapse.

You may be familiar with the concept of a governor on an automobile, riding lawn mower, or go-cart; there is a mechanism on the engine that prevents the engine from being damaged by exceeding its design capabilities. In the same vein, our presynaptic neuron has receptors that act in a negative feedback loop to limit the release of neurotransmitter into the synapse.

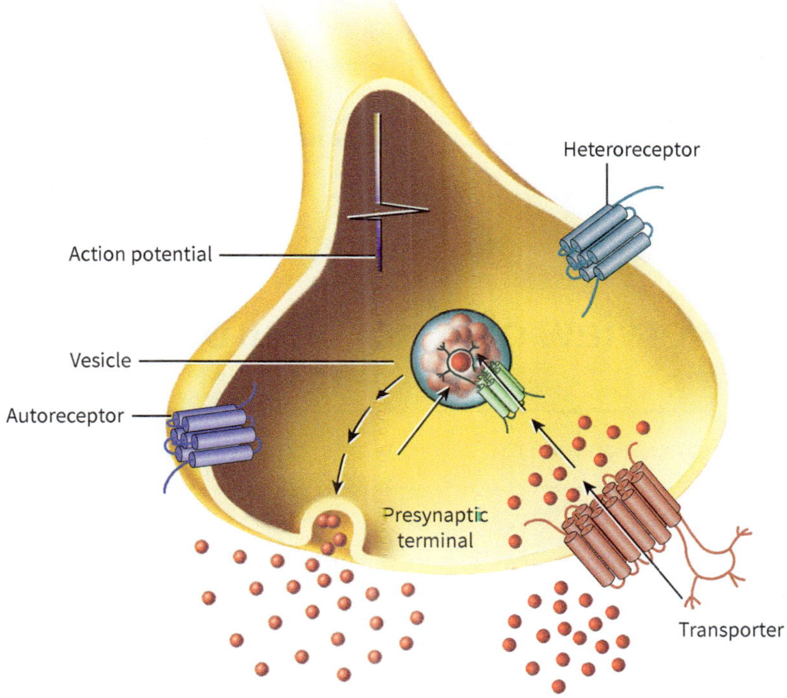

FIGURE 3.6 Autoreceptors and heteroreceptors on the presynaptic neuron Autoreceptors in the presynaptic neuron regulate neurotransmitter availability only for neurotransmitters released by that neuron. Heteroreceptors perform the same function but also respond to neurotransmitters released by other neurons.

The function of a negative feedback loop is to keep a system in balance (homeostasis) by preventing the system from increasing activity until it is out of control. Specialized receptors in the presynaptic neuron provide this function by regulating internal processes in the neuron, for example, monitoring the amount of neurotransmitter available in the presynaptic vesicles to inhibit neurotransmitter availability.

These presynaptic receptor types come in two flavors, autoreceptors and heteroreceptors. **Autoreceptors** (from the Greek for self) are only sensitive to neurotransmitters and hormones released by that individual neuron itself. Autoreceptors are G protein-coupled receptors, so they act as second messengers. If too many neurotransmitters are being released, autoreceptors inhibit neurotransmitter synthesis until a suitable balance is regained. **Heteroreceptors** (from the Greek for other) respond to neurotransmitters and hormones released by other neurons (**Figure 3.6**). These presynaptic receptors can be located anywhere in the neuron, including the dendrite, cell body, axon, or terminal button. These presynaptic receptors are yet another regulatory mechanism in the transmission of information throughout our nervous system.

3.2 Before You Go On

Section Summary

When a neurotransmitter is released into the synaptic cleft, it must dock at a receptor for information transmission to occur. When ionotropic receptors fire, they create a signal that propagates down the axon to another neuron, whereas metabotropic, or G protein-coupled, receptors are second messengers that create many metabolic changes within the postsynaptic neuron. Receptors may be down- or upregulated by decreasing or increasing their number in response to over- or understimulation. Autoreceptors only allow a specific neurotransmitter to dock, whereas heteroreceptors respond to several neurotransmitters.

Comprehension Check

1. An important difference between ionotropic and metabotropic receptors is that ionotropic receptors _____, whereas metabotropic receptors _____.

- **a.** have a direct effect between neurons; work indirectly within a neuron
- **b.** have a very important impact on neurons; do not
- **c.** are more influenced by drugs and disease; are less so
- **d.** are not autoreceptors; are autoreceptors

2. When a postsynaptic neuron is flooded with neurotransmitters, what may it do?

   a. upregulate
   b. become a heteroreceptor
   c. become an autoreceptor
   d. downregulate

3. An autoreceptor is like a _____, whereas a heteroreceptor is like a _____.

   a. bicycle; automobile
   b. can opener; knife sharpener
   c. lamp; table
   d. private club; public restaurant

3.3 An Overview of Major Neurotransmitters

LO 3.3 **Identify** the most common neurotransmitters, explaining the functions of each.

You have figured out by now in your studies that neurotransmitters are part of the complex mechanism that sustains life as well as creates consciousness. Imagine for a moment that we are creating a device to act in a complex and purposeful way, and this device requires chemical messengers to act. What would we want from these messengers to create a device that can survive in a complex world? We would need to create a system that could deal with a seemingly infinite number of situations.

We would need some messengers to act very quickly when a threat appears, such as suddenly encountering a cliff or a predator, and we would want those messengers to draw on every possible resource to save the machine. But rapid responses that engage every mechanism are costly in terms of resources, so we would also want other messengers that respond in a less costly, more gradual fashion to local conditions that change slowly. We would need some messengers that are like the gas pedal on a car to get us where we need to go, but we would also need a brake to stop a process once a goal has been achieved.

Finally, because our imagined machine is as complex as we are, we would need different messengers for many different purposes. Some messengers would control body movement, some would help the machine think and solve problems, and, because this is a very complex machine, some would even create emotions that help the machine stay motivated to succeed.

What you will discover in this next section is that you and I and most other organisms have developed this complex set of messengers known as neurotransmitters. Neurotransmitters are **endogenous** chemicals; that is, they are created in our own bodies. There are over 100 neurotransmitters discovered so far, and there are probably more to be discovered. For our purposes, we will focus on those we know the most about and those that seem to play the most significant role in behavior.

Before we do, keep the following two ideas in mind to help you organize your understanding of neurotransmitters. First, there are six basic attributes that are essential to human behavior—mood or attitude, pleasure or reward, movement, learning and attention, excitation, and inhibition—and there are neurotransmitters dedicated to each of these attributes. Second, neurotransmitters are not randomly squirted anywhere into your brain but are created in specific locations and then travel through pathways that deliver them to specific locations. If you think of neurotransmitter pathways as rivers throughout the brain, you are on the right track. With these ideas in the foreground, we now turn to a discussion of neurotransmitters.

Four Key Neurotransmitter Systems MCAT

Here we will explore in more detail four key neurotransmitters: serotonin, dopamine, acetylcholine, and noradrenaline. Each of these neurotransmitters affects distinct aspects of human behavior (Table 3.3). Together, the sources, pathways, and effects of each neurotransmitter can be considered as a system: the serotonergic, dopaminergic, cholinergic, and noradrenergic systems.

TABLE 3.3 Behavioral Effects of Major Neurotransmitters

Neurotransmitter	System	Behavioral effects
Serotonin (SE or 5-HT)	Serotonergic	Mood, satiety, body temperature, sleep, pain perception
Dopamine (DA)	Dopaminergic	Arousal, reward
Acetylcholine (ACh)	Cholinergic	Memory, learning, arousal, reward
Noradrenaline (NE)	Noradrenergic	Arousal, alertness

The serotonergic system Serotonin (SE or 5-HT) is derived from tryptophan, an amino acid that is found in many protein-rich foods but is especially plentiful in chocolate, oats, milk, red meat, eggs, poultry, bananas, and peanuts. Serotonin is secreted principally from the Raphe nuclei located in the brainstem. From there it travels to the cerebellum and through the entire cortex, as illustrated in Figure 3.7. The serotonergic pathway permits SE to wield a powerful influence on many behavioral systems located throughout the CNS.

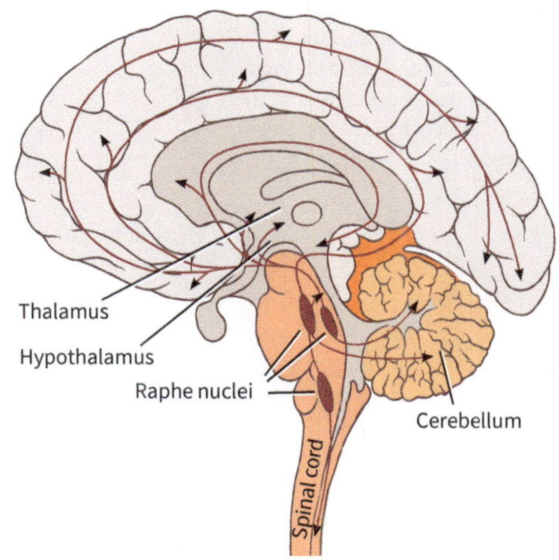

FIGURE 3.7 The serotonergic pathway The serotonergic pathway originates in the brainstem and extends throughout the brain where it affects mood and satiety.

Serotonin is one of the most widely studied neurotransmitters because it plays a role in mood, anxiety, depression, and general cognition. Your quality of life is in many ways a function of your mood; feeling upbeat and optimistic is more likely to lead to successful behavior than feeling depressed and anxious. SE plays a large role in determining how you respond to life's everyday challenges. SE has also been implicated as an active agent in psychedelic drugs such as LSD, psilocybin, and mescaline. For this reason, researchers have investigated the role SE might play in psychotic behavior. We will discuss the role of SE in psychiatric disorders in detail in a later chapter.

There are serotonin receptors in the CNS, and your gut is also richly provided with receptors for serotonin. For this reason, you may not be surprised that serotonin plays an important role in regulating satiety, or a feeling of fullness. When SE is released in the gut, this causes the gut to contract around the food, giving a feeling of fullness. One of the many protective mechanisms we have in our body is a response to foods that irritate the gut lining; we release more SE until we either vomit the noxious food out or experience diarrhea. Conversely, drugs that block the effect of SE in the gut are used to treat nausea and vomiting.

The dopaminergic system Dopamine (DA) is a neurotransmitter that is involved in a number of important functions, but most importantly reward and movement. Think for a moment about how pleasurable experiences motivate or reward you to behave in certain ways. A good grade on an exam motivates you to study for your next exam, a pay raise or promotion rewards your hard work, and quality time spent with family and friends enhances life. The dopamine system is involved in some way in every type of reward that has been studied, not all of which are positive or healthy. You will not be surprised to discover that addictive drugs such as amphetamines, methamphetamines, and cocaine amplify the action of DA and by doing so, reward their continued use.

Dopamine is generated by perhaps as few as 400,000 DA-producing neurons that are confined to a few small areas, but the axons of these neurons project through three major pathways in the brain to exert a powerful effect on behavior. The three dopaminergic pathways in the brain are the mesocortical, mesolimbic, and nigrostriatal pathways (Figure 3.8). The prefix *meso-* comes from the Greek for middle, which is a clue that the first two pathways originate in the middle region of the brain. The mesocortical pathway terminates in the cortex, and the mesolimbic pathway sends dopamine to the limbic system. The nigrostriatal pathway originates in the substantia nigra and terminates in the striatum.

Some of the more important dopamine-secreting areas are the substantia nigra, the ventral tegmental area, and the posterior hypothalamus. We will discuss these regions and others

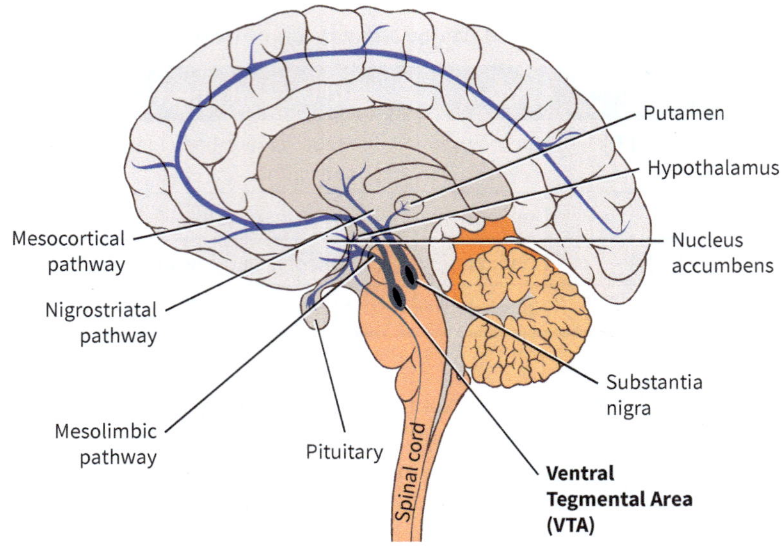

FIGURE 3.8 The major dopaminergic pathways Dopamine originates in the brainstem and travels to various cortical and subcortical structures where it affects reward and movement.

in detail in the chapter on the architecture of the nervous system as well as in discussions of the biological basis of reward, addictions, mood, memory, and other behaviors, but a short discussion of the control of movement may illustrate the importance of dopamine.

Consider for a moment what is involved in reaching your hand out in front of you. You have to coordinate several muscle groups and scan the visual world to decide if there is room in front of you to stretch your hand out. You may even activate those brain regions responsible for judgment as you ponder whether stretching your hand out in front of you is appropriate given where you are right now. There is a collection of nuclei known as the basal ganglia that lie deep underneath the cortex and are connected to many structures, because these decisions and others are typically involved in motor movements. The basal ganglia are essential to the initiation of movements and their control; damage to the dopamine-secreting regions of the basal ganglia leads to serious neurological disorders such as Parkinson's disease and other motor disorders. You may be familiar with a condition known as restless legs syndrome (RLS), a malady in which a person's sleep is disrupted by uncontrolled leg movements. RLS has also been shown to be associated with a decrease in dopamine. Sleepy sickness, or encephalitis lethargica, was characterized by extreme lethargy, rigidity, and an inability to initiate movement similar to Parkinson's disease (see **Check It Out: The Sleepy Sickness Epidemic**).

Check It Out | The Sleepy Sickness Epidemic

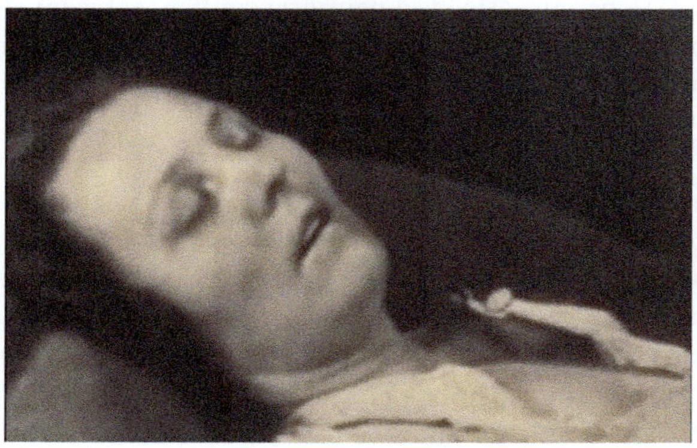

Months or even years after their flu symptoms receded, some survivors of the 1918 flu became very sleepy, often falling asleep in the middle of everyday activities, which explains the common name for this disease, the sleepy sickness. One person fell asleep during their own wedding, and another fell asleep during childbirth. Watch the video **Encephalitis Lethargica**, and answer the following questions:

1. Search using the term "encephalitis lethargica." What area of the brain was implicated in this disorder?
2. Given your answer to the prior question, what would you suggest as a treatment for encephalitis lethargica?

The ventral tegmental area (VTA) in the brainstem is the largest of the dopamine-producing nuclei, although it is still a relatively small structure. The VTA is involved in reward and motivation and coordinates with other structures to plan motor movements.

The cholinergic system The neurotransmitter acetylcholine (ACh) is found throughout the CNS, especially in regions such as the cortex and frontal lobe that create cognition. In the PNS, ACh activates muscles and transmits signals that regulate our blood vessels, our internal organs, and even pupil dilation. Neurons that respond to acetylcholine are known as cholinergic neurons.

We saw with DA and SE that there are physical pathways in which these two neurotransmitters flow throughout the brain, and ACh is no different. Knowing the course of the cholinergic pathway (**Figure 3.9**) helps us to see that this pathway serves many functions.

ACh originates in a region of the brainstem known as the pedunculopontine and interpeduncular nuclei. As intimidating as these names might appear, it is simply the space between the peduncle and the pontine, two brainstem nuclei. A second source of ACh is the basal nucleus of Meynert and the medial septal nucleus. Figure 3.9 shows that, once synthesized, ACh travels up through the frontal lobe and into the cortex and services the thalamus and the occipital lobe. Because the frontal lobe is responsible for planning and judgment and because humans are very visual, it makes sense that the neurotransmitter that supports learning and memory would service these areas.

ACh has been found to be closely involved in higher cognitive functions such as learning, memory, and attention. As a consequence, drugs that increase the action of ACh may hold promise for enhancing learning and memory in some individuals. There is some reason to believe that a decrease in the level of ACh at the synapse contributes to Alzheimer's disease (AD). For this reason, one form of therapy for AD is to inhibit acetylcholinesterase from breaking down ACh, thereby increasing the supply of ACh at the receptor site. Sadly, clinical studies of the effects of ACh inhibitors on the cognitive performance of patients with AD show mixed results, with some individuals improving but many not. Why that might be will be discussed in greater detail in the chapter on neurological disorders.

Acetylcholine receptors come in two flavors, or variants: muscarinic and nicotinic. These two ACh receptor types are so named because the former reacts to muscarine and the latter reacts to nicotine. The difference between the two is the nature of their action. Muscarinic receptors are characterized as metabotropic receptors that provide a slow response mediated by a second messenger system. In contrast, nicotinic receptors are ionotropic, consisting of a ligand-gated ion channel that responds quickly to activation. The combination of a slow and gradual response with a fast response provides the ACh system with considerable flexibility in function.

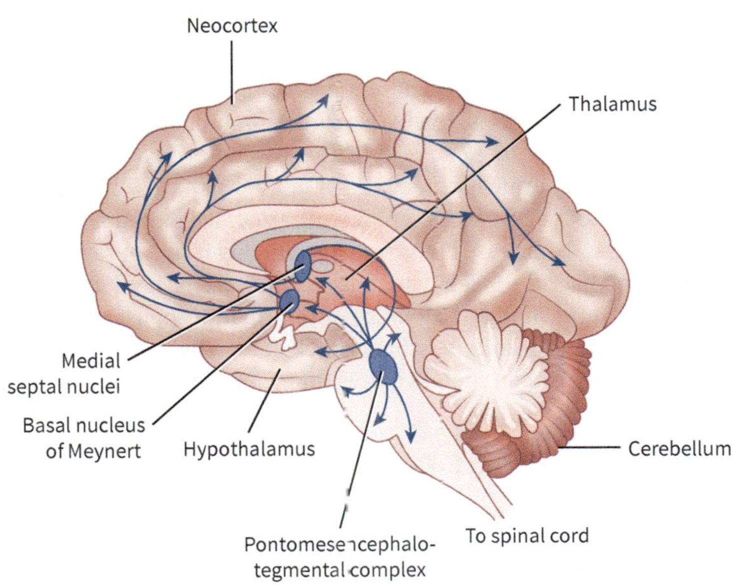

FIGURE 3.9 The cholinergic pathway Acetylcholine is secreted by nuclei in the brainstem and conveyed throughout the brain where it affects memory and attention.

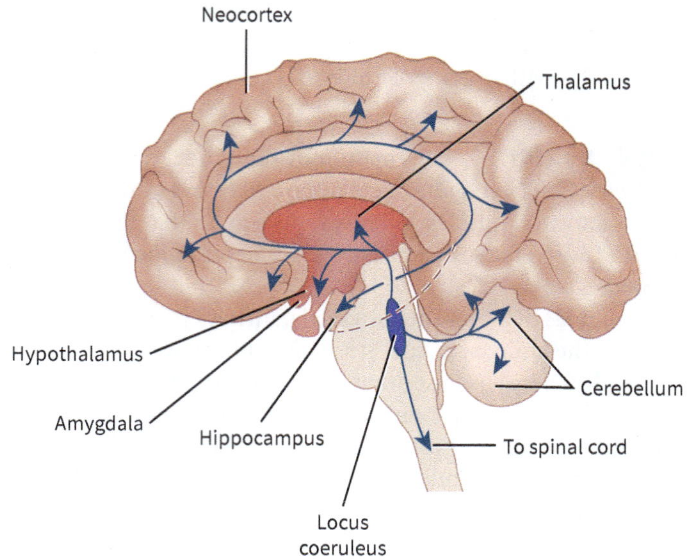

FIGURE 3.10 The noradrenergic pathway Noradrenalin is secreted in the brainstem and then travels throughout the brain to affect alertness, arousal, and response to stress.

The noradrenergic system Noradrenaline (NA), also called norepinephrine (NE), is a chemical messenger that serves as both a hormone and a neurotransmitter. NA is released in the brain by several small nuclei in the brainstem; principal among these is the locus coeruleus, where perhaps only a few thousand neurons are responsible for NA secretion. Structures and processes that are part of the NA system are known as the noradrenergic pathway, which is portrayed in **Figure 3.10**.

NA affects both the CNS and, in the PNS, the sympathetic nervous system. In the brain, NA increases general cortical activation, which results either in greater alertness or anxiety. As previously mentioned, NA is also a hormone, and it is released into the bloodstream by the adrenal glands, which sit atop the kidneys. In the body, NA boosts blood pressure by causing blood vessels to constrict (vasoconstriction). It increases heart rate and respiration and also stimulates the release of glucose from the liver. All of these actions increase the supply of blood, oxygen, and fuel to your brain and muscles. As such, noradrenaline is part of the body's fight-or-flight mechanism.

In our discussion of neurotransmitters and their associated receptors, details that are not essential to conveying the big picture are sometimes deferred to keep the amount of new information to a reasonable level. You will learn as you continue your studies of behavioral neuroscience that a receptor for a given neurotransmitter often has several variants, each with a slightly different effect, and this is true for NA. For example, various NA receptors affect many organs and processes outside of the brain, and two examples are worth mentioning. NA stimulates what are known as alpha-receptors (α-adrenoreceptors) throughout the body. When activated, α-adrenoreceptors constrict blood vessels and quiet the gastrointestinal tract. In contrast, activation of beta-receptors (β-adrenoreceptors) brings about the release of glycogen (a source of fuel for your muscles), increases cardiac output, and dilates passages in your lungs known as bronchi, leading to greater oxygen availability.

Noradrenaline/norepinephrine is related to but not the same as adrenaline/epinephrine, a hormone secreted by the adrenal glands. Epinephrine does act as a minor neurotransmitter, but its major importance is to increase blood pressure and heart rate in response to stress. You may know someone who has an EpiPen, which is an autoinjector used to deliver a measured dose of epinephrine and as such is part of your fight-or-flight response in case of emergency. An EpiPen can be used to treat a serious asthma attack or anaphylaxis, a rapid and potentially fatal allergic reaction to an allergen such as a bee sting, foods such as peanuts, or a medicine.

Other Neurotransmitters of Note MCAT

As previously noted, there a more than 100 known neurotransmitters at this time, so to discuss each in detail would not enhance your overall understanding of how neurotransmitters work in the synapse. However, we will briefly discuss a few that you will encounter later in your exploration of the human brain.

Nitric oxide Nitric oxide (NO) is a gaseous neurotransmitter that affects the inner layer, or endothelium, of blood vessels by signaling them to relax or dilate, thereby regulating blood flow and blood pressure. Several popular drugs prescribed to treat erectile dysfunction (ED) do so by dilating blood vessels, thereby increasing blood flow to a penile structure known as the corpus cavernosum. NO is not to be confused with nitrous oxide (N_2O), a common anesthetic, or nitrogen dioxide (NO_2), a common air pollutant.

Glutamate Glutamate (Glu), or glutamic acid, is the most common excitatory neurotransmitter in the vertebrate nervous system. Glutamate is synthesized and stored in the presynaptic vesicles. It is then released into the synapse, where it may bind to glutamate receptors on the postsynaptic cell.

You already know that the difference between the internal and external electrical gradients creates an action potential that is the basis for neural signal transmission and that a neuron is more likely to fire when it has more positive ions inside. Glutamate acts on both ionotropic and metabotropic receptors to increase the number of positive ions inside the cell, thereby making it more likely that an action potential will be created.

In addition, when we learn something new, a physical change in the synapse known as long-term potentiation (LTP) occurs to strengthen the response of the receptor to stimulation. This capacity for neural change is known as **synaptic plasticity** and is the foundation of our ability to create and store memories. Glu receptors are prominent in areas of the brain such as the hippocampus, which are known to play an essential role in learning and retrieving information. Given that we have a neurotransmitter that is designed to excite a response, you might imagine that nature would demand a corresponding inhibitory mechanism to keep our neurons in check, and that neurotransmitter is GABA.

GABA Gamma-aminobutyric acid (GABA) is the inhibitory partner to glutamate. GABA is found in mammalian nervous systems, and where glutamate has an excitatory role in the nervous system, GABA inhibits neural systems. Without GABA, nervous systems would have a tendency to run out of control, just as a car without brakes would be impossible to drive.

GABA is one of the more interesting neurotransmitters because it changes its mode of action over the course of human development. In the developing nervous system, stem cells become neurons, and as these neurons develop, they find their way into the structures of the CNS. Why a given cell becomes a white matter tract reaching from the corpus callosum to the cortex and not part of the amygdala is of great interest for those who study neural development. It is clear that GABA influences the migration and differentiation of developing neural structures and acts as the major excitatory neurotransmitter prior to the maturation of glutamic neurons. In the mature human brain, GABA is inhibitory.

Like many neurotransmitters, GABA also has important functions outside the CNS. For instance, GABA is a part of the insulin–glucagon feedback loop in that GABA is secreted in conjunction with insulin to convert glucose in your blood into glycogen for storage. GABAminergic receptors in the soft tissues (epithelium) of your airway (trachea) may play a role in asthma.

One way to visualize the glutamate and GABA mechanisms is to think about the electrochemical gradient that must be overcome for the sodium-potassium pump to create an action potential. If you think of that gradient as a hill to be overcome, glutamate levels off the slope somewhat, making it easier to get over the top. In contrast, GABA makes that hill steeper.

Drugs and disorders that bring about an excess of glutamate result in insomnia, anxiety, poor concentration, a low threshold for pain, and similar symptoms. In extreme situations, glutamate toxicity overexcites cells, leading to their death. GABA is inhibitory, so too little GABA leaves the neurons in the CNS in an overexcited state that can cause epilepsy or mood disorders. In contrast, too much of the inhibitory effects of GABA can lead to drowsiness and sleep disorders.

You will see throughout your study of behavioral neuroscience that physiological mechanisms often come in pairs that balance each other to maintain equilibrium. There is an easy way to remember which is inhibitory and which is excitatory. GABA comes ahead of glutamate when the two are alphabetized, and the red light is always at the top or far left and the green light is always at the bottom or far right of a stoplight. Therefore, GABA is

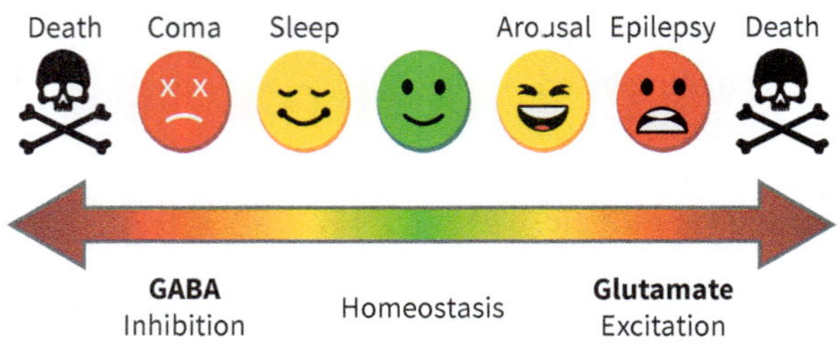

FIGURE 3.11 GABA and glutamate on a continuum of excitation and inhibition GABA and glutamate are paired to activate or inhibit neural functions throughout the nervous system. A balance of these neurotransmitters contributes to overall health, whereas too much of either can disrupt health or even be fatal.

inhibitory (red light), and glutamate is excitatory (green light). Too much of either GABA or glutamate can have fatal consequences (Figure 3.11).

So far we have discussed neurotransmitters that control basic functions such as alertness, memory, movement, mood, and blood pressure. Another important aspect of humanity is our close interpersonal connections between romantic partners or between parent and child. These connections are part of what makes us human, and as such, you might predict that nature would not leave them up to chance. In that light, we now turn to oxytocin.

Oxytocin Sometimes referred to as the love hormone, oxytocin is a molecule that is both a neurotransmitter and a hormone (we will discuss its role as a hormone when we cover the endocrine system), and it enjoys signaling properties and receptors in the CNS and elsewhere. Oxytocin is created in the hypothalamus and then secreted into the bloodstream through the posterior pituitary, a structure that is connected to the hypothalamus and descends from it. Animal research involving voles (small, mouse-like rodents) and rats indicates that exposure to oxytocin leads to an increase in social behavior and caregiving to offspring. In addition, oxytocin plays a very important part in stimulating uterine contractions during labor. We will discuss oxytocin in some detail when we turn our attention to neurodevelopment and the biological basis of sexual behavior.

Substance P Substance P is a neurotransmitter that is released quickly in response to noxious or damaging stimuli. It should not surprise you that organisms would evolve a mechanism to respond quickly and powerfully to tissue damage. Specialized structures known as nociceptors, or pain receptors, quickly signal when physical damage occurs. If you are familiar with the meaning of the word noxious (physically harmful), you can easily remember the function of nociceptors. In response to physical damage to bodily tissue such as a cut, abrasion, sunburn, or even an infection, glutamate (an excitatory neurotransmitter) is released along with substance P. Substance P appears to signal to the brain that damage has occurred. Some individuals experience constant pain due to disease or physical damage, so medical researchers are interested in understanding more about how our nervous system transmits the perception of pain. The topic of pain perception is important and will be discussed in greater detail elsewhere.

Before we move on from our introduction to neurotransmitters, take a moment to consider the overview presented in Table 3.4. You have already seen that physiological systems such as neurotransmitters are complex but have a logic to them. There are opposing pairs (excitatory and inhibitory) that act as gas and brake to maintain an equilibrium. Some neurotransmitters exert their influence in a rapid and abrupt fashion, whereas others act in a slower and more gradual fashion. Both systems acting together work to our advantage.

TABLE 3.4 Notable Neurotransmitters and Their Behavioral Effects

Neurotransmitter	Behavioral effects
Nitric oxide (NO)	Relaxation of blood vessels throughout the body
Glutamate	Excitation of neural activity
GABA	Inhibition of neural activity
Oxytocin	Creates feeling of empathy, affection and parental bonding; in body, stimulates uterine contractions
Substance P	Pain perception

3.3 Before You Go On

Section Summary

Neurotransmitters are the chemical messengers responsible for much of behavior. Among the most significant, serotonin affects mood and satiety. Dopamine is associated with reward and movement, acetylcholine with memory, and norepinephrine with arousal and reward. GABA and glutamate are an inhibitory and excitatory pair, oxytocin is associated with empathy and attachment, and substance P provides a mechanism for experiencing pain.

Comprehension Check

1. Mood disorders such as depression are often treated by drugs that increase the level of which neurotransmitter?

- **a.** dopamine
- **b.** serotonin
- **c.** oxytocin
- **d.** norepinephrine

2. Damage to the substantia nigra often results in a loss of neurons that secrete which neurotransmitter?

- **a.** dopamine
- **b.** serotonin
- **c.** norepinephrine
- **d.** acetylcholine

3. Research into the causes of Alzheimer's disease and other dementias has often focused on which neurotransmitter?

- **a.** dopamine
- **b.** serotonin
- **c.** norepinephrine
- **d.** acetylcholine

4. Researchers investigating the biological basis of romantic and parental attachments have focused on which neurotransmitter?

- **a.** nitrous oxide
- **b.** nitric oxide
- **c.** oxytocin
- **d.** substance P

5. GABA and glutamate work together similar to the way _____ work in a car.

- **a.** gas and brake pedals
- **b.** brake and gas pedals
- **c.** left and right turn signals
- **d.** high beams and low beams

3.4 Disorders and Treatments Related to Neurotransmission

LO 3.4 Integrate your understanding of synaptic processes to understand related disorders and their treatments.

Some students may feel that the details of neural transmission are not relevant to their interests, but allow me to argue against that view and propose instead that a grasp of the basics of neural transmission is valuable. In fact, what you have learned about the neuron, the synapse, and neurotransmitters will help you understand rare diseases but also common psychiatric and neurological disorders such as schizophrenia, depression, obsessive-compulsive disorder (OCD), anxiety and panic disorders, addiction, and many other fascinating topics that we will discuss in later chapters.

Synaptopathies

As you have seen in this chapter, the synapse is key to the transfer of signals between neurons. In this context, you will not be surprised to hear that many in the scientific and medical community believe that disease or damage that affects processing at the synapse (termed **synaptopathy**) is at the root of many disorders such as fragile X syndrome, Down syndrome, epilepsy, and autism spectrum, degenerative disorders such as Alzheimer's and Parkinson's, and even psychiatric disorders such as schizophrenia. You can learn more about the promise of synaptopathology research in **Advances in Behavioral Neuroscience: Dimitri Kullmann and the Silent Synapse**.

Advances in Behavioral Neuroscience | Dimitri Kullmann and the Silent Synapse

In your study of how the brain manages all our bodily functions and creates the experience of consciousness, you learned quite a bit about the complex mechanisms involved when one neuron speaks (so to speak) to another. One critical component of that conversation is the synapse, or the gap between one neuron and another. Neurotransmitters flow back and forth, receptors open and close, the number of receptors is increased or decreased, and so on. Clearly, the synapse is a mechanism involving many moving parts, without which you and I would not exist.

In light of the synapse's importance, it seems reasonable to investigate synaptic dysfunction as a cause of neural and psychiatric disorders, and that is exactly what the emerging discipline of synaptopathology is designed to do. Dr. Dimitri Kullmann of the University College of London's (UCL) Queens Square Institute of Neurology heads an initiative at UCL to look into the role that synaptic dysfunction plays in a wide variety of disorders such as epilepsy, migraines, and movement disorders. Dr. Kullmann's research and that of his colleagues at the Institute looks for genetic markers that signify a dysfunction in the synapse. The goal is to identify genes responsible for synaptic dysfunction, then search for ways to treat that dysfunction through genetic engineering, and eventually treat the disease successfully.

Kullmann and his colleagues are at the forefront of research into the mechanisms operating at the level of the synapse, often on the scale of nanometers with a duration of only milliseconds. One of Kullmann's important discoveries is the silent synapse, a glutaminergic synapse that would periodically go silent, or unresponsive to stimulation, during the course of development. The purpose of this periodic silence during neurodevelopment is unclear, but the possibility of pathology in these receptors may explain neural disorders such as epilepsy and migraines. Investigating the genetics behind synapse formation may lead to a treatment and possibly a cure for these and other maladies.

Questions

1. **Search on the term "Epilepsy frequency in humans" to answer this question. Out of 100 people you know, what are the odds that you know a person with epilepsy?**
2. **If you do not know a person with epilepsy, why might epileptics refrain from disclosing their condition?**
3. **Search on the term "Migraine frequency in humans" to answer this question. Out of 100 people you know, what are the odds that one of those people suffers from migraines?**
4. **If you know a person who suffers from migraines, ask them to describe a typical migraine and then summarize their response. If you do not know such a person, search the internet to find the typical symptoms and summarize them here.**

Some synaptopathies result from deficiency in a given neurotransmitter. For example, tyrosine hydroxylase deficiency is a rare pediatric genetic disorder in which a lack of tyrosine hydroxylase, a precursor to dopamine synthesis, results in a loss of DA throughout the nervous system. Because DA is both a neurotransmitter and a precursor to ACh and SE, this deficiency results in a wide range of motor and intellectual deficits. Understanding that a neurotransmitter must be created in the neuron helps you to understand maladies such as this.

Neurotransmitter transport is essential to normal synaptic functioning, and widespread damage to transport structures is implicated in several neurological disorders such as Parkinson and Alzheimer's diseases as well as amyotrophic lateral sclerosis (ALS), a neurodegenerative disease that took the lives of baseball player Lou Gehrig, astrophysicist Stephen Hawking, and Stephen Hillenberg, the creator of *SpongeBob SquarePants*. Knowing that a neurotransmitter is useless unless it can be transported down into the terminal button where it is available to the synapse means that researchers would look at these structures as they try to unravel the cause of a deadly disorder.

Toxins can also impair synaptic function. You may have heard of nerve agents such as Novichok which, although illegal, are sometimes used in war and to kill or disable individuals who are politically inconvenient. Novichok blocks the degradation of ACh at the point where cholinergic neurons synapse upon muscles, resulting in a breakdown of cardiac and respiratory function and death if not immediately treated. Realizing that preventing the normal degradation and reuptake of a neurotransmitter can be deadly points the way to drugs and treatments that would reverse the process and save a life.

Disorders can also arise from problems related to neurotransmitter regulation. The *Stories of the Brain* that opened this chapter featured two individuals with disorders that affected their movement. In fact, both of these disorders were related to the same neurotransmitter, dopamine. Kurt developed Parkinson disease, which impaired his movement as a result of too little dopamine. Jean Marie's tardive dyskinesia, which caused constant movement, likely resulted from a supersensitivity to dopamine after many years of taking medication for schizophrenia that altered her dopamine receptors.

Alcohol, Drugs, and Addiction

We will discuss addictions in much more detail in future chapters, but it is worth a moment to consider them within the context of neurotransmitters and their receptors. Many effects of alcohol and drugs can be explained at the level of neurotransmission, as can the symptoms of addiction.

For example, alcohol increases the effect of the inhibitory neurotransmitter (GABA), which leads to the clumsy movements and slurred speech typical of the tipsy individual. Alcohol also inhibits the excitatory neurotransmitter (glutamate), leading to slower reaction times and diminished intellectual function and judgment. Finally, alcohol also increases the levels of dopamine, a neurotransmitter that produces, among other effects, a pleasant feeling of euphoria. The person who drinks more than is wise may appear to others to exhibit slurred speech, clumsy movements, poor judgment, and a happy mood. You now know that the effects of alcohol are due to an increase in GABA and dopamine and a decrease of glutamate.

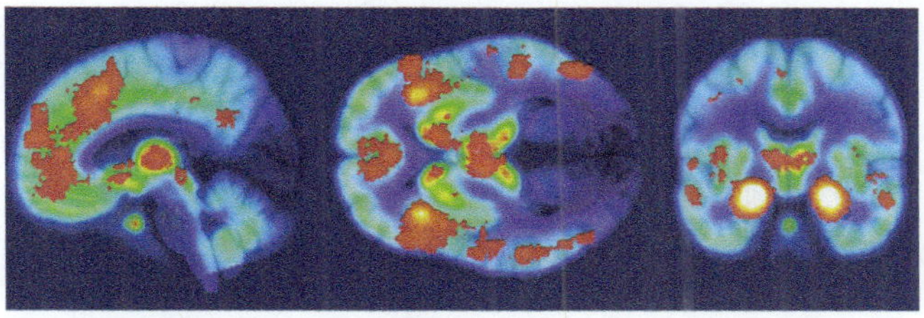

FIGURE 3.12 Overlap between the human emotion circuit and the opioid receptor system Participants were stimulated to experience emotions, and regions of the brain that showed an increase in activation were recorded (in green). The same individuals were then given an opioid with a radioactive tag, so that when that opioid reached its receptors in the brain, their locations could be imaged (in red). When the regions activated by each procedure are superimposed, opioid receptors are seen clustered in the same regions activated by emotions.

In a second example, opioid addiction can be explained by examining the effects of opioids on neurotransmission. Opioids are neurotransmitters that form the basis for much of human emotion, motivation, attachment behavior, reactions to pain, and food-intake regulation. Opioids act on receptors found mostly in the amygdala, thalamus, hypothalamus, and caudate nucleus, which are regions associated with emotion, feelings of well-being, and associative learning (the acquisition of habits) (Figure 3.12). Have you ever wondered about the neural mechanism that makes you feel happy when you ace an exam, receive a raise or promotion at work, or otherwise reach some goal that you have worked toward for some time? The answer is that we generate an opioid that binds to our opioid receptors and in doing so, creates the perception of well-being and happiness.

Opioids generated in the brain are known as endogenous opioids to distinguish them from the **exogenous** opioids created outside our bodies by plants or chemists. Exogenous opioids such as morphine and codeine are widely used as surgical anesthetics and as analgesics to reduce pain following an injury or surgery. They are also abused, and misuse of prescription opioids as well as unregulated opioids such as heroin and fentanyl is responsible for around 50,000 overdose deaths in the United States each year.

When stimulated with low doses, opioid receptors create the experience of euphoria. In higher doses, opioids create nausea, itchiness, constipation, and drowsiness. Higher doses are often fatal because opioids can depress respiration to the point of stopping it altogether. Continued use of opioids creates tolerance, which is a need for ever-larger doses to achieve the same level of intoxication or analgesia. Because they are so involved in regions of the brain associated with emotions and learning, repeatedly stimulating opioid receptors can lead to an addiction that is highly resistant to breaking. We will return to the link between endogenous opioid receptors and behavior in discussions of addiction and social behaviors.

Drug Treatments Targeting Neurotransmission

Given the role of neurotransmitters in psychiatric and other disorders, drug treatments targeting neurotransmission are a promising path toward treatments for many diseases we currently struggle to treat. In subsequent discussions of these and other topics of interest, we will often be talking about ways to increase or decrease the amount of neurotransmitter delivered to the synapse, factors that increase or decrease the number of receptors at the synaptic cleft, and ways to inhibit the reuptake of neurotransmitter at the synaptic cleft so that a neurotransmitter lingers in the cleft and so has a greater opportunity to bind to a receptor. Most psychoactive drugs that are used to treat anxiety, psychosis, and depression work by affecting one or more of the steps of neurotransmission that we outlined in this chapter. Therefore, understanding those steps is the foundation for understanding much of mental illness and our tools for treating those illnesses.

Many treatments involve employing agonist and antagonist drugs. An **agonist** is chemical that is so similar to a given neurotransmitter that it can bind and activate receptors for that neurotransmitter like a second key for the same lock. In nature there is always symmetry, so if there is a second key, there must also be a substance that blocks the receptor. That substance is known as an **antagonist**. The word agonist comes from the Greek for champion, but if you struggle to remember which is which, it's easy to see that an antagonist is a blocker, so an agonist must be a substance that binds to a receptor and activates it. These two mechanisms are depicted in **Figure 3.13**.

Researchers have identified exogenous chemicals that serve as agonists and antagonists for many key neurotransmitters, both from natural extracts and synthetic molecules designed in a lab. Using such chemicals, doctors can prescribe drugs that either increase or block transmission of a targeted neurotransmitter. You can see that to address disorders related to neurotransmission, sometimes we would like to increase the supply of a neurotransmitter at the synapse, whereas other times we would like to block the receptor in an effort to decrease the activity of that neurotransmitter system.

A few examples will help drive the point home. For instance, we know that a particular variant of the adrenalin receptor, the α_1 receptor, stimulates narrowing of the blood vessels (vasoconstriction) and thereby increases blood pressure. Individuals suffering from low blood pressure (hypotension) may benefit from an α_1 agonist. On the other hand, an α_1 receptor antagonist decreases vasoconstriction in the body, thereby decreasing blood pressure. These drugs are often recommended for treating high blood pressure (hypertension).

Similarly, a plant product called curare inhibits the receptor for acetylcholine (Ach) at the junction where a neuron activates a muscle, causing paralysis. For this reason, curare or similar drugs are often administered by anesthesiologists during surgery to keep the patient immobile. Without respiratory assistance, curare can lead to death by asphyxiation.

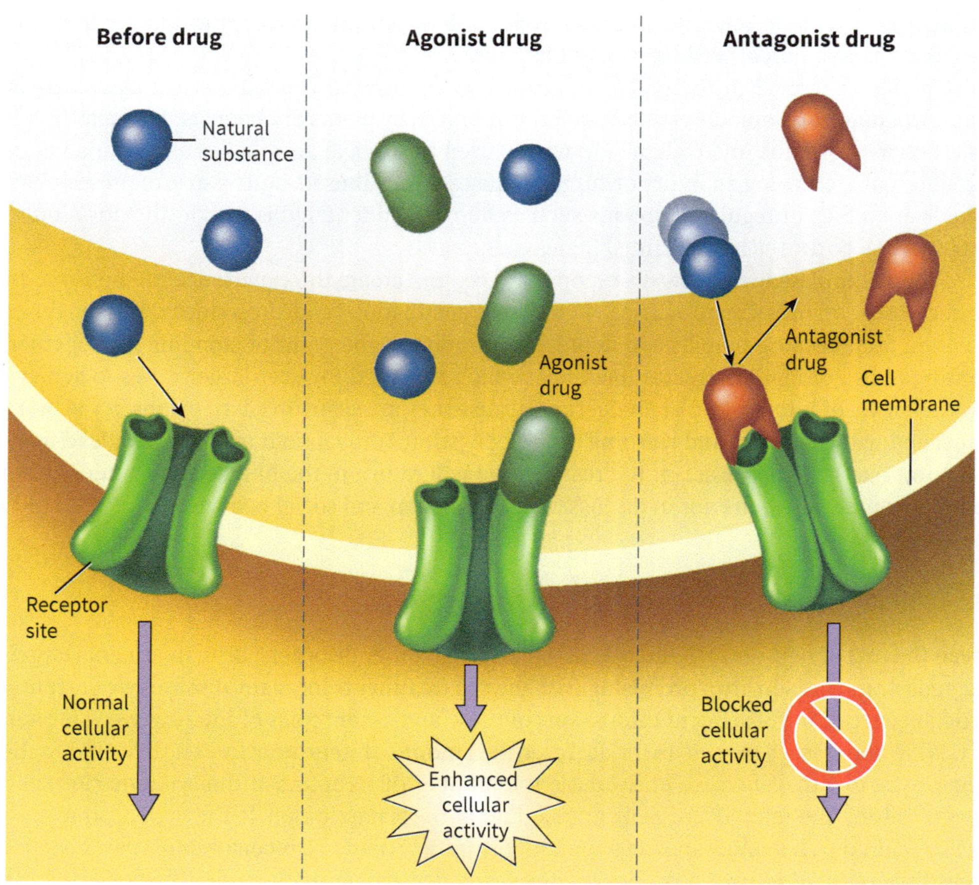

FIGURE 3.13 Agonists and antagonists at the receptor An agonist docks to a receptor and activates it, whereas an antagonist blocks the receptor from activating.

3.4 Before You Go On

Section Summary

Synaptopathies, disorders related to synapse function, include a wide range of degenerative, psychiatric, and other disorders. Understanding how neurotransmitters function also helps us to understand the addictive properties of drugs. Agonists and antagonists are often used to treat psychiatric and neuropsychological disorders by either increasing or reducing the availability of a neurotransmitter in the synaptic cleft.

Comprehension Check

1. Because one of the effects of consuming alcohol is to inhibit production of glutamate, alcohol consumption is known to have what effect?

- **a.** a positive effect on mood
- **b.** a change in fluid absorption in the gut
- **c.** a depressive effect on the brain
- **d.** improved reaction time and reflexes

2. When you feel elated after you receive good news about your grades, money, or your social life, that feeling of euphoria is likely due to what?

- **a.** the excitatory effects of glutamate
- **b.** an increase in GABA
- **c.** an increase in serotonin
- **d.** the release of endogenous opioids

3. What is the effect of an agonist on a receptor?

- **a.** disabling the receptor
- **b.** causing it to switch from an autoreceptor to a heteroreceptor
- **c.** facilitating the action of the neurotransmitter at the receptor
- **d.** causing the receptor to fire only occasionally

4. If a given neurotransmitter is required for you to fall asleep and I administer an antagonist for that neurotransmitter, what outcome would you predict?

- **a.** You will fall asleep quickly.
- **b.** You will struggle to fall asleep.
- **c.** You will crave more of that neurotransmitter.
- **d.** The antagonist will have no effect on sleep.

Conclusion

We began this chapter with the stories of Kurt and Jean Marie, two individuals whose lives were altered by having too much or too little of the neurotransmitter dopamine. Their stories illustrate the delicate balance that our nervous system strives and sometimes fails to maintain. Ideally, our nervous system carefully regulates the levels of more than a hundred neurotransmitters, and, in doing so, keeps us alive and creates our conscious awareness of the world. Understanding the relationships between neurotransmitters and the synapse and the many mechanisms involved such as upregulation and downregulation and agonists and antagonists will give you insights into both normal and disordered behavior.

Critical Thinking Questions

1. Your computer is a digital device that transmits information through signals that can be either a zero or a one. In contrast, your brain is an analog device whose components (the synapse, neurotransmitter, and receptor) can signal in many different states. Why might our analog system be more useful than a digital system in creating complex behaviors?
2. Throughout your study of how the nervous system works, you have seen that nature has placed opposing mechanisms in place to regulate behavior, and upregulation and down regulation are examples. Pick one of these processes and describe how it functions to maintain appropriate behavior.
3. Neurotransmitters are involved in every form of behavior including disorders such as depression, schizophrenia, and Parkinson's Disease. One might think that we can cure these and other disorders by simply taking some neurotransmitter away or adding more, depending upon the disorder, and yet when we do so, improvements in behavior are sometimes transitory. In light of your understanding of synaptic mechanisms, why might this make sense?
4. Propranolol is a drug that blocks receptors in the sympathetic nervous system; the behavioral effect is to reduce heart rate, 'jitters' or trembling of the limbs, and feelings of anxiety when a person is excited or feeling under pressure. Individuals who perform demanding and delicate tasks such as surgeons or individuals who defuse bombs may take propranolol prior to engaging in their task. Do you think this is a good or a bad idea? Explain your answer.

Dig Deeper into Behavioral Neuroscience

3.1 The Synapse

The Dana Foundation is heavily involved in neuroscience research and education, and they provide a comprehensive online overview of the synapse.

Dana Foundation. *What happens at the synapse?*

Many degenerative disorders such as Alzheimer's disease involve degeneration of the synapse. If a neuroprotective agent could be discovered that reduced or eliminated synaptic damage, such diseases could be successfully treated.

Cong, Q., Soteros, B. M., Wollet, M., Kim, J. H., & Sia, G.-M. (2020). The endogenous neuronal complement inhibitor SRPX2 protects against complement-mediated synapse elimination during development. *Nature Neuroscience*. https://doi.org/10.1038/s41593-020-0672-0

3.2 Receptors and Postsynaptic Effects

Solomon Snyder is one of the world's most influential neuroscientists. The following is an overview of the complex interplay between neurotransmitters and receptors.

Snyder, S. (2009, October 14). Neurotransmitters, receptors and second messengers galore in 40 years. *Journal of Neuroscience*, *29*(41), 12717–12721. https://doi.org/10.1523/JNEUROSCI.3670-09.2009

*Atrial fibrillation (A-fib) is a condition that affects millions of individuals each year and is a leading contributor to death in hundreds of thousands. The AF Association is a clearing house for information about A-fib; their website is https://www.heartrhythmalliance.org/afa/us

You can learn more about the role of electrical synapses and A-fib at NIH. https://pubmed.ncbi.nlm.nih.gov/12062332

3.3 An Overview of Major Neurotransmitters

The Dana Foundation provides a global overview of neurotransmitters in an approachable fashion. https://dana.org/article/neurotransmitters/

Over a century ago, a worldwide pandemic killed over a half million people. Although some of those who survived developed serious neurological complications, by and large, the epidemic and these unfortunate survivors were largely forgotten until neurologist Oliver Sacks wrote about them. A technical review of what we know about encephalitis lethargica is available here.

Berger, J. R., & Vilensky, J. A. (2014). Encephalitis lethargica (von Economo's encephalitis). *Handbook of Clinical Neurology*, *123*, 745–761. https://doi.org/10.1016/B978-0-444-53488-0.00036-5. ISBN 9780444534880. PMID: 25015515.

A popular book about these patients was written by the neurologist Oliver Sacks; the book describes their symptoms and how a young neurologist (Sacks) discovered a treatment that, although not a cure, increased our understanding of the basal ganglia and movement. The book was made into a movie with the same name.

Sacks, O. (1990). *Awakenings*. New York: Harper Perennial.

3.4 Disorders and Treatments Related to Neurotransmission

The MSD manual provides a comprehensive review of the many disorders that are attributed to neurotransmission dysfunction.

https://www.msdmanuals.com/professional/multimedia/table/v1031997

Chapter 3 Before You Go On Answers

3.1 Before You Go On Answers

1. a. cytoplasm; synaptic vesicles
2. b. synaptic cleft
3. d. returns the neurotransmitter to the presynaptic neuron for reuse
4. c. operate much faster

3.2 Before You Go On Answers

1. a. have a direct effect between neurons; work indirectly within a neuron
2. d. downregulate
3. d. private club; public restaurant

3.3 Before You Go On Answers

1. b. serotonin
2. a. dopamine
3. d. acetylcholine
4. c. oxytocin
5. b. brake and gas pedals

3.4 Before You Go On Answers

1. c. a depressive effect on the brain
2. d. the release of endogenous opioids
3. c. facilitating the action of the neurotransmitter at the receptor
4. b. You will struggle to fall asleep.

Chapter 3 Glossary

agonist A substance that binds to the receptor and facilitates its response.

antagonist A substance that binds to the receptor and disrupts or prevents its response.

autoreceptor A type of receptor that is only sensitive to the neurotransmitter released by that neuron.

downregulation A decrease in the number of receptors on a target cell that results in lower sensitivity.

endogenous Created within the organism.

exogenous Created outside the organism.

G protein One of a family of proteins that act as a molecular switch inside a cell to transmit signals from outside the cell to inside.

gap junction A specialized rapid-acting direct connection between two neurons.

heteroreceptor A type of receptor that is sensitive to more than one type of neurotransmitter.

ionotropic receptor A receptor characterized by a chemical gate that opens in response to the presence of the neurotransmitter; also known as a ligand-gated ion channel.

metabotropic receptor A receptor that acts as a second messenger, using G proteins to activate intracellular events; also known as a G protein-coupled receptor.

neurotransmitter A molecule released by a neuron that causes the transfer of a signal elsewhere in the body.

reuptake pump A protein in the cell membrane of a presynaptic neuron that is involved in bringing neurotransmitters from the synaptic cleft back into the cell.

second messenger An intracellular signaling molecule released by the cell in response to an outside signal from a first messenger.

synapse The functional junction between two neurons where information transmission occurs.

synaptic cleft The small gap between two neurons.

synaptic plasticity The capacity to increase or decrease synaptic function in response to increases or decreases in its activity.

synaptic vesicle A small bag-like structure within a neuron that holds neurotransmitters.

synaptopathy Any disease of the brain, spinal cord, or peripheral nervous system that is associated with a dysfunction of the synapse.

transport protein A protein that returns neurotransmitters back into the presynaptic neuron; sometimes called a transporter.

upregulation An increase in the number of receptors on a target cell that results in greater sensitivity.

CHAPTER 4

The Nervous System

Hybrid Images/Image Source/Getty Images

Just as a conductor organizes the individual performers into an orchestra, our nervous system organizes the structures that keep us alive, create consciousness, and direct purposeful actions.

CHAPTER OUTLINE

4.1 An Overview of the Nervous System
- Divisions of the Nervous System
- Gross Anatomy of the Brain

  Check It Out: The Gross Anatomy of the Brain
- The Brainstem
- The Spinal Cord

  Advances in Behavioral Neuroscience: Nicholas Hatsopoulos and Spinal Cord Injury Repair

4.2 Major Structures and Support Systems of the CNS
- The Cerebral Cortex
- The Cerebellum

  Check It Out: Comparing Sheep and Human Brains
- Essential Support Systems of the CNS

  Check It Out: The Ventricular and Glymphatic Systems

4.3 Major Subcortical Processing Circuits
- The Corpus Callosum: Information Transfer
- The Thalamus and Hypothalamus: Integration and Regulation
- The Limbic System: Emotion and Memory
- The Basal Ganglia: Initiation and Control of Movement

  Check It Out: Delgado's Bull

4.4 The Peripheral Nervous System
- Divisions of the Peripheral Nervous System
- The Neuromuscular Junction

  Think About It: Myasthenia Gravis and the Neuromuscular Junction
- The Cranial Nerves

LEARNING OBJECTIVES

LO 4.1 **Understand** the major divisions of the nervous system and the landmarks that define them.

LO 4.2 **Identify** the major structures and support systems of the CNS and their functions.

LO 4.3 **Integrate** the major subcortical circuits into the larger functions of the CNS.

LO 4.4 **Describe** the components of the peripheral nervous system and their functions.

Stories of the Brain

Quite tall and fit, Christopher Reeve was a stage and soap opera actor with little in the way of film experience when he was cast as Hollywood's Superman, a superhero previously known only from comic books. The film was a great success, and Reeve's input into the development of the character of Superman was key to the movie's appeal. His Superman was more than just a person endowed with superpowers; he was kind, romantic, and at times troubled and even unsure of himself.

After starring in four Superman movies, Reeve moved on from the role that made him famous and continued his career as an actor, director, and author. He became an enthusiastic equestrian and trained hard to develop the skills necessary to compete at a high level. While riding in an equestrian event in 1995, Reeve was thrown over a fence when his horse balked at a jump. Landing headfirst, he suffered a spinal cord injury that left him with no sensation below the neck and needing a ventilator to breathe.

After his accident, Reeve became an advocate and spokesperson for those with disabilities. Through daily therapy and practice, he eventually learned to breathe on his own for up to 90 minutes. Reeve's therapy included electrical stimulation of his muscles to retard atrophy, and, to the astonishment of his doctors, over some years he began to regain a small amount of feeling and muscular control. Sadly, Christopher Reeve died of an infection related to his condition in 2004. His story demonstrates the complex interactions among the components of the nervous system and the devastating impact of injury to one of those components.

This chapter covers one of nature's most complex creations, the human nervous system. Even something as complex as the human nervous system is made up of parts, however, and our goal is to understand the parts so we can comprehend the system as a whole. We will begin our exploration of the nervous system by describing those features of the central nervous system that are visible to the naked eye, and then we will delve under the surface to explore important structures residing inside the brain. Finally, we will transition to the peripheral nervous system, which is the complex collection of nerves and other mechanisms that allow the brain to create movement and action. By the end of this chapter, we will come to understand how your desire to wave your hand results in motion and why Christopher Reeve's identical intention did not. We now turn to a discussion of the human nervous system that is responsible for creating your behavior.

4.1 An Overview of the Nervous System

LO 4.1 Understand the major divisions of the nervous system and the landmarks that define them.

Most people think of the brain as the driving force behind all behavior, but the reality is that behavior is orchestrated through interactions at many levels, including at the level of the neuron, the brain and spinal cord, and even glands scattered throughout the body. If we wanted to understand a symphony orchestra, we would consider individual instruments such as the bassoon and tympani, we would look at the role that a large structure such as the string or woodwind section plays in the performance, and we would step back at times to listen to the music produced by the symphony as a whole. By the same token, to understand the biological basis of behavior, we must think about not only neurons and neurotransmitters but also the nervous system that creates and coordinates our behavior.

FIGURE 4.1 The human nervous system **a.** The human nervous system is comprised of two major divisions, the CNS and PNS. **b.** The CNS includes the brain and spinal cord, and the PNS can be divided into the somatic and autonomic systems, the latter being further divided into the sympathetic and parasympathetic systems.

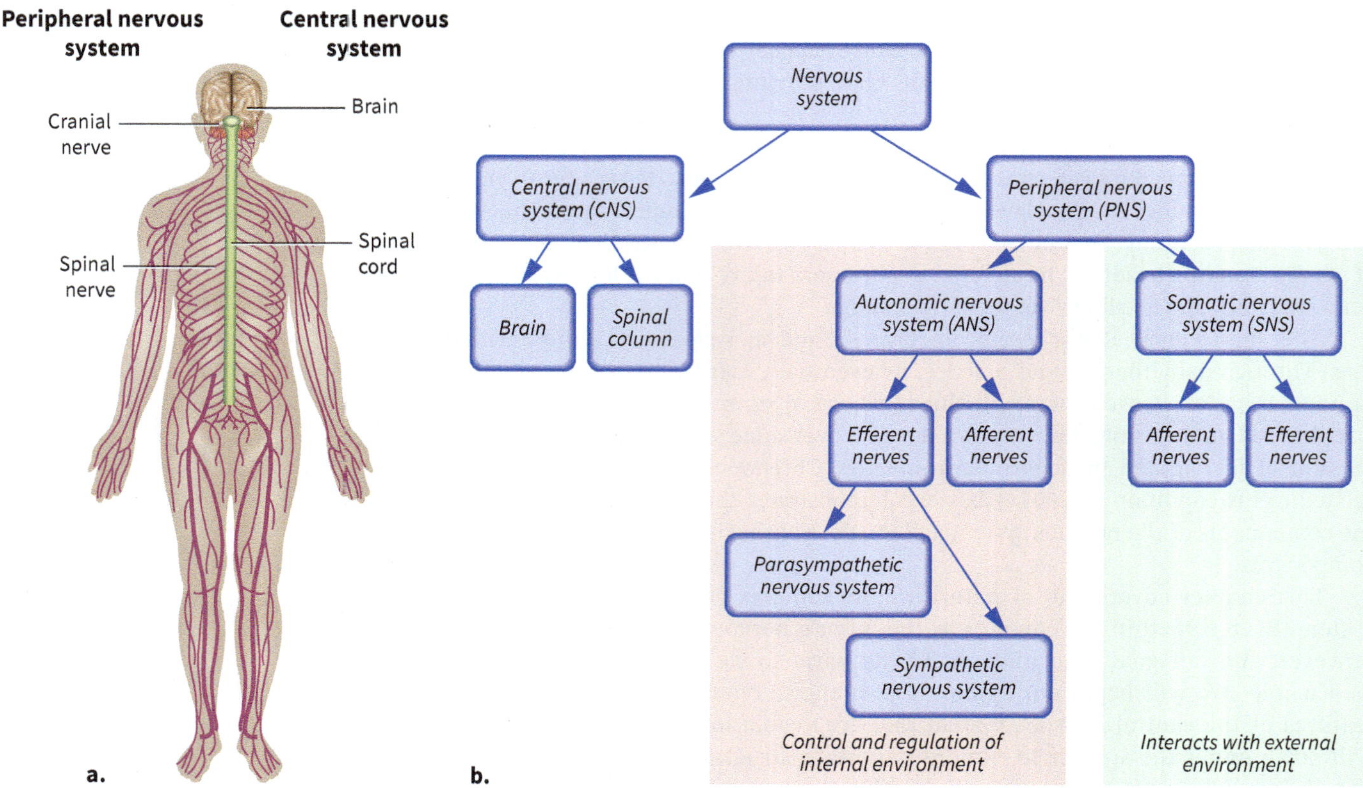

Divisions of the Nervous System MCAT

The vertebrate nervous system has two basic components: the **central nervous system (CNS)** and the **peripheral nervous system (PNS)**. The CNS consists of the brain and the spinal cord, and these structures serve as the central command-and-control system for the body. The PNS consists of the complex peripheral web of nerves as well as junctions known as ganglia that connect the CNS to the body (Figure 4.1).

The PNS itself is divided into two major systems, the somatic nervous system (SNS), which interacts with the external environment, and the autonomic system nervous system (ANS), which controls and regulates the internal environment. The ANS is further divided into two complementary divisions, the sympathetic and parasympathetic systems. In this chapter we will discuss each of these systems in turn, but as we do, keep in mind where the structures being discussed are located in the overall scheme of the nervous system.

Common anatomical terms If you were going to travel to another country with a different language and culture, you would be wise to take a moment at the outset to learn the basic phrases that you would need to ask for and understand directions. The same is true in anatomy, where specialized terms indicate locations in the body. You should familiarize yourself with the directional terms depicted in Figure 4.2 to orient your understanding of structures in the nervous system.

The development of the nervous system We can also understand the structures of the nervous system by tracing their embryonic development. The adult human nervous system develops out of four major structures that appear by the fourth week after

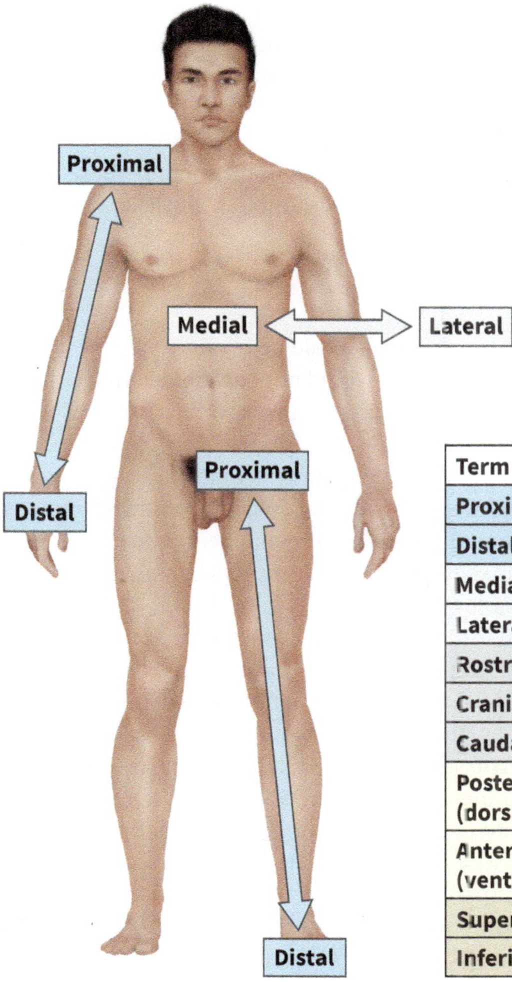

Term	Orientation
Proximal	Near
Distal	Distant
Medial	Near the middle
Lateral	To the side
Rostral	Near the front or nose
Cranial	In the head
Caudal	Near the end or tail
Posterior (dorsal)	Behind
Anterior (ventral)	In front
Superior	On top
Inferior	On the bottom

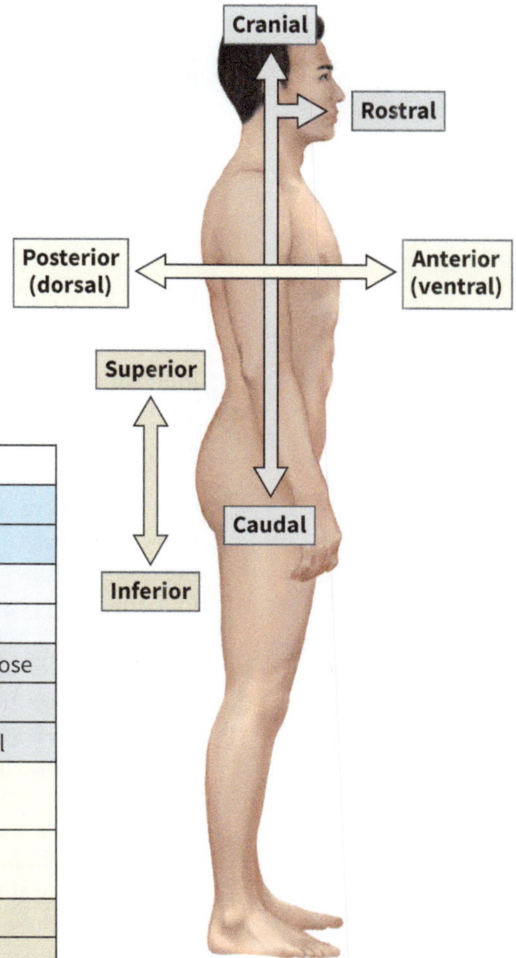

FIGURE 4.2 **Anatomical positional and directional terms** Knowing the meanings of terms that describe position and direction in the body will help orient you as you learn the structures in the nervous system.

conception: the prosencephalon (forebrain), the mesencephalon (midbrain), the rhombencephalon (hindbrain), and the spinal cord (Figure 4.3).

As the embryo develops, each of these structures further differentiate into the complex organ we recognize as a human brain. The forebrain develops into the **cerebrum**, the largest part of the brain, where much of our conscious experience is created, and into important

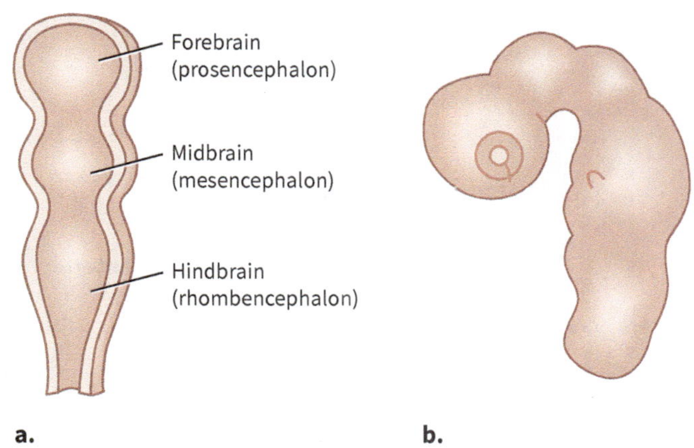

FIGURE 4.3 **The four major divisions of the developing brain** **a.** The embryonic human nervous system consists of the prosencephalon, mesencephalon, rhombencephalon, and spinal cord and is depicted here as a lengthwise section of a straight tube to better illustrate their divisions. **b.** This three-dimensional rendering shows the actual shape of the embryonic nervous system, in which the basic curved structure of the brain is already apparent.

structures that are essential for experiencing information from our senses. The midbrain and hindbrain develop into the **brainstem** and spinal cord, structures that are essential for motor (muscle) coordination, the proper routing of signals from our senses, and basic life-support functions such as respiration, heart rate, sleep and wakefulness, and hunger. The **spinal cord** transmits signals from the brain into the body and from the body back to the brain. In this introduction to the nervous system, we will present each of these structures in turn before exploring the details of the CNS and PNS in the rest of the chapter.

Gross Anatomy of the Brain MCAT

If you held a human brain in your hand, those features that would be readily apparent to your inquisitive eye are the gross, or macroscopic, features. You would notice that the brain consists of two large hemispheres (halves) that appear identical and that the external surface is wrinkled with what might be described as ridges and furrows.

When physicians and scientists first began to study the brain systematically, they realized that its outward appearance suggested the presence of several major anatomical regions, known as lobes (**Figure 4.4**). These are the frontal lobe, the parietal lobe, the temporal lobe, the occipital lobe, and the insula. Each lobe is separated from the others by either a ridge or fold called a **gyrus** (plural *gyri*) or a fissure or valley called a **sulcus** (plural *sulci*). To remember which is which, recall that when you sulk, you feel low, so a sulcus is a valley. A specific gyrus or sulcus serves as a reference point or landmark to describe a location on the cortex. For example, the central sulcus is a prominent ridge that separates the frontal and parietal lobes.

Each lobe is specialized for specific functions, and knowing the functions of each lobe is key to understanding how your brain works. The **frontal lobe** is responsible for higher-level cognitive processes such as conscious thought, planning, and judgment. When you are paying attention to an in-class discussion, planning your path to your degree, or deciding whether attending a party the night before a big exam is really the wisest choice, you are using your frontal lobe.

The **parietal lobe** is the place where information from your many senses as well as spatial information is integrated and then associated with the proper response of your muscles. When you reach out to catch a ball, you are using your parietal lobe, and when you are finding your way around your neighborhood or putting together a piece of furniture that came unassembled, your parietal lobe is engaged.

Your **temporal lobe** has much of the responsibility for language comprehension and production. When you are chatting with a friend about classes and life, your temporal lobe is busy comprehending your friend's speech and creating your verbal responses. The **occipital lobe** creates your visual world. Finally, there is a fifth lobe that is somewhat hidden in the lateral

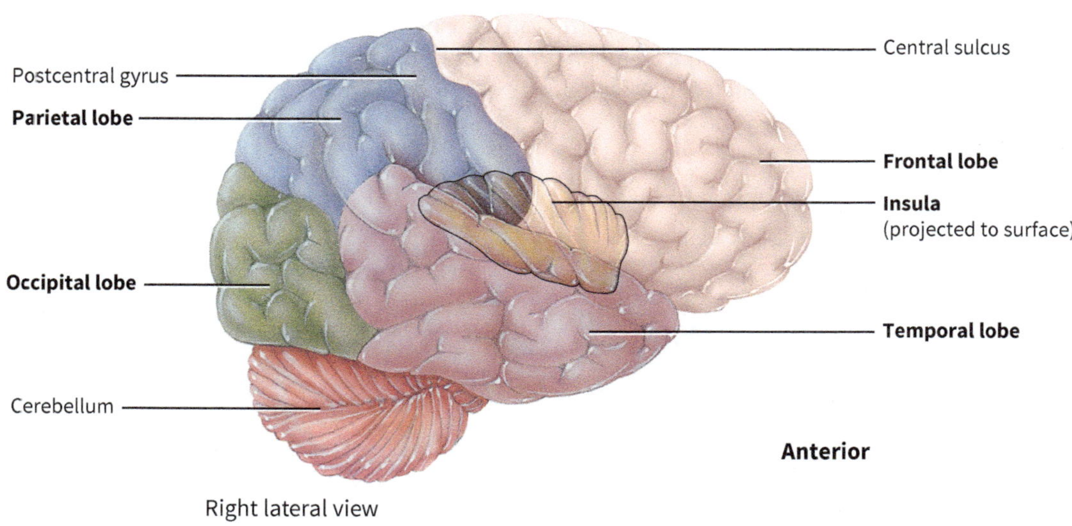

FIGURE 4.4 **Major lobes of the brain and associated anatomical landmarks** The frontal, parietal, temporal, and occipital lobes of the brain are readily visible in this lateral view, as are the sulci and gyri that define them. The insula is hidden underneath the temporal and parietal lobes.

fissure between the frontal, parietal, and temporal lobes and is called the **insula**, or insular cortex. The insula is involved with sensory processing, emotions, taste and pain perception, and maintaining the basic housekeeping functions that keep us alive. We will see in a subsequent chapter that the insula is implicated in several psychiatric disorders.

Just as there is specialized vocabulary to explain the orientation of structures in the body, there are terms to help orient us within the brain. Brains are commonly divided or sectioned into one of three different views or planes, depending on what features we want to display (**Figure 4.5**). A sagittal, or longitudinal, plane separates the brain along a left–right axis. A transverse, or horizontal, plane sections the brain into superior and inferior components. Finally, a coronal, or frontal, plane is a cut separating the brain along a rostral–caudal axis.

In **Check It Out: Gross Anatomy of the Brain** you can check your knowledge of the external features of the brain.

FIGURE 4.5 **Planes of the Brain** **a.** A median or sagittal plane separates the brain along a left-right axis. **b.** A transverse plane separates the brain along a dorsal-ventral or top-to-bottom axis. **c.** A coronal plane separates the brain along a caudal-rostral or front-to-back axis.

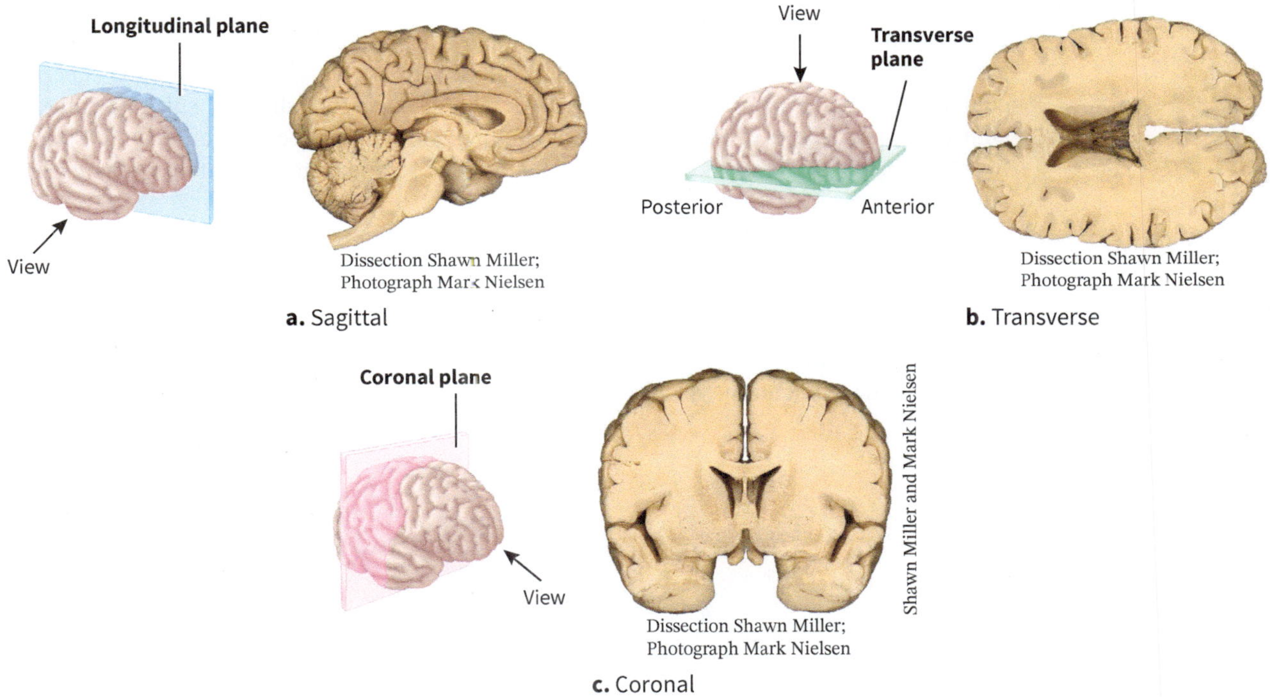

Check It Out | Gross Anatomy of the Brain

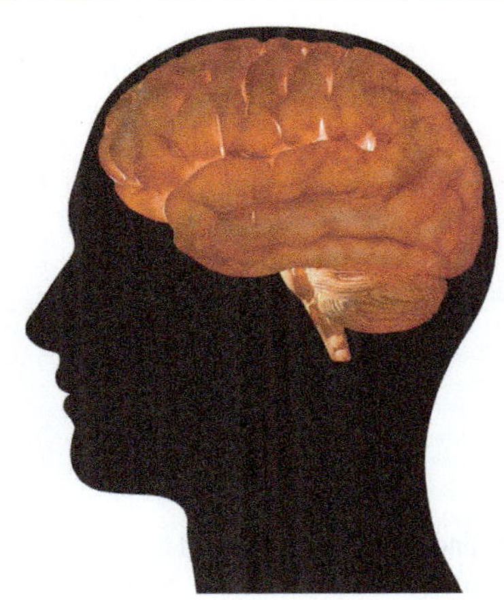

tose/Getty Images

Each lobe of the brain has a specific function, and knowing those functions is essential to your understanding of behavioral neuroscience and neurological disorders. Watch the video **Gross Anatomy of the Brain**, and then answer these questions.

1. Why might a blow to the back of the head leave an individual with a serious visual deficit?
2. A person suffers a wound to the left side of their head, above their ear. What cognitive ability might be affected by such a wound?
3. Why is it useful to label the various sulci and gyri on the surface of the brain?

The Brainstem MCAT

The brainstem is found beneath the cerebrum, where it sits atop the spinal cord (Figure 4.6). All the sensory and motor connections between the CNS and the rest of the body pass through the brainstem. In addition, the brainstem is the site of many of the brain's housekeeping functions. Respiration, cardiac function, alertness, and sleep, as well as other aspects of homeostasis (maintaining physiological balance) such as hunger and thirst, are all centered in the brainstem. Think of the brainstem as the site of the less exotic but essential housekeeping chores that keep you alive entirely without your conscious awareness. Major structures of the brainstem are the midbrain, the pons, and the medulla oblongata.

The midbrain The midbrain (also known as the mesencephalon) is located at the top of the brainstem and is divided into the tectum, which is the dorsal section, and the tegmentum, which is the ventral section. The midbrain is involved in vision, hearing, temperature regulation, pain perception, motor control, alertness, and wakefulness, so damage to the midbrain can be catastrophic. Let us take a moment to look at this important structure, starting with the tectum.

The tectum (Latin for roof) is an important part of the brain's mechanism for tracking or focusing on visual and auditory signals. The corpora quadragemini (Latin for the structure of four twins) consists of two sets of mound-like masses of gray matter, or nuclei: the superior colliculi and the inferior colliculi (from the Latin *colliculus* for hill). The superior colliculi control several aspects of eye movement, including maintaining focus on a target and tracking that target as it moves. The inferior colliculi are an important way station in the brain's auditory processing stream.

As you progress down from the tectum toward the pons, you encounter two stalk-like structures that attach the cerebrum to the brainstem called the cerebral peduncles, which consist of the tegmentum and other structures. Ascending and descending motor nerve tracts are found in the peduncles. The tegmentum (from the Latin for covering) is the ventral part of the brainstem and is involved in homeostasis, pain perception, and also inhibition of movements. Damage to the tegmentum typically results in coma.

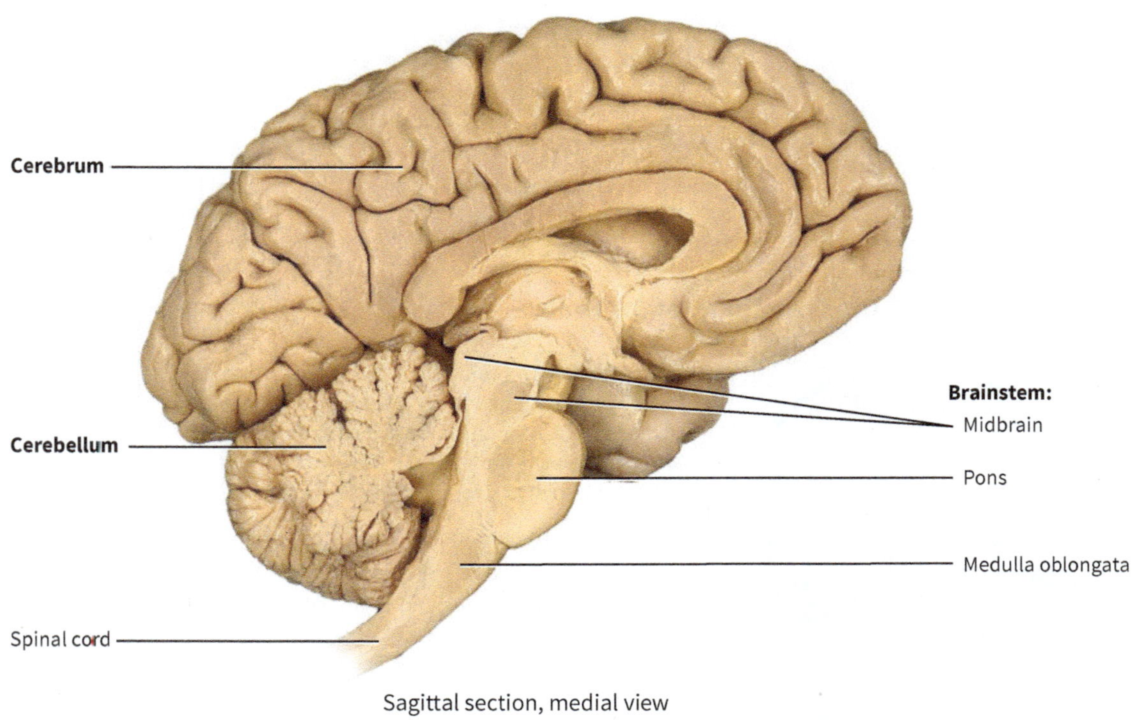

FIGURE 4.6 The brainstem The brainstem is located between the cerebrum and the spinal cord, and it consists of the midbrain, the pons, and the medulla oblongata. The brainstem is responsible for maintaining consciousness, as well as basic functions such as respiration, metabolism, and cardiac function.

The pons The **pons** (bridge) is so named because it bridges the midbrain and the medulla oblongata. Like other brainstem structures, it has several functions. Sensory data from the face and mouth are processed here, so damage to the pons can lead to difficulties in chewing and swallowing. Nerve tracts responsible for balance and the vestibular sense also pass through here, so damage can lead to balance problems as well.

One interesting function of the pons is to protect you from hurting yourself while you are dreaming. If not for the pons, when you dreamed that you were surfing a wave at Waikiki Beach or fighting a dragon attacking the Shire, your body would act out the physical actions associated with your dream, and you could easily hurt yourself or others around you. Your brain has a clever protective mechanism that comes into play when you are dreaming. Essentially, your pons disconnects your body from your brain to prevent you from thrashing about while dreaming. Occasionally, the pons reestablishes the connection when you are dreaming, with unfortunate results. The opposite also happens occasionally when someone is awakened unexpectedly and finds for a moment that they feel paralyzed. This can be terrifying, but in a few seconds the pons reestablishes the connection and normal life resumes. Serious damage to the pons can cause locked-in syndrome. In this agonizing condition, the individual is unable to move any voluntary muscles except for the eyes, all while perfectly aware of their surroundings and able to think.

The medulla oblongata The **medulla oblongata**, or simply the medulla, is at the bottom of the brainstem and is a center for the control of respiratory, digestive, and cardiac functions. Damage to the medulla typically occurs from the trauma of an automobile accident or violent attack; extended hypoxia (oxygen deprivation) will also damage the medulla. Damage to the medulla can lead to the inability to control heart rate, blood pressure, and respiration as well as the kidneys' removal of toxins from the blood. Significant damage to the medulla is generally fatal.

The Spinal Cord MCAT

The spinal cord is a physical connection between the central and peripheral nervous systems, so you will not be surprised to discover that nature has gone to some lengths to protect it. That protective structure is the spine. Your spine consists of a series of irregularly shaped bones called vertebrae with a channel in the middle where the spinal cord and associated nervous tissue are securely located (Figure 4.7). The spine is divided into the cervical, thoracic, lumbar, and sacral sections, and spinal segments are individually labeled by their region (C, T, L, S) and numbered according to their order in that region. For example, the first cervical vertebra is C1, the last of the thoracic vertebrae is T12, and the second lumbar vertebra is L2.

The function of the spinal cord is to transmit information from the brain down into the body and to transmit sensory information from the body back to the brain. It also plays a role in reflexes that are processed at the level of the spinal cord and do not involve the brain. At each spinal vertebra there is a spinal segment, or tract, that contains a motor (efferent) pathway for controlling muscles and organs and a sensory (afferent) pathway for information from the body to reach the brain (Figure 4.8). Signals from the brain travel down the spinal cord and from there into bundles of nerve fibers known as nerve tracts or simply nerves that carry the information to **ganglia** (singular *ganglion*), which are groups of nerve cells that function as a unit.

For reflexive responses, the entire action can occur in a short loop at the level of the spinal segment. For example, if you step on a sharp nail, you do not want to take the time for a signal to travel from your foot, up your spinal cord and brainstem, and all the way to your frontal lobe before deciding to pull your foot back and then sending a signal back to the foot along the same reverse path; you just want to pull your foot back as quickly as possible. Such simple responses are handled at the level of the spinal ganglion, which is why you might retract a foot or hand from a nail or hot surface before you are even aware of the danger. For more complex processes, the loop can continue all the way up to the brain. Through this combination of

FIGURE 4.7 The spinal cord and spine **a.** The spinal cord is situated in a channel of the spine where it is protected from damage. **b.** The spine consists of a series of spinal vertebrae that are divided into five regions, and each individual vertebra is identified by its region and position within that region.

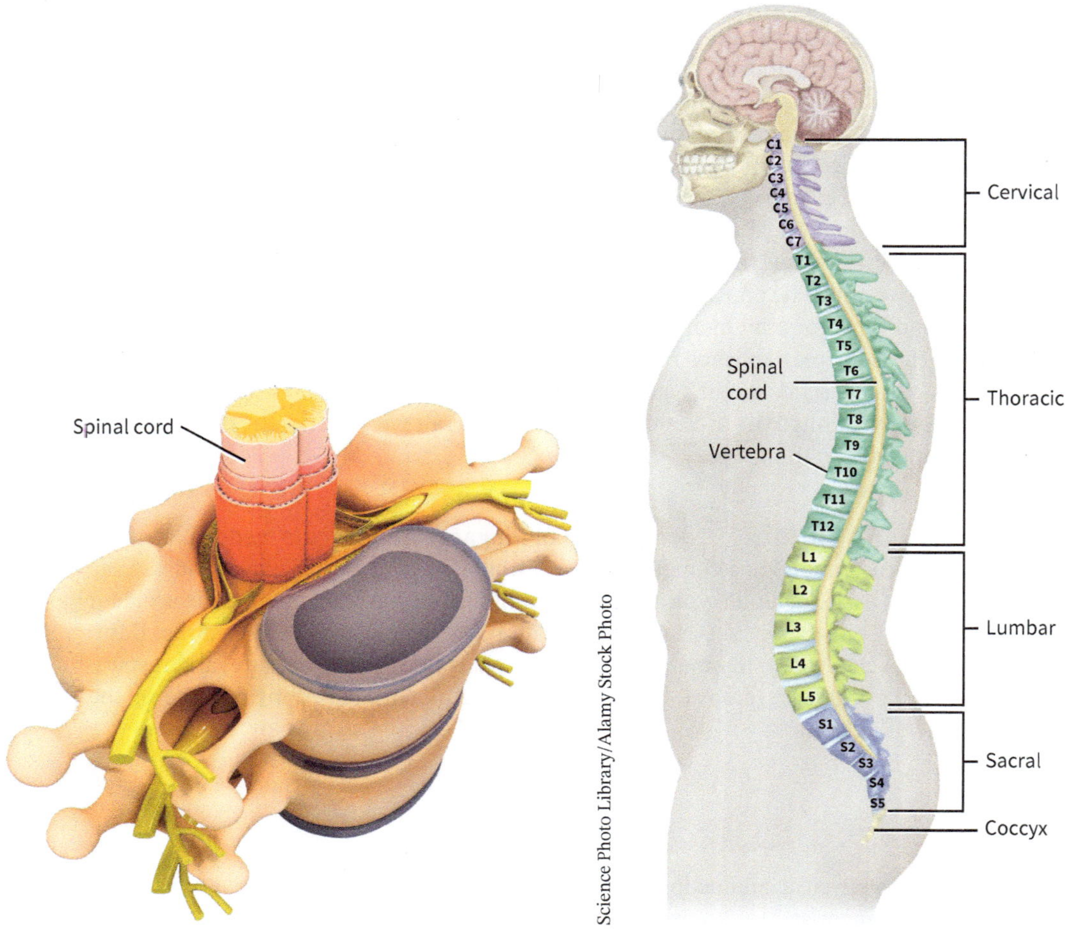

a. Section of spinal cord

Science Photo Library/Alamy Stock Photo

b. Spinal vertebrae

Source: Tonya Hines, 2019 / Mayfield Brain & Spine.

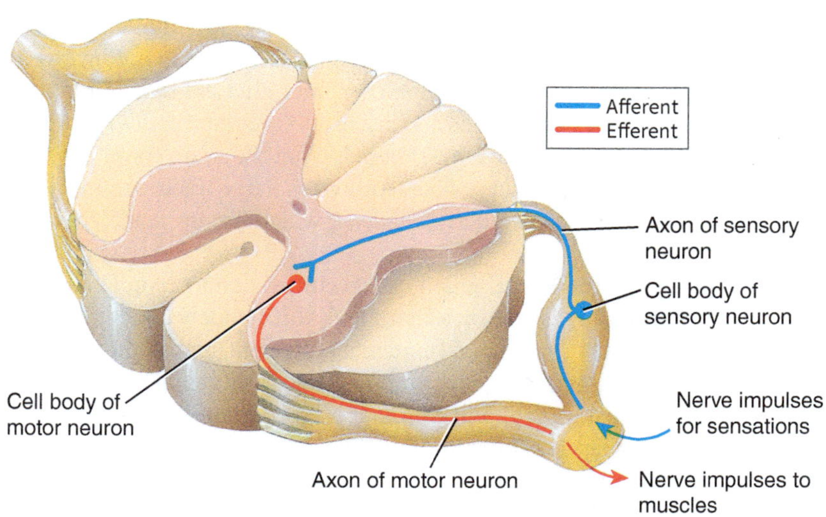

FIGURE 4.8 A spinal cord tract Each vertebra of the spine contains a spinal segment that carries communication between the brain and specific muscles or organs in the body through ascending and descending neural tracts. Efferent pathways (in red) transmit information from the brain to the body, and afferent pathways (in blue) carry sensory information from the body back to the brain.

local control and interaction with the brain, our spinal cord manages to simultaneously control intentional acts and also react instinctively to certain types of stimuli. Our spinal cord is an essential partner to the brain as it transmits commands from the CNS down into the body and serves as a communication link for sensory and other information from the body to travel back to the brain.

Each segment of the spine has a connection both to specific muscles and also to sensory receptors that communicate with that segment. For example, the nerves at cervical vertebrae C1–C6 control the muscles that flex the neck, the nerves at C6 control muscles that rotate the arm, and the nerves at vertebrae L6–S2 extend the hip and flex the foot. When spinal injury occurs, the location of the injury determines which parts of the body lose motor control or sensation. The higher the spinal segment damaged, the more global the deficit. For example, if the spinal cord is cut at C1–C2, the individual will lose all sensation and motor control below the neck and will need a ventilator to breathe. Damage to the spinal cord lower down, such as between vertebrae T6 and T12, will preserve the function of the chest and other muscles necessary for breathing but will affect abdominal and back muscles used to maintain balance. Damage to the lumbar sections of the spinal cord will preserve the function of all segments but will likely result in weakness of the legs as well as loss of bladder and bowel control.

We began this chapter with the story of how Superman (really the actor who played him) came to be paralyzed. In light of what you have learned, you can diagnose the location of the damage that Christopher Reeve suffered. Because he lost all feeling and motor control below his neck, the damage had to be above the thoracic vertebrates. In fact, Reeve fractured the two highest cervical vertebrates, C1 and C2, so he was completely paralyzed. Once damaged, nerve cells do not heal, so spinal cord damage like that experienced by Christopher Reeve can lead to lifelong disability. Some researchers are looking for a way to turn on a genetic or molecular switch that would allow neural tissue to regrow, and other scientists are attempting to use technology to bypass the damaged tissue and reroute these signals to their proper destination. You can read about one of these researchers in **Advances in Behavioral Neuroscience: Nicholas Hatsopoulos**.

Advances in Behavioral Neuroscience | Nicholas Hatsopoulos and Spinal Cord Injury Repair

Dr. Nicholas Hatsopoulos

In this chapter, you became acquainted with the complexity of the system that initiates, coordinates, and then executes motor functions and does so with feedback from the body. To create a device that can operate like a functional hand, arm, or leg has been the dream of researchers, and Dr. Nicholas Hatsopoulos is one of the foremost researchers working to make this dream a reality.

His research focuses on how our neural system coordinates motor learning with the goal of understanding how signals from associated areas of the brain relate to specific motor movements. This information is essential to creating an interface that can take signals from the brain or from surviving descending nerve tracts and use them to control a truly neural prosthetic device that will improve the lives of individuals with spinal cord injuries, amputations, and other disorders that result in a loss of muscular control over peripheral structures.

After graduating with a physics major from Williams College, Dr. Hatsopoulos earned a Ph.D. in cognitive science at Brown University where he studied motor behavior and computational models of movement. Eventually, he began to focus his professional interests on the neurophysiology of movement control. In addition to being an expert in the neuroscience of movement, he is an enthusiastic guitar and bouzouki player as well as a fan of Greek folk dancing. He finds that these personal interests inform his research into the acquisition and control of complex motor skills. Dr. Hatsopoulos' path in science demonstrates that a breadth of interests, rather than being a distraction, often enhances professional excellence.

Questions

1. **Search the internet to obtain an estimate of the number of spinal cord injuries that occur each year either nationally or globally. Then search to obtain an estimate of the number of limb amputations each year. What were those two numbers, and were they higher or lower than you thought they would be? Based on your answer, what would be the societal impact of a remedy for either?**
2. **Search the internet to discover the common causes of spinal injuries and of amputations. Which cause would you be most likely to experience, and what steps could you take to protect yourself?**
3. **Dr. Hatsopoulos has expertise in psychology, physics, computer science, and the neurology of motor movement. Briefly describe how each of these fits into his research.**

4.1 Before You Go On

Section Summary

The nervous system includes a central nervous system (CNS) and a peripheral nervous system (PNS). The CNS consists of the brain and spinal cord. Gyri and sulci divide the brain into five lobes, each of which is specialized in function. The brainstem controls the housekeeping functions that keep us alive and forms the basis for our conscious awareness of the world around us. The spinal cord carries information from the brain into the body and from receptors in the body back to the brain for evaluation.

Comprehension Check

1. The CNS consists of which structures?
   a. brain and peripheral nerves
   b. spinal cord and autonomic nervous system
   c. brain and spinal cord
   d. brain and sympathetic nervous system
2. Your adult brain developed out of which simpler developmental structures?
   a. forebrain, lateral brain, and mesobrain
   b. hindbrain, forebrain, and midbrain
   c. rostroencephalon, caudalencephalon, and mesoencephalon
   d. sagittal, horizontal, and coronal encephalons
3. Damage to the frontal lobe would likely result in loss of what?
   a. vision
   b. language
   c. movement
   d. problem solving and planning
4. Which of these is a function of the midbrain?
   a. sense of hearing
   b. coordinating the muscles of the mouth and tongue
   c. recognizing facial expressions
   d. alertness and arousal
5. What are three major divisions of the spinal cord?
   a. lumbar, thoracic, and cervical
   b. cranial, thoracic, and cervical
   c. cranial, corticospinal, and coccyx
   d. cranial, lumbar, and corticospinal

4.2 Major Structures and Support Systems of the CNS

LO 4.2 Identify the major structures and support systems of the CNS and their functions.

Now that you are oriented in the basic components of the nervous system, we will take a closer look at the major structures and support systems of the CNS.

The Cerebral Cortex MCAT

The outermost structure of the brain is the **cerebral cortex**. A cortex (plural *cortices*), from the Latin for bark, is the outermost layer of an organ, and the cerebral cortex is a large sheet of tissue that covers the brain. Your cerebral cortex consists of gray matter arranged in a layer between 2 and 4 mm thick (Figure 4.9). The cortex is comprised of cell bodies, and there are about 20 billion neurons in your cortex. Underneath the cortex is white matter, so named because it consists of bundles of myelinated axons that act as communication channels throughout the brain. If you think of the skin of an orange as the cortex, and the pulp underneath as white matter, you have grasped the basic architecture of the brain.

Perhaps the most prominent structural difference that distinguishes human brains from the brains of other organisms is the large size of the cerebral cortex. If unfolded, the cortex would be about the size of a large pizza or a small umbrella. To fit this large sheet into our skull, it must be folded over in the same way that a sheet of paper can fit within your palm if you crumple it up. Similarly, folding of the cortical surface into gyri (outward folds) and sulci (inward folds) allows a larger cortex to fit within the skull.

On the surface of the cortex are two strips, the somatosensory cortex and the primary motor cortex, which are specialized for the control of movement (Figure 4.10). Considered

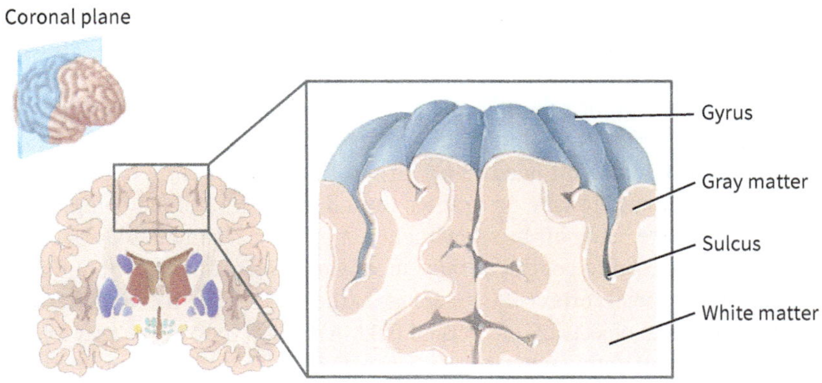

FIGURE 4.9 Folds in the human cerebral cortex A coronal section of the brain shows that the cerebral cortex is comprised of gray matter folded into gyri, or hills, and sulci, or valleys. Beneath the cortex is white matter comprised of axons and support cells such as glia.

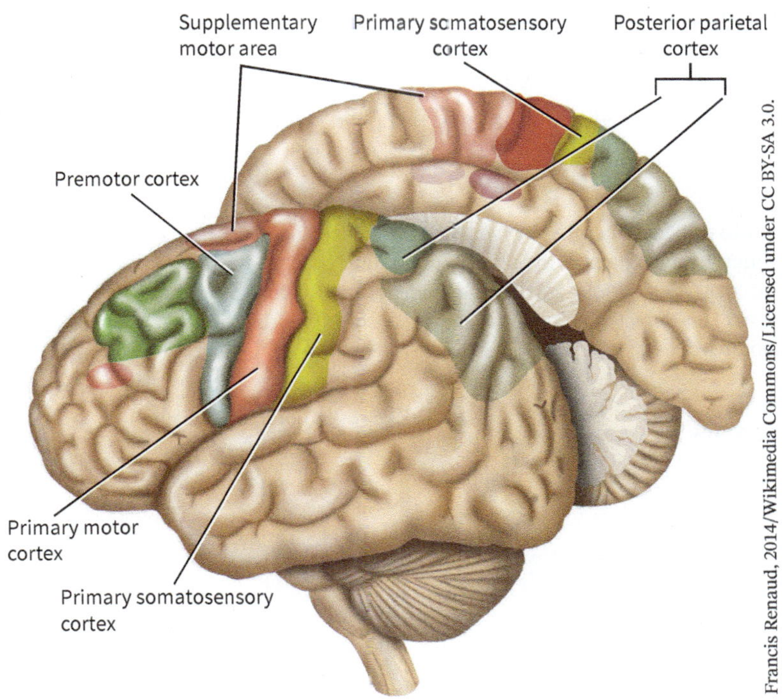

FIGURE 4.10 The human motor cortex There are several cortical areas dedicated to the control of movement. The primary somatosensory cortex receives sensory input from the body, whereas the primary motor, premotor, and supplementary motor cortices organize and initiate motor responses in the body.

together, the somatosensory and motor cortices are known as the **sensorimotor cortex**. The primary somatosensory cortex (PSC), or simply the **somatosensory cortex**, is rostral to, or immediately behind, the central sulcus, where it receives input from sensory receptors throughout the body and interprets them. When a mosquito lands on your arm, for example, signals from touch receptors in your skin travel up through your spinal cord and into the somatosensory cortex; from there, a connection to the frontal lobe allows you to choose a response (to swat it away), and that signal then goes to the motor cortex to initiate the movement of your arm. The motor response to that pesky mosquito invokes several cortical regions, including the primary motor cortex, the premotor cortex, and the supplementary motor area (SMA).

The **primary motor cortex (PMC)** is located just anterior to the central sulcus and posterior to the precentral gyrus, landmarks that divide the cerebral hemispheres into anterior and posterior portions. When you perform almost any voluntary action, from standing up from your seat, to scratching your nose, to playing a note on a musical instrument, your PMC is involved. The PMC contains Betz cells, which are the largest cells in the CNS. The PMC generates the neural signals that control muscular movement throughout the body. Signals emanating from the left hemisphere cross over to control skeletal muscles on the right side of the body, whereas signals generated in the right have a corresponding effect on the left side.

Where the PMC controls effectors such as arms, legs, and hands, the premotor cortex controls large trunk muscles. The role of the premotor cortex is not fully understood, but there is compelling evidence that it is involved in planning motor movements; namely, individuals who have experienced damage to the premotor cortex struggle to complete complex skilled motor actions while simple movements remain intact.

The SMA has an important role in sequencing or planning complex movements into a single action. The SMA arranges the individual components of the complex sequence, much as you might lay out the tools you need to perform a household maintenance task prior to beginning the repair. It is easy to take for granted our ability to coordinate complex muscular movements on tasks where being off by a few inches or a half a second could spell disaster.

No discussion of the cortical control of movement would be complete without mentioning the role of the posterior parietal cortex (PPC). Damage to the PPC results in deficits in reaching for objects and also attention to objects in one's visual field. Although the PPC plays a role in movement, that role is more about noting what is in the visual field and less about coordination of muscles, so we will discuss the parietal lobe's involvement in object recognition in a different chapter.

One important observation about these sensorimotor and motor cortices is that they are functionally aligned; for example, the region of the sensorimotor cortex that receives input from your left index finger is adjacent to the area of the motor cortex that organizes movement of that same finger. This means that sensory feedback from your finger does not have far to go to affect the motor cortex that controls that same finger. **Figure 4.11** depicts this arrangement in a diagram known as the sensorimotor homunculus. Figure 4.11a depicts the

FIGURE 4.11 The sensorimotor homunculus The sensorimotor homunculus depicts the location and amount of cortex dedicated to different structures on the sensory and motor cortices. Different structures may have more or less space on each cortex depending on whether motor control or sensation is more important to their purpose. **a.** This depicts the functions of the sensory strip. **b.** This depicts the corresponding functions of the motor strip.

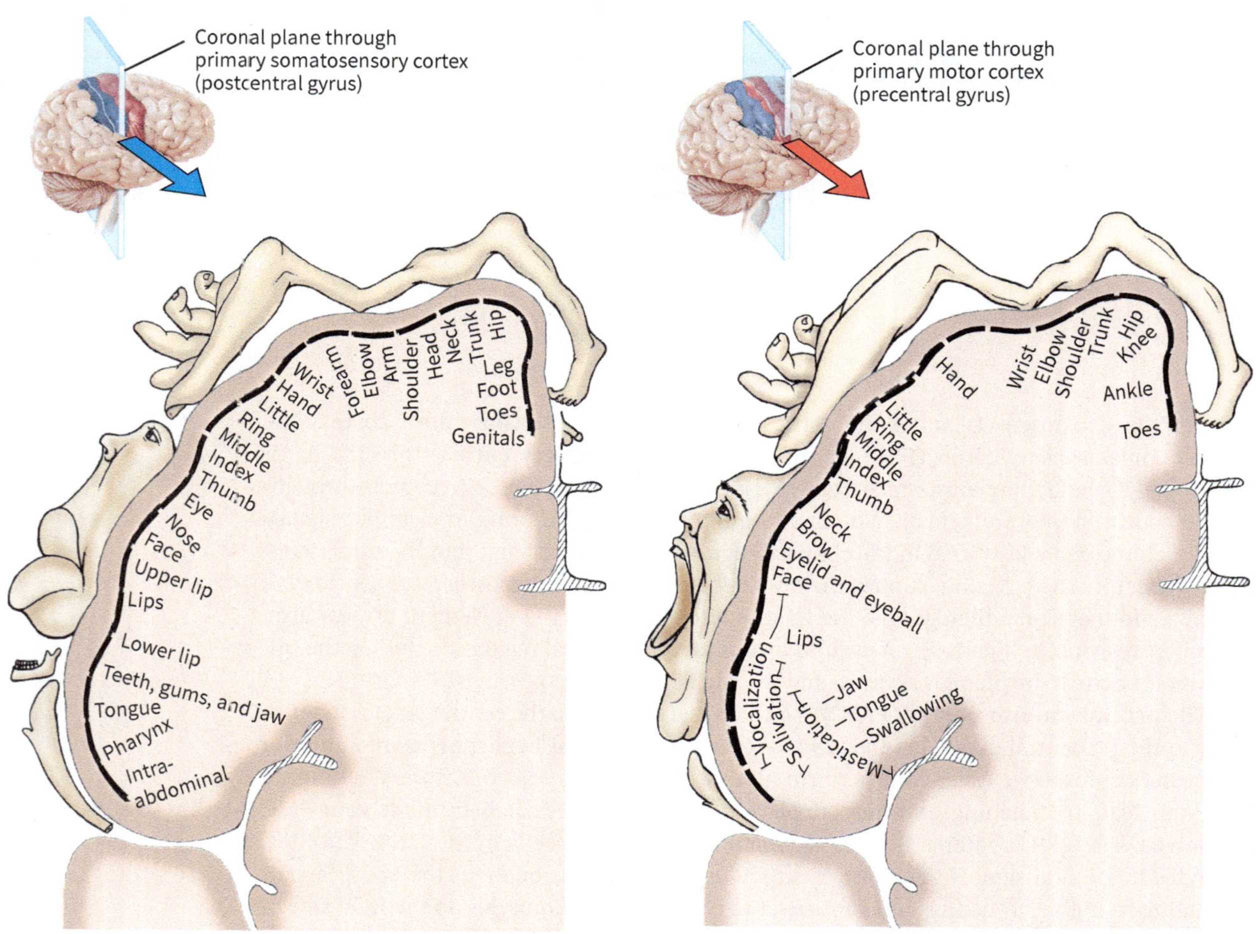

a. Coronal section of primary somatosensory cortex in right cerebral hemisphere

b. Coronal section of primary motor cortex in right cerebral hemisphere functions of the sensory strip and Figure 4.11b represents the corresponding functions of the motor strip. Note that although sensory and motor areas for specific structures are generally aligned, the amount of sensory and motor cortex devoted to each structure might differ depending on its function. For example, considerably more sensory cortex is devoted to receiving sensation from the lips than motor cortex is devoted to their muscular control, and the opposite is true for the hand. You can think about how you use your hands and lips to understand why this is the case.

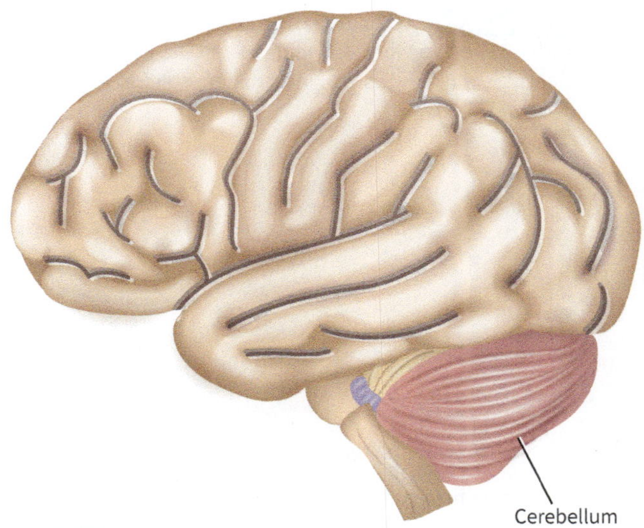

FIGURE 4.12 **The cerebellum** The cerebellum is a part of the hindbrain that is located underneath the occipital lobe and next to the brainstem.

The Cerebellum MCAT

The **cerebellum** (from the Latin for little brain; plural *cerebella*) is situated at the rear, or caudal, end of the brain and tucked under the cerebrum (**Figure 4.12**). Its main function is the coordination of motor activity throughout the body. The cerebellum does not initiate movement, but it does control coordination and timing of complex actions such as balance, reaching for an object, or walking.

The cerebellum is a complex structure in its own right and perhaps earns its name of little brain. Although it is only 2.4 in. by 2 in. by 4 in. (6 cm by 5 cm by 10 cm) and about the size of a slightly asymmetric clementine, it is comprised of many tightly folded layers of gray matter known as folia, which, if unfolded, would be a strip about a yard long (~1 meter) and 2 in. (5 cm) wide. Underneath this layer is a second structure known as the arbor vitae, or tree of life, so named for its tree-like appearance. The arbor vitae is comprised of myelinated fibers.

Cerebellar neurons The cerebellum is packed with around 100 billion neurons, as opposed to the cerebral cortex, which contains perhaps 20 billion neurons. The cerebellum is a complex structure consisting of several types of cells arranged in tightly packed layers (**Figure 4.13**). There are two primary types of neurons in the cerebellum, and these are Purkinje cells and granule cells. Purkinje cells are large (second only to Betz cells located in the gray matter), and their arboreal, or tree-like, structure is striking, with perhaps as many as 200,000 spines per dendrite. These dendrites permit one cell to influence many surrounding cells. Dendritic spines are involved in storing habits and memories, so they are important in

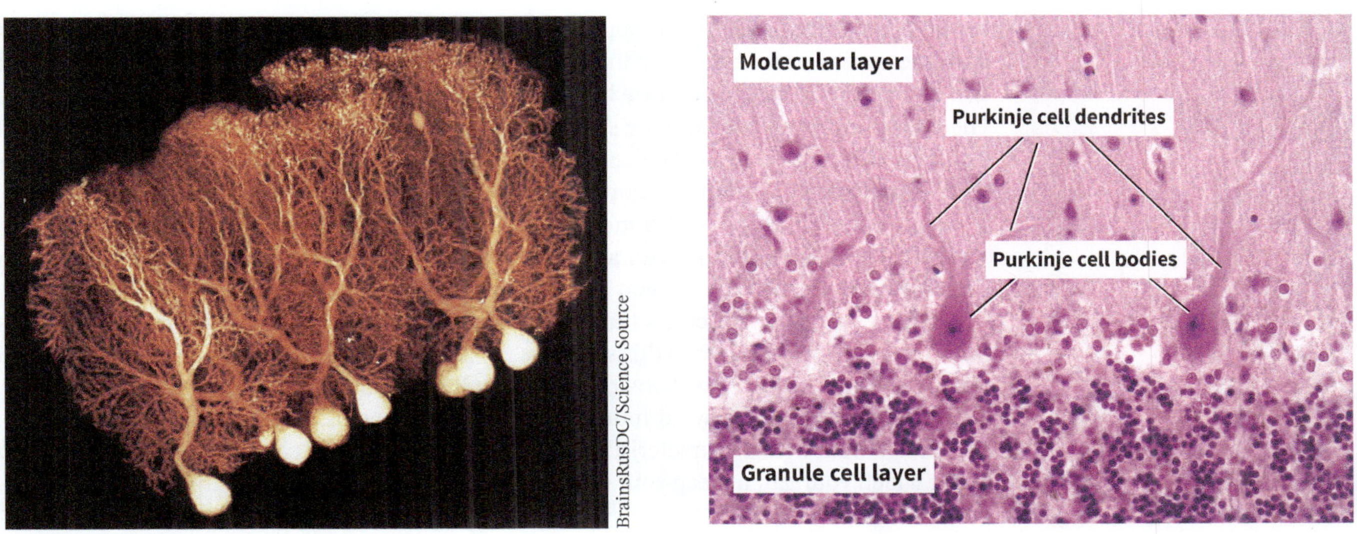

FIGURE 4.13 **Cerebellar Purkinje and granular cells** The cerebellum consists of several tightly packed and interconnected layers of specialized cells such as Purkinje and granular cells. Smaller and with fewer dendrites, granular cells interact with Purkinje cells, most likely to coordinate their action.

FIGURE 4.14 The cerebellum in superior and inferior views The cerebellum can be divided into two basic regions: the midline and the hemispheres. **a.** The midline consists of the vermis and flocculonodular lobes. **b.** The two hemispheres are on each side of the midline.

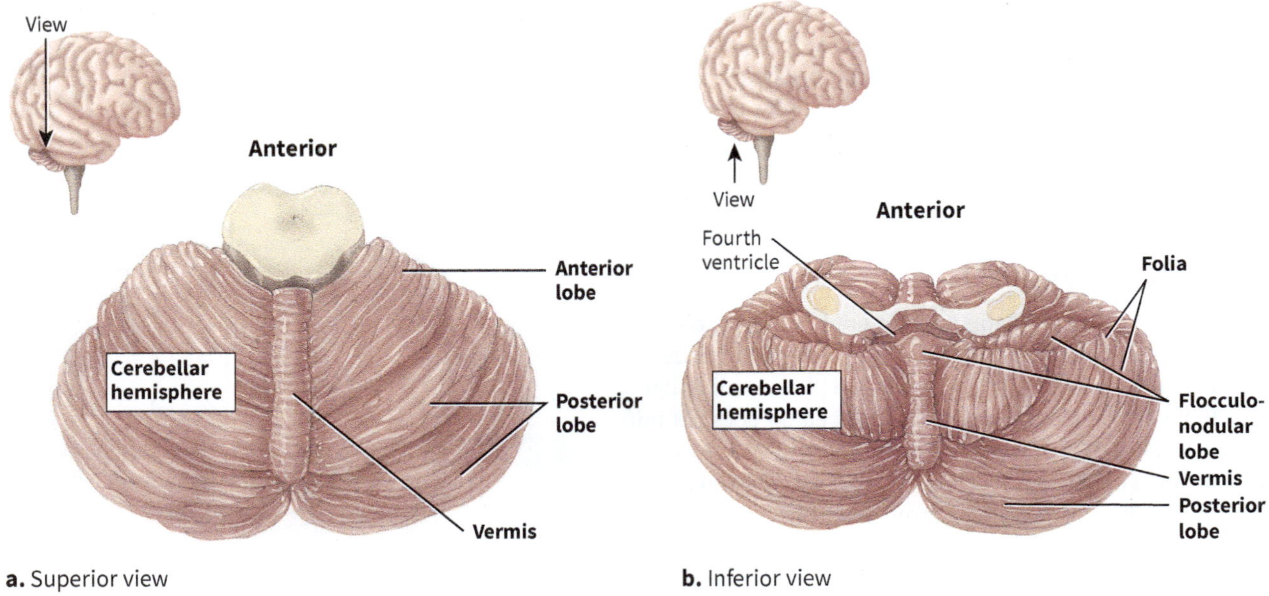

the establishment of skilled movements. Purkinje cells release the inhibitory neurotransmitter GABA, so they are thought to inhibit neuromuscular activity in a way that regulates and coordinates movement.

A second cortical cell type is the granular cell. In contrast to the large Purkinje cells, cerebellar granular cells are among the smallest of the neurons in the human brain and have few dendrites with few dendritic spines. Based on their relationship with other cerebellar cells, researchers theorize that granular cells are dedicated to making very fine adjustments to cerebellar response. What these granular cells lose in size, they more than make up in number; it is estimated that there are perhaps 50 million cerebellar granular cells, more than any other cell type in the brain.

Cerebellar lobes Moving from the cellular level to a larger field of focus, we now see the cerebellum can be divided into the anterior lobe, the posterior lobe, and the flocculonodular lobe (Figure 4.14). In addition, a stripe down the middle is known as the vermis (Latin for worm) because the tissue has the appearance of worms. The flocculonodular lobe accepts input from the vestibular system in the inner ear and uses that information to maintain balance. It also accepts input from a structure known as the superior colliculus to guide eye movements. The vermis accepts input from the spinal cord and from our visual and auditory systems, and its function is to regulate movements of the trunk and the limbs. The anterior and posterior lobes participate in planning motions.

Animals such as sheep also must move about, so they also have a cerebellum. You can compare a sheep cerebellum with a human cerebellum and reflect on cerebellar function in **Check It Out: Comparing Sheep and Human Brains**.

What exactly does the cerebellum do? The answer, based on clinical data from individuals suffering damage or disease to this structure, is that the cerebellum controls the coordination of motor output to the body. If you can catch a ball, snap your fingers, play a pentatonic scale, or deftly move your fork toward your mouth, thank your cerebellum. Individuals with cerebellar damage walk with a characteristically wide gait. If this seems confusing, imagine how you might widen your gait to keep your balance while walking on a swaying train or boat.

Essential Support Systems of the CNS MCAT

Just as an army includes support people who provide the soldiers with food, shelter, and repair equipment essential to their mission, our brain has structures that protect it, feed it, and

Check It Out | Comparing Sheep and Human Brains

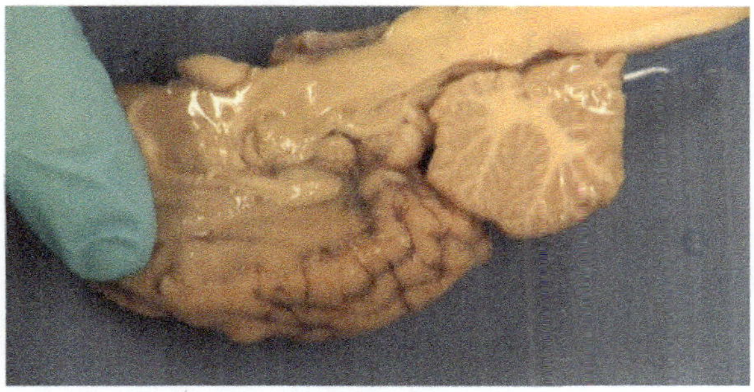

An image in a textbook does not always convey what an anatomical structure looks like in reality, and this is certainly true of the cerebellum. The cerebellum is modest in size, but it is tightly packed with a variety of cells. Watch the video **The Cerebellum**, and answer these questions.

1. Compare the relative size of the human and sheep cortices and also the human and sheep cerebella. What does that cortical comparison suggest about the intellectual and motor demands of a sheep's life compared with those of a human?
2. Why might the very dense organization of the cerebellum confer an advantage in organizing and coordinating movements over the more widely dispersed arrangement of the cortex?
3. Would a more efficient cerebellum give you an advantage at some activities such as sports? If we had a way to identify those individuals whose cerebellar function was above average, should they be encouraged to participate in sports? Explain your answer.

cleanse it of metabolic by-products. We now turn our attention to these important structures and systems.

The meninges Your brain is a very delicate organ, so protecting it against the trials and tribulations of everyday life is a priority. The human brain is protected by the skull, but two additional contributors to the brain's protection are the meninges and the cerebrospinal fluid. If you were to gently peel the scalp back and then cut through the skull to create a coronal view, you would see something like the presentation in **Figure 4.15**.

Inside the skull, the brain is wrapped in three additional layers of protection known as the **meninges**. These three layers are the **dura mater** (tough mother in Latin), the **arachnoid layer**

FIGURE 4.15 The meninges The meninges surround and cushion the brain and provide a framework for blood vessels that serve the brain.

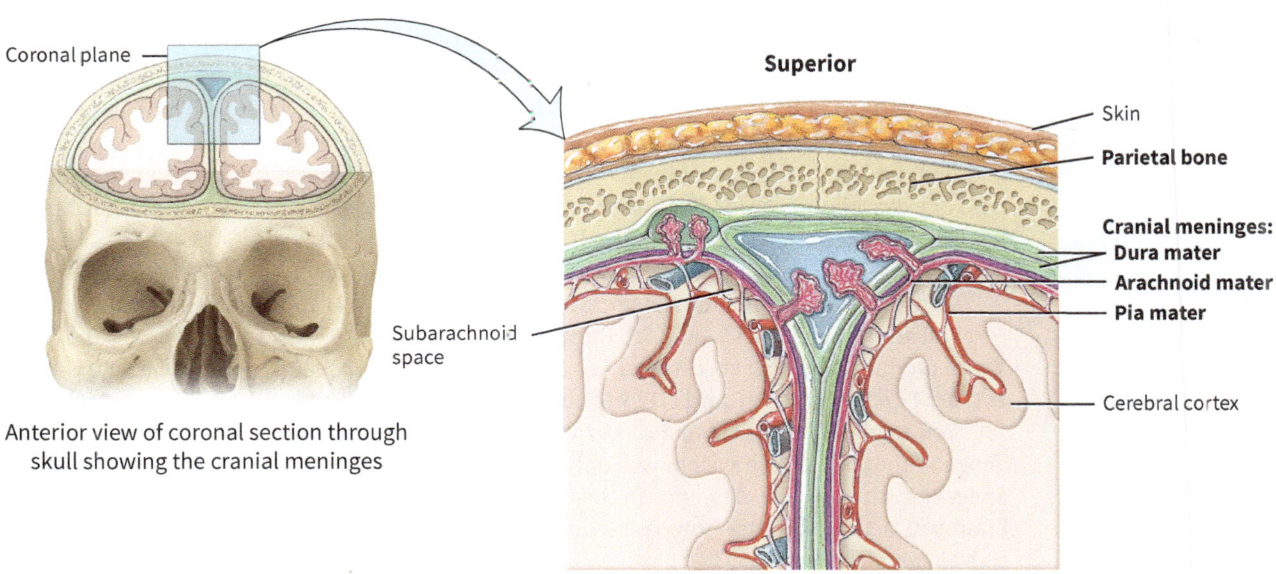

FIGURE 4.16 **The human ventricular system** CSF is created in the lateral ventricles and courses through the ventricles, around the brain, and then down into the spinal column. CSF serves to cushion the brain and remove metabolic waste products.

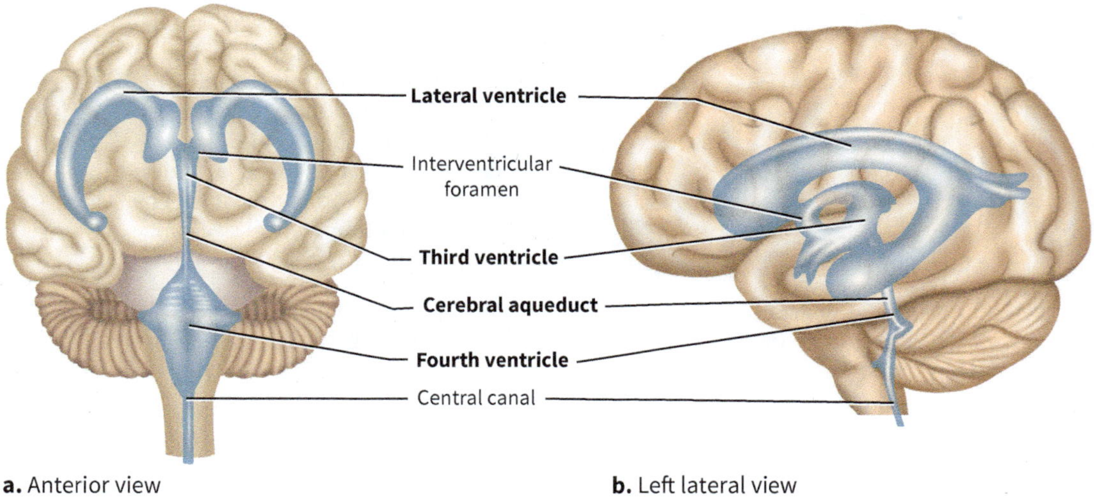

(named for its spiderweb appearance), and the **pia mater** (soft mother in Latin). The dura and arachnoid layers comprise an outer layer; the pia is the inside layer, and it adheres closely to the cortex. In between the outer and inner layers is the fluid-filled subarachnoid space, so named because it lies underneath the arachnoid layer.

The ventricular system The meninges form a protective cover over the delicate brain, but you might have also wondered how it is that your brain does not chafe against the base of the skull as you go through your daily activities. The answer is that the brain is protected by a thin layer of **cerebrospinal fluid (CSF)**. In essence, our brain floats within our skull. CSF is secreted by specialized tissue known as the choroid plexus. We generate a little more than two and a half cups (~600 ml) of CSF each day. Suspending the brain in fluid decreases its weight and absorbs shock.

In addition to its protective function, the CSF also aids in equalizing the chemical environment throughout the brain and in transporting metabolic waste. CSF circulates around the brain in the space between the dura and pia mater, through what is known as the ventricular system (Figure 4.16). Inside your brain are a series of four cisterns, or spaces, known as **ventricles**. The two lateral ventricles are situated one in each hemisphere, and below them and on the midline of the brain are the third and fourth ventricles. CSF is produced in the lateral ventricles and then passes through the interventricular foramen into the third and then fourth ventricles, around the outside of the brain, and finally down the central canal of the spinal cord.

The glymphatic system Metabolic processes consume energy and oxygen and create waste by-products. In the body, the lymphatic system collects extracellular waste products and fluid and then disposes of them. Your brain has no lymphatic system, so how does it dispose of extracellular waste? The answer is the **glymphatic system**, which provides metabolic by-product clearance within the brain. It is called the glymphatic system because it is similar to the lymphatic system but involves waste removal from the glia.

Through the glymphatic system, CSF enters the brain via tiny channels that surround cerebral arteries and then is pumped through the brain, carrying waste products with the flow. Waste products then exit via cerebral veins using channels associated with those veins (**Check It Out: The Ventricular and Glymphatic Systems**).

You might be surprised to discover that although CSF was known in antiquity, clear evidence of the glymphatic system was only achieved in 2012. Researchers knew there must be a mechanism for waste clearance in the CNS, but that mechanism only functioned when the brain was active, so it was impossible to observe in the living brain until new visualization techniques were created.

Check It Out | The Ventricular and Glymphatic Systems

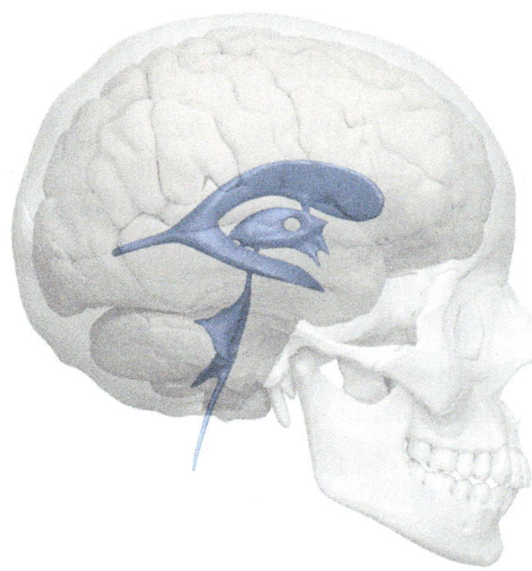

Courtesy of DBCLS (Data Base Center for Life Sciences)

The ventricular system cushions the brain, provides buoyancy to reduce its weight on the bottom of the skull, and, along with the glymphatic system, helps regulate the chemical environment and remove metabolic waste. View the short video on the **Ventricular and Glymphatic Systems**, and then answer these questions.

1. Search the internet using the phrase "disorders of the ventricular system," and identify and briefly describe one such disorder.
2. Some researchers hypothesize that disruption of the glymphatic system might be the cause of Alzheimer's disease. Why might that make sense?

4.2 Before You Go On

Section Summary

The somatosensory cortex and the motor cortex are essential to the control of movement. The cerebellum at the base of the brain is a highly organized structure that coordinates overall motor function. The meninges cushion and protect the brain. Cerebrospinal fluid circulates through the ventricular system and around the brain, removing waste and providing a supportive and protective layer, whereas the glymphatic system removes metabolic by-products from the brain.

Comprehension Check

1. A gyrus is like a __________, and a sulcus is like a __________.
   - **a.** ridge; valley
   - **b.** valley; ridge
   - **c.** small ridge; large ridge
   - **d.** small valley; large valley
2. The somatosensory cortex __________, and the motor cortex __________.
   - **a.** directs voluntary muscles; directs involuntary muscles
   - **b.** sends signals to the senses ; accepts signals from muscles
   - **c.** accepts signals from the senses; sends signals to muscles
   - **d.** is much larger; is smaller
3. Damage to either of the lobes of the cerebellum would likely result in what?
   - **a.** gait and balance problems
   - **b.** mental confusion
   - **c.** loss of vision
   - **d.** inability to grasp a cup
4. What is the function of the glymphatic system?
   - **a.** regulating metabolic processes throughout the CNS
   - **b.** removing excess CSF from the brain
   - **c.** removing metabolic by-products from the brain
   - **d.** supplying the brain with oxygen and glucose

4.3 Major Subcortical Processing Circuits

LO 4.3 Integrate the major subcortical circuits into the larger functions of the CNS.

We began our travels through the CNS by looking at its most obvious external features. We now dip under the surface of the cortex to see the most important and interesting subcortical structures. These subcortical processing structures are responsible for the rapid and efficient transfer of information throughout the brain, for integrating the senses and regulating your internal environment, for creating and maintaining emotions and memories, and for initiating and controlling movement. We begin our exploration of circuits in the brain that create complex behaviors with the corpus callosum.

The Corpus Callosum: Information Transfer MCAT

The **corpus callosum** (Latin for hard body) is a wide bundle of white matter called a commissure that crosses the midline of the brain to connect the left and right hemispheres (Figure 4.17). There are several commissures in the human brain, but with roughly 250 million axonal projections from one hemisphere to the other, the corpus callosum is the largest white matter tract in the entire brain.

This structure carries much of the communication between the two hemispheres, so it plays an important role in the integration and coordination of processing that involves multiple streams. For example, cognitive tasks that have both verbal and spatial components depend on the corpus callosum to permit the cross talk among cortical processing centers at which humans are so adept.

FIGURE 4.17 The corpus callosum **a.** In this MRI scan, the corpus callosum appears as a thin semicircle of dense myelinated tissue in the middle of the brain. **b.** Using diffusion tensor imaging, we see that the corpus callosum connects white matter tracts that service the entire cerebrum. The corpus callosum is more accurately described as a complex structure that links the left and right cortical hemispheres and subcortical structures into a powerful thinking machine.

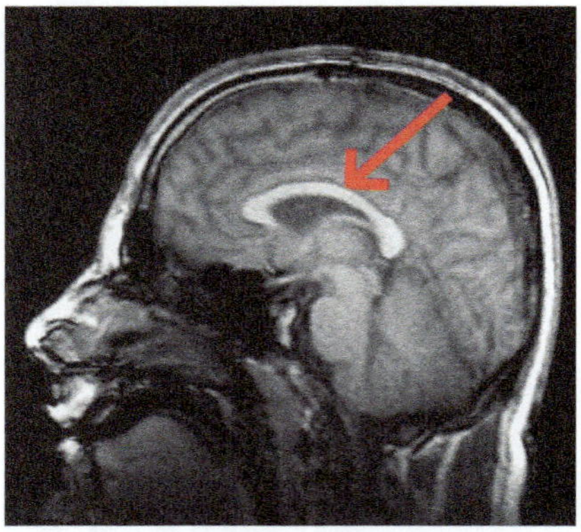

a.

Courtesy Nevit Dilman

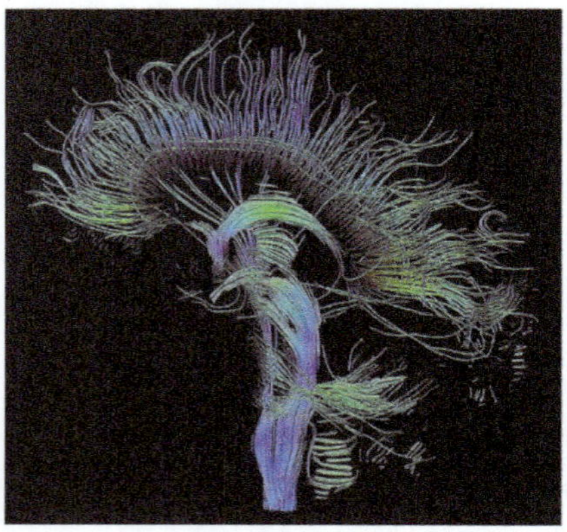

b.

Courtesy Dr. Thomas Schultz

The Thalamus and Hypothalamus: Integration and Regulation MCAT

Beneath the cortex and perched atop the brainstem is the **thalamus** (inner chamber), a bilateral structure that sits on the midline of the brain (Figure 4.18). The **hypothalamus** is just beneath the thalamus, hence the name (from the Greek *hypo* for below). These two structures are responsible for integrating sensory information and regulating the bodily process that help us maintain homeostasis.

The thalamus consists of two identical walnut-sized bodies between the brainstem and the cortex. The fact that it enjoys a prime piece of real estate in the brain suggests that it plays an important role, and research as well as clinical observations support that view. The thalamus is a hub for information from the different sensory systems. It also plays a role in regulating arousal, wakefulness, and sleep. Damage to the thalamus can result in transitory or permanent coma.

The hypothalamus plays an important role in the brain's control of many bodily functions including hunger and satiety, thermoregulation, fatigue and sleep, and circadian rhythms. It secretes hormones and communicates with the endocrine system to regulate homeostasis. We will consider this important structure again in a later chapter when we discuss the regulation of internal states, feeding, and satiety. As you might well imagine, researchers interested in either stimulating eating or suppressing hunger find the hypothalamus to be a fascinating topic.

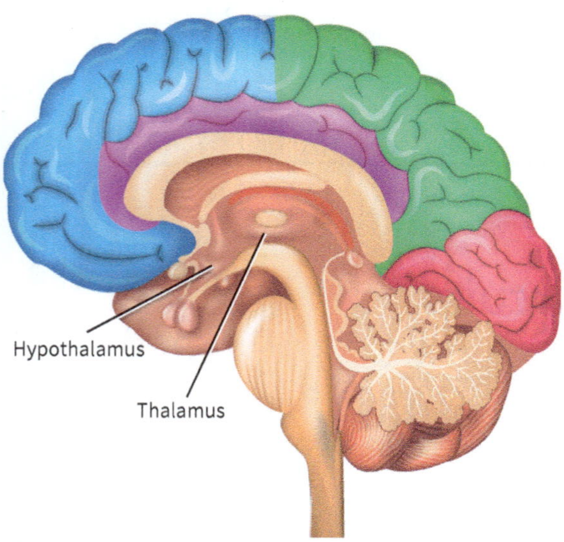

Tefi/Shutterstock

FIGURE 4.18 The thalamus and hypothalamus The thalamus sits atop the brainstem, and the hypothalamus sits below the thalamus; both structures are essential to numerous brain functions.

The Limbic System: Emotion and Memory MCAT

In the late 1930s, neuroanatomist James Papez was investigating the impact of rabies on the brain with the goal of understanding brain–behavior relationships. Rabid animals are characteristically aggressive, and Papez found that hippocampal damage was common in these animals. This led Papez to propose that a group of subcortical structures support and control emotions. Subsequent work by James McLean and others led to the current accepted model known as the **limbic system**. The word limbic comes from the Latin for border because this system forms an edge of the cortex. The limbic system is a collection of subcortical structures responsible for the expression of our internal emotional state. It is also involved in motivation and reward as well as memory.

The limbic system is generally considered to be comprised of the amygdala, cingulate cortex, septal nuclei, hippocampus, mammillary bodies, and fornix (Figure 4.19). Other structures such as the frontal

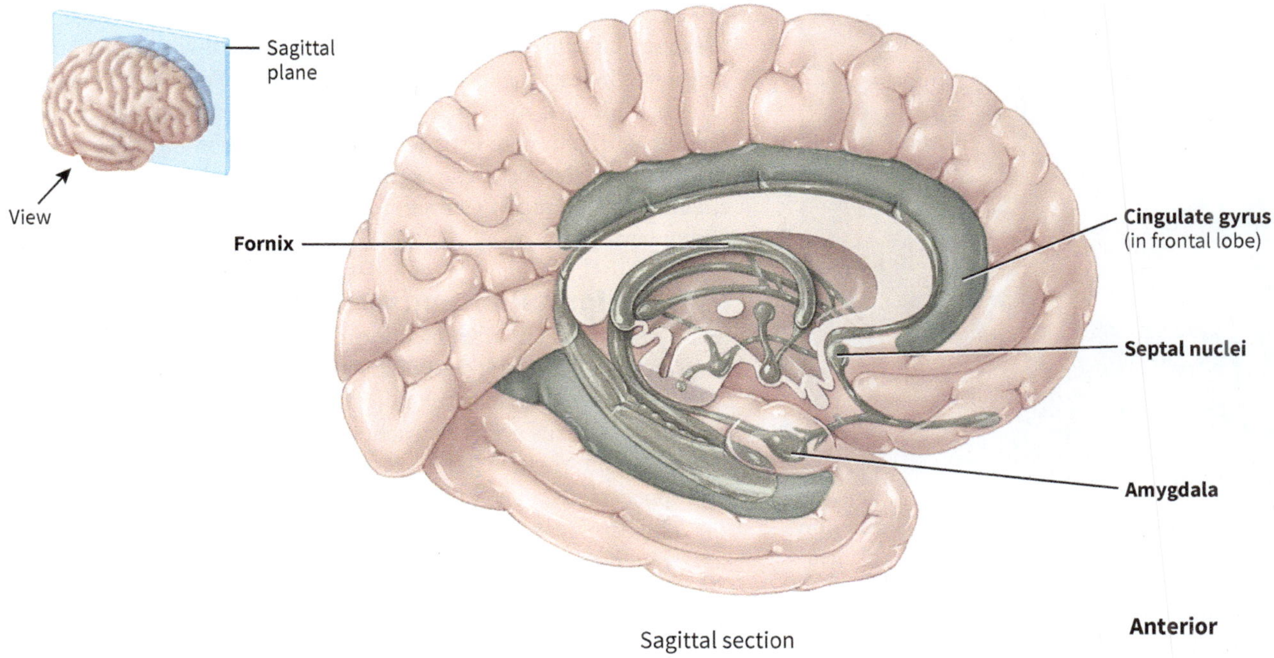

FIGURE 4.19 **The limbic system** The limbic system is a collection of interconnected subcortical structures that communicate with the cortex to create emotions and facilitate the storage of memories.

lobe have direct connections to members of this circuit. Unlike other brain structures that can be clearly delineated, there is some debate about which structures should be counted as part of the limbic system and even whether the concept itself is useful. With that ambiguity in mind, we turn to a discussion of this interconnected set of structures.

Emotional response

The **amygdala** (from the Latin for almond because of their shape) are small structures bilaterally situated deep within the temporal lobes. The amygdala play a critical role in emotional responses and also are intimately involved in memory. If a certain movie from your youth or a specific song that is linked to an emotional time in your life elicits memories of events long past, you are experiencing the link between emotions and memory that the limbic system creates. This link is often a good thing; who would reject the warm memories of Grandma's Sunday dinners that flood into your mind when you are served one of her signature dishes? Sadly, this same mechanism linking emotions to memories long forgotten can work against a person, as when combat veterans are emotionally triggered by watching intense war movies such as *Blackhawk Down* or *Saving Private Ryan.*

The cingulate cortex is superior to the corpus callosum, takes input from the thalamus, and sends its output to the temporal lobe. The cingulate cortex is involved with emotional processing and especially the connection among emotions, your sense of smell, and learning. For example, you may have a memory of wonderful aromas that bring you back to holiday dinners with family, and that sensory memory helps you to remember events that occurred in that setting. By the same token, odors that are associated with unpleasant experiences can have significant emotional impact. Soldiers in World War I were sometimes exposed to mustard gas, which forms large blisters on the skin and inside the lungs with potentially fatal consequences. To disguise the natural odor of mustard gas, chemists mixed it with xylyl bromide, which smells like lilacs. When the war was over, soldiers walking through a park would flee in terror when a springtime breeze brought the scent of a lilac bush to their nose (and to the cingulate cortex).

The **septal nuclei** are a small group of nuclei that connect to the hippocampus, hypothalamus, and amygdala and are positioned below the corpus callosum. In the 1950s, James Olds and his colleagues reported that they had discovered a reward center in the rat brain. Olds implanted an electrode in the septum of a rat that had been taught to press a bar to get a food reward; after the operation, the rat would receive a small electrical stimulation to the septum with each bar press. Figure 4.20 depicts a rat in Olds' experiment.

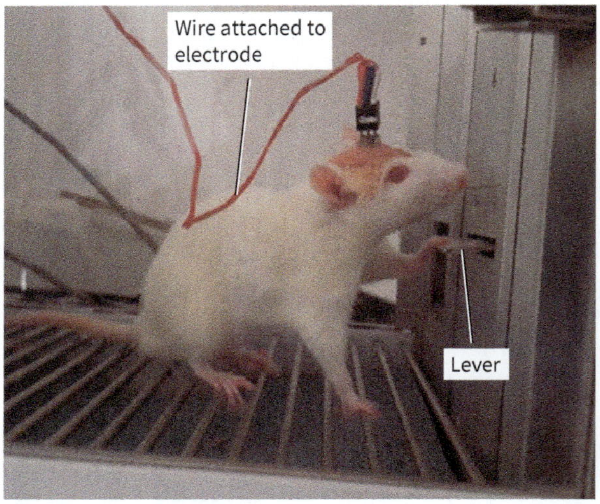

FIGURE 4.20 Rat pressing a bar to receive stimulation in the septal region An electrode is surgically implanted in the septal nuclei of a rat that was trained to receive a reward in the form of stimulation to the brain's reward center every time the bar was pressed.

Olds found that rats receiving septal stimulation would bar press to exhaustion and even to death. They concluded that the septal nuclei play a role in learning by reinforcing responses. There are other structures besides these that are a part of the limbic system, but given the richness of the human experience, you will not be surprised to learn that many of these structures perform in more than one behavioral circuit. We now turn to structures that are involved in learning and remembering.

Memory formation and retrieval

Our ability to learn and retain enormous amounts of information is a hallmark of humans. Not surprisingly, a great deal of attention has been devoted to identifying and understanding the brain structures that enable this important function. Many cortical structures are involved in the storage and retrieval of information, but three structures essential for memory are the hippocampus, the mammillary bodies, and the fornix.

The **hippocampus** is a small subcortical structure whose name comes from the Latin for seahorse because of its similarity in appearance. It is a curved tube located in the medial temporal lobe of the brain. The hippocampus is an important structure in short-term and long-term memory consolidation as well as in spatial memory, so bilateral damage to the hippocampus results in devastating memory loss. We will return to this structure when we discuss the biological basis of memory and then extend our discussion into such topics as amnesia, Alzheimer's disease, and related disorders.

The **mammillary bodies** are two small bilateral structures located at the tip of the fornix. They are generally described as part of the hippocampus, but some consider them to be more connected to the hypothalamus. Significant damage to the mammillary bodies typically results in anterograde amnesia, or the inability to learn new information.

The **fornix** is a neural tract that carries information from the hippocampus to the hypothalamus. A complete understanding of the fornix has not yet been realized, but we do know that damage to the fornix results in memory deficits, especially recall as opposed to recognition memory.

We now turn from a discussion of structures in the brain that create emotions and memories to a critical subcortical circuit of structures that create our ability to move about in the world, the basal ganglia.

The Basal Ganglia: Initiation and Control of Movement MCAT

The **basal ganglia** are a collection of subcortical structures that act as a unit to control purposeful action. The basal ganglia consist of the globus pallidus (or pale globe), the subthalamic nucleus, the nucleus accumbens, the substantia nigra (black substance), and the caudate nucleus and putamen (Latin for nutshell), which together are referred to as the striatum (Figure 4.21).

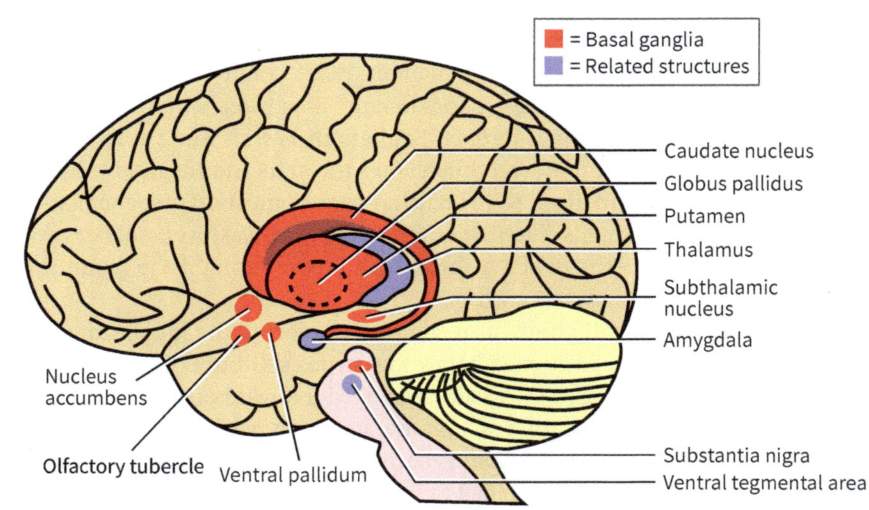

FIGURE 4.21 The basal ganglia and related structures The basal ganglia consist of a collection of nuclei that, acting together, initiate and control movement.

The basal ganglia are involved in both the initiation and inhibition of movement, as well as smoothing out and coordinating movements. Perhaps a good metaphor for the basal ganglia would be the control surfaces of an airplane. The pilot flies the plane by intentionally directing the elevator, ailerons, and rudder, but in addition, constant small adjustments to these surfaces are made to inhibit unwanted drifting or yawing out of stable flight. In the same way, the basal ganglia play a key role in intentional movement as well as inhibiting unwanted muscular tremors. Without these structures, you could not initiate muscular movement, and without their proper functioning, you would struggle to complete even simple movements.

The globus pallidus is involved in the regulation of voluntary movements. It serves as an inhibitory partner to the cerebellum, which initiates voluntary movements. When you decide to reach out your hand to scratch your nose, it is the complementary action of the cerebellum and the globus pallidus that creates that smooth arc of your hand to its target.

The subthalamic nucleus is a small, lens-shaped nucleus directly underneath the thalamus, as its name implies. Its complete function is at present unknown, but it does appear to play a role in suppressing involuntary movements.

The nucleus accumbens is a center involved in generating feelings of reward. Its role in the basal ganglia is thought to be related to selection of an action based on the potential reward for that action.

The **substantia nigra** secretes dopamine, a neurotransmitter essential for the proper functioning of the basal ganglia. One way to learn how a system works is to observe it when it is broken. Damage to the substantia nigra leads to Parkinson's disease, and damage to the striatum leads to Huntington's chorea. We will discuss both these unfortunate maladies elsewhere.

The striatum is a nucleus in the basal ganglia that is also involved in the initiation of complex movements, but its role goes beyond that to include executive function and working memory. The striatum receives input from the cortex and transmits information to the thalamus and other regions. Recent research has implicated the striatum in two complex behaviors: the reward circuit and response to novel or unfamiliar situations. For this reason, the striatum is a focus of research for those studying addictive behaviors such as substance abuse and gambling, as well as response to uncertainty and risk.

The caudate nucleus consists of bilateral structures that are a part of the striatum, which in turn is a part of the basal ganglia. The caudate nucleus is involved in creating and controlling motor activity but is also associated with procedural learning and other forms of memory. José Delgado, a pioneer in the area of brain stimulation, found that he could change behaviors through brain stimulation. In perhaps the most famous of his demonstrations, Delgado entered an arena with a bull that had an electrode implanted in the caudate nucleus. As the bull charged Delgado, he pressed a button on the radio transmitter connected to the implanted electrode. The bull stopped in its tracks and reversed course (see **Check It Out: Delgado's Bull**).

Check It Out | Delgado's Bull

Yale Events and Activities Photographs (RU 690). Manuscripts and Archives, Yale University Library

By stimulating the caudate nucleus, Delgado interrupted the bull's charge. Search the internet for "Delgado's bull" to watch a video of Delgado's original experiment with the bull, and then answer these questions:

1. If it were possible to implant a device to blunt uncontrollable violent urges, should such a procedure be considered for violent individuals who choose the procedure over jail?
2. Would your answer change if they were unwilling to have the surgery? Explain your answer.

Finally, the putamen helps with the selection of movement and as such has been implicated in Tourette's and Parkinson's, two disorders we will discuss in detail in the chapter on neurological disorders.

Here we conclude our comprehensive overview of the central nervous system: the cortex and its lobes that create consciousness; the various subcortical structures that create movement, memory, and internal regulation; and the spinal cord that sends information down to the body and receives information back in return. How does the nervous system in the body function as the partner to the CNS? In the next section, we begin to address that question.

4.3 Before You Go On

Section Summary

Subcortical structures are essential to how the brain creates behavior. The corpus callosum is the largest white matter tract in the brain, and it serves as an information highway, linking the cortex and other structures throughout the brain. Underneath the cortex are several functional units that maintain basic housekeeping functions, create emotions and also memories, and even initiate movements.

Comprehension Check

1. What is the function of the corpus callosum?
   a. coordinating muscle activity throughout the body
   b. transferring information between hemispheres
   c. enhancing memory of past events
   d. transferring information from sensory systems
2. The hypothalamus is important because __________.
   a. without it, we lose our sense of vision
   b. it coordinates many basic metabolic functions
   c. it is responsible for emotions
   d. it coordinates the actions of the cortex and the cerebellum
3. The amygdala is a part of a system that is responsible for regulating what?
   a. body temperature
   b. blood pressure and circulation
   c. emotions
   d. muscle coordination
4. Damage to the hippocampus and mamillary bodies would likely result in what?
   a. difficulties in maintaining body temperature
   b. an immediate loss of consciousness
   c. visual difficulties
   d. memory problems
5. Damage to the basal ganglia would likely result in what?
   a. a movement disorder
   b. loss of memory for recent events
   c. inappropriate emotional responses
   d. loss of vision

4.4 The Peripheral Nervous System

LO 4.4 Describe the components of the peripheral nervous system and their functions.

In this section we expand on your understanding of the nervous system from the brain and spine into the PNS, which extends throughout your body (**Figure 4.22**). When you decide to take a drink, the signal from your brain to move your hand travels down the PNS to activate the muscle. It will be our task to illuminate the various structures that allow you to take a refreshing sip from your cup.

Before we begin, it is useful to have an organizing theme in mind that will help you understand the mechanisms involved in the PNS. Remember that your brain is the organ that creates behavior, but it needs the PNS to carry out its commands in the body. Therefore, the PNS needs to have efferent communication channels from the brain into the body to initiate movement, and it needs corresponding afferent communication channels from the body back to the brain to provide feedback on those movements. Some of these communication channels must react rapidly and some more slowly. Some must have specific targets, whereas others must be systemic, or general, and affect many organs or muscles. Finally, the

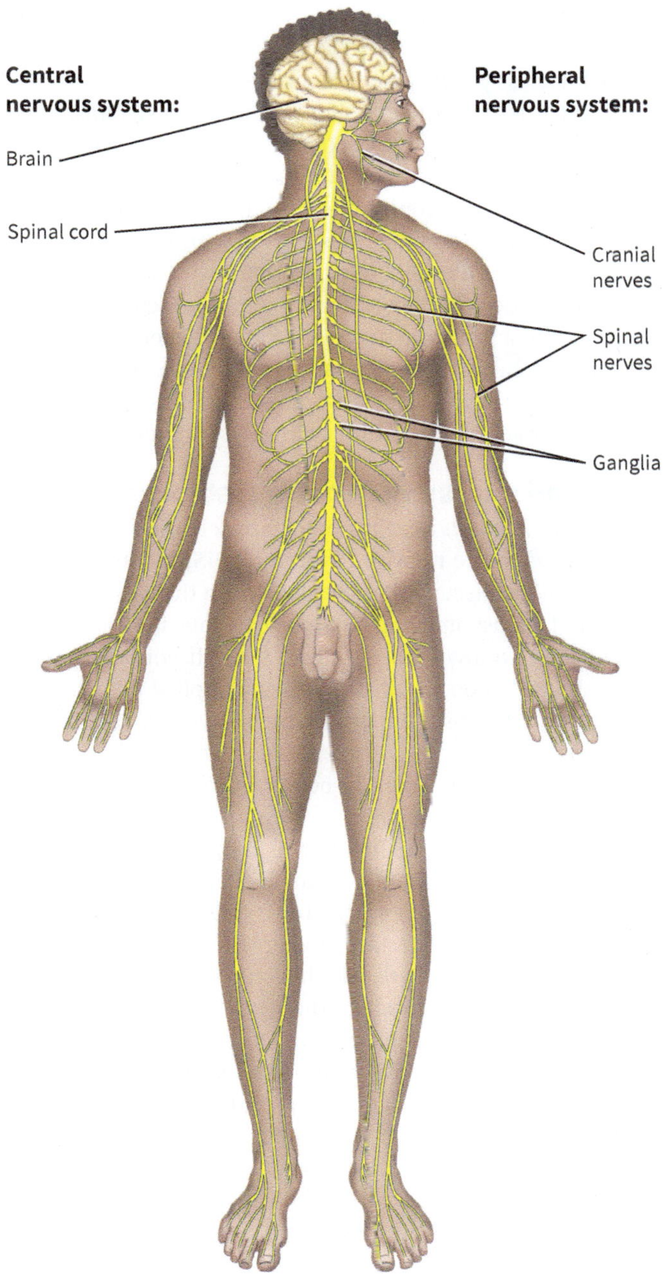

FIGURE 4.22 **The peripheral nervous system** The PNS consists of the cranial nerves, spinal nerves, and ganglia that connect the brain to the body. Without the PNS, the brain could not control the muscles and internal organs in your body.

PNS needs excitatory mechanisms to initiate behavior and corresponding inhibitory mechanisms to stop behavior when necessary. The principle that biological mechanisms form complementary pairs like this—efferent/afferent, rapid/slow, specific/systemic, and excitatory/inhibitory—is one that you have already seen in your studies of neurons, neurotransmitters, and the brain. Now let us begin to look closely at how the PNS illustrates this principle.

Divisions of the Peripheral Nervous System MCAT

The PNS is divided into two different systems, the somatic nervous system and the autonomic nervous system. The autonomic system can be further divided into the sympathetic and parasympathetic systems.

The somatic and autonomic nervous systems

The **somatic nervous system** controls voluntary movements. When you turn the page, move your mouse, sing a song, or go for a stroll, your brain is employing the somatic system to stimulate muscle contractions to achieve its goals. While the somatic nervous system controls voluntary muscle movements, the

autonomic nervous system (ANS) controls involuntary muscular actions as well as the functions of your viscera, or organs. The ANS controls the many housekeeping functions that run beneath our consciousness yet are essential for life.

Perhaps your computer is a useful metaphor for these two systems. When we open a document on our computer, we are engaging the computer's somatic system in that we are choosing to open the document. At the same time, dozens if not hundreds of programs are running constantly in the background; they continuously poll the keyboard looking for a response, refresh the screen, check to see if an email has arrived, and keep the hard drive spinning at precisely the right speed. We don't see these applications running, but without them your computer would be useless. These housekeeping tasks are your computer's autonomic system. We are generally unaware of autonomic functions, yet we depend on them to live. Examples of autonomic functions are heart rate, respiration, salivation, digestion, pupillary dilation, perspiration, and sexual arousal. In some cases, muscle groups controlled by the autonomic system can also come under voluntary control such as in swallowing and breathing.

The sympathetic and parasympathetic systems

Within the ANS, there are two systems relevant for our interests as behavioral neuroscientists, the **sympathetic nervous system** and the **parasympathetic nervous system (PNS)**. A third system of the ANS, the enteric nervous system, will be discussed in the chapter on the endocrine system.

The sympathetic and parasympathetic systems operate independently to control ANS functions. What might the two divisions of the ANS do? If you answered that one initiates action and the other inhibits action, you have correctly applied the principle of complementary mechanisms. The sympathetic system has traditionally been described as a fight-or-flight mechanism and the parasympathetic as a rest-and-digest or feed-and-breed counterpart. More recent research suggests that the sympathetic is better described as a rapid initiation system and the parasympathetic as a slower-acting dampening system. It is easy to think of the two systems as being in conflict—that we are either in flight-or-flight mode or rest-and-recover mode—but the reality is that in many situations the two systems cooperate quite nicely. For example, when you stand up after some time at rest, your sympathetic system increases your blood pressure to prevent you from fainting. Once a satisfactory blood pressure is attained, the parasympathetic inhibits further increases in blood pressure, which would endanger your health.

A second example that demonstrates both the mode of action and the different response rates may be familiar to you. Many of us have had a close call in our lives, perhaps while driving. You may have just missed a deer that ran in front of you, or a car came out of nowhere and almost hit you. You skillfully avoid catastrophe and carefully pull over to the side of the road into a safe location, and as you do, you realize for the first time that you are shaking like a leaf and your knees are weak. Given what you know about the ANS, can you describe your physiological response?

The answer is that your sympathetic system responded rapidly to aid your survival by increasing your heart rate and blood pressure and secreting adrenaline into your blood stream. You were ready to flee or fight for your life, but now as you collect your thoughts on the side of the road, the danger has passed. It takes awhile for the parasympathetic system to bring those visceral and muscular systems back to their normal state, and during that interval you may continue to feel that your heart is racing, your breathing is shallow, and your hands are trembling. Understanding the interplay between the sympathetic and parasympathetic systems explains what you experienced.

Parasympathetic activation of a muscle extends from a section of the spinal cord, through a ganglion, and then to the target muscle or gland via the neurotransmitter acetylcholine (ACh). Sympathetic activation of the muscle or gland also extends from a section of the spinal cord, synapses on a ganglion, and then activates a muscle or gland via the neurotransmitter norepinephrine (NE). The contrasting action of these two systems is presented in Table 4.1.

The Neuromuscular Junction MCAT

Commands from the CNS travel down the PNS and then cause a muscle to either contract or relax. Muscle is soft tissue that, when stimulated, contracts to produce force. Muscle comes in three varieties: smooth, striate, and cardiac. Smooth muscle is found in organs such as your esophagus and digestive system; striate, or skeletal, muscle is the muscle attached to your skeleton that allows you

TABLE 4.1 The Sympathetic and Parasympathetic Systems

Sympathetic	Parasympathetic
Dilates pupils	Constricts pupils
Inhibits salivation	Stimulates salivation
Opens airways	Constricts airways
Speeds heartbeat	Slows heartbeat
Diverts blood to striate muscles (e.g., legs and arms)	Diverts blood to smooth muscles (e.g., intestines)
Stimulates glucose release	Glucose levels unaffected
Inhibits digestive system	Stimulates digestive system
Stimulates release of adrenaline	Stimulates release of acetylcholine

to move and create force. Cardiac muscle is the muscle of your heart. Your nervous system connects to muscles through the **neuromuscular junction** (Figure 4.23). At the neuromuscular junction, the presynaptic neuron releases acetylcholine (ACh), and a corresponding receptor in the muscle fiber may then fire, thereby causing the muscle to contract.

Disease and toxins can affect the function of the neuromuscular junction. For example, the bacterium *Clostridium botulinum* creates a lethal neurotoxin that prevents the release of acetylcholine from the presynaptic neuron. This inactivates the neuromuscular junction and causes muscular paralysis, leaving the victim unable to breathe. Eating improperly preserved food tainted with infinitesimally small amounts of the toxin can be fatal. The same toxin, known commercially as Botox©, is commonly used cosmetically to smooth wrinkles by paralyzing facial muscles. Snakes and other venomous animals can also use a neurotoxin to affect the neuromuscular junction and immobilize their prey. Some snake neurotoxins disable the ACh receptor at the neuromuscular junction, leading to muscular paralysis and possibly death. The bite of a blue-ringed octopus injects tetrodotoxin, a neurotoxin that disables sodium channels, thereby inhibiting the firing of neurons throughout the body and causing complete muscular paralysis in minutes. In a final example, myasthenia gravis is a disorder in which the autoimmune system destroys the ACh receptors at the neuromuscular junction, leading to flaccid muscle tone. Myasthenia gravis affects the muscles that control the eyes, face, and throat, causing drooping eyelids, double vision due to strabismus (an inability to align both eyes), and other symptoms (**Think About It: Myasthenia Gravis and the Neuromuscular Junction**).

You might have asked yourself why myasthenia gravis affects the muscles that control the eye before it affects any other muscles, and the answer to that question will become apparent as we discuss the cranial nerves. As you learn about them, keep a mental list of the number of cranial nerves that are dedicated to the eyes and coordinating their movement to create your visual world. A system that requires many nerves to function is a system that can be easily disrupted.

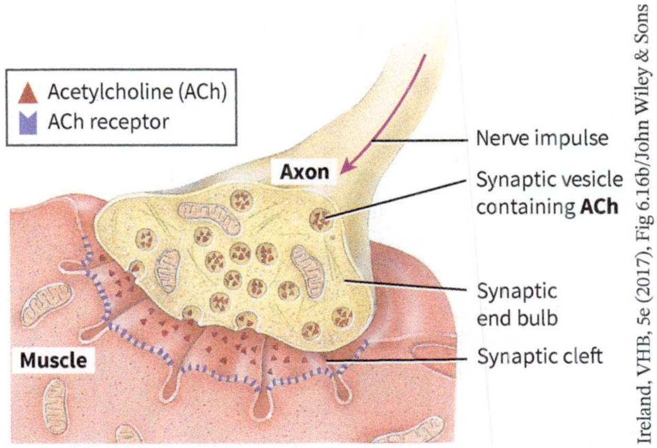

FIGURE 4.23 The neuromuscular junction At the neuromuscular junction, the nervous system completes the circuit between intention and action. An action potential travels down the axon of the presynaptic neuron of the peripheral nervous system until it reaches the neuromuscular junction, where it releases ACh into the synapse. This stimulates ACh receptors on the postsynaptic muscle fiber to fire, causing the muscle to contract.

The Cranial Nerves MCAT

With the exception of the optic nerve, the olfactory nerve and trigeminal nerve whose axons extend into the CNS, the **cranial nerves** are considered to be a part of the PNS because their axons extend beyond the CNS. There are a dozen cranial nerves that handle sensory input, control motor output, or do both. For instance, the optic nerve transmits signals from the retina to the brain; the oculomotor nerve controls the muscles that rotate the eyes, and the facial nerve directs motor output to facial muscles while also receiving sensory input from the rear portion of the tongue. Figure 4.24 illustrates the location and function of the cranial nerves.

Think About It | Myasthenia Gravis and the Neuromuscular Junction

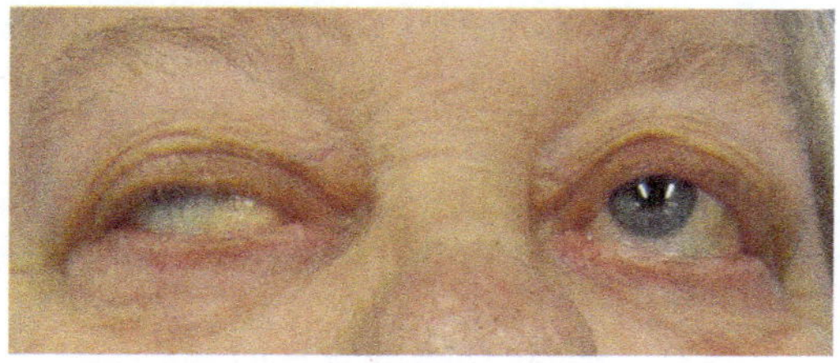

Muscle weakness and strabismus in an individual with myasthenia gravis

The path from your initial intention to make a movement to the actual movement is quite complex. When you decide to scratch your head, a signal is created in your brain that travels through several subcortical structures and down through the spinal cord until it reaches the spinal ganglion at the appropriate level. It then travels out to the neuromuscular junction that motivates the appropriate muscles.

Unfortunately, for some individuals that signal never arrives at the muscle due to damage at the neuromuscular junction. You might think of the path from the CNS to your muscle as mirroring the postal system; a letter that is created by a writer travels through a complex series of postal centers to reach you. If the postal worker does not place the letter in your mailbox, however, the message never gets to you. In the same way, a signal that began in the CNS and traveled through the PNS never reaches its muscle target if the neuromuscular junction is damaged.

The neuromuscular junction depends on the neurotransmitter acetylcholine to operate, and when that receptor at the junction between the neuron and the muscle is damaged, myasthenia gravis may be the cause. Initial symptoms typically are a weakness in facial muscles, especially around the eyes, and also diplopia or double vision. In more severe cases, the fatigue and weakness that characterize myasthenia gravis can spread to other muscles throughout the body.

Questions

1. **Search the internet using the phrase "frequency of myasthenia gravis." How common is this disorder? Name a risk factor for this disorder.**
2. **What class of drugs is commonly used to treat myasthenia gravis, and why does this treatment make sense?**

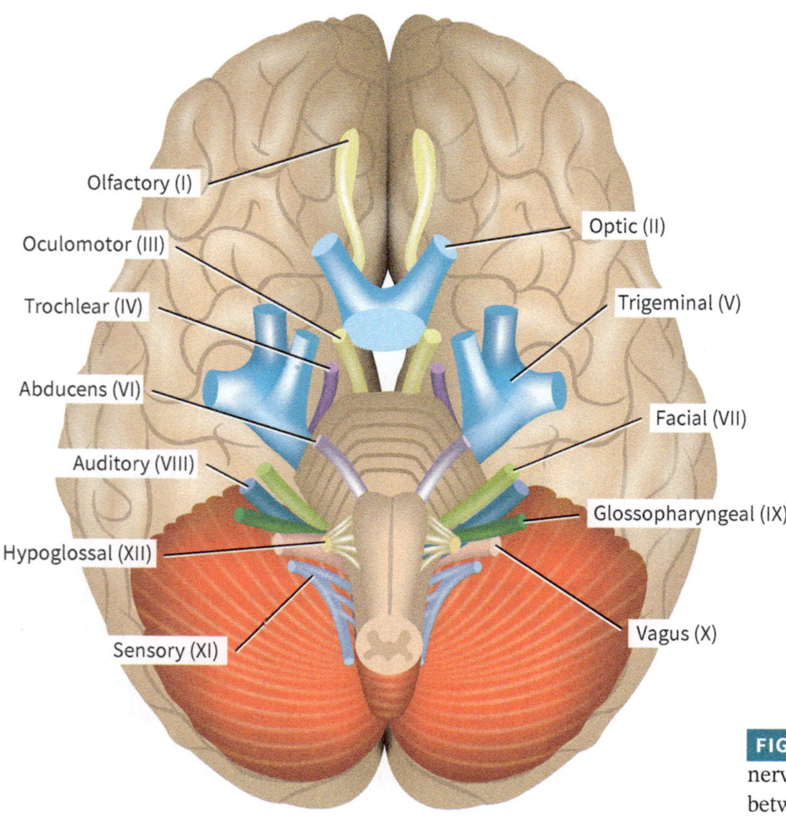

Nerve	Function
I Olfactory	Smell
II Optic	Vision
III Oculomotor	Muscles that move the eye
IV Trochlear	Muscles that move the eye
V Trigeminal	Face
VI Abducens	Muscles that move the eye
VII Facial	Muscles of face
VIII Auditory (vestibulocochlear)	Hearing
IX Glossopharyngeal	Muscles of mouth and throat
X Vagus	Heart and other internal organs
XI Sensory (accessory)	Senses and muscles of the head and neck
XII Hypoglossal	Muscles of tongue

FIGURE 4.24 **The cranial nerves** The 12 cranial nerves are the portals that allow communication between the brain and the body.

Students of anatomy, physiology, and neuroscience have always struggled to remember the cranial nerves and their functions. Some of these 12 nerves are sensory (afferent) only in that they take information from the senses into the CNS; others are motor (efferent) only as they convey the commands of the CNS down into the PNS; and some are both afferent and efferent. How to remember the names of these nerves and whether they are sensory, motor, or both? A common mnemonic provides a framework to recall the names of the cranial nerves: **O**n **O**ld **O**lympus' **T**owering **T**op **A** **F**inn **A**nd **G**erman **V**iewed **S**ome **H**ops. Can you come up with your own sentence to help you remember the cranial nerves?

We conclude our discussion of the nervous system with a brief summary of the neuromuscular loop that may help to tie it all together (**Figure 4.25**). While sitting in class and

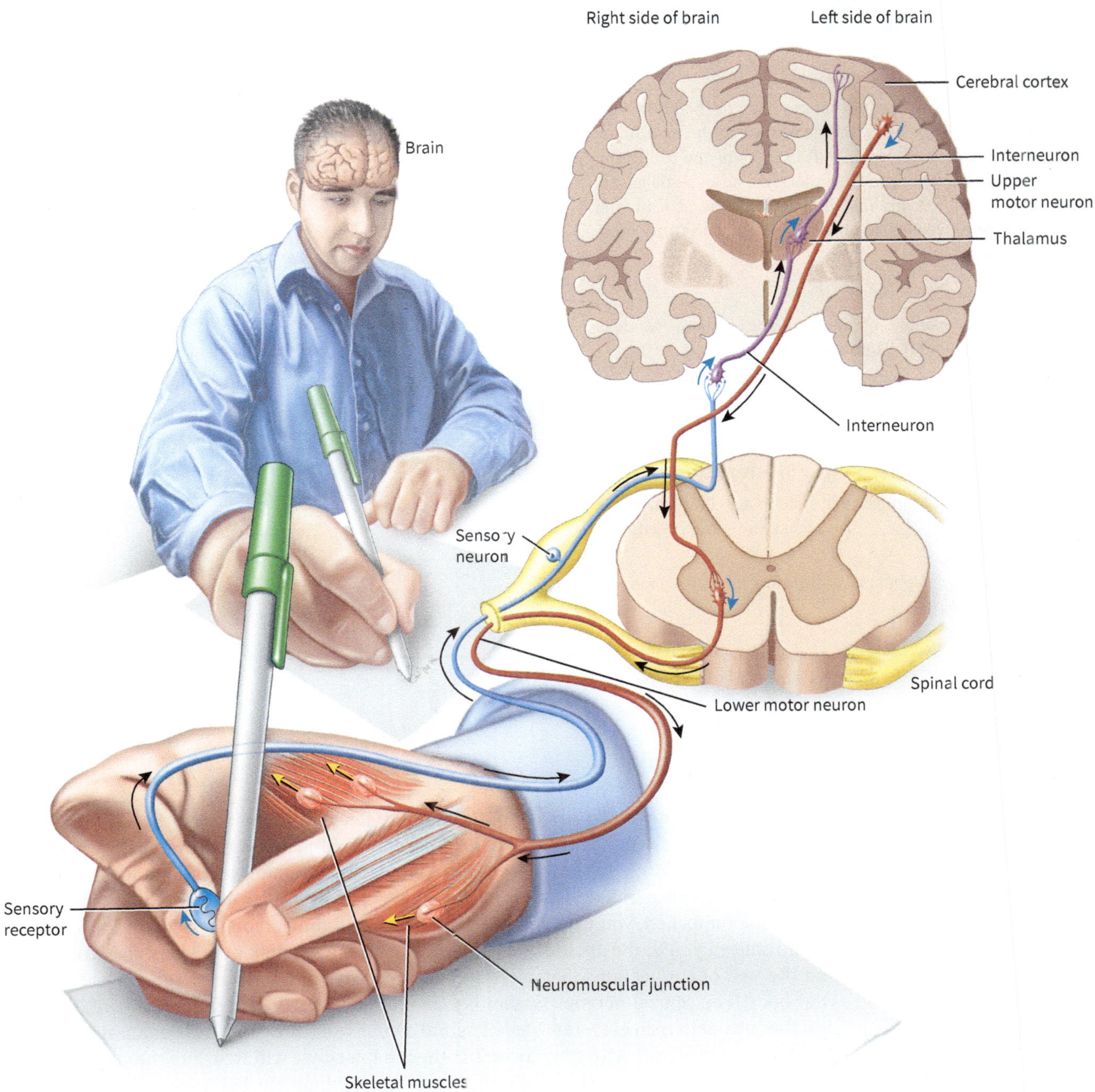

FIGURE 4.25 The neuromuscular loop An intention that originates in the motor strip of the central nervous system travels down the spinal cord and into the peripheral nervous system, where it moves the muscles holding the pen. Sensory receptors in the fingers and hand provide essential feedback, which then travels back up to the sensory strip in the cortex to maintain performance.

taking notes as your instructor lectures, you decide to write down an important point. That intention creates a command in the motor strip that travels down motor neurons, passing through the thalamus and down the spinal cord into a ganglion. This impulse connects to a neuromuscular junction that motivates your hand and fingers to record your thought. At the same time, a complementary sensory system provides important feedback about the position of your pen and sends that signal back up through the spinal cord where it terminates in the sensory strip of the cortex. These two systems, working in concert, create the skilled motion that enables you to succeed in class.

4.4 Before You Go On

Section Summary

The peripheral nervous system connects the body to the brain. The connection between the PNS and a muscle is made at the neuromuscular junction. An understanding of the PNS explains the paralyzing action of neurotoxins like Botox and disorders like myasthenia gravis. The cranial nerves provide communication between the CNS and the body, including sensory systems, the head, shoulder, and trunk muscles, and some internal organs.

Comprehension Check

1. __________ is an example of the somatic system, whereas __________ is an example of the autonomic system.

- **a.** Reaching for a pencil; regulating blood pressure
- **b.** Regulating blood pressure; digesting a meal
- **c.** Reaching for a pencil; typing on a keyboard
- **d.** Regulating blood pressure; reaching for a pencil

2. The sympathetic and parasympathetic systems are complementary in that the sympathetic __________, whereas the parasympathetic __________.

- **a.** controls voluntary responses; controls involuntary responses
- **b.** is excitatory; is inhibitory
- **c.** controls emotional responses; controls intellectual responses
- **d.** is inhibitory; controls voluntary responses

3. Neurotoxins such as botulinus as well as some snake and octopus venoms influence the __________ and __________.

- **a.** basal ganglia; paralyze muscles
- **b.** primary motor cortex; deplete dopamine
- **c.** spinobulbar tract; paralyze face and throat muscles
- **d.** neuromuscular junctions; paralyze muscles

4. What is the primary function of the cranial nerves?

- **a.** connecting the muscles of the head and neck with the brain
- **b.** allowing the spine to communicate with the brain
- **c.** communicating between the senses and the brain
- **d.** directing the muscles that are essential for life support

Conclusion

We began this chapter with the story of Christopher Reeve, an actor so healthy and vibrant he was chosen to represent the mythical hero Superman in film, and yet an unfortunate accident broke the connection between his central nervous system where intentions for movement are created, and his peripheral nervous system, which carries out those intentions. To understand what happened, you need to have an awareness of the nervous system. In this chapter, we introduced you to the anatomy of the nervous system, its major divisions and structures, and the web of connections that speed information back and forth between your brain and your body. The central nervous system is a computational device beyond compare, and the peripheral system is an elaborate web of neural connections that provides your brain with the power that comes from active behavior. In other chapters, we will return to the structures we discussed here with the goal of understanding in depth how our brain learns new material, remembers what it has already learned, chooses to eat or not eat, and feels attraction to one person but not to another.

Critical Thinking Questions

1. Some people say that bigger is better, but size alone may not tell the whole story about what a nervous system can do. Sperm whales, elephants, and bottle-nosed dolphins all have a bigger brain than you and I, yet humans with their smaller brains have created cities and culture and have traveled into outer space. If intelligence is not determined by brain size alone, what aspect of brain structure might better explain human intellectual success?

2. Your cerebrum consists of gray matter, which is considered to be computational or thinking tissue; white matter, which permits rapid communication among the many structures in the brain; and centers like the cerebellum that are essential for coordinating movements. It's very likely that there are individual differences among us in all these and other cerebral structures. If it were possible for you to know if you had a natural advantage at some aspect of brain function, would you want to know? Which aspect would you choose to have an advantage in if you could? Explain your answers.

3. Your limbic system is responsible for emotional expression and interpretation and is also involved in creating and storing emotional aspects of memory. Damage to structures of the limbic system result in a wide range of maladies, ranging from psychiatric disorders such as depression to neurological disorders such as Alzheimer's. Why might limbic dysfunction create such a wide range of symptoms?

4. In your study of the peripheral nervous system, you learned that the sympathetic nervous system prepares you for fight or flight, whereas the parasympathetic nervous system serves as a brake on the sympathetic system. Many of us experience anxiety at some point in our lives while delivering a speech or a class report. How might your understanding of these systems help you deal with such anxieties?

Dig Deeper into Behavioral Neuroscience

4.1 An Overview of the Nervous System

The human brain is quite complicated. Rita Carter's book is an easily readable overview.

Carter, R. (2019). *The Human Brain Book: An Illustrated Guide to Its Structure, Function, and Disorders*. DK Illustrated. ISBN-13: 978-146547954A product very popular with medical and neuroanatomy students is the neuroanatomy coloring book, which teaches neuroanatomy through active learning.

Anatomy Academy. (2020). *Neuroanatomy coloring book*. Muze Publishing,. ISBN-13: 978-1838188610

Anatomical terminology can be confusing. Here is one of several online tutorials regarding anatomical terminology that present anatomical terms and directions in an accessible fashion.

https://open.oregonstate.education/aandp/chapter/1-4-anatomical-terminology/

Each year in the United States, roughly 18,000 individuals experience a spinal cord injury (SCI), and at this time, about 300,000 Americans live with an SCI. Many researchers are working toward a cure, and searching using the phrase "breakthroughs in treating spinal cord injuries" will lead to sources such as these.

https://www.scientificamerican.com/article/repairing-the-damaged-spinal-cord/

https://www.sciencedaily.com/releases/2018/08/180830102535.htm

The National Institute of Neurological Disorders and Stroke (NINDS) of the NIH is a clearinghouse for information about current research about new treatments.

https://www.ninds.nih.gov/Disorders/All-Disorders/Spinal-Cord-Injury-Information-Page

4.2 Major Structures and Support Systems of the CNS

Our ventricular system delivers essential nutrients to the CNS and removes waste products. You can learn more about what happens when this system is dysfunctional at the National Institute of Neurological Disorders and Stroke.

https://www.ninds.nih.gov/Disorders/Patient-Caregiver-Education/Fact-Sheets/Hydrocephalus-Fact-Sheet

It is surprising to see that we are still discovering new structures in the brain, and the glymphatic system is an example.

Jessen, N. A., Munk, A. S., Lundgaard, I., & Nedergaard, M. (2015, December). The glymphatic system: A beginner's guide. *Neurochemical Research*, *40*(12), 2583–2599. https://doi.org/10.1007/s11064-015-1581-6.

4.3 Major Subcortical Processing Circuits

The limbic system is responsible for the creation and interpretation of emotions. As such, it has attracted a great deal of interest from neuroscientists as well as clinical researchers.

Catani, M., Dell'Acqua, F., & Thiebaut de Schotten, M. (2013). A revised limbic system model for memory, emotion and behavior. *Neuroscience & Biobehavioral Reviews*, *37*(8), 1724–1737.
https://doi.org/10.1016/j.neubiorev.2013.07.001

James Olds and Peter Milner conducted the seminal study that alerted behavioral neuroscientists to the reward function of the limbic system.

Olds, J., & Milner, P. (1954). Positive reinforcement produced by electrical stimulation of the septal area and other regions of rat brain. *Journal of Comparative Physiology and Psychology*, *47*, 419–427.

José Delgado's demonstration of behavioral management via an implanted electrode was startling; there is more to the story than is commonly reported. Here is an interesting and accessible read from *Scientific American*.

https://blogs.scientificamerican.com/cross-check/tribute-to-jose-delgado-legendary-and-slightly-scary-pioneer-of-mind-control/

4.4 The Peripheral Nervous System

Without the PNS, our brain would be useless. The following chapter is a very thorough overview of the PNS.

Wilkinson,J.L.(1992).Peripheralnervoussystem.InJ.L.Wilkinson(Ed.,), *Neuroanatomy for Medical Students* (2nd Ed.), 32–51. Butterworth-Heinemann. ISBN: 9780750614474.
https://doi.org/10.1016/B978-0-7506-1447-4.50007-4.

In this chapter, you learned about the neuromuscular junction and also that some organisms have learned to use neurotoxins that interrupt its function as a defensive mechanism. The blue-ringed octopus is a most beautiful and deadly organism whose toxin affects the neuromuscular junction.

Sheumack, D. D., Howden, M. E., Spence, I., & Quinn, R. J. (1978). Maculotoxin: A neurotoxin from the venom glands of the octopus *Hapalochlaena maculosa* identified as tetrodotoxin. *Science, 199* (4325), 188–189.
https://doi.org/10.1126/science.619451. PMID: 619451.

Other animals use tetrodotoxin as a defensive mechanism to deter predators. Search using the term "tetrodotoxin" to learn more about this agent.

Chapter 4 Before You Go On Answers

4.1 Before You Go On Answers

1. c. brain and spinal cord
2. b. hindbrain, forebrain, and midbrain
3. d. problem solving and planning
4. d. alertness and arousal
5. a. lumbar, thoracic, and cervical

4.2 Before You Go On Answers

1. a. ridge; valley
2. c. accepts signals from the senses; sends signals to muscles
3. a. gait and balance problems
4. c. removing metabolic by-products from the brain

4.3 Before You Go On Answers

1. b. transferring information between hemispheres
2. b. it coordinates so many basic metabolic functions
3. c. emotions
4. d. memory problems
5. a. a movement disorder

4.4 Before You Go On Answers

1. a. Reaching for a pencil; regulating blood pressure
2. b. is excitatory; is inhibitory
3. d. neuromuscular junctions; paralyze muscles
4. c. communicating between the senses and the brain

Chapter 4 Glossary

amygdala A structure in the temporal lobe that is a part of the limbic system and is believed to play a role in emotions, memory, and decision-making.

arachnoid layer The spiderweb-like middle layer of the meninges.

autonomic nervous system (ANS) A division of the peripheral nervous system that consists of the sympathetic and parasympathetic systems and is responsible for controlling smooth muscle, glands, and the fight-or-flight response.

basal ganglia A collection of subcortical nuclei that regulate voluntary movements.

brainstem A distal part of the brain that is made up of the midbrain, pons, and medulla oblongata.

central nervous system (CNS) That part of the nervous system consisting of the brain and spinal cord.

cerebellum A small structure at the rear and underneath the cerebral hemispheres that controls movement on the ipsilateral side of the body and is involved in motor learning.

cerebral cortex The outer layer of the cerebrum consisting of two hemispheres and made of gray matter.

cerebrospinal fluid (CSF) A clear fluid that circulates through the four ventricles of the brain and the spinal cord, providing cushioning and waste removal for the brain.

cerebrum The largest part of the brain, consisting of the cerebral hemispheres and major subcortical structures.

corpus callosum A wide, flat bundle of neural fibers that connects the two hemispheres and is the largest white matter structure in the brain.

cranial nerve One of a set of twelve pairs of nerves that transmit data to and from the brain and the body.

dura mater The thick outermost layer of the meninges.

fornix A small structure that accepts output from the hippocampus as part of the circuit that supports memory.

frontal lobe The front part of the brain that controls problem solving and other executive functions.

ganglia A group of nerve cells that function as a unit.

glymphatic system A waste-removal system in the brain that uses CSF.

gyrus A ridge or hillock on the cerebral cortex.

hippocampus A bilateral subcortical structure that is part of the limbic system and is responsible for consolidation of short-term memories.

hypothalamus A structure in the brain that is comprised of several nuclei and is responsible for controlling bodily functions such as hunger, body temperature, and thirst.

insula A part of the cortex that is folded into the lateral sulcus and is involved in consciousness and pain perception; also known as the insular cortex.

limbic system A collection of subcortical structures that are associated with emotions, motivation, and long-term memory.

mammillary body One of two small round bodies that are associated with memory, bilaterally located, and are part of the limbic system.

medulla oblongata Part of the brainstem that controls cardiac and respiratory function.

meninges The protective layers that surround the brain.

neuromuscular junction The synaptic contact between a neuron and a muscle fiber.

occipital lobe The most posterior lobe of the brain that is responsible for much of visual processing.

parasympathetic nervous system (PNS) That inhibitory part of the autonomic nervous system that is responsible for feed-and-breed or rest-and-digest behaviors.

parietal lobe The lobe of the brain involved in integrating sensory information and also in the processing of visuospatial information.

peripheral nervous system (PNS) That part of the nervous system that connects the CNS to organs, muscles, and nerves outside the CNS.

pia mater The innermost of the meninges, which keeps CSF in place.

pons A part of the brainstem that is responsible for sleep, respiration, swallowing, and bladder control.

primary motor cortex (PMC) A principal area of the cerebral cortex that generates the commands that control movement.

sensorimotor cortex The area of the brain that contains the primary sensory and motor areas of the brain.

septal nucleus One of a set of structures beneath the corpus callosum that play a role in reward.

somatic nervous system That part of the peripheral nervous system that controls voluntary movements through skeletal muscles.

somatosensory cortex The structure in the cerebral cortex that receives sensory input from the body for interpretation and analysis.

spinal cord A long tubular section of neural tissue that extends from the medulla down into the lumbar region.

substantia nigra A structure in the basal ganglia that is involved in reward and movement.

sulcus A crevice on the surface of the brain.

sympathetic nervous system The excitatory part of the autonomic nervous system that is responsible for fight-or-flight behaviors.

temporal lobe A part of the brain located beneath the lateral fissure on the cortex and associated with language comprehension and production along with visual memory.

thalamus A structure at the midline of the brain that connects the cortex with sensory and motor areas.

ventricle One of four interconnected chambers in the brain that create and circulate cerebrospinal fluid.

CHAPTER 5

Vision

IvanMikhaylov/Getty Images

We perceive the vibrant colors of a sunset when physical input to our eyes is translated to sensory data in our brains.

CHAPTER OUTLINE

5.1 An Introduction to Vision

- The Physics of Light
- The Architecture of the Human Eye
- How the Focal Plane Affects Our Vision
- Measuring Visual Acuity and Contrast Sensitivity

  Try It Out: The Landolt C Task
- The Visual Processing Stream

5.2 The Retina

- Photoreceptors
- Retinal Processing Beyond Rods and Cones

  Check It Out: Peeking into the Retina Through OCT

5.3 The Subcortical Centers

- The Geniculostriate Pathway

  Check It Out: The Eye of the Batter
- The Tectopulvinar Pathway
- Damage to the Visual Stream

5.4 The Visual Cortex

- The Primary Visual Cortex
- Processing in the Visual Cortex

  Advances in Behavioral Neuroscience: Margaret Wong-Riley and the PVC
- The Extrastriate Cortex

  Check It Out: Motion After effects
- The Ventral and Dorsal Streams

5.5 Color Vision

- The Trichromatic Theory
- Color Vision Abnormalities
- Color Sensitivity Across the Visual Spectrum

  Think About It: Safety and Color Vision
- The Opponent Process Theory

  Try It Out: The Checkerboard Illusion

LEARNING OBJECTIVES

LO 5.1 **Describe** the perceptual characteristics of light and the basic anatomy of the human eye.

LO 5.2 **Identify** the major components of the retina and their functions.

LO 5.3 **Identify** the major subcortical structures in the visual processing stream and their functions.

LO 5.4 **Describe** the organization of the visual cortex and its role in visual perception.

LO 5.5 **Describe** the mechanisms that create color vision.

Stories of the Brain

Jonathan was a successful painter whose works were proudly displayed in galleries and in private homes. Like many men in his mid-60s, Jonathan was an early riser. However, one day his wife was surprised to find him still asleep at midmorning. After breakfast, Jonathan went out to his car and discovered that there was a large dent in the passenger side. Assuming someone had backed into his car in the night, he went off to his studio to work. There he encountered his second surprise: lying on his desk was a police report of a traffic accident from the day before. Jonathan had apparently been blindsided by a truck, given a description of the event to a police officer, and signed the report. However, he had no recollection of the events. Even more bizarre was that he could make neither heads nor tails of the writing. It was as if the report were written in some odd alphabet that, although vaguely familiar, was incomprehensible.

Fearing that he had suffered a stroke, Jonathan made his way to a physician. The doctor found some loss of color vision and a severe difficulty in recognizing letters. There was also evidence of a concussion but no life-threatening conditions, so Jonathan was told to go home and see how he felt the next day. Jonathan returned home with the hope that he would be good as new in the morning.

Unfortunately, the next day Jonathan awoke to a new world. Before his very eyes, the world of color to which he had devoted his talents was taken away from him. Over time, he regained his ability to recognize words, but his complete inability to see colors remained. He now saw the world only in black, white, and shades of gray.

Before the accident, he had marveled at a bowl of fruit, savoring the rich and complex colors. Now he was disgusted by the unreal appearance of these same apples, oranges, and bananas. Clouds were almost indistinguishable from the blue sky behind them. Food appeared repulsive; ketchup was black, as were tomatoes. Although before the accident he enjoyed painting people, now they seemed like ghastly statues, with gray and ratlike skin. Jonathan now found the sight of people revolting. Unable to get these images out of his mind, he found sexual relations with his wife almost impossible.

At first, Jonathan felt that his professional life was over. How could a person paint without color, and where would he find inspiration when what he saw repulsed him? Although all color in his world was now gone, in its place was an unparalleled ability to distinguish small details in shades of gray. After a time, Jonathan learned to paint again, but now his compositions were free-form, abstract, and composed completely in black, white, and shades of gray (see photos).

What changed in Jonathan's brain so that suddenly he couldn't see colors? How did a change in his vision affect so many different aspects of his life? This chapter will answer these questions and more as we explore one of our most essential senses, vision, including its mechanics, its disorders, and all the surprising ways it affects our thinking.

millefloreimages/Adobe Stock

millefloreimages/Adobe Stock

Paintings before and after the accident

5.1 An Introduction to Vision

LO 5.1 Describe the perceptual characteristics of light and the basic anatomy of the human eye.

Take a break from your studies for just a moment and look around you. When you gaze out the window, you may see city streets lined with shops and filled with busy people or mountains on a distant horizon. When you look up, your skies may be blue and sunny, or filled with clouds that threaten rain, or dark and starry. Whatever you are looking at right now, one thing is certain: you are swimming in a sea of visual information. Color, texture, and distance can all be determined with a glance as you look about your world. Humans are visually prepotent organisms, which means that vision is our primary sense. Our senses are designed to provide important information about objects at a distance to create an internal model of the world around us. How does the raw energy of the physical world become the meaningful and orderly visual world that we perceive? Let's begin with a short review of light itself.

The Physics of Light

What we think of as light is actually the name we give to the very small portion of the **electromagnetic spectrum** that humans can see. Other organisms are receptive to, or see, different portions of this same spectrum such as infrared or ultraviolet. Another way of defining light from the human perspective is to say that our sensors are only receptive to about 0.0035% of the overall range of electromagnetic radiation depicted in **Figure 5.1a**. Your entire human visual experience sits within that tiny slice of radiation.

Light can be described as a particle—a photon—or as a wave. The properties of a light wave inform our perception. When the magnitude of the wave increases, we perceive that physical change as a greater intensity of brightness (**Figure 5.1b**). The frequency of a light wave, or the number of full wavelengths that pass by a given point in space every second, is perceived as color. Lower frequencies are perceived as violet or blue, and higher frequencies are perceived as orange or red. Recall for a moment the chapter's opening image of a sunset. What the natural world presents to us as billions of electromagnetic waves, varying in their magnitude and frequency, our brain translates into a brilliant sunset.

Light is the physical input that forms the basis for our visual perception, and our eyes are the collection device for receiving that input. Given that our eye reflects our world, let us begin our tour of the visual system with the eye itself.

FIGURE 5.1 The electromagnetic spectrum a. Humans can only sense a small segment of the electromagnetic spectrum. **b.** The magnitude of a wave is perceived as brightness, whereas the frequency of a wave is perceived as color.

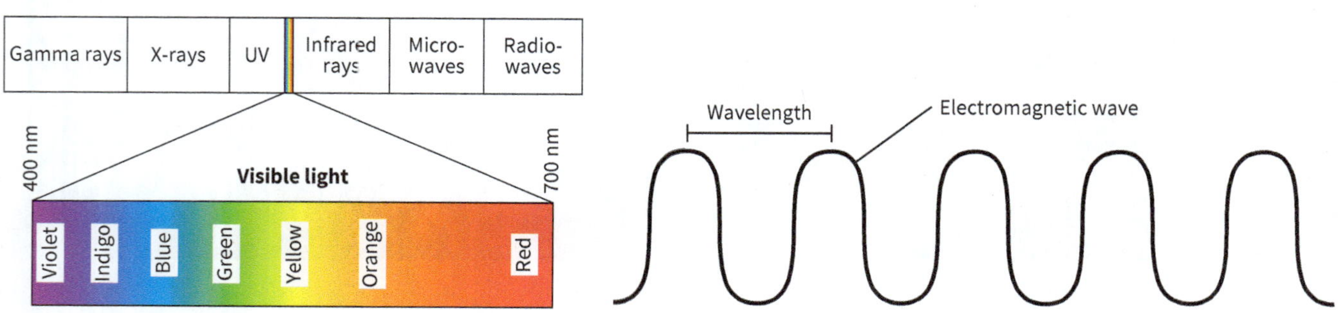

a. Electromagnetic spectrum

b. An electromagnetic wave

The Architecture of the Human Eye MCAT

There are many different kinds of eyes. Given that some organisms live in low-light environments and others spend their lives underwater or in the desert, it is not surprising that organisms from different habitats have eyes that are adapted to the demands of their world. You and I are no different; our eyes are designed to help us succeed in our world.

To understand the workings of the human eye, first you need to remember that every evolutionary adaptation represents a trade-off among many possibilities. You cannot have the sharp vision of a hawk that easily detects a slight rustle in the grass as it circles high above its unsuspecting prey and also have the low-light vision of a fox that leaves its burrow to hunt in the dead of night.

Neither hawks nor foxes get to choose their visual capabilities. Instead, they have what vision their genes allow. Those organisms whose visual system provides them with an advantage in their environment at locating food and avoiding predators are more likely to survive and reproduce. How does this endless competition relate to you and me? Archeological evidence suggests that our ancestors hunted during the day and not in the dark, when they would be at a disadvantage against animals with excellent night vision. Therefore, it is not surprising that the human eye is adapted for daytime conditions when there is ample light.

The anatomy of the human eye reflects millennia of adaptation to optimize our visual system for our environment. The human eye (Figure 5.2) is a slightly oblong sphere that consists of a tough, dark inner layer called the **choroid**, which is sheathed in a whitish, fibrous protective layer called the **sclera**. The inner chamber of the eye is filled with a jelly-like substance called vitreous humor. Several muscles that connect the eyeball to the skull allow us to rotate our eye toward what we want to see.

At the front of the eye are structures dedicated to protecting the eye as well as collecting and focusing light. A transparent dome called the **cornea** covers and protects the opening to the eye as well as the lens. A circular structure known as the **iris** gives our eyes their characteristic color. The opening in the iris to the inner eye is called the pupil. The size of the pupil is determined by the pupillary sphincter muscle, which can open it wide in response to low levels of illumination to let in more light, or close it down tight in response to bright light.

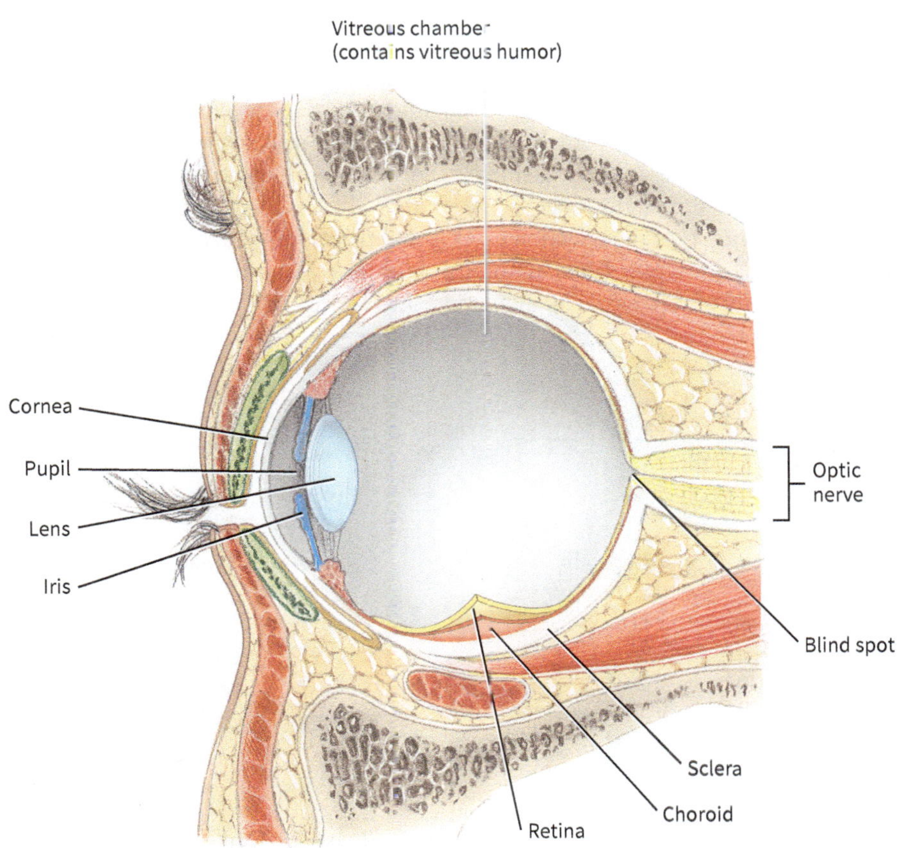

FIGURE 5.2 Gross anatomy of the human eye In this cross section of the human eye, light enters from the left, goes through the cornea and iris and is focused by the lens, and then follows the visual axis to the retina, which contains photosensitive cells. Signals from the retina leave the eye at the optic nerve.

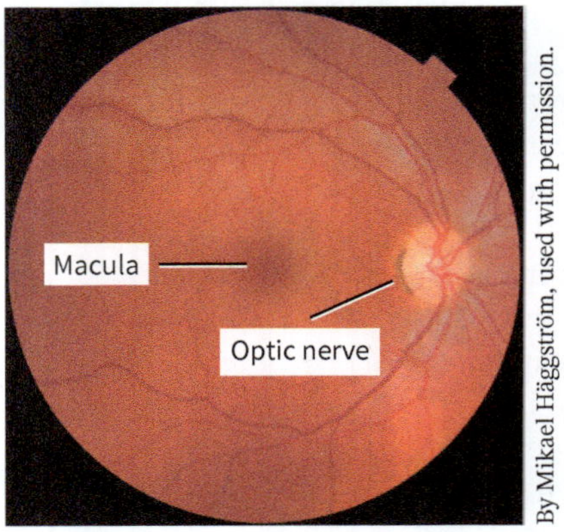

FIGURE 5.3 Interior view of a human eye The dark center of the eye is the macula, the region with the greatest number of photoreceptors. On the right side is the optic nerve, where information created by photoreceptors throughout the retina leaves the eye. The blood vessels that service the eye stretch across the fundus in a spidery web.

The tiny ciliary muscles on each side of the lens change the focus of the eye by either stretching or squeezing the lens. Changing the shape of the lens focuses light on the retina, a process known as refraction.

After light has entered through the pupil and has been focused by the lens, it hits a layer of cells at the back of the eye called the **retina**. These specialized neural cells are photoreceptors, and it is here that the physical energy of light begins its transformation into the psychological sense we call vision. Information created in the retina leaves the eye via the **optic nerve**, and because there are no photoreceptors where the nerve leaves the eye, this point is called the blind spot. The region of the retina directly at the back of the eye, called the **macula**, is especially rich in receptor cells, and at the center of the macula is the **fovea**. There are about 4,000 photoreceptors in the fovea, which is 1.5 mm in diameter, or roughly the width of a grain of rice. Images that fall on the fovea are perceived in rich detail because the fovea has a dense array of photoreceptors that translate the fine details of the image entering the eye into a complex signal. Figure 5.3 shows the **fundus**, or the interior of the eye opposite the lens, and is a view you would see if you were looking into a person's eye with an ophthalmoscope.

Now that you are familiar with how the key structures in the eye collect and focus light, you should have a sense of what causes the vision differences that lead many of us to wear glasses or contact lenses, as we detail in the next section.

How the Focal Plane Affects Our Vision

Recall from our discussion of the anatomy of the eye that the light reflected from an object enters our eye and is refracted by the cornea and lens. The lens of the eye can change shape to focus the entering light on the fovea at the back of the eye. A physicist would say that in normal vision, also known as **emmetropia**, the focal plane of the image falls directly on the retina so that the image is perfectly focused (Figure 5.4a).

When the lens cannot properly focus light on the retina, the focal plane may fall either in front of or behind the retina so that the resulting image is blurry. In **myopia**, commonly called nearsightedness, the focal plane of the image falls in front of the retina, and the individual can see objects clearly up close but not far away (Figure 5.4b). Individuals with **hyperopia**, commonly called farsightedness, have the opposite problem: the focal plane falls behind the retina so that they can see distant objects fairly well but need help to see objects nearby (Figure 5.4c). Some of us are born hyperopic, but almost all of us develop farsightedness during middle age due to a loss of flexibility in the lens. When a person's hyperopia is due to aging, it is called **presbyopia**. Glasses or contact lenses correct far- or nearsightedness by moving the image onto the fovea from an improper focal plane.

One last common type of refraction error is **astigmatism**, which generally occurs when the cornea or lens is not symmetrical. In this case, the focal plane is skewed by falling partly on the retina and partly off the retina as shown in Figure 5.4d. This mismatch creates an image

FIGURE 5.4 Vision and the focal plane The cornea and lens together refract, or bend, light entering the eye, and the clarity of the image depends on where the focal plane lands on the retina. **a.** In emmetropia, the focal plane falls directly on the retina. **b.** In myopia, the focal plane falls in front of the retina. **c.** In hyperopia, it falls behind the retina. **d.** In astigmatism, the focal plane falls in several places on the retina.

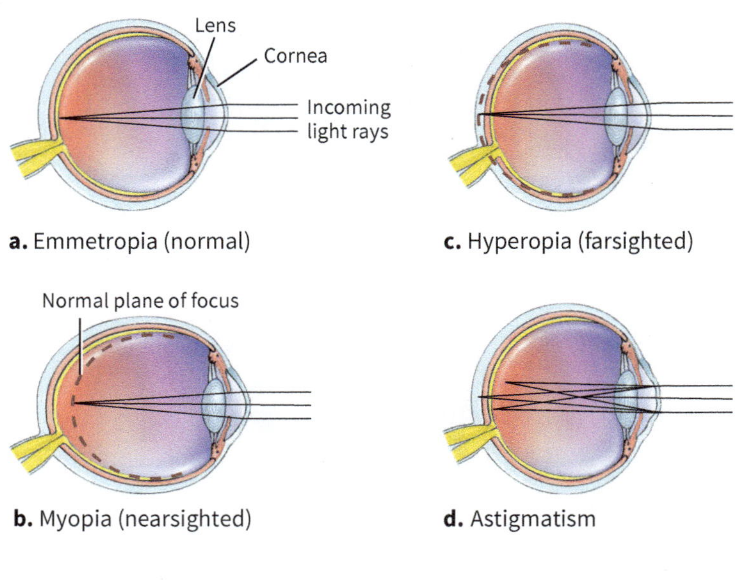

in which one part is in focus and another part is not. Astigmatism is not rare, occurring in perhaps one in three children. It can be corrected with an asymmetrical lens that is carefully designed to correct the individual's astigmatism. In order to discuss and correct vision problems such as refraction errors, scientists have developed means of quantifying and measuring visual capabilities, the topic of our next section.

Measuring Visual Acuity and Contrast Sensitivity MCAT

Children may experience difficulties reading the board in class, elderly drivers may have trouble reading road signs, and some of us struggle to read text or see objects without the help of corrective lenses. For a variety of reasons, it is useful to accurately measure acuity, or sharpness, of vision. One standard measure of acuity is the Snellen chart, which can be found on the wall of almost every physician's office. The Snellen chart presents letters of diminishing size to the viewer, who tries to identify the letters from a standard distance. Perfect vision is characterized as 20/20 (or 6/6 in metric countries), which means the individual sees at 20 feet (6 meters) what a person with normal vision sees at the same distance. By the same measure, an individual with 20/40 vision sees at 20 feet (6 meters) what a person with normal vision sees at 40 feet (12 meters).

The Snellen chart is used as a clinical screening tool to diagnose common vision problems, but more sensitive measures also exist. For research purposes, scientists often prefer the Landolt C task (sometimes called the Landolt ring) as a measure of visual acuity because it provides a finer level of analysis. In this test, a computer delivers a series of visual stimuli in the form of rings with a segment missing. On each trial, the observer is asked to specify the direction of the missing segment using the numeric keypad on a computer. With each correct response, the stimulus reduces in size until the observer can no longer detect the missing segment (see **Try It Out: The Landolt C Task**).

Acuity is one useful measure of visual capabilities, but it is not the only measure or even the only important measure. Another important measure is **contrast sensitivity**. Contrast is the difference between the intensity of a stimulus and its background (**Figure 5.5**). Contrast allows us to distinguish an object from its surroundings, known as the **figure–ground organization**. If you think distinguishing an object from its surroundings is easy, it is only because our ancestors have been working on it for eons. You may remember the story of the princess and the pea from your childhood in which the sensitive princess could detect even the smallest pea under her bedding. In like fashion, an individual with high contrast sensitivity can detect a stimulus that differs only slightly in intensity from its background.

Contrast sensitivity comes into play in many real-world tasks. Older drivers who test very successfully on the static Snellen test may perform poorly on measures of contrast sensitivity. Their ability to perceive the road at night or in foggy conditions is considerably impaired when compared to that of a typical younger driver.

Try It Out | The Landolt C Task

You can test your own vision using a Landolt C task by downloading the Freiburg vision and contrast test (FrACT), which is available from your instructor. Watch the brief video Freiburg Visual Acuity Task, collect your own measure of visual acuity, interpret it in light of the worksheet for this activity, and then answer these questions.

1. What advantages might there be to the Landolt C task over typical paper tests such as the Snellen chart? What are the disadvantages?
2. Visual acuity is very important to your success. Search the internet using the phrase "How to protect my vision" and identify two risks to your vision and what you can do to protect your vision from them.

FIGURE 5.5 Differences in contrast affecting visual perception In this image, the contrast between objects and background has been reduced on the right side of the image. The loss of detail leads to a loss of information in the image.

The Visual Processing Stream

Now that you have a grasp of the major anatomical structures of the human eye, you are ready to see how our visual processing stream takes raw energy from photons striking receptors in the retina and converts that signal into our complex visual world. We can break that process down into three stages, which will be the focus of the next three sections of the chapter: the retina, the subcortical centers, and finally the visual cortex (**Figure 5.6**).

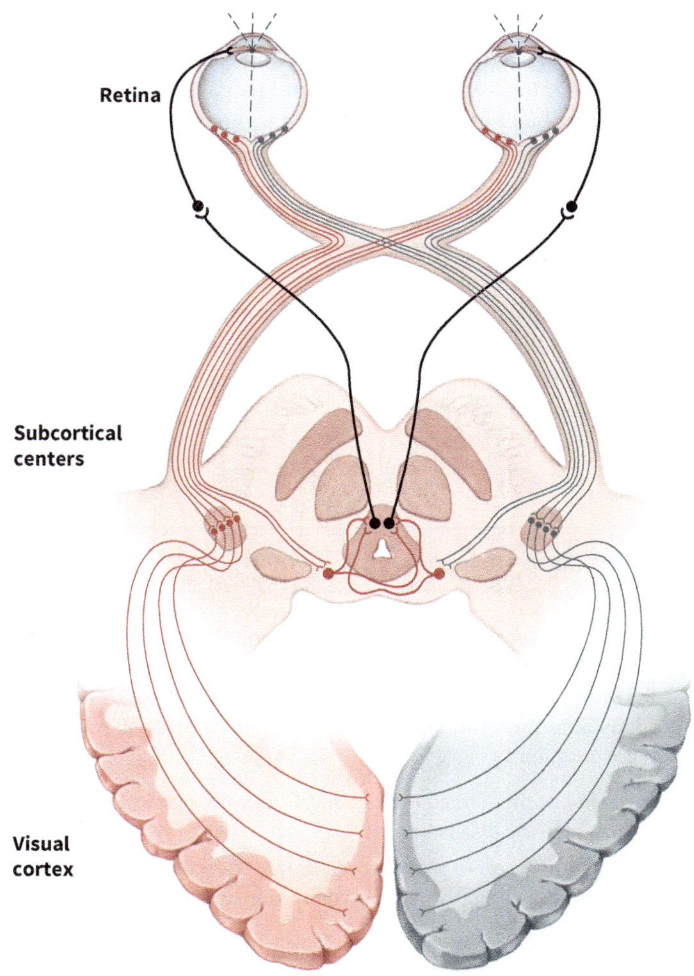

FIGURE 5.6 The visual processing stream Information enters the visual stream through the retina, is transmitted through several subcortical centers, and ends its journey in the visual cortex. At each stage, processing occurs that enhances the original signal quality. The visual stream takes in sensation and creates perception.

At each stage of the visual processing stream, raw sensory data is further refined until it forms a meaningful mental image. You might compare your visual system to a computer, in which a bit is an elementary unit of a signal. Your visual system is a marvelous device that takes in as many as ten million bits of visual data per second at the level of the retina. Along with the useful bits of data that will form our visual perception is a lot of background noise, or random fluctuations that don't actually contribute to the picture. At each processing stage in the visual stream, the noise is reduced, and the information-laden signal is enhanced. We start with raw physical energy in the form of light, and after many steps in various neural structures, the process culminates in an appreciation of something as visually nuanced as Leonardo da Vinci's *Mona Lisa*.

We will begin our discussion by looking at the retina, that very important cellular layer at the back of the eye.

5.1 Before You Go On

Section Summary

The physical characteristics of visible light waves are translated into brightness and color by our perceptual system. Light enters the eye, and the image of the world is refracted by the cornea and lens onto the retina. When the focal plane does not fall correctly on the retina, visual difficulties such as myopia and hyperopia occur. Visual acuity and contrast sensitivity are two measures of vision. As light enters the eye, it is first processed at the retina, then in the subcortical centers, and finally at the visual cortex.

Comprehension Check

1. Visible light, infrared light, microwaves, X-rays, and gamma waves __________.

- **a.** differ mainly in their intensity
- **b.** are all part of the electromagnetic spectrum
- **c.** differ in that the first two are light and the last three are radiation
- **d.** are all at different frequencies of the visible spectrum

2. The fibrous structure that protects the eye from physical damage is the __________.

- **a.** sclera
- **b.** choroid
- **c.** retina
- **d.** dura

3. The region of the retina with the densest collection of photoreceptors is the __________.

- **a.** fundus
- **b.** iris
- **c.** optic nerve
- **d.** fovea

4. When the focal plane, or image, falls in front of the retina, that condition is known as __________.

- **a.** emmetropia
- **b.** hyperopia
- **c.** myopia
- **d.** diplopia

5. Select the task that would be difficult if you had poor contrast sensitivity.

- **a.** separating knives from a pile of forks and spoons
- **b.** reading the numbers on a gauge in a brightly lit room
- **c.** telling one person from another as you enter a room
- **d.** walking down a flight of stairs in the dark

5.2 The Retina

LO 5.2 Identify the major components of the retina and their functions.

At the heart of every sensory system is the point where the physical energy of the world is translated into an electrochemical signal that can be transmitted by neurons through the nervous system. In the visual system, the retina is the site of this transformation and the neurons that create it are the photoreceptor cells.

Photoreceptors MCAT

You learned in the chapter on neurons that although types of neurons differ in many ways, almost all are triggered by a stimulus that reaches a specific threshold to propagate a signal along

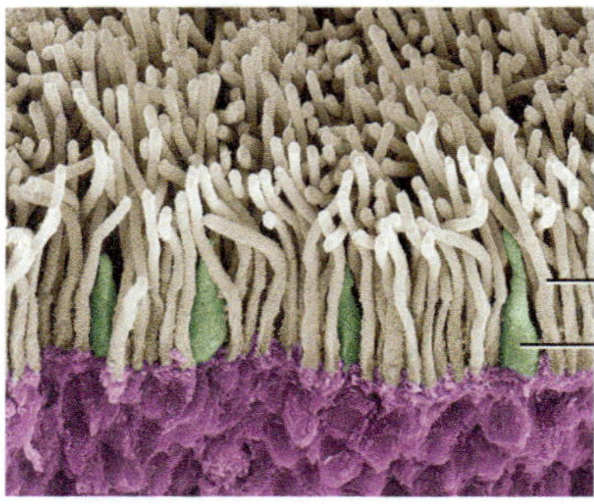

FIGURE 5.7 Rods and cones *in situ* Rods are responsible for vision in dim light and are sensitive to movement. Rods are larger than cones, which are responsible for color vision as well as acuity.

their axon. At the retina, the energy contained in a photon is received by a neuron called a photoreceptor, where it is transformed into a signal that then travels along the axon toward the brain. When a photoreceptor cell is stimulated by light at or beyond its threshold, that receptor cell fires. If the level of stimulation does not reach that cell's threshold, the receptor will not fire, and no sensation will be passed farther down the visual processing stream. The typical mammalian visual receptor cells resemble either rods or cones, our first topic.

Rods and cones

When scientists first looked through a microscope at the retina, they noticed an obvious difference in the shapes of rods and cones that implied that each does something different (Figure 5.7). Research has since shown that cones are responsible for acuity, or sharpness, of vision, but they need a considerable amount of light to operate. The longer rods do not provide information about color, but they require relatively little light to produce an image, and they are also quite sensitive to motion. The shorter cones generally report to one downstream retinal cell, whereas rods send their signals to many. This explains why cones provide an image that is sharp and detailed.

When light falls on a rod, a chemical inside the rod called rhodopsin breaks down into a molecule called retinal and light-sensitive proteins called **opsins**. The energy released by this reaction causes the cell to fire, thereby creating the sensation of light. Cones also contain retinal and one of several opsins that are each sensitive to a specific wavelength of light. This fact will be important to understanding color vision later in the chapter. That rods and cones differ in appearance and also contain different opsins suggests that each type of receptor has a different purpose. Table 5.1 summarizes the principal functional differences between these receptor types.

Another observation about rods and cones when viewed with a microscope is that they are not present in equal numbers, and they are distributed differently on the retina. There are about 100 to 120 million rods in the human retina and only about 7 million cones. Even more interesting is that cones exist almost entirely in the middle of the retina where an image would naturally fall when you look straight at an object, whereas rods have the opposite distribution (Figure 5.8). This means that when you look directly at an object, the image will fall directly on the cones, whereas the image of objects in your peripheral vision will fall on the rods.

TABLE 5.1 Functional Characteristics of Rods and Cones

Function	Rods	Cones
Acuity	Low	High
Threshold	Dim light	Bright light
Color sensitivity	None	Good color discrimination
Specificity of output	Stimulate many retinal cells	Stimulate one retinal cell
Dark adaptation	30 minutes	5 to 6 minutes

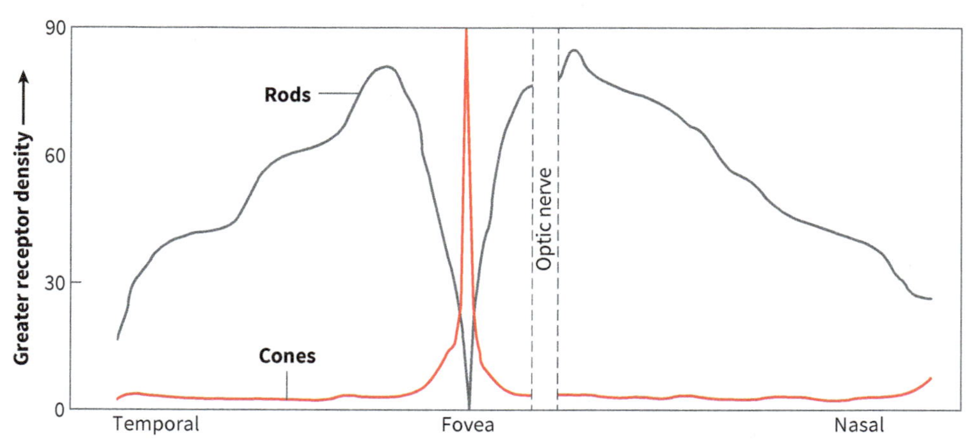

FIGURE 5.8 Spatial organization of photoreceptors in the human eye Cones create a color image in sharp detail, and they are located almost entirely in the fovea. In contrast, rods do not exist in the fovea but are distributed across the retina. Because there are no rods or cones where the optic nerve exits the eye, any image that falls on that spot (the blind spot) cannot be seen.

Let's take a moment to consider the implications of Figure 5.8. The human visual system provides a detailed image in color when the level of illumination is high and the viewer is looking right at the object so that the image falls directly on the cones of the fovea. When you are viewing a scene in dim light, the level of stimulation falls near the threshold of the cones, so not all of them fire. Therefore, in dim light, you maintain an image of the object but begin to lose accurate color vision. When the level of illumination dips below the threshold of the cones, we see very little directly in front of us because only our rods are signaling. However, our eye is very sensitive to objects approaching us from the side. Our evolutionary history has provided us with an eye that favors visual acuity over low-light vision; animals that hunt at night or live in the ocean would have different visual capabilities to service their needs.

The ipRGC For a long time, scientists thought rods and cones were the only two types of photoreceptor, but we now know that a third photoreceptor services the rods and cones, known as the intrinsically photosensitive retinal ganglion cell (ipRGC).

The story of the discovery of a third photoreceptor demonstrates that science sometimes moves slowly but inexorably toward discovery. In 1923, geneticist Clyde Keeler accidently bred mice with no rods or cones. He assumed these mice were blind but was astonished to find they showed a pupillary response to light. This could only be explained by the existence of a third photoreceptor. It was not until 2005 that this receptor was identified in mice as a ganglion cell in the retina that was photosensitive. What purpose does this third photoreceptor serve? The answer appears to be that it responds in a slow-motion fashion to seasonal changes in illumination. These cells report to the hypothalamus, a region implicated in determining the circadian rhythms that govern physiological patterns over a 24-hour cycle.

Retinal Processing Beyond Rod and Cones MCAT

Where does the output of photoreceptors go? Photoreceptors communicate with other neuronal cells in the retina that also play an important role in vision. The retina is actually composed of layers of cells beyond rods and cones, including horizontal, bipolar, amacrine, and ganglion cells (see **Check It Out: Peeking into the Retina Through OCT**). These cells poll the responses of the rods and cones they serve and begin the process of making sense out of the activity of the individual photoreceptors in the same way that political pollsters make sense out of their surveys with individual voters.

Students often are puzzled when they see for the first time that humans have an inverted retina; that is, light actually passes through several layers of these support cells *before* reaching the photoreceptors that create the initial visual signal. The explanation for this arrangement is that these receptors operate best with a generous blood supply and those blood vessels are

Check It Out | Peeking into the Retina Through OCT

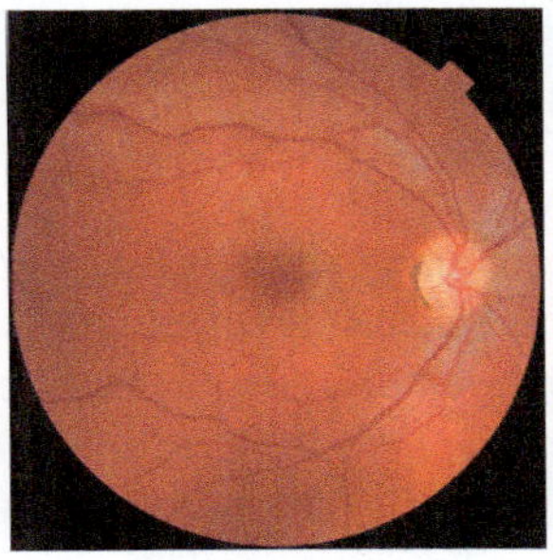

By Mikael Häggström, used with permission.

A technique known as optical coherence tomography (OCT) allows us to see the actual cellular layers of the retina. Take two minutes to view the video OCT: A Window into the Retina, and then answer these questions.

1. At the beginning of the video, you saw the thin tracings of the blood supply to the retinal cells. Why do you think the retina is richly serviced by your eye's blood supply?
2. You saw in the video that the thinnest part of the retina is the macula, which consists almost entirely of photoreceptors. Why do you suppose that the support cells are swept to the side at the macula?

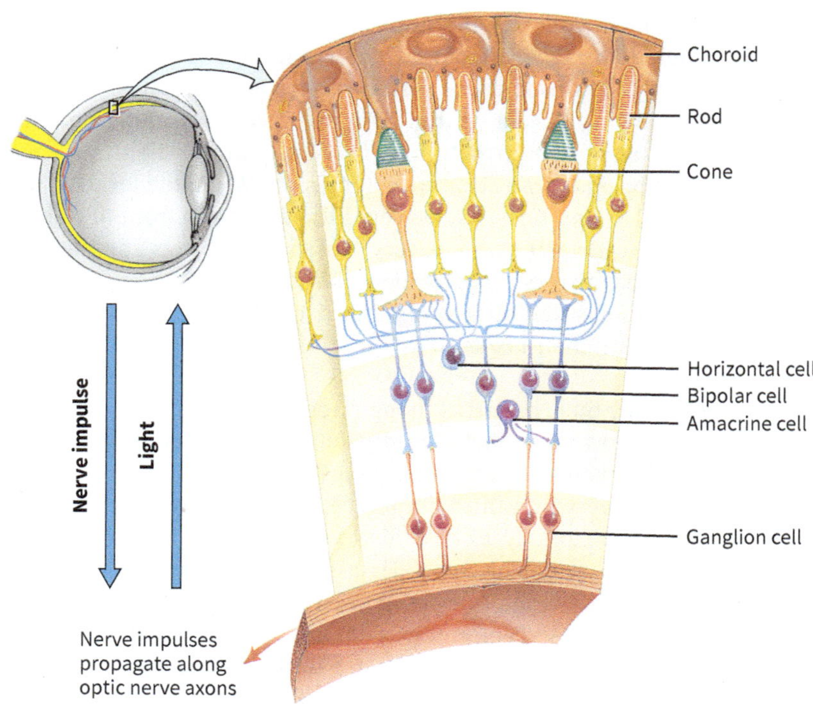

FIGURE 5.9 The human retina in cross section Light enters the eye, passes through several retinal layers, and then either stimulates a photoreceptor or is absorbed by the choroid. A signal from a rod or cone travels back from the receptor for processing in the bipolar and ganglion layers and then to the optic nerve for further processing in the visual stream.

in the choroid, which is adjacent to the receptors. Let's revisit what happens in the retina within the context of these additional cells: light enters the eye, passes through several layers of the retina prior to contacting the rods and cones, and is absorbed by the choroid (**Figure 5.9**). When these photoreceptors fire in response to light, the impulse travels down the axon to the next retinal layer, which consists of horizontal cells and bipolar cells. Horizontal cells are so called because they are horizontally arranged in the retina in order to simultaneously poll several rods. Bipolar cells are characterized by a cell body with a process such as a dendrite or axon at each end. Bipolar cells convey retinal activity farther into the retina. Many rods share a horizontal or bipolar cell, but cones enjoy the luxury of a more exclusive synaptic relationship with bipolar cells. That arrangement explains why cones are less sensitive to low levels of stimulation and yet provide greater detail.

Complicating the picture somewhat are amacrine cells, which are the small cell bodies you see sharing space with the bipolar cell bodies in Figure 5.9. The role of amacrine cells is poorly understood, although it is clear that they contribute to vertical communication among retinal layers as well as motion detection, and it is thought that they play an inhibitory role in visual processing at the retinal level.

Finally, specialized cells known as ganglion cells are represented by the four large cell bodies near the bottom of Figure 5.9. Each ganglion cell receives input from an average of 100 photoreceptors, although this number varies greatly. In the fovea, a ganglion cell may only connect to four or five cones, but in the retinal periphery, a ganglion cell may connect to thousands of rods. The number of receptors reporting to a ganglion cell controls the level of detail that each ganglion cell passes to processing centers downstream.

Ganglion cells help convert simple data from the retina into a complex image. An important insight into how the nervous system creates our complex perceptual world is that stimuli enter the processing stream through a sensory organ, and then at each subsequent stage the signal is refined and information is added. This concept is clearly demonstrated by **receptive fields** in the visual processing stream. The receptive field (from the Latin *recipio* for receive) of a neuron in the visual system is that region of the neuron's visual field that can initiate or inhibit a response. A simple but familiar example of a receptive field is a motion-activated security light; it only reacts if an object enters a specific region. Sensory receptors throughout the body have receptive fields with different characteristics; for example, the skin on your lips is more sensitive than the skin on your lower back. In the visual processing system, we have a

FIGURE 5.10 **Ganglion receptive fields** **a.** A center-on receptive field is stimulated by a light represented by a yellow dot. When that light falls on the center of the receptive field, the cell fires rapidly. When a light falls within the receptive field but not in the center, the cell inhibits firing. **b.** For a surround-on receptive field, when the light falls on the center of the receptive field, the cell inhibits firing, but when the light falls in the periphery of the receptive field, the cell fires.

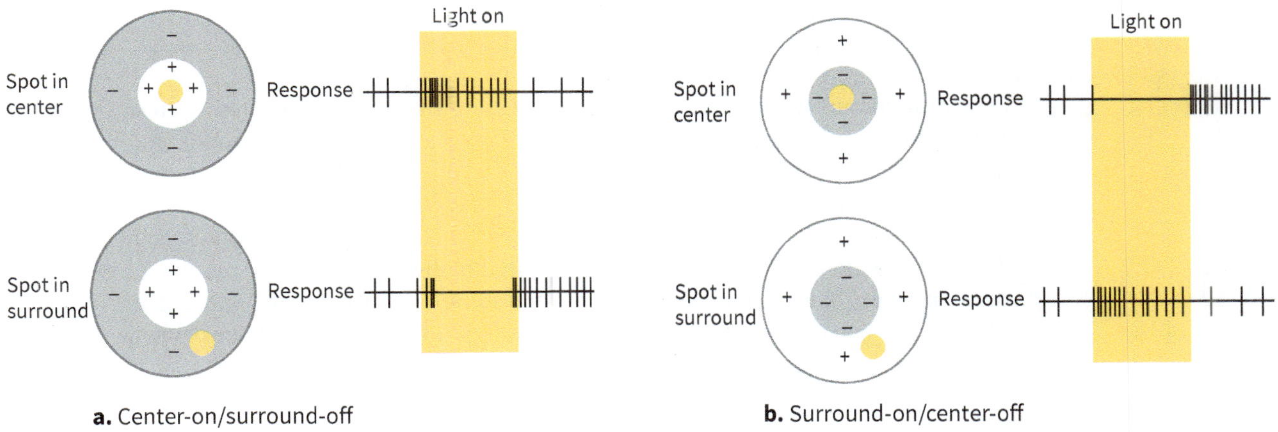

series of receptive fields that provide additional information about edges and other characteristics of objects in our visual world. You will see the concept of a receptive field in the retina, in the subcortical centers, and finally in the visual cortex.

Remember that neurons are limited in what they can do. A neuron can fire or not fire, and it can fire at a faster or slower rate. It would be easy to assume that photoreceptors, once stimulated by light, simply signal on or off, but our retina provides more than a simple yes/no response to light. Photoreceptors typically signal on or off but at the retinal ganglion level, some cells (center-on/surround-off cells) only fire when the center of the receptive field is lit and the surround is dark, whereas other ganglionic cells (surround-on/center-off cells) respond in the opposite fashion (**Figure 5.10**). As the edge of an object moves across the various center-on and surround-on fields, the sum of the many signals becomes the first step in perceiving the edge of an object.

From raw data, our visual system begins to build a representation of the external world. We will introduce a more complex understanding of receptor fields in a moment, but for now, we continue to follow the path of processing deeper into the brain.

5.2 Before You Go On

Section Summary

The retina is the place where the physical energy of light is transformed into a neural signal. There are three types of photoreceptor cells: cones for color, rods for dim-light environments, and a specialized receptor called ipRGC for seasonal effects. Horizontal, bipolar, amacrine, and ganglion cells help create an information-rich signal. Ganglion receptive fields are an important first step in creating the edges and lines that our visual system turns into perceptual objects.

Comprehension Check

1. Animals that are active mostly at night would ________ animals that are active mostly in the day.
   a. rely on cones more than
   b. rely on rods more than
   c. have an equal number of rods and cones as
   d. have fewer cones than

2. If your eyes only had rods, you might struggle ________.
   a. to tell the difference between the green and red signals of a traffic light
   b. to see deer on the side of the road at dusk
   c. to perceive a ball being thrown at you from the side
   d. with tasks that require a quick response

3. Retinal ganglion cells provide the first level of information about what?
   a. the color of objects
   b. the size of objects
   c. the distance of the object from the observer
   d. lines and edges

5.3 The Subcortical Centers

LO 5.3 Identify the major subcortical structures in the visual processing stream and their functions.

Returning to our overview of the visual processing stream, we have now seen how light is collected by the eye and translated into an electrochemical signal in the retina. In this section we will follow this signal out of the retina and into the subcortical centers of the brain where it will be further processed before ending in the visual cortex.

There are two major conduits for the output of the retina, each with a specific purpose: the geniculostriate pathway and the tectopulvinar pathway. We will first discuss the geniculostriate pathway, the primary pathway for visual stimuli to travel to the brain, which is responsible for our perception of objects. Then we will turn to the tectopulvinar pathway, which is responsible for the spatial aspects of visual perception.

The Geniculostriate Pathway MCAT

The **geniculostriate pathway** is the primary route for visual stimuli to travel from the eye to the brain. It begins at the retina and goes through the **lateral geniculate nucleus (LGN)** and then into a cortical region known as the **primary visual cortex (PVC)**, or the striate cortex. This pathway is thought to handle close to 90% of all the output of the retinal ganglion cells. We begin our discussion of this important path with the LGN itself.

The LGN is an important sensory relay system located in the thalamus of the brain that provides a basis for the analysis of form, color, and movement. There are two LGN, one on each side of the thalamus. Each is roughly the size of a small bird's egg, and if it were unrolled and flattened, it would be about the size of a credit card. The lateral geniculate nucleus is *lateral* because there are two nuclei in this region of the thalamus, and the LGN is the one that is lateral, or closer to the outside. Its neighbor, the medial geniculate nucleus, is medial, or closer to the midline of the brain. The word *geniculate* comes from the Latin *genu* for knee because these nuclei are curved like a bent knee. The LGN is a complex and important structure, so we will first discuss its relationship with the retina and then its anatomical structure.

The output of the retina leaves the eye via the optic nerve and proceeds to the **optic chiasm** (pronounced *KAI-as-um*), the junction where the nerve fibers cross on their way to the brain. The term chiasm comes from the Greek letter *chi*, or X, because of the X shape formed by the decussation, or crossing, of the output of each eye as it is directed to the separate brain hemispheres. Each eye divides its retinal output into a nasal (near the nose) and a temporal (near the temple) **hemifield**, or half an image. From the chiasm, the output from each field travels along the optic tract to the brain. The rule is simple for both eyes: the temporal hemifield goes to the ipsilateral (same) hemisphere, whereas the nasal hemifield for each eye is directed to the contralateral (opposite) hemisphere. This means that each eye sends its output to both left and right cortical hemispheres, ensuring distributed cortical processing of the input from both eyes.

The principle of integrating input is maintained in the LGN, where the left visual field consists of input from the left temporal and right nasal retinal fields, and the right visual field consists of input from the right temporal and left nasal retinal fields. The left and right eyes both send their temporal and nasal hemifield outputs to the optic chiasm. From the optic chiasm, the temporal signal is sent to the LGN on the same (ipsilateral) side of the brain, while the nasal signal is sent to the opposite (contralateral) LGN (**Figure 5.11a**). In this way, each LGN receives input from both eyes.

Once the signal reaches the LGN, we can be even more specific about where the input from each hemifield is received. The LGN consists of six layers, which differ by cell type (M-, P-, and K-cells, to be discussed presently). Information coming to layers 5, 3, and 2 originates in the ipsilateral (same) eye, and information coming to layers 6, 4, and 1 originates in

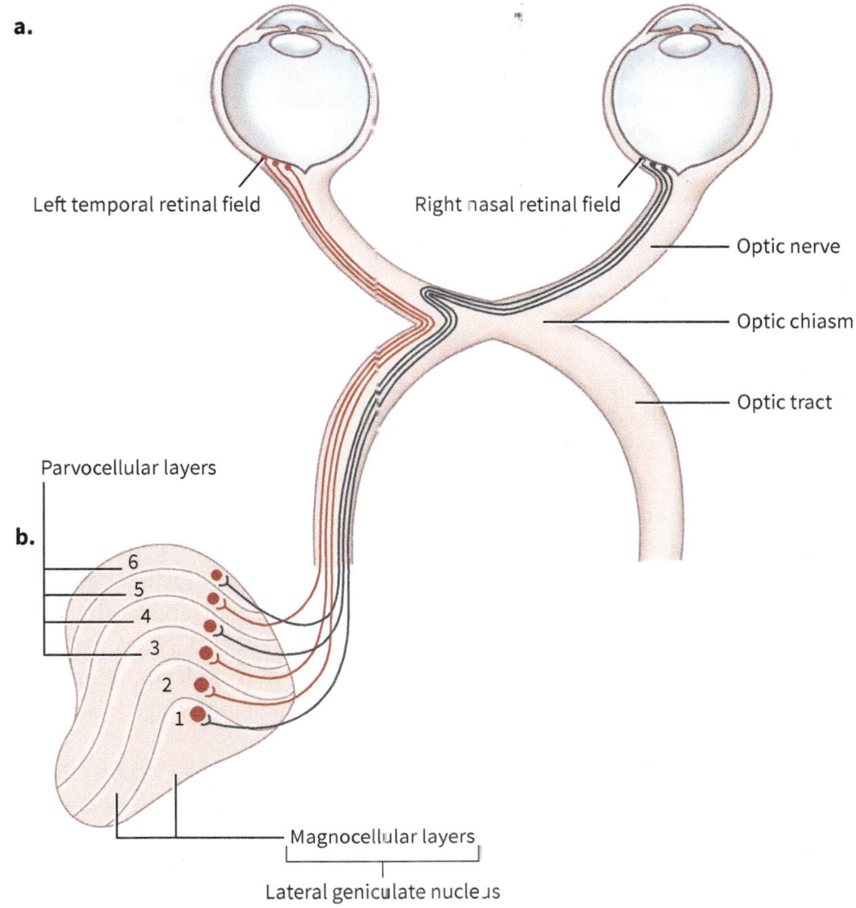

FIGURE 5.11 Visual input and layer organization of the LGN **a.** Each eye sends the output of both retinal fields down the optic nerve. At the optic chiasm, the left temporal field continues to the left LGN, and the right nasal field crosses over to the left LGN. In this way, the left LGN integrates the hemifields to form a complete image with data from both eyes. **b.** Within the left LGN, the right nasal retinal field reports to layers 6, 4, and 1, whereas the left temporal field reports to layers 5, 3, and 2. In the LGN, temporal and nasal streams are divided between M-cell and P-cell layers.

the contralateral (opposite) eye (Figure 5.11b). Note that retinal input to layers 1 and 2 goes directly to M-cells while input to layers 3 through 6 is directed to P-cells.

The unique structure of the LGN provides clues to its function in our visual processing system. A toothpick driven through the six layers of the LGN would affect the various processing operations for the same point in your visual world for both eyes. This general arrangement of our visual system in which a two-dimensional representation of the visual world is maintained by spatial relations among processing units is known as **retinotopic mapping**.

The six layers of the LGN differ in the retinal source of their input, and they also differ anatomically. Layers 1 and 2 consist of relatively large **magnocellular cells (M-cells)**, so they are labeled M1 and M2. The outermost four layers (P3 to P6) are much smaller cells known as **parvocellular cells (P-cells)**. These names are derived from the Latin for large (*magnus*) and small (*parvus*). If you think of the six layers of the LGN as layers in a cake, there should be a thin layer of icing between each layer. In between each of the six M- and P-cell layers is a thin layer of a third type of cell, known as a **koniocellular cell (K-cell)**. The root word for *koniocellular* is the Greek word for dust because K-cells are lightly dusted on the ventral surface of the M- and P-cell layers.

Each of the three distinctly different cell types in the LGN has a specific purpose (Table 5.2). The large M-cells are fast acting and transient in response. They appear to be responsible for rapidly creating coarse or rough outlines of objects due to their large receptive fields. In contrast, the action of the P-cells is slower and more sustained. They appear to provide both color vision and the fine level of detail we enjoy. Earlier, we mentioned that receptive fields exist at three levels in the visual processing stream; the first is at the retinal level, and the second is at the level of the LGN, where M-cells and P-cells add another level of analysis to the identification of objects in our visual world. We will turn to the third level later in the chapter.

Finally, the K-cells provide input from cones that result in sensitivity to short wavelengths such as blue. K-cells likely provide other important visual functions such as the detection of

TABLE 5.2 Attributes of the Cells of the LGN

	Magnocellular	Parvocellular	Koniocellular
Location	Layers 1 and 2	Layers 3 to 6	Between M and P layers
Input	Rods	Cones	Throughout retina
Speed	Fast acting	Slow acting	Slow acting
Function	Movement and shapes	Small details	Varied

moving objects in our visual field. Damage to the LGN can result in any number of visual problems. For example, lesions of the P-cells result in a loss of fine detail and color vision because they accept input from cones, whereas lesions of the M-cells disrupt the detection of fast-moving objects in the visual field because their input comes from rods.

The LGN has been an object of intense interest to neuroscientists due to its importance in the visual processing stream. Like the retina, the LGN also has receptive fields that are center-on/surround-off as well as center-off/surround-on. The LGN further enhances the signal from the retina, adding binocular information about color, detail, and motion. The original primitive signals from the retinal ganglia, now enhanced by the LGN, are passed onto the visual cortex for further processing. A concept essential to your understanding of the visual processing stream is that at each stage, the total amount of data in the signal is reduced while its information content is increased.

Before you proceed, take a moment to consider the importance of the LGN as a major processing structure in the human visual stream (see **Check It Out: The Eye of the Batter**). Raw data from the photoreceptors in both eyes arrives at the LGN, where the beginnings of a three-dimensional representation are created from the output of three types of cells, each of which makes a unique contribution to the whole. That stereoscopic view and the resulting relationship among different objects in the visual field represent an important step in creating our complex visual world.

With this overview of the geniculostriate pathway completed, we now turn to the tectopulvinar pathway, which also plays a significant role in our visual processing.

The Tectopulvinar Pathway MCAT

While the geniculostriate pathway is dedicated to identifying objects in our visual world, the task of the **tectopulvinar pathway** is to orient us toward the source of those same objects.

Check It Out | The Eye of the Batter

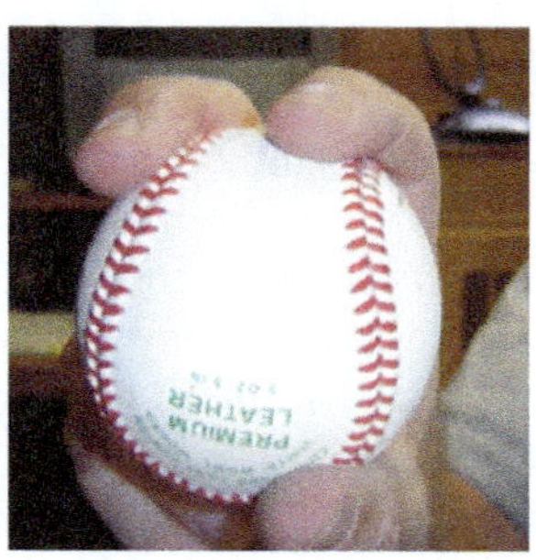

When we think of the importance of visual acuity, we typically consider only static acuity tasks such as reading letters on an eye chart. In the real world, dynamic visual acuity is much more important. Dynamic acuity allows us to determine the identity and path of moving objects, which helps us drive safely, for example. In this activity, you will see how dynamic visual acuity allows us to make subtle judgements about objects moving toward us at over 90 miles per hour (145 kph). Watch the video Eye of the Batter: Skilled Visual Acuity, and then answer these questions.

1. When a varsity or professional pitcher throws the ball, how long does the batter's visual system have to analyze the rotational pattern of the pitch?
2. List two tasks that a person might perform in everyday life or on a job where reasonable dynamic visual acuity would be a requirement. Explain your answer.

Knowing what an object is and where it is in relation to you might seem to be two ways of saying the same thing, but they differ. From an evolutionary perspective, we might wonder if the first system to evolve would be a rudimentary sensory system that simply orients the organism to an object in their immediate environment such as a predator or prey. This is supported by the fact that systems like the tectopulvinar path exist even in simple organisms that lack more sophisticated visual processing skills.

More relevant to the topic is this: the geniculostriate pathway provides data on *what* is out there, and the tectopulvinar pathway gives us information about *where* the object is. This suggests that if one system is damaged and the other intact, it should be possible for a patient to be aware of an object's location without knowing it is present. As odd as that sounds, we will find out later that this is a real condition known as blindsight. Before we discuss these and other related maladies, however, we need to resume our discussion of the tectopulvinar pathway.

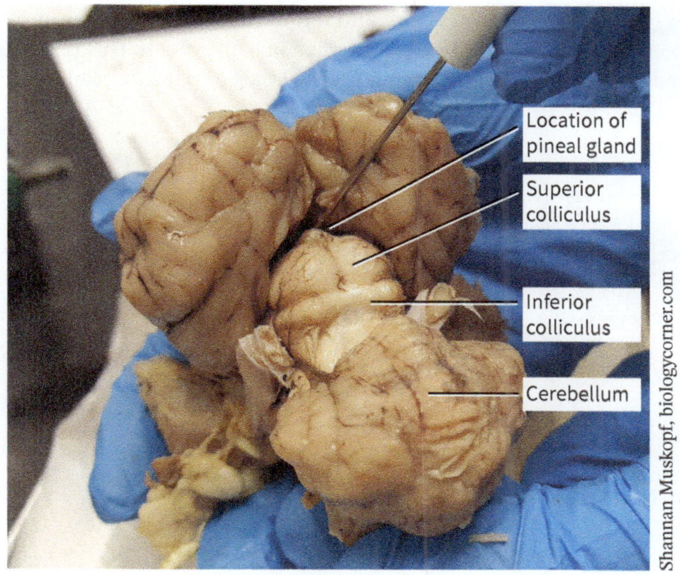

FIGURE 5.12 View of the superior colliculus The superior colliculus is located on top of the midbrain, just above a structure called the inferior colliculus and below the pineal gland. The superior colliculus is an essential part of the circuit that orients our visual attention in response to events in the visual world.

Most of the output of the optic nerve is channeled to the visual cortex via the LGN, but around 10% of the retinal ganglia connect to the superior colliculus and also to the pulvinar and lateral nuclei. This pipeline for visual processing then continues on to the visual cortex. We will consider the role of each structure in turn.

We saw in our discussion of the CNS that there is a region in the midbrain known as the tectum (or *roof*) located atop the midbrain. The tectum contains the superior colliculus, which has an important function in our visual system and is depicted in its position on the top of the brainstem in **Figure 5.12**. The role of the superior colliculus appears to be directing the eyes toward a specific region in the visual field in response to external stimuli. Given the function of the superior colliculus, it might not surprise you that the inferior colliculus, which is situated next to the superior colliculus, performs a similar function by orienting us toward incoming auditory stimuli. Having the two structures in close physical proximity aids us by employing visual and auditory cues to guide us in orienting our eyes toward objects in our environment.

Another center in the tectopulvinar pathway is the pulvinar nucleus, a small structure located near the superior colliculus that appears to play a role in directing and maintaining attention toward an object. In response to external stimuli or feedback from the cortex, the tectopulvinar pathway directs your attention to objects in your visual field, thereby directing your gaze to a specific location.

Damage to the Visual Stream

By looking at the visual processing stream in its entirety, you can understand how damage to one part of the stream can affect overall visual perception. Each retina is divided into half of a visual field, or hemifield. The outside (temporal) hemifield is one stream, and the inside (nasal) hemifield is a separate stream. If you follow the temporal and nasal hemifields out of the eye and into the LGN of the thalamus and then to the primary visual cortex of the brain, it is apparent that each eye provides a data stream to both hemispheres. You can easily see why this arrangement is more advantageous than having the left eye report only to the left hemisphere; otherwise, the loss of one eye would hinder visual processing entirely in that hemisphere.

The implications of an arrangement where each eye reports to both hemispheres in a specific fashion becomes clear in a patient with damage to one side of the occipital lobe or the other, creating what is known as a **hemianopsia**, from the Greek *hemi* for half and *anopsia* for loss of vision. Individuals experiencing a hemianopsia typically lose either the left or the right half of their visual field (**Figure 5.13**).

FIGURE 5.13 **Normal vision and left hemianopsia** **a.** This image shows a view of Paris as though seen by an individual with normal vision in both eyes. **b.** This is the same view as though seen by an individual with damage to the right visual tract or right occipital lobe.

a. Normal vision **b.** Left hemianopsia

5.3 Before You Go On

Section Summary

The primary pathway in the visual processing stream is the geniculostriate pathway, which passes through the LGN, where processing of form, detail, and color occur. A secondary and simpler system, the tectopulvinar system, directs our visual attention to significant objects in the environment. Damage to these streams can result in a loss of vision in one part of the visual field.

Comprehension Check

1. Identify the one statement about the lateral geniculate nucleus that is false.

- **a.** The magnocellular layers are responsible for detecting shapes.
- **b.** The parvocellular layers are responsible for detecting details.
- **c.** The koniocellular layers are responsible for contrast sensitivity.
- **d.** The magnocellular layers are all located in the right LGN, and the parvocellular layers are in the left LGN.

2. An important distinction between the geniculostriate and tectopulvinar pathways is that the geniculostriate pathway ________, and the tectopulvinar pathway ________.

- **a.** is fast acting; responds slowly
- **b.** responds to shapes; responds to color
- **c.** detects what we see; detects where we see it
- **d.** detects locations; detects forms

3. One important function of the superior colliculus is to ________.

- **a.** direct our eyes toward a moving object
- **b.** guide both eyes to focus on the same object
- **c.** aid in our critical ability to focus on distant objects
- **d.** control the size of the pupil to let in the right amount of light

4. The fact that a person may lose their vision only on the left side of their visual field is evidence of what?

- **a.** The right eye is dominant over the left eye.
- **b.** The output of each eye is divided between the left and right hemispheres.
- **c.** Damage to the LGN does not necessarily lead to blindness.
- **d.** Damage to the optic nerve generally causes total blindness.

5.4 The Visual Cortex

LO 5.4 **Describe** the organization of the visual cortex and its role in visual perception.

What you have seen in our journey though the visual processing system is that raw data enters the retina and travels through several subcortical processing centers before arriving at the visual cortex. As the signal is passed along, it becomes more complex. In a sense, the signal loses some of the characteristics of the original signal but becomes more useful to the organism; its informational properties are enhanced.

You might think that with so many areas in your eye and brain devoted to visual processing, there would not be anything left for your cortex to do, but that would be a mistake. The visual cortex is essential to interpreting our visual world, and we will see later in our discussion that cortical damage leads to some very interesting clinical problems.

Our goal here is to explain how the visual cortex analyzes input from upstream sources to create our visual world, but the truth is, scientists still do not understand the whole picture. Vision scientists are discovering more about this system every day, and as they do, new pieces of the puzzle fall into place. Furthermore, many of the techniques necessary for investigation are invasive and not appropriate for humans, so much of what we do know comes from work on zebrafish, mice, cats, hamsters, and monkeys. For these reasons, you should expect that our understanding of the human visual system will be continuously coming into clearer focus in the years ahead. With that caveat in mind, we will focus our attention initially on the primary visual cortex.

The Primary Visual Cortex MCAT

At the rear of the brain lies the occipital lobe. The rearmost portion of the occipital lobe is known as the primary visual cortex (PVC), or area V1. It is also known as the striate cortex due to a stripe of myelinated cells that is readily apparent in a cross section of V1. The PVC receives its input from the LGN via a band of tissue known as the optic radiations. The path from the retina through the LGN and superior colliculus to the primary visual cortex is presented in **Figure 5.14**.

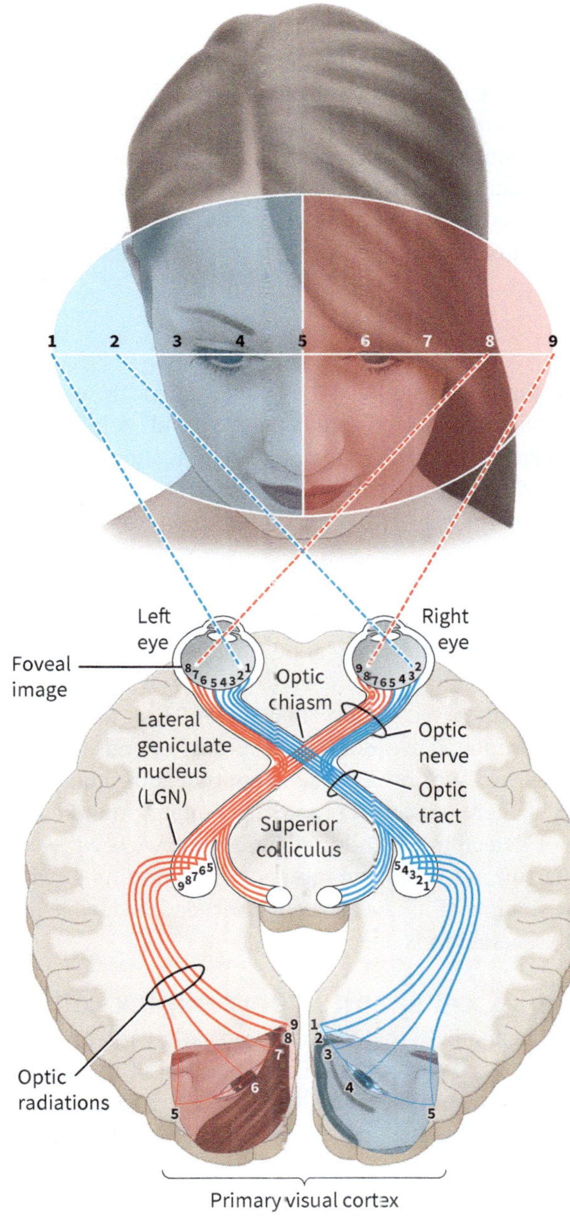

FIGURE 5.14 The pathway from retina to PVC Each retinal hemifield creates a signal that is sent to the superior colliculus and passes through the LGN or superior colliculus and finally arrives at the primary, or striate, visual cortex via the optic radiations. The PVC is the seat of complex visual processing tasks.

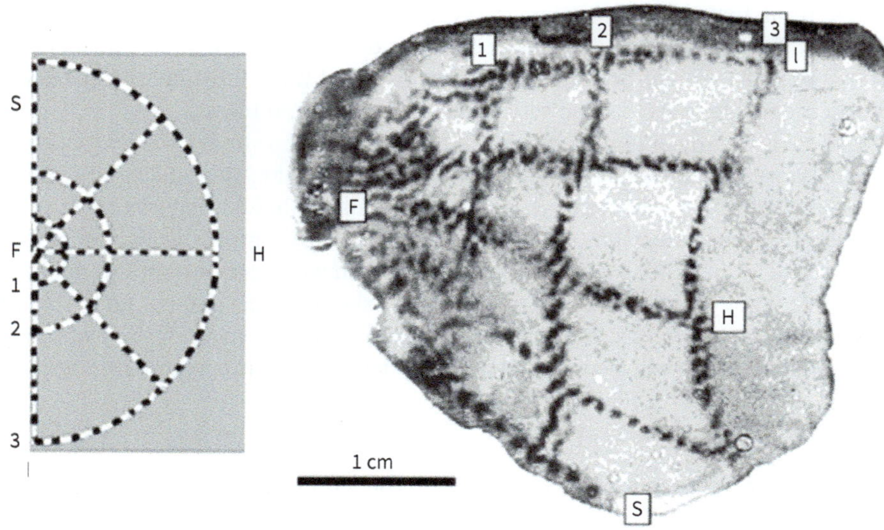

From Tootell, RB, et al., (1982), American Association for the Advancement of Science.

FIGURE 5.15 Cortical retinotopic mapping The geometric pattern of the stimulus was re-created in area V1 by retinotopic mapping. In area V1, your visual processing system creates a two-dimensional representation of the three-dimensional world around us. Note that the general pattern of the labels S, F, 1, 2, 3, I, and H in the original stimulus on the left has been maintained in the pattern of cortical activation on the right.

Activation in area V1 creates a retinotopic map; that is, the cortex creates a two-dimensional image of the three-dimensional object being perceived. This was shown in a study where a radioactive tracer was used to identify the cortical regions activated by visual processing and then an image was shown to a monkey; the pattern of activation on the cortex roughly mirrored the appearance of the original object being viewed. The stimulus and the pattern of activation in area V1 are illustrated in **Figure 5.15**. V1 is organized in such a way that specialized processing modules for color, form, and movement are dedicated to each region of the retina. The data from left and right eyes is processed at adjacent regions. The beauty of V1 is that it provides a complex analysis of the visual world while maintaining the retinotopic representation necessary for integrating objects in relation to each other.

Processing in the Visual Cortex

How does our visual processing stream take the simple signals from receptors and create the edges that allow us to separate an object from its background? A mental image may help you understand how complexity is created from simple signals. Imagine an athletic field where the ground is equipped with a tightly packed array of motion detectors, the same sort that you might find on a garage or porch. These detectors face upward and signal when an object passes overhead.

Now imagine that a hawk flies over the field and its shadow falls on these motion detectors. Individual detectors signal a change in illumination to let us know something is out there, but that alone doesn't tell us anything about the type or direction of motion. This level corresponds to the action of retinal photoreceptors. A few feet underneath the field, some detectors are wired together into a second level of detector. These detectors only fire if multiple detectors in a line are all stimulated in order. This level corresponds to the concept of receptive fields that we discussed earlier. We now have a system that adds motion to the signal. Finally, in our imaginary example, there is a control room deep underground where the output of the first and second levels of detectors are collected, and there we see the edges combined with their movement. The person in the control room sees this outline moving across the screen and recognizes it as the shadow of a hawk soaring overhead. Evidence indicates that cortical receptors similarly have several levels of complexity.

Simple cells, complex cells, and hypercomplex cells In addition to the receptive fields in the retina and LGN that we discussed earlier, researchers have found several processing modules in the visual cortex that take the output of these structures and create a more complex analysis of the incoming visual data. The quest to understand how the human visual system works led two individuals all the way to a Nobel Prize in Physiology or Medicine in 1981. David Hubel and Torsten Wiesel suspected that receptive fields might exist in the cortex that add information to the signal from the retina, so they devised the following

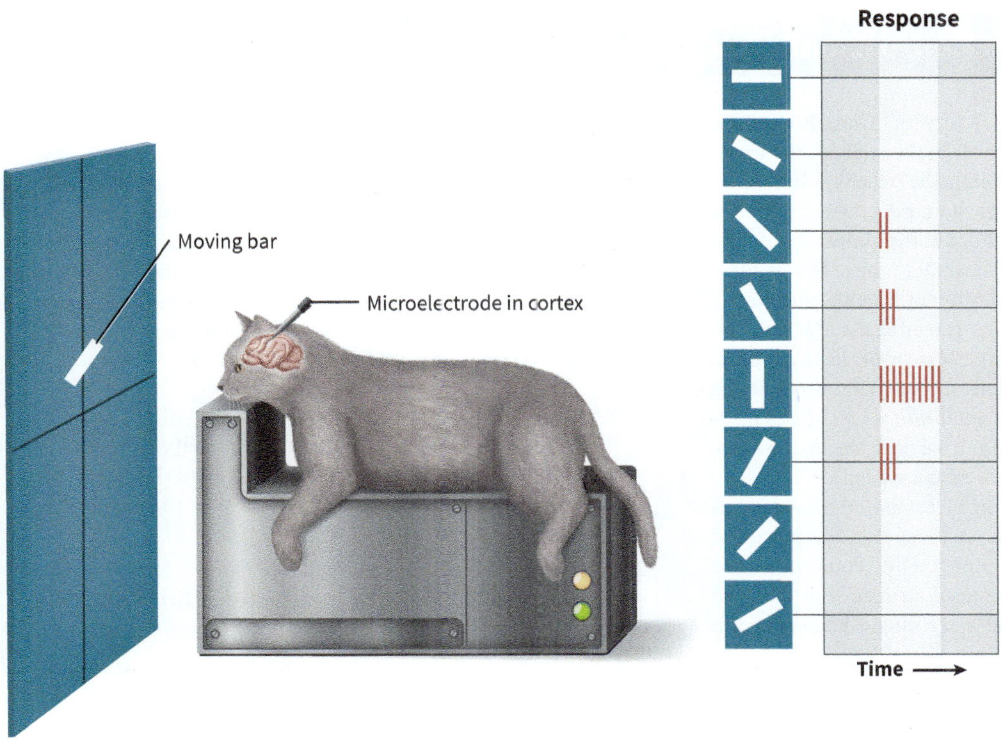

FIGURE 5.16 The Hubel and Wiesel single-cell recording paradigm After anesthetizing the cat and inserting a recording electrode into the primary visual cortex, Hubel and Wiesel presented the cat with a lit bar that moved across the cat's visual field in a series of orientations. They discovered that neurons in V1 respond selectively to specific orientations that define that cell's receptive field.

experiment. A microelectrode was surgically implanted in a single cell in a cat's visual cortex, and then the cat was positioned facing a projection screen. A stimulus in the form of a bar was then presented on the screen. This arrangement is portrayed in **Figure 5.16**.

Hubel and Wiesel expected that the visual stimulus would send a signal to the cortex and the microelectrode would detect that signal, but this was not what they initially found. The cells did not fire until the stimulus bar moved, and to their great surprise, they discovered that the cortical cells collecting data from the retina have receptive fields that are tuned to code in very specific ways. For example, some cells increase their firing rate when a line moves across the receptor from upper left to lower right, but only when that line moves across that receptor field in that specific direction. As the direction of the movement of the line diverges from the optimal angle for that cell, the rate of fire slows and eventually stops entirely. These cortical receptive fields are an essential first step in detecting the lines and edges of objects in our visual world.

Because every cell is tuned to respond to a specific orientation, movement of the edge of a visual stimulus across the retina will stimulate some receptors but not others. Hubel and Wiesel described these structures as simple cells, complex cells, and hypercomplex cells. Explaining how different cells all operate to turn the firing of a photoreceptor into the image of an apple can be very confusing, so perhaps a simple model can get the main points across. Photoreceptors and supporting cells in the retina send their output to the LGN, which then analyzes and processes the data and transmits that signal to area V1. Simple cells fire in response to a stimulus that falls on their receptive field, and they fire optimally if the orientation of the stimulus matches their excitatory orientation. When the stimulus moves off that region, the cells inhibit their firing. In this way, simple cells elicit a brief response to a stationary object. Complex cells, on the other hand, collect the output of several simple cells and provide information about the movement of that object across the larger visual field. Hypercomplex cells, like complex cells, respond to movement in a particular orientation but also provide information about the length of the object.

Research that followed up on the work of Hubel and Wiesel appears to show that even simple and complex cells in the PVC can show the end-stopping property that was originally ascribed only to hypercomplex cells. Rather than descend into a jungle of details, we can conclude that in the PVC there is a hierarchy of processing modules that interact to create the edges and characteristics such as the size of objects in our visual world. This allows us identify objects and also note their movement in our visual field. As complex as these processing modules are, our understanding of how our brain creates a visual image of the world was further expanded by the discovery of the columnar organization of the visual cortex.

Advances in Behavioral Neuroscience | Margaret Wong-Riley and the PVC

Dr. Margaret Wong-Riley is a pioneer in the study of the visual system. She received her Ph.D. from Stanford and completed a post-doc at the University of Wisconsin. After time at the National Institutes of Health, she had a stellar career at the Medical College of Wisconsin in Milwaukee, where she was editor-in-chief of *Eye and Brain*, a prestigious medical journal dedicated to the study of the visual system.

Dr. Wong-Riley was a pioneer in investigating how the primate visual system is organized. She realized that the puzzle of how cortical organization reflected functionality could be solved by finding a way to measure activation of structures in the PVC in relation to function. She devised a technique known as the cytochrome *c* oxidase (COX) stain to illuminate structures in the PVC at a very granular level. The enzyme cytochrome oxidase reflects metabolic activity, thereby identifying areas of greater and lesser activity. Her work identified tiny structures known as blobs that are sensitive to color. These blobs are surrounded by regions known as interblobs that are sensitive to the orientation of a stimulus rather than its color. Blobs receive input from the P-cell, or parvocellular, regions of the LGN and report to area V2 of the visual system.

Questions

1. **Why would the ability to measure activity in a part of the nervous system (especially your visual system) be useful in our quest to understand how the brain creates behavior?**
2. **Name a visual task that you feel is very important to you and would be worth investigating. Briefly explain why you chose that task.**

The columnar organization of the visual cortex Before we continue, let's take a moment to consider the hierarchy of processing that occurs in the visual processing stream. Receptive fields in retinal cells provide a first level of signal processing, which then is augmented by receptive fields in the LGN. At the level of the visual cortex, simple, complex and hypercomplex cells further enhance the signal. To appreciate the computational power of your brain, we now turn to an exploration of the columnar organization of the visual cortex.

Researchers believe that area V1 consists of approximately 5,000 specialized processing clusters known as **hypercolumns** (sometimes also called cortical modules or macrocolumns), each about 0.5 mm across. Anatomists discerned that V1 consists of at least six distinct layers, but their unique function was a mystery until research by Margaret Wong-Riley and colleagues differentiated the layers into more and less active regions including blobs and interblobs (see **ABN: Margaret Wong-Riley and the PVC**). It appears that blobs contribute to the perception of color, whereas the interblob regions are devoted to form and motion analysis.

The work of Wong-Riley and others revealed the functional organization of the visual cortex, best described as a collection of thousands of well-organized vertical columns (**Figure 5.17**). Your visual cortex consists of six separate layers that are situated between subcortical white matter and the pia mater. There are several different processing modules that we have identified so far in the visual cortex.

Orientation columns identify the spatial orientation of features such as lines and edges in your visual world. If you inspect the lines at the top of the orientation columns in Figure 5.17, you will see that they indicate that the sensitivity of each orientation column is both specific and also represents a continuum. For example, the leftmost orientation column is sensitive to a line whose top is tilted to the right, whereas the corresponding rightmost column is sensitive to a line whose top is tilted to the left.

Orientation columns are organized into sections that accept input from the left and right eyes, known as ocular dominance columns. Blobs are interspersed among the orientation/dominance columns, where they receive input from the P-cells on the LGN (for example, layers 6 and 5), and in that way, they receive information about color and fine detail. Finally, the regions between blobs, interblobs, appear to be sensitive to orientation. Before we move on, take one last look at Figure 5.17 to notice the feedback and feedforward mechanisms in the PVC. Layer 5 sends feedback to the superior colliculus, while layer 6 sends its output back to the LGN. Layers 2 and 3 send their output on to the extrastriate cortex.

Hypercolumns are the basic processing modules of the PVC, so we might ask if early life experience enhances the subsequent performance of these structures. Just as encouragement

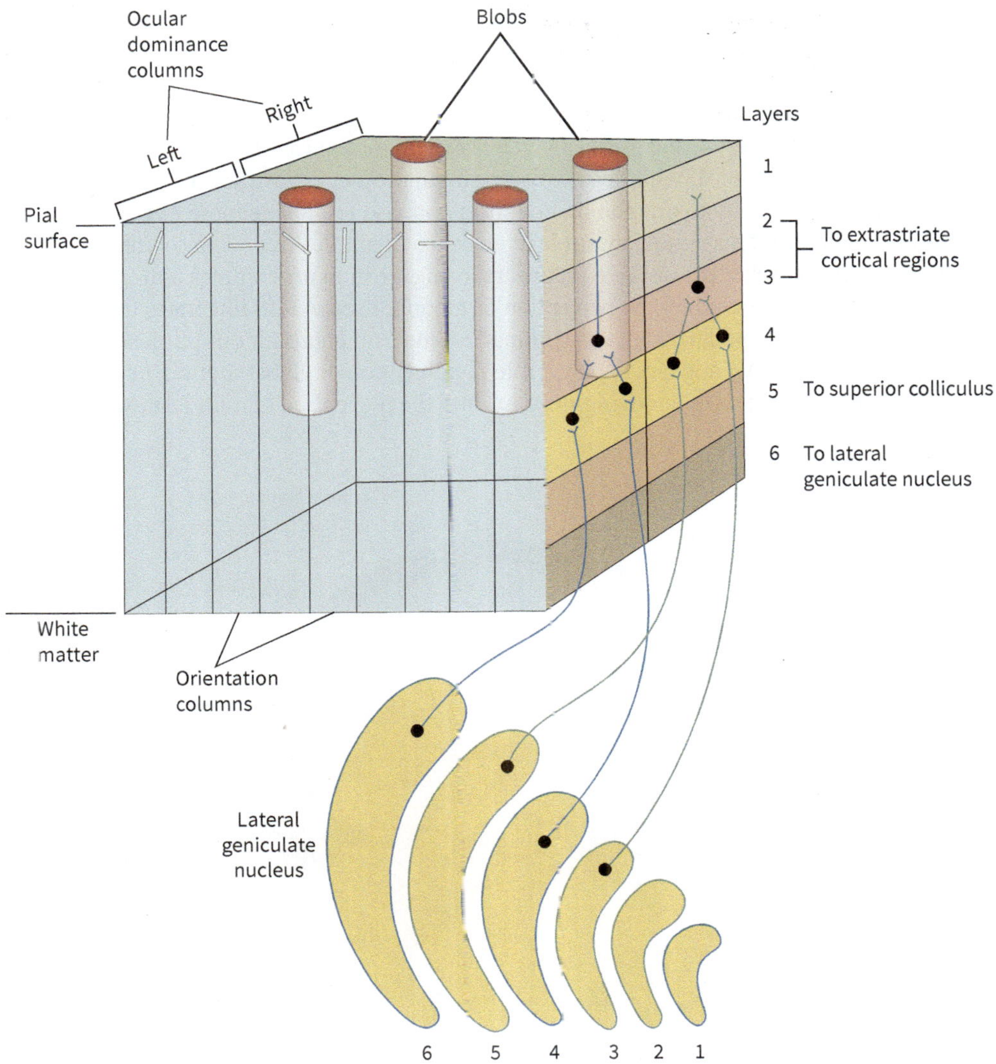

FIGURE 5.17 Structure of hypercolumns in V1 Area V1 consists of many cortical columns, each of which is a single processing element that, in concert with its adjacent columns, creates a complex visual representation of the word that includes color, orientation of edges and lines, and fine details. Area V1 receives input from the LGN and transmits information farther along the processing stream while sending feedback to subcortical centers.

to exercise from an early age will likely lead to greater lifelong fitness, early visual experience plays a critical role in how hypercolumns develop and create useful processing modules. There are cases where injury or disease to the cornea or lens allowed light to enter a person's eye but not normal visual detail. The cortical organization that experience normally provides is missing in these individuals. If their vision is restored in adulthood, they are typically never able to acquire normal perception, and they struggle to recognize everyday objects.

To summarize the elegant processing mechanism of V1, the input from both eyes that came from the LGN is sent to the visual cortex, which in turn sends some of the signal on to other cortical centers for further processing, returns some of the signal to subcortical processing centers as feedback, and does its own processing of color and orientation. The primary visual cortex makes the most advanced video chips look primitive by comparison.

V1 is not the only area of interest in the occipital cortex; nearby are other regions that differ in structure, which have been designated V2, V3, V4, V5, and so on. These additional regions are the topic of our next section.

The Extrastriate Cortex MCAT

As you gaze upon a beautiful sunset like the one at the beginning of the chapter, the activity at V1 is not the end of the line for a process that began some milliseconds ago when a photon from that sunset struck a photoreceptor in your retina. When V1 is activated, areas V2 and beyond also engage. Because regions V2, V3, V4, and V5 are outside V1 (the striate cortex), they are referred to as the **extrastriate cortex**. Activation of the extrastriate cortex can be demonstrated by studies where a slightly radioactive tracer is coupled with either oxygen or glucose. When a cortical region is activated, it takes up more oxygen and glucose. The result is that when metabolic activity increases, detectors positioned over the skull pick up the emission emanating from any region that is activated by the task.

We do not yet know the whole story of how our brain creates our visual world, but we do know that regions V2, V3, V4, and especially V5 play an important role. Region V2 receives input from V1 (and feeds back to V1) while also providing input to other extrastriate regions. V2 appears to share with V1 sensitivity to orientation and form and also to provide more elaborate figure–ground information. V2 may also play a role in object recognition. Area V5 (also known as MT) is of great interest to scientists because it appears to engage in many complex and higher-order processing operations, including the perception of motion. **Figure 5.18** illustrates the locations of areas V1 through V5 on the cortex and summarizes their functions for our purposes.

In particular, area V5 appears to play a role in perception of global motion, or the linking of several disparate aspects of a moving stimulus into the perception of a single moving object.

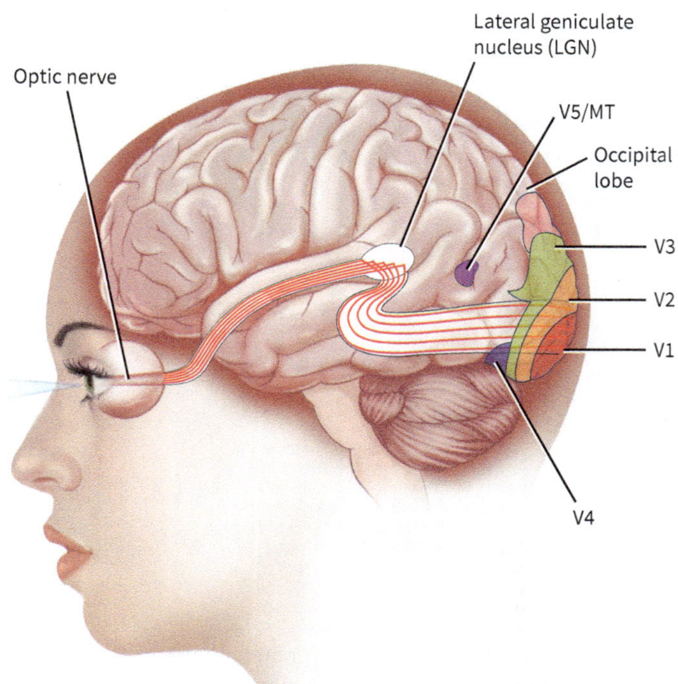

Structure	Location	Task	Connections
V1	Striate cortex	Edge detection	Upstream from superior colliculus Downstream to V2 and beyond
V2	Pre-striate cortex	Creates detail, color, orientation, and figure-ground	Upstream from V1 Downstream to V2 through V5
V3	Next to V2	Motion, global awareness of large pattern	Upstream from V1 Downstream to parietal cortex
V4	Extrastriate cortex	Determining salience of stimulus	Upstream from V1 Downstream to in-ferotemporal cortex
V5	Extrastriate cortex next to V4	Motion and complex perception	Upstream from V1 through V3 and K-cell of LGN Downstream to specialized cortical area

FIGURE 5.18 **Regions V1 through V5 of the visual cortex** The signal that began at the retina and was further processed in the subcortical centers ends its journey in the cortex where the most sophisticated processing occurs. Starting with area V1, an increasingly complex set of analyses in these cortical areas creates our detailed visual world.

Check It Out | Motion Aftereffects

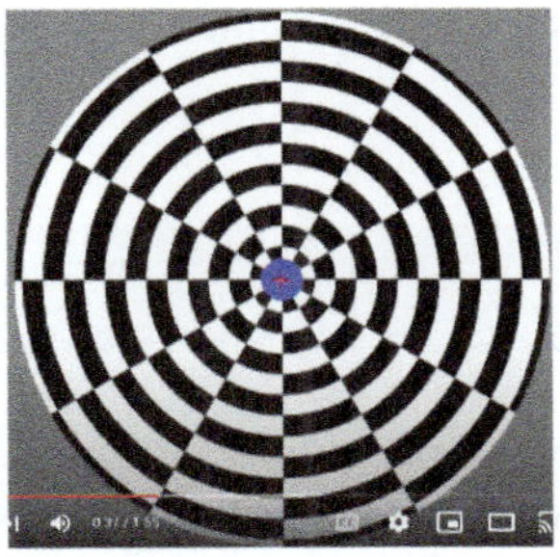

To experience the phenomenon of motion aftereffect, watch the brief video Motion Aftereffects, and then answer these questions.

1. Do you think that the functioning of V5 is innate or develops with experience in the visual world? Briefly explain your answer.
2. What are two situations where akinetopsia would put someone at risk? Briefly explain why you chose those situations.

Area V5 takes a series of individual images and creates a sort of movie of the world out of those individual images so that they appear to be continuous. The importance of this function becomes clear when it doesn't work properly. Some people experience a deficit in motion detection known as akinetopsia, which means they experience the visual world as a series of individual still images. The phenomenon of motion aftereffect (MAE) demonstrates how your visual system accommodates a moving field. When the motion of the field stops, an aftereffect occurs (see **Check It Out: Motion Aftereffects**).

The Ventral and Dorsal Streams MCAT

After input to V1 is processed, the output is passed to two distinct streams (**Figure 5.19**). The **dorsal stream** goes from V1 to areas V2 and V5 and then to the parietal lobe. This pathway is also known as the *where pathway* because damage to these regions often results in an inability to perceive *where* the object is in relation to the self. The **ventral stream** originates in V1 and then proceeds to V2, V4, and then the temporal lobe. This path is known as the *what pathway* because damage here may lead to an inability to recognize *what* an object is. This separation of what and where streams is important for understanding how our visual system works as well as understanding symptoms when it is damaged. We will discuss in detail the general topic of object perception and these two processing streams, as well as fascinating clinical disorders such as agnosia and blindsight, in a subsequent chapter.

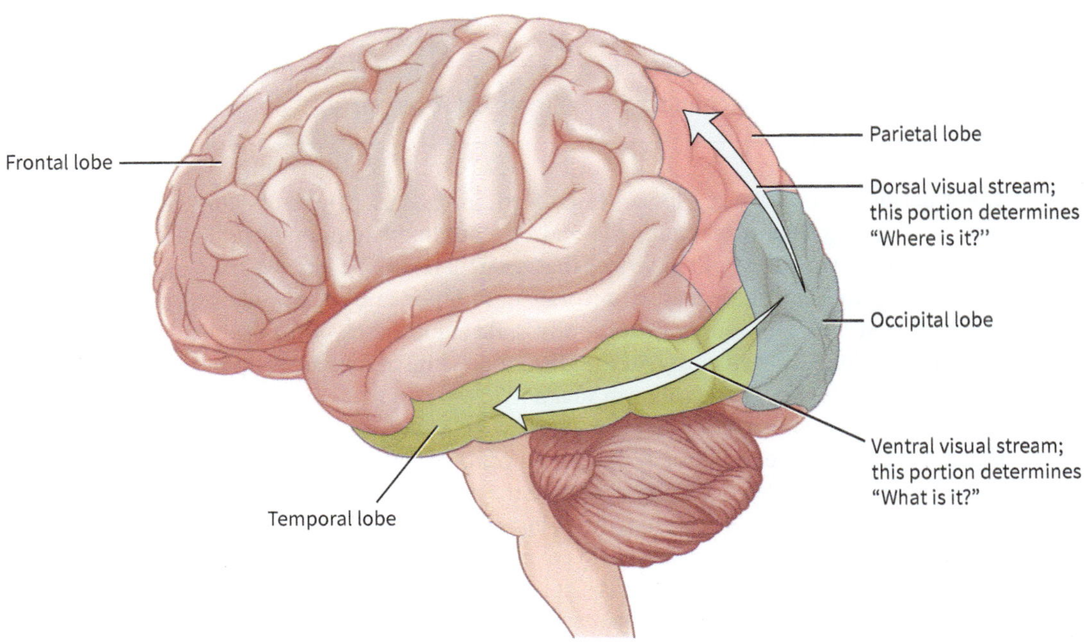

FIGURE 5.19 The ventral and dorsal processing streams The ventral stream is dedicated to deciding what is out there, and the dorsal stream is dedicated to analyzing where that object is in space.

Our journey through the visual processing stream, which began at the retina and passed through multiple subcortical centers on its way to the cortex, shows how we are able to rapidly analyze information in a dynamic visual world. At a subsequent point, we will discuss how your brain takes the output of these many centers and combines them to recognize objects. Let us now turn to a discussion of color vision.

5.4 Before You Go On

Section Summary

The visual cortex is the seat of complex visual processing. It consists of a collection of anatomical areas, each with a specific processing role. These areas communicate with each other to create our complex visual world. Damage to the visual cortex typically results in a significant loss of visual function.

Comprehension Check

1. An important characteristic of the primary visual cortex (PVC) is that ________.

- **a.** it creates vision almost entirely on its own
- **b.** it is a single very large processing unit, unlike the LGN
- **c.** it consists of a collection of interconnected processors
- **d.** damage to the PVC generally results in loss of color vision

2. The discovery of the columnar organization of the visual cortex was important because it demonstrated that ________.

- **a.** color vision is created in the LGN
- **b.** these columns connect the tectopulvinar and geniculostriate pathways
- **c.** blobs and interblobs create vision in dim light
- **d.** millions of small, discrete processors create a complex visual world

3. Which statement is true about the extrastriate cortex?

- **a.** The extrastriate cortex is so named because it consists of additional, or extra, striated cortex.
- **b.** Input from the LGN goes to the extrastriate cortex from where it is passed to the PVC.
- **c.** Damage to the extrastriate cortex typically leads to minimal vision loss.
- **d.** It upholds the principle that complexity is added at each stage in the visual processing stream.

4. The dorsal visual processing stream is known as the ________ pathway, whereas the ventral stream is known as the ________ pathway.

- **a.** where; what
- **b.** what; where
- **c.** color; black-and-white
- **d.** dim-light; bright-light

5.5 Color Vision

LO 5.5 Describe the mechanisms that create color vision.

Color permeates our world. Flowers burst forth in nature's palette of colors to announce the arrival of spring. Rainbows mark the end of a summer thunderstorm, and cool grays and whites combine to create winter scenes that have their own beauty. We express ourselves with color in our clothing, in the way we decorate our homes, and even by altering the color of our lips, hair, and skin. How do we turn the physical characteristic of the frequency of light waves into the psychological perception of the glistening white beach, the turquoise tropical sea, and the azure summer sky of a Caribbean vacation?

Many organisms get along just fine without color vision, so what advantage does chromatic sensitivity confer? One obvious advantage that color vision provides is a greater ability to distinguish food that is spoiled or unripe from food that is edible (Figure 5.20). Other organisms use their sense of smell to make this judgment, but humans rely on vision, so we need a visual cue.

A second hypothetical advantage to color vision is subtler. For those organisms that come into sexual receptivity on a cyclical basis, it is advantageous for the female to signal when she is fertile and receptive and equally advantageous for the male to know when his romantic interest is likely to be met with success or failure. Many female mammals and especially primates experience a change in color somewhere on their body that signals ovulation. One hypothesis is that the evolution of color vision is at least partially due to our ancestors using color as a signal

FIGURE 5.20 **Color vision as an evolutionary advantage** **a.** The colorful fruit stands out from the leaves in the background. **b.** Without color, the decrease in contrast makes it more difficult to distinguish the figure from the ground, which could prove to be an evolutionary disadvantage.

for sexual receptivity. You might well respond to this hypothesis by noting that humans do not appear to change color when they are in the mood to mate. However, in a subsequent chapter, we will discuss the evidence that humans may respond to subtle perceptual cues that signal ovulation. We now turn to a discussion of how our sensory system creates color perception.

The Trichromatic Theory MCAT

We began our discussion of vision at the level of the retina, where we introduced the two main varieties of photoreceptors: rods and cones. It was apparent to early anatomists that cones are responsible for visual acuity as well as color vision. Once scientists had identified cones as the source of color vision, they then had to understand how these cells generate a response to specific frequencies of light.

Types of cones Chemical analysis of the pigments within cones and other techniques led to the realization that humans are **trichromats**, with three different types of cones. One cone is most sensitive to long-wavelength light (the L cone), a second to middle wavelengths (the M cone), and a third to short wavelengths (the S cone). It became clear that at one level at least, color vision results from the overall pattern of excitation of the three cones in response to light falling on the retina. Many textbooks refer to these as the blue, green, and red cones, but Figure 5.21 shows that the spectral sensitivity peaks for the three cones actually correspond to the psychological experiences we call dark blue, greenish-yellow, and orange-red.

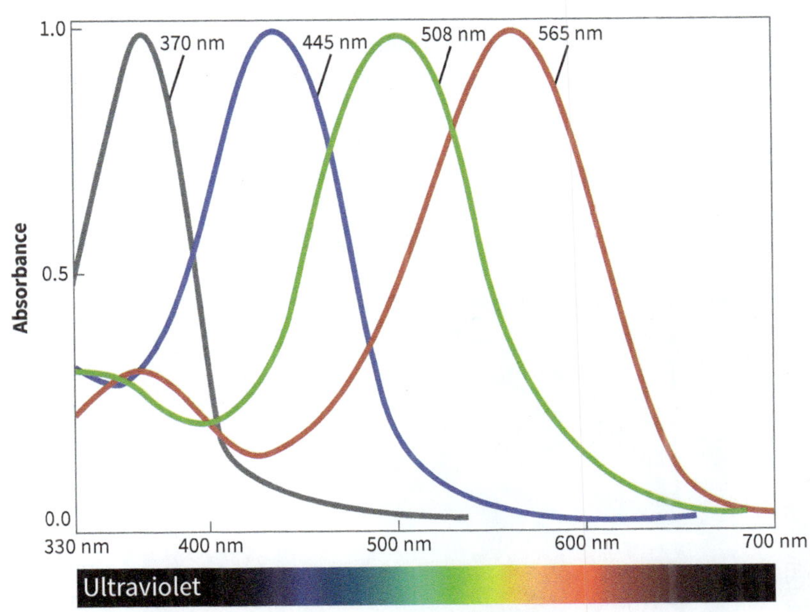

FIGURE 5.21 **Retinal receptor sensitivity as a function of wavelength** The graph represents the sensitivity of each of the three cones and also of rods to the frequency of the light falling on the receptors. The short-wavelength cone has a sensitivity peak at 437 nm, the middle-wavelength cone peaks at 533 nm, and the long-wavelength cone peaks at 564 nm. The second curve from the left depicts the response of the rods, which are maximally sensitive at 500 nm.

Distribution of cones across the fovea By now you know that your retina has cones and that these are concentrated in a central region known as the fovea. Interestingly, the S, M, and L cones are not equally distributed across the

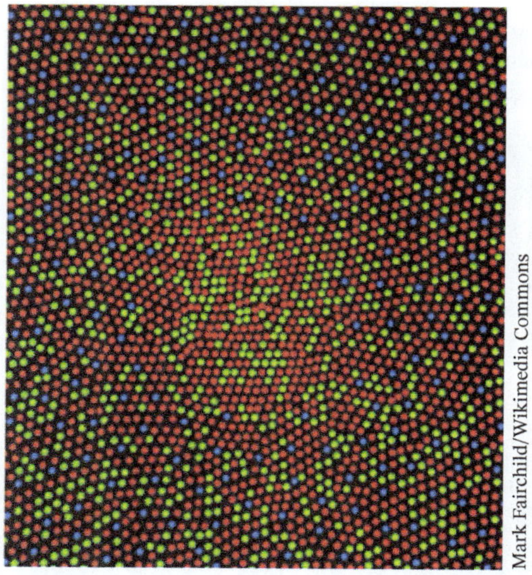

FIGURE 5.22 **Distribution of S, M, and L cones in the retina** There are three varieties of cones, each with a different sensitivity function. In this image, the S cones are blue, the M cones are green, and the L cones are red. You can see that the three types are not distributed in equal numbers across the retina.

retina. There are many more M and L cones than S cones in the human retina (Figure 5.22). It is hypothesized that the S cone evolved first because it would permit distinguishing a ripe red or orange fruit from the green of surrounding foliage.

As you now see, colors in the visible spectrum are created by the activity of some combination of the three cones. Some colors are the result of the firing of a single cone. For instance, deep violet is created by the firing of the S cone alone, and deep red is the result of the L cone reacting. However, our perception of most colors stems from two or more cones firing simultaneously. When there is a problem with the cones, it leads to color vision abnormalities, as we will see in the next section.

Color Vision Abnormalities MCAT

Most people have normal color vision, but some do not, and you may be surprised to find that mildly color-blind individuals your own age are sometimes blissfully unaware of their perceptual difference. When my roommate landed his first job after college, he bought several suits, ties, and dress shirts to make an impression on his potential brokerage clients. One day, his boss took him aside and told him that he needed to sharpen up his appearance. That night his girlfriend asked to see some of the rejected outfits and agreed that the colors looked terrible together. When asked to sort all his ties by color, he placed the green, blue, and purple ties in one pile he called blue. Evidently, he was colorblind. After his girlfriend sewed small tags into his clothing indicating their colors, he was able to create ensembles that would not clash using a grid that he posted on his wall.

Color deficiency is a more accurate descriptive term for what is commonly referred to as color blindness. How do you think color vision deficiencies occur? Think for a moment before you proceed. What physiological process would explain color vision deficiencies? Would your answer predict more than one type of color vision abnormality? By now you have probably hypothesized that the simplest explanation for color vision deficiencies is a problem with the cones.

Anomalous trichromats Some individuals have all three cone systems, but one or more of the cones has an abnormal sensitivity function. To understand what this means, look at the sensitivity function for the three receptor types provided in Figure 5.23 that represents normal color vision.

Imagine that the sensitivity function for the S, M, or L cones is shifted to the left or right of its normal position or that the sensitivity of one or more of the receptors is lowered when compared with normal sensitivity. These individuals, known collectively as anomalous trichromats, experience mild to moderate color deficiency. The language they use to classify colors is normal, but the assignment of specific colors to primary groups is a bit skewed.

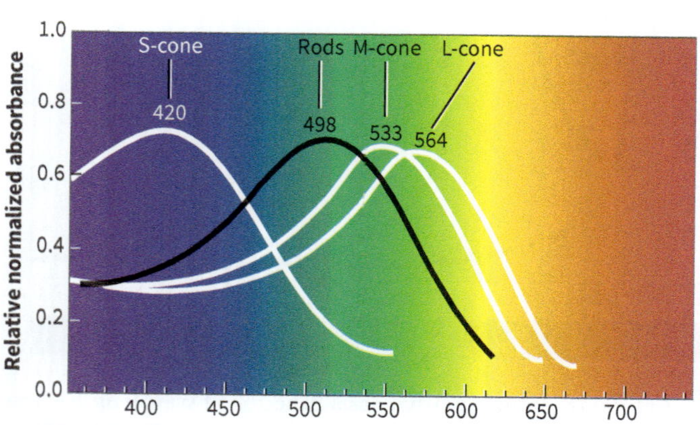

FIGURE 5.23 **Normal response functions for S, M, and L cones** Every photoreceptor has a unique sensitivity function. When a cone is deficient or missing, a person's entire color perception changes.

Variations in the number of cones A second cause for color vision deficiency comes about when an individual is entirely missing one or two of the three cone systems. **Dichromats**, who have two cones, are relatively rare and experience a much narrower range of color vision than individuals with three working cones. Dichromats come in three varieties: protanopes are missing the L cone, deuteranopes the M cone, and tritanopes the S cone (Figure 5.24).

Finally, there are the rare individuals who have only one cone system. It takes at least two cones to have color vision, so these **monochromats** have no color vision at all. They see everything in monochromatic shades.

Most mammals have some color vision, although not necessarily the same as ours. Cats and dogs are dichromats and

a. Normal-color-vision

b. Simulation-of-protanopia

c. Simulation-of-deuteranopia

d. Simulation-of-tritanopia

FIGURE 5.24 **Normal vision and simulated color deficiencies** The first image represents what a person with normal color vision would experience. The others are simulations of what an individual missing the L, M, and S cones respectively would perceive.

appear to have their greatest sensitivity in the blue-green region. Nonhuman primates' color vision is closest to that of humans. You might be surprised to learn that many birds and insects have a wider range of color perception than humans, and this is because they have four cones and are classified as **tetrachromats**.

Is it possible that there are humans with better than normal color vision because they have a fourth cone? The answer is unclear. From an evolutionary perspective, one would predict that a mutation for tetrachromacy would be rare because there does not appear to be a meaningful advantage to being able to perceive finer differences among colors. Because the genes for color vision are on the X chromosome, we might predict that if a tetrachromat were found, that person would be a female. In fact, Dr. Gabriel Gordon in the UK has located a tetrachromat, and this woman has both a wider range of sensitivity than normal and can distinguish very small differences in hue that appear identical to you and me. In fact, she may perceive thousands of subtly different hues that we cannot.

Incidence of color vision abnormalities

How common are these color vision deficiencies? Because the genetic information for cones is carried on the X chromosome and males have only one X, males are much more likely to experience color vision deficiencies. Females have two X chromosomes, so information missing on one X is often supplied by the other X. About 8% of males have some form of color blindness, typically red–green confusion, whereas only about 0.5% of females have this problem. Only about 2% of males are dichromats, and most of these are protanopes or deuteranopes. Tritanopia is exceedingly rare and is found in only one individual in 20,000. A true monochromat is even rarer; perhaps one person in 30,000 has no color vision whatsoever and sees the world in shades of gray.

Color deficiency is the result of a recessive trait, so it is uncommon in large and diverse populations. Given that, we would only expect to see a large proportion of color-deficient individuals in a small population that is geographically isolated so that a random mutation would be spread throughout the population. One such population exists. The atoll of Pingelap in the Pacific nation of Micronesia is less than one square mile in total area with a population of about 250 people. As much as 10% of the residents have **achromatopsia**; that is, they have no working cones in their eyes. It is believed that the unusual incidence of achromatopsia on Pingelap is due to the combination of two events. First, a typhoon swept over the island in 1775, leaving only about 20 survivors. Second, it is thought that the ruler at that time, Doahkaesa Mwanenihsed, carried the gene for achromatopsia, which was passed through the generations that descendended from that small and isolated gene pool.

Consequences of color vision deficiencies Most people with color vision aberrations live completely normal lives, although there are a few occupations and situations where color perception really matters. For example, in the early days of automobiles, there was no standardization of the placement of red and green lights in traffic signals. Drivers with protanopia were at great risk for confusing stop and go lights. Eventually, the federal government standardized the placement of red and green lights in traffic signals. If you have normal color vision, you may have to think for a moment to remember what the placement of red and green lights is, but if you have red–green confusion, you already know that red is always at the top of vertical stoplights and at the left of horizontal stoplights.

Negotiating your way through intersections is not the only place where color vision is important. Some obvious examples of careers where color vision plays a critical role include artist and fashion designer. Some occupations that you might not realize have a need for color vision are chef and microbiology lab technician. Professionals in the hospitality industry often have to use color to determine whether meat or other perishables are beginning to spoil. The consequences of accidentally serving spoiled food include illness and even death in the case of infants or immune-compromised individuals. Laboratory pathologists rely on color; after a blood or tissue sample is collected, it is sent to a laboratory where the cells are cultured and typically chemically stained in a way that helps identify whether the cells are cancerous or perhaps representative of a specific bacterial infection. The laboratory technician or pathologist who examines the stained cells must often make a decision based on the color of the tissue or cells after they are stained; a color vision deficit could lead to a misdiagnosis resulting in nontreatment or mistreatment, which could lead to illness or death of the patient.

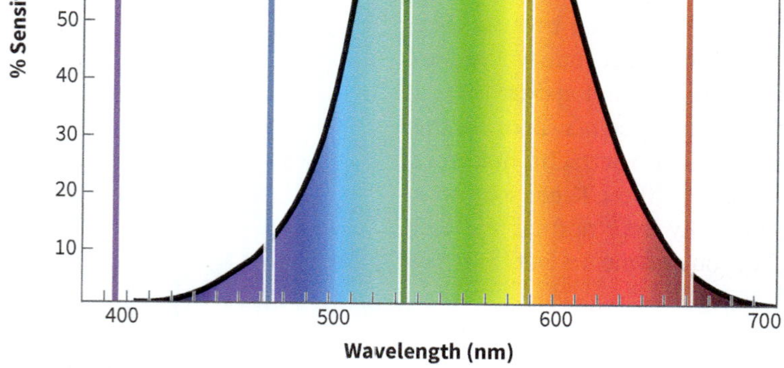

FIGURE 5.25 Chromatic sensitivity function for humans We are not equally sensitive to all parts of the visible spectrum, even when they are presented at the same level of illumination. If two lights were presented dimly and then their brightness was slowly increased, one light at 555 nm and one at 650 nm, you would perceive the 555-nm light long before the 650-nm light.

Color Sensitivity Across the Visual Spectrum

Figure 5.1 provided a view of the visible spectrum as a part of the electromagnetic spectrum, and you know from your basic knowledge of science that our eyes are sensitive to a segment of that spectrum that ranges from around 400 to 700 nm. It would be easy to assume that within that frequency band, we are equally sensitive to all frequencies, but in fact the human eye is not equally sensitive to all frequencies (Figure 5.25).

Imagine a field shrouded in a dense fog; at some distance from you are a hundred individuals, each holding a light that is shining with one of the colors in Figure 5.25. The people and their lights are arranged to reflect the spectrum; those on the far left are purple and blue,

Think About It | Safety and Color Vision

Vesnaandjic/E+/Getty Images

Oregon Department of Transportation/
Wikimedia commons/CC BY 2.0

Your author was asked some years ago by the police chief of a major city to provide expertise in the choice of jackets worn by SWAT teams who might be asked to enter a building in a hostage situation. The police were planning to use tear gas to bring out any terrorists, so the question they asked was, "What color jackets should our people be wearing as people emerge from the cloud of tear gas so that a sniper would immediately identify that person as police and not accidently shoot them?" Before you read on, stop and look again at Figure 5.25 and answer the question based on your knowledge of human psychophysics.

My response was that they should order jackets that are fluorescent lime green, or as close to 555 nm as possible. Take a moment to think of an occupation or object that we would wish to be highly visible and an occupation or object that we would wish to be unobtrusive. What color or colors would you assign to each? Do you see this awareness of the relationship of frequency sensitivity applied to everyday objects, or do you see culture or custom being more important?

In the end they ordered dark blue, because *police wear dark blue*.

Questions

1. **Do you think their choice of a dark blue jacket was wise? Explain your answer.**
2. **What color do you think emergency vehicles and fire trucks should be if the goal is to make the vehicle most visible in a dark or foggy situation?**

and those on far right are orange and then red. Each light is exactly 10 watts. At your command, they begin to move toward you in a perfectly straight line. The first person you will see is holding a light that shines at around 555 nm. We are most sensitive to light at 555 nm, a color most people describe as a fluorescent lime green. Understanding our sensitivity to different colors can have important safety implications, as is demonstrated in **Think About It: Safety and Color Vision**.

The Opponent Process Theory MCAT

Sometimes, the way to tell a story is to keep it simple at first and later let the listener know that things are a bit more complicated. The story behind color vision is a good example, and one problem that has been hidden from you until now is that the trichromatic theory explains much of our color vision, but it doesn't explain all the phenomena we see. For example, a vexing observation for trichromatic theory is that of afterimages. Stare fixedly at the black cross in the top of **Figure 5.26** while you slowly count to 30, and then look at the second cross at the bottom. What did you see?

If you are like most people, you saw a reverse afterimage when you fixated on the second cross. That is, the blue and yellow sides reversed. A similar afterimage occurs with red and green. Trichromatic theory doesn't easily explain what you just experienced. Why are afterimages often in reverse colors? Why is it that some color combinations seem almost to vibrate at their juncture?

Think of the many colors you have seen and the many mixtures such as the reddish-orange of a sunset or the blue-green of the sea. Would you believe me if I said that I know two colors that you have never seen

FIGURE 5.26 Afterimages Trichromatic color theory explains a great deal of our color vision experience, but it does not explain phenomena such as color afterimages. Afterimages can be explained by the opponent process theory.

and that, in fact, these are two color combinations that you can't even visualize? If you doubt me, prepare to be convinced; the two colors you cannot even imagine are a greenish-red and a bluish-yellow. Now that you see that trichromatic theory cannot explain all of the perception of color vision, let us briefly discuss some of what we know about this opponent process.

Color opponents Scientists realized that the color pairs of blue–yellow and red–green are not capable of being simultaneously perceived, and the traditional three cone theory doesn't explain why these colors act as if they are opponents. That is, when one color of the pair is perceived, the other half of the pair cannot be. In the early 1900s, physiologist Ewald Hering realized that some visual phenomena such as the perception of these unique color pairs were best explained by assuming retinal receptor cells acted as if they were either/or. That is, a blue–yellow receptor could fire for either blue or yellow but not both simultaneously, or the red–green receptor could signal red or green, but not both together.

Hering termed his theory of color vision the opponent process theory, and for some time, the trichromatic and opponent process theories competed for primacy. Which is correct? We now know that both are correct but at different levels. The trichromatic theory explains what is happening at the cone (receptor) level, but it appears that as the data from cones are combined at a higher level of the retina, the opponent process theory dominates.

Now that you know a bit about the opponent process theory of color vision, perhaps you can explain why the line between two opponent colors often appears to move ever so slightly. The answer is simple. Our eye is never completely still. It moves slightly even when you have it trained on that line between blue and yellow. The receptors that fire either blue or yellow are therefore constantly in a state of flux, firing either one way or the other. That pattern of firing, when combined with the small motions of the eyeball, create the perception of a line that is ragged and uneven rather than straight.

Color constancy You might well ask how the trichromatic and opponent process theories fit together and why we have an opponent process at all, when three receptors seem sufficient to explain color vision. A simple answer is that the three cones provide a view of color perception at the level of the retina, but higher up in the visual system, a second process involving groups of cells uses information from larger areas to create what is known as color constancy. The opponent process is responsible for this valuable aspect of color vision.

Here is an example of color constancy from your everyday life. You probably have some brightly colored object at hand, perhaps a piece of fruit that you stuck in a pocket for a snack when you left for class. Place that colored object in a bright light; now place it in shadow (but not darkness). Has the brightly colored object changed color? The answer is no; objects do not change color in response to fairly substantial changes in the level of ambient illumination. A bright yellow banana still appears bright yellow, even when you pull it off the shelf in a dimly lit dorm room as you try to get out the door for an early class without waking your roommate. Although it is an illusion that the banana is bright yellow in the dim light, that is a useful illusion; without it, we would have trouble recognizing objects across a wide range of circumstances. A dramatic illustration of the power of the visual system to create the useful illusion of color constancy across different levels of illumination is provided by the checkerboard illusion of Edward Adelson (**Try It Out: The Checkerboard Illusion**).

Try It Out | The Checkerboard Illusion

JunCTionS / Wikimedia / Public Domain

Although colors appear stable during changes in the level of illumination, this is actually an illusion that helps us navigate through the world. Take a moment to view the short video The Checkerboard Illusion, and then answer these questions.

1. One theme in behavioral neuroscience is that our brain can create an illusion that is better than reality. How does color constancy reflect that theme?
2. If you lost the ability to maintain color constancy, how might that affect you?

In the *Stories of the Brain* at the beginning of this chapter, you read about an artist who lost all color vision after a blow to the head. You now know quite a bit about where in the brain color vision is created. Take a moment to review the visual processing stream and make a guess as to where you think the problem might have occurred. You can read more about this individual in "The Case of the Color-Blind Painter" in Sacks (1995); the reference is listed in the Additional Reading section.

5.5 Before You Go On

Section Summary

According to the trichromatic theory of color vision, three separate types of cones, each sensitive to specific wavelengths of light, combine to create a wide palette of colors. Some individuals are either missing a cone or have a cone that operates at a very low level, and this leads to color vision abnormalities. The opponent process theory explains phenomena such as afterimages and color constancy.

Comprehension Check

1. What are the three types of cones most humans have?
   - **a.** short-wavelength, medium-wavelength, and long-wavelength
   - **b.** blue, red, and yellow
   - **c.** small, medium, and large
   - **d.** wavelength, amplitude, and frequency
2. Individuals who are missing a cone or have a cone that does not function normally will ________.
   - **a.** see colors normally but at a much lower intensity
   - **b.** have color vision but will not see differences in color
   - **c.** have trouble telling the difference between red and blue
   - **d.** often be male because males are prone to color vision abnormalities
3. Which color would first become visible in dim light or fog?
   - **a.** reddish orange
   - **b.** lime green
   - **c.** light blue
   - **d.** purple
4. Which of these is an example of color constancy?
   - **a.** A red sweater stays red after many washings.
   - **b.** After staring at a color, you continue to see that color after you look away.
   - **c.** You perceive an apple to be red in dim or bright light.
   - **d.** You cannot perceive red and green at the same time.

Conclusion

In this chapter you learned that your sense of vision is truly wondrous. It provides you with a considerable amount of information about the world. We learned that vision creates a model of the world that may be reliable from day to day and even generally reliable from person to person but that the model we perceive is different from that perceived by some other people and certainly differs from that experienced by other organisms.

As you progress in your study of the human brain, you will find that nature is much more interesting than you might have imagined. If you continue on into clinical neurology, neuropsychology, or cognitive neuroscience, you may encounter individuals who know something is out in their visual world but cannot tell you what that object is, even when they can accurately draw the object. You may encounter individuals who will tell you they are unaware of the presence of objects placed in their visual field yet when asked to guess and point in the direction of an unseen object, they can do so. Finally, you may meet individuals who are unaware of objects in their left visual field. Such odd combinations of symptoms might have appeared incomprehensible to you prior to reading this chapter, but now these deficits should begin to make a little sense. After you digest subsequent chapters, such patients should make even better sense to you.

Critical Thinking Questions

1. Cones provide sharpness of vision, and rods provide notice of objects approaching from the side. Which of these two systems do you think is more primitive and may have developed first in our evolutionary past? Briefly explain your answer.

2. Some snakes can sense infrared (IR) electromagnetic waves, and some insects can see in ultraviolet (UV). IR sensitivity allows snakes to detect heat from other organisms, and seeing in UV helps insects spot predators or prey against a snowy background. Can you think of an advantage or disadvantage for humans to being able to perceive light in IR or UV? Describe a specific situation in your response.
3. You may have noticed while driving at night that the eyes of some animals such as deer and raccoons have a very reflective choroid, whereas humans, hawks, and lions do not. Why might that be?
4. Since the time of Plato and Aristotle, philosophers have pondered whether our senses provide an accurate image of the world or create an imperfect reflection of reality. After reading this chapter, what is your opinion?

Dig Deeper into Behavioral Neuroscience

5.1 An Introduction to Vision

Here is a nontechnical but accurate discussion of the physics of light.

Watzke, M. (2015). *Light: The visible spectrum and beyond.* Black Dog and Leventhal. ISBN-13: 978-1631910067

Our understanding of our sense of vision has dramatically changed over the centuries.

Keele, K. D. (1965, May). Leonardo da Vinci on vision. *Proceedings of the Royal Society of Medicine, 48*(5), 384–390. https://www.ncbi.nlm.nih.gov/pmc/articles/PMC1918888/?page=1

A more detailed overview of the anatomy of the human eye is available here.

Oyster, C. (1999). *The human eye: Structure and function.* Cary, NC: Oxford Press.

5.2 The Retina

These sources are useful to those who want to read more about the retina.

Osterberg, G. (1935). Topography of the layer of rods and cones in the human retina. *Acta Ophthalmologica,* Supplement 6, 1–103.

Kolb, H. (1995). Simple anatomy of the retina. *Webvision.* https://pubmed.ncbi.nlm.nih.gov/21413391/. PMID: 21413391.

People are interested in seasonal effects upon mood, so the ipRGC is currently of great interest both to scientists and clinicians.

Coghlan, A. (2007). How blind people see sunrise and sunset. *New Scientist, 196*(2635–2636), 9. https://doi.org/10.1016/S0262-4079(07)63172-8

5.3 The Subcortical Centers

To read more about the LGN and other subcortical structures, the following is a good source.

Goodale, M., & Milner, D. (2004). *Sight unseen.* New York: Oxford University Press.

5.4 The Visual Cortex

Getting a handle on receptive fields can be confusing. Seeing what Hubel and Wiesel saw in their lab can make the concept easier to understand. Watching the video Cortical Receptive Fields from the original experiment can help you to more easily understand receptive fields.

Our visual cortex is where a great deal of visual perception occurs, especially the more complex aspects.

Mather, G. (2015). The Visual Cortex. School of Life Sciences, University of Sussex.

It is surprising that our perception of what is out there is handled by a different part of the brain that decides where the object is.

Goodale, M. A., & Milner, A. D. (1992). Separate visual pathways for perception and action. *Trends in Neuroscience, 15*(1), 20–25. https://doi.org/10.1016/0166-2236(92)90344-8. PMID: 1374953.

If you were blind until adulthood and then suddenly had your sight restored, would you quickly recognize simple shapes and objects in your surroundings, or would you struggle to comprehend your visual world?

Kurson, R. (2008). *Crashing through: The extraordinary true story of the man who dared to see.* Random House. ISBN-13: 978-0812973686

5.5 Color Vision

The chapter opened with a story about an artist who lost all color vision after a head injury. How that affected the artist's personal life and artistic expression is described in some detail by Sacks, and the inclusion of pictures of art created both before and after the injury is very provocative.

Sacks, O. (1995). The case of the color-blind painter. In *An anthropologist on Mars: Seven paradoxical tales.* New York: Knopf.

Chapter 5 Before You Go On Answers

5.1 Before You Go On

1. b. are all part of the electromagnetic spectrum
2. a. sclera
3. d. fovea
4. c. myopia
5. d. walking down a flight of stairs in the dark

5.2 Before You Go On

1. b. rely on rods more than
2. a. to tell the difference between the green and red signals of a traffic light
3. d. lines and edges

5.3 Before You Go On

1. d. The magnocellular layers are all located in the right LGN, and the parvocellular layers are in the left LGN.
2. c. detects what we see; detects where we see it
3. a. direct our eyes toward a moving object
4. b. The output of each eye is divided between the left and right hemispheres.

5.4 Before You Go On

1. c. it consists of a collection of interconnected processors
2. d. millions of small, discrete processors create a complex visual world
3. d. It upholds the principle that complexity is added at each stage in the visual processing stream.
4. a. where; what

5.5 Before You Go On

1. a. short-wavelength, medium-wavelength, and long-wavelength
2. d. often be male because males are prone to color vision abnormalities
3. b. lime green
4. c. You perceive an apple to be red in dim or bright light.

Chapter 5 Glossary

achromatopsia The total inability to perceive differences in color.

astigmatism A visual defect characterized by blurry vision caused by imperfections in the curvature of the eye or cornea.

choroid The layer of the eye between the retina and the sclera that contains the blood supply.

contrast sensitivity The ability to distinguish small differences in the color and brightness of an object.

cornea The transparent covering for the lens, iris, and pupil.

dichromat An organism whose color vision is created by two receptors with different sensitivity functions.

dorsal stream A cortical pathway in the visual system that is involved in recognizing where objects are in space.

electromagnetic spectrum The range of frequencies of radiation from less than the width of an atomic nucleus to thousands of kilometers.

emmetropia Normal vision, in which the focal plane of an image falls directly on the retina.

extrastriate cortex The region of the occipital lobe next to the primary visual cortex.

figure–ground organization The ability to detect an object as separate from its background.

fovea The region of the retina at the very back of the eye that is responsible for visual acuity.

fundus In the eye, the part at the back of the eye farthest from the pupil.

geniculostriate pathway The main visual pathway that takes retinal input and sends it through the lateral geniculate nucleus and then through the thalamus to the striate cortex of the occipital lobe.

hemianopsia A defect in the visual field characterized by a loss of vision, most commonly only one side of the visual midline.

hemifield One half of the visual field.

hypercolumn The structure formed when ocular dominance columns and orientation columns combine into a larger functional unit.

hyperopia The condition of only being able to see distant objects clearly; also known as far-sightedness.

iris The circular structure responsible for controlling the amount of light entering the eye.

koniocellular cell (K-cell) A type of cell in the LGN whose function is not well understood but that may provide information about brightness, color, and tracking of objects.

lateral geniculate nucleus (LGN) A nucleus of the thalamus that is a relay station in the visual pathway that receives information from the retina and then transmits the processed information to the occipital lobe.

macula A region of the retina at the very back of the eye that is richly supplied with cones.

magnocellular cell (M-cell) A type of cell in the LGN that provides information about motion and edge detection.

monochromat An organism that perceives light in terms of intensity only, without respect to frequency (color).

myopia The condition of only being able to see nearby objects clearly; also known as nearsightedness.

opsin One of a class of light-sensitive proteins responsible for converting the energy in a photon into an electrochemical signal.

optic chiasm The place in the brain where the two optic nerves cross so that each eye sends its signal to both hemispheres.

optic nerve A bundle of axons that transmit information from the retina to the brain.

parvocellular cell (P-cell) A type of cell in the LGN that is sensitive to color and very fine details.

presbyopia Hyperopia caused by aging.

primary visual cortex (PVC) A critical region for visual perception located at the very rear of the occipital lobe; also known as area V1 or the striate cortex.

receptive field A neural structure that creates a maximal response according to the characteristics of the stimulus.

retina A layer at the back of the eye that contains cells sensitive to light, that, when triggered, create a nerve impulse via the optic nerve that goes into the brain.

retinotopic mapping The mapping of visual stimuli to the cortical regions that process those same stimuli in a fashion that preserves the spatial organization of the original stimulus.

sclera The tough, fibrous layer of the eye; also known as the white of the eye.

tectopulvinar pathway A smaller visual pathway that takes retinal input and sends it through the superior colliculus of the tectum and then through the pulvinar nucleus, ending at the occipital lobe.

tetrachromat An organism whose color vision is created by four receptors with different sensitivity functions.

trichromat An organism whose color vision is created by three receptors with different sensitivity functions.

ventral stream A cortical pathway in the visual processing stream that is responsible for identifying what an object is.

CHAPTER 6

Audition, Chemosenses, and Cutaneous Senses

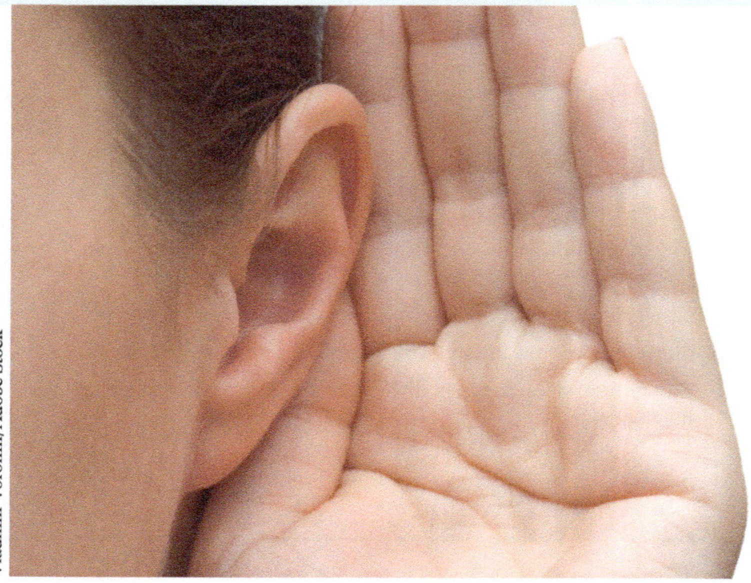
Vladimir Voronin/Adobe Stock

Our hearing, chemosenses, and cutaneous senses, as well as our vestibular system, inform us about our surroundings.

CHAPTER OUTLINE

6.1 Audition: Hearing and the Ear
- The Physics of Sound
  Try It Out: Test the Limits of Your Hearing
- The Anatomy of the Ear
- Transmission of Sound Waves Through the Ear
- Addressing Hearing Deficits
  Check It Out: Experience a Hearing Deficit and a Cochlear Implant

6.2 Auditory Processing and the Vestibular System
- Auditory Processing in the Brain
- Amplitude and Frequency Coding
  Think About It: Auditory Sensitivity and Age
- Auditory Localization
  Try It Out: Experience Auditory Localization
- The Vestibular System

6.3 The Chemosenses: Olfaction
- The Anatomy and Physiology of Olfaction
- Olfactory Receptors
- Individual Differences in Olfaction
  Check It Out: Conversation with an Anosmic

6.4 The Chemosenses: Gustation
- The Anatomy and Physiology of Taste
- Taste Receptors
  Try It Out: Are You a Picky Eater?
- Individual Differences in Taste
  Advances in Behavioral Neuroscience: Linda Bartoshuk and Supertasters

6.5 The Cutaneous Senses
- Pressure and Vibration
  Check It Out: Global Kinesthetic Deficit
- Pain

LEARNING OBJECTIVES

LO 6.1 **Describe** how structures in the ear collect and transfer the energy from sound waves into the auditory system.

LO 6.2 **Describe** how auditory stimuli are transduced into neural signals and explain the vestibular system.

LO 6.3 **Describe** how molecules in the air are interpreted by your olfactory system.

LO 6.4 **Describe** how taste receptors create our gustatory experience.

LO 6.5 **Understand** how receptors in the skin provide important information about the world.

Stories of the Brain

Meredith was a happy child, generally described as having a sunny personality, who was active in sports and loved school. As the oldest child in her family, she often helped her parents out with the care of her younger siblings. Like any child, she was sometimes puzzled by events around her and would ask questions to try to understand how the world works. Although she was bright, there was one thing about the world that she could never understand, and that was when people described what they smelled. When her family or friends talked about the aroma of a freshly baked pie or a holiday turkey roasting in the oven, she was perplexed as she did not know what they were describing. When others spoke about the smell of new-mown grass, a fall bonfire, or the hot smell of an automobile engine that was overheating, Meredith tried hard to associate what she perceived with their description, but somehow it never made sense to her. She wondered if she was missing something in her perceptual world that others took for granted, but when she mentioned this to her parents, they thought that was impossible.

One day while changing a soiled diaper, she had a flash of insight. She put her nose an inch from the contents of the diaper and said to her mother, "I don't smell anything! Now do you believe me?" After a series of tests by a neurologist, Meredith's intuition about her situation was confirmed. She was a congenital anosmic, unable since birth to experience the sensory world of smells that the rest of us travel in every day. Her neurologist can find no sign of damage or tumor in her brain; there is just something missing in the olfactory system that typically provides us with our sense of smell.

Vision may be our most important sense, but our other senses are also invaluable to our survival. Our auditory sense underlies spoken language and provides clues about the identity and location of objects in our local environment. The chemosenses of smell and taste guide us in selecting food that is healthy and avoiding that which is spoiled. Consider the hazards Meredith could face because she cannot smell smoke, a gas leak, or spoiled milk. Finally, the cutaneous senses embedded in our skin provide essential information that helps us avoid damage from objects that are too hot, cold, or sharp. With the understanding that our auditory system as well as our chemosenses and cutaneous senses are valuable assets, we can begin our exploration of these three fascinating sensory systems.

6.1 Audition: Hearing and the Ear

LO 6.1 **Describe** how structures in the ear collect and transfer the energy from sound waves into the auditory system.

If vision is our primary sense, then audition is a close second in importance. We use hearing to detect the stealthy approach of a predator creeping toward us, the click that lets us know we locked the car door, and the relaxing sounds of nature that refresh us as we walk through a park. Hearing also supports language and music. How does our auditory system take physical vibrations in the air and transform them into meaningful sounds?

The Physics of Sound

Sound waves begin as pressure waves, and our sense of hearing is really just a sophisticated way of detecting and analyzing small changes in atmospheric pressure. All the sounds you have ever heard in your life, complex and varied though they might be, can be described at the most basic level as having the characteristics of amplitude, frequency, and timbre.

Pressure waves that are larger in amplitude are perceived as louder, whereas waves of lesser amplitude are perceived as being fainter. Perceived loudness is measured on the decibel (dB) scale, and Table 6.1 provides some examples to help you understand and evaluate

TABLE 6.1 The Decibel Levels of Common Sounds

Decibel level (dB)	Sound
6	Whisper in a library
30	Normal conversation at 3 feet
90	Jackhammer at 50 feet
107	Power lawn mower at 3 feet
140	Jet engine at 100 feet

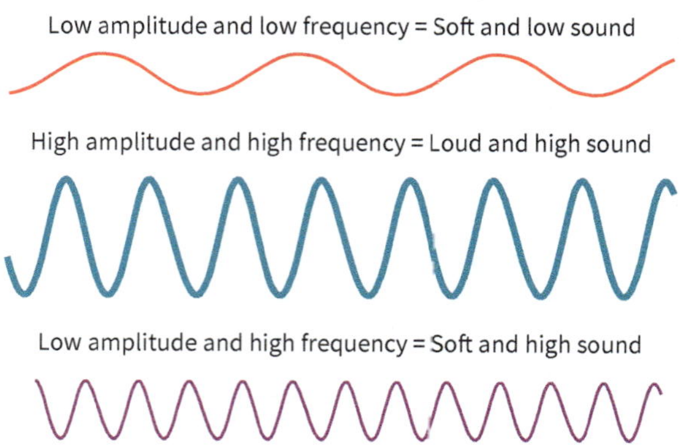

FIGURE 6.1 Amplitude and frequency Amplitude is a measure of the height of a wave, whereas frequency is how many times the wave repeats per second. These three waves illustrate the properties of amplitude and frequency as they relate to sound.

decibel levels in relation to everyday experiences. Sustained exposure to sounds over 90 dB leads to hearing loss, and even short-term exposure to sounds over 140 dB will likely cause permanent hearing loss. For example, musicians who play loud music may begin to lose their hearing in midlife.

While the amplitude of a pressure wave is interpreted as loudness, its frequency is coded as tone or pitch. Frequency is defined as the number of times per second that a sound wave repeats. Waves of greater frequency are perceived as being higher pitched, whereas those of lower frequency are perceived as bass sounds (**Figure 6.1**). Frequency is measured in terms of cycles per second (cps) or more commonly, Hertz (Hz).

The optimal frequency range for human hearing is 20 to 20,000 Hz (20 kHz), but very few people can detect sounds above 15 kHz, and most adults over the age of 60 cannot reliably hear sounds above 8 kHz. You can test your own auditory range with this **Try It Out: Test the Limits of Your Hearing**.

Amplitude and frequency describe many but not all of the perceptual characteristics of sound. Timbre (pronounced *tam-bur*) is defined as the tone color or quality of a sound. You understand timbre if you recognize that the same musical note at the same volume sounds different when played on a piano, flute, and guitar.

With an understanding that our sensory world is built on the foundations of the physical world, we can turn to an overview of how our auditory mechanism transforms those physical vibrations into meaningful sounds.

The Anatomy of the Ear MCAT

The human mechanism for collecting sound stimuli is the ear. The human ear consists of three basic regions: the outer ear, middle ear, and inner ear. The outer ear collects sound waves in the air and channels them into the middle ear, which amplifies these vibrations and transmits them to the inner ear, where sophisticated analysis of the sound waves begins. With

Try It Out | Test the Limits of Your Hearing

You can test your own high-frequency hearing by playing the video **Auditory Sensitivity and Frequency**, which presents you with a series of tones all created at the same physical intensity but at six different frequencies. Then answer the following questions.

1. Were you able to hear all the tones? If so, did you perceive all six as equally loud?
2. Ask someone over age 40 to watch the same video. How were their results the same or different?
3. What are some situations in everyday life when a decreased range of auditory sensitivity would be problematic?

that overview in mind, we can turn to a more detailed description of each of these three regions. The basic structures of the human ear are presented in **Figure 6.2** in more detail.

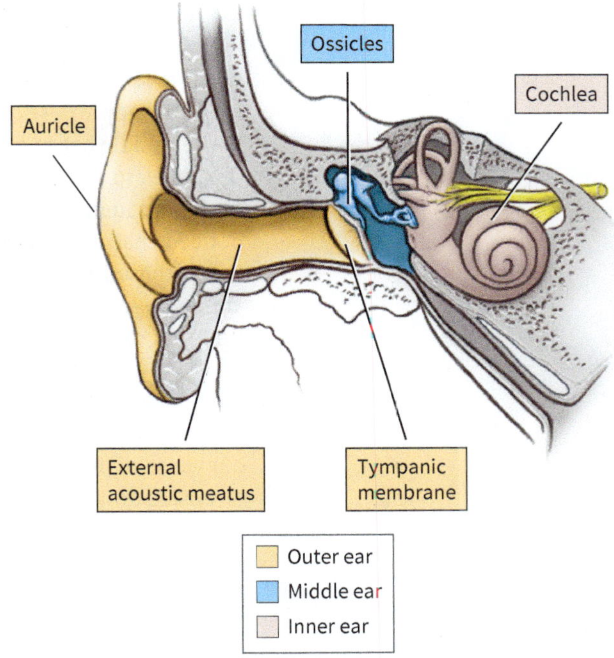

FIGURE 6.2 **The human ear** The outer ear consists of the auricle, external acoustic meatus, and tympanic membrane. The middle ear consists of the ossicles, and the inner ear consists of the cochlea.

The outer ear

The outer ear starts at the external structure commonly referred to as the ear but technically known as the **auricle**; an alternate term for the auricle, the pinna, is also correct but generally only used in reference to animals such as dogs, cats, and foxes. The outer ear ends at the **tympanic membrane**, commonly known as the eardrum. Sound waves traveling through the air touch the auricle first, which is cupped to reflect sound waves into the **external acoustic meatus** (also known as the auditory canal) in the same way that a scoop collects water. Sound waves travel down the auditory canal and strike the tympanic membrane, causing the membrane to move in response to incoming sound waves.

The middle ear

The middle ear consists of the **ossicles**, three bones named after their shape: the **malleus** (hammer), **incus** (anvil), and **stapes** (stirrup, pronounced *stay-peez*). These bones are quite small; the largest is the malleus, which is only 8 mm (0.3 in.) long, whereas the smallest is the stapes at 3.5 mm (0.13 in.). Movement of the tympanic membrane transfers energy to the malleus, which connects to the incus. The incus then transmits its movements to the stapes, which is physically connected to the cochlea. The ossicles are shown in **Figure 6.3**.

What function do the ossicles provide? Why not have the eardrum connect directly to the auditory organ of sensation? The answer is that these three bones mechanically amplify sound by transmitting the movement of the tympani onto a smaller surface area. You understand the principle of mechanical amplification if you know why you would prefer to have someone step on your foot while wearing an athletic shoe over the same person stepping on your foot with a spiked heel. Force is amplified when applied over a smaller area.

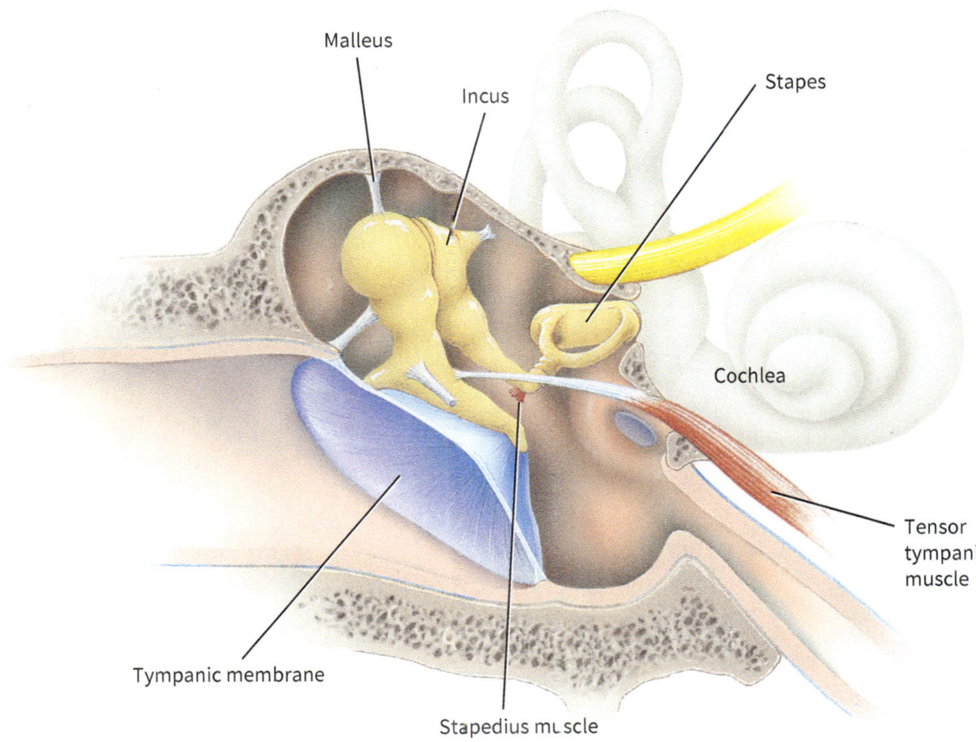

FIGURE 6.3 **The ossicles** The ossicles transmit sound waves from the outer to inner ear, mechanically amplifying them in the process. Movement of the tympanic membrane transfers energy to the malleus, which connects to the incus. The incus then transmits its movements to the stapes, which connects to the cochlea.

Given that the ossicles act as a mechanical amplifier of the auditory signal, we might imagine that there is a mechanism for protecting the middle and inner ears from damage when a very loud sound occurs, and in fact, nature has provided us with the auditory reflex for just that purpose. Excessively loud noises cause the stapedius and tensor tympani muscles in the middle ear to contract, thereby stabilizing the ossicular chain. It takes about 100 ms for this reflex to stabilize the ossicles, so it cannot protect your hearing from the damage resulting from a sudden noise such as a nearby thunderclap, explosion, or gunshot. If you have ever noticed that you found it harder to hear for a while after attending a loud concert, you can thank these two muscles for doing what they can to preserve your hearing. If these muscles are damaged due to disease, the dampening of the stapes is reduced, and the individual may find that many sounds are now uncomfortably loud, a condition known as hyperacusis.

The inner ear In the inner ear are the semicircular canals and the cochlea. The **semicircular canals** sit on top of the cochlea and are not involved in hearing. They are responsible for our sense of orientation in space, including whether we are standing upright, tilted to the side, or upside down. We will discuss them as a part of the vestibular system later in this chapter.

In our discussion of the visual system, the point of transduction from an external physical event into a format the nervous system can process was at the photoreceptors of the retina. For audition, that point where physical energy out in the world enters our nervous system is the **cochlea** (pronounced *coke-lee-ah*). The basic structure of the cochlea is presented in Figure 6.4.

The cochlea is a complex device consisting of many interconnected parts, so to fit it into the skull, nature has folded it into a compact coil that consists of a hollow spiral chamber that is divided into three separate chambers, or scalae: the **scala tympani**, or tympanic chamber; the **scala vestibuli**, or vestibular chamber; and the **scala media**, or middle chamber, also known as the cochlear duct. At the end of the cochlea, the scala tympani and scala vestibuli join together at their tip (known as the **helicotrema**). These two chambers are filled with perilymph, a fluid rich in sodium ions, whereas the scala media contains endolymph, which is rich in potassium ions. There are two membrane-covered openings from the middle ear into the cochlea: the **oval window** leads into the cochlea from the middle ear, and the **round window** is where the energy created by sound waves is dissipated at the end of its journey through the cochlea.

FIGURE 6.4 The cochlea **a.** If the cochlea were uncoiled, you would see the stapes connecting at the oval window, which leads into the scala vestibuli. At the helicotrema, the scala vestibuli meets the scala tympani, which leads to the round window. The organ of Corti runs through the middle. **b.** In cross section, you would see the two fluid-filled chambers with the organ of Corti nestled in between.

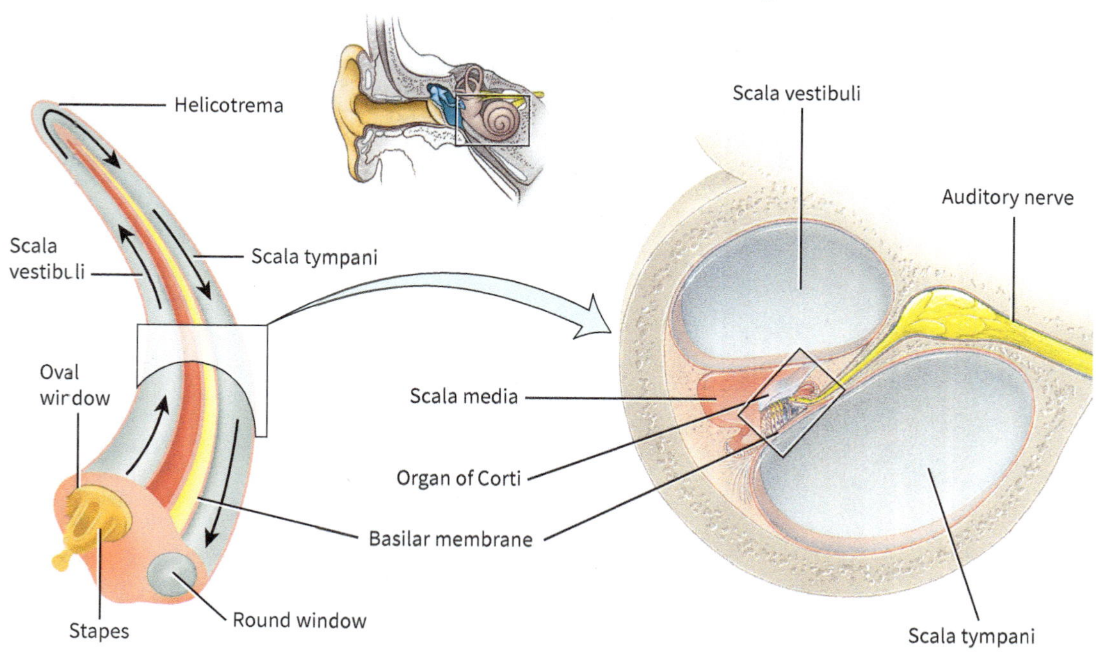

a. Cochlea shown uncoiled

b. Cochlea shown in cross section

The **organ of Corti** exists inside the scala media and consists of the **basilar membrane**, the **tectorial membrane**, and unique receptors known as **hair cells**. These structures are the point where the physical energy that entered the auditory system via the ear becomes a neural impulse that exits the ear via the auditory nerve also known as cranial nerve VIII. Because the auditory nerve consists of the vestibular and the cochlear nerves, it is also sometimes referred to as the vestibulocochlear nerve.

Transmission of Sound Waves Through the Ear MCAT

The human ear is a sophisticated signal-analysis system that converts the physical energy of vibrations in the air into speech, music, and the many sounds that make up our auditory world. The point where physical energy becomes sound is the cochlea, and we now turn to a discussion of this elegant mechanism (**Figure 6.5**).

Sound waves are collected by the auricle and travel down the external acoustic meatus, causing the tympanic membrane to vibrate (**step 1**). From here, these vibrations are transferred to the ossicles in the middle ear (**step 2**). The third ossicle, the stapes, is in direct contact with the cochlea via the oval window. When the stapes oscillates in response to sound, that oscillation creates a wave in the cochlear fluid (**step 3**). As this fluid wave travels from the oval window down the scala vestibuli, it moves the basilar membrane where the hair cells reside (**step 4**). When this wave reaches the helicotrema, it reverses and continues back up the scala tympani (**step 5**). Finally, the energy created by the action of the stapes against the oval window is dissipated against the round window (**step 6**). Picture that when the membrane of the oval window is pushed in by an energy wave, the membrane of the round window is pushed out. This coordinated vibration of the two membranes in opposite phase allows the fluid inside the cochlea to move the basilar membrane, creating the perception of sound.

Having now established how sound waves are transmitted through the ear, we will take a closer look at what happens inside the cochlea (**Figure 6.6**). Hair cells on the basilar membrane are an essential part of the process of transduction from a physical force to a neural signal. The basilar membrane has hair cells on the dorsal surface that extend outward toward the tectorial membrane. The tectorial membrane is less of a membrane and more of a gelatinous

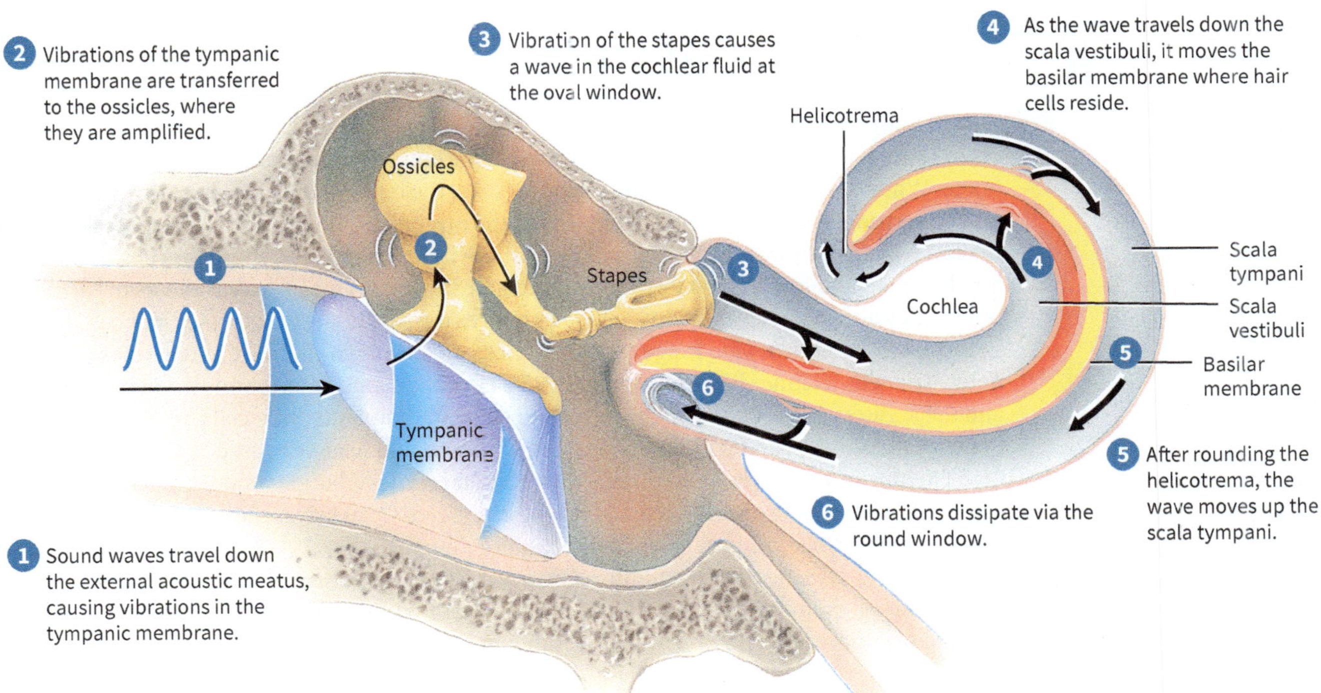

FIGURE 6.5 Transmission of sound waves through the ear The movement of pressure waves is converted into the perceptual experience of sound in the cochlea.

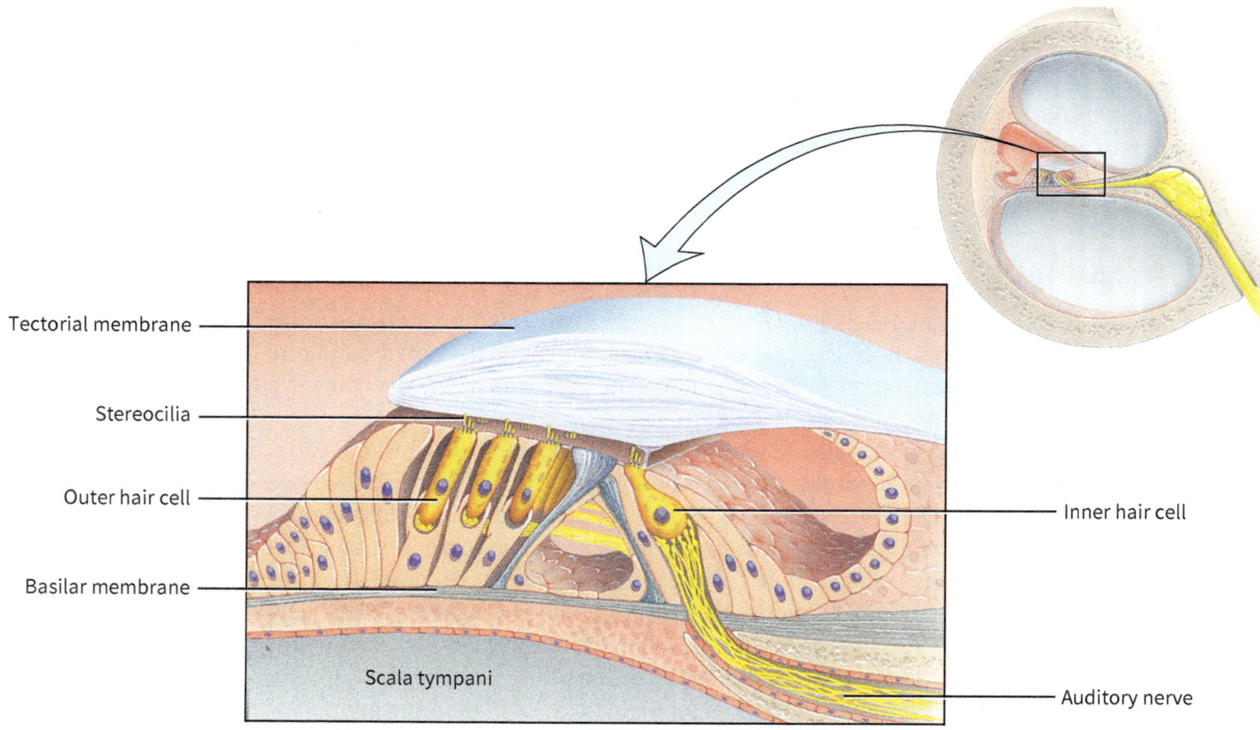

FIGURE 6.6 **The organ of Corti** A cross section of the organ of Corti is expanded to show the details of the arrangement of the basilar and tectorial membranes and hair cells. Hair cells on the basilar membrane move in response to movement of the endolymph. As they do, the stereocilia atop the hair cells stimulate the tectorial membrane, creating the signal our brain interprets as sound.

construction that is anchored at only one end to the basilar membrane. Imagine an underwater cave with kelp that stretches from floor to ceiling. As waves reach into the cave, the kelp gently moves to and fro. That image gives you a good idea of what is occurring in your cochlea with every sound you hear. These fluid oscillations are the physical energy that is converted into sound inside the cochlea.

There are two varieties of hair cells, outer and inner, and they are arranged in three rows of outer and one row of inner hair cells. Hair cells are embedded in the basilar membrane and topped with thin, finger-like projections called **stereocilia** (singular *stereocilium*). The tallest of the outer hair cells have the tips of their stereocilia embedded in the overhead tectorial membrane. When stimulated by a passing pressure wave, the outer hair cells respond by shortening and elongating, thereby physically amplifying the original motion of the basilar membrane to the tectorial membrane, which brushes against the inner hair cells. Outer hair cells selectively amplify quiet and low-frequency sounds. When the outer cells move against the tectorial membrane, they alter its natural rigidity, which enhances the sensitivity of the inner hair cells. Inner hair cells respond to a wide range of frequencies, and although there are three times as many outer hair cells as inner hair cells, roughly 95% of the synaptic connections between hair cells and the auditory nerve belong to the inner hair cells. Without outer hair cells we would struggle to hear faint sounds, and without inner hair cells we would hear very little at all.

The role of the tectorial membrane in audition is not well understood, but its presence is essential to normal hearing. The signal from the organ of Corti leaves the cochlea and provides input to higher processing centers via the auditory nerve.

Just how sensitive is our auditory system? A pressure wave that moves the tympanic membrane by no more than the diameter of an atom will create a signal in the brain. If our hearing was more sensitive than that, the random motion of molecules on our eardrum would bathe us in a constant murmur of noise. The difference between the faintest sound we can detect and the loudest sound we can tolerate is a range of more than a million in amplitude.

Our ancestors evolved in a much quieter world, where the ability to detect a rustle of prey or predator in the tall grass was literally a matter of life or death. By contrast, our modern world is a cacophony of stimulation. Car horns, jackhammers, and ear buds offer more input than our hair cells and their stereocilia were ever designed to handle. Unfortunately, hair cells damaged by excessive sound or age do not grow back, although modern technologies provide ways to cope with hearing deficits.

Addressing Hearing Deficits MCAT

Now that you understand the complexity of the structures within the ear, you can imagine the variety of ways for some failure of this mechanism to cause hearing problems. According to the National Institutes of Health, about 15% of American adults report some trouble hearing. Hearing loss can occur as a result of damage at any stage of the path that sound waves take from the auricle to the cochlea or at the auditory nerve. For adults, hearing loss typically results from continued exposure to loud noises from lawn mowers, power tools, leaf blowers, snow blowers, and music, be it via headphones or in live venues. Gradual damage to the hair cells in the cochlea is a natural consequence of aging. Congenital hearing deficits (those present from birth) are more rare, and addressing these presents special challenges.

Most people imagine that having a hearing disability means that the sounds around you are not loud enough. If that were the case, then simply amplifying the signal would cure many if not most hearing problems. Having read this chapter, you probably are able to see that the situation is generally more complex than that. For many individuals suffering with a hearing loss, the cause is a loss of sensitivity to some frequencies but not others, or a difficulty separating the signal from background noise. Turning up the volume of the signal is not helpful in such cases. You will see that your understanding of auditory anatomy informs the discussion of remedies for hearing deficits.

Hearing aids are the most common way to address hearing deficits. Although advances in technology have led to better performance and also better aesthetics, these devices are inherently limited for many individuals because their mode of operation is to amplify the original auditory signal outside the ear. As we have discussed, for many auditory problems, this is not a satisfactory solution.

Cochlear implants are devices that bypass the outer and middle ears and feed a signal directly into the cochlea. These devices consist of an external microphone that is attached to the skull and a processor to analyze the signal, which is then transmitted through the skull to a receiver and stimulator attached to bone (**Figure 6.7**). Finally, the sounds hit an array of electrodes that are woven into the cochlear nerve. You might immediately recognize that more electrodes translate into better resolution, especially of speech sounds. Given the size of the cochlea and the difficulty of inserting multiple electrodes into the nerve, 22 electrodes is the current limit.

Many who receive an implant report that the quality of their lives is significantly enhanced, especially for speech comprehension. Because cochlear implants are specifically designed for speech, however, wearers often report that music perception is not improved and may actually be disturbed. Many of us have an inaccurate understanding of what a person with a hearing deficit experiences and also overestimate the improvement that a cochlear implant provides. You can get a more accurate understanding of both issues in this **Check It Out: Experience a Hearing Deficit and a Cochlear Implant.**

Not every hearing-impaired individual is a good candidate for a cochlear implant. The age of implantation is a significant factor for patients who are congenitally (from birth) hearing impaired. The developing brain benefits from early exposure to complex auditory stimulation, so early intervention leads to considerably better success than late intervention. Individuals who are profoundly deaf and receive an implant as adults often report that the device is disruptive. This is likely due to a lack of experience with basic speech sounds (known as phonemes) while the brain was still in a plastic state.

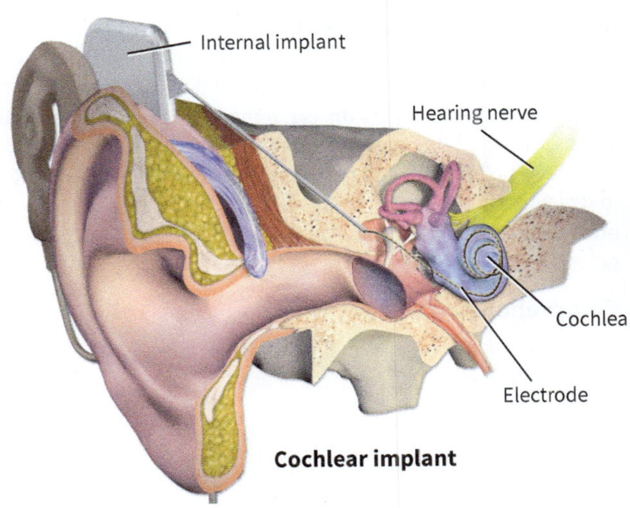

FIGURE 6.7 Cochlear implant A cochlear implant consists of a microphone and sound processor mounted to the outside of the skull and an internal processor implanted in the head that sends its signal to an electrode array implanted in the cochlear nerve.

Check It Out | Experience a Hearing Deficit and a Cochlear Implant

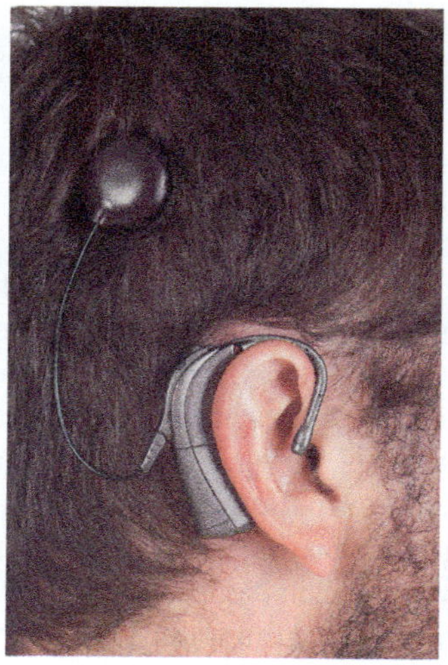

ELizabethHoffmann/iStock/Getty Images

For those of us fortunate enough to have good hearing, it can be difficult to imagine the reality that others face. Watch the video **Understanding Hearing Deficits**, and then answer these questions.

1. The first example contrasted normal speech perception with what an individual with a common frequency deficit might experience. Would simply speaking louder result in better speech perception for an individual with a significant hearing deficit? Explain your answer.
2. In the second example, which simulated auditory perception via a cochlear implant, you experienced the enhancement that additional electrodes add by increasing the spectrum of sound frequencies sampled. At what point were you able to reasonably comprehend the speech? Explain why no one has created an electrode implant with 100 electrodes.
3. In the final example, you heard a simulation of music perception through a cochlear implant. Which of these musical genres (Norwegian thrash death metal, hip-hop, and John Philip Sousa marches) would a person with cochlear implants be most likely to appreciate and why? You may need to search for a sample of each genre to answer the question.

You now have an overview of how a physical event in the world creates a pressure wave that is converted into a neural signal through a chain of events that started with the tympanic membrane and ended with the organ of Corti. Where does this signal travel from here, and how does it provide us with information about loudness, tone, and location? The answer is quite interesting.

6.1 Before You Go On

Section Summary

What we experience as sound begins as a pressure wave in the air that can be characterized by its amplitude (perceived as loudness) and frequency (perceived as pitch). The ear can be divided into the outer ear, the middle ear, and the inner ear, which includes the cochlea. When a sound wave enters the ear, it moves the tympanic membrane and ossicles, and this movement is transferred to the cochlea, where stimulation of the hair cells creates the perception of sound. Damage to structures in the ear can lead to hearing loss, and cochlear implants provide an avenue for some individuals to regain partial auditory perception.

Comprehension Check

1. The amplitude of a sound wave determines _____, and its frequency determines _____.
   - **a.** loudness; pitch
   - **b.** pitch; loudness
   - **c.** timbre; loudness
   - **d.** pitch; timbre
2. The external ear consists of which structures?
   - **a.** auricle, external acoustic meatus, and tympanic membrane
   - **b.** auricle, osculum, and eardrum
   - **c.** auricle, scala media, and tympanic membrane
   - **d.** pinna, auditory meatus, and auditory nerve
3. What are the three bones of the middle ear?
   - **a.** handle, hammer, and ring
   - **b.** hammer, anvil, and stirrup
   - **c.** ring, hammer, and anvil
   - **d.** handle, bridle, and ring
4. What is the place in the auditory system where a physical action becomes a sensory or psychological reality?
   - **a.** ossicles
   - **b.** auditory nerve
   - **c.** helicotrema
   - **d.** cochlea
5. One important advantage of a cochlear implant over traditional hearing aids is that it _____.
   - **a.** is not noticeable by others
   - **b.** is more economical than a battery-powered device
   - **c.** sends its signal directly to the auditory nerve
   - **d.** services only one ear, thereby leaving the other ear free

6.2 Auditory Processing and the Vestibular System

LO 6.2 Describe how auditory stimuli are transduced into neural signals and explain the vestibular system.

So far we have discussed how sounds waves move through structures in the ear to reach auditory receptors, but you know from our consideration of the visual system that there is much more to the creation of perception than just the stimulation of a sensory receptor. In audition just as with vision, a pathway consisting of several processing structures begins at the sensory receptor and ends in a region of the cortex dedicated to the analysis of that sensory modality. In this next section we trace the progress of the auditory signal from the cochlea into the brain. We end the section with a brief discussion of a related sensory system, the vestibular system.

Auditory Processing in the Brain MCAT

Just as visual processing begins at receptors in the retina and proceeds through a series of processing centers in the brainstem, midbrain, and thalamus before it arrives at the visual cortex for elaborate processing, quite a bit of auditory processing occurs upstream from the auditory nerve.

The path to comprehension of sound proceeds from the auditory nerve into the cochlear nuclei in the medulla, superior olivary nucleus, and the lateral lemniscus in the brainstem, the inferior colliculus in the midbrain, the medial geniculate body in the thalamus, and finally the primary auditory cortex in the temporal lobe (**Figure 6.8**). Each of these centers in the auditory

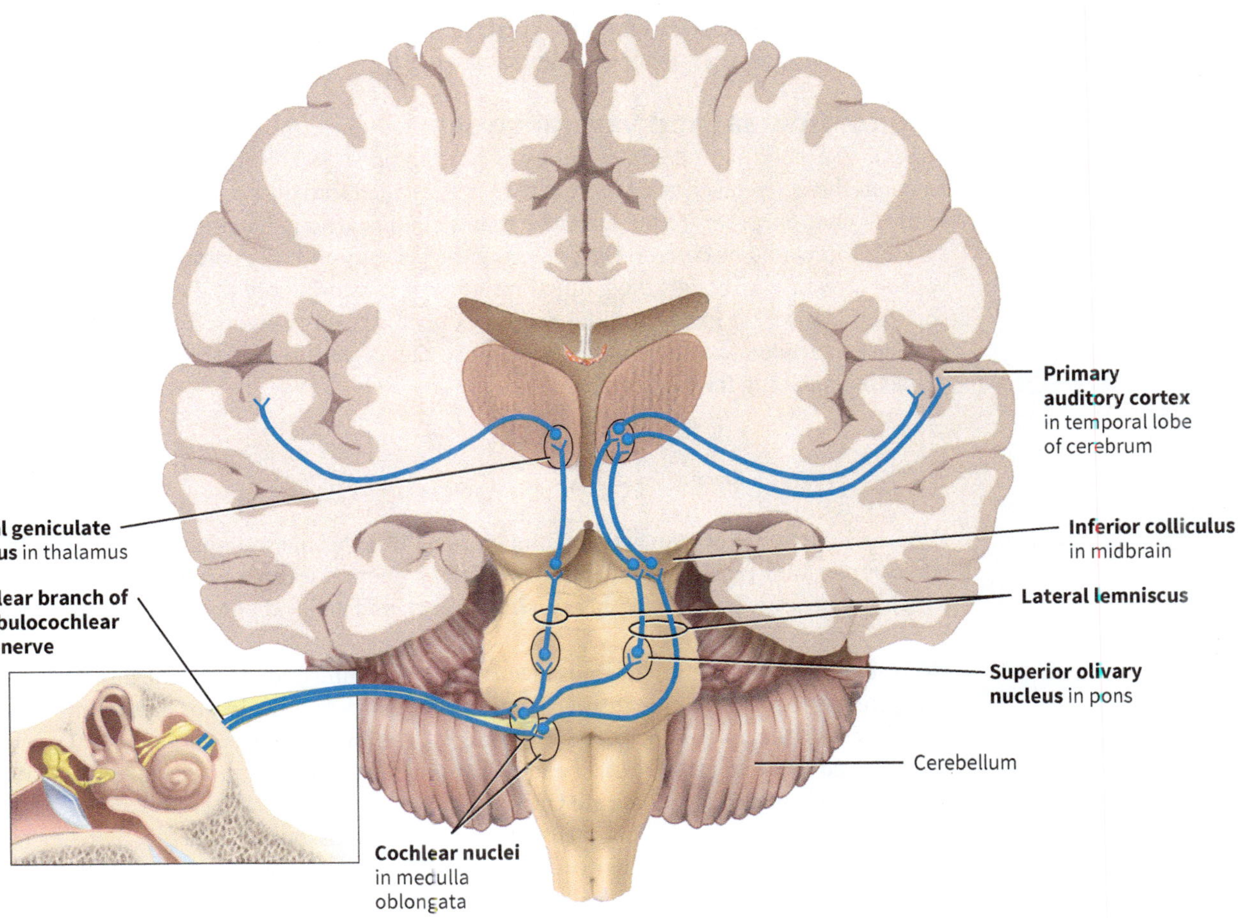

FIGURE 6.8 The auditory pathway The signal from the cochlear half of cranial nerve VIII travels up through a series of nuclei in the brainstem, midbrain, and thalamus until it reaches the primary auditory cortex.

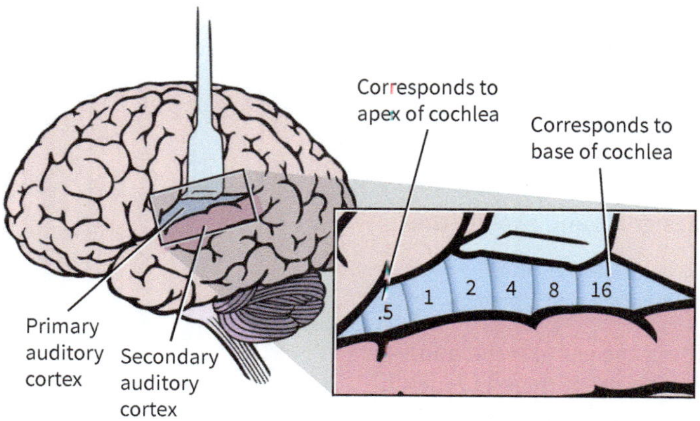

FIGURE 6.9 The primary auditory cortex with tonotopic map The primary auditory cortex is organized tonotopically with lower frequencies being processed toward the rostral side and higher frequencies processed toward the caudal side. Frequencies are shown in kHz.

path between the organ of Corti and the cortex performs some processing operation that cleans up the raw signal and then hands off a signal that contains more information and less noise to the next stage in the auditory processing stream. It is worth a moment to reflect on the fact that the auditory processing stream has obvious similarities to its visual counterpart. In both vision and audition, the path from receptor upstream passes through the brainstem. Vision engages the superior colliculus and then the lateral geniculate nucleus, whereas auditory information passes through the inferior colliculus and then the medial geniculate nucleus. In both cases, the processing pathway ends in a specialized region of the cortex. With that concept in mind, let us proceed to the auditory counterpart to the occipital lobe, the primary auditory cortex.

The primary auditory cortex

In audition as with vision and other senses, perception is not complete until the input from the receptors and the upstream processing centers reaches a specialized region of the cerebral cortex. For audition, this region is the **primary auditory cortex (PAC)**. We will discuss the PAC in more detail in another chapter dedicated to the neural basis of language and music comprehension. For our current purposes, it is enough to say that the output of the auditory processing stream terminates in the superior aspect of the temporal lobe.

One point about the PAC worth mentioning is that there is a physical relationship between location on the PAC and tone perception. Signals from the base of the cochlea report to posterior regions of the PAC and are perceived as high notes, whereas signals from the apex of the cochlea report to the anterior PAC and are experienced as low notes (Figure 6.9). This physical correspondence between the frequency of the tone and the location of processing on the auditory cortex is known as a tonotopic map.

Auditory dorsal and ventral streams

In our discussion of the human visual system, you saw that a complex series of processing steps culminated in the arrival of the signal at the occipital cortex, where it then was divided into dorsal and ventral streams, each with its own purpose. If you surmised that a similar arrangement exists with the auditory system, you would be correct. Data entering the primary auditory cortex exits into dorsal and ventral streams. Although their functions are not as thoroughly understood as their visual counterparts, we do know that the auditory dorsal stream is involved in sound localization and speech articulation, whereas the ventral stream is involved in recognition of sounds and extracting meaning from speech.

Amplitude and Frequency Coding MCAT

How does our sense of audition turn the motion of the basilar membrane and hair cells into our complex and rich world of sound we enjoy? There are three proposed mechanisms by which frequency is coded: rate theory, place theory, and volley theory.

Rate theory

The most obvious mechanism for coding frequency is the direct translation of the frequency of oscillation into a perceptual frequency or tone. In this model, movement of the hair cells 25 cps would be perceived as a 25-Hz or low-bass tone. Similarly, 800 cps would be perceived as a tone of 800 Hz. The idea that the rate of fire of hair cells directly creates the perception of tone is known as **rate theory**.

Rate theory explains our ability to perceive some sounds, but the idea that there is a one-to-one correspondence between hair cells firing and frequency perception is problematic. This is because the neurons serving hair cells cannot fire faster than about 1,000 times a second. If rate of fire created the totality of our sense of hearing, then we could not experience any sound over 1,000 Hz. This would be a problem because human speech typically ranges from 125 to 8,000 Hz.

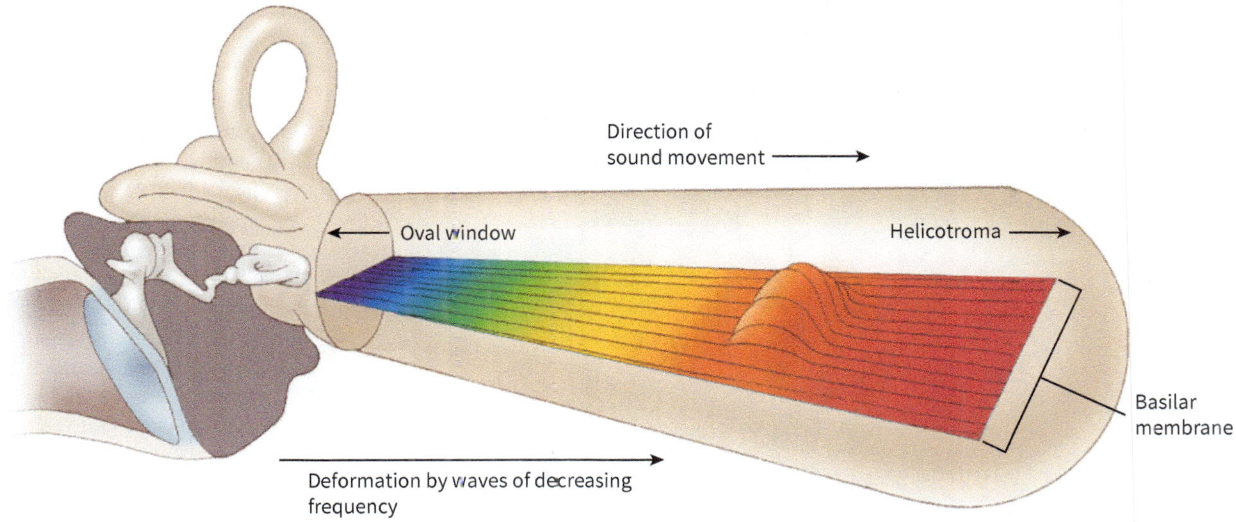

FIGURE 6.10 The place theory of audition The action of the stapes against the oval window causes movement in the cochlear fluid that is transferred to the basilar membrane and causes it to deform. The location of the deformation of the basilar membrane is coded by our nervous system as frequency, where stimulation closer to the narrow base of the basilar membrane near the oval window is coded as a higher tone and stimulation closer to the wider apex near the helicotrema is coded as a lower tone.

Place theory Research into the auditory system has led us to believe that there is another mechanism for creating our auditory world that provides sound perception of stimuli above 1,000 Hz. **Place theory** proposes that there is a general relationship between the point of stimulation on the basilar membrane and sensory experience.

The basilar membrane deforms in response to the frequency of the sound, much like the wake of a motorboat changes according to the speed of the boat. You may have noticed that a slow-moving boat creates a wake that projects off the stern of the boat in a narrow V, but as the boat picks up speed, the wake expands out into a much wider V shape. In a similar fashion, the vibration corresponding to a higher-frequency sound enters the fluid in the cochlea and creates what is known as a standing wave on the basilar membrane. Lower-pitched sounds create this wave near the wider end of the basilar membrane, whereas higher-pitched sounds create a deformation closer to the tip (Figure 6.10). Amplitude is also coded by deformation of the basilar membrane in that a larger wave is interpreted as a louder sound.

Volley theory There is one last perspective on how receptors on the basilar membrane create our complex world, and that is the **volley theory**, which proposes that receptors firing slightly out of phase with each other can respond to a stimulus above the maximum firing rate of any one receptor (Figure 6.11). Rate theory proposed that perception of a tone below 1,000 Hz was created by receptors firing at that rate. Imagine for a moment that you are an auditory receptor and you fire in response to a stimulus by clapping your hands. You are able to clap up to 1,000 times a second, but no faster. How then would you signal a sound of 2,000 Hz? You could convince the receptor next to you to clap just after you do. The two of you together can now accommodate sounds up to 2,000 Hz. With two such accomplices, each one clapping just a bit later than the previous one, you could theoretically accommodate sounds up to 3,000 Hz. You now see the principle of volley theory.

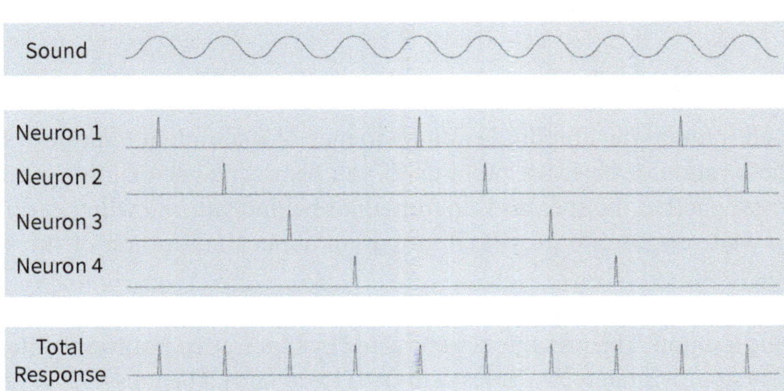

FIGURE 6.11 The volley theory of audition Phased coordination of firing by auditory receptors can provide a rate of response that exceeds the maximum firing rate of any single receptor.

Think About It | Auditory Sensitivity and Age

Drazen/Adobe Stock

As we age, our ability to detect high-frequency sounds declines. This fact of life has been applied to the real world in a very creative way. Store owners sometimes complain that teenagers loiter around the front door of an establishment without actually buying anything, and that their presence can drive away older customers. How do they move the teenagers to another place without calling the police and making a scene? One company has turned the superior auditory capabilities of the youth into a disadvantage. They market a device known as the Mosquito that constantly emits a loud tone at 17.4 kHz. Because of differences in the range of their hearing, teenagers hear a piercing screech that is unbearably annoying, whereas older patrons hear nothing at all. Broadcasting this tone outside a business should prevent teenagers from congregating there.

They say that turnabout is fair play, so it is only reasonable that the same concept used to the disadvantage of the youth should be turned against the older generation. Some adolescents have learned to use a tone about 15 kHz as a ringtone for their cellphones in places such as the classroom or workplace when cellphone use is discouraged or not permitted. A ringtone at this frequency would be perceived by the young student but not by the older teacher or administrator due to their age-related loss of high-frequency response.

Questions

1. **Do you think the use of a high-frequency tone to deter adolescents from loitering near a business is an ethical or unethical use of technology? Briefly explain your answer.**
2. **Do you think the high-frequency ringtone is an ethical use of technology? Is it different from use of the Mosquito? Briefly explain your answer.**
3. **What other situations can you think of where age-related differences in hearing could be used to the advantage of one group or another?**

We have now discussed three possible mechanisms for coding frequency: receptors all firing in concert with a specific frequency (rate theory), the membrane being deformed at a specific point by a wave traveling down its surface (place theory), or multiple receptors firing slightly out of phase (volley theory). Which theory is correct? The answer is that all three are correct, but each theory describes what happens in a different frequency range. Rate theory operates between 20 and 1,000 Hz, volley theory between 500 and 5,000 Hz, and place theory contributes a response to sounds above 5,000 Hz. These different mechanisms provide us with a wide range of sensitivity, optimally between 20 and 20,000 Hz (or 20 kHz), but age and exposure to loud sounds can degrade that range, especially at the higher frequencies (**Think About It: Auditory Sensitivity and Age**).

Auditory Localization

In addition to interpreting the amplitude and frequency of a sound, our auditory system can inform us of the location of objects around us. If you hear someone calling your name, you instinctively know whether the speaker is in front of or behind you and whether on your right or left. How do we use sound to provide information about the location of the source of a sound in relation to the perceiver?

The answer is quite interesting. **Figure 6.12** depicts an object that is off to one side of the perceiver creating a sound. The pressure wave created by that source is traveling literally at the speed of sound, or approximately 770 miles per hour (1,240 kph). The pressure waves traveling from the source arrive at the perceiver's right ear just milliseconds before they reach the left ear because path X_R is slightly shorter than path X_L. The left and right superior olivary nuclei (which were presented in Figure 6.8) compare the arrival times of the two signals, which may be only a few milliseconds apart, create an auditory image of the location of the source in relation to the perceiver.

Although we are generally quite accurate at identifying the direction from which a sound originated, there are exceptions to this rule, and if you consider Figure 6.12 for just a moment, you will see the circumstances that lead our superior olivary nuclei to be confused. Sounds that originate directly in front of and directly behind the perceiver reach both ears at the same time, so distinguishing between these two origins is difficult. You can explore our ability to localize sounds for yourself in **Try It Out: Experience Auditory Localization**.

Auditory localization is so effective that individuals who are blind or severely visually impaired can develop the ability to navigate by sound alone. They generate a sound such as a vocal click or a tap with their cane and then use the echo from objects in their surroundings to create an auditory image of the world. Through echolocation, such individuals may be able to distinguish chairs from tables, a car from a tree, and so on. Furthermore, when recordings that simulate what each ear of an echolocator was hearing as they navigated through an environment were played through headphones to other blind echolocators, those participants perceived the same objects in the environment as the original participants. MRI studies on human echolocators suggest that in blind individuals, the areas of the brain normally reserved for analyzing visual stimuli are enlisted to create an auditory scene.

By this time you have likely developed an appreciation for the complexity and the elegance of the auditory system. Pressure waves in the air are processed by a complex mechanism that converts a physical force into the sounds that we experience as loudness, pitch, and even direction. Before we move on to the chemosenses, we need to cover another system that shares real estate in the ear and brain with the auditory system.

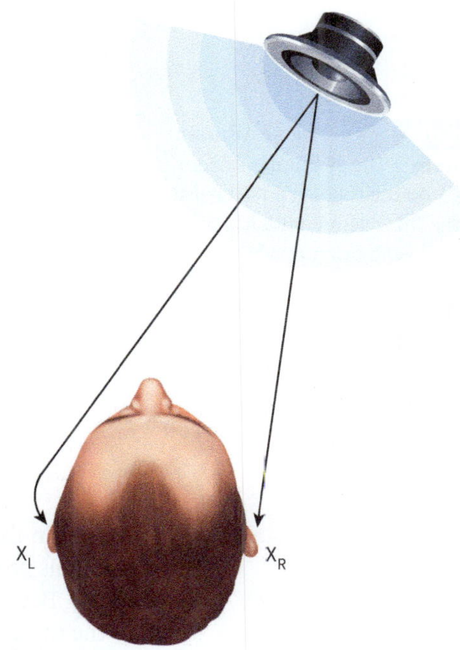

FIGURE 6.12 Sound localization Sound from a loudspeaker reaches a person's ears at slightly different times if the source of the sound is not directly ahead or behind. This offset in stimulus arrival time is used by the two superior olivary nuclei to calculate the location of the source.

The Vestibular System MCAT

Although not related to hearing, the **vestibular system** is physically attached to the cochlea in the inner ear. Its function is to detect motion, the position of your head with respect to your body, and the spatial orientation of your body, so it is essential in maintaining balance and posture.

The vestibular system consists of two interconnected systems—the three semicircular canals and two otolith organs known as the utricle and saccule—and these two systems have complementary functions. The semicircular canals (also known as the vestibular labyrinth) are three semicircles or loops filled with endolymph. Each canal is oriented in one of the three axes along which your head can move: up and down, back and forth, and side to side. At the base of each of the three canals is a small bulbous structure called the ampulla, which contains endolymph, and the crista, which contains hair cells topped with stereocilia (**Figure 6.13**). When your head moves, that movement is transmitted to the endolymph in the ampulla and the hair

Try It Out | Experience Auditory Localization

You can demonstrate our remarkable ability to localize sounds by asking a friend to sit down and close their eyes; you stand in front and create an auditory signal by snapping your fingers or clicking your fingernails together. Ask your friend to point to the sound source and then move the source around their head as they point to the source of the sound. Then answer these questions.

1. How accurate was the individual at pointing to the source of the sound?
2. A crime is committed in which a gunshot is heard by several witnesses. Some witnesses swear the gunshot come from directly in front, and others are equally sure it came from behind. Were some of these "earwitnesses" lying, or could there be another explanation? Explain your answer.

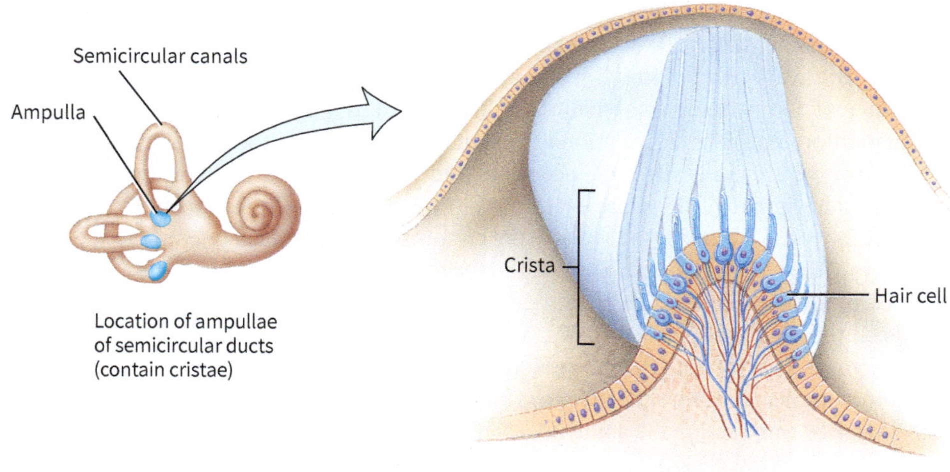

FIGURE 6.13 The vestibular system The vestibular system sits atop the cochlea and consists of the three semicircular canals, each oriented along a specific axis. At the base of each canal is an ampulla and crista with hair cells. Movement of endolymph stimulates the hair cells and thereby provides information about your balance and orientation in space.

cells in the crista fire, creating a signal. The sum of the responses from the three semicircular canals provides an accurate picture of the orientation of your head in relation to the world.

Between the cochlea and the semicircular canals of each ear are the two otolithic organs: the **utricle** and **saccule**, two endolymph-filled chambers lined with hair cells that are embedded in the otolithic membrane (**Figure 6.14**). These hair cells are topped with tiny particles known as **otoliths**. Movement in a linear plane creates a corresponding movement of the endolymph, thereby stimulating the hair cells, which we then perceive as forward or backward motion. A simple image will make this mechanism east to understand. Imagine for a moment that you are holding upright a flexible rod that is topped with a small weight. As you move back and forth, the weight at the end of the rod causes the rod tip to amplify the movements. When you feel acceleration or deceleration in a car, airplane, or roller coaster, it is your otoliths that are providing you with that information.

The viscosity of the endolymph is such that the hair cells lining the semicircular canals and the otoliths are precisely balanced against the constant pull of gravity, but there are circumstances where disruptions to the signaling can occur with unfortunate results. In everyday life, the information from our vestibular system and our eyes matches up, but the jerky and unpredictable motions of amusement park rides such as roller coasters can lead to dizziness and nausea. When astronauts travel into space and are released from the gravitational pull that their vestibular systems expect, this sends contradictory information into the brain. The result is nausea and vomiting, which, in a spacesuit, would be fatal. Finally, when an Earth-bound individual imbibes more alcohol than is wise, that alcohol is carried by the circulatory system throughout the body, and some of it diffuses into the vestibular system, thereby changing the viscosity of the endolymph and causing the hair cells to send confusing and contradictory messages. The resulting dizziness, known colloquially as the spins or the whirlybeds, typically leads to nausea and vomiting.

The output of the vestibular system is the vestibular component of the auditory nerve (which you may recall is also referred to as the vestibulocochlear nerve). Like the cochlear portion, the output of this system then travels up into the brain for additional processing.

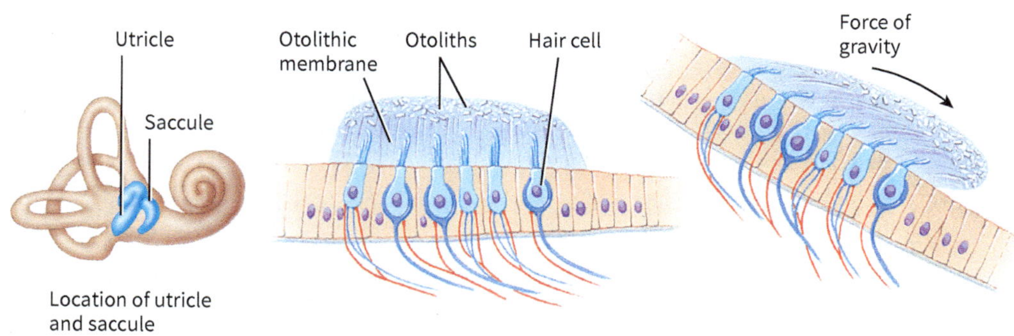

FIGURE 6.14 The utricle and saccule Hair cells in the utricle and saccule respond to the movement of the endolymph and thereby provide information about acceleration and deceleration.

6.2 Before You Go On

Section Summary

Auditory signals generated in the cochlea are processed in the primary auditory cortex. The rate of receptor firing in addition to movement of the basilar membrane creates our ability to hear sounds over a wide range of amplitudes and frequencies. Our brain interprets subtle differences in the arrival times of sounds at the two ears as information about the position of the source. The vestibular system uses the movement of fluid in the inner ear to generate information about orientation, acceleration, and deceleration.

Comprehension Check

1. What is the path of auditory processing in order?

- **a.** from the cochlea to the thalamus to the frontal lobe
- **b.** from the cochlea to the superior colliculus to the parietal lobe
- **c.** from the cochlea to the medulla to the orbitofrontal lobe
- **d.** from the cochlea to the brainstem to the temporal lobe

2. Perception of a stimulus at 400 Hz is created by the _____, whereas perception of a stimulus at 8,000 Hz is created by the _____.

- **a.** basilar membrane; hair cells
- **b.** tympanic membrane; basilar membrane
- **c.** secondary auditory cortex; organ of Corti
- **d.** hair cells; movement of the basilar membrane

3. We are generally very accurate at detecting the direction of a sound, but one exception is sounds that _____.

- **a.** originate directly ahead or behind us
- **b.** originate 45 degrees to the side
- **c.** are too loud to properly analyze
- **d.** are speech, which are processed differently

4. If I lean over to one side or the other, what structure is providing me with information about my position?

- **a.** utricle
- **b.** saccule
- **c.** semicircular canal
- **d.** cochlea

5. If you are a passenger in an automobile moving at 30 mph (50 kph) and your eyes are closed, what structure will alert you when the car begins to decelerate?

- **a.** semicircular canal
- **b.** otolith
- **c.** cochlea
- **d.** basilar membrane

6.3 The Chemosenses: Olfaction

LO 6.3 Describe how molecules in the air are interpreted by your olfactory system.

The chemosenses play an important if often unrecognized role in dealing with the world. Unlike our sense of vision and hearing, which allow us to perceive objects at a distance, our chemosenses requite direct physical contact with a molecule that creates a perceptual experience. We have two chemosenses: olfaction (smell) and gustation (taste). We will discuss each in turn, but as we do, be aware that they are related.

Compared to our visual and auditory senses, our sense of smell is not well understood. Part of the problem has to do with the mechanical difficulties of doing research on olfaction. It is much easier to present visual and auditory stimuli at specific intensities and durations than olfactory stimuli. In addition, the location of the olfactory mechanism makes it somewhat problematic to investigate. That being said, researchers have uncovered some of nature's mysteries of olfaction.

The Anatomy and Physiology of Olfaction MCAT

You have already seen in your exploration of vision and audition that the signal path for a sense begins at a sensory organ when receptors are stimulated to create an action potential, which then passes through a series of intermediate processing centers and finally terminates in the cortex. For olfaction, this sensory organ is the nose.

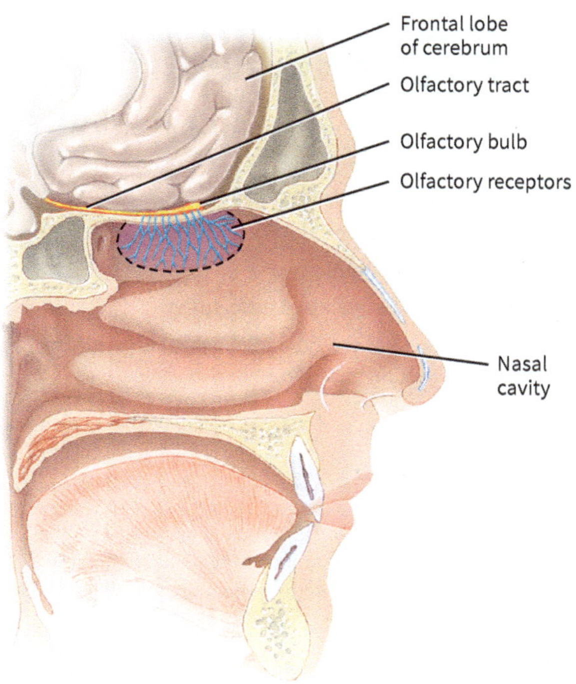

FIGURE 6.15 The human olfactory system Your sense of smell begins with receptors located at the top of the nasal cavity and just beneath the frontal lobe. These receptors extend from the olfactory bulb through tiny perforations in the cribriform plate at the base of the skull. The olfactory receptors sample the air for odorants and transmit the resulting signal to the olfactory tract, which transmits the signal into the brain.

Our sense of smell begins in the nasal cavity. Inside this space are a series of ridges known as the turbinates that tumble the air for better reception by the olfactory system. The **olfactory bulb** sits at the base of the skull just below eyebrow level and underneath the frontal region of the brain. Olfactory receptors extend through the tiny perforations in the **cribriform plate** and into the nasal cavity. These receptors are embedded in the nasal mucosa, and they are constantly sampling the air for molecules called odorants, which are defined as any substance that stimulates the olfactory system. Finally, olfactory receptors transmit the resulting signal to the olfactory tract, which sends the signal into the brain. These key structures of the olfactory system are presented in Figure 6.15.

A look at the path that olfactory data travels provides important clues to the importance of our sense of smell (Figure 6.16). Odorants traveling into the nasal cavity reach olfactory receptors embedded in the olfactory epithelium (**step 1**). Signals from the olfactory bulb travel up to a cortical region known as the piriform cortex that is dedicated to olfaction (**step 2**). From there the signal diverges into multiple paths: to the amygdala (involved with emotions), the entorhinal cortex and hippocampus (associated with memory), and the thalamus (where the different senses are integrated) (**step 3**). The signals finally terminate in the frontal lobe (specifically the orbitofrontal cortex) and the insula, where sophisticated attentional and emotional decisions are created (**step 4**).

The take-home message is that although olfaction is less important than either vision or audition (as evidenced by it receiving less

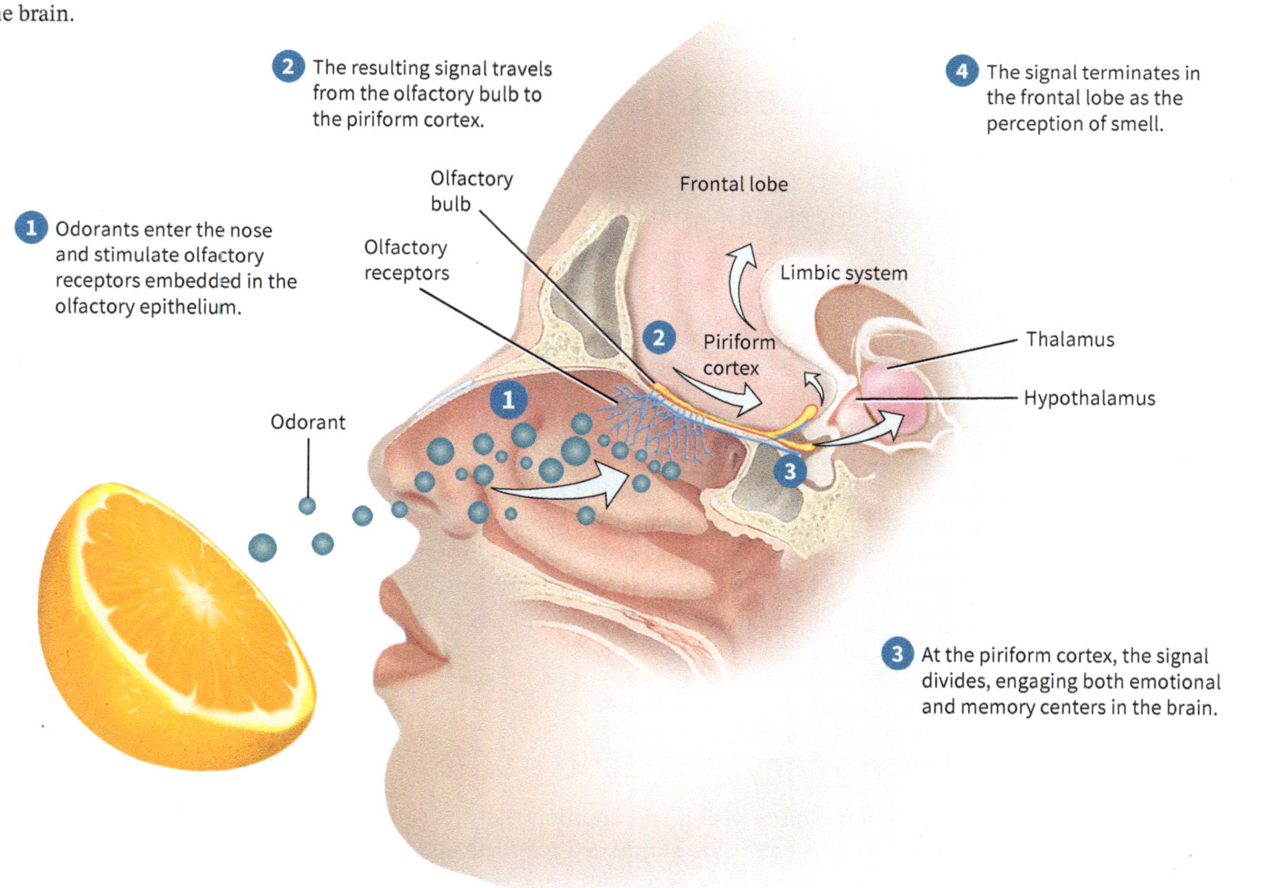

FIGURE 6.16 The olfactory circuit The olfactory circuit begins in the nasal cavity and ends in the brain.

cortical real estate), the olfactory circuit combines the perception of odor with memory and emotions. If an odor you experienced as a child has ever brought you back to a happy memory from your childhood, you see the power of the olfactory circuit.

Olfactory Receptors MCAT

It appears that humans have receptors for approximately 400 different molecules. Out of this alphabet of odorants, we create complex smells in much the same way that a symphony creates a complex sound out of the combination of many different instruments playing at once. Humans can recognize 10,000 different odorants, a remarkable feat.

How do olfactory receptors work? We always wish for a neat and tidy story in science, and sometimes nature cooperates. Unfortunately, that is not the case with olfaction. The simple answer to our question would be a lock-and-key model, where odorant molecules fit into specific receptor sites and cause them to fire. If that were the case, then molecules with similar configurations should express similar odors, and dissimilar molecules should not smell the same; however, the evidence does not support this model. For example, beta-cyclocitral ($C_{10}H_{13}O$) and *p*-anisaldehyde ($C_8H_8O_2$) have dissimilar molecular shapes yet both are described as having a minty odor. Thymol and menthol have almost identical molecular structures, yet the former has the aroma of the herb thyme, whereas the latter smells like common cough drops.

Two other theories are often invoked to explain olfaction. The odotope theory, or weak shape theory, proposes that rather than a single lock-and-key mechanism, olfactory receptors respond to small but individual physical features of a molecule so that a single odorant might stimulate several receptors simultaneously. This model explains the data better than the simple lock-and-key model, but it fails to fully predict the odorant–odor relationship.

A more recent and intriguing theory of olfaction changes the focus of perception from molecular shape to the vibrational frequency of the atoms themselves. The evidence to support vibration theory is contradictory, however. Enantiomers have the same molecular formula, but their structures are mirror images of each other, much as your left and right hands are. According to vibration theory, enantiomers should smell the same, and although some do, others do not. For instance, the right-handed enantiomer of carvone smells like spearmint, which flavors gum and mouthwash, whereas its mirror image smells like caraway, an ingredient in rye bread (**Figure 6.17**).

The olfactory system may be described as an elegant but mysterious mechanism; elegant because it is complicated but mysterious because we do not quite understand what makes a molecule an odorant.

Individual Differences in Olfaction MCAT

We have verbal descriptors of our sensory experience that permit us to communicate our internal perceptual state to another person. We could describe a sunset as blood red and a conversation as screeching. Terms that are often used to convey olfactory perceptions to another person

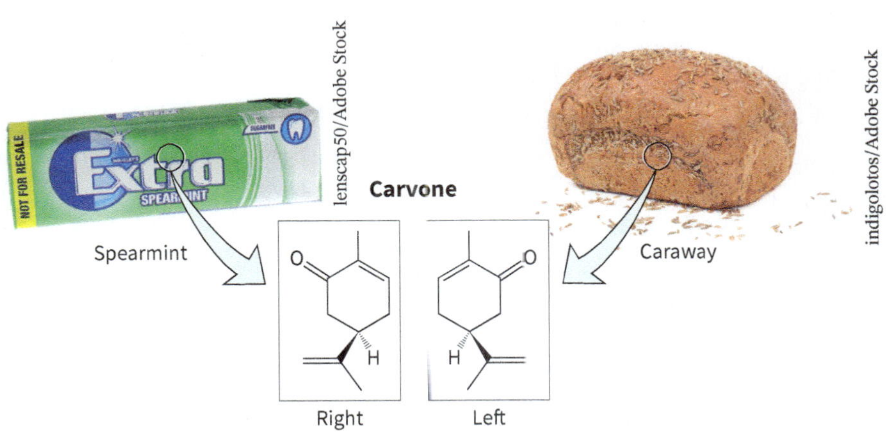

FIGURE 6.17 Right and left enantiomers of carvone Carvone comes in two versions that are mirror images of each other. Despite the chemical and structural similarity between the two, the right-handed molecule smells like spearmint and the left smells like caraway.

include musky, pungent, mothball, ethereal (a chemical smell), floral, and minty. When we share our responses to smells with other people, we soon find that not everyone experiences smell the same way.

Genetics and olfaction Genetic differences are one explanation for individual differences in olfaction. In 2004, neuroscientists Linda Buck and Richard Axel were awarded the Nobel Prize in Physiology and Medicine for their identification of genes that control olfaction. Their work shed new light on how genes can affect even our sense of smell. We now believe that each of us smells the world in a slightly different way. Most people detect almost every odorant but not necessarily at the same sensitivity, and perhaps each of us is unable to smell one or two of the roughly 10,000 known odorants. For example, androsterone (a substance produced in boar testicles and in some people's perspiration) is not detected by half the population, is powerfully aversive for 35%, and has a flowery, musky odor for 15%. Not everyone has the opportunity to be exposed to a boar's odor (and you can be thankful for that), but some percentage of us have noticed that our urine has an unpleasant odor after eating asparagus. In both cases, olfactory sensitivity to the odorant has a genetic basis.

Disorders of the olfactory system Flu and colds can temporarily disrupt our sense of smell, but some individuals experience a significant change in olfaction. Damage to the olfactory receptors can result in hyposmia, a decreased sense of smell, or even **anosmia**, a complete loss of this important sense. Anosmia is infrequent but not unknown; perhaps 3% of adults over the age of 40 are anosmic. There are several causes of anosmia: infection, damage to the olfactory receptors or the olfactory nerve, or chronic exposure to toxic chemicals. In rare cases, individuals are born without a sense of smell, as was Meredith in this chapter's *Stories of the Brain*. You can hear from Meredith herself what it is like to never have had a sense of smell in this **Check It Out: Conversation with an Anosmic**.

In rare cases, damage to olfactory receptors or cortical regions associated with olfaction may lead to dysosmia, a distortion in the sense of smell. Such individuals may perceive unpleasant odors instead of their normal pleasant response, leading to difficulties in eating.

Given that the olfactory receptors and the olfactory bulb are literally touching the brain, it seems reasonable to wonder if changes in olfaction might be a useful predictor of disease onset or subsequent disease severity. In fact, changes in olfaction may be a predictor of the course of diseases such as multiple sclerosis and dementia. One of the symptoms of COVID-19 is a sudden decrease or even loss of the sense of smell, and this symptom may be a very useful predictor of infection prior to the individual experiencing more overt symptoms such as a fever or respiratory distress.

Check It Out | Conversation with an Anosmic

As a child, Meredith was puzzled when family and friends mentioned the aroma of a fine meal or the smell of a scented candle, but she did not give the matter much thought until an incident brought it home to all that she was different. Hear from Meredith what it is like to never have experienced the world of smells in the video **Conversation with an Anosmic**, and then answer these questions.

1. If you were a congenital anosmic, what sensory experience would you miss the most, and why?
2. What incident finally convinced Meredith's family that she was different?
3. Think of two situations in which an anosmic would be a risk due to their sensory deficit, and explain why that person is at greater risk in that situation.

6.3 Before You Go On

Section Summary

The olfactory system is located at the apex of the nasal cavity, and its connections pass through the skull and into the brain. Humans have olfactory receptors for more than 400 different molecules. Individual differences in olfaction include having a diminished, absent, or distorted sense of smell.

Comprehension Check

1. Where are olfactory receptors located?
   - **a.** attached to the turbinate bones
   - **b.** extending from the olfactory bulb into the nasal cavity
   - **c.** wrapped around the olfactory canal
   - **d.** within the piriform cortex
2. The path from the olfactory receptors upstream into the brain _____.
   - **a.** is unique because it is very short
   - **b.** can easily be disrupted by disease or physical injury
   - **c.** involves the parietal lobe
   - **d.** includes connections to memory and emotion centers
3. If two molecules have the same molecular structure, _____.
   - **a.** they will smell the same
   - **b.** they will smell very similar
   - **c.** they will smell completely opposite
   - **d.** we cannot confidently predict their odors
4. An individual with anosmia _____.
   - **a.** has lost some of their sense of smell
   - **b.** has a distorted sense of smell
   - **c.** has no sense of smell
   - **d.** is repelled by strong odors

6.4 The Chemosenses: Gustation

LO 6.4 Describe how taste receptors create our gustatory experience.

Our olfactory and gustatory systems are linked; one enhances the other. You may have noticed that when your nasal passages are congested by a cold, your enjoyment of food is diminished. There is a reasonable case to be made that these chemosenses made an early appearance in the history of life on Earth. An organism that could accept or reject a potential food based on some sensory characteristic prior to fully ingesting it would have an advantage over an organism that had to ingest the food and deal with the consequences. How does our sense of taste work?

The Anatomy and Physiology of Taste MCAT

To maintain our distinction between sensory structure and site of coding into the nervous system, the eye, ear, nose, and tongue are all sensory structures, whereas the retina, basilar membrane, olfactory receptor cells, and taste buds are sites where sensory coding takes place.

Gustatory sense organs The human tongue is an organ consisting of muscle and tissue. It is unique in that its musculature is only connected at one end. Although there are taste receptors in the roof of the mouth and cheeks and even the top of the esophagus, the tongue is the primary organ of gustation, or taste. The surface of the tongue is covered with **papillae**, tiny hillocks that protrude above the surface of the tongue. Each papilla contains a pore, or opening, in the tongue's epithelium that leads down to a receptor known as a **taste bud**. Each papilla contains hundreds of taste buds. There are typically between 8,000 and 10,000 papillae of four varieties distributed across the tongue (Figure 6.18).

Taste buds are continuously being replaced. If you slightly burn your tongue with a hot beverage, you will lose some taste sensitivity for a few days, but it will return. Severe burns permanently damage the tongue epithelium, and in that case, there is a permanent loss of taste sensitivity.

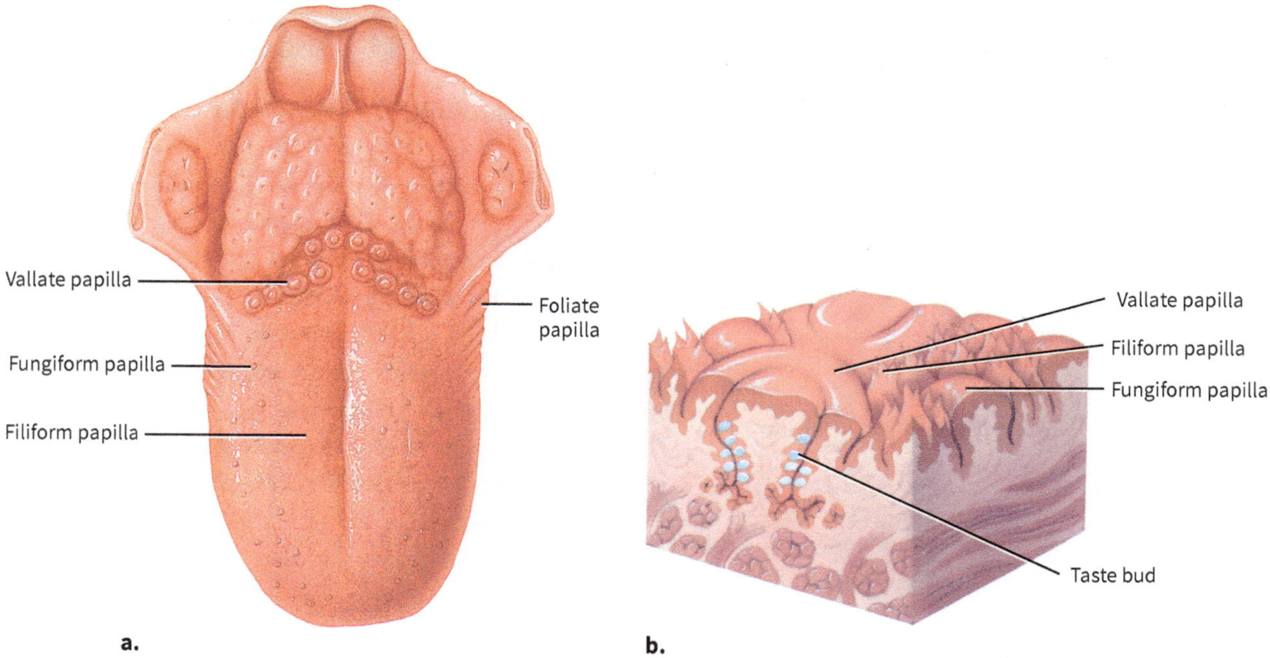

FIGURE 6.18 **The tongue, papillae, and taste buds** **a.** The tongue contains several varieties of papillae, which are distributed on the dorsal surface of the tongue in a systematic fashion. **b.** Around these papillae are taste buds, which contain the receptors for taste.

The four types of papillae are fungiform, filiform, foliate, and vallate (Table 6.2). It would be neat and tidy if each type of papillae were responsible for a specific taste dimension and if each papilla variety was assigned its own patch of real estate on the tongue. In fact, a paper published in 1901 made this claim, and this was then taught to students for generations. In the 1970s, scientists realized that the tongue map that so many students had been taught was incorrect.

Cortical representation of taste Just as with other senses, the gustatory pathway leads from the receptors through the brainstem and midbrain before terminating in the cortex. Information from the taste receptors is carried by the facial (VII), glossopharyngeal (IX), and vagus (X) nerves to centers in the medulla and thalamus. These centers then connect to regions in the insula and frontal operculum that are known as the gustatory cortex. Studies in which a tiny probe delivered electrical stimulation to the insula have reported that such stimulation can produce taste sensations. Although we know much less about the gustatory cortex than we know of its visual and auditory counterparts, it is thought that this structure is responsible for signaling taste familiarity. Familiar foods are foods that have been safely eaten in the past, so a preference for such foods provides a survival advantage.

TABLE 6.2 **The Four Types of Papillae**

Type	Shape	Location	Function
Fungiform	Mushroom shaped	On the tip and sides of the tongue	Their receptors distinguish the five basic taste dimensions. They are the most common papillae and provide the sensation of touch.
Filiform	Thin with hair-like structures on top	In the middle of the tongue's dorsal surface	They act mechanically like a grater to abrade food into smaller particles.
Foliate	Vertical folds	On the edges of the tongue especially toward the rear	They contribute to the sensation of taste.
Vallate (or circumvallate)	Dome shaped	At the back of the tongue	They are more sensitive to bitter tastes. They are the least common type. Most people have only 10 to14 in total.

Taste Receptors MCAT

Just as sight has the attribute of color and hearing has the characteristic of pitch or tone, taste has its own underlying dimensions. The five basic tastes are thought to be sweet, sour, bitter, salty, and umami (savory). The umami receptor responds to glutamate, a substance present in savory foods such as meats, meat broths, and fermented sauces such as soy sauce.

How do taste receptors work? The answer is complicated and not completely understood, but it appears that there are two different actors in the gustatory system. Taste buds themselves provide a narrow-field response; that is, they appear to be mostly tuned to one of the basic taste dimensions and respond weakly or not at all to the others. However, there are presynaptic cells nestled in the papillae adjacent to the receptors that integrate the response of many receptor cells. These cells provide a wide-field response to all of the basic taste dimensions. In this way, our gustatory system is both narrow field and wide field and is capable of creating the complex perceptual experience. The sophistication of this system may explain why humans expend so much energy creating cuisine. This also explains why artificial sweeteners such as saccharine and aspartame have both a sweet and a slightly bitter characteristic; finding a molecule that has no calories and stimulates only the sweet receptors is more difficult than you might think.

In an earlier chapter, you learned that receptors at the postsynaptic neuron can be blocked. Taste receptors can also be blocked with surprising results that illuminate how these receptors work. For example, eating the Indian herb *Gymnema sylvestre*, known as sugar destroyer, will totally block your sweet receptors for as long as an hour. The West African plant *Synsepalum dulcificum*, known as the miracle fruit, contains a molecule known as miraculin that binds to sweet receptors. At neutral pH, miraculin acts as an antagonist at the sweet receptor site, but at lower pH values (that is, sour foods), miraculin changes from an antagonist to an agonist. For about 30 minutes after eating miraculin, sour foods taste sweet.

There are few behaviors more central to life than eating, and humans have developed a mind-boggling array of cuisines. Most of us enjoy sampling novel foods, but some find novelty quite aversive. This aversion to trying foods outside your comfort zone is known as food neophobia. There is some survival value to avoiding novel foods unless they have been shown to be safe. You can explore this issue in **Try It Out: Are You a Picky Eater?**

Individual Differences in Taste MCAT

Just as with hearing and smell, taste can vary among individuals based on damage or genetics.

Try It Out | Are You a Picky Eater?

Svitlana/Adobe Stock

Most of us are willing to eat a wide variety of foods, but some of us are very skeptical of new foods. Complete the questionnaire **Are You a Picky Eater?** Then answer these questions.

1. Consider the number of statements that you checked yes to on the questionnaire. Would you consider yourself to be a picky eater or a person with a wide range of tastes? Which (environment or genetics) was most important in guiding your food choices?
2. Is there a food most people enjoy that you avoid whenever possible? If so, name that food or foods. Can you explain the source of your dislike? Is it the taste, the texture, or an unpleasant experience that accompanied an exposure to the food?

Disorders of gustation A diminished sense of taste is called hypogeusia and can be the result of infections or various diseases. Some individuals experience dysgeusia, a condition where the receptors produce an incorrect or distorted response. For example, certain medical conditions and pharmaceuticals such as Lunesta, Flagyl, and some anticancer agents can produce a metallic taste in the mouth. More serious is **ageusia**, which is the complete lack of a sense of taste. Ageusic patients are at risk for food-borne illnesses because they cannot detect the flavor of spoiled food.

Genetic influences on taste We do not all experience the same perceptual world, and your genes play a role in what you experience. Our papillae code for tastes based on the proteins that are expressed in that taste bud, and our genes control the expression of those proteins, leading to a great deal of individual difference in taste perception. For example, cilantro is an herb often used in Asian, Indian, Mexican, Mediterranean, and other cuisines. Cilantro has an appealing taste to many people but an unpleasant soapy taste to others. Research has shown that some people have a gene that makes them extremely sensitive to the aldehydes in cilantro.

An early insight into the link between genes and gustation was the discovery that the chemical PTC, which is naturally found in cabbage and broccoli, is perceived as bitter to about two-thirds of tasters and tasteless to the rest. The related chemical PROP is used in research and demonstrations due to its lower toxicity. The gene *TAS2R38* expresses a protein that provides us with the ability to perceive certain substances such as PTC as bitter tasting. Different genotypes of this gene lead to varying abilities to perceive bitter substances. Individuals with some variants are quite sensitive to bitter, whereas other genotypes attenuate the experience. A heightened sensitivity to bitterness also means that ethanol is experienced as being bitter, so PROP-sensitive individuals find the taste of alcohol unpleasant and are therefore less likely to enjoy the consumption of alcohol.

Some individuals experience tastes at a greater intensity than is usual, with perhaps an extreme sensitivity to bitter tastes, and this innate difference may explain why some individuals are very selective in their food preferences. The term supertaster was coined by Linda Bartoshuk, a scientist whose career was devoted to investigating how genes shape the perception of taste (see **Advances in Behavioral Neuroscience: Linda Barkoshuk and Supertasters**).

Advances in Behavioral Neuroscience | Linda Bartoshuk and Supertasters

Linda Bartoshuk/GoodLayers

French painter Maurice de Vlaminck said that "Good art is like good cooking; taste can be experienced but not explained." He might have been correct about art, but Professor Linda Bartoshuk would argue that taste is not beyond the tools of science. Dr. Bartoshuk has spent her life exploring our sense of taste, and her discoveries have changed the way we view what was once thought of as a subject outside the realm of science.

Linda Bartoshuk was born in rural South Dakota and fell in love with science at an early age, reading every science book she could get her hands on. In high school, she was often the only female in science and math courses, where she earned the top grades and was eventually awarded a college scholarship. She originally intended to be an astronomer, but after being told that women were not allowed to use the big telescopes, Bartoshuk switched to psychophysics, the study of how physical stimuli create mental phenomena. In her graduate study at Brown University, her Ph.D. adviser told her that women were less desirable as students because "they are always crying and washing their hair."

Despite these instances of discrimination, Linda Bartoshuk persisted in her studies and is currently the Bushnell Professor of Food Science and Human Nutrition and the Director of Psychophysical Research at the University of Florida. In her research, Dr. Bartoshuk asked a simple but profound question: how does the response of gustatory receptors become the complex world of taste? Barkoshuk and her colleagues found individuals whose sensitivity to tastes was unusually high. For example, supertasters find PROP to be considerably more bitter than the rest of us. Bartoshuk reasoned that supertasters might differ in criterion or in actual perception; that is, they might experience the same taste perception as you and I but have different criteria for judging tastes, or they might actually perceive tastes with greater intensity than is typical.

According to Bartoshuk's theory, supertasters should differ from most individuals in some anatomical or physiological aspect of gustation. Confirmatory evidence has now shown that supertasters have many more taste buds than normal, and this abundance of receptors certainly plays a role in the phenomenon.

Today Dr. Bartoshuk is one of the world's foremost authorities on how the brain creates the subjective experience of taste. Her work on genetic influences on taste perception has changed our view of how we experience taste, and her discovery of supertasters has made her known worldwide.

Questions

1. **Search online for "Linda Bartoshuk research" and read a bit about her work. Pick one project she has been involved in and briefly summarize it and why you thought it was important.**
2. **Are you a supertaster? Test yourself and see. Complete the worksheet Are You a Supertaster? to find out how you can self-administer a supertaster test with materials you can obtain at any grocery store.**

6.4 Before You Go On

Section Summary

Our sense of taste helps us to avoid noxious foods and consume foods high in caloric value. Taste receptors located mainly on the tongue but also in the mouth code for sweet, sour, bitter, salty, and savory. Just as there are individual differences in height, weight, and complexion, there are individuals who have deficient or unusually sensitive senses of taste.

Comprehension Check

1. Structures on the tongue known as _____ contain receptors known as _____.
   a. papillae; taste buds
   b. taste buds; papillae
   c. fungiform vacuoles; taste buds
   d. ventricular receptors; taste buds
2. What are the taste dimensions?
   a. sweet, bitter, hot, umami, and salty
   b. sweet, umami, bitter, and salty
   c. sweet, salty, bitter, sour, and umami
   d. sweet, salty, spicy, umami, and juicy
3. Research on the biological basis of taste suggests what?
   a. We all taste the world in the same way.
   b. Some people can learn to tolerate bitter foods.
   c. There is no genetic influence on taste.
   d. We all taste the world in our own unique way.
4. One thing we know about supertasters is that they _____.
   a. are picky people who are not easy to satisfy
   b. do not enjoy eating most foods
   c. have a great preference for sweets
   d. have taste receptors that work differently

6.5 The Cutaneous Senses

LO 6.5 Understand how receptors in the skin provide important information about the world.

Can you name the largest organ in your body? You might be surprised to know that your skin (or dermis) covers more surface area than any other organ. Your skin accounts for roughly 16% of your body weight and, if removed from the average adult, would cover roughly 20 square feet (2 m^2). Each square inch of your skin contains 19 million cells, 19 feet of blood vessels, and 1,000 sensory cells.

Our skin functions as a barrier from the hostile environment, protecting us from injury, dangerous pathogens, and extremes of hot and cold. Our skin helps cool us through evaporation and maintain the proper balance of water in our bodies. It also is essential in the production of vitamin D. In addition, our skin contains several types of sensory receptors that help us survive. These sensors are part of your cutaneous senses, and they provide information about pressure and vibration, temperature, and pain to your nervous system.

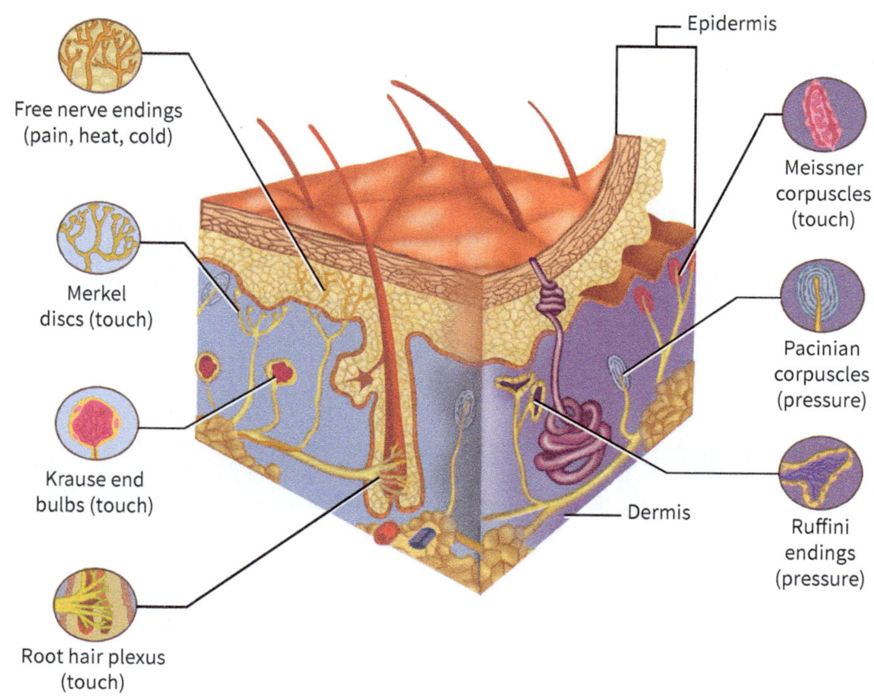

FIGURE 6.19 Sensory receptors of the skin Your skin contains several types of receptors that convey information about pressure, pain, and temperature from the skin up to the central nervous system for interpretation.

Pressure and Vibration MCAT

Like your chemosenses and unlike vision and audition, your cutaneous receptors require direct contact with the stimulus. Pressure and vibration are signaled by mechanoreceptors in the skin. Mechanoreceptors respond to mechanical or physical pressure applied to four different types of receptors in the skin (Figure 6.19).

Mechanoreceptors can be classified in terms of their adaptation rate and the size of their receptive field. Adaptation rate is the time it takes for continuous stimulation to be ignored by your nervous system. Although you are likely to be wearing clothes as you read this sentence and the fabrics are pressing on your skin, you have adapted to that stimulus so that you can ignore it and better attend to this chapter. You were introduced to the topic of receptive fields in our discussion of the visual system, and receptive fields also exist in the skin; your fingertips have very narrow receptive fields, so they can detect tiny differences in pressure and texture, whereas the skin on your back has wide receptive fields and is much less sensitive to small differences in sensation. Table 6.3 presents the response characteristics of these four receptor types.

Merkel disks provide fine-grain information on pressure, position, and the shape and edges of objects. You employ these receptors when you are feeling the texture of a fabric, reading Braille, or trying to determine whether a screw is Phillips-head or standard by touch.

Ruffini's endings are slow-adapting mechanoreceptors that respond to pressure and torsional forces (stretching) of the skin. They are most frequently found in the fingertips where they can detect movement of an object in your grip. These receptors help you know if your grip on a hammer or a fork is sufficient for the task.

TABLE 6.3 Cutaneous Receptors and Their Characteristics

Receptor type	Sensitive to	Adaptation rate	Receptive field
Merkel disk	Sustained touch, pressure, and edges	Slow	Narrow
Ruffini's ending	Sustained pressure	Slow	Wide
Meissner's corpuscle	Texture and slow vibration	Fast	Narrow
Pacinian corpuscle	Deep pressure and fast vibration	Fast	Wide

Meissner's corpuscles are sensitive to light touch, and they rapidly adapt to stimulation. They respond to the low-frequency vibrations that occur when an object is sliding across your skin. They would alert you to the fact that the glass you have picked up is sliding out of your hand due to insufficient pressure.

Pacinian corpuscles respond quickly to sudden changes in pressure and also vibration. They likely play an important role in detecting surface texture. They fire when your skin first comes into contact with an object. At the microscopic level, these encounters consist of a number of high-frequency collisions between you and the object. These receptors alert you that your hand is now in contact with your phone.

Testing of sensitivity to pressure and vibration is often a part of a neurological examination because damage to an individual sensory or spinal tract often leads to a loss of feeling at a corresponding body region. Therefore, damage to a spinal tract in a specific vertebra can often be localized by a neuropsychologist or neurologist by stimulating the cutaneous receptors in specific areas of the skin and asking the patient if they can detect that stimulation.

Not all touch receptors are in the skin. Some are buried in muscles and tendons and fire when the muscles stretch. Kinesthetic receptors tell us where our body parts are, even when we cannot see them. There is evidence that those among us with a more acute kinesthetic sense excel at sports that require eye–hand coordination and at athletics in general.

Nerve damage can cause an individual to have diminished kinesthetic response for a body part, but it is exceedingly rare for an individual to lose all kinesthetic feeling. If you can imagine not having any idea where your legs or arms are unless you are looking directly at them in a given moment, then you can imagine the difficulties such an individual would face. You can consider their plight in this **Check It Out: Global Kinesthetic Deficit**.

Pain MCAT

At some point, we have all experienced the perception of pain from an injury of one sort or another. A skinned knee, a cut from a knife, a sunburn, and the discomfort of a pulled muscle are all common examples of painful situations. Perhaps you have wished at some point that you were not able to feel pain. Would that be a good idea? Before we answer that question, we should first ask how the perception of pain is generated.

There is a great deal of scientific interest in understanding pain perception because the experience of pain is the most common event that brings a patient into a physician's office. Pain can signal a serious internal injury, be so unrelenting that it destroys a person's quality of life, and even be experienced from a limb that was amputated. Understanding pain is both a complicated scientific problem and an issue with significant personal impact for many.

Check It Out | Global Kinesthetic Deficit

Dragon Images/Shutterstock.com

When he was 19 years old, Ian Waterman lost all kinesthetic feedback due to a viral infection. He is one of the few patients in the world with a global kinesthetic deficit who has learned to walk again. Search the internet using the term "Ian Waterman" and you will find several short videos that describe him and his unusual success at rehabilitating himself. Then answer these questions.

1. How would you describe his problem to someone who has not seen the videos?
2. What would you say are the greatest problems he faces in his everyday life? Name a task and a career he would be wise to avoid.

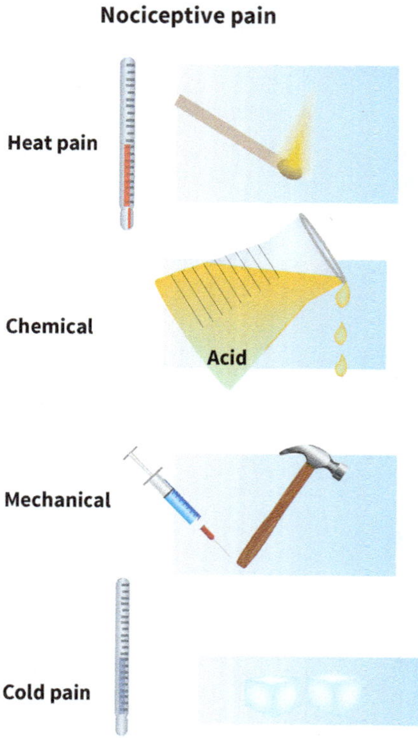

FIGURE 6.20 Four varieties of nociceptive receptors Our pain receptors respond to tissue damage from excessive heat or cold, chemicals, and mechanical tearing or cuts.

In the context of understanding the link between the nervous system and behavior, we can simplify our task by focusing on **nociceptive pain**, or the pain that is caused by stimulation of nerve fibers. Typical sources of nociceptive pain are thermal (heat or cold), mechanical (tearing, cutting, or crushing), and chemical (irritation from substances such as acids or spices) as illustrated in Figure 6.20.

Pain that results from stimulation of nerve fibers in the skin is typically described as sharp, and the location of the injury is generally easy to specify. In contrast, pain that results from stimulation of fibers deep in the body is known as visceral pain, and it is characterized as a dull ache whose source is often difficult to locate.

Researchers have identified two types of fibers that transmit the sensation of pain: **C fibers** and **A-delta (Aδ) fibers**. C fibers are unmyelinated, so their conduction speed is slow, on the order of 0.5–2 m/s. C fibers are responsible for the perception of deep and nonlocalized pain. C-fiber axons are clumped together in what are known as Remak bundles. It has been observed that when a nerve tract is damaged, nearby C fibers spontaneously increase their activity. There is some evidence to suggest that when these C fibers adjacent to an injured nerve fiber are activated, they release substances that aid in the regeneration of the damaged nerve. These bundles are currently of interest to pain researchers because harnessing the regenerative properties of Remak bundles would help individuals suffering nerve damage.

Unlike C fibers, Aδ fibers are thicker and myelinated. As a result, signal transmission by the Aδ fiber is fast, on the order of 5 to 30 m/s. Aδ fibers are responsible for what is described as bright, sharp pain, whereas C fibers provide the experience of dull, aching pain. In terms of temperature, the C fibers are thought to be responsible for the perception of warm, whereas the Aδ fibers are thought to carry the perception of cold. These two receptor types are not segregated but typically occur in conjunction with each other.

In very rare cases, individuals are born with a congenital insensitivity to pain (CIP) and are totally unable to perceive pain. This might seem like a good problem to have, but in fact the ability to experience pain is a protective mechanism that helps us avoid bodily damage. Pain motivates us to pull our hand back from a hot stove and stay off a bruised ankle until it heals. Pain researchers would like to better understand the mechanism behind this lack of pain perception, because it might result in a treatment for chronic pain that does not involve massive doses of opioids.

To demonstrate the consequences of not feeling pain, consider the case of Dr. Letizia Marsili, who was skiing in the Italian Alps when she fell and crashed hard into the slope, shoulder first. She thought nothing of it and continued on with her day, but at dinner than night, she was startled to see that her knife and fork kept falling from her hand. A visit to the doctor the next day revealed that her shoulder was broken. All through her life, cuts, burns, and bruises resulted in a momentary feeling of discomfort, which then immediately went away. Other members of her family expressed this trait in varying degrees, and the fact that about 40 individuals living in a small village in Sweden all have CIP suggests that there is some genetic basis for the disorder.

Cases of CIP are quite rare, but evidence that we do not all perceive pain equally may be as close at hand as your nearest redheaded friend. Redheads appear to require more anesthesia for sedation, but may be more responsive to opioids and so require a lower dose to achieve pain control. Finally, redheads may also be more sensitive to the cold. The evidence suggests that the same gene that codes for their fair skin and hair color also changes the sensitivity of some cutaneous receptors. Clearly, we still have a lot to learn about pain.

6.5 Before You Go On

Section Summary

Your skin senses provide information about touch, temperature, and especially pain. There are several different cutaneous receptors, and each has its own characteristics. Nociceptive pain can be chemical, thermal, or mechanical. Unmyelinated C fibers are responsible for deep, nonlocalized pain, whereas myelinated Aδ fibers are responsible for sharp, localized pain.

Comprehension Check

1. Pacinian and Meissner's corpuscles _____, whereas Ruffini's endings and Merkel disks _____.

- **a.** are pressure receptors; are temperature receptors
- **b.** do not develop until adolescence; are present at birth
- **c.** are vibration receptors; signal both pressure and vibration
- **d.** are fast acting; are slow acting

2. Which of these is NOT a source of nociceptive pain?
   a. scalding your mouth with hot coffee
   b. a stress headache
   c. a paper cut
   d. stubbing your toe

3. A key difference between C fibers and Aδ fibers is that C fibers _____ and Aδ fibers _____.
   a. are myelinated; are unmyelinated
   b. are isolated; form Remak bundles
   c. cause dull, aching pain; cause sharp, local pain
   d. transmit signals quickly; transmit signals slowly

Conclusion

In this chapter, we explored your auditory sense, chemosenses, and cutaneous senses, as well as your vestibular system. Although we humans rely more on vision than these other senses, the information they provide helps us survive and thrive. Audition, like vision, allows us to sense objects at a distance; sound can warn us of an approaching predator and alert us to the presence of a stream. Audition is also a powerful platform for communication. Our vestibular sense provides essential information about our position in space.

In contrast, our chemosenses require more direct contact with the stimulus. Taste and smell are dependent on foods and odorants entering our bodies and acting on receptors. Both taste and smell are designed to encourage us to consume foods rich in calories or nutrients and avoid those that might be dangerous. Smell also alerts us to dangers such as a predator or a fire.

Finally, the most direct contact of all our senses are the cutaneous senses that provide the information we need to grasp and hold objects. Although pain is aversive, it serves to motivate us to retreat from and avoid dangers.

The goal of the chapter was to describe the anatomy of these senses and, at a deeper level, to show how each of our sensory modalities follows a common plan: a receptor that communicates with centers in the brainstem, the thalamus, and finally terminates in a region of the cortex for complex interpretation.

Critical Thinking Questions

1. There is no air on the Moon or in space. The atmosphere on the surface of Mars is only 7% as dense as on the Earth's surface, and water is roughly 830 times as dense as air. A sound is made in each of these locations. Using your knowledge of how the auditory system operates, explain why there are large differences in how far a sound would travel in each of these environments.
2. The hair cells in your cochlea are easily damaged; sustained exposure to sounds over 70 dB can cause damage, and any exposure to sounds over 120 dB can cause immediate damage. What situations that you encounter in your daily life might put your hearing at risk if you were not wearing ear protection?
3. Sofia developed an ear infection in her childhood that left her with significant hearing loss in one ear. She easily understands conversations, TV shows, and class discussions but often errs when trying to identify the source or direction of a sound that occurs when she is outside. Explain.
4. When people enter their 80s and beyond, they often begin to lose olfactory and gustatory receptors. These same people are more at risk for food poisoning that can lead to nausea and leave them dangerously dehydrated. Explain why loss of these receptors puts the elderly at risk.
5. We have more receptors for cold in our face and ears than in other parts of the body. From an evolutionary perspective, why does it make sense that the face has an abundance of cold receptors?

Dig Deeper into Behavioral Neuroscience

6.1 Audition: Hearing and the Ear

Our auditory system is both complicated and fascinating. These sources provide a more detailed description.

Emmanual, D. C., & Letowski, T. (2012). *Hearing Science*. Lippencott, Williams and Wilkins.

Schnupp, J., Nelken, I., & King, A. (2011). *Auditory Neuroscience*. MIT Press.

6.2 Auditory Processing and the Vestibular System

For some people, motion sickness is a temporary consequence of a flu or a drug, but for others it can be a constant, debilitating condition. There is much we know about motion sickness, as this article demonstrates.

Lackner, J. R. (2014). Motion sickness: More than nausea and vomiting. *Experimental Brain Research, 232*(8), 2493–2510. https://doi/org/10.1007/s00221-014-4008-8. PMC: 4112051.

Why do some people enjoy roller coasters and boat rides while others suffer from nausea? Understanding why some people's vestibular systems are disrupted and others' are not may help us understand the vestibular system.

https://www.theatlantic.com/health/archive/2015/02/the-mysterious-science-of-motion-sickness/385469/

6.3 The Chemosenses: Olfaction

We know a great deal about the chemosenses and cutaneous senses, but there is still much we do not know. These texts provide a comprehensive overview.

Wolfe, J. M., Kluender, K. R., & Levi, D. M. (2011). *Sensation and Perception* (3rd ed.) Sinauer Associates.
Goldstein, E. B. (2009). *Sensation and Perception*. Cengage.

Understanding that some of our perceptual experience is guided by our genetic heritage can be a thought-provoking concept, and these sources are worth the read.

Markt, S. C., Nutall, E., et al. (2016, December 13). Sniffing out significant "pee values": Genome wide association study of asparagus anosmia. *BMJ, 355*, i6071. https://doi.org/10.1136/bmj.i6071. PMC: 5154975.

Wysocki, C. J., & Beauchamp, G. K. (1984, August). Ability to smell androsterone is genetically determined. *Proceedings of the National Academy of Sciences USA, 81*(15), 4899–4902. https://doi.org/10.1073/pnas.81.15.4899

6.4 The Chemosenses: Gustation

Our understanding of how our nervous system creates the conscious experience is always evolving, and this is true for our understanding of gustation. The recent discovery of a taste receptor in mice that responds to sweet, sour, bitter, and umami but not salty has caused scientists to look for evidence of this receptor in other mammals, including humans.

Banak, D. D., Bentley, E. D., et al. (2020). A subset of broadly responsive Type II taste cells contribute to the detection of bitter, sweet, and umami stimuli. *PLOS Genetics, 16*(8), e1008925. https://doi.org/10.1371/journal.pgen.1008925

6.5 The Cutaneous Senses

This text explores the loss of proprioception:

Sachs, O. (1970). The disembodied lady. In Sachs, O., *The Man Who Mistook His Wife for a Hat and Other Clinical Tales*. New York: Harper and Row.

Despite the importance of physical pain, there is much we do not know, especially how to reduce it when warranted.

Melzack, R. (1973). *The Puzzle of Pain*. New York: Basic Books.

Redheaded individuals have a lower pain threshold than others, and this has implications for everyday life but especially for situations where anesthesia is required.

Camann, W. (2004, August 16). Redheads need more anesthesia! *Medscape*. http://www.medscape.com/viewarticle/487261

Liem, E. B., Lin, C-M., Suleman, M-I., D'Oufas, A. G., Gregg, R. G., Veauthier, J. M., Loyd, G., & Sessler, D. I. (2004, August). Anesthetic requirement is increased in redheads. *Anesthesiology, 101*(2), 279–283. http://www.ncbi.nlm.nih.gov/pmc/articles/PMC1362956/

*What would it be like to never have experienced pain? A family in Norrbotten, Sweden, has a genetic mutation that prevents them from feeling pain.

Minde, J. (2006). Norrbottnian congenital insensitivity to pain. *Acta Orthopaedica. Supplementum, 77*(321), 2–32.
http://umu.diva-portal.org/smash/record.jsf?pid=diva2:144411

In this article, you can read about Ashlyn Blocker, a woman who has never experienced pain.

http://www.nytimes.com/2012/11/18/magazine/ashlyn-blocker-feels-no-pain.html?_r=0

Chapter 6 Before You Go On Answers

6.1 Before You Go On Answers

1. a. loudness; pitch
2. a. auricle, external acoustic meatus, and tympanic membrane
3. b. hammer, anvil, and stirrup
4. d. cochlea
5. c. sends its signal directly to the auditory nerve

6.2 Before You Go On Answers

1. d. from the cochlea to the brainstem to the temporal lobe
2. d. hair cells; movement of the basilar membrane
3. a. originate directly ahead or behind us
4. c. semicircular canal
5. b. otolith

6.3 Before You Go On Answers

1. b. extending from the olfactory bulb into the nasal cavity
2. d. includes connections to memory and emotion centers
3. c. they will smell completely opposite
4. c. has no sense of smell

6.4 Before You Go On Answers

1. a. papillae; taste buds
2. c. sweet, salty, bitter, sour, and umami
3. d. We all taste the world in our own unique way.
4. d. have taste receptors that work differently

6.5 Before You Go On Answers

1. d. are fast acting; are slow acting
2. b. a stress headache
3. c. cause dull, aching pain; cause sharp, local pain

Chapter 6 Glossary

A-delta (Aδ) fiber A small, fast-acting, myelinated fiber that conducts pain signals.

ageusia The complete loss of the sense of taste.

anosmia The complete loss of the sense of smell.

auricle The technical term for the external part of the ear.

basilar membrane One of two membranes in the cochlea (the other being the tectorial membrane) that is essential for auditory perception.

C fiber An unmyelinated afferent nerve fiber with a low conduction speed.

cochlea A part of the inner ear involved in hearing that contains the organ of Corti.

cribriform plate A bone at the roof of the mouth that is perforated by olfactory receptors.

external acoustic meatus The passageway that leads from the outside of the head to the tympanic membrane; also known as the auditory canal.

hair cell An auditory sensory receptor located on the basilar membrane in the organ of Corti.

helicotrema The tip of the cochlea where the scala tympani and scala vestibuli meet.

incus The middle bone of the ossicles; also known as the anvil.

malleus The outermost bone of the ossicles; also known as the hammer.

Meissner's corpuscle A tactile mechanoreceptor in the skin that is sensitive to light touch.

Merkel disk A receptor in the skin that is sensitive to pressure.

nociceptive pain Pain caused by stimulation of nerves by noxious stimuli such as heat or cold, crushing, intense chemicals, or cutting.

olfactory bulb A neural structure in the skull above the cribriform plate that receives information from olfactory receptors and sends olfactory information upstream to the brainstem and eventually to the olfactory cortex.

organ of Corti The receptor organ for sound that is the site of transduction of physical signals into an action potential that is encoded as sound.

ossicle One of the three bones of the middle ear consisting of the malleus, incus, and stapes.

otolith A particle at the end of a hair cell in the saccule and utricle.

oval window The membrane-covered opening from the middle ear to the inner ear

Pacinian corpuscle A skin receptors responsible for sensitivity to vibration and pressure.

papilla A small, nipple-like structure on the upper surface of the tongue that give it its rough texture.

place theory The theory that our perception of high-frequency sound results from the location of stimulation on the basilar membrane.

primary auditory cortex (PAC) The part of the temporal lobe that processes sound.

rate theory The theory that auditory perception of low-frequency sound depends on the rate of fire of hair cells on the basilar membrane.

round window One of two openings in the middle ear, which moves in opposite phase from the oval window.

Ruffini's ending A skin receptor that responds to sustained pressure and shows little adaptation.

saccule A structure that, when paired with the utricle, provides information about motion and balance.

scala media The endolymph-filled chamber within the cochlea that houses the organ of Corti; also known as the cochlear duct.

scala tympani The perilymph-filled chamber within the cochlea that is inferior to, or below, the scala media.

scala vestibuli The perilymph-filled chamber within the cochlea that is superior to, or above, the scala media.

semicircular canal One of three semicircular interconnected tubes in the inner ear that provide sensory input regarding rotary movement.

stapes The innermost bone of the ossicles; also known as the stirrup.

stereocilium A mechanosensitive organelle in the inner ear that responds to fluid motion in the cochlea.

taste bud A taste receptor cell in the tongue and oral cavity.

tectorial membrane One of two membranes in the cochlea (the other being the basilar membrane) that is essential for normal hearing.

tympanic membrane A thin membrane that separates the outer ear from the middle ear; also known as the eardrum.

utricle A structure that, when paired with the saccule, provides information about movement and balance.

vestibular system A sensory system responsible for providing our brain with information about movement and orientation in space.

volley theory A theory that the firing of multiple neurons in the auditory system when taken together codes for perception of frequencies above the firing rate of an individual neuron.

CHAPTER 7

Homeostatic Regulation of the Internal Environment

NASA/Public Domain

A spacesuit allows a fragile human to survive in outer space by maintaining the conditions essential for life. In a similar way, our brain is constantly monitoring and adjusting our internal environment to keep us alive in our own world.

CHAPTER OUTLINE

7.1 An Overview of Homeostasis
- Balance and Change in the Body
- Feedback Systems

  Try It Out: Negative Feedback Systems in Nature and Technology

7.2 Homeostatic Regulation of Temperature
- Ectothermic versus Endothermic Thermoregulation
- The Neural Basis of Thermoregulation
- Hyperthermia and Hypothermia

  Think About It: Avoiding Exercise-Induced Hyperthermia

7.3 Homeostatic Regulation of Fluids Through Thirst
- Volumetric Thirst and the Renin–Angiotensin System
- Osmotic Thirst and Cellular Sodium Concentration
- Disorders Related to Fluid Imbalance

  Try It Out: Recognizing the Dangers of Fluid Imbalance

7.4 Homeostatic Regulation of Hunger and Satiety
- The Gastrointestinal Tract
- The Neural Basis of Hunger and Satiety

  Advances in Behavioral Neuroscience: Philip Teitelbaum and the Motivation to Eat
- Hormonal Regulation of Hunger
- Hormonal Regulation of Satiety

7.5 Disorders of Hunger and Satiety
- Obesity

  Try It Out: Measure Your BMI
- Prader-Willi Syndrome
- Anorexia Nervosa
- The Challenges of Treating Disordered Eating

LEARNING OBJECTIVES

LO 7.1 **Explain** the importance of homeostasis and the role of negative feedback in regulating the internal environment.

LO 7.2 **Identify** the processes that regulate internal temperature.

LO 7.3 **Explain** and distinguish the two mechanisms that control thirst.

LO 7.4 **Evaluate** critically the multiple mechanisms that regulate hunger and satiety.

LO 7.5 **Analyze** the causes and treatments of feeding disorders in light of the mechanisms that regulate feeding.

Stories of the Brain

Lester was born after a difficult pregnancy, and at birth he was lethargic and struggled to nurse. His parents soon noticed his poor muscle tone and floppy limbs, which contrasted to the energetic movements of his older sister at that age. As time went on, he slept constantly and missed almost every cognitive and motor developmental milestone. When their pediatrician examined Lester, she noted that he exhibited strabismus, a condition in which one's eyes do not naturally align while gazing at an object. Their doctor advised them to keep a watchful eye on Lester's development and report anything out of the norm, especially any unusual eating behavior.

Around his fourth birthday, Lester's eating changed dramatically, and it seemed no amount of food would satisfy his hunger. One night at dinner, his mother decided to see if there was a limit to how much their four-year-old son could eat, so she provided Lester with a full adult meal. After cleaning his plate, he begged for more food. His worried parents took him to a pediatrician who noted Lester's slow intellectual development, obesity, and insatiable hunger. A genetic test confirmed the diagnosis: Lester had Prader-Willi syndrome, a rare genetic disorder characterized by a constant need to eat without ever feeling satisfied. Lester is now 10 years old, and his parents have to padlock the refrigerator and pantry to prevent him from eating everything in sight. There is no cure for Prader-Willi, but regular exercise to combat obesity and a structured environment have helped Lester lead a more normal life.

Lester's behavior raises several interesting points. There must be a mechanism in the nervous system that regulates hunger and also satiety, and in Lester's case, that mechanism appears to be dysfunctional. Understanding the processes that control feeding can provide insights into Lester's behavior but also help us to understand why some of us struggle to maintain a healthy weight. Finally, the specific question of how feeding is regulated opens the door to considering this chapter's larger questions: how does the nervous system regulate our internal environment, and what happens when those mechanisms go awry?

In this chapter, we will first discuss feedback mechanisms that control homeostasis and then consider the biological mechanisms that regulate our internal temperature, fluid intake, and calorie consumption. With an understanding of how these mechanisms normally work, we will briefly discuss what happens when they are dysfunctional. Just as a spacesuit keeps an astronaut alive in the hostile environment of outer space, our homeostatic mechanisms provide us with a safe internal environment.

7.1 An Overview of Homeostasis

LO 7.1 Explain the importance of homeostasis and the role of negative feedback in regulating the internal environment.

Every living organism must meet certain requirements to survive, and humans are no exception. We must maintain a body temperature within a certain range, neither too hot nor too cold. We need enough fluids so that our blood can carry glucose, oxygen, and other nutrients throughout our body. We need enough fuel in the form of glucose to power our cells. Wherever we go and whatever we do, our nervous system is working constantly behind the scenes to ensure that our essential biological systems can function despite changing conditions around us.

Balance and Change in the Body MCAT

In 1926, physiologist Walter Cannon coined the term **homeostasis** to describe physiological processes that only function within a specific range of values. The word homeostasis comes from the Greek for same (*homeo*) and standing still (*stasis*), and homeostasis is defined as a regulated internal state that supports life. In prior chapters you have already seen how

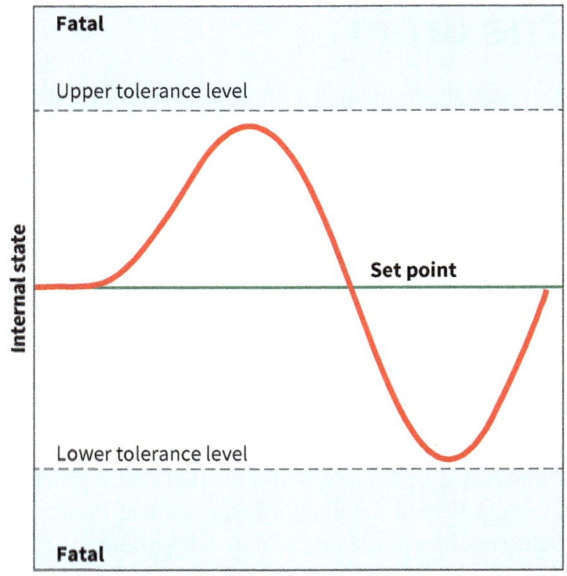

FIGURE 7.1 Set point Biologically, the set point is the sweet spot in which organisms survive and thrive. Too high above or too low below the set point can be dangerous and even fatal.

imbalances in certain ions or neurotransmitters can have serious effects on our health and can even be fatal. Other factors that must stay in balance include blood pressure, oxygen levels, temperature, and pH. Intuitively, you can see that humans must remain within a certain range in all of these parameters to survive and thrive (Figure 7.1). The point at which systems tend to stabilize is called the **set point**. If your blood pressure significantly exceeds the set point, you may have a stroke, but if it falls too far below this same set point, you will pass out. Too far in either extreme is fatal.

Cannon's perspective that our nervous system defends set points through homeostatic processes has been a valuable tool for understanding the biological basis of behavior, but in 1988, neuroscientist Peter Sterling and his colleague Joseph Eyer proposed a modification of homeostasis they termed **allostasis**. In contrast with homeostatic mechanisms that assume the existence of an unchanging set point, Sterling and Eyer suggested that experience can modify these set points. They believe that allostasis, from the Greek for stability through change, better describes how our nervous system regulates many physiological functions. That our nervous system may adjust set points or targets in response to stress and other factors has gained traction with those trying to understand eating disorders, addictive behaviors, and even pharmacological treatment of psychiatric disorders. Table 7.1 summarizes the major differences between homeostatic and allostatic perspectives.

Various bodily systems are involved in maintaining homeostasis, and the nervous system is primary among them. Our nervous system is responsible for maintaining the proper internal temperature, regulating fluid intake, and obtaining sufficient calories to power our cells. In the next chapter, we will discuss how the endocrine system contributes to homeostasis in other important ways.

You have seen time and again in your study of the nervous system that there are opposing mechanisms that operate to regulate behavior, and homeostasis is no different. Before we begin to discuss the neural basis of homeostasis, we will take a moment to discuss the properties of feedback systems, as you will encounter these systems throughout your study of behavioral neuroscience.

TABLE 7.1 Comparing Homeostatic and Allostatic Perspectives

Homeostasis	Allostasis
Unchanging set point	Changing set point
Trends toward stability	Trends toward change
No set point adjustment based on history	Set point changes in response to history

Feedback Systems MCAT

There are several biological mechanisms that work to keep metabolic processes at an optimal level, and the most common is known as a **negative feedback system**, sometimes called a negative feedback loop. A negative feedback system maintains homeostasis through a three-part mechanism that includes a sensor, a control mechanism, and a brake or effector. The sensor is constantly monitoring the level of a stimulus. When it detects a deviation from the set point, the control mechanism activates the effector or brake to reduce the difference between the current state and the set point until the difference is eliminated. This may be accomplished by an effector, which increases or activates something, or a brake, which decreases or deactivates something.

A home cooling system is a familiar example of a negative feedback system. When the system is engaged in the summer and a desired temperature is set, a device in the thermostat called a thermocouple begins to monitor the ambient temperature. When the room temperature exceeds that set point by a given amount, the thermostat activates the compressor and fan in the air conditioner, and cold air begins to cool the room. This continues until the thermocouple senses that the room temperature has cooled to the set point, and then the thermostat turns the unit off (Figure 7.2).

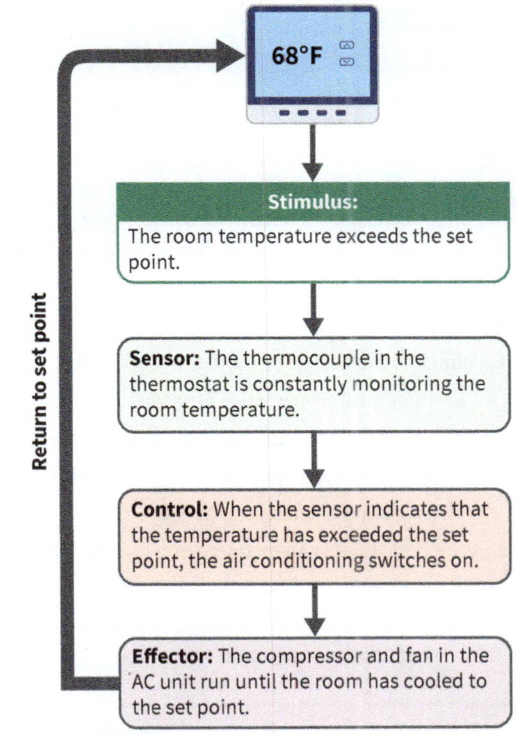

FIGURE 7.2 A negative feedback system When the sensor detects that the system is out of equilibrium, it activates a control system that engages the effector to bring the system back into balance. When the sensor detects that the system is back in balance, it signals the control system to deactivate the effector.

In this chapter, you will see that negative feedback systems abound in homeostatic mechanisms and elsewhere in the neural mechanisms that control behavior. You may have already learned in a previous chapter about autoreceptors and heteroreceptors that help to regulate neurotransmitter release, and you will discover in a later chapter that the basic principle of excitatory and inhibitory mechanisms that work together to regulate mood is essential to developing a grasp of psychiatric disorders and their treatment.

In light of our emphasis on negative feedback systems, it is only natural to ask if positive feedback systems play any role in regulating biological functions, and the answer is they do exist. Unlike negative feedback systems, which operate to reduce change, positive feedback systems enhance change, and some examples may clarify this distinction. Contractions of the uterine wall during childbirth stimulate the release of hormones that stimulate additional contractions. After childbirth, when a mother lactates (secretes milk), the infant feeds, and that stimulates the production and release of more milk. The reason that negative feedback systems are more common in nature than positive ones is that negative feedback systems tend toward stability, whereas positive feedback systems tend toward instability. Negative feedback systems occur throughout our everyday life, and you can practice identifying them in this Try It Out: Negative Feedback Systems in Nature and Technology.

Try It Out | Negative Feedback Systems in Nature and Technology

AlexMax/Getty Images

As you learn how the brain and body work together, you see time and again that negative feedback loops are an efficient mechanism for maintaining equilibrium. A regulatory system that is dormant unless some aspect veers out of limits is so useful that we see them in many situations. Download the worksheet Negative Feedback Systems in Nature and Technology and explain how a negative feedback loop functions to maintain an equilibrium in each situation.

7.1 Before You Go On

Section Summary

Our nervous system works to maintains homeostatic levels of basic functions such as temperature, fluids, and calories. It does so through negative feedback systems that sense when a system deviates sufficiently from the set point and then returns the system to homeostasis or equilibrium.

Comprehension Check

1. Which is the best definition of homeostasis?
   - **a.** maintaining a positive attitude
   - **b.** breathing in oxygen and releasing carbon dioxide
   - **c.** maintaining a constant internal environment
   - **d.** disposing of metabolic waste
2. Which of these is an example of a homeostatic system?
   - **a.** a skier skiing down a steep slope
   - **b.** a bank account whose balance changes as checks are written and deposits are made
   - **c.** a tightrope walker balancing on a tightrope
   - **d.** the word count in a document you are writing
3. The difference between homeostasis and allostasis is that homeostasis ______, and allostasis ______.
   - **a.** refers only to men; refers to men and women
   - **b.** involves a static set point; involves a dynamic set point
   - **c.** involves a negative feedback loop; involves a positive feedback loop
   - **d.** explains feeding and thirst; only explains feeding
4. Which of these is an example of a negative feedback system?
   - **a.** the mute button on a TV remote
   - **b.** the soap dispenser in a dish-washing machine
   - **c.** different cooking options on a microwave oven
   - **d.** cruise control in a car

7.2 Homeostatic Regulation of Temperature

LO 7.2 Identify the processes that regulate internal temperature.

Most living organisms share the need to regulate their internal temperature, and humans are no different. We need to maintain a relatively stable internal temperature in order for the physiological processes that support life to function properly.

Ectothermic versus Endothermic Thermoregulation MCAT

Different animals have evolved different mechanisms for regulating body temperature. **Ectotherms** (from the Greek for outside heat) such as snakes and lizards regulate their internal temperature through environmental means; when you see a snake sunning itself on a rock, it is doing so to raise its body temperature to that of the environment. Humans and other mammals are **endotherms** (from the Greek for inside heat), organisms that use the heat by-product of their metabolism to keep their body temperature at the desired point.

Given the importance of thermoregulation and the diversity of life, you will not be surprised to find that a wide array of behaviors are employed to maintain thermoregulation (Figure 7.3). For instance, the toucan is an endotherm that lives in the jungle and because of its natural habitat is challenged to dissipate heat. Its colorful bill, which can be a third of the length of its body, is filled with blood vessels. By adjusting the blood flow in its bill, the toucan can either conserve or dissipate heat, changing its temperature by as much as 18 °F (10 °C) in just a few minutes. In contrast, a turtle sunning itself on a rock or log is a familiar example of ectothermic regulation. The turtle maintains its body temperature by moving in and out of the warming rays of the sun.

The Neural Basis of Thermoregulation MCAT

How does our brain regulate our internal temperature? It should not surprise you to learn that there is a negative feedback system to keep our body temperature within safe bounds. The hypothalamus is the control center for temperature regulation, and it does so through

FIGURE 7.3 Two animals with different adaptations for thermoregulation a. The toucan is an endotherm that uses its bill to dissipate heat. b. The turtle is an ectotherm that uses the sun to achieve a proper internal temperature.

a. Toucan in jungle

b. Turtle sunning on rock

the anterior hypothalamic nucleus (AH) and the adjacent preoptic nucleus, also known as the preoptic area (PO or POA). The AH and the PO are constantly monitoring temperature through thermoreceptors in the skin, body organs, and also in the hypothalamus. The body's thermoregulatory set point is around 98.6 °F (37 °C), with variations at different times of day and at different life stages. One important effector mechanism in thermoregulation involves the diameter of blood vessels. Blood vessels can be opened wider (dilated) through vasodilation or narrowed (constricted) through vasoconstriction.

When thermoreceptors report to the hypothalamus that the body's core and skin are too cold, the hypothalamus initiates vasoconstriction, which keeps blood away from the skin and concentrated in the core, thereby conserving heat. This is why you can warm up your cold hands by tucking them into your armpits. In addition, your brain signals your muscles to go into spasms that we call shivers, and these have the effect of warming the muscles and thereby maintaining core body temperature. Shivering of muscles in your jaw may cause your teeth to chatter. Conversely, when the hypothalamus detects that the body and skin are too hot, it initiates vasodilation to bring warm blood to the skin where evaporative cooling through sweating can occur (Figure 7.4).

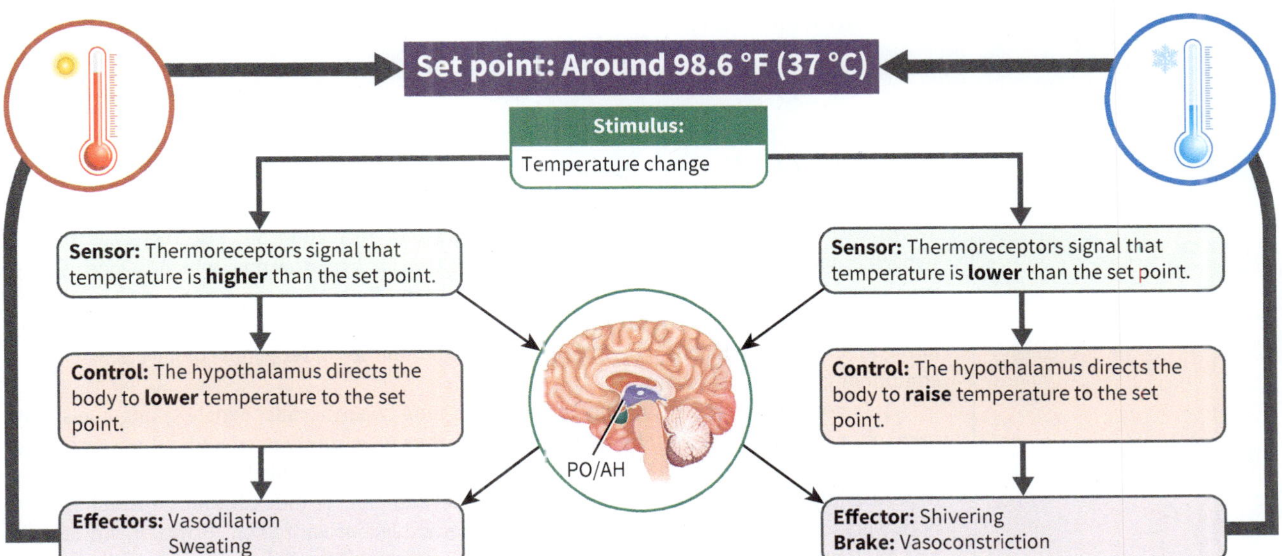

FIGURE 7.4 The brain and temperature regulation The preoptic nucleus and the anterior hypothalamic nucleus monitor our internal temperature and, in response to extremes of heat or cold, initiate responses that return our internal temperature to a safe level.

Hyperthermia and Hypothermia MCAT

We all occasionally sweat in the summer and shiver in the winter, but exposure to extreme temperatures creates a challenge to homeostasis. Dysfunction occurs when environmental conditions outpace the ability of our thermoregulatory systems to adapt. If we overheat, we suffer from **hyperthermia**, and if we get too cold, we develop **hypothermia**. Internal temperatures that are either too high or too low can lead to death.

Hyperthermia When we become overheated, hyperthermia causes vasodilation, which brings the blood to the surface of the skin where it can be cooled by the evaporation created by sweating. Hyperthermia also results in a rapid but weak pulse and also rapid breathing, which dissipates heat through the lungs. When temperatures exceed your body's ability to cool itself, heat exhaustion occurs. In extreme cases of hyperthermia known as **heat stroke**, the individual may become delirious or hostile. Hyperthermia is classified as any core body temperature above 101 °F (38 °C), but death can result when core temperatures above 104 °F (40 °C) occur. For individuals of any age, exercising in the heat can be dangerous (see **Think About It: Avoiding Exercise-Induced Hyperthermia**). Treatment for hyperthermia consists of removing the individual from the hot environment and into a cool or shady place, administering fluids (ideally cold), and seeking medical attention.

You might well ask if a fever counts as hyperthermia and therefore should be treated in the same way. The answer is no to both questions because the underlying causes differ: a fever is the body's controlled response to internal conditions (such as infection), whereas hyperthermia occurs when external temperatures exceed the body's ability to cool itself. **Table 7.2** summarizes the differences between a fever and hyperthermia. The fact that fever is an internally generated state is also reflected in the presence of a diurnal (daily) cycle for fever. Because the hypothalamus is responding to multiple internal signals, a person's fever ebbs and flows, typically lowest in the morning and highest at night. Because hyperthermia is caused by external events, no such cycle occurs.

Our bodies strive to maintain critical biological functions within safe boundaries, and bacteria and viruses must do the same. Raising your internal temperature beyond what these infectious agents can withstand is a part of your body's defensive mechanism against them. During an immune response, your hypothalamus raises the set point for your internal temperature, much like raising the setting on your thermostat to turn the heat up in your house. This response highlights an important difference between a fever and hyperthermia: in a fever, your hypothalamus is responding to an internal signal by raising body temperature, whereas in hyperthermia, your hypothalamus is responding to an external stimulus by trying to lower body temperature.

Think About It | Avoiding Exercise-Induced Hyperthermia

Warm weather provides ample opportunity for outdoor recreation but also increased risk of hyperthermia. Hot, humid weather puts people at risk for hyperthermia, which is increasingly problematic as temperatures reach historic highs and deadly heat waves become more common. Older people are at greater risk of hyperthermia because their bodies are less efficient at regulating temperature, which is why it is especially important for older people to have access to air conditioning during heat waves.

It is not necessary to be an elite athlete to develop hyperthermia; people taking a leisurely hike in a park have died from hyperthermia when they were unprepared and did not recognize the signs. You may have heard about the young couple and their small child and dog who died of hyperthermia just 1.5 miles from their car at the tail end of a 7.7-mile hike near Yosemite. They both were experienced hikers and had an 84-oz. water bladder with them. However, although it was 75 °F when they set out, the temperature climbed to 107 °F, and there was no shade. Serious elevation change meant that the hike required vigorous exercise. As this tragic example demonstrates, hyperthermia can be rapid and deadly, so it is essential to recognize signs and address them immediately, as shown in this chart from the Centers for Disease Control (CDC).

HEAT-RELATED ILLNESSES

WHAT TO LOOK FOR	WHAT TO DO
HEAT STROKE	
• High body temperature (103°F or higher) • Hot, red, dry, or damp skin • Fast, strong pulse • Headache • Dizziness • Nausea • Confusion • Losing consciousness (passing out)	• Call 911 right away-heat stroke is a medical emergency • Move the person to a cooler place • Help lower the person's temperature with cool cloths or a cool bath • Do not give the person anything to drink
HEAT EXHAUSTION	
• Heavy sweating • Cold, pale, and clammy skin • Fast, weak pulse • Nausea or vomiting • Muscle cramps • Tiredness or weakness • Dizziness • Headache • Fainting (passing out)	• Move to a cool place • Loosen your clothes • Put cool, wet cloths on your body or take a cool bath • Sip water **Get medical help right away if:** • You are throwing up • Your symptoms get worse • Your symptoms last longer than 1 hour
HEAT CRAMPS	
• Heavy sweating during intense exercise • Muscle pain or spasms	• Stop physical activity and move to a cool place • Drink water or a sports drink • Wait for cramps to go away before you do any more physical activity **Get medical help right away if:** • Cramps last longer than 1 hour • You're on a low-sodium diet • You have heart problems
SUNBURN	
• Painful, red, and warm skin • Blisters on the skin	• Stay out of the sun until your sunburn heals • Put cool cloths on sunburned areas or take a cool bath • Put moisturizing lotion on sunburned areas • Do not break blisters
HEAT RASH	
• Red clusters of small blisters that look like pimples on the skin (usually on the neck, chest, groin, or in elbow creases)	• Stay in a cool, dry place • Keep the rash dry • Use powder (like baby powder) to soothe the rash

CDC

Questions

1. Search online for the question "How many people worldwide die from hyperthermia each year?" What was the estimate, and was it higher or lower than you expected?
2. Search online for the question "Who is at risk for heat stroke?" and identify three characteristics of individuals that place them at greater risk for heat-related illnesses.
3. What responsibility should schools, coaches, or trainers bear for protecting students from heat stroke?

TABLE 7.2 Comparing Fever and Hyperthermia

Fever	Hyperthermia
Initiated by hypothalamus	Not initiated by hypothalamus
Can be dangerous if not treated	Can be rapidly fatal
Rarely exceeds 104 °F (41 °C)	Can exceed 104 °F (41 °C)
Diurnal variation	No diurnal variation
Treated by antipyretics (acetaminophen) and in extreme cases, cooling cloths and baths	Antipyretics not effective; cold fluids, cool cloths to speed evaporative cooling, removal from situation and immediate medical care

Hypothermia When core body temperature dips below 95 °F (35 °C), we experience hypothermia. Common circumstances leading to hypothermia are exposure to the elements and water immersion. An interesting phenomenon associated with extreme hypothermia is paradoxical undressing, in which a person in the late stages of hypothermia takes off their clothes, thereby accelerating heat loss. Paradoxical undressing occurs in 20% to 50% of hypothermia deaths.

You may be familiar with the tragedy that occurred on Mount Everest in May 1996 when eight climbers died in a blizzard during their summit attempt. One of the guides was Scott Fischer, a world-class climber and a professional guide who collapsed after reaching the summit. Fischer was an expert at surviving in what is known as the death zone, and yet when his body was found, he had taken off his mittens and unzipped and removed part of his down suit in temperatures of −30 °F (−34 °C). What might be the mechanism behind paradoxical undressing? In the early stages of hypothermia, your brain directs blood vessels in your hands, arms, feet, and legs to vasoconstrict and thereby maintain your heat in the core. In the final stages of hypothermia, however, your hypothalamus might lose control and flood your periphery with warm blood that stimulates thermoreceptors in your skin, making you feel hot.

Treatment for hypothermia consists of rewarming the individual by providing warm and dry clothing, offering warm drinks or food, and placing them in a warm room or bath. Medical professionals may administer warmed IV fluids.

7.2 Before You Go On

Section Summary

Endotherms generate their own heat, whereas ectotherms use heat from the environment. Proper regulation of internal temperature occurs through the action of the hypothalamus, which receives input from thermoreceptors in the skin and organs throughout the body. In response to a body temperature that is too high (hyperthermia) or too low (hypothermia), your nervous system engages vasodilation or vasoconstriction of blood vessels to dissipate or retain heat, respectively.

Comprehension Check

1. Ectotherms and endotherms differ in that ectotherms _____, and endotherms _____.

- **a.** rely on the environment to regulate heat; regulate heat through internal means
- **b.** require more calories to live; require fewer calories
- **c.** cannot survive in cold climates; can survive in cold climates
- **d.** have a hypothalamus; do not have a hypothalamus

2. Which region of the nervous system is most involved in temperature regulation?

- **a.** frontal lobe
- **b.** thermoregulatory nucleus
- **c.** hypothalamus
- **d.** spinal cord

3. The difference between hyperthermia and hypothermia is that hyperthermia results from _____ and hypothermia results from _____.

- **a.** dangerously low body temperature; dangerously high body temperature
- **b.** dangerously high body temperature; dangerously low body temperature
- **c.** renal dysfunction; circulatory dysfunction
- **d.** an inability to direct blood supply to the skin; keeping blood in the body core

4. Overheating normally causes your blood vessels to _____, whereas hypothermia causes those same blood vessels to _____.

- **a.** constrict; dilate
- **b.** dilate; collapse
- **c.** constrict; constrict
- **d.** dilate; constrict

7.3 Homeostatic Regulation of Fluids Through Thirst

LO 7.3 Explain and distinguish the two mechanisms that control thirst.

Water is essential to life, and that truth is reflected in common wisdom. People say, "You can live without love, but you can't live without water," and "When the well is dry, you discover the value of water." Some individuals have lived without food for three weeks, but most of us would die within three to four days without water. Our bodies are 60% water, so you would not be surprised to hear that we have developed a mechanism to ensure we stay hydrated.

As is so often the case, when life depends on a process, nature provides redundancy. There are two mechanisms for detecting a lack of water in your body, and they operate in completely different ways to create the perception of thirst. **Volumetric thirst** responds to a decrease in blood volume, whereas **osmotic thirst** reacts to an accumulation of sodium within the extracellular space as a result of a loss of fluid. In each case, the mechanism operates through a negative feedback system. When a sensor signals that a minimum threshold has been exceeded, the sensor triggers a mechanism that responds until the receptor signals that the system is back within acceptable limits. Unlike thermoregulation, which happens to some extent without your conscious awareness, you are an active participant in managing your fluid levels by seeking out liquid and taking a drink.

Volumetric Thirst and the Renin–Angiotensin System MCAT

As we have already said, volumetric thirst is tied to blood volume. Volumetric thirst is caused by a loss of blood volume through blood loss or uncontrolled vomiting or diarrhea. The renin–angiotensin system (RAS) is a mechanism designed to maintain proper blood pressure and fluid balance. A loss of fluid in the body (known as **hypovolemia** or a loss of blood volume) triggers receptors in the kidneys; in response to that signal, a three-part chain of responses leads to the perception of thirst. Having too much fluid in your body (hypervolemia) is typically a consequence of kidney failure or heart disease. Because this condition is related to disease rather than mechanisms that regulate thirst, we will not discuss it here.

When the receptors in the kidneys signal a drop in blood volume, the kidneys secrete the protein renin into the bloodstream. Renin then combines with the peptide angiotensinogen, which is secreted by the liver. Renin breaks angiotensinogen into angiotensin I, which is then converted into angiotensin II, a very powerful vasoconstrictor. When blood vessels constrict, that forces the heart to work harder against the increased pressure, thereby raising blood pressure in response to the original signal of low blood volume. Angiotensin II also signals the posterior hypothalamus and the adrenal cortex to stimulate the kidneys to increase reabsorption of sodium, which also raises blood pressure.

This might seem a bit complicated, but the basic theme of volumetric thirst is that when the total amount of blood in our circulatory system falls below a set point, a signal from receptors in our kidneys leads to vasoconstriction that raises blood pressure and stimulates intracellular sodium retention (Figure 7.5). This process creates the perception of thirst in an effort to increase our blood supply by prompting us to drink. Note the similarity of this process to the general depiction of a negative feedback system in Figure 7.5, which you encountered at the beginning of this chapter.

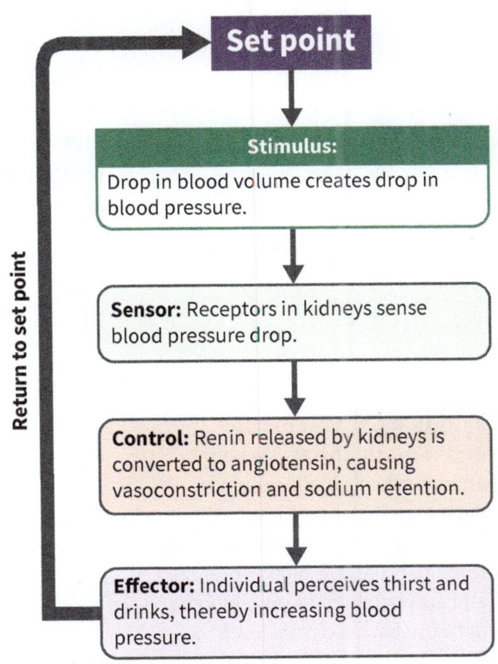

FIGURE 7.5 Volumetric thirst controlled by a negative feedback loop When blood pressure drops due to a loss of blood volume, this triggers a three-step process that causes the perception of thirst, prompting us to drink.

Some scientists consider volumetric thirst to be extracellular thirst because this mechanism responds to changes in the extracellular fluid balance. Given a mechanism for monitoring extracellular fluid balance, you might imagine that there is a corresponding mechanism monitoring the fluid balance within our cells, and you would be right. The intracellular fluid balance mechanism regulates osmolality, or the concentration of sodium within our cells. We now turn to an overview of that mechanism.

Osmotic Thirst and Cellular Sodium Concentration MCAT

The second mechanism that motivates a person to drink is that of osmotic thirst. Osmolality is a measure of the number of dissolved particles in a fluid, and in this context, osmolality refers to the monitoring of dissolved sodium in the blood. When the concentration of sodium within a cell is too high, water migrates from within the cell to the extracellular space, leading the cell to become dehydrated and shrink.

Whereas volumetric thirst is a response to a decrease in the extracellular fluid (that is, the fluid *outside* the cells, also known as extracellular fluid), osmotic thirst is created when the concentration of sodium in the intracellular environment (that is, the fluid *within* a cell) exceeds a safe level. One easy way to remember that osmotic thirst involves intracellular fluid and volumetric thirst involves extracellular fluid is to note that, alphabetically, O comes before V and inside comes before outside.

Figure 7.6 illustrates the differences between these two systems. Blood volume is represented by the height of the fluid in the column, and sodium concentration is portrayed by the number of dots in a column. When a person is in homeostasis, their intracellular and extracellular fluid levels are at an appropriate set point. When blood volume decreases below an acceptable level, the individual perceives volumetric thirst. Lower blood supply means lower blood pressure, so individuals experiencing volumetric thirst often crave salts, which cause vasoconstriction and raise blood pressure. In contrast, when the concentration of sodium within the cell becomes too high (shown in the right column, which has 60 sodium dots as opposed to 42 in homeostasis), the individual perceives osmotic thirst and drinks to dilute the sodium concentration within the cell.

Like volumetric thirst and other homeostatic process, osmotic thirst is a negative feedback system. Osmoreceptors in the anterior hypothalamus are continuously monitoring osmolality, but in addition, subcortical structures known as the **circumventricular organs (CVOs)** are uniquely equipped to detect changes in osmolality. The circumventricular organs are so named because they are located around or circling the ventricles. Two such regions, the vascular organ of lamina terminalis (VOLT) and the subfornical area (SFO), which are located below the fornix, are the location of sensors for changes in intracellular fluid. You know that brain structures are protected from many blood-borne toxins by the selectively permeable blood–brain barrier. Certain circumventricular organs are unique in that the capillaries that service them are highly permeable, which allows them to detect angiotensin released in response to volumetric thirst. In response to signals from the VOLT and SFO that originate when the intracellular fluid compartment becomes dehydrated, we experience thirst and

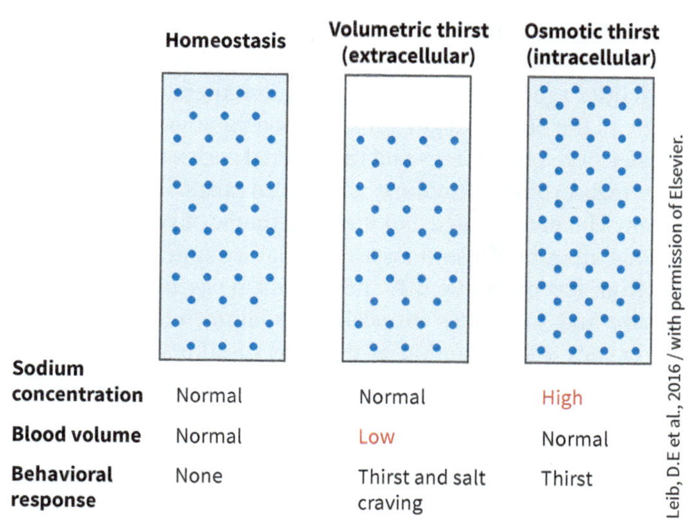

FIGURE 7.6 Volumetric and osmotic thirst When an individual's blood volume decreases below safe levels, the individual experiences volumetric thirst to replenish blood volume and a salt craving to raise blood pressure back to a safe level. When sodium levels within a cell become concentrated due to dehydration, the individual experiences osmotic thirst as a way to dilute the intracellular environment.

are prompted to drink (Figure 7.7). It is worth noting that the volumetric and osmotic mechanisms often work together to motivate us to drink.

Once our thirst is quenched, we stop drinking. You may wonder how it is that drinking in response to thirst quenches that thirst so quickly when it takes at least 5 minutes for your gut to begin to move water into the circulatory system and perhaps as much as 15 minutes for half of what you drank to make that transition. Logically, we should continue to feel thirsty and overhydrate until our two fluid-maintenance mechanisms have time to catch up. The answer is that stimulation from receptors in the mouth, throat, and esophagus signal for drinking to cease prior to the point where the individual would overhydrate. It is also worth pointing out that not all beverages are equal when it comes to replenishing our fluid reserves. Substances that remove water from the body are known as diuretics, and alcohol has significant diuretic properties as anyone who has ever consumed even a moderate amount of alcohol knows. Coffee and soda are mild diuretics, and many teas contain a small amount of caffeine and are therefore very mild diuretics. Many people rely on sport drinks to replenish their fluids lost through exercise, and although many of these drinks contain carbohydrates to provide energy, the best fluid replacement is water.

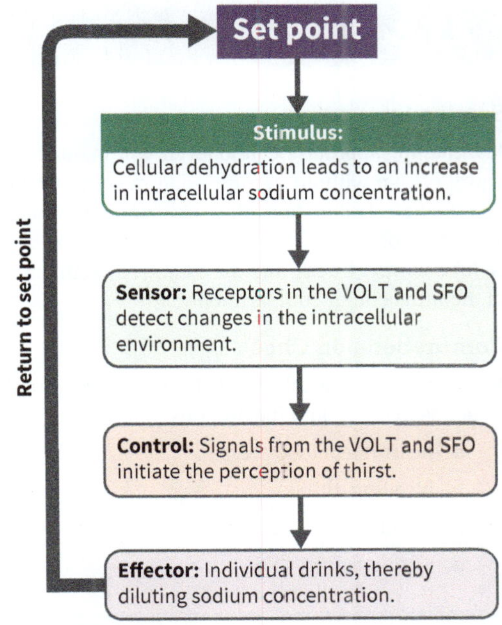

FIGURE 7.7 Osmotic thirst and the circumventricular organs When receptors in the brain detect an abnormal concentration of sodium in the blood, the individual perceives thirst and is motivated to drink, thereby diluting the blood's sodium concentration.

Disorders Related to Fluid Imbalance MCAT

Our body is designed to maintain a given volume of blood and fluid in the intracellular and extracellular environments and a specific level of electrolytes such as sodium and potassium. In earlier chapters, you saw the importance of Na^+ and K^+ ions for proper neural functioning. If our bodily fluids become too dilute or too concentrated, the concentration of ions necessary for neural function is not present, which can result in cardiac arrest and death. We will end our discussion of the mechanisms that maintain fluid homeostasis by briefly touching on maladies resulting from dysfunction of the system.

Polydipsia, or excessive thirst, is a symptom of several diseases. Polydipsia occurs in diabetes because the increase in blood glucose concentration leads the body to crave fluids in an effort to remove excess glucose from the blood and regain homeostasis. Diseases such as cholera, typhoid, and dysentery can deplete the body of fluids so rapidly due to vomiting and diarrhea that if untreated, death may ensue. As is so often the case, too much of a good thing can be just as dangerous as too little. Excess fluid in bodily tissues creates additional stress on the heart and kidneys and can lead to serious consequences.

Injuries (external or internal) that lead to a significant loss of blood volume are also a threat to life. The average person has between 1.2 and 1.5 gallons (4.5–5.7 L) of blood in their circulatory system; the loss of more than 0.5 gallon (1.9 L) of blood leads to hemorrhagic or hypovolemic shock due to the circulatory system's inability to transport oxygen and glucose. Without swift intervention, coma and death typically follow. In this Try It Out: Recognizing the Dangers of Fluid Imbalance, you can see how large deviations from fluid homeostasis have serious consequences.

Try It Out | Recognizing the Dangers of Fluid Imbalance

btrenkel/Getty Images

In this last section, you learned about the two mechanisms that keep our internal fluids at the right volume and at an appropriate electrolyte level. In this activity, you can apply what you have learned about neuroscience and thirst to situations you may encounter in the real world by completing the worksheet: Recognizing the Dangers of Fluid Imbalance.

7.3 Before You Go On

Section Summary

Maintaining appropriate fluid levels within our cells and in the circulatory system without diluting sodium concentration is essential for life. We have two systems that are designed to maintain proper fluid level: the volumetric mechanism maintains the balance in the extracellular compartment, and the osmotic system maintains the balance in the intracellular compartment.

Comprehension Check

1. There are two mechanisms for regulating blood fluid levels; one responds to _____ and the other to _____.

- **a.** heart rate; blood pressure
- **b.** bleeding; thirst
- **c.** blood volume; cellular sodium concentration
- **d.** angiotensin; CCK

2. Our volumetric system operates by the kidneys releasing _____, which causes _____.

- **a.** renin; vasoconstriction
- **b.** renin; vasodilation
- **c.** angiotensinogen; vasoconstriction
- **d.** renin and angiotensinogen; vasodilation

3. Our osmotic system operates by detectors in the _____ detecting changes in _____.

- **a.** circumventricular organs; blood volume
- **b.** circumventricular organs; cellular sodium
- **c.** posterior hippocampus; angiotensinogen
- **d.** mouth; moisture levels

7.4 Homeostatic Regulation of Hunger and Satiety

LO 7.4 Evaluate critically the multiple mechanisms that regulate hunger and satiety.

Regulation of temperature and thirst occurs through relatively simple mechanisms, so disorders are rare. We do not see people spontaneously bursting into flames or developing icicles on their limbs. Similarly, we do not generally see people spontaneously drinking gallons of water or deciding that drinking is something they no longer choose to do. If caloric regulation were as straightforward as these other homeostatic mechanisms, we would rarely see individuals who are obese or anorexic. Yet we frequently do see large departures from a healthy weight level.

Why might the regulation of feeding be so different from thermoregulation and thirst? The answer may be found in our distant past. Our ancestors evolved in an environment where food scarcity was an everyday challenge, and we would predict that multiple mechanisms that stimulate hunger would confer an advantage. Today, however, many of us live in an environment of food abundance that would have been unimaginable to our ancestors.

With that perspective in hand, we now turn to an overview of how the brain regulates feeding. We will begin by describing the digestive system and the role it plays in feeding and then proceed to a discussion of the complex web of hypothalamic centers and hormones that regulate feeding and **satiety**, or the feeling of fullness. We will conclude this discussion in the following section by applying what we learned to understand dysregulation of feeding and satiety.

The Gastrointestinal Tract MCAT

How does our body motivate us to feed? To understand the process, we begin with a quick tour of your digestive system, known as the **gastrointestinal tract (GI tract)**, which refers to the tract that leads all the way from the mouth to anus (Figure 7.8). Food enters the mouth, where it is chewed and mixed with digestive juices secreted by the salivary glands, and then travels down the esophagus and into the stomach. The contents of the stomach consist of ground up food as well as digestive juices, and this semi-liquid mixture is known as **chyme**. The stomach

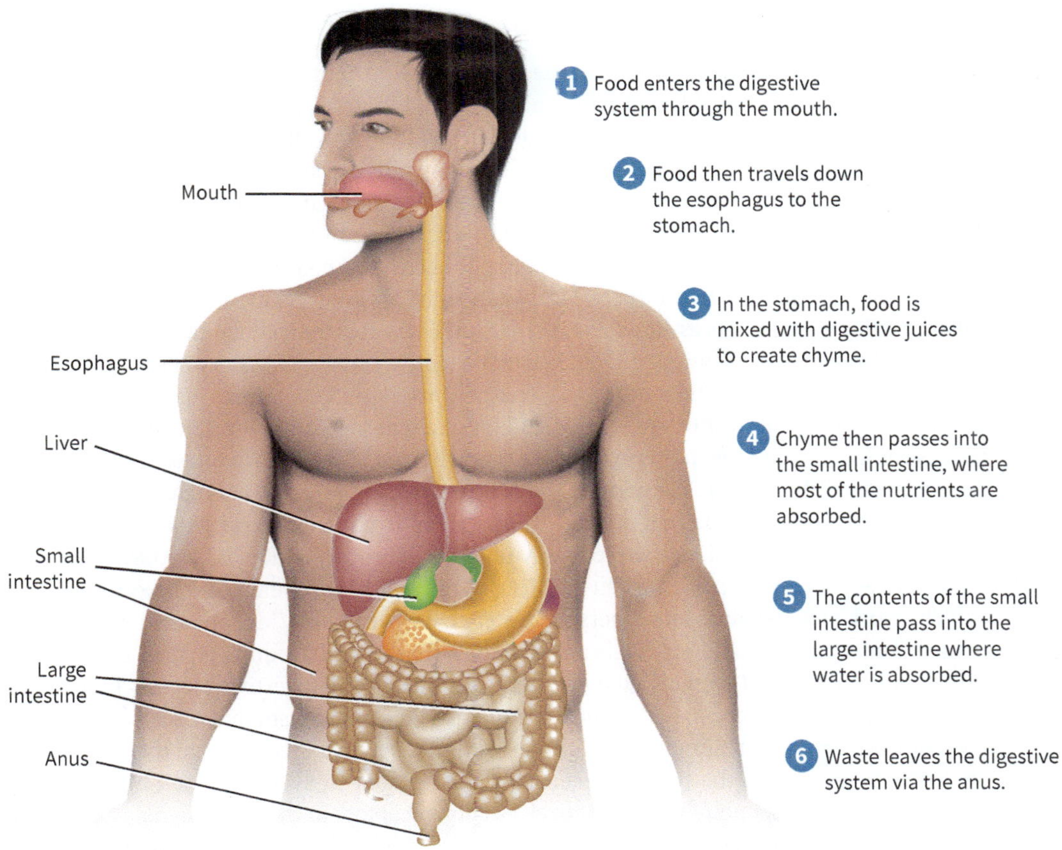

FIGURE 7.8 **The human gastrointestinal system** Food enters the digestive system through the mouth, passes through the stomach and small and large intestines, and exits via the anus.

expands as it begins to extract nutrients such as glucose, vitamins, and minerals from the chyme. After about one hour, the contents of the stomach pass into the small intestine where most of the digestive process occurs. Secretions from the liver, gall bladder, and pancreas enter the small intestine and break down complex molecules into soluble forms. Eventually, the chyme is passed to the large intestine, where water in the chyme is reabsorbed into the bloodstream over the next 2 to 3 days. Finally, feces exit the anus roughly 50 hours after the meal.

Some control mechanisms of hunger and satiety are located in the digestive system itself. An empty stomach creates muscular contractions that we call hunger pangs, and these motivate us to eat. When we eat a meal, our stomach stretches, and our nervous system interprets that expansion as a signal of satiety. Could maintaining caloric balance be as simple as your brain interpreting a certain set of signals from your stomach? In fact, there are receptors in the GI tract that signal via the vagus nerve (also known as cranial nerve X) that the stomach is distended, but there is so much more to satiety than mere stomach distension.

In addition to a simple signal from the GI tract indicating stomach distension, a hormonal signal emanates from the duodenum the upper portion of the small intestine, which occurs when the stomach releases chyme into the small intestine. This hormone, known as **cholecystokinin (CCK)**, stimulates the release of enzymes into the small intestine that aid the digestion of fats and proteins. This slows the process of emptying the stomach contents into the small intestine, thereby maintaining a feeling of fullness. In this way, CCK acts as an appetite suppressant by signaling satiety.

You might think that CCK is the answer to every overweight person's concerns about weight; just take a supplement of CCK and your desire for food will diminish and the pounds will melt away. The reality is that our body's drive for homeostasis is so effective that when an individual is continuously administered CCK as an appetite suppressant, the initial benefits soon disappear as the body adapts and the individual develops resistance to its effects. Intermittent administration of CCK as an appetite suppressant does not lead to long-term effectiveness either, because then the individual desires smaller meals but more of them.

The Neural Basis of Hunger and Satiety MCAT

In the preceding section, we discussed the physical path that food follows in your digestive system and the signals the GI tract sends to initiate and limit feeding. This simple model does not explain how morbidly obese individuals can eat the equivalent of several meals at one sitting and still experience hunger; nor does it explain why some people find food so unappetizing that they become emaciated. To answer these and other questions, we will turn to the brain for a more complete understanding of the neural basis of hunger and satiety.

It is easy to take our current understanding of behavioral neuroscience for granted, but in reality, quite a bit of our current knowledge was acquired within the lifetime of those scientists still alive today. Researchers assumed that there must be centers in the brain that were regulating hunger and satiety, but identifying them was not a simple matter. Our story starts in the 1950s in the laboratory of Elliot Stellar at Johns Hopkins University. Stellar spent his academic career studying the biological basis of motivation, and his graduate student Philip Teitelbaum extended that research effort into the question of how we are motivated to eat (**Advances in Behavioral Neuroscience: Philip Teitelbaum and the Motivation to Eat**). In his research on rats, Teitelbaum was able to determine that damage to the lateral hypothalamic nucleus (LH) interfered with feeding so that the animals wasted away unless fed through a tube. In contrast, damage to the ventromedial hypothalamic nucleus (VMH) led the animals to overeat and become obese.

The work of Teitelbaum and others suggested that the LH and VMH operated in tandem to regulate hunger and satiety. You likely already know that our cells require glucose as an energy source and our GI tract extracts glucose from food during the digestive process. In a simple world, hunger and satiety can both be easily explained as a mechanism that reacts to changes in blood glucose levels. This simple model proposed that when glucoreceptors in the LH detected low levels of glucose in the blood, the LH signaled hunger. Conversely, when blood glucose levels exceeded that set point in the feedback system, receptors in the VMH signaled satiety. In both cases, the mechanism operates as a negative feedback loop; when a sensor detects sufficient deviation from a set point, a control system activates to return the system to its original state. As you have likely surmised, the reality of calorie regulation is more complex.

Advances in Behavioral Neuroscience | Philip Teitelbaum and the Motivation to Eat

Philipfest

Dr. Philip Teitelbaum is a behavioral neuroscientist whose work has shaped our understanding of hypothalamic function in feeding and satiety as well as recovery from neurological injury. Two lessons to learn from his work are that questioning dogma is always worthwhile, and creativity in experimental design will lead to a deeper understanding of neural function.

Teitelbaum was born in Brooklyn, New York and earned his undergraduate degree at City College of New York and his Ph.D. at Johns Hopkins; academic posts at Harvard, the University of Pennsylvania, and the University of Illinois followed. Teitelbaum was interested in understanding the neural basis of hunger and satiety, so he focused on the lateral hypothalamus (LH). The common research strategy was to lesion (injure) a neural center in a rat and see what function was disturbed, so Teitelbaum and his colleagues lesioned the LH in rats and found that they would no longer eat or drink and had to be maintained with a gastric tube. The thinking was that removing the LH from the brain's homeostatic circuitry eliminated the ability to feel hunger.

That might have been the end of the story except that Teitelbaum went further and wondered if these LH-lesioned rats would in fact eat if provided with highly palatable foods. Chocolate chips were introduced, and the rats began to eat; later a mash of milk, chocolate, and cereal was introduced, and after a period of time, feeding resumed, albeit with some residual deficits. Teitelbaum's work showed that neural plasticity was possible in such situations, and more importantly, that simple models of hypothalamic function were inadequate to explain complex functions such as hunger and satiety. The LH and other neural centers were much more complicated than previously thought. His work on the neural basis of motivation has been an important foundation of the field of affective neuroscience, the neural basis of emotions, personality, and mood.

Dr. Teitelbaum received the distinguished scientific contribution award from the American Psychological Association (APA) and has served as the president of the Division of Comparative and Physiological Psychology of the APA. He is a member of the National Academy of Sciences and the Society of Experimental Psychologists.

Questions

1. **Earlier you learned that stomach distension is a signal for satiety. Why then have other signaling mechanisms for satiety and hunger?**
2. **As you consider your own eating habits, do you think they are more a result of your genetics, your environment, or your own personality? Briefly explain your answer.**

Damage to the VMH in laboratory rats did lead to **hyperphagia**, or excessing eating, but not because it directly stimulated eating. A primary role of the hypothalamus is energy regulation, and damage to the VMH causes the animal to overeat in response to a misperception of low energy. As a result, its metabolism converts its food to stored fat instead of glucose. Paradoxically, this excessive storage of fat leaves the animal low on glucose, so it eats more to compensate. If this is confusing, imagine that as a car's gas gauge nears empty, the driver fills a gas can but stores the surplus fuel in the garage. The available fuel supply has increased, but not the amount accessible by the car. Our understanding of the role of the LH was expanded when it was recognized that LH lesions produced animals that were less responsive to the environment. The LH was less of a hunger center than a center that promotes wakefulness and arousal, which of course affects feeding. As researchers continued to investigate the neural basis of feeding, they realized that the hypothalamus was coordinating an array of hormones. We now turn to a discussion of these important chemical messengers.

Hormonal Regulation of Hunger MCAT

We began our discussion of the biological basis of hunger with the simple idea that the LH initiated feeding when blood levels of glucose fell too far, and the VMH initiated satiety when blood glucose levels were restored. The reality is not so simple. Hunger and satiety are perceptual experiences that are created by a complex web of interactions among hypothalamic centers in the CNS, receptors in the GI tract, and hormones that are secreted in the CNS and in the GI tract. Because the overall picture is complicated, we will begin by discussing the role of these important hormones, which are produced throughout the body, and then relate these to hypothalamic mechanisms.

Ghrelin

One hormone that appears to stimulate hunger is **ghrelin**. Ghrelin is produced in the GI tract as well as in regions of the hypothalamus such as the arcuate nucleus (AN). In the GI tract, ghrelin stimulates movement of chyme through the digestive tract, whereas in the CNS it appears to stimulate the perception of hunger. In fact, animal studies that involved injecting ghrelin directly into the brain and abdomen led to increased feeding and did so in a dose-dependent relationship (that is, more ghrelin led to more feeding). Clinical research with humans reinforced the hypothesis that ghrelin acts as an appetite initiator. Measurements of ghrelin concentration in the liquid portion of blood (known as plasma) demonstrate that ghrelin levels rise precipitously prior to a meal and begin to drop as soon as feeding commences (Figure 7.9).

That ghrelin administration in animals leads to increased feeding, food foraging, and food hoarding is compelling evidence that ghrelin plays a role in the mediation of hunger for you and me, but the situation is more complex than it appears at first. For instance, individuals who elect gastric bypass surgery to control their weight lose a considerable proportion of the tissue that secretes ghrelin, yet they continue to experience hunger. Similarly, mice that are bred to not produce ghrelin still feed at a normal rate. This evidence suggests that our nervous system does not rely on ghrelin alone to stimulate hunger.

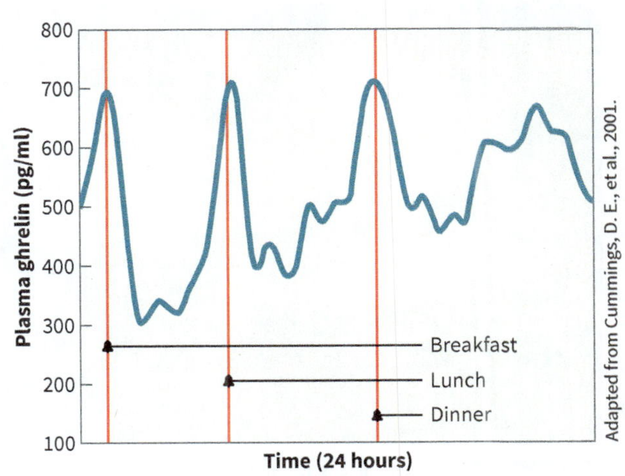

FIGURE 7.9 Plasma ghrelin levels in relation to mealtimes in humans Ghrelin levels in the blood begin to rise prior to the initiation of feeding, and this is accompanied by the perception of hunger.

Neuropeptide Y

The role of ghrelin is well established, but nature seems to leave nothing to chance, so you should not be surprised to find another actor involved in stimulating hunger. **Neuropeptide Y (NPY)** is secreted by cells in the AN of the hypothalamus as well as the periventricular nucleus (PVN), and our current understanding is that ghrelin release in the AN stimulates NPY release. Studies using a rat model have shown that increases in NPY in the brain result in increased appetite, whereas blocking the action of NPY receptors leads to a corresponding decrease in feeding. One interesting observation about the effect of NPY on hunger is that NPY appears to specifically increase appetite for carbohydrates. Carbohydrates are a compact source of energy, so they confer an advantage for individuals leading a very active life such as our hunter-gatherer ancestors. For people today who do not have to expend a great deal of energy to obtain food, overreliance on carbohydrates leads to weight gain.

Orexin Another hormone involved in appetite is **orexin**, which is primarily secreted by cells in the LH. Orexin appears to increase appetite while also suppressing post-meal feedback from the digestive system that would normally act to signal satiety. You saw earlier that the LH is more than just a satiety center; it stimulates feeding as part of a general regulation of arousal. This explains why orexin-deficient animals lose weight despite orexin acting to stimulate appetite; when an animal becomes lethargic and uninterested in feeding, it loses weight. Orexin receptors appear to be influenced by blood glucose levels, in that chronic hyperglycemia (high glucose) leads to a reduction in orexin-receptor activity. This link between orexin, appetite, and glucoregulation has caused researchers to consider orexin as a target for treatment of obesity.

Agouti-related peptide There is one final hormone of note that contributes to the perception of hunger. **Agouti-related peptide (AgRP)** is a neuropeptide produced in the AN of the hypothalamus. AgRP increases appetite and decreases metabolism, thereby leading to weight gain and is thought to play a role in obesity.

You now can see that we have multiple mechanisms to create the perception of hunger so that we feed. Ghrelin, NPY, orexin, and AgRP are all seen to influence feeding behavior in some fashion, but to fully understand feeding behavior, we need to understand its complementary process, satiety. We therefore turn our attention to leptin.

Hormonal Regulation of Satiety MCAT

You have seen again and again that the brain creates mechanisms so that for every gas pedal there is a brake, so you will not be surprised to learn that there are satiety hormones paired with ghrelin. Earlier in our discussion of the GI tract, you learned that CCK is secreted during feeding and its effect is to produce the feeling of fullness during a meal. CCK is created in the periphery, so we would expect there to be a satiety messenger created in the CNS that corresponds to ghrelin. One line of research into feeding and obesity was to breed mice for obesity and then look for differences between their physiology and that of normal mice. In doing so, researchers identified **leptin**, a hormone secreted by the small intestine that inhibits feeding.

Leptin (which comes from the Greek meaning slender or thin) acts on receptors in the arcuate nucleus of the hypothalamus. The arcuate nucleus is adjacent to the ventromedial nucleus, another hypothalamic nucleus associated with satiety. Leptin is contained in our fat cells (also known as **adipose tissue**) as well as cells in the small intestine. Keeping in mind that body fat is a form of stored energy, leptin monitors body fat reserves and ideally acts through a negative feedback loop as an appetite suppressor when fat reserves become too large. Animal studies suggest that large fat stores result in increased leptin levels, which then lead to decreased appetite and increased physical activity to burn those fat stores. Rats bred to generate no leptin act as if they are starving; they will eat continuously and move very little in order to conserve energy (**Figure 7.10**).

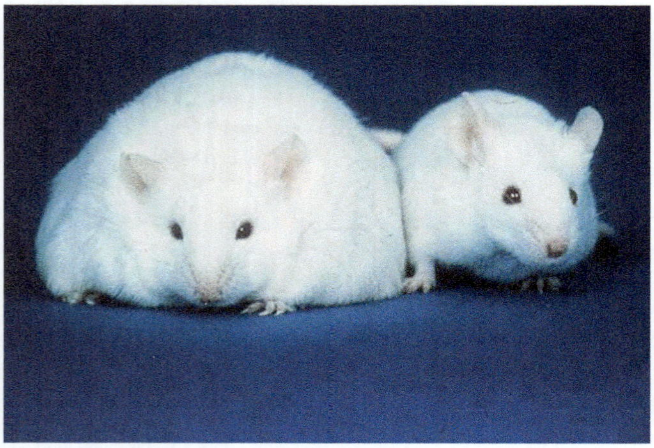

Sunholm/Wikimedia Commons/Public Domain

FIGURE 7.10 Mouse bred for leptin deficiency compared to normal mouse The mouse on the left side of the image has been bred to be deficient in leptin. In contrast to the normal mouse on the right side of the image, the leptin-deficient mouse eats more than normal and is obese.

We began this section with the prediction that the mechanism that motivates us to feed through the perception of hunger and satiety would be complex, and the evidence supports that prediction. The digestive tract is part of the signaling system, and the hypothalamus secretes both excitatory and inhibitory messengers (**Table 7.3**). These messengers react in an interactive fashion to ensure adequate calorie intake.

As researchers began to understand the homeostatic mechanisms that regulated hunger and satiety, their hope, to paraphrase a popular song of the 1960s, was that "One pill would make you larger, and one pill would make you small." We now know quite a bit about the underlying neurological and hormonal mechanisms that are in play here, but an easy solution to weight management is not apparent.

TABLE 7.3 Mechanisms That Regulate Hunger and Satiety

Mechanism	Location/Source	Action
Blood glucose levels	Circulatory system	Hunger and satiety
Ghrelin	GI tract and arcuate nucleus of hypothalamus	Hunger
Neuropeptide Y (NPY)	Arcuate nucleus of hypothalamus	Hunger
Orexin	Lateral hypothalamus	Hunger
Agouti-related peptide (AgRP)	Arcuate nucleus of hypothalamus	Hunger
Leptin	Fat cells and cells of the small intestine in GI tract	Satiety
Stomach distension	Stomach of GI tract	Satiety
Cholecystokinin (CCK)	Small intestine	Satiety

7.4 Before You Go On

Section Summary

Our need to consume calories in order to survive is reflected in a number of interconnected mechanisms that motivate us to eat and then cause us to feel satisfied. Some signals come from physical changes to the digestive system such as stomach distension. Others involve hormones and other chemical messengers such as ghrelin, NPY, orexin, leptin, and AgRP.

Comprehension Check

1. When you eat, food passes through your gastrointestinal tract in which order?
   - **a.** large intestine, stomach, small intestine
   - **b.** esophagus, stomach, large intestine, small intestine
   - **c.** esophagus, small intestine, stomach, large intestine
   - **d.** esophagus, stomach, small intestine, large intestine
2. One way the digestive system participates in the regulation of satiety is through _____.
   - **a.** secretion of bile from the gall bladder
   - **b.** suppression of CCK
   - **c.** release of CCK
   - **d.** release of ghrelin into the bloodstream
3. When you feel hungry, it is likely that _____ levels are high and _____ levels are low.
   - **a.** ghrelin, NPY, AgRP, and orexin; leptin
   - **b.** leptin and AgRP; ghrelin, NPY, and orexin
   - **c.** NPY and ghrelin; leptin, AgRP, and orexin
   - **d.** leptin and orexin; NPY and ghrelin
4. When you feel satisfied after a meal _____ levels are high and _____ levels are low.
   - **a.** ghrelin; leptin
   - **b.** ghrelin; leptin and CCK
   - **c.** CCK and ghrelin; other
   - **d.** ghrelin; CCK and leptin

7.5 Disorders of Hunger and Satiety

LO 7.5 Analyze the causes and treatments of feeding disorders in light of the mechanisms that regulate feeding

Any conversation about how we take on calories to fuel our bodies must also consider some vexing problems. For example, why aren't healthy food tasty and unhealthy foods unappetizing? Why do we sometimes eat more than is good for us, and why do some people eat until they weigh 500 pounds (226 kg) and more? Finally, how it is that some people simply choose not to eat and may waste away? In this section we will discuss what happens when feeding mechanisms go awry.

Try It Out | Measure Your BMI

There is a considerable body of evidence that supports the claim that BMI is related to health and longevity. Download the worksheet **Measure Your BMI**, calculate your BMI, and then learn about how it is used, its limitations, as well as other measures of body mass.

Obesity MCAT

How do we define **obesity**? One answer is through the **body mass index (BMI)**, which is defined as one's weight divided by the square of one's height. A healthy BMI is considered to be between 18.5 and 24.9, and those with a BMI of 25 to 29.9 are considered to be overweight. Individuals with BMI over 30 are considered to be obese, and those with a BMI over 40 are considered to be morbidly obese, a condition in which the weight represents a clear and immediate danger to one's health. Obese individuals are at significant risk for hypertension, coronary heart disease, stroke, type 2 diabetes, and other illnesses, so treating obesity is a public health issue. Learn about the BMI and its limitations in **Try It Out: Measure Your BMI**.

One of the many appealing aspects of behavioral neuroscience is that it consistently connects science to real-world problems, and understanding the homeostatic mechanisms that underlie feeding is an example. Take a moment to examine **Figure 7.11**, which illustrates the epidemic of obesity (a) across the world and (b) in the United States.

Obesity is a worldwide health problem, but one that is especially common in developed nations. More than 40% of adults in the United States are obese according to the WHO criteria. Obesity is driven by biological and cultural factors, as can be deduced by the regional trends in U.S. obesity incidence. There are multiple reasons why obesity is a problem in developed countries. The availability of unlimited amounts of food, especially foods heavy in sugars and fats that we are predisposed to favor, is a major factor in the obesity epidemic. In addition, there is compelling evidence that there is a genetic component to obesity; obese twins raised apart are very likely to grow up with similar weights and distribution of body fat rather than reflecting the social and dietary environment in which they grew up.

What you have learned so far is that we have a constellation of mechanisms that are all designed to initiate feeding but fewer opposing mechanisms to keep our caloric intake in check. Our body weight is maintained by a set point that has a genetic basis and is regulated by a complex circuit involving the hypothalamus and other regulatory mechanisms in the body. However, evidence indicates that consistent and continued overeating creates an allostatic response that leads to an upward drift in the hypothalamic set point. In a sense, we are the victims of our own success as a species; we have learned to provide ourselves with more food than is healthy. Obesity is linked to a number of serious health risks, so one of the challenges of behavioral neuroscience going forward is to find ways to alter feeding behavior and its underlying mechanisms in a healthy fashion.

Causes of obesity

Now that you understand the roles of various chemical messengers in caloric regulation, you should be able to come up with several hypotheses for how levels of these messengers that are too high or too low might contribute to obesity.

One obvious hypothesis involves leptin, which stimulates feelings of satiety. Obesity might be caused by a lack of leptin, but we know that obese individuals have as much or more leptin than their thin counterparts. It might also be that obese people have fewer leptin receptors, but again, the evidence is that obese individuals do not meaningfully differ in the number of leptin receptors. The best evidence at this time is that a dysfunction in the post-receptor pathway explains why obese individuals have a sufficient supply of leptin and leptin receptors but do not appear to respond to it by feeling satiety. One thought that explains the tenuous relationship between leptin and satiety is that leptin is not a simple satiety switch to prevent overeating but may actually be a starvation-avoidance mechanism that prompts the organism to eat until enough fat is stored to survive a lean time. In this model, obese individuals may be storing fat in preparation for a famine in spite of abundant food availability.

FIGURE 7.11 **Obesity across the world and the United States** **a.** Obesity is a problem afflicting the more developed nations in the world. **b.** In the majority of the 50 U.S. states, obesity is a serious health issue. Regions with the highest levels of obesity are in red, followed by orange, then tan, then yellow, and finally, green.

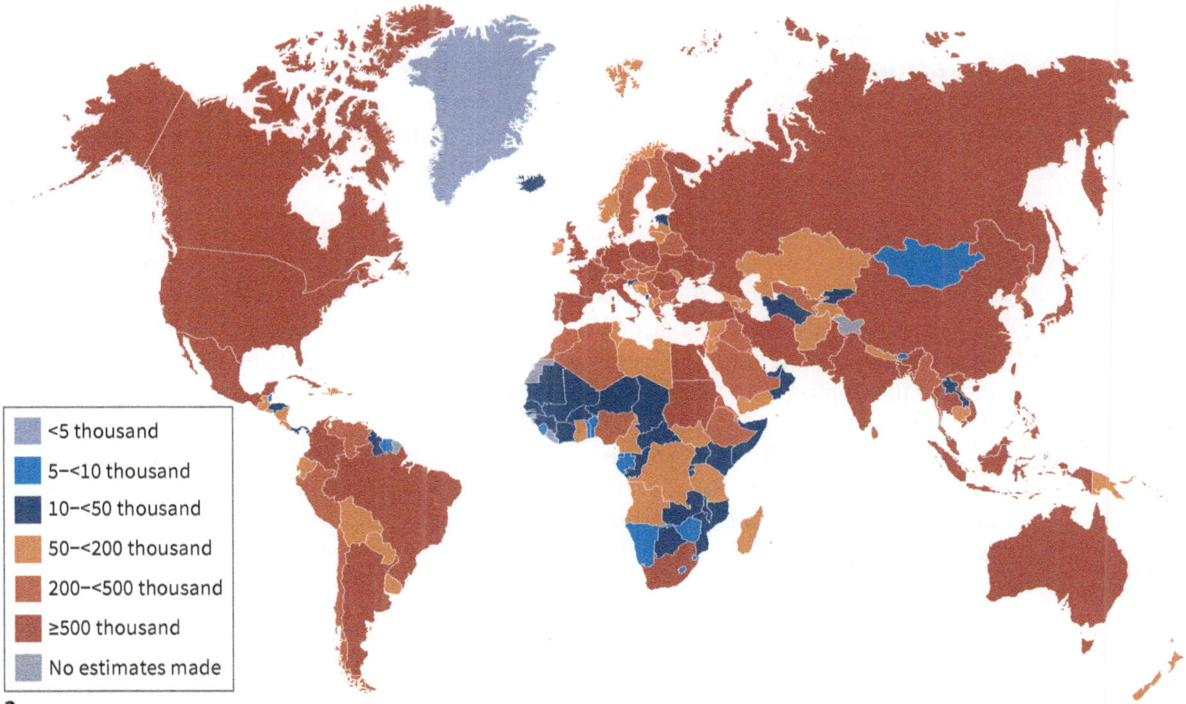

a.

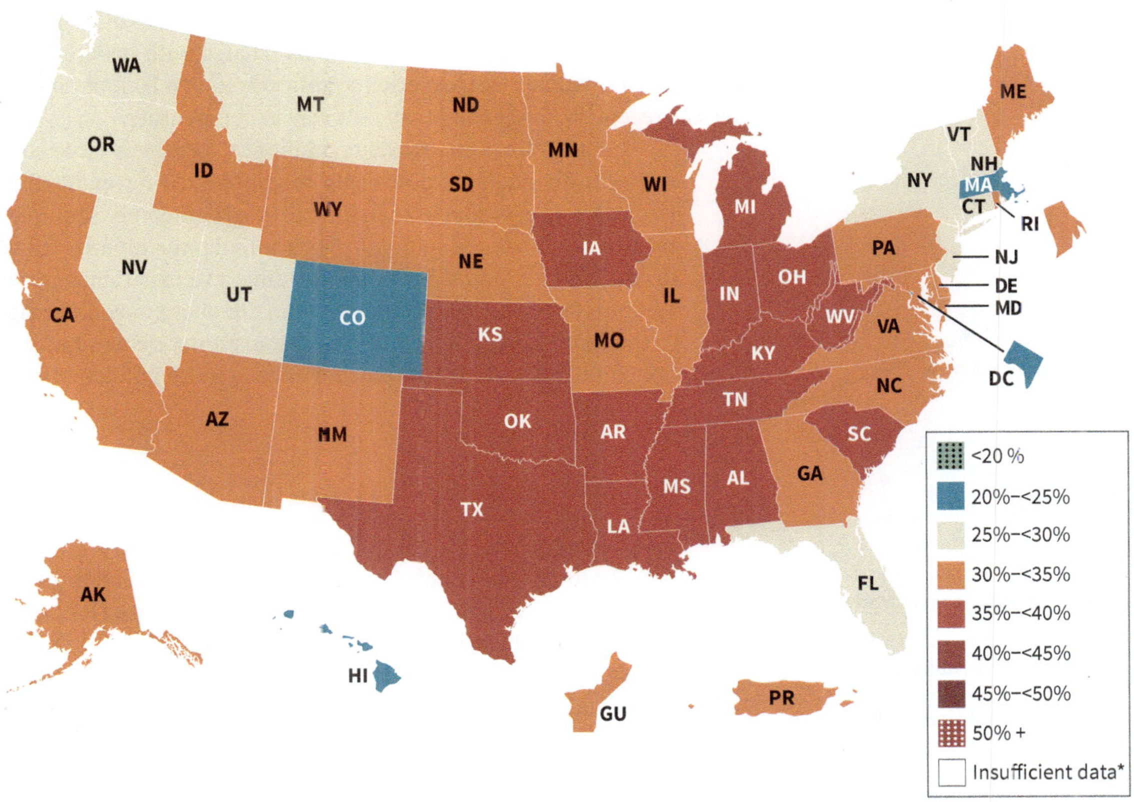

b.

A second obvious hypothesis is that obese individuals have an overabundance of ghrelin, which stimulates appetite. Again, the evidence does not support that theory. In fact, obese individuals appear to have less ghrelin in their bloodstream than do individuals of normal weight. In large part, the causes of obesity remain a mystery and are a topic of great research interest.

Treatment of obesity You might have thought before you read this chapter that treating obesity is simple; perhaps just telling people to eat less or take a pill that interrupts the mechanism that creates hunger. By now you realize that we are endowed with multiple mechanisms that are designed to motivate us to eat, so treating morbid obesity is not a simple matter.

Take a moment to consider what you have just learned in relation to other topics you have explored so far. Essential functions typically have more than just one control mechanism to ensure your continued survival, and hunger and satiety are no exception. If a feeling of fullness were the only satiety signal, then there are some predictions one could make about treating obesity. One such prediction is that drinking a glass of water before a meal would distend the stomach and thereby reduce hunger; in fact, this strategy does lead to weight loss if followed regularly. The second is that bariatric surgery (removing a part of the stomach to make it smaller or placing a band around the stomach to restrict its ability to expand) should lead to a healthy weight. You may be familiar with individuals such as actors Rosie O'Donnell and Star Jones, musician Randy Jackson, meteorologist Al Roker, author Anne Rice, and politician Chris Christie, who have all publicly disclosed their decision to have bariatric surgery to reduce the size of their stomach in an attempt to lose weight. In fact, bariatric surgery does result in significant weight loss in most cases as the individual experiences fullness with less food consumed. However, a feeling of fullness is only one part of the body's caloric regulation, so solutions that only address stomach distension are unlikely to be a simple answer to our natural propensity to eat more food than we need.

Just in case you felt that understanding the root causes of obesity were complicated enough, there is more. One way to view overeating is to consider it as addictive behavior, and you may already know that dopamine (DA) receptors are implicated in the function of the limbic system and the perception of pleasure. A comparison of DA receptors in a sample of obese and normal weight controls found that those individuals with higher BMI levels also had fewer DA receptors in the striatum, a part of the brain associated with reward (**Figure 7.12**). This suggests that obese individuals may experience anhedonia, or a lack of pleasurable feelings, due to fewer DA receptors, and eating may be a way to compensate.

Clearly, the homeostatic regulation of feeding is quite complex, and telling individuals who struggle to control their weight to simply eat less will not lead to a successful treatment outcome. Multiple chemical messengers from the gut and the hypothalamus interact to motivate us to experience hunger to initiate feeding and experience satiety to stop feeding. The fact that individuals who are obese may differ significantly in the functioning of a critical neurotransmitter system makes the development of effective treatments for obesity more problematic.

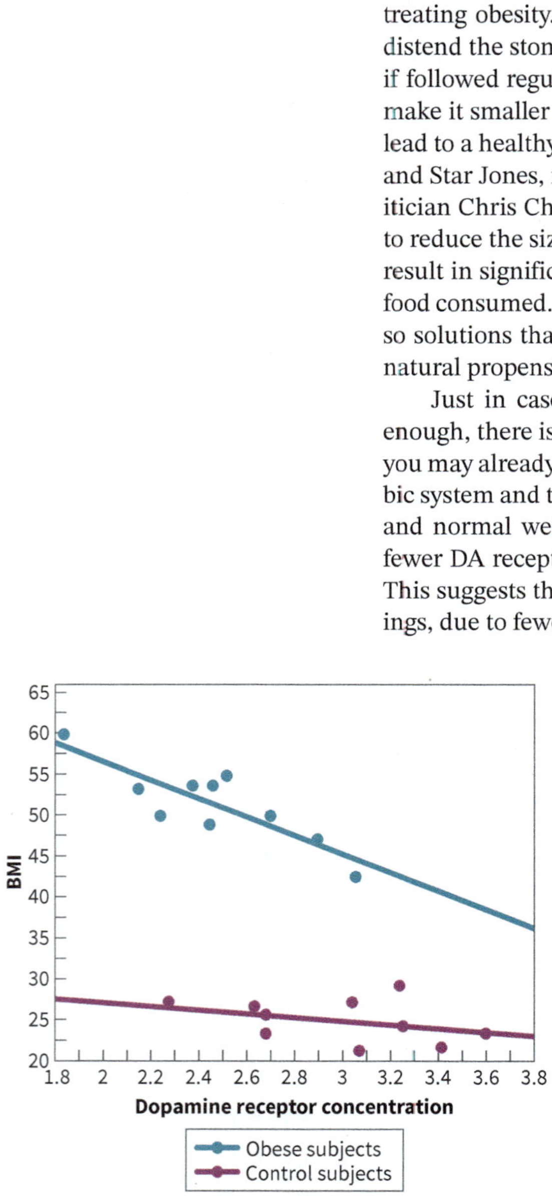

FIGURE 7.12 DA-receptor concentration and obesity Individuals of normal weight showed very little variation in DA receptor concentration in the striatum. Obese individuals whose BMI ranged from >40 to 60 demonstrated an inverse relationship; the higher the BMI, the fewer DA receptors in the striatum.

Prader-Willi Syndrome MCAT

Investigating cases of dysfunction is one way to gain insight into the function itself. A congenital disorder related to dysfunction of the hunger and satiety mechanisms is **Prader-Willi syndrome (PWS)**, which was mentioned in this chapter's *Stories of the Brain*. PWS is a rare genetic disorder that affects perhaps one out of every 15,000 births each year. PWS equally affects males and females as well as different ethnic groups. It is caused by loss of a specific region of genetic material on chromosome 15. Typically, the male and female parents each contribute one copy of this chromosome to their offspring, but in this specific case, only the male copy is active. For PWS to occur, either the information on the male copy is damaged or the offspring receives two female copies.

Prader-Willi children exhibit mild intellectual impairments, weak muscle tone, and difficulties breast-feeding that can be serious enough to require tube feeding to avoid a failure to thrive. These symptoms are serious on their own, but our interest is in a defining symptom of PWS called hyperphagia. Hyperphagia is defined as an abnormal appetite and interest in food. PWS individuals do not experience satiety and are compelled to eat beyond all normal limits (**Figure 7.13**).

At this point in our discussion, you should be able to make some predictions about possible causes of PWS. One clue comes from studies that compared the volume of the hypothalamus and pituitary of individuals with PWS with controls. Those with PWS were reported to have a lowered connectivity between the hypothalamus and the rest of the brain, and also to have a much smaller pituitary gland. These two observations suggest that at least part of the cause of PWS is anatomical and therefore not easily remedied. The genetic basis for Prader-Willi syndrome does imply that genetics plays a part in weight management of the rest of us.

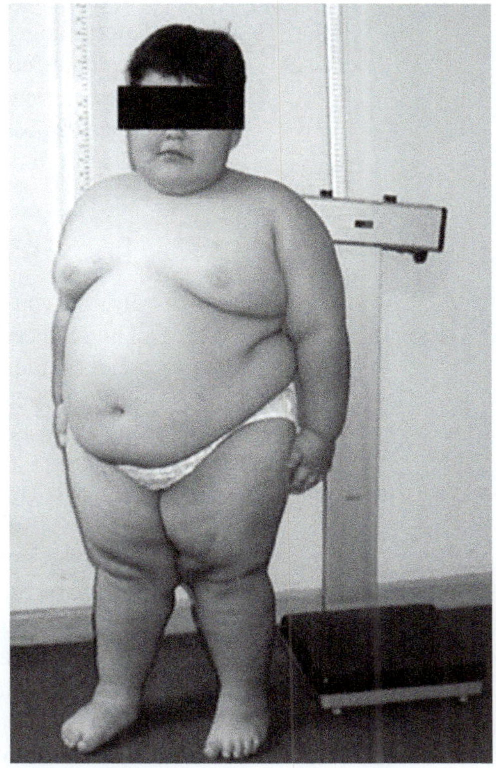

FIGURE 7.13 Prader-Willi syndrome Individuals with Prader-Willi syndrome struggle to control their weight. They also exhibit intellectual deficits and weak muscle tone.

Anorexia Nervosa MCAT

Throughout our discussion of homeostasis, we have seen that extremes in either direction are dangerous. Therefore, if morbid obesity represents the danger of too many calories, there should be a parallel extreme on the side of too few calories. **Anorexia nervosa** is an eating disorder characterized by extreme low weight and a poor appetite resulting from a highly restrictive diet and/or extreme exercise. A BMI of 17 is considered to be mild anorexia, and a BMI of 15 or less is severe. To put these numbers in context, a person who is 6 feet tall (1.8 m) and 110 pounds (50 kg) would have a BMI of 15. Anorexia leads to various nutritional disorders and can be fatal. The lifetime incidence of anorexia is low, perhaps just 0.5% of the population, but that estimate does not tell the whole story. Males rarely exhibit anorexia (~0.05%), whereas as much as 5% of females reach the diagnostic criteria, and some studies have found as many as 13% of female teenagers in the United States experience an episode of anorexia (**Figure 7.14**).

Causes of anorexia

The causes of anorexia are complex, but our interest in the biology of behavior leads us to an understanding of some aspects of the disorder. Neuroimaging studies that compare the brains of anorexics and neurotypical individuals suggest that there are several underlying differences that may contribute to the emergence of the disorder. Anorexics typically show a significant disturbance in body image. They may literally be wasting away, yet they see themselves as grotesquely obese, an example of a dysfunction of self-perception. Other research suggests that anorexics receive less stimulation in the reward system from eating and that, in some way not yet clearly understood, serotonin dysregulation is part of the problem.

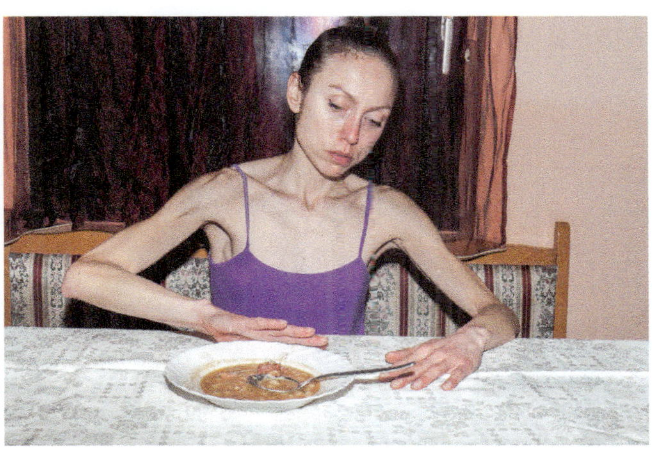

FIGURE 7.14 An anorexic individual struggling to eat Anorexia is characterized by low body weight, poor appetite, and a perception of one's own body that is not in accord with reality.

Treatment of anorexia Treatment for anorexia involves a multimodal approach of diet change, psychotherapy that involves the individual's family, and often hospitalization to regulate food intake. Despite the research suggesting serotonin's involvement in anorexia, medications that affect serotonin levels have not been particularly successful. Research investigating possible differences in ghrelin blood levels or receptor sensitivity has not provided a clear understanding that leads to a reliable treatment plan for anorexia.

The Challenges of Treating Disordered Eating MCAT

By now you have seen how complicated are the mechanisms that regulate hunger and satiety. The science behind homeostasis is fascinating on its own, but for many, the goal is that the science informs treatment. If hunger and satiety were simple mechanisms, we would have the answer to caloric homeostasis problems such as obesity and anorexia. Theoretically, delivering agonists of ghrelin, NPY, or orexin on the one hand or blocking leptin production in the body or leptin receptors in the hypothalamus on the other hand should stimulate hunger in those who need to gain weight. The opposite course of delivering agonists of leptin or blocking receptors for ghrelin, NPY, and orexin should be a silver bullet for obese individuals needing to shed weight. Unfortunately, injecting a hormone such as ghrelin or leptin to the third ventricle of a rodent (a common research technique) may be justified to learn about homeostasis but is not a practical course of treatment for humans.

As you examine **Figure 7.15**, you can easily understand why altering our feeding behavior is complicated. Hypothalamic centers such as the VMH, LH, and AN and associated hormones such as ghrelin, NPY, orexin, and leptin all interact dynamically and do so in communication with receptors in both the CNS and the periphery.

Finally, while our focus in this chapter is on the neurology underlying behavior, humans enjoy the luxury of a sophisticated cortex that also influences our behavior. In discussions of biological mechanism such as homeostasis, it is easy to forget the unique contributions that

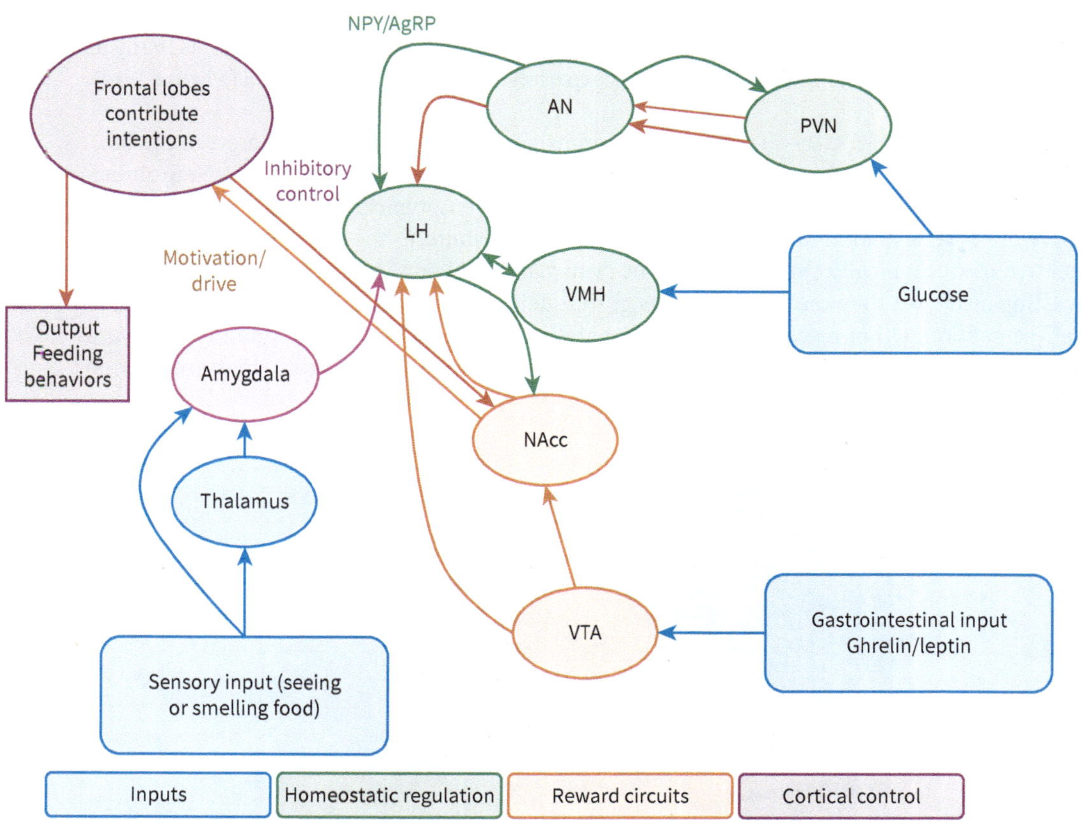

FIGURE 7.15 Homeostatic reward and cortical circuits involved in hunger and satiety The neurological basis of feeding behavior involves multiple signals, multiple sensors throughout the body and brain, and cortical control through intentions and plans.

our cortex adds to the problems associated with feeding regulation. Social media and culture bombard us with messages about our appearance and weight and create expectations that are unreasonable and may lead to cycles of binging and purging that have their own disruptive impact on our regulatory systems. Finally, when faced with disappointment, our nervous system may guide us to fill the empty space with food. Our nervous system is dynamic and adaptive, so when we actively adjust the functioning of one of these systems, others react to maintain the status quo. If the mechanisms that control hunger and satiety were simple, feeding disorders would already be a thing of the past.

7.5 Before You Go On

Section Summary

The availability of food that most people enjoy has led to a worldwide epidemic of obesity. The genetic basis for Prader-Willi syndrome, a disorder that involves insatiable eating, demonstrates the role of genes in caloric regulation. Less frequent than obesity but more deadly is anorexia, an eating disorder in which an individual refuses to eat, leading to disease and even death. Treating disordered eating is extremely challenging because of the complexity of this homeostatic system.

Comprehension Check

1. Which statement is true about the United States and other developed countries?

   a. Obesity is common.
   b. Obesity is not a serious health problem.
   c. Obesity exists but is relatively rare.
   d. There are no countries or states where obesity is a serious health issue.

2. What is the optimal BMI for young adults?

   a. 15 for females and 19 for males
   b. between 18 and 25 for both males and females
   c. 25 for females and 28 for males
   d. 28 for females and 30 for males

3. Individuals with Prader-Willi syndrome exhibit what?

   a. anorexia associated with low levels of ghrelin and NPY
   b. low levels of the satiety hormone leptin
   c. leptin-receptor insensitivity
   d. hyperphagia associated with hypothalamic dysfunction

4. When an anorexic looks at their own body, how do they typically see themselves?

   a. very thin and unhealthy
   b. very thin but in a healthy way
   c. overweight and unhealthy
   d. overweight but in a healthy way

Conclusion

We began our discussion of homeostasis with an iconic image of an astronaut who is being kept alive in an inhospitable environment by a spacesuit. In a sense, we have a similar set of systems that keep us at a safe temperature and supplied with sufficient fluid and fuel to maintain our body's essential processes. Given that even relatively small departures from our normal state will lead to dysfunction or death, you should not be surprised that these mechanisms often enjoy multiple redundancies to ensure our survival. Finally, despite the complexity of these systems, they do occasionally break, and understanding how they are designed to work provides a window to understand and treat dysfunction.

Critical Thinking Questions

1. Homeostatic regulation of temperature, thirst, and feeding are typically discussed within the context of negative feedback systems, but many wonder if an allostatic mechanism involving a dynamic set point might be a better model. Do you see any evidence for an allostatic model for temperature, thirst, or feeding regulation? Briefly explain.
2. Ectotherms require the environment as a source of heat, whereas endotherms generate their own heat. Given this, what relationship would you predict between latitude (distance from the equator) and the prevalence of ectothermic and endothermic species?
3. When we look at the homeostatic regulation of feeding, we see dysregulation at both ends of the spectrum (obesity and anorexia),

but when we look at homeostatic regulation of thirst, we only see polydipsia. Why might that be?

4. Hunger and satiety are controlled by the GI tract, hypothalamic centers, hormonal messengers, and genetics. Pick two of these and propose a treatment that might be effective for obesity.

5. It is easy to assign a biological explanation for every phenomenon we encounter in behavioral neuroscience, but humans enjoy the advantages of a cortex that enables complex thought. Although anorexia is quite rare, it is about five times more prevalent among females. Suggest an explanation for this difference that does not rely on a biological factor.

Dig Deeper into Behavioral Neuroscience

7.1 An Overview of Homeostasis

Two classic papers that defined homeostasis and allostasis for a generation of scientists were the work of Walter Cannon and also of Sterling and Eyer.

Cannon, W. B. (1926). Physiological regulation of normal states: some tentative postulates concerning biological homeostatics. In A. Pettit (Ed.), *A Charles Richet: ses amis, ses collègues, ses élèves* (in French). Paris: Les Éditions Médicales.

Sterling, P., & Eyer, J. (1988). Allostasis: A new paradigm to explain arousal pathology. In S. Fisher and J. T. Reason (Eds.), *Handbook of life stress, cognition, and health*. Chicester, NY: Wiley. ISBN: 9780471912699. OCLC: 17234042.

7.2 Homeostatic Regulation of Temperature

More than 1,000 people each year die of hyperthermia; victims are primarily male or over 65 years of age.

Tansey, E. A., & Johnson, C. D. (2015, September). Recent advances in thermoregulation. *Advances in Physiology Education*, *39*(3), 139–148. https://doi.org/10.1152/advan.00126.2014

In the United States each year, roughly 1,500 ER visits and about 600 deaths result from hypothermia.

Turk, E. E. (2010). Hypothermia. *Forensic Science, Medicine and Pathology, 6,* 106–115. https://doi.org/10.1007/s12024-010-9142-4

The Centers for Disease Control maintain online resources for identifying and treating hypothermia and hyperthermia. Search on "CDC and (hypothermia/hyperthermia)" for the latest information.

7.3 Homeostatic Regulation of Fluids Through Thirst

How does our brain regulate thirst and fluid intake?

Zimmerman, C. A., Leib, D. E., & Knight, Z. A. (2017, August). Neural circuits underlying thirst and fluid homeostasis. *Nature Reviews: Neuroscience*, *18*(8), 459–469. https://doi.org/10.1038/nrn.2017.71

7.4 Homeostatic Regulation of Hunger and Satiety

These sources explain much of our current understanding of the mechanism behind eating and drinking.

Amin, T., & Mercer, J. G. (2016). Hunger and satiety mechanisms and their potential exploitation in the regulation of food intake. *Current Obesity Reports*, *5*, 106–112. https://doi.org/10.1007/s13679-015-0184-5

Ghrelin is called the hunger hormone, and it certainly plays a role in maintaining an appetite. Does it have any therapeutic value?

Garin, M. C., Burns, C. M. Kaul, S., & Cappola, A. R. (2013, May). The human experience with ghrelin administration. *Journal of Clinical Endocrinology & Metabolism*, *98*(5), 1825–1837. https://doi.org/10.1210/jc.2012-4247

7.5 Disorders of Hunger and Satiety

Why isn't leptin a cure for morbid obesity? These sources provide additional information.

Izquierdo, A. G., Crujeiras, A. B., Casanueva, F. F., & Carreira, M. C. (2019). Leptin, obesity, and leptin resistance: Where are we 25 years later? *Nutrients*, *11*(11), 2704. https://doi.org/10.3390/nu11112704

Could obesity be a result of underlying differences in neurology? An obvious place to look is in the hypothalamus.

Thomas, K., Beyer, F., Lewe, G., et al. (2019). Higher body mass index is linked to altered hypothalamic microstructure. *Scientific Reports, 9*, 17373. https://doi.org/10.1038/s41598-019-53578-4

Individual neural structures are important, but how components interact as systems is equally important. This study investigates the role of communication among centers in feeding regulation.

Contreras-Rodríguez, O., Vilar-López, R., Andrews, Z. B., Navas, J. F., Soriano-Mas, C., & Verdejo-García, A. (2017). Altered cross-talk between the hypothalamus and non-homeostatic regions linked to obesity and difficulty to lose weight. *Scientific Reports, 7*(1), 9951. https://doi.org/10.1038/s41598-017-09874

*One fascinating line of enquiry investigates the role of the reward circuit in obesity. Eating is a pleasurable activity for most, but for some, a dysfunctional reward circuit may lead to overeating.

Wang, G. J., et al. (2001, February 3). Brain dopamine and obesity. *Lancet*, *357*(9253), 354–357.
https://doi.org/10.1016/s0140-6736(00)03643-6. PMID: 11210998.

Chapter 7 Before You Go On Answers

7.1 Before You Go On Answers

1. c. maintaining a constant internal environment

2. c. a tightrope walker balancing on a tightrope

3. b. involves a static set point; involves a dynamic set point

4. d. cruise control in a car

7.2 Before You Go On Answers

1. a. rely on the environment to regulate heat; regulate heat through internal means

2. c. hypothalamus

3. b. dangerously high body temperature; dangerously low body temperature

4. d. dilate; constrict

7.3 Before You Go On Answers

1. c. blood volume; cellular sodium concentration

2. a. renin; vasoconstriction

3. b. circumventricular organs; cellular sodium

7.4 Before You Go On Answers

1. d. esophagus, stomach, small intestine, large intestine

2. c. release of CCK

3. a. ghrelin, NPY, AgRP, and orexin; leptin

4. d. ghrelin; CCK and leptin

7.5 Before You Go On Answers

1. a. Obesity is common.

2. b. between 18 and 25 for both males and females

3. d. hyperphagia associated with hypothalamic dysfunction

4. c. overweight and unhealthy

Chapter 7 Glossary

adipose tissue Body fat.

agouti-related peptide (AgRP) A peptide that increases appetite.

allostasis A model of regulation of internal processes that assumes a dynamic set point.

anorexia nervosa An eating disorder characterized by extreme weight loss, fear of gaining weight, and body dysmorphia.

body mass index (BMI) A measure of body mass used to categorize a person as under- or overweight.

cholecystokinin (CCK) A hormone essential to the digestion of fats and protein.

chyme Partially digested food, water, and digestive juices that is created in the stomach and passes into the small intestine.

circumventricular organ (CVO) One of the structures surrounding the third and fourth ventricles that sense changes in osmolality.

ectotherm An animal dependent on external sources of body heat.

endotherm An animal dependent on or capable of internal generation of body heat.

gastrointestinal tract (GI tract) The entire digestive system from mouth to anus.

ghrelin A hormone secreted by the GI tract that stimulates hunger.

heat stroke A life-threatening illness caused by the body's temperature rising higher than 104 °F (40 °C).

homeostasis The state of being in balance or equilibrium among several independent factors, especially as applied to biological systems.

hyperphagia A constant compulsion to eat, thought to be associated with hypothalamic damage.

hyperthermia The state of body core temperature being greater than can be controlled by thermoregulation, resulting in rapid breathing, weak pulse, loss of consciousness, and eventually death; also known as heatstroke.

hypothermia Body core temperature below 95 °F (35 °C), which results in confusion, shivering, and eventually death.

hypovolemia A decrease in the body's blood volume.

leptin A hormone created by adipose tissue that inhibits hunger.

negative feedback system A control process that stabilizes system output; also known as a negative feedback loop.

neuropeptide Y (NPY) A hormone that appears to stimulate hunger.

obesity The state of having too much body fat, with a BMI of 30 or greater.

orexin A neuropeptide that stimulates appetite.

osmotic thirst Thirst that occurs in response to intracellular dehydration caused by high sodium concentration.

polydipsia Excessive thirst as a symptom of a disease or other disturbance.

Prader-Willi syndrome A genetic disorder characterized by weak muscles, slow development, and an insatiable appetite.

satiety The state of feeling full after eating.

set point In the context of homeostatic functions, a desired target value for a variable such as weight, temperature, or glucose.

volumetric thirst Thirst that occurs in response to a loss of blood volume.

CHAPTER **8**

The Endocrine System

damircudic / Getty Images

Our endocrine system helps us properly regulate our internal state and respond to stress.

CHAPTER OUTLINE

8.1 An Introduction to the Endocrine System
- The Discovery of Hormones as Messengers
- Comparing the Endocrine and Nervous Systems
  Try It Out: Fast- and Slow-Response Systems
- Glands and Hormones
- Negative Feedback and Endocrine Function

8.2 Endocrine Glands in the CNS
- The Hypothalamus, Pituitary Gland, and Pineal Gland
- Pituitary Hormones
- Endocrine Regulation of Growth
  Think About It: Synthetic Growth Hormone Use and Abuse

8.3 Endocrine Glands in the Periphery
- The Thymus and Thyroid
- The Pancreas and Glucoregulation
  Check It Out: Glucoregulation
  Think About It: Preventing Type 2 Diabetes
- The Adrenal Glands and Stress
- The Gonads and Sexual Development

8.4 The Stress Response
- The HPA Axis
- The General Adaptational Syndrome
- Post-Traumatic Stress Disorder
  Advances in Behavioral Neuroscience: Carrie Hill Kennedy and PTSD

LEARNING OBJECTIVES

LO 8.1 **Understand** the advantages of a slow-acting, generalized control system.

LO 8.2 **Describe** the endocrine glands of the CNS, their purposes, and associated hormones.

LO 8.3 **Describe** the endocrine glands of the periphery, their purposes, and associated hormones.

LO 8.4 **Understand** how the endocrine system regulates our response to stress.

Stories of the Brain

The sun was slowly sinking toward the horizon as Madeline glanced out her car window and marveled at the blues and pinks reflected in the clouds. Her radio was quietly playing, and her attention was focused on the events of the day, conversations with friends and professors, assignments pending for the coming week, and her plans for the weekend. The road in front of her took a gentle curve to the right where a stand of trees bordered the edge of the road.

Without warning, a car burst out of a hidden intersection on the right, its horn blaring. As she glimpsed the other driver's terrified face, Madeline yanked the steering wheel hard to the left in an effort to avoid a collision. Her car crossed both lanes and dove into the ditch on the left side of the road, coming to a stop amid a cloud of dust. Luckily, she never failed to wear her seatbelt.

Getting out of the car to check for damage, she realized that she had narrowly avoided a potentially fatal accident. Her knees suddenly begin to wobble, her heart was racing, and she felt a bit light-headed. As she leaned against her car for support and began to settle down, she wondered why she felt in control at the moment of greatest danger, but once the danger had passed, her body reacted with a racing heart and shaking limbs. Why did her body react as it did?

In fact, Madeline's response to a life-threatening situation evolved with early humans. Although our ancestors did not have to deal with distracted drivers suddenly appearing in their path, they were threatened by human and animal predators, food scarcity, inclement weather, and injuries and diseases that they were ill-equipped to treat. Stress has always been a part of the human condition.

We begin our discussion of the endocrine system with the discovery more than a century ago that chemical messengers in the bloodstream have profound effects on behavior. We will then describe the many structures that make up the endocrine system in the brain and elsewhere in the body. With that basic knowledge in hand, we will take a closer look at how the endocrine system controls our response to stress.

8.1 An Introduction to the Endocrine System

LO 8.1 Understand the advantages of a slow-acting, generalized control system.

In our discussion of the biological basis of human behavior, it would be easy to assume that our nervous system controls everything that happens in our bodies, but that would not be accurate. It may come as a surprise to you to discover that there is a second network that is responsible for a different set of functions such as stimulating growth during development, initiating sexual behavior, regulating blood pressure, maintaining blood glucose levels, and preparing you to fight or flee. That second and equally important network is the **endocrine system**. We begin our exploration of the endocrine system with a brief history of its discovery.

The Discovery of Hormones as Messengers MCAT

The first medical practitioners did not have the luxury of MRI or blood work to help them diagnose and treat maladies, but they were astute observers of their patients. Aristotle (4th century BCE) noted that castration of an animal or a person, although a relatively small change, led to significant changes in behavior. Hippocrates (a contemporary of Aristotle) proposed that human behavior was dictated by the proportion of four humors, or vital bodily fluids, in the blood and that these were responsible for the proper functioning of the body (**Figure 8.1**). The theory that disease was caused by a humor imbalance in the blood was eventually discarded when microscopes allowed scientists to visualize germs. Nonetheless, the idea that there were

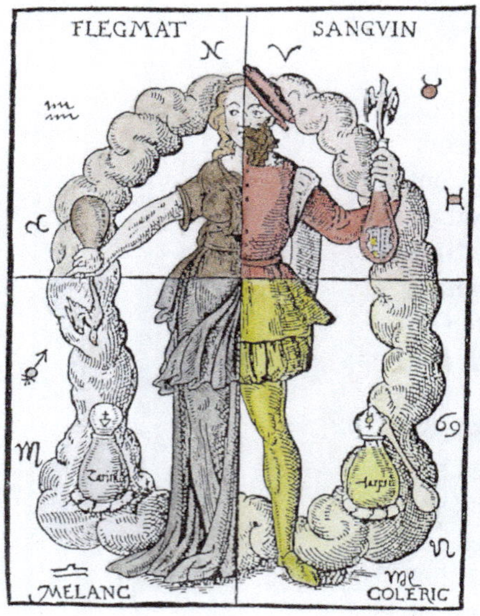

Mary Evans Library/News/Adobe Stock

FIGURE 8.1 The four humors Hippocrates proposed that human temperament was determined by four humors that traveled through the body. Although this theory was disproven, the idea that chemicals in the blood influence behavior was prescient.

chemical messengers in the blood was a valuable insight. Although the four humors are no longer a part of our scientific theories, their original names form the basis for the character traits we refer to as phlegmatic (unemotional), choleric (hot-tempered), melancholic (depressed), and sanguine (optimistic).

In 1849, German physician Arnold Berthold performed a series of experiments on roosters that demonstrated the existence of chemical messengers outside the nervous system. Berthold castrated roosters and noted that they lost typical rooster behaviors such as crowing, mating, and defending their territory. These changes suggested to him that the testes were directly responsible for such behaviors. He then transplanted a testis into the abdomen of a castrated rooster, and that rooster resumed the typical behaviors that were lost through castration. From these results, Berthold deduced that the testis must be releasing some agent into the blood that was responsible for these behaviors.

The system that Berthold had discovered was the endocrine system, and the chemical messengers responsible for the changes in the roosters' behavior were hormones. We now turn our attention to an exploration of the many attributes of the endocrine system and how its unique role complements the function of the nervous system.

Comparing the Endocrine and Nervous Systems MCAT

You might wonder why we need the endocrine system when we have a nervous system, and the answer is that they perform very different functions. Your nervous system responds to a stimulus quickly, often in less than one hundredth of a second, and it does so at the level of the synapse. In contrast, your endocrine system responds gradually through chemical messengers carried in the blood, and it exerts an effect over a wide range of targets. The two systems working together provide us with a neurological and endocrine-behavioral response mechanism that is elegant and subtle. The differences between the two systems are summarized in Table 8.1.

You might question why we need a slow-responding system to complement the rapid response of the nervous system. If you have ever walked on a slick surface such as ice, you

TABLE 8.1 Comparing the Nervous System and Endocrine System

Characteristic	Nervous system	Endocrine system
Mechanism	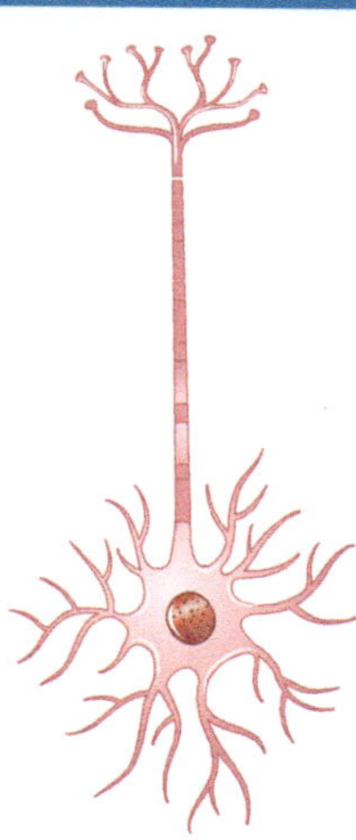 Neural transmission at the level of the synapse	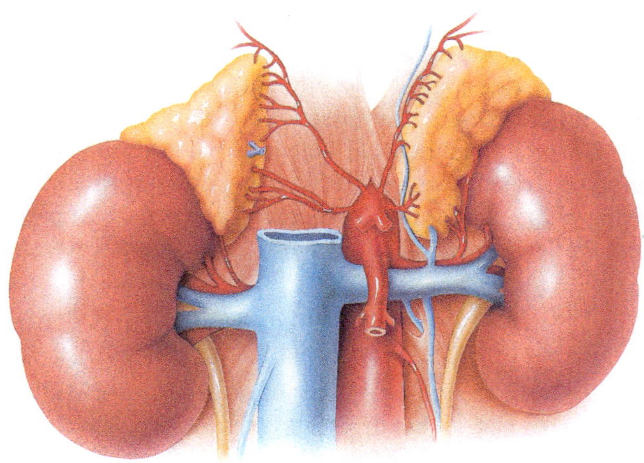 Chemical messangers produced from glands
Response time	Fast acting but transient	Slow and sustained acting
Effects	Acts on one or only a few cells	Acts throughout the body

Try It Out | Fast- and Slow-Response Systems

GK Hart/Vikki Hart/Stone/Getty Images

In this activity, you are asked to consider other systems where success may depend on either a fast or a slow response, where fast responding is defined as a system that responds at full strength to the first possible sign of a signal and slow responding is defined as a system that samples signal strength over time before responding. Complete the worksheet **Fast- and Slow-Response Systems**.

understand that, at times, a slow and gradual response is advantageous. The same is true for the brain's control of the body. Everyday metabolic processes must be constantly monitored and adjusted; our bodies must be guided to develop from infancy, through childhood and adolescence, into adulthood and beyond, and processes such as sexual maturation and response require a careful orchestration of events. Consider the advantages of a mechanism like the endocrine system in this **Try It Out: Fast- and Slow-Response Systems**.

As you study behavioral neuroscience, you will see again and again that biological systems are complex and commonly contain complementary and opposing mechanisms such as excitatory versus inhibitory processes, afferent versus efferent paths, and fast-acting versus slow-acting mechanisms. The interactions among these several processes can provide a subtle behavioral response to environmental challenges. Your nervous system is designed to respond immediately, whereas your endocrine system creates a slow and gradual response that is just as important. The combination of the two systems creates a powerful tool for humans to survive and thrive.

We began our exploration of the nervous system with the neuron, and therefore we begin our consideration of the endocrine system with the gland.

Glands and Hormones MCAT

The endocrine system acts in a slow and gradual way because of the particular way its chemical messages are communicated. You know from your study of nervous system that neurotransmitters are formed within neurons and then secreted at the synapse to deliver a chemical message to adjacent neurons. In the endocrine system, the chemical messenger is a **hormone**, which is secreted into the bloodstream by glands and then travels through the body. The effect of neural transmission is immediate and transitory, whereas the action of a hormone is gradual and longer lasting.

Glands A gland is a structure in the body that synthesizes and then releases a substance. Glands are either exocrine glands or endocrine glands. **Exocrine glands** release their secretions into ducts or channels; examples of ducted glands are sweat glands and sebaceous glands in the skin, mammary glands, and the various ducted glands in the digestive system whose secretions help us digest our food. **Endocrine glands** differ from exocrine glands in that they are ductless. Their secretions enter the capillary system or the extracellular fluid and are carried throughout the body by the circulatory system. If the distinction between exocrine and endocrine glands seems murky, a simple metaphor can help you see the difference. A garden hose is like an exocrine gland in that it delivers water through a duct (the hose) to one location, whereas a rainstorm resembles an endocrine gland because it also delivers water but does so over the entire area. **Figure 8.2** gives an overview of the glands that make up the endocrine system.

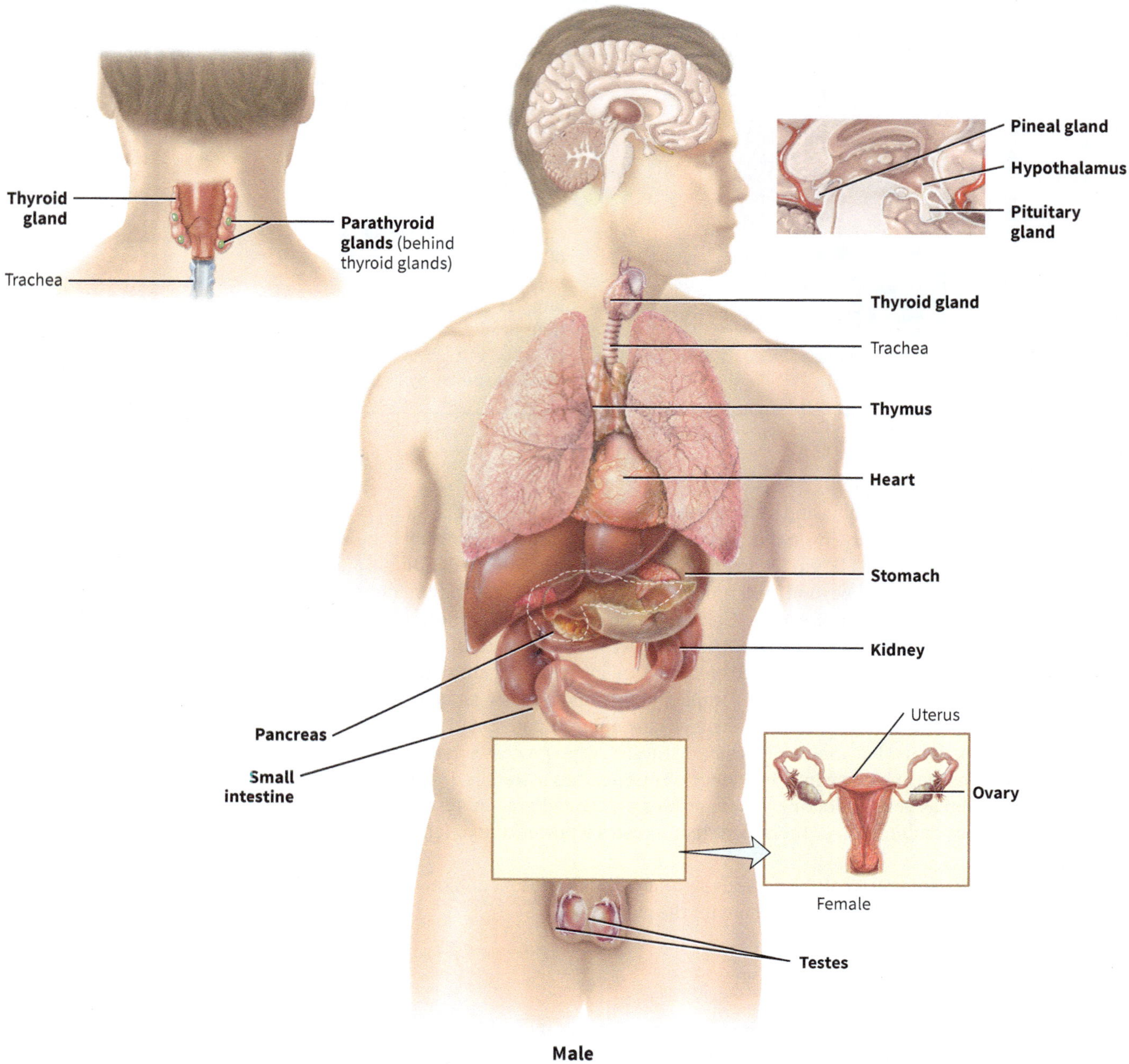

FIGURE 8.2 The endocrine glands The hypothalamus, the pituitary, and the pineal gland are the part of the endocrine system that is located in the brain. The thyroid and parathyroid glands, thymus, adrenal glands, pancreas, and gonads are located outside the brain. The endocrine glands regulate a wide range of metabolic and behavioral functions.

Hormones The secretions of the endocrine glands are hormones. Hormones are chemical messengers that travel through the blood to effect a change in either physiology or behavior. Hormones affect covert behaviors such as homeostasis and metabolism, as well as overt behaviors such as wakefulness, sexual behavior, and mood swings.

Although neurons and endocrine glands both carry information, they do so in very different ways. Neurons employ neurotransmitters to carry a signal, whereas an endocrine gland uses a hormone for that same purpose. Neurons transfer that signal across a specific synapse and therefore have a very localized effect, whereas glands secrete a hormone into the bloodstream where it can travel throughout the body, generating a global effect.

As hormones are carried through the body, they signal target cells to alter their function; for example, a hormone may affect growth, activity level, or even the function of the target cell. Just as synapses have receptors specific to each neurotransmitter, there are receptors specific to each hormone produced by the endocrine system. A hormone transports a signal from one cell to another cell, but only if that target cell has the receptor for that specific hormone, as illustrated in Figure 8.3.

The endocrine system is responsible for many of the background maintenance process that keep us functioning, so it is important that these be kept in balance. Like the homeostatic systems we covered in a prior chapter, the endocrine system relies heavily on negative feedback systems to regulate function, so a brief review of their properties is in order.

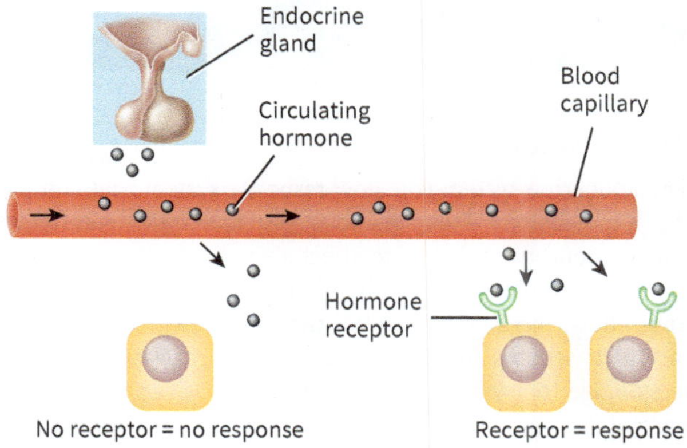

FIGURE 8.3 Hormones traveling through the blood The endocrine system uses the circulatory system to affect change throughout the body. Only cells with the receptor for that hormone are influenced.

Negative Feedback and Endocrine Function MCAT

One important aspect of the endocrine system is that it operates by a **negative feedback system**. The key to understanding a negative feedback system is that it does not spring into action unless the system is out of balance, and only then does it respond to reduce the difference between the current state and a goal state until equilibrium is restored.

A familiar example of a negative feedback system is the toilet (Figure 8.4). A common version of a residential toilet consists of a reservoir or tank. When the toilet is not in use, the ball float keeps the valve closed and the water level unchanged (**step 1**). This water level is the set point of the system. When the toilet is flushed, the water level drops, and the ball float sinks as the water rushes out (**step 2**). When the tank is empty, the ball float opens the valve wide, letting water rush into the tank (**step 3**). As the tank fills with water, the ball float rises and slowly closes the valve. When the set point is reached, flow into the tank ceases and the system is returned back to its initial state.

The reason this type of feedback system is negative is because as long as the system is in equilibrium, nothing happens. It is only when the system gets significantly out of equilibrium that action takes place, and that action will continue until equilibrium is restored. Recall from the discussion of homeostasis in an earlier chapter that the key components of a negative feedback system are a sensor, a control mechanism, and an effector or brake that can make a change in the system. In our everyday example of the toilet, the float serves as a sensor to identify when the water level in the tank falls below the set point. The valve serves as the control mechanism in that it can open to let water into the tank and close once equilibrium is reached. The flush of the toilet, which empties the tank, serves as the effector.

In our discussion of the endocrine system, you will see that negative feedback systems are often used to keep our everyday conditions in balance. This includes growth, glucoregulation, and the stress response. With this background in hand, we can proceed in the next section to an overview of the major endocrine glands that affect behavior.

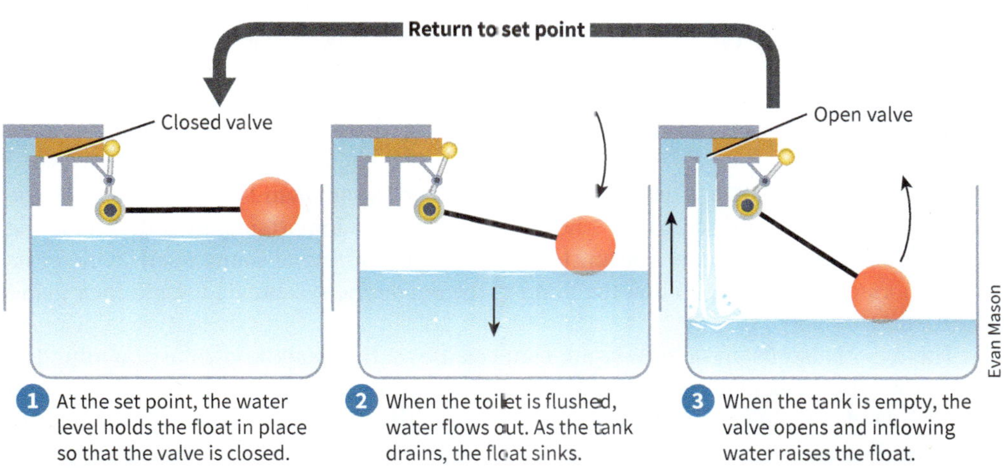

FIGURE 8.4 A negative feedback system The mechanism that drains and refills a toilet tank is an example of a negative feedback system.

8.1 Before You Go On

Section Summary

The endocrine system is a slow-response system that complements the rapid action of our nervous system. Unlike neurons, which employ neurotransmitters to affect a single neuron, glands use hormones distributed through the bloodstream to affect global activity. Your endocrine system operates through negative feedback systems in which sensors monitor conditions and a control center works to stabilize the system through effectors if it varies too much from normal.

Comprehension Check

1. An early approach to understanding how glands affect behavior was stated by __________, who proposed that behavior was controlled by __________.
   - **a.** Hippocrates; the four humors
   - **b.** Aristotle; the brain
   - **c.** Archimedes; what we eat
   - **d.** Hippocrates; ideas
2. One thing that distinguishes the endocrine system from the nervous system is that the endocrine system __________.
   - **a.** controls few bodily functions
   - **b.** uses neurotransmitters but for different purposes
   - **c.** responds to conditions in a slow and gradual fashion
   - **d.** does not begin to operate until adulthood
3. Which statement is true of the endocrine system?
   - **a.** It allows an internal environment to fluctuate greatly.
   - **b.** It is not typical in vertebrates.
   - **c.** It maintains a relatively stable environment through a feedback mechanism.
   - **d.** It is a relatively recent evolutionary adaptation.
4. Glands create hormones, which __________.
   - **a.** are transmitted across the synapse to affect behavior
   - **b.** begin to be secreted as a person reaches adolescence
   - **c.** directly alter the action of neurotransmitters to affect behavior
   - **d.** enter the bloodstream to affect systems throughout the body
5. Which statement is true of a negative feedback system?
   - **a.** It maintains a set point through both positive and negative forces.
   - **b.** It prevents small deviations from a set point from getting larger.
   - **c.** It amplifies a process until it is brought back under control.
   - **d.** It maintains a set point by creating negative consequences.

8.2 Endocrine Glands in the CNS

LO 8.2 Describe the endocrine glands of the CNS, their purposes, and associated hormones.

There is a lot of ground to cover in the glands of the endocrine system, so we will first discuss endocrine glands in the CNS and then those in the rest of the body.

The Hypothalamus, Pituitary Gland, and Pineal Gland MCAT

The portion of the endocrine system that is in the central nervous system includes the hypothalamus, the pituitary gland, and the pineal gland.

The hypothalamus You were introduced to the hypothalamus in the chapter on the nervous system. We now discuss this structure in its role as an important part of the endocrine system. The **hypothalamus** is a subcortical structure consisting of many nuclei. One of its most important functions is to link the brain to the endocrine system via the pituitary gland (Figure 8.5).

The hypothalamus and pituitary are connected by a thin stalk known as the infundibulum. A key to understanding the endocrine system is to see that your hypothalamus and pituitary gland combine to create the **hypothalamic–pituitary (HP) axis**, a command center for regulating a wide range of metabolic functions. As capillaries carry blood through the hypothalamus, there are gaps in their walls that allow large proteins and other molecules to filter in. In this way, the hypothalamus is constantly monitoring critical metabolic functions.

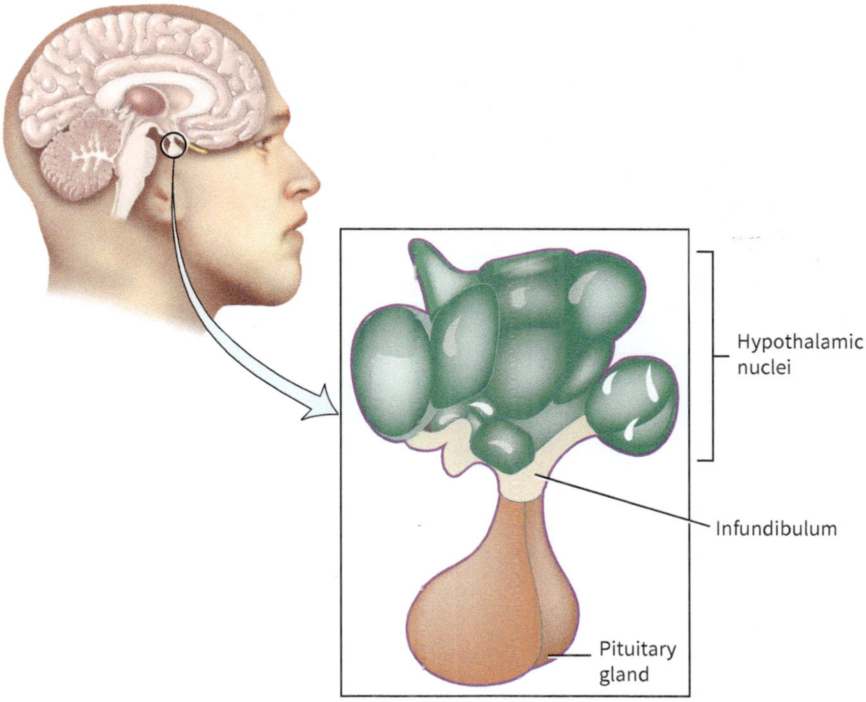

FIGURE 8.5 The hypothalamus and pituitary The hypothalamus sits deep inside the brain just above the midbrain and below the thalamus. The pituitary is connected to the hypothalamus by the infundibulum.

The hypothalamus responds to metabolic changes in two ways: directly and indirectly. It responds directly by stimulating the pituitary to release hormones that affect tissues throughout the body and indirectly by stimulating specific endocrine glands that have their own regulatory functions. The HP axis is a complex mechanism that coordinates behaviors such as metabolic level, hunger and thirst, sexual development and behavior, and also arousal, sleep, and biorhythms. The hypothalamus also assists in regulating attachment and emotion via the hormone oxytocin. Last but not least, the hypothalamus in conjunction with the pituitary and the adrenal cortex serves a regulatory role in dealing with stress. We will return to all of these examples in greater detail in this and other chapters, but any discussion of the endocrine function of the hypothalamus must include the pituitary, so we now turn to this tiny but powerful gland.

The pituitary gland

The **pituitary gland** is a small gland the size of a bean that functions in close physical connection to the hypothalamus and whose influence extends to almost every part of the body. Your pituitary is called the master gland because it controls peripheral glands, including the thyroid, adrenal glands, and gonads. The pituitary itself is divided into an anterior and a posterior lobe, each of which produces a variety of significant hormones that we will discuss in the next section.

The pituitary is involved in many processes essential to survival, so you will not be surprised that nature has provided this structure with considerable protection in the form of a bony shield known as the sella turcica (from the Latin for Turkish saddle). Your pituitary sits in this bony cup, well insulated from worldly hazards (**Figure 8.6**).

The pineal gland

The **pineal gland** (so named because it is shaped like a pine cone) is about the size of a grain of rice and is located in a groove on the midline of the two lobes of the thalamus. It secretes melatonin, a hormone that aids in the regulation of the sleep–wake cycle, circadian rhythms, and seasonal changes in metabolism and affect. The pineal receives input from the retina about light stimuli via the hypothalamus and the suprachiasmatic nuclei. It is believed that through this pathway, the pineal regulates metabolic processes in response to seasonal changes.

Ideally, the pineal gland responds to seasonal changes in daylight by altering metabolic processes to reflect the different demands of the summer and winter. Because it controls your daily biological rhythms, damage to the pineal gland often leads to sleep disorders. For this

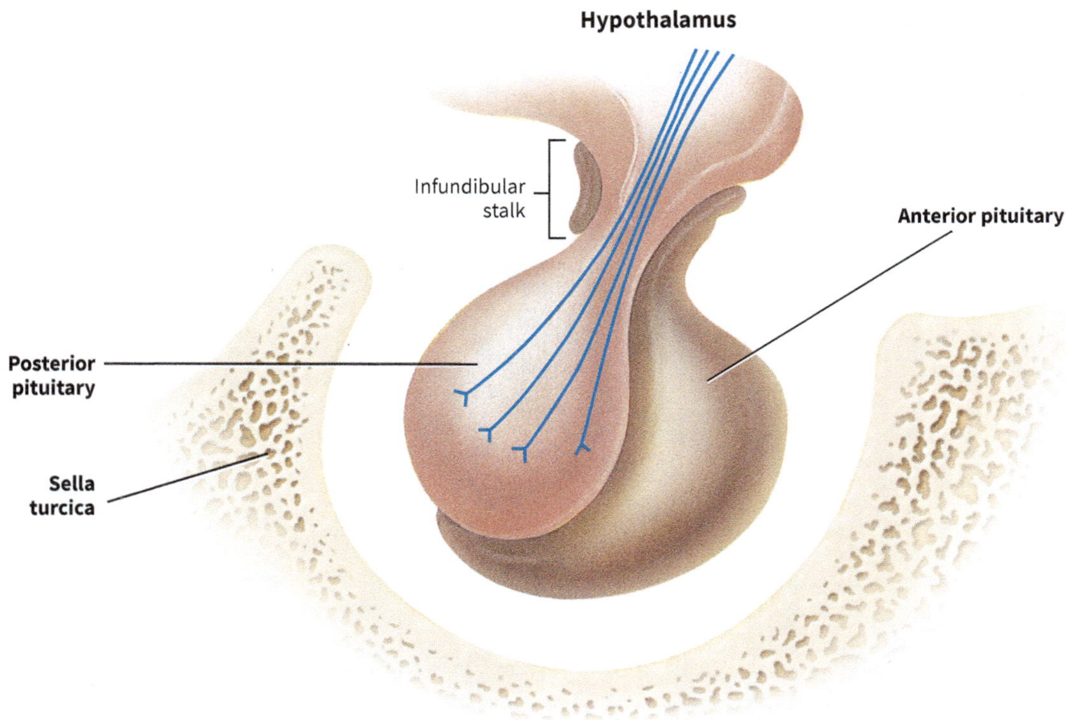

FIGURE 8.6 **The pituitary in the sella turcica** The pituitary gland is located at the base of the skull, and it is protected by its location in the sella turcica.

reason, the pineal gland is a suspect in disruptions of circadian rhythms, insomnia, jetlag, and other disorders of the sleep–wake cycle such as seasonal affective disorder (SAD), now reclassified as major depressive disorder (MDD) with seasonal pattern.

Hyper- and hypopineal activity have also been linked to precocious puberty and delayed puberty, respectively. Finally, you will not be surprised to hear that pineal dysfunction may play a role in some cases of major depression. These topics are discussed in greater detail in other chapters.

Pituitary Hormones MCAT

The pituitary secretes hormones that regulate your growth, blood pressure, and even reproductive functions. The pituitary plays a central role in bodily processes in that it is a vehicle through which the hypothalamus directs many functions (**Figure 8.7**).

Hormones from the anterior pituitary The anterior pituitary lobe secretes seven different hormones that regulate many different bodily functions. Growth hormone (GH) stimulates the formation of bones and muscles, especially during childhood. We will return to the role of GH in regulating growth later in this section.

Thyroid-stimulating hormone (TSH), not surprisingly, stimulates the activity of the thyroid, a gland we will discuss shortly. Adrenocorticotrophic hormone (ACTH) stimulates the adrenal glands to produce cortisol, a hormone associated with the stress response. The anterior pituitary also secretes prolactin, which stimulates lactation (milk production), and also follicle-stimulating hormone (FSH) and luteinizing hormone (LH), which are necessary for the development and release of ova in the female reproductive cycle and for testosterone and sperm production in males. As a result, dysregulation of the anterior pituitary can cause reproductive difficulties.

Hormones from the posterior pituitary The posterior pituitary lobe plays a slightly different role. Unlike the anterior pituitary, the posterior lobe does not itself synthesize hormones but rather receives and stores the hormones oxytocin and antidiuretic hormone (ADH) that are created within the hypothalamus.

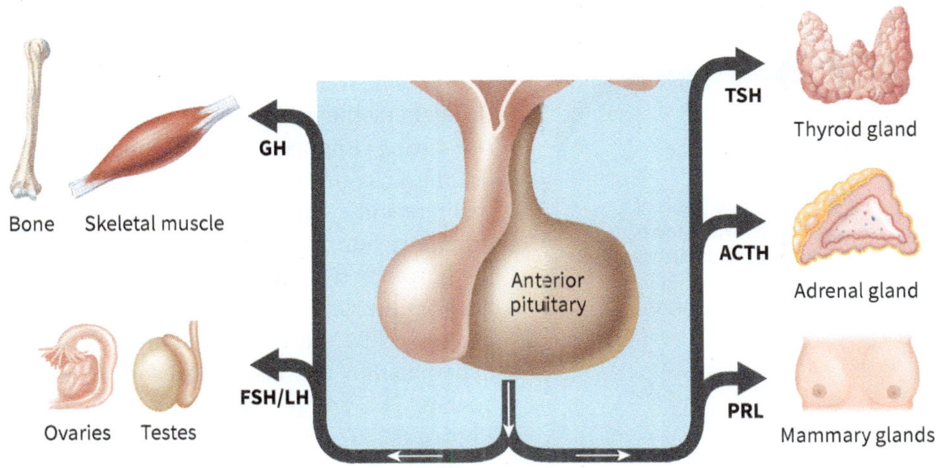

FIGURE 8.7 The influence of the pituitary throughout the body The anterior pituitary acts as an intermediary between the hypothalamus and several endocrine glands. When stimulated by the hypothalamus, the anterior pituitary secretes hormones that regulate functions such as growth, metabolic rate, response to stress, and reproductive functions.

Oxytocin is a hormone that plays a role in maternal behavior as well as sexual response, and current social neuroscience research suggests that it may be associated with feelings of trust and empathy. Oxytocin is a part of the normal human sexual response cycle that leads to orgasm and is also a part of the normal let-down reflex that causes milk to be released from the mammary glands. Oxytocin is also involved in uterine contractions during labor.

ADH, also known as vasopressin, regulates water retention and also blood pressure. There is evidence that it also plays a part in memory consolidation and possibly some aspects of social behavior.

The list of pituitary hormones and their effects in Table 8.2 should convince you that your pituitary puts in a full day's work every day (and that pituitary dysfunction is a serious problem).

It should be clear by now that the hypothalamus and the pituitary exert a powerful and wide-ranging effect on the body, and an example might help to make that clear. There are few things more important during the course of development than regulating growth.

Endocrine Regulation of Growth MCAT

In childhood and adolescence, the hypothalamus secretes growth hormone-releasing hormone (GHRH), which is aptly named as it stimulates the anterior lobe of the pituitary to release growth hormone (GH). GH (also known as somatotropin from the Greek for body responder) is secreted by the anterior lobe of the pituitary, and its effect is to stimulate bone and tissue growth throughout the body. GH is released daily; its levels increase during childhood, peak during the puberty growth spurt, and then slowly decline.

In our discussion of the endocrine system, you have seen numerous examples where a hormone acts directly on target cells throughout the body but also stimulates secretion by

TABLE 8.2 Pituitary Hormonal Messengers and Their Effects

Pituitary lobe	Associated hormones	Effects
Anterior	Growth hormone (GH)	Promotes growth of body tissue
	Prolactin (PRL)	Promotes lactation in the mammary glands
	Adrenocorticotropic hormone	Stimulates adrenal cortex to release hormones
	Thyroid-stimulating hormone (TSH)	Stimulates thyroid to release hormones
	Follicle-stimulating hormone (FSH), luteinizing hormone (LH)	Stimulates gamete (ova and sperm) production by the gonads
Posterior	Oxytocin	Stimulates uterine contractions during childbirth, relaxation
	Antidiuretic hormone (ADH)	Stimulates water absorption by kidneys

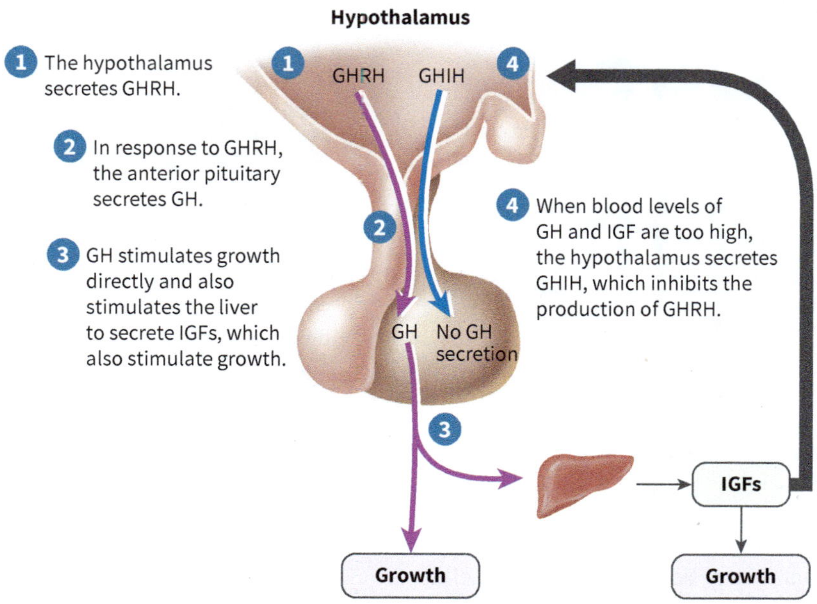

FIGURE 8.8 **Endocrine regulation of growth** The HA axis regulates growth in a negative feedback system.

another organ. In the same way, GH stimulates the liver to secrete insulin-like growth factors (IGFs), which act like GH to stimulate growth throughout the body. Also like GH, IGF levels rise during childhood, peak in puberty, and then decline. Of course, GH and IGF act as a gas pedal, so there must be a corresponding brake. That inhibitory hormone is growth hormone–inhibiting hormone (GHIH), also known as somatostatin from the Greek for body unchanging. GHIH is released by the ventromedial nucleus of the hypothalamus and also by cells in the digestive system. This complex circuit is summarized in Figure 8.8. The hypothalamus releases GHRH (**step 1**), which causes the anterior pituitary to release GH (**step 2**), which causes the liver to release IGFs (**step 3**). GHIH acts as an inhibitor of GHRH to keep the system in check (**step 4**).

Hormones and other substances that affect metabolic level, muscle development, and stamina are sometimes used therapeutically. They are also sometimes abused, which can have dangerous consequences, as you will discover in **Think About It: Synthetic Growth Hormone Use and Abuse**.

Think About It | Synthetic Growth Hormone Use and Abuse

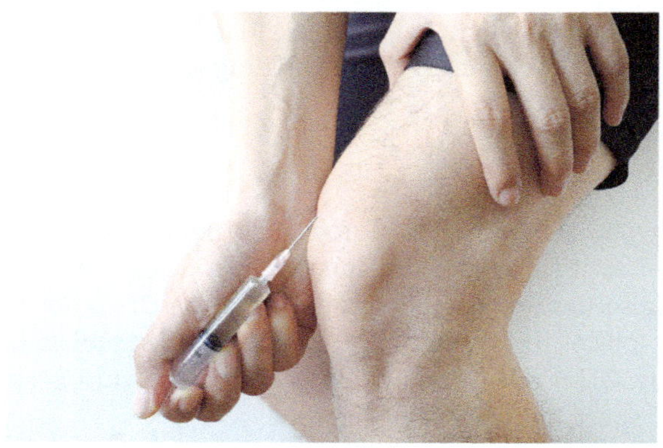

Seasontime/Shutterstock

You probably are aware that synthetic forms of hormones such as growth hormone (GH, also referred to as human growth hormone, or HGH) are used to stimulate growth in children who are unusually undersized and to supplement hormones in adults who have a deficiency. Although there are legitimate medical uses for GH, due to the damage they can create when misused, GH is a restricted drug that is only available by prescription.

Some athletes use GH along with anabolic steroids to increase muscle mass to provide an unfair advantage. In fact, although GH may increase the size of muscles, there is no compelling evidence that it increases strength. Over the long term, GH use may even damage muscles, including those in the heart. Some individuals take GH because they believe it will slow aging, although there is also no evidence that this is the case.

GH is a controlled substance, legally available only by prescription and only for specific uses such as treating poor growth in children and adults or loss of muscle mass due to a wasting disease and is not legal for recreational use.

Questions

1. **Search using the terms "health risks for using GH for recreational use." List five possible side effects of recreational use of GH.**
2. **Do you think that people should be able to use GH and other chemical aids to increase muscle mass beyond what their body can do naturally? Why or why not?**
3. **Do professional sports organizations, the NCAA, schools, or coaches have a responsibility to educate players about GH abuse or test for its use, or should the decision to use GH rest with the individual?**

FIGURE 8.9 **Acromegaly and pituitary dwarfism** Andre Roussimoff (a) and Lavinia Warren (b) were individuals whose stature was affected by their pituitary dysfunction.

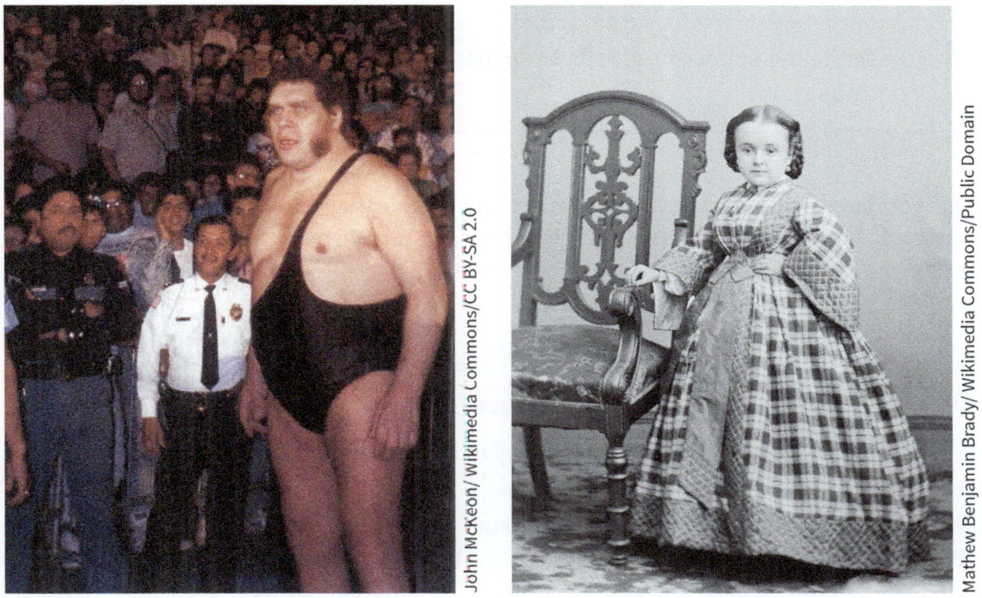

a. b.

Given the role of the pituitary gland in regulating growth, pituitary dysfunction can lead to too much or too little growth. Oversecretion of GH leads to a condition known as acromegaly, or gigantism. These individuals are tall and have unusually large hands, feet, and facial features. You may be familiar with the actors Richard Kiel or Ted Cassidy or the wrestler Andre Roussimoff, whose professional name was Andre the Giant. All three are examples of acromegaly. Just as oversecretion of GH leads to problems, undersecretion causes its own difficulties. Too little GH leads to hypopituitarism, which is characterized by small stature, constant fatigue, and loss of body hair. The results of too much and too little pituitary function are illustrated in **Figure 8.9**.

8.2 Before You Go On

Section Summary

Endocrine glands within the CNS include the hypothalamus, the pituitary gland, and the pineal gland. The pineal gland is associated with sleep–wake and seasonal cycles. In partnership with the hypothalamus, the pituitary gland secretes hormones that govern numerous homeostatic processes, including growth, sexual development, and the stress response.

Comprehension Check

1. Which are examples of behaviors under hypothalamic control?

- **a.** hunger and satiety
- **b.** temperature regulation
- **c.** sleep and biorhythms
- **d.** all of these

2. One of the many important functions of the pituitary is to secrete hormones that stimulate __________.

- **a.** the development of the visual system
- **b.** growth (as in height)
- **c.** your appetite and control satiety
- **d.** you to wake in the morning

3. Your pineal gland has been implicated in what important behavior?

- **a.** sleep–wake cycles and circadian rhythms
- **b.** the ability to digest lactose
- **c.** the ability to digest gluten
- **d.** the development of external genitalia

4. One individual is extremely small in stature, whereas another is unusually large, with large hands and feet, a large jaw, and protruding forehead. As different as they are, they might share which trait?

- **a.** dysfunction of the HPA axis
- **b.** dysfunction of the adrenal glands
- **c.** a pituitary dysfunction
- **d.** an insensitivity to human growth hormone

8.3 Endocrine Glands in the Periphery

LO 8.3 Describe the endocrine glands of the periphery, their purposes, and associated hormones.

So far you have seen how the hypothalamus and pituitary serve a central role in governing homeostatic processes throughout the endocrine system. Now we will take a closer look at these peripheral glands, including the thymus, thyroid, pancreas, and gonads. Each has a critical role to play in the endocrine system.

The Thymus and Thyroid MCAT

We begin our survey of peripheral endocrine glands with the thymus and thyroid.

The thymus

The **thymus** is a small gland comprised of two lobes located above the heart and under your sternum, or breastbone. The thymus has an important role in stimulating the development of the immune system in children and adolescents, but typically shrinks as the individual becomes an adult. Your thymus is critical to the maintenance of your immune system in that without it, an individual is extremely susceptible to infections. A disruption of the thymus can cause the immune system to attack the cholinergic (ACh) receptors at the neuromuscular junction, leading to myasthenia gravis, an autoimmune disorder characterized by muscular weakness, lethargy, and even death.

The thyroid and parathyroid glands

The **thyroid** is a butterfly-shaped gland located in the neck, just above the trachea and adjacent to the pharynx, the opening at the top of the esophagus and trachea (Figure 8.10). It has two lobes, one on each side of the trachea. The prefix *para-* means beside, and the parathyroids consist of four glands, two on each side of the thyroid. These glands are responsible for maintaining the blood's calcium and phosphate balance, which is essential for homeostasis.

The thyroid secretes the hormones thyroxine (T_4) and triiodothyronine (T_3), which are critical for the regulation of metabolic activity throughout the body. Although T_4 and T_3 differ somewhat, for our purposes we can refer to them collectively as thyroxine. Thyroxine increases energy production at the cellular level throughout the body, so it is essential for normal physical health. Digestion, heart and muscle function, bone cell replacement, and even neural function are all affected by your thyroid.

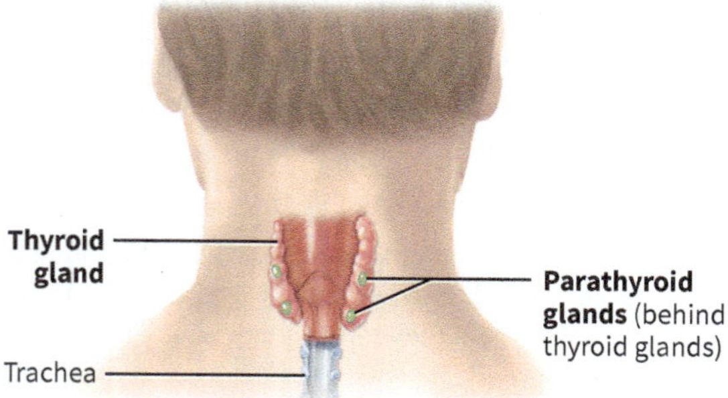

FIGURE 8.10 The thyroid and parathyroid glands The thyroid sits at the top of the trachea; on each side are the two parathyroid glands. The thyroid secretes thyroid hormones that regulate your metabolism. The parathyroids regulate calcium in the blood and bones.

Hyper- and hypothyroidism

Your thyroid manages your metabolism, so over- and underactivity of this gland can lead to serious consequences. When an individual's thyroid is overactive, that is known as **hyperthyroidism**, and it causes weight loss, anxiety, and fatigue. There are many causes for hyperthyroidism, but the most common is Graves' disease, an autoimmune disorder in which your thyroid responds to the immune system by overproducing hormones. A classic symptom of Graves' disease is bulging eyes (Figure 8.11). Famous individuals with Graves' disease include former U.S. President George H. W. Bush and his spouse, Barbara Bush, musician Missy Elliott, and actor Maggie Smith of Harry Potter fame.

Hypothyroidism is defined as an abnormally low level of thyroid activity, and it has serious consequences, including fatigue, loss of hair, weight gain, and sensitivity to cold. If you imagine that your metabolic activity is a symphony of many processes, you might think of the thyroid as the conductor of the orchestra. When the conductor is less energetic at wielding their baton, trouble ensues, as your author personally

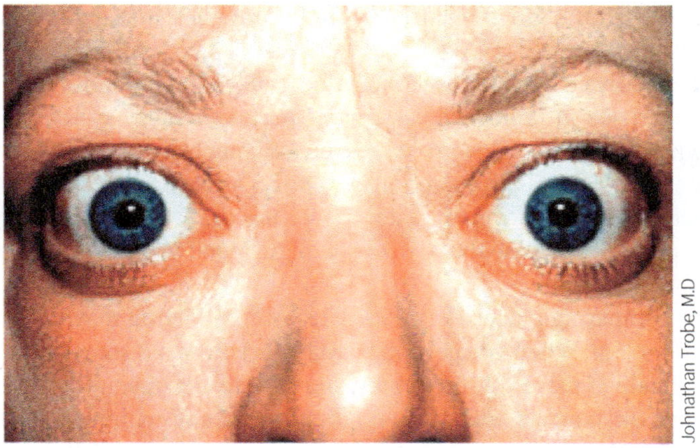

FIGURE 8.11 An individual with Graves' disease Bulging eyes are a common symptom of hyperthyroidism.

discovered. A few years ago, I became easily fatigued and found it hard to concentrate in faculty meetings (your instructor will sympathize). In my yearly physical, I discovered I was hypothyroidic, and with a thyroxin supplement, my energy and focus returned.

The Pancreas and Glucoregulation MCAT

Earlier in this chapter you learned that glands can be exocrine (ducted) or endocrine (ductless). Some glands are both, and the **pancreas** is an example. The pancreas secretes digestive juices via ducts, but given that our interest is in endocrine function, we will focus on the role of the pancreas in maintaining the consistent level of glucose needed to power our cells, called **glucoregulation**.

Glucoregulation

The details of glucoregulation can be complex, but the basic framework is straightforward and demonstrates again how a negative feedback system maintains balance (Figure 8.12).

When you eat, your body converts carbohydrates into glucose, a simple sugar (monosaccharide) that is the primary source of energy for your cells. When the digestive system releases glucose into your bloodstream, your blood glucose level rises above the set point, a state known as **hyperglycemia**. The overabundance of glucose stimulates beta cells in a region of the pancreas known as the islets of Langerhans to release insulin. As the insulin is transported by your circulatory system to cells throughout your body, it enables your cells to

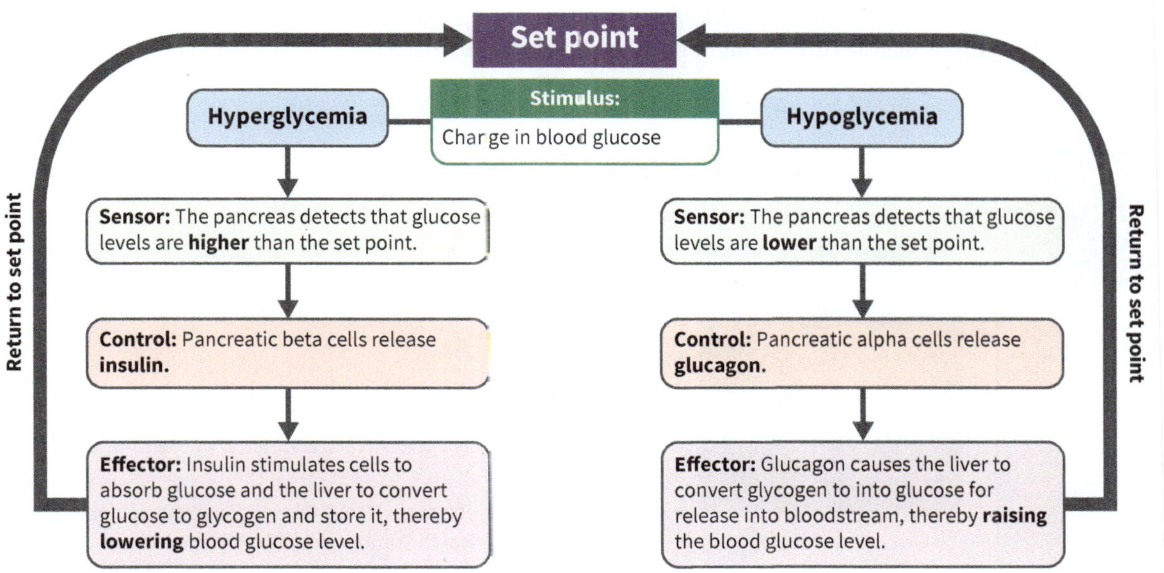

FIGURE 8.12 Glucoregulation The pancreas and the liver work cooperatively after feeding to store excess glucose as glycogen and then convert glycogen back to glucose when needed.

use the glucose. Not all the glucose is immediately converted into energy, however, and the excess is stored in your liver as glycogen. Storing excess glucose in the liver allows the glucoregulatory system to save it for later use when glucose levels are too low.

As you so often see in biological systems, a complementary system is needed to raise glucose levels when they are too low. When we go too long between meals, our blood glucose levels sink lower than the set point, a condition known as **hypoglycemia**. In response, alpha cells in the pancreas secrete glucagon. Glucagon causes the liver to convert its store of glycogen back into glucose and release that glucose into the blood where cells convert glucose into energy.

A metaphor comparing glucose regulation to your budgeting may help to make the system easier to understand. When you get your paycheck, you spend some of the money on what you need immediately. If there is extra money, you can store it in your bank account until you need it later. When an expense arises, you go to the ATM and withdraw the funds needed to cover the expense. In this metaphor, glucose is money, the activities that money enables is the metabolic energy released by glucose, your liver is the bank, and glucagon is the ATM card.

You can easily see that the pancreas is constantly cycling between glucagon release and insulin release to maintain the proper level of blood glucose, and this cycle is a response to our consuming calories and then expending those same calories through maintaining our metabolism, supplying muscles with energy, and fueling our brain. The continuous regulation of blood glucose through the action of the pancreas is demonstrated in this **Check It Out: Glucoregulation**.

Diabetes Maintaining the proper glucose level is essential to the functioning of your brain and body. What happens if that delicate balance is disrupted? When the pancreas is congenitally (from birth) unable to secrete sufficient insulin to regulate glucose levels, that individual has type 1 **diabetes** (also known as juvenile-onset diabetes). Type 1 diabetes is genetic and generally appears sometime in adolescence. Treatment requires that the individual carefully monitor carbohydrate intake and blood glucose levels and then inject insulin in the correct amount necessary for digestion.

Type 2 diabetes (also known as adult-onset diabetes) develops in adulthood and is characterized by insulin resistance as opposed to insulin absence. In type 2 diabetes, the pancreas is secreting sufficient insulin, but the surface receptors in the body's cells are less responsive so that glucose is not absorbed. The pancreas continues to secrete insulin in an effort to regulate blood glucose.

Recall that every cell in your body requires glucose as fuel; you can then easily see that providing your cells with too much or too little will lead to a wide range of disorders, and this is why diabetes is such a health concern. Diabetes damages the retinas, kidneys, and other organs. In addition, excess glucose damages the walls of blood vessels throughout the brain and body, making them less elastic and unable to properly function. Damage to cerebral blood vessels leads to diffuse damage in the brain, and damage to peripheral blood vessels typically

Check It Out | Glucoregulation

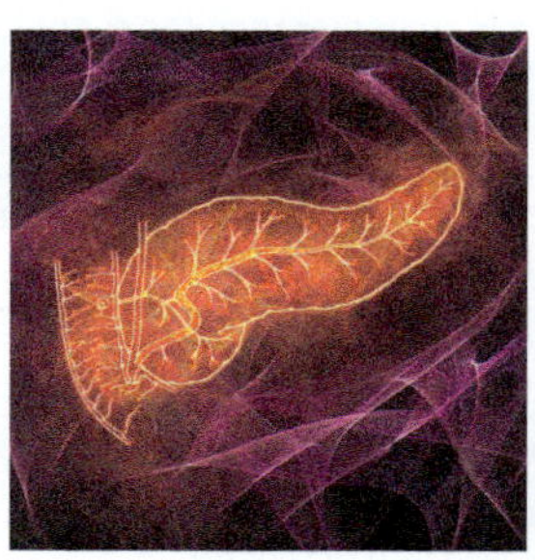

MEHAU KULYK/Science Photo Library/Getty Images

Watch the brief video **The Pancreas and Glucoregulation** that depicts how our pancreas continuously adjusts blood glucose levels throughout the day, and then answer these questions.

1. What do you think the effect is on your pancreas of consuming a large amount of carbohydrates such as bread, cookies and cakes, cereals, and rice at one sitting? Briefly describe.
2. What can happen if the brain does not get enough glucose? Search online for "hypoglycemia and cognition" and briefly describe the mental symptoms of hypoglycemia.

first emerges at the feet and legs. In the most severe cases, amputation of toes, feet, and even legs may be required.

Diabetes also has important effects on the brain. Your brain uses glucose as an energy source, and it requires a supply that matches with demand. Your automobile engine will stall if the carburetor or fuel injectors deliver too much or too little fuel to the cylinders. Similarly, when more glucose is available in the blood than organs can use, the excess can damage tissue, including the nervous system. Individuals with uncontrolled diabetes may cycle through hypo- and hyperglycemia, and each cycle may cause damage to cerebral arteries. Diabetics experience a significant increase in risk for early dementia, and when hypoglycemic, diabetics may experience mental confusion and generalized difficulties with problem solving and rational thought. Explore the personal and societal costs of diabetes in **Think About It: Preventing Type 2 Diabetes**.

Think About It | Preventing Type 2 Diabetes

moodboard/Adobe Stock

It is hard to overestimate the extent to which diabetes affects modern life in the developed world. Diabetes affects about 10% of the U.S. population and one of four people over the age of 65, so although we may not be aware of their condition, each of us every day encounters many people who are diabetic.

Type 1 diabetes is a chronic condition that generally appears in adolescence and likely has some genetic component. Type 2 diabetes can develop as a result of life choices such as a poor diet and smoking. Managing diabetes of either type involves maintaining a healthy lifestyle. Uncontrolled diabetes can have devastating personal costs, including amputation, dementia, and death. Diabetes also has a tremendous societal cost. According to the Centers for Disease Control (CDC), $1 out of every $4 of U.S. healthcare spending goes toward people with diabetes.

Where does individual responsibility end and societal responsibility begin when it comes to diabetes? Some municipalities have imposed a so-called sugar tax on sugary drinks to offset the high cost of diabetes and other health impacts of obesity. Some employers offer incentives for employees to lose weight. On the other hand, it is very hard to influence the lifestyle choices of individuals on a broad scale.

Your author knew a type 2 diabetic who refused to take her disease seriously. She lived on a diet high in carbohydrates, smoked heavily, and had a BMI above 45. Her blood glucose levels were constantly out of control, vacillating between very low and very high. She began to have episodes where she would become confused and even pass out. We had a frank chat in which I told her that she was damaging her cerebral blood vessels, and it would not be long before she began to have strokes. She declined to change her lifestyle, diet, or begin even a mild exercise program. Her response was that the enjoyment she derived from eating a six-pack of muffins for breakfast was worth the early demise. After a series of mini strokes that led to hospitalizations, she suffered a massive fatal stroke at age 74.

Questions

1. **Search using the terms "risk factors for diabetes" and identify three risk factors for type 2 diabetes.**
2. **Search using the terms "What can I do to prevent diabetes" and identify five things you can do to avoid developing type 2 diabetes.**
3. **What could governments, employers, or insurers do to reduce the prevalence of diabetes? Make an argument for or against such interventions.**

The Adrenal Glands and Stress MCAT

Perched atop each kidney are the **adrenal glands**. Their function is to mobilize the body to respond to stress, both physical and emotional. The adrenal glands consist of an inner part known as the adrenal medulla and an outer part known as the adrenal cortex, each with its own unique function (**Figure 8.13**).

The adrenal glands secrete hormones that prepare you for fight or flight. The adrenal medulla secretes epinephrine (adrenalin) and norepinephrine (noradrenaline). These chemical messengers act as neurotransmitters inside the CNS and as hormones in the periphery. If a lion suddenly appeared in front of you, what would your body need to do for its survival? If you answered increase blood flow to the large muscles (say, of the legs and arms) and increase respiration and heart rate, glucose metabolism, and blood flow to the brain, you'd be on the right track. That response is exactly what epinephrine and norepinephrine do for you. Norepinephrine also stimulates the amygdala and appears to sharpen attentional focus, which also aid your chances of survival.

The adrenal cortex secretes estrogens and androgens in small amounts. More important for our purposes are cortisol and aldosterone. Cortisol is a glucocorticoid hormone produced by the adrenal cortex that plays several important roles in the body. It helps control the body's use of fats, proteins, and carbohydrates; suppresses inflammation; regulates blood pressure; increases blood glucose; and can also decrease bone formation. This hormone also controls the sleep–wake cycle. It is released during times of stress to help your body get an energy boost and better handle emergency situations. We will encounter cortisol later in this chapter when we discuss your body's response to stress.

Aldosterone is a hormone secreted by the cortex of the adrenal glands. One function of your kidneys is to manage sodium and potassium levels in the blood. Aldosterone controls the rate at which sodium and potassium are retained by the kidneys, which indirectly

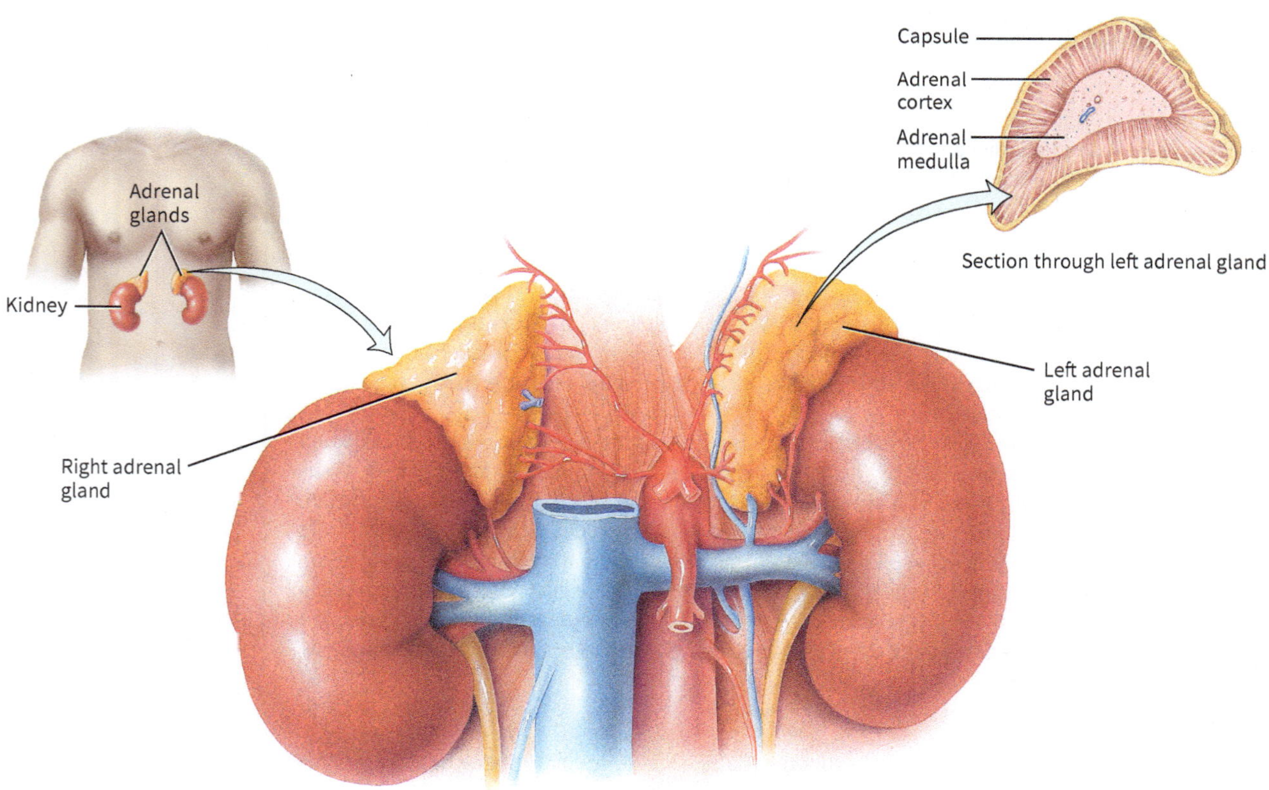

FIGURE 8.13 **The adrenal glands** The adrenal glands sit atop the kidneys and secrete hormones that help you respond to stress.

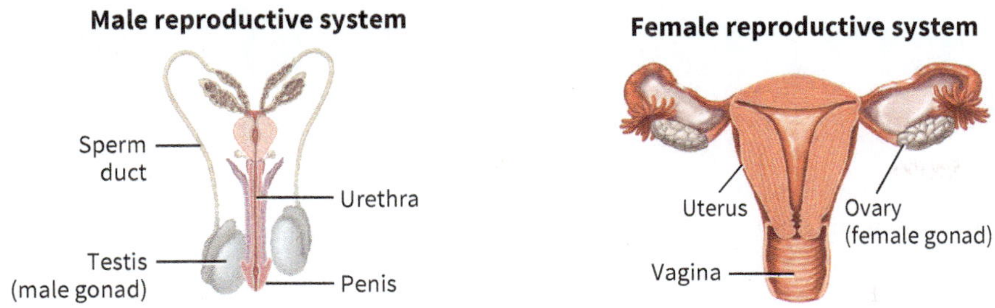

FIGURE 8.14 The male and female reproductive systems The male gonads consist of the testes, whereas the female gonads consist of the ovaries.

affects water retention, which in turn affects blood pressure. Because aldosterone secreted by the adrenal cortex affects renal (kidney) function, adrenal disorders often lead to renal disorders.

The Gonads and Sexual Development MCAT

A **gonad** is an organ that creates gametes; in humans, these are egg and sperm cells. The location of the gonads in the body is depicted in Figure 8.14. Female gonads are ovaries, and male gonads are testes. There are two ovaries (one on each side of the pelvis) in females and two testes in males.

Your gonads secrete sex hormones, of which the two main varieties are androgens and estrogens. **Androgens** (from the Greek root for man) regulate the development of primary sexual characteristics such as male genitalia as well as secondary sexual characteristics such as bodily and facial hair and muscular development. Androgens are also essential for sexual function, and they affect sexual interest (libido) in males and females. Females also secrete androgens, although in much smaller amounts. Androgens are also secreted by the adrenal glands and the ovaries. The primary androgen is testosterone, although in your studies you will encounter others such as dihydrotestosterone (DHT) and androstenedione.

Estrogens are a class of hormones that, like androgens, are responsible for the development of primary sexual characteristics such as a vagina as well as secondary sexual characteristics such as lack of facial hair, development of breasts, and widening of hips. Estrogens are secreted by ovaries as well as the adrenal glands, and although females secrete more estrogens than males, males also secrete estrogens. The primary estrogens are estrone (E1), estradiol (E2), and estriol (E3). Of these, estradiol is the most potent. That being said, it is not uncommon for people to use the term estrogen to refer to female hormones in a general fashion.

The ovaries are internal, and they release estrogen, testosterone, and progesterone. Estrogen plays a central role in egg production and maturation and also drives the development of female secondary sex characteristics such as a lack of facial hair, breast development, and deposition of subcutaneous fat. Progesterone mostly influences the menstrual cycle as well as egg implantation and gestation during pregnancy.

The testes are located in the scrotum, and they secrete testosterone. Testosterone stimulates the maturation of sperm and the development of male secondary sex characteristics such as facial and body hair, a deep voice, and larger skeletal muscles. Both estrogen and testosterone have an impact on some working memory processes and are purported to play a role in certain social behaviors.

Males and females normally express both sets of hormones, but at very different levels. Androgens and estrogens do more than stimulate sexual differentiation; they also play a critical role in the menstrual cycle and sexual behavior. The timing and amount of these hormones during prenatal development has an important part in the development of sexual orientation, and their effect on the subsequent development of adult cognitive processes has been extensively investigated. These topics will be discussed in other chapters.

8.3 Before You Go On

Section Summary

The thymus plays a role in immune system function. The thyroid produces the hormone thyroxine, which plays a critical role in maintaining metabolic function. The pancreas cycles between production of insulin and glucagon to regulate blood glucose. The adrenal glands play a role in the stress response. The gonads (testes in males and ovaries in females) produce androgens and estrogens, which are responsible for sexual development.

Comprehension Check

1. What is the function of thyroxine?
   - a. regulating your metabolism
   - b. stimulating the pancreas to release insulin
   - c. causing your blood vessels to dilate in response to stress
   - d. regulating blood glucose levels
2. The medical term for low blood glucose is _____, and the corresponding term for high blood glucose is _____.
   - a. glucoremia; glucosemia
   - b. pancreatitis; pancreatosis
   - c. hyperglycemia; hypoglycemia
   - d. hypoglycemia; hyperglycemia
3. Diabetes is a dysfunction of which gland(s)?
   - a. adrenal glands
   - b. pancreas
   - c. pituitary
   - d. HPA axis
4. Your adrenal glands are involved in regulating what?
   - a. sexual behavior
   - b. fight-or-flight responses
   - c. the development of facial hair
   - d. social awareness
5. Which of these statements is true about androgens and estrogens?
   - a. Estrogens are much more powerful than androgens.
   - b. Androgens are secreted only from the testes.
   - c. Males and females secrete both male and female sex hormones.
   - d. Androgens and estrogens are both hormones and neurotransmitters.

8.4 The Stress Response

LO 8.4 Understand how the endocrine system regulates our response to stress.

This chapter's *Stories of the Brain* was an anecdote about a person whose physiological response to a narrow escape from an auto accident was probably familiar to us all. Now that you have an overview of the glands of the endocrine system, we will take a closer look at how and why that stress response occurs.

Modern life is complicated, and complications create stress. We navigate social, family, and romantic entanglements, worry about finances, and juggle the demands of jobs and school. Our early ancestors dealt with stress as they tried to avoid predators while securing food and shelter for themselves and their clan. You are unlikely to spend part of today hiding from a prowling lion or pack of howling hyenas, but your life has its own stressors, and they are no less real and no less dangerous.

What exactly is stress, and how does our body respond to it? Stress is your body's response to external threats, be they physical, biological, or psychological. When a hostile army appears at the gate of a city, the defenders mobilize their forces at weak spots on the wall and prepare to fight. In the same way, your endocrine system mobilizes its resources to keep you alive. We now turn to the story of how your endocrine system deals with stress through a circuit known as the hypothalamic–pituitary–adrenal axis.

The HPA Axis MCAT

In prior sections we have already mentioned that the hypothalamus, the pituitary gland, and the adrenal cortex each play a role in the stress response. These three glands working together are known as the **hypothalamic–pituitary–adrenal (HPA) axis**. The HPA axis is part of a larger system for responding to stress, and understanding how our endocrine system deals with stress is an important topic in behavioral neuroscience. Consider that there are two forms of stress: immediate stress and continuing stress. All too familiar examples of each are the immediate stress of an unexpected pop quiz and the continuing stress that weekly reading assignments, papers, and exams create in a student's life (**Figure 8.15**).

These two types of stress are managed by your nervous system and endocrine system in different ways and so we now turn to the first of these systems for managing stress: the fight-or-flight response.

The fight-or-flight response Our first response to an emergency is the **fight-or-flight response**. When faced with impending danger, our CNS engages the sympathetic nervous system to signal the adrenal medulla to immediately flood the blood supply with epinephrine and norepinephrine (also called adrenalin and noradrenaline). You may remember from a previous chapter that epinephrine and norepinephrine act in the nervous system as a neurotransmitter by conducting messages over the synapse, but outside the CNS, they act as a hormone by conducting messages through the bloodstream. This release causes an almost instant increase in heart rate, blood pressure, and respiration.

The release of these hormones also causes an increase in the conversion of glycogen to glucose, thereby adding to the energy store available to cells and tissues. Additionally, blood is shunted from peripheral vessels to the major limbs, and the coronary (heart) blood vessels dilate. This sudden surge in blood circulation enhances the body's ability to fight or flee. The fight-or-flight response is an example of coordination between the nervous system and the endocrine system.

Recall that the autonomic nervous system consists of two components: the sympathetic and the parasympathetic systems. The sympathetic system acts like a gas pedal while the parasympathetic operates as the brake. When a stressor stimulates the HPA axis, the adrenal glands release epinephrine, which rapidly initiates a systemic response. If you think for a moment about this chapter's *Stories of the Brain*, Madeline's response to narrowly avoiding a danger while driving illustrated the action of the HPA axis. Her system responded instantly to a threat by raising her blood pressure, heart rate, and respiration. Then, when she was safe from danger, the parasympathetic brake responded, lagging behind the action of the HPA axis and the sympathetic nervous system. From an evolutionary perspective, this apparent mismatch makes sense. If you suddenly found yourself face to face with a lion, even when

Gallo Images-Dave Hamman / Getty Images

Maskot / Adobe Stock

FIGURE 8.15 **Immediate stress and continuing stress** **a.** Our ancestors often had to deal with stressors that were immediate. **b.** Our lives are stressful, but often the causes are continuous.

you retreated to a safe spot, it would be advantageous to be in a heightened state of readiness until you were confident the threat had passed. Now you know how your endocrine system cooperates with your nervous system to respond to the stress of immediate danger, but how does it respond to everyday stressors?

Resistance to continuing stress It is a good thing that we do not encounter an unexpected quiz, a hungry lion, or a car speeding toward us out of control every day, but stress is a part of life. We have academic demands, work and family pressures, as well as concerns about how best to arrange our personal relationships, careers, and life. The second system for managing stress is concerned with what might be called chronic or continuous stress.

Continuous stress stimulates the hypothalamus to engage the pituitary, which then reaches out into the body to create a gradual but lasting response to stress through the adrenal cortex, liver, and thyroid. The pituitary releases the hormone ACTH (adrenocorticotropic hormone), which stimulates the adrenal cortex to release cortisol, which in turn stimulates the release of glucose and anti-inflammatory substances. The pituitary also releases growth hormone (GH or HGH), which then stimulates the liver to increase the conversion of glycogen to glucose and release that glucose into the bloodstream. Finally, the pituitary releases the hormones FSH and LH (follicle stimulating hormone and luteinizing hormone), which then stimulates the thyroid to release thyroid hormones, which increase general metabolic rate.

There are two important takeaways from what you have just learned. The first is that in contrast to the flight-or-flight system, which prepares you to deal with an immediate but transitory danger, this second, slower-responding system creates ongoing resistance to stress by mobilizing resources throughout your body. The second takeaway is less obvious but no less important; just as increasing your spending without increasing your income eventually leads to financial distress, experiencing continued stress without a period of recovery eventually leads to the depletion of resources. We will discuss how this depletion leads to disease in a subsequent section when we consider the general adaptational syndrome.

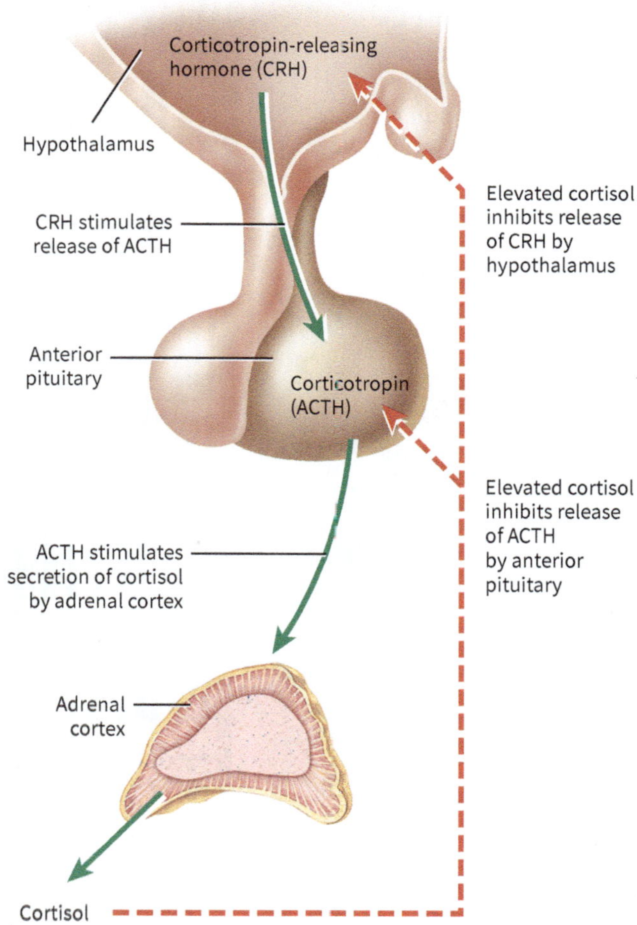

FIGURE 8.16 The HPA axis and cortisol released in response to stress Stress stimulates the hypothalamus to release the hormone CRH, which stimulates the pituitary to release ACTH, which in turn stimulates the adrenal cortex to release cortisol.

Cortisol The fight-or flight response gives us an immediate boost to respond to danger through epinephrine and norepinephrine, but that response cannot be maintained for long. A longer-term response to stress is generated by the hormone cortisol, also regulated by the HPA axis in a negative feedback system (**Figure 8.16**).

In response to a stressor, the hypothalamus (the sensor) stimulates the anterior pituitary via the hormone CRH (corticotropin-releasing hormone); the anterior pituitary (the control) in turn secretes the hormone ACTH (adrenocorticotropin hormone), which stimulates the adrenal cortex. The adrenal cortex (the effector) then secretes cortisol, which increases glucose metabolism, giving your body an energy boost. Cortisol also boosts the body's immune response and acts as an anti-inflammatory, all functions that are valuable when you are stressed. Given that cortisol is an integral part of your fight-or-flight response, you will not be surprised to know that cortisol release suppresses nonessential functions such as the digestive and reproductive systems.

As you would expect in a negative feedback system, there is also a braking mechanism to bring the system back to homeostasis once the threat has passed. Higher levels of cortisol inhibit the release of CRH by the hypothalamus and of ACTH by the pituitary, which returns the system to homeostasis.

This release of cortisol provides a boost of energy that is helpful in responding to a temporary stressor, but chronic exposure to high levels of cortisol is not healthy. Soldiers in war zones or refugees fleeing natural or social disasters are exposed to high levels of stress hormones such as cortisol over periods of months or even years. One does not have to be in combat or fleeing a catastrophe to be exposed to continuous and excessive stress; work, friends, family, and school can create enough stress to overwhelm the resistance circuit in a way that leads to disease. In that light, there is evidence that the HPA axis is involved in post-traumatic stress disorder (PTSD), discussed later in the chapter.

Our ancestors evolved in a very different environment than you and I navigate daily. They had to provide food and shelter each day while avoiding predators and possibly hostile and competing tribes, but their concerns were answered with direct physical response. Hunger pangs were addressed by seeking food, and this might entail a daily journey of several miles of checking traps and snares and gathering root and berries to be returned to the camp for processing and eventual consumption; menacing predators of the two- and four-footed varieties were met with a vigorous defense. These stressors activated the HPA axis, leading to hormonal release that activated the body's resources. Our ancestors dissipated these hormones by roaming for food, capturing prey, avoiding predators, and even occasionally fighting with hostile tribes. Today, our stressors are no less serious, but in the modern world, you cannot react to the stresses of exams, work, and family through a six-mile (10-km) walk or by lifting your club over your head and charging into battle. We know that stress hormones such as epinephrine prepare the body for fight or flight, but modern life does not lend itself to such a response to our stressors. As a result, some of us experience chronic stress, which brings its own health concerns. Our understanding of the HPA axis and hormonal release in response to stress leads us to consider how the body deals with chronic stress

The General Adaptational Syndrome MCAT

Now that you have an idea of your body's response to stress, we will return for a moment to our battle metaphor of stress in which your forces mobilized to defend the city from an external threat. At first, your forces are fresh and repulse the attack furiously, but as the enemy mounts successive assaults on the city's defenses, your forces become fatigued and eventually fail to defend the city walls. In the same way, your response to stress becomes fatigued if stressors are continuous. Without a period to rebuild and refresh one's resources, the available defenses can be overwhelmed.

Perhaps there have been times in your academic career when you were sleeping very little, cramming for exams, and trying to cover your hours at a job while living a full social life. You finished your final exams, and then immediately came down with a flu or other illness. This scenario is an example of the **general adaptational syndrome (GAS)** that was identified by Hungarian endocrinologist Hans Selye (1907–1982).

Early in his medical career, Selye observed that patients who were experiencing pressures of many different types all exhibited a single set of symptoms, including depression, sleep disruption, and cardiovascular disease. Subsequent laboratory work with rodents that were repeatedly exposed to stressors established that chronic stress without a recovery period leads to these health problems. The GAS response consists of three phases: alarm, resistance, and exhaustion (Figure 8.17).

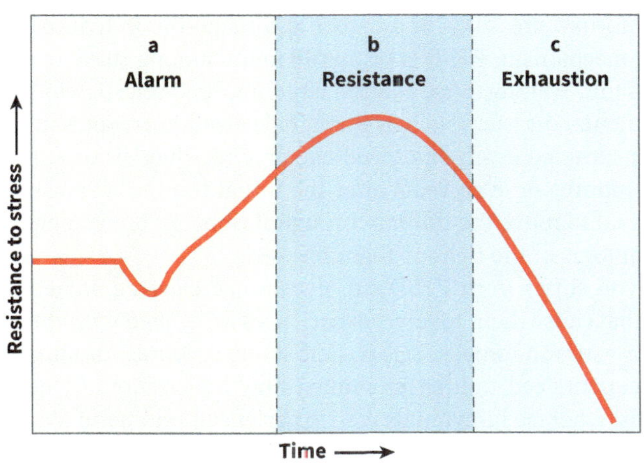

FIGURE 8.17 The GAS response **a.** When stress challenges the body, we respond by increasing our metabolic resources to provide resistance. **b.** Over time, our ability to respond to stressors peaks and then begins to decline. **c.** If stress continues unabated without a period of relief, our ability to resist declines.

FIGURE 8.18 Bungee jumping Individuals define stress in unique ways. For some, jumping off a platform hundreds of feet in the air with only a bungee cord to protect you is a hobby; others would find this experience terrifying.

In the alarm phase, the body mounts a response to the stressor through mobilizing the sympathetic nervous system. An initial dip in resistance to stress is followed by a ramping up of resistance. Epinephrine is released into the bloodstream, and heart rate, blood pressure, and respiration all increase to meet the challenge.

In the resistance phase, the body has ramped up its defenses and is meeting the challenge. Additional glucose is released into the bloodstream, as is cortisol, a steroid secreted by the adrenal glands that aids in the release and use of glucose.

In the recovery phase, the threat is countered with sufficient resources until the stressor is eliminated entirely, in which case the body returns to its pre-threat level. Should the body's ability to respond to threat be diminished due to being overly taxed, the individual enters the exhaustion phase, in which their defenses can be overwhelmed. Exhaustion will eventually lead to illness or even death.

GAS does not fully explain our experience of stress; for example, some people thrive on stress and actively seek out activities and careers that many of us would consider to be stressful. People define stress in different ways; someone might find bungee jumping off a bridge to be great fun (**Figure 8.18**) but be so stressed about giving a talk in front of a group that even the thought of it makes them feel faint. No matter how individuals experience stress, the general opinion is that without some regular period of relaxation and release from stress, our mental and physical health is at risk.

Post-Traumatic Stress Disorder MCAT

When your fight-or-flight response is triggered, your endocrine system releases hormones to prepare you to flee or fight for your life. After the threat is removed, most people are able to put aside the memory of the events and move on, but for others, events in their daily lives trigger the fight-or-flight response again and again. **Post-traumatic stress disorder (PTSD)** is a serious psychological injury characterized by flashbacks, anxiety attacks, and obtrusive thoughts.

PTSD can affect family, romantic, and social relationships; it can disrupt a person's ability to cope with the normal pressures of a career and make the individual prone to substance abuse as a coping mechanism. PTSD is primarily found among survivors of traumatic events such as sexual assault, war, terrorist attacks, child abuse, domestic violence, vehicular accidents, or environmental disasters such as earthquakes and hurricanes in which there is great loss of life. The unexpected death of a loved one or a health crisis can also precipitate PTSD. PTSD may occur months or even years after the initial trauma when some triggering event causes the individual to relive the trauma through flashbacks or even hallucinations, accompanied by the symptoms of the fight-or-flight response.

Individuals who suffer from PTSD can experience extreme anxiety if they experience stimuli or events that cause them to relive experiences associated with their trauma. Common triggers are sensory stimuli (smells, sights, and sounds), words, or situations. For example, individuals who experienced trauma in combat may be triggered by car backfires or loud, unexpected sounds, whereas individuals who experienced trauma in a crowd fleeing a threat may be triggered when finding themselves in a peaceful crowd. One consequence of PTSD is that the individual may be hypervigilant, that is, have a heightened level of sympathetic arousal and so be easily triggered, even when removed from the stressful environment. Therapeutic intervention using counseling techniques designed to either bring the stress response under control or replace triggering memories with other memories have been found to be useful. Modest success has also occurred by coupling therapeutic drugs known as SSRIs or benzodiazepines with therapy.

Not everyone who experiences horrific or gruesome events develops PTSD, so we might wonder if there are risk factors that help us to both predict who might experience PTSD and also understand the mechanism behind it. We do know that if one identical twin experienced PTSD after serving in combat, the other was significantly more likely than a fraternal twin to also experience PTSD. This suggests a genetic component to risk factors for PTSD.

Heightened startle response has also been associated with a greater propensity for developing PTSD. The startle response is your body's reaction to a sudden and unexpected event such as a loud noise (80 dB or above) or a sharp and disconcerting movement. Your eyes widen, your pupils dilate, your body moves almost imperceptibly into a crouch, and you are ready to fight or flee. The startle response is mediated through the brainstem and begins to occur in as little as 40 ms.

Finally, your knowledge of the HPA axis may help you understand one of the more intuitive theories about the neurophysiology of PTSD. The HPA axis is involved in maintaining fear, and hormonal analysis of individuals suffering from PTSD suggests a dysfunction in the inhibitory mechanism of the HPA axis in PTSD. In other words, the negative feedback system fails to return to homeostasis.

Because combat often puts people into life-threatening situations, military psychologists and researchers are heavily invested in understanding and treating PTSD. Meet one of these researchers in **Advances in Behavioral Neuroscience: Carrie Hill Kennedy and PTSD**.

Advances in Behavioral Neuroscience | Carrie Hill Kennedy and PTSD

Carrie Hill Kennedy

Dr. Carrie Hill Kennedy is in a unique position to understand the implications of living through a traumatic event because she is not only a research scientist and a clinical neuropsychologist; she is also an officer on active duty in the U.S. military who has been deployed to combat zones. Because military personnel are more likely to experience severe stress while in combat, Dr. Kennedy has had a unique opportunity to investigate risk factors that may make someone more likely to develop PTSD as well as how best to treat it after the fact.

Members of the military are not the only individuals who develop PTSD, but they provide a laboratory for understanding the causes of PTSD and the best treatment modes. Our best estimate is that roughly 7% of us will experience PTSD at some point in our lives, with the incidence for women being slightly higher. Individuals experiencing natural disasters such as hurricanes or earthquakes as well as individuals who are victims of terrorist attacks may develop PTSD. In addition, firefighters, police, EMT, and rescue workers, as well as those who witness such events up close, are at risk for PTSD. The symptoms of PTSD may linger or emerge long after the event.

PTSD has been hypothesized to be an acquired or maladaptive response of the HPA axis to fearful stimuli. Other mechanisms may operate to permanently fix unpleasant events in long-term memory, thereby leading to flashbacks and distressing dreams. There is much we know about PTSD, but much we still have to learn. For example, are some personality types more resistant to PTSD than others? Do men and women respond the same to therapies, or are there gender differences in therapeutic outcomes? What is the best way to intervene in individuals at risk for PTSD? Is PTSD overdiagnosed? Is there something we can do to inoculate (so to speak) people at risk for PTSD, so they are less likely to develop it? These and other questions are the object of Dr. Kennedy's work, and her insights are widely applicable.

Questions

1. **Identify a traumatic experience that someone in your cohort (age group) might experience. If someone developed PTSD after experiencing this trauma, what everyday events might trigger stress? How might this impact their daily lives?**
2. **Search online for the term PTSD and identify one fact about PTSD that you did not know previously.**

8.4 Before You Go On

Section Summary

When we are stressed, our endocrine system responds by secreting hormones that mobilize resources throughout the body. After a stress is met, the system needs time to reset and repair any cellular damage that occurred in response to the challenge, and when there is no opportunity for refreshment, chronic stress can lead to loss of resilience and negative effects on health, as explained by the GAS. Experiencing a trauma can lead to PTSD, a debilitating disorder in which a person experiences stress responses for months and even years after the traumatic event.

Comprehension Check

1. The HPA axis is a mechanism that regulates the ________.
   a. reaction to stress
   b. sleep–wake cycle
   c. digestion and metabolism of nutrients
   d. expression of sexual behavior
2. One thing we know about stress is that ________.
   a. it is a modern concept that our ancestors rarely experienced
   b. people who experience stress are generally physically weak
   c. every time you are stressed, you become stronger
   d. repeated stress can damage bodily systems
3. What are the stages of stress response according to the GAS theory?
   a. alarm, response, and exhaustion
   b. alarm, recovery, and response
   c. alarm, refreshment, and recovery
   d. alerting, alarm, and response
4. Which is true of post-traumatic stress disorder?
   a. It is an immediate response to a traumatic event.
   b. It is a delayed response to a traumatic event.
   c. It rarely happens if you only witnessed the traumatic event.
   d. It is caused by a disorder of the hypothalamus.

Conclusion

We began our discussion of the endocrine system with the story of Madeline, whose brush with a threat caused her endocrine systems to prepare her to fight for her survival if need be. You now have a comprehensive perspective on the endocrine system. In contrast to the rapid and very localized response of a postsynaptic receptor to a neurotransmitter, the endocrine system provides a mechanism for gradual, systemic response to an ever-changing environment. The endocrine system demonstrates that a chemical messenger such as a hormone can have an important effect at a receptor far away from the gland itself. Many endocrine disorders cause fatigue, muscle weakness, and a general metabolic disturbance. These common symptoms demonstrate both the importance of the endocrine system as well as the difficulty a medical professional faces in diagnosing these conditions.

Understanding how hormones help regulate your body systems in a changing environment will inform your understanding of many topics in behavioral neuroscience including sexuality, development, disorders, and more. Finally, our endocrine system defends us from stressors, but only until its resources are depleted.

Critical Thinking Questions

1. The human nervous system activity is characterized by rapid and local effect, whereas endocrine system activity is gradual and global. Assume both systems operated both as rapid/local or both as gradual/global. How would each scenario change human behavior and survivability?
2. The hypothalamus is located deep in the brain, where it is somewhat protected from damage. In rare cases, a tumor on the hypothalamus can disrupt its function. Why would you not be surprised that hypothalamic damage leads to a broad range of disruptions?
3. The most common endocrine disorder is diabetes. Some think that our modern lifestyle encourages us to behave in ways that increase our risk of developing diabetes. Choose at least two behaviors or aspects of modern life, and describe how they amplify the risk of developing type 2 diabetes.

4. During final exams, students often skip meals and go without sleep. Sometimes immediately after a student's last exam, they develop a cold or flu. How might the GAS (general adaptational syndrome) explain this phenomenon?

5. The term *endocrine system* refers to a collection of glands that release hormones into the bloodstream that affect target cells in organs throughout the body; however, many behavioral neuroscientists, medical researchers, and practitioners use the term *neuroendocrine system* or *neuroendocrinology* to refer specifically to the circuit consisting of the hypothalamus and pituitary that interacts more directly with the nervous system. Do you think this distinction between the general endocrine system and the more specific neuroendocrine system is useful? Briefly explain.

Dig Deeper into Behavioral Neuroscience

8.1 An Introduction to the Endocrine System

Our endocrine system regulates almost every aspect of our daily lives. General metabolic function, food digestion, glucose regulation, the development of our gonads, and their expression of sexual behaviors are all a part of the endocrine system. You can read more about this system here.

https://openstax.org/books/anatomy-and-physiology/pages/17-1-an-overview-of-the-endocrine-system

8.2 and 8.3 Glands of the Endocrine System

There is much more to the endocrine glands than can be covered in the context of behavioral neuroscience. The *Merck Manual* is an important reference for medical information that physicians keep close at hand; a consumer version can be found online that provides additional information about the glands that make up the endocrine system.

https://www.merckmanuals.com/home/hormonal-and-metabolic-disorders/biology-of-the-endocrine-system/endocrine-glands

As important as our endocrine system is, you should not be surprised that endocrine dysfunction has serious consequences. You can read more about neuroendocrine disorders here.

https://www.niddk.nih.gov/health-information/endocrine-diseases

8.4 The Stress Response

We all experience stress in our own way, so we all need to know how to recognize its negative consequences and how to successfully deal with it.

https://www.nimh.nih.gov/health/publications/stress/index.shtml

PTSD results in significant and negative consequences for the life of the individual. These two sources can show you more about PTSD both in civilian life and in the military.

https://www.nimh.nih.gov/health/topics/post-traumatic-stress-disorder-ptsd/index.shtml

Kennedy, C. H. (2020). Basic truths about military stress, trauma and post-traumatic stress disorder. In C. H. Kennedy, *Military Stress Reactions: Rethinking Trauma and PTSD*. New York City: Guilford Press. ISBN-13: 978-1462542949

Chapter 8 Before You Go On Answers

8.1 Before You Go On Answers

1. (a) Hippocrates; the four humors
2. (c) responds to conditions in a slow and gradual fashion
3. (c) It maintains a relatively stable environment through a feedback mechanism.
4. (d) enter the bloodstream to affect systems throughout the body
5. (b) It prevents small deviations from a set point from getting larger.

8.2 Before You Go On Answers

1. (d) all of these
2. (b) growth (as in height)
3. (a) sleep–wake cycles and circadian rhythms
4. (c) a pituitary dysfunction

8.3 Before You Go On Answers

1. (a) regulating your metabolism
2. (d) hypoglycemia; hyperglycemia
3. (b) pancreas
4. (b) fight-or-flight responses
5. (c) Males and females secrete both male and female sex hormones.

8.4 Before You Go On Answers

1. (a) reaction to stress
2. (d) repeated stress can damage bodily systems
3. (a) alarm, response, and exhaustion
4. (b) It is a delayed response to a traumatic event.

Chapter 8 Glossary

adrenal gland One of two small glands atop the kidneys that secrete hormones that regulate metabolism, immune response, and other functions.

androgen A type of hormone that regulates the development and maintenance of male characteristics in males and contributes to growth and reproduction in males and females.

diabetes A disorder involving how the body regulates blood glucose.

endocrine gland A ductless gland that releases hormones into the bloodstream; examples are the pancreas, hypothalamus, and adrenal glands.

endocrine system A messenger system that regulates a wide range of behavioral and metabolic functions by means of glands secreting hormones directly into the circulatory system.

estrogen A type of hormone that regulates the development and maintenance of female characteristics in females and contributes to growth and development in males and females.

exocrine gland A gland that releases its secretions by way of a duct; examples are sweat, salivary, and mammary glands.

fight-or-flight response A collection of physiological responses to a perceived threat or attack.

general adaptational syndrome (GAS) A theory that describes the body's response to continued stress.

glucoregulation The body's system for regulating blood glucose.

gonad An organ (testes in males and ovaries in females) that produces the gametes or sex hormones.

hormone Any of a class of signaling molecules that are released into the bloodstream, travel to distant organs, and regulate physiology or behavior.

hyperglycemia A state of high blood glucose.

hyperthyroidism The condition of having an overactive thyroid or an overabundance of thyroid hormones.

hypoglycemia A state of low blood glucose.

hypothalamic–pituitary (HP) axis An important circuit in the endocrine system in which the hypothalamus and pituitary interact with other endocrine glands such as the adrenals, the gonads, and the thyroid to regulate essential functions.

hypothalamic–pituitary–adrenal (HPA) axis A circuit in the body that mediates the effects of stressors by regulating physiological processes.

hypothalamus A structure in the brain that is comprised of several nuclei and is responsible for controlling bodily functions such as hunger, body temperature, and thirst.

hypothyroidism The condition of having an underactive thyroid or an undersupply of thyroid hormones.

negative feedback system A control process that stabilizes system output; also known as a negative feedback loop.

pancreas An organ in the abdomen that is essential for converting food into glucose.

pineal gland A small gland in the brain that is essential in regulating sleep and circadian rhythms.

pituitary gland A small gland in the brain that regulates growth and also signals many other endocrine glands.

post-traumatic stress disorder (PTSD) A mental health disorder created by experiencing or witnessing a traumatic event that is triggered by subsequent events; symptoms include nightmares, flashbacks, and severe anxiety.

thymus A gland in the upper chest that contributes to the immune response.

thyroid A gland in the upper chest that plays a major role in regulating metabolic functions.

CHAPTER 9

Kobby Dagan / Shutterstock

The same brain mechanisms that reward us for adaptive behavior can reinforce substance abuse and addictive behaviors like gambling.

Reward, Addiction, and Psychopharmacology

CHAPTER OUTLINE

9.1 The Biological Basis for Reward

- Discovering the Biological Basis for Reward
  - Think About It: The Ethics of Deep Brain Stimulation
- Reward Centers in the Brain
- The Mesolimbic and Mesocortical Pathways
- The Reward Circuit and Human Behavior

9.2 Addiction as Dysfunction of the Reward Circuit

- The Genetic Basis for Addiction
- Substance Abuse Disorders
  - Try It Out: Diagnosing Addictive Behavior
- Risk Factors for Addiction

9.3 The Essentials of Psychopharmacology

- How Drugs Enter the Body
- Pharmacodynamics
  - Think About It: Behavioral Tolerance and Drug Overdoses
- Pharmacokinetics

9.4 Intoxicants

- Alcohol
  - Check It Out: Effects of Alcohol on Visual Search
- Opioids
  - Advances in Behavioral Neuroscience: Candace Pert and Opioid Receptors
- Marijuana

9.5 Stimulants and Other Commonly Abused Drugs

- Nicotine
- Caffeine
  - Think About It: Do We Abuse Caffeine?
- Other Commonly Abused Drugs

LEARNING OBJECTIVES

LO 9.1 **Understand** the importance of reward in motivating goal-directed behavior.

LO 9.2 **Explain** how the reward circuit establishes and maintains addictions.

LO 9.3 **Describe** aspects of psychopharmacology relevant to behavioral neuroscience.

LO 9.4 **Describe** the pharmacology and behavioral effects of alcohol, opioids, and marijuana.

LO 9.5 **Describe** the pharmacology and behavioral effects of common stimulants, depressants, hallucinogens, and dissociatives.

Stories of the Brain

Roy was the last person you would imagine to be an addict. He was considered to be a stable individual with a good job, a wife, and a family. Unbeknownst to anyone but his spouse and closest friends, however, Roy had a secret: he liked to gamble. At first it was just neighborhood card games, but one day he wandered into a casino, and it felt to him as if he had come home. The sights and sounds of the casino excited him more than anything he had ever known. He began to gamble at a casino once a month, but soon that became once a week, and within a year, he was heading to the casino after work every day.

At first, Roy was able to hide his behavior from his family, but as his gambling compulsion increased, so did his debts, and within a few years, Roy no longer owned his own house. On the day his dad died, Roy had no money to pay for a headstone for the grave, so he asked family members to cover the cost. When the check came, Roy stared at it for hours, then cashed it and went to a casino, where he lost every cent. Despite his shame at what he had done, he continued to gamble. Roy justified his behavior by saying that when his luck turned, he would win all his money back and then some, but what he admitted to himself was that the excitement of gambling was more important than whether he actually won or lost.

After his marriage broke up and he lost his home and had to move in with his daughter, Roy finally admitted to himself that he was addicted to gambling. Roy still goes to a casino occasionally to gamble, but because he registered himself on a national list of gambling addicts, he cannot take any winnings. Even now when it is literally impossible to win, Roy still gambles just to experience the thrill.

Addiction can involve more than drugs; some individuals like Roy are addicted to gambling, whereas others are addicted to exercise, shopping, engaging in social media, and even sex. In the past, we thought addiction was caused by a character flaw, but we now believe it to be a disorder of a neural reward mechanism. There must be a very powerful mechanism that supports addictive behavior in the face of overwhelming negative consequences, and in this chapter, we will begin by exploring this reward mechanism and then consider how it is involved in addiction. Then we will outline the types of drugs that are commonly abused and describe the mechanism by which drugs affect the body and behavior. We begin our discussion by highlighting the importance of reward in shaping and maintaining behavior.

9.1 The Biological Basis for Reward

LO 9.1 Understand the importance of reward in motivating goal-directed behavior.

One of the most interesting human characteristics is our ability to work toward a distant goal. We have all had the experience of deferring or delaying an immediate reward in order to achieve a larger goal in the future. We try not to spend our entire paycheck and instead budget to save up for a rainy day, we decide not to party and instead catch up on schoolwork in order to earn a high grade that might eventually lead to a good job, and we hit the gym or decline that dessert today in an effort to be healthier in a distant tomorrow. Humans are unique in their ability to sacrifice now in hopes of achieving a distant and uncertain goal; no tiger chooses to let some antelopes live in order to have a sufficient supply of them in the future.

A constant theme in our exploration of behavioral neuroscience is that behavior that is essential for us to survive and thrive will not be left to chance; a mechanism to support that behavior will be hard-wired into the organism. We see this theme again with the reward circuit that motivates behavior. This circuit consists of individual components that are linked together into two separate but cooperative pathways. We begin with the discovery of reward centers in the brain.

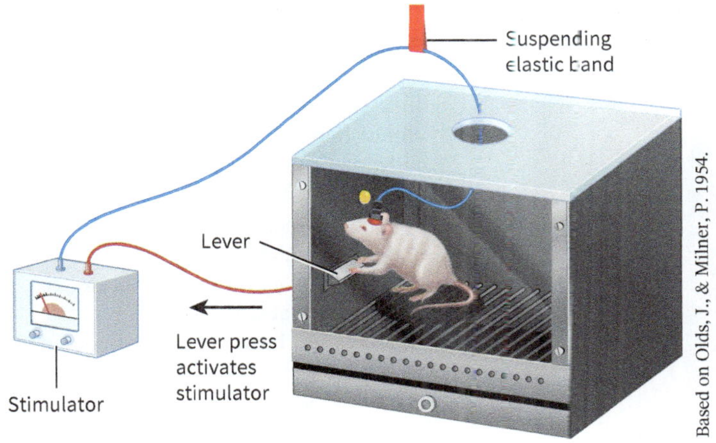

Based on Olds, J., & Milner, P. 1954.

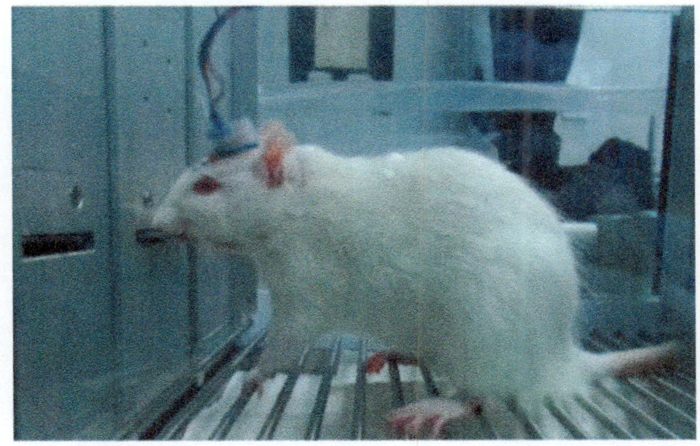

Chin Su Koh, Ph.D.

FIGURE 9.1 **The Olds and Milner brain-stimulation experiment** A rat was trained to press a bar for a reward and then had an electrode implanted in the septal region. When the rat pressed the bar, a tiny electrical pulse was delivered to the septum. These rats would press the bar continuously until they were exhausted.

Discovering the Biological Basis for Reward

In the 1950s, James Olds and Peter Milner—two pioneers in modern-day behavioral neuroscience—were investigating the nature of reward in a rat's brain. Their approach was to insert an electrode into one of various subcortical nuclei of a rat and then stimulate that region with a tiny pulse of current. Olds and Milner stimulated the brain whenever the rat happened to walk into a specific corner of their cage. Their prediction was that the stimulation would be aversive, and the rodent would avoid that corner. To their surprise, they found that when the stimulated region was the septal nuclei, the rat preferred that corner. When the experimenters allowed the rats to control the stimulation themselves by pressing a bar in their cage to deliver stimulation, the rats would bar press as much as 700 times per hour (**Figure 9.1**).

Olds and Milner concluded that they had discovered a reward center in the brain and that this center played a major role in motivating behavior. At first it was thought that reward was just a matter of stimulating the septal nuclei, but as we see time and again, the truth is more complex and more interesting, as we will see in the following sections.

You may wonder whether the electrode-implantation procedure pioneered by Olds and Milner has ever been applied to a human. In fact, a few individuals have had an electrode implanted in the septal region, and they described the feeling that accompanied stimulation as pleasurable and protested when the stimulating unit was removed. In one case, an individual who was given the ability to self-stimulate at home become so obsessive at stimulating that she developed an open sore on her finger from constantly adjusting the current. Her obsession with stimulation was only ended by the removal of the unit. Although providing individuals with the ability to stimulate a reward center at will is not on the horizon, there are circumstances where direct stimulation of a subcortical structure does have therapeutic value (**Think About It: The Ethics of Deep Brain Stimulation**).

Reward Centers in the Brain MCAT

Since the experiments by Olds and Milner, subsequent work has expanded our original understanding of a reward center in several ways. For example, stimulation may lead to an experience closer to craving than to pleasure. We also know that the reward system involves several subcortical centers that cooperatively interact to create motivation and even pleasurable sensations: the nucleus accumbens, the ventral tegmental area, and the medial forebrain bundle.

Think About It | The Ethics of Deep Brain Stimulation

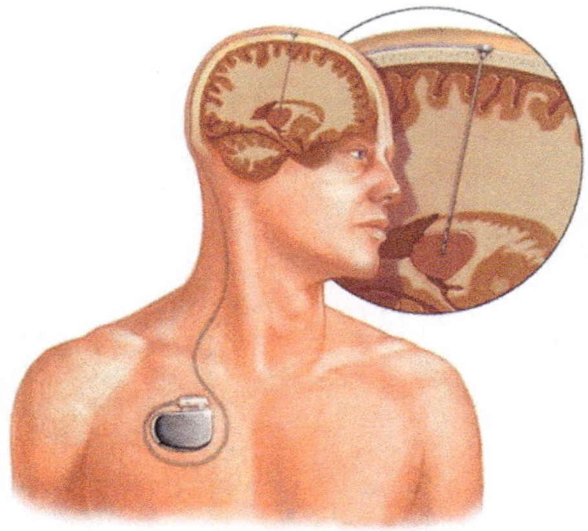

In a procedure known as deep brain stimulation (DBS), an electrode is carefully implanted in a specific subcortical structure and the individual is provided with an external unit known as a neurostimulator, which is a pacemaker that, at the press of a button, delivers an electrical pulse to that structure.

DBS has been found to be helpful in reducing the symptoms of motor disorders such as Parkinson's and essential tremor, and also reducing the frequency of epileptic seizures. Some individuals suffer from intractable pain due to cancer, and no amount of pain medication can blunt their discomfort. DBS has been suggested as a way to help such individuals by providing pain relief without drug side effects as they approach the end of their lives.

As our understanding of the neuroanatomical basis of psychiatric disorders has improved, some have proposed that DBS might also be useful in the treatment of disorders such as treatment-resistant depression, obsessive-compulsive disorder (OCD), anxiety disorders, Tourette's syndrome, and possibly even Alzheimer's disease and addictions. Technology has a way of evolving without regard to its original purpose, so the potential exists for DBS to be used for other purposes as well. We all might want to be happier, focus more, or feel more satisfied in our relationships with the help of DBS.

Implanting an electrode in the brain to reduce the tremors of Parkinson's seems very reasonable, but when the question turns to self-stimulation to alter mood or for pleasure alone, the ethics become less clear.

Questions

1. **Feeling happy is an appropriate goal state that we typically achieve through some effort. Would it be a good idea if people could choose to have an appliance installed that allowed them to self-stimulate a reward center at will? Explain.**
2. **Some behaviors are considered to be antisocial and thereby disruptive to society. Should DBS be used to reward positive behavior and punish antisocial behavior in individuals who have been judged to be highly resistant to remediation? Explain.**

The nucleus accumbens The **nucleus accumbens (NAcc)** is a small bilateral structure at the head of the caudate nucleus and is one of several structures at the top of the brainstem. The NAcc can be divided into two different structures: a core and a shell that surrounds the core. Both structural and behavioral characteristics differentiate the core from the shell. Extensive research with animal models indicates that the core stimulates motor responses to stimuli, whereas the shell integrates an emotional connection to the stimulus. One way to view the dual role of the NAcc is that the shell creates an emotional tag to the stimulus, and the core initiates a motor response appropriate to the emotional tag.

Now the picture gets even more interesting. For example, we know that chronic smokers typically have a routine they follow as they prepare to light a cigarette. Is it possible that mere exposure to an image linked to an addiction will trigger a neural response in the brain of an addict? Breiter and colleagues (2001) used imaging techniques to examine the relationship between addictive stimuli and activation of the NAcc. They found that the NAcc was activated not only by a stimulus but also by an image of a stimulus. That is, even seeing a cigarette, a deck of cards, or a needle can cause the NAcc of an addict to fire and thereby create a craving (**Figure 9.2**). Now you can easily understand the influence that commercials, movies, or videos that feature images of addictive substances or behaviors would have on a person struggling with an addiction.

The ventral tegmental area The **ventral tegmental area (VTA)** is so named because the structure is a covering (from the Latin *tegmentum*), and it constitutes the floor, or ventral surface, of the midbrain. The VTA enjoys reciprocal relationships with many subcortical regions, particularly the amygdala, the NAcc, and the prefrontal cortex. Research investigating activation of the VTA suggests that this region becomes activated when the individual experiences a reward. The romantics among us will not be surprised then that fMRI research has indicated the VTA is activated when a person experiences feelings of intense romantic attachment. The VTA also communicates with the orbitofrontal cortex (OFC), one of the least understood regions of the brain. The OFC is believed to evaluate rewards versus punishments for actions being contemplated. As such, disruption of the normal function of the OFC can lead to disinhibition with regard to antisocial behaviors.

Mike Cullen / Shutterstock

FIGURE 9.2 **Images of addictive behaviors** Images of addictive substances or behaviors can create powerful cravings that work to maintain the addiction.

The medial forebrain bundle The **medial forebrain bundle (MFB)** is an important white matter tract that connects the septal nuclei and the ventral tegmental area (VTA) with the hypothalamus. The NAcc, VTA, and MFB belong to two pathways that involve the neurotransmitter dopamine (DA): the mesolimbic and the mesocortical pathways. Each pathway has a unique and complementary function in the reward circuit. We now broaden our focus from the individual components of the reward system to the circuits that explain how humans can work for weeks, months, and even years to achieve a cherished goal.

The Mesolimbic and Mesocortical Pathways MCAT

If you were designing an organism from scratch, one of the attributes that you would include would be a mechanism for motivating the organism to move toward food, a mate, or environmental conditions more conducive to success. In addition, you would need a supervisory system to decide which stimuli are appealing and which are aversive. Both a mechanism that creates an emotional experience of pleasure and a second system that makes decisions about how to achieve that pleasurable reward are present in each of us.

If we think of a neural pathway as an information highway, then we have a useful metaphor to help us comprehend the reward circuit. A highway has origins and destinations, and it also has vehicles that travel along the highway. The reward circuit highway consists of the mesolimbic and mesocortical pathways, and the vehicle that conveys information along these pathways is DA.

The **mesolimbic pathway** provides us with motivation to achieve goals and control over response to stimuli. This makes sense because the limbic system includes structures such as the amygdala that create a strong emotional response when stimulated. The second circuit is the **mesocortical pathway**, which connects the VTA with the cortex, particularly the frontal and prefrontal areas that are known to be the seat of judgment and planning. The combination of a mechanism that creates a feeling of reward for engaging in a behavior and a second mechanism that creates a plan for obtaining that feeling is powerful indeed. Figure 9.3 illustrates several of the reward centers along with the two pathways.

We began our discussion of reward and addiction with the *Stories of the Brain* about Roy's gambling addiction, so it is perhaps appropriate that we use the related schema of card playing to help you see the big picture of the reward circuit. If you've ever played card games such as hearts, spades, or gin rummy, you already have a model of how the reward system works by integrating a collection of individual skills into two overarching mechanisms. Individual skills such as counting cards, observing emotions on the faces of others (known as a tell), and recalling whether the other players take risks or play conservatively are individual entities similar to the NAcc, VTA, and MFB. At a higher level, the successful card player takes pleasure in winning and creates strategies to ensure winning. In the same way, the mesolimbic pathway provides the sensation of reward at access to the substance or the completion of the behavior,

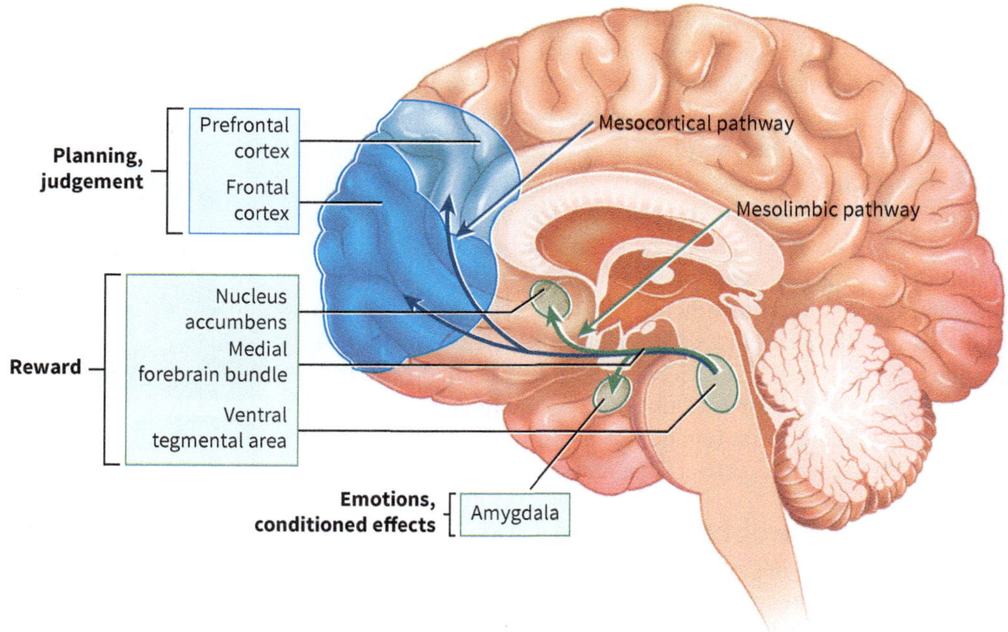

FIGURE 9.3 The reward circuit The reward circuit consists of two pathways that are organized around three functions: emotions, reward, and planning. The mesolimbic pathway connects the limbic system with the VTA and NAcc, whereas the mesocortical pathway connects the VTA with the frontal and prefrontal regions. The medial forebrain bundle connects the two pathways so they can interact.

whereas the mesocortical pathway generates a plan to obtain the reward and even possibly a rationale for doing so. The two pathways together are referred to as the mesocorticolimbic pathway.

As is so often the case in behavioral neuroscience, there is a subtle and nuanced interplay between cerebral processing structures and neurotransmitters. Dopamine is the primary neurotransmitter in the reward circuit, and it acts to initiate feelings of reward, so you would expect that there would be an inhibitory partner to the excitatory role of DA. That brake to DA's gas pedal is serotonin (SE). DA levels in the brain correlate with the expression of impulsive behavior, and the relationship between acting impulsively and acquiring an addiction is clear. SE levels have an inhibitory effect on impulsivity by inhibiting DA release in the mesocorticolimbic pathways. Our understanding of the complementary functions of DA and SE in the reward circuit have led clinicians to use drugs that increase SE levels in the brain (which normally are prescribed for psychiatric disorders such as schizophrenia) to treat addictive behaviors, with some success.

The Reward Circuit and Human Behavior MCAT

It makes perfect sense that our brain would motivate us to act in a way that is in our best interest. When our ancestors ran down an antelope, started a fire, or saw their children grow into adults, a feeling of accomplishment and pleasure that accompanied the behavior would lead to better odds of survival for the individual and the entire clan.

Although the details of our lives have changed over the millennia, our overarching needs have not. We no longer have to hunt and gather all our food; we show up at a dining hall or visit the local market when we are hungry. Instead of gathering downed tree limbs and struggling to start a fire, we touch the thermostat when we are cold. Finally, we still strive to reach long-term goals such as studying for a test or accumulating credits towards graduation, and it is essential to our survival as individuals and as a species that behaving adaptively would be accompanied by a positive feeling.

We can also apply what we know about the mesolimbic circuit to romantic and filial love and infatuation. Feelings of anticipation, excitement, and sadness during absences that are associated with our relationship with partners and family are the glue that holds society together. If you have ever been infatuated with someone, you may have felt giddy and excited in their presence, you looked forward to their arrival with great anticipation and when they were absent, you felt a longing that was physically palpable. Perhaps a person you have been

FIGURE 9.4 Human pair bonding The reward circuit supports the development of a behavioral and physiological bond that underlies our success as a species.

Aruta Images / Shutterstock.com

a.

Kim Ruoff / Adobe Stock

b.

in love with has ended the relationship, or you have lost a family member or close friend, and the grief you experienced was so strong as to almost overwhelm you.

Love is also the foundation of attachment, the bonding of a parent and child. Examples of strong interpersonal bonds that are at the basis of human happiness are depicted in **Figure 9.4**.

Whenever a stimulus, be it a thought or a behavior, is paired with the release of DA in the reward circuit, the chances of that behavior being emitted in the future are increased. Unfortunately, not every behavior that stimulates the reward system is adaptive, and the powerful motivation it provides can be highjacked to achieve a short-term high with long-term destructive consequences in the form of addiction. Addiction can take many forms: a person can be addicted to a substance such as a drug, but addictions also occur to behaviors such as eating, exercise, video-game playing, and even sex. There are serious concerns today that some people are addicted to social media and spend untold hours reading or posting comments online in conversations with people they do not know and are unlikely to ever meet. Understanding how the reward circuit enforces certain behaviors can help us recognize and combat destructive addictions.

You now have an insight into the mechanism and also the power of addiction, the subject of our next section.

9.1 Before You Go On

Section Summary

Humans have evolved a mechanism for rewarding behaviors that are likely to increase survival. The mechanism consists of the nucleus accumbens and the ventral tegmental area, which are linked by the medial forebrain bundle. There are two circuits associated with reward: the mesolimbic pathway involves attaching an emotion to reinforce the behavior, whereas the mesocortical pathway involves planning or evaluating the behavior.

Comprehension Check

1. What is the function of our reward system?
   - **a.** motivating us to find shelter
   - **b.** directing us to locate food
   - **c.** helping us to form social groups that make us more successful
   - **d.** all of the these

2. What are three subcortical structures involved in the reward circuit?
   - **a.** the lateral hypothalamus, the hippocampus, and the nucleus accumbens
   - **b.** the MFB, the VTA, and the nucleus accumbens
   - **c.** the lateral hypothalamus, the MFB, and the VTA
   - **d.** the lateral hypothalamus, the medial hypothalamus, and the cerebrum

3. The mesolimbic circuit provides _____, and the mesocortical circuit provides _____.
   - **a.** a link between emotion and behavior; planning for reward
   - **b.** reinforcement for positive behaviors; reinforcement for negative behaviors
   - **c.** reward for motor behaviors; reward for emotional behaviors
   - **d.** reinforcement for food choices; reinforcement for emotional responses

9.2 Addiction as Dysfunction of the Reward Circuit

LO 9.2 Explain how the reward circuit establishes and maintains addictions.

You have just learned how effective the reward circuit is at reinforcing behavior. Experiments with both animals and humans have shown that direct stimulation of these structures is highly reinforcing, to the point of obsession. In addition to promoting adaptive behaviors, however, this reward circuit can also lead to unhealthy and even fatal behaviors, including those associated with addiction.

What you have seen so far is that a complex circuit involving several subcortical structures that communicate through DA and SE with the frontal lobe and the limbic system motivate or incentivize us to behave in a specific way. How does the release of DA in the mesocortico-limbic system become associated with enjoying a good book, excessive gambling, taking an illegal drug, or being unable to resist commenting on social media? The answer is that we share with most living organisms that ability to associate a behavior with a reward through a process called conditioning. We will discuss conditioning at length in a different chapter, but you understand the process of linking a behavior to a stimulus if you've ever thought of spicy foods and found yourself salivating or felt anxious as you were about to receive an injection.

In the prior section, you read about powerful feelings associated with romantic love: infatuation, anticipation of events, the high of presence, sadness at separation, and despair at loss. These emotions are also associated with addictive behavior. A drug addict experiences craving that builds over time. When they score the drug, there is an excitement that precedes the actual use of the drug, and a feeling of euphoria when they take the drug. When they do without, their body rebels, they become ill, and there is a long period before the person is not troubled with cravings. Romantic and familial love is an adaptive function of the mesolimbic circuit, whereas addiction is not.

To understand what goes wrong in the reward system that leads to addiction, we must first establish the genetic component of reward.

The Genetic Basis for Addiction MCAT

Recall from our prior discussion of genetics that genes create proteins and proteins do a great deal of the work inside a cell that leads to behavior. One such gene, the **FosB gene**, plays an important role in everyday cell operations. During the process of gene expression, the occasional alternative splice results in a variant that, although similar to the original, codes for novel proteins. This principle can result in an adaptive advantage or a disorder that has significant implications. The FosB gene at times replicates itself as a spliced variant, and one such variant is the delta FosB gene.

The **delta FosB gene** (or ΔFosB) is integral in the reward system as well as in establishing and maintaining addictive behavior. The nucleus accumbens (NAcc) is essential to creating a feeling of satisfaction at the completion of a task. It does so by stimulating the production of several genes in the Fos family, including c-Fos, FosB, and delta FosB. As delta FosB production is stimulated, dopamine levels at the synaptic level are increased in the NAcc. As with any switch, it is important to be able to turn it off as well as on. Delta FosB acts as a sustained molecular switch in that it maintains the activation of dopaminergic neurons in the NAcc for long periods of time, in contrast to c-Fos and FosB levels, which decline over time.

Because of its role in the mesolimbic system, delta FosB is involved in the maintenance of any number of healthy behaviors such as exercising regularly, enjoying appealing food, and experiencing the delightful presence of friends and family as well as sexual interest. However, healthy behaviors such as exercise, eating, and sex can also develop into addictions. When they do, it is likely that delta FosB is involved. Delta FosB is also stimulated by addictive drugs such as opioids and alcohol or by behaviors such as gambling.

FIGURE 9.5 Delta FosB expression in the nucleus accumbens **a.** Administration of an addictive drug causes a sudden and short-lived increase in the expression of the c-Fos gene followed by a lesser and longer-lasting increase in activity of the delta FosB gene, which stimulates the NAcc to signal reward and pleasure. **b.** With repeated exposure to the addictive substance, the baseline levels of delta FosB increase, leading to greater feelings of craving and withdrawal.

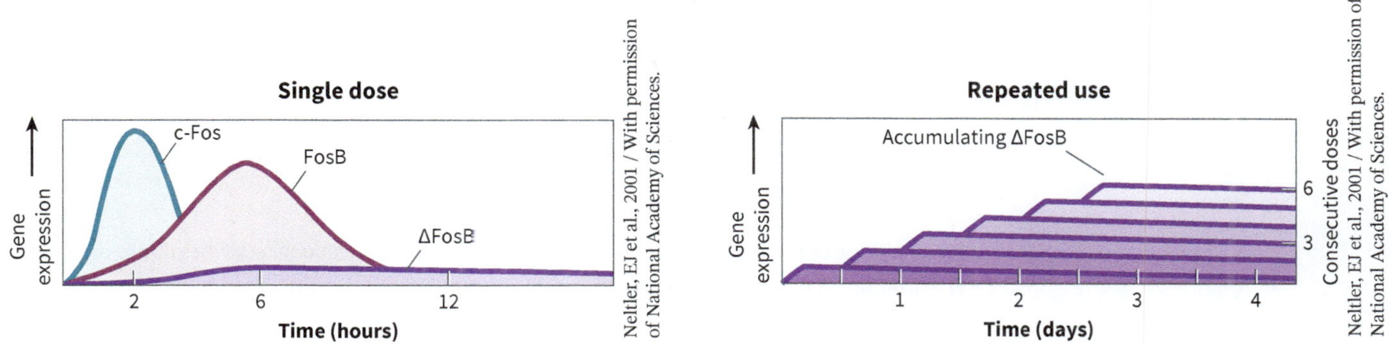

a. b.

The increase in dopamine by delta FosB leads to the establishment and maintenance of addictive behaviors by initiating structural changes in the NAcc. The single, or acute, administration of an addictive substance causes a spike in the expression of the c-Fos gene; this response is followed by a smaller but more sustained increase in expression of the delta FosB gene and other variants (**Figure 9.5a**). You already know that delta FosB activity is related to the development of craving for an addictive substance. Ideally, delta FosB levels would fall after a time in the way that c-Fos and FosB levels do, and craving would diminish. However, with repeated exposure, the expression of delta FosB stairsteps, or incrementally increases activity, so that with each additional exposure, delta FosB expression is ratcheted up to ever higher levels (**Figure 9.5b**).

What you have just read should startle you: exposure to an addictive stimulus actually rewrites your genetic code, rewiring your reward circuit to increase the intensity of the craving for the stimulus, be it heroin, gambling, or social media. How this plays out in the real world is unfortunate; with each exposure to the addictive drug, the addictive properties of that drug are enhanced, and the new and higher level of craving becomes a baseline for future enhancement. Is it any wonder that drug addiction is so resistant to therapeutic intervention?

Before we go further, we should delineate where casual use becomes addiction.

Substance Abuse Disorders MCAT

The neuroscience of addiction can explain the physiology behind addictive behavior, but we also need a clinical definition of substance-related and addictive behavior. Exercise can be healthy, but compulsive exercise can be an addiction. Many of us drink alcohol but are not addicted to it. At what point does a behavior become a disorder? How do dependence, abuse, and addiction differ?

You may see these three terms used almost interchangeably, but they differ in their meaning when used in a professional setting. If an individual experiences distress when they are denied access to the substance or the behavior, that feeling of **withdrawal** is a defining characteristic of **dependence**. For example, people are often prescribed opioids to blunt pain after an injury or a medical procedure. In light of what you know about the reward circuit, you can see that the mesocorticolimbic circuit will adjust its function to account for the presence of this drug on opioid receptors. As a result of these changes, removing the opioid can result in severe withdrawal symptoms, including anxiety, sweating, vomiting, and diarrhea. Withdrawal is not limited to opioids and other drugs; if you feel significant distress when you are unable to drink alcohol, smoke or vape, or interact with social media, you are experiencing dependence.

When a person's behavior leads to harm to themselves or to others in their lives, they are on a continuum between abuse and addiction with the difference being one of severity. When using the drug or engaging in the behavior begins to interfere with one's life or the individual substitutes the addiction for normal activities such as eating, sleeping, or socializing,

TABLE 9.1 ***DSM-5-TR*** **Diagnostic Criteria for Substance Abuse Disorder**

1.	Continued use of the drug despite negative personal consequences.
2.	Repeatedly unable to carry out important obligations at school, work, or home.
3.	Recurring use of the drug in dangerous circumstances.
4.	Continued use of the drug despite interpersonal conflicts resulting from use.
5.	Tolerance defined as the need to take more of the drug to achieve a desired effect.
6.	Withdrawal symptoms when the drug is unavailable.
7.	Increasing use of the drug over time.
8.	Persistent desire or unsuccessful attempts to take less of the drug.
9.	Devoting a lot of time to obtain, use, or recover from drug use.
10.	Stopping or reducing important social, work-related, or recreational activities due to drug use.
11.	Continued use of the drug despite an awareness of persistent or recurrent problems due to the use of the drug.
12.	Craving or a strong desire for the drug.

that satisfies the definition of **abuse**. When the harm escalates to loss of job, relationship problems, or even legal consequences, that individual has crossed the line from abuse to **addiction**. Although the difference between abuse and addiction may feel subjective, it is the responsibility of the mental health profession to quantify the difference as best as possible to guide medical and mental health professionals.

The *Diagnostic and Statistical Manual of Mental Disorders* (*DSM*) provides mental health professionals with clear diagnostic criteria for the classification of mental disorders, including addiction. The fifth edition of the *DSM* (*DSM-5-TR*) has combined the categories of substance abuse and substance dependence that were distinguished in earlier versions of the *DSM* into a single continuum from mild to severe. In this view, an individual exhibiting two or three of the criteria is classified as exhibiting mild substance abuse, whereas four or five indicates moderate substance abuse, and six or seven would be diagnosed as severe substance abuse. These criteria are reproduced in **Table 9.1**.

You can apply what you know about substance abuse disorder to evaluate three individuals in this **Try It Out: Diagnosing Addictive Behavior**.

Risk Factors for Addiction MCAT

Clinical psychologists and addiction professionals know quite a bit about traits that predict the development of addictive behaviors, so we know that there are psychological, environmental, and genetic risk factors for addiction.

Try It Out | Diagnosing Addictive Behavior

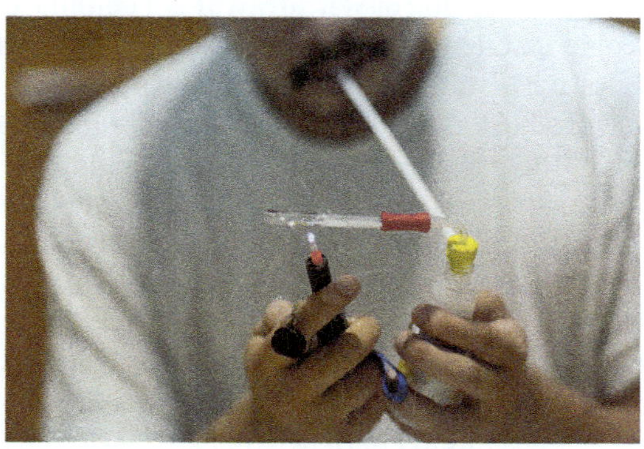

kunanon / Shutterstock.com

To apply what you have learned about diagnosing substance abuse disorder, evaluate three scenarios on the worksheet **Diagnosing Addictive Behavior** according to the criteria from the *DSM*.

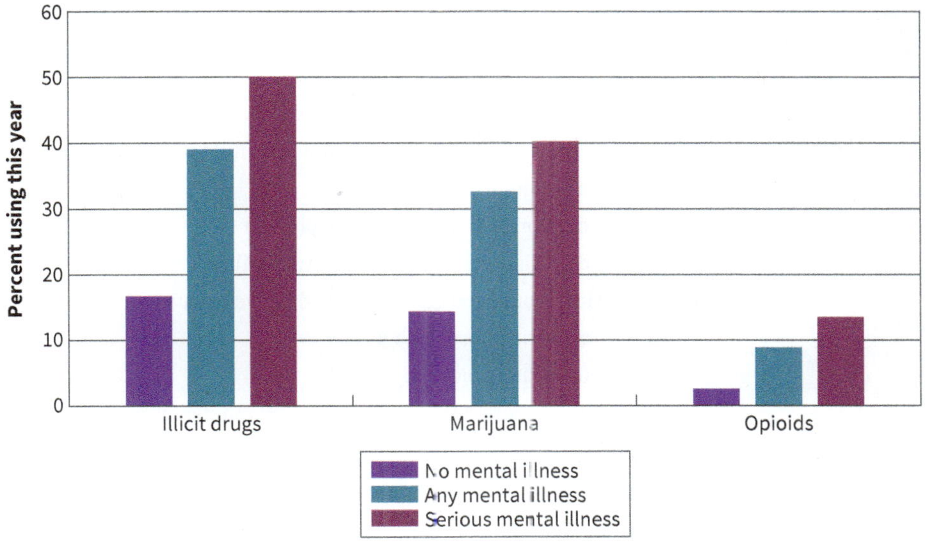

FIGURE 9.6 Mental illness and substance use According to a 2019 survey by the U.S. Substance Abuse and Mental Health Services Administration (SAMSA), adults aged 18 or older who experience mental illness are more likely to have used substances like illicit drugs or marijuana in the past year. Those with serious mental illness are even more likely.

Psychological factors include stress, impulsivity, and sensation-seeking. Individuals already being treated for depression or anxiety also have a heightened level of risk for addiction. Figure 9.6 shows how substance use by adults correlates with mental illness.

Environmental factors include exposure to physical, emotional, or sexual abuse; a history of substance abuse in the family or in peer group; and living in poverty. Educational level is negatively correlated with drug use, but that may be more a function of income. Individuals whose family use drugs or children whose parents are uninvolved in their lives are more at risk for developing an addiction. Early use is also a predictor; individuals who start using alcohol or drugs in their early teens are more likely to develop addictions than those whose substance use begins later. Males are perhaps twice as likely as females to develop an addictive behavior.

Genetically, there is evidence that some of us have more delta FosB than others, and that has obvious implications for developing compulsive behaviors. Finally, it seems reasonable to assume that if some of us have an excess of delta FosB, others might have almost none. Clearly, there is much to be learned about the genetics behind addiction.

The type of drug used and the method and length of administration are also factors in developing additions. Those who use hard drugs such as heroin, cocaine, methamphetamine, or prescription opioids are more at risk then those who use alcohol or marijuana. Those who have been using an addictive substance for some time are more likely to transition from use into dependance, abuse, and addiction. Finally, those who inject drugs or smoke them are more likely to develop an addiction because they deliver a more immediate and intense high. Risk factors for developing an addiction are summarized in Table 9.2.

We will touch on basic principles of psychopharmacology related to behavioral neuroscience in the next section.

TABLE 9.2 Risk Factors for Addiction

Psychology	Environment	Genetics	Drug and method of administration
Mental health comorbidity such as anxiety or depression Stress Impulsivity Sensation-seeking	Social norms Peer pressure Educational level Poverty Exposure to abuse	Family history of addiction Gender More delta Fos gene	Route of administration, with injecting and inhaling posing the greatest risk Type of drug Early age of first use Length of use Access to the addictive substance or behavior

9.2 Before You Go On

Section Summary

The reward circuit that supports healthy, goal-directed behavior is also the motivating force behind addictions. When the delta FosB gene, which enhances goal-directed reward, becomes associated with an unhealthy behavior, it maintains and even accelerates the addiction. The *DSM* provides a framework for diagnosing addictive behaviors on a continuum from mild to severe. There are psychological, environmental, and genetic risk factors for addiction.

Comprehension Check

1. Which statement is true of the delta FosB gene?
   - **a.** It prevents a person from acquiring an addiction.
   - **b.** It is produced in response to opioids but not alcohol.
   - **c.** It maintains addictions in response to repeated exposure.
   - **d.** It is only expressed for unhealthy addictions.
2. The *DSM-5-TR* considers addiction to be what?
   - **a.** a disorder on a continuum from mild to severe
   - **b.** a moral failure
   - **c.** more serious than schizophrenia
   - **d.** a disorder best treated by tranquilizers
3. How would you characterize addictive behaviors?
   - **a.** entirely psychological
   - **b.** entirely biological
   - **c.** supported by psychological, environmental, and genetic factors
   - **d.** none of these

9.3 The Essentials of Psychopharmacology

LO 9.3 Describe aspects of psychopharmacology relevant to behavioral neuroscience.

As is the case with so many topics in behavioral neuroscience, to gain a clearer picture of addiction requires familiarity with multiple levels of focus. In the previous section we discussed the neurobiology of addiction. In this section we will discuss the principles of **psychopharmacology**, the study of how drugs affect cognition, mood, and other aspects of behavior. The same principles will help you understand both drug abuse and the action of therapeutic drugs in the treatment of psychiatric disorders, a topic you will encounter in a subsequent chapter. We will summarize the essentials of psychopharmacology in three sections: methods of entrance, pharmacodynamics, and pharmacokinetics.

How Drugs Enter the Body MCAT

Drugs can enter the body via several routes. Typical routes are injection, oral administration, inhalation, and transdermal administration. Table 9.3 offers a comparison of these methods.

Drugs can be injected into a vein (IV), into a muscle (IM), or just under the skin or subcutaneous layer (SC). Drugs injected IV cause a rapid response because they travel through the right side of the heart, then to the lungs, and finally up to the brain. However, sometimes gradual absorption is desirable, so diabetics inject their insulin subcutaneously into the abdomen in order to slow the absorption of insulin into the bloodstream and thereby mimic the natural operation of the pancreas.

Drugs ingested via the mouth as a pill or a liquid must first travel through the stomach, where they are mixed with gastric juices. They then travel into the small intestine, where many drugs are absorbed by the small blood vessels that line the walls of the intestines. Drugs that have an affinity with water (hydrophilic drugs) may be absorbed and excreted differently than drugs that have an affinity with fats (lipophilic drugs).

Drugs that are inhaled through the mouth enter the bloodstream via the lungs and are quickly carried up to the brain. An even more effective route is through the nasal cavity. Drugs

TABLE 9.3 Drug Delivery Methods

Intravenous (IV) injection 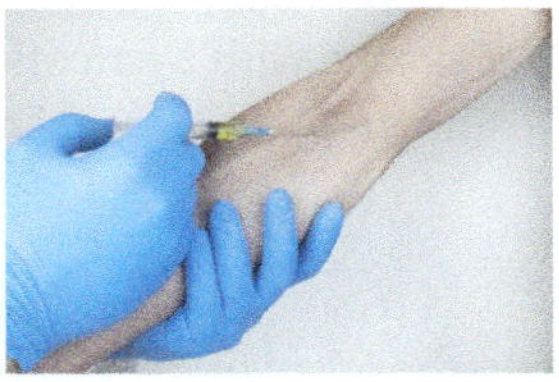 Alona Siniehina / Shutterstock.com	**Speed of action**: 30 to 60 seconds to reach the brain **Advantages**: Precise control over dose; rapid entry into bloodstream **Disadvantages**: Requires trained staff for legal use; risk of infection if needles are shared **Examples**: Heroin; chemotherapy, some antibiotics
Intramuscular (IM) injection 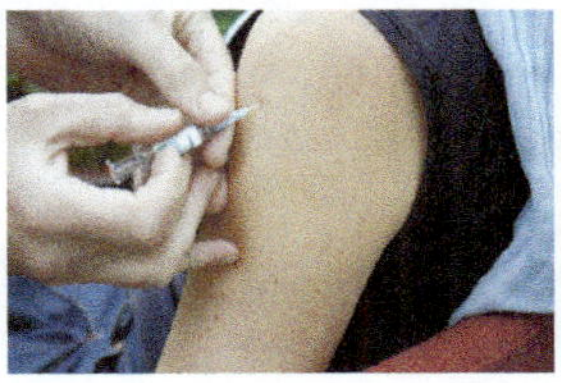 Photopat medical healthcare / Alamy Stock Photo	**Speed of action**: 3 to 5 minutes to enter the bloodstream **Advantages**: Low cost if performed by patient as opposed to medical professional; better patient compliance **Disadvantages**: Variation in dose if administered by nonprofessional **Examples**: Vaccines, vitamin B, some blood thinners such as Lovenox, and some multiple sclerosis or rheumatoid arthritis drugs
Subcutaneous (SC) injection 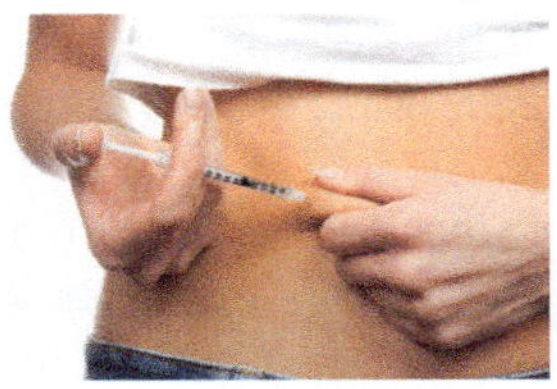 Dmitry Lobanov / Adobe Stock	**Speed of action**: Slow and gradual **Advantages**: Highly effective; slow entry into the bloodstream when desirable **Disadvantages**: Site bruising, skin infections, slow entry into the bloodstream **Example**: Insulin
Oral administration 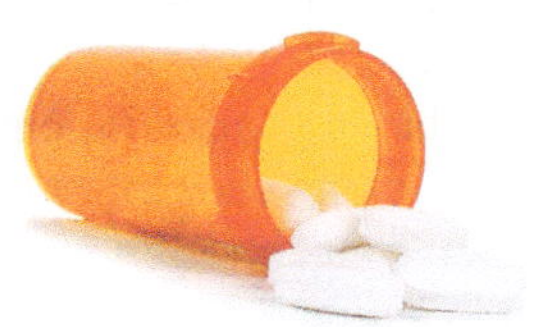 Mike Cullen / Shutterstock	**Speed of action**: 20 to 30 minutes to reach the brain **Advantages**: Consistent dosage, ease of administration and storage **Disadvantages**: Older patients often miss or double their dose **Examples**: Most pharmaceuticals, alcohol
Inhalation  Mike Cullen / Shutterstock	**Speed of action**: 10 seconds to reach the brain **Advantages**: Rapid delivery to brain and bloodstream **Disadvantages**: Can be a risk factor for lung disease **Examples**: Cigarettes, marijuana, vaping, asthma inhalers
Intranasal 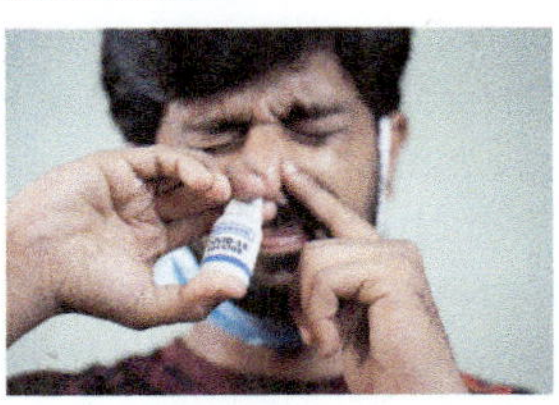 WESTOCK / Adobe Stock	**Speed of action**: Rapid onset (minutes) **Benefits**: Lower cost and self administration **Disadvantages** Dose not controlled **Examples**: Migraine drugs; naloxone (Narcan), nicotine replacement, some hormone treatments
Transdermal administration 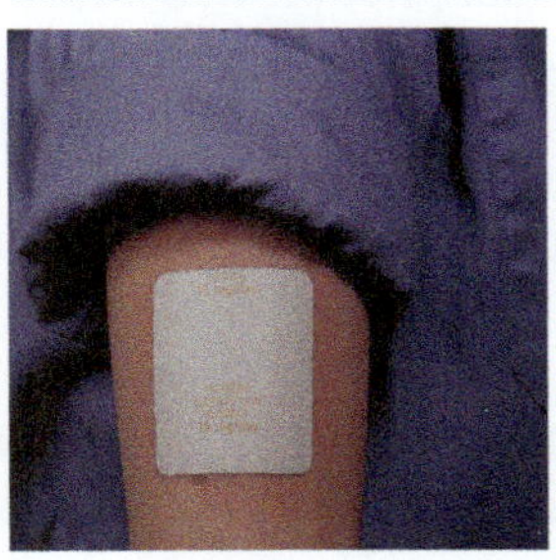 Spike Mafford / Getty Images	**Speed of action**: Slow and gradual **Advantages**: Known dosage and rate of release; no expertise required to apply **Disadvantages**: Skin rashes; may be inconvenient **Examples**: Nicotine (patch), hormones, treatment of motion sickness, pain medication

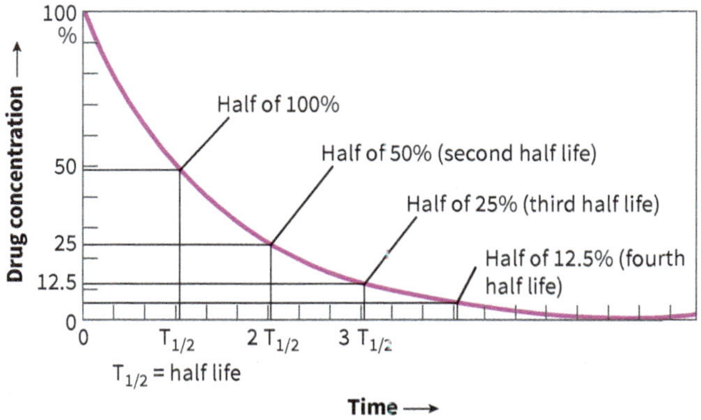

FIGURE 9.7 **Half-life of an active drug** Assume that this drug has a half-life of 1 hour. At the time of administration, the drug is at 100%. In the first hour, only 50% of the drug will remain in the system. By hour two, only 25% will remain, and so on. By hour five, only 3.13% of the active components of the drug will remain in the system.

administered intranasally may enter the brain through the nasal tissue and thereby bypass the -brain barrier. Intranasal application is also useful for pediatric situations where a needle might be aversive, or where speed of entry into the brain is of paramount importance such as treating a patient overdosing with a narcotic.

Pharmacodynamics MCAT

The means of ingestion is one variable in the effect of drugs, but their method of action is another. **Pharmacodynamics** is the study of how a drug interacts with the body, and there are several concepts that will help you to better understand how drugs affect behavior: half-life, potentiation, tolerance, and sensitivity.

Half-life

You may be familiar with the concept of a **half-life** in physics; it is the time required for a radioactive compound to lose half its radioactivity. Similarly, as time passes after administration and drugs gradually work their way out of the system through the cleansing action of the kidneys and to a lesser extent the liver, the therapeutic effect of the drug diminishes. Drugs with a longer half-life remain in the system longer. Figure 9.7 illustrates the typical half-life function of a drug.

Understanding how a drug's half-life affects its therapeutic action involves more than just being able to read a half-life graph. With advanced age or with disease, a person's kidneys operate less efficiently, so a drug may take longer to be cleared. Other factors such as sex and stomach contents also may affect pharmacodynamic half-life. Now that you understand the concept of a drug's half-life, you are ready to see the larger picture of the pharmacodynamics of a drug.

A drug only delivers a therapeutic effect within a certain range of blood concentrations, known as the therapeutic range. When a drug first enters the bloodstream, its concentration typically rises quickly to a point where it begins to exert its effect. Blood concentration peaks and then slowly declines as the liver and kidneys perform their clearance functions. At some point, the concentration of the drug in the blood falls below the floor of the therapeutic range and the drug ceases to be effective, although it remains in the bloodstream. Finally, a point is reached where the drug is completely cleared from the system. Pharmacological action over time varies depending on the drug in question, but one example is shown in Figure 9.8.

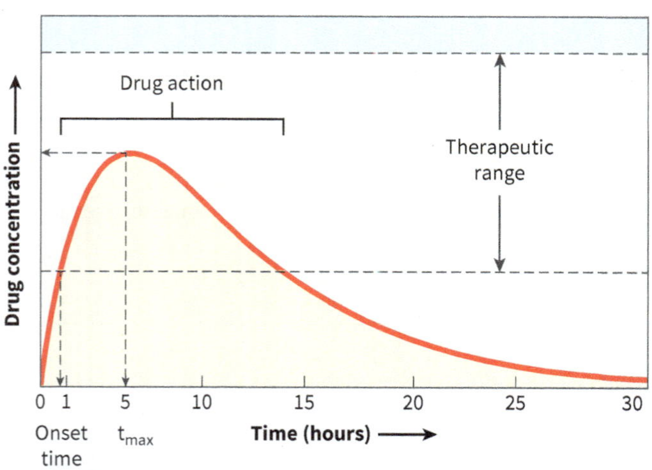

Mehrotra, N., et al., 2007 / with permission of Springer Nature.

FIGURE 9.8 **Pharmacological action of a drug over time** About one hour after initial administration, the drug reaches the therapeutic range and begins to have an effect. It peaks at five hours, begins to decline, and around 15 hours after administration, the concentration falls below the floor of the therapeutic range and the drug ceases to be effective.

Potentiation

Another concept of note in pharmacodynamics is **potentiation** (also known as synergy). A drug that potentiates another drug amplifies the action of the first drug, and this amplification can have either positive or negative effects. For example, NSAIDs (nonsteroidal anti-inflammatory drugs) such as ibuprofen potentiate opioids, that is, enhance their effect. Similarly, diazepam, which is commonly prescribed to treat anxiety, potentiates the effects of alcohol.

The implications of potentiation are important. An individual who is struggling with depression may find that although a given dose of drug A does not provide a therapeutic effect, adding even a small amount of a potentiating drug B creates a meaningful improvement in symptoms. Conversely, a drug user who has habituated to a given drug dose may overdose when they take their typical dose of drug A with a small amount of a second, potentiating drug.

Sensitivity and tolerance

At the synaptic level, a drug normally activates its receptor to create a signal that travels up the postsynaptic neuron. However, in some circumstances, receptor response can become more or less sensitive. You are already familiar with the concepts of up- and downregulation from the chapter on neurotransmitters. **Sensitivity** occurs when a drug's function is upregulated. For instance, a specific malady may be treated by administering a drug that blocks receptors (an agonist drug). Because our systems are always striving for balance, the postsynaptic neuron may respond to receptor blockage by creating additional receptor sites. When this happens, receptor upregulation makes the postsynaptic neuron more sensitive to the drug and, in doing so, defeats the original therapeutic purpose of the receptor blocker.

Conversely, **tolerance** occurs when long-term exposure to a drug leads to downregulation, in which the postsynaptic neuron absorbs receptors in an effort to regulate activity back to prior levels. Figure 9.9 demonstrates how a drug normally generates a neural signal at the synapse, along with sensitivity and tolerance.

You may be confused as to why blockading receptors with a drug leads to sensitivity, whereas increasing the amount of the drug at the synapse leads to tolerance considering that both involve increased activity at the synapse. The answer is that different conditions (reducing available receptors versus overabundance of drug) lead to different mechanisms.

In our everyday lives, we are more likely to experience tolerance at the synaptic level than increased sensitivity. Many students, professors, and adults in other walks of life that require sustained attention have found that they have developed a tolerance for caffeine whether it is delivered in a cup of coffee, caffeinated soda, or caffeinated energy drink. Over time, more of the drink is required to maintain the same level of alertness. When a person has their first alcoholic drink, they often feel quite lightheaded and tipsy. If that same person continues to drink alcohol on a regular basis, it is very likely that they will need to have more than one drink to achieve the same level of intoxication.

Continual exposure to greater and greater amounts of the drug will generally lead to an increased tolerance and that tolerance can be fatal. John Bonham, the drummer for the band Led Zeppelin, was a heavy drinker. On the day he died, he became so intoxicated that he passed out, aspirated his own vomit, and then choked to death without waking. The coroner later reported that he had the equivalent of 40 shots of vodka in his bloodstream (1 shot = 1.5 fl. oz. or 44.4 ml). Bonham's tolerance to alcohol, developed over several years of heavy drinking, allowed him to consume enough alcohol to suppress the normal survival mechanisms of his brainstem that would have roused him as he was choking. Individuals who regularly take substances such as alcohol, cannabinoids, or opioids develop tolerance over time and therefore require an increased dose to achieve the same response.

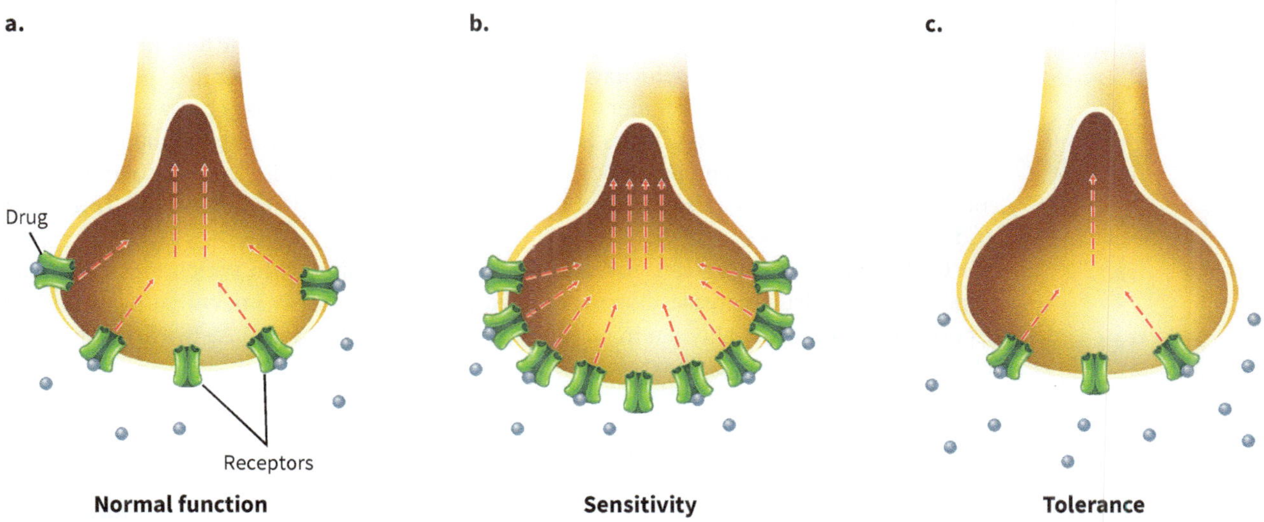

FIGURE 9.9 **Receptor sensitivity and tolerance** **a.** Normally, a drug stimulates a receptor at the synapse, which leads to a response. **b.** Continued use of an agonist may lead to increased expression of receptor sites, causing increased sensitivity. **c.** Continued high availability of the drug can reduce the number of receptor sites, leading to tolerance.

Think About It | Behavioral Tolerance and Drug Overdoses

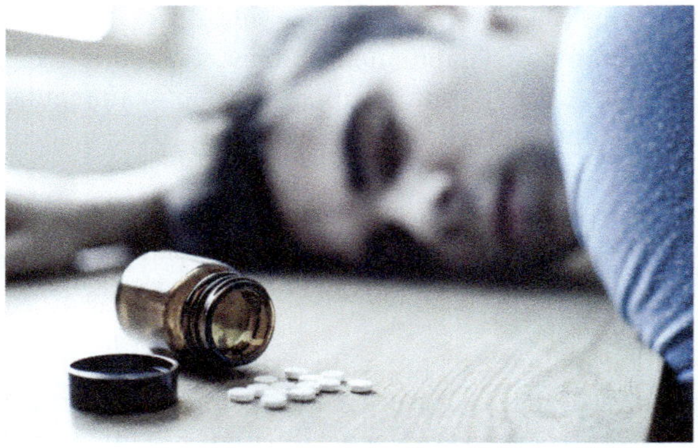

SpeedKingz / Shutterstock

You may have observed this effect in your own life if you are in the habit of patronizing a particular tavern or house where in the past you have enjoyed yourself while having a drink or two and socializing. Experience has conditioned you to tolerate the effects of alcohol *in that context*. Now you visit a different place and consume the exact same amount of alcohol as you are used to consuming in the previous location, and to your surprise, the effect of the alcohol is stronger than you expected. This is due to behavioral tolerance.

In a seminal set of studies, Shepard Siegal illustrated the power of behavioral tolerance. Two groups of rats were given increasingly higher doses of heroin until both groups developed a tolerance for a dose that would be fatal for a first-time user. Both groups were then given a dose that would be fatal to 100% of rats that had not developed a tolerance; half were injected in their normal living quarters, and the other half received the injection in an unfamiliar room where 60 dB white noise was present. The rats receiving the overdose in the context to which they had been conditioned to the drug showed a 32% fatality rate, whereas those rodents who had developed the same considerable tolerance to heroin but were injected in a novel environment were twice as likely (64%) to die.

Questions

1. **A habitual user with a significant tolerance has been in the habit of injecting drugs in the basement of their house. When the user takes a dose equal to their normal heavy dose at a new acquaintance's house, they stop breathing and go into cardiac arrest. Explain why this happened.**
2. **When a person develops tolerance to one intoxicant such as alcohol, opioids, or cocaine through repeated use, they may assume that this tolerance will naturally transfer to other intoxicants. Based on your understanding of tolerance, do you see any risks associated with this perspective? Explain.**

Understanding the concept of tolerance is essential to understanding how drug overdoses can occur among individuals who have become habituated to a drug over long periods of use. This principle draws on your understanding not only of pharmacology but also of learning and conditioning phenomena. There is strong evidence that an individual is conditioned not only to the physiological and psychological effects of the drug but also to their environmental context, which is known as **behavioral tolerance** (**Think About It: Behavioral Tolerance and Drug Overdoses**).

Chronic use of a drug can also lead to increased sensitivity, in which an individual's response to that drug increases due to repeated exposure. You may know that a person first exposed to urushiol, the allergen in poison ivy, may have a small response the first time but a larger response at subsequent exposures.

Pharmacokinetics MCAT

In contrast to pharmacodynamics, which is the effect of the drug on the body, **pharmacokinetics** is the effect that the body has on the action of drugs, including the way drugs are

Blood alcohol content (BAC) for males (M) and females (F)										
Number of drinks		Body weight in pounds								Driving condition
		100	120	140	160	180	200	220	240	
0	M	.00	.00	.00	.00	.00	.00	.00	.00	Only safe driving limit
	F	.00	.00	.00	.00	.00	.00	.00	.00	
1	M	.06	.05	.04	.04	.03	.03	.03	.02	Driving skills impaired
	F	.07	.06	.05	.04	.04	.03	.03	.03	
2	M	.12	.10	.09	.07	.07	.06	.05	.05	
	F	.13	.11	.09	.08	.07	.07	.06	.06	
3	M	.18	.15	.13	.11	.10	.09	.08	.07	
	F	.20	.17	.14	.12	.11	.10	.09	.08	Legally intoxicated
4	M	.24	.20	.17	.15	.13	.12	.11	.10	
	F	.26	.22	.19	.17	.15	.13	.12	.11	
5	M	.30	.25	.21	.19	.17	.15	.14	.12	
	F	.33	.28	.24	.21	.18	.17	.15	.14	

FIGURE 9.10 **Intoxication as a function of alcohol consumed and body weight** Notice that a 160-pound female will be legally intoxicated after two drinks, whereas a male of the same weight will be legally intoxicated after three drinks.

absorbed and metabolized. Examples of factors that affect pharmacokinetics are mode of delivery, age, sex, and health.

Drugs that are orally administered enter the bloodstream via the hepatic portal system. That is, they are absorbed into the small intestine, but that blood is directed to the liver where a significant amount of the drug is likely to be eliminated. Only a small fraction of the original dose then enters the bloodstream. This filtering of orally administrated drugs through the liver is known as first-pass metabolism.

For this reason, individuals whose liver function is below par will see a larger effect of a drug, and that effect will last longer as their hepatic system struggles to clear the drug from their bloodstream. The key role of the liver in clearing substances from the bloodstream can lead to damage and disease when substances are abused. For example, excessive alcohol use over time causes cirrhosis of the liver, which can be fatal, and overuse of pain relievers such as acetaminophen or NSAIDs can also cause damage. Some prescription drugs when overused may also lead to liver disease.

Elderly individuals who are experiencing the normal age-related decrease in liver function will experience a greater impact of a drug; this explains why it is dangerous to share prescribed medication, especially with the elderly. For example, Percocet prescribed at a dosage appropriate for a young adult to minimize discomfort after a surgical or dental procedure could have serious and possibly fatal consequences when taken by an elderly individual due to their reduced liver function.

Finally, males and females metabolize drugs differently for many reasons. On average, females weigh less than males and have more adipose (fatty) tissue. As a result, lipophilic drugs act differently for males than for females. Figure 9.10 shows how alcohol intoxication differs based on sex and body weight.

In a simple world, drug effects are linear; that is, a drug creates an effect, and more drug creates more of the same effect. However, our brains follow the rules of nature and not our wishes, so some drugs exhibit a **biphasic effect**; that is, they create one effect at a low dose and the opposite effect at a high dose. Perhaps the most familiar example of a biphasic effect is alcohol; at low doses it creates mild euphoria, facilitates social behavior, and stimulates appetite. At high doses, alcohol creates nausea, loss of consciousness, and even death. How a body metabolizes that alcohol at varying concentrations is explained by pharmacokinetics, whereas the biphasic action on mood would fall under pharmacodynamics.

With this background in the neurobiology of addiction and the principles of pharmacology, we next turn to a brief discussion of drugs that are commonly abused. We begin with intoxicants, which create a feeling of euphoria.

9.3 Before You Go On

Section Summary

Drugs enter the body via different routes, and the chosen route determines how quickly the drug enters the brain. Drugs have a half-life in the body and are eliminated through the liver and kidneys. Repeated use can lead to increased tolerance, which requires the individual to take larger amounts to obtain the same effect. Some drugs potentiate the effects of other drugs, and this can be an asset or a danger. The action of a drug can vary based on the age, sex, and health of the user.

Comprehension Check

1. Put the routes of entry in order according to how quickly a drug enters the bloodstream from fastest to slowest.
   - **a.** lungs, veins, mouth, skin
   - **b.** nose, lungs, veins, mouth, skin
   - **c.** veins, mouth, lungs, skin
   - **d.** veins, lungs, mouth, skin
2. What is one of the major advantages of delivering a drug via a transdermal patch?
   - **a.** It avoids the stomach so that nausea isn't a problem.
   - **b.** It cannot affect the lungs.
   - **c.** It is appropriate for children and the elderly.
   - **d.** The release of the drug is spread out over time.
3. If I take a drug at noon that has a half-life of two hours, then at _____ percent is available and at _____ percent is available.
   - **a.** 2 p.m., 75; 4 p.m., 50
   - **b.** 4 p.m., 50; 8 p.m., 25
   - **c.** 2 p.m., 50; 4 p.m., 25
   - **d.** 4 p.m., 75; 8 p.m., 50
4. John used to drink one drink each evening, but recently he realized he is drinking much more to achieve the same level of effect. What phenomenon explains this?
   - **a.** half-life
   - **b.** tolerance
   - **c.** potentiation
   - **d.** sensitivity
5. A drug that has a biphasic effect _____.
   - **a.** has different effects at low and high doses
   - **b.** is metabolized in two stages
   - **c.** is not addictive
   - **d.** has different effects after the first use

9.4 Intoxicants

LO 9.4 Describe the pharmacology and behavioral effects of alcohol, opioids, and marijuana.

Intoxicants generate a feeling of euphoria at low to moderate doses; at high doses they typically produce negative or unpleasant feelings. Common intoxicants are alcohol, opioids, and THC, the active agent in marijuana.

Alcohol

In most western societies, alcohol use is legal and forms a part of many social situations, where it is used in moderation as a social lubricant to relax people. Unfortunately, some individuals drink excessively, which can have serious consequences for health and well-being. One such behavior is **binge drinking**, defined as consuming five or more drinks for men or four or more drinks for women in two hours or less. According to the National Institutes of Health (NIH), in 2019, 4.2 million adolescents had binged on alcohol in the past month and over 800,000 had binged on five or more days in the past month. Figure 9.11 portrays adolescent alcohol use by age and sex in the United States.

Pharmacology of alcohol Alcohol is soluble in fats and in water, so it is easily absorbed by the gastrointestinal tract. Initially, alcohol acts as an irritant to the stomach, stimulating the flow of digestive juices and thereby aiding digestion. Therefore, in small amounts, alcohol use can be conducive to digesting a meal, but in large amounts, alcohol damages the

FIGURE 9.11 Adolescent alcohol use **a.** By ages 12–13, some adolescents have begun to drink alcohol. **b.** By ages 16–17, more than 10% have binged in the past month.

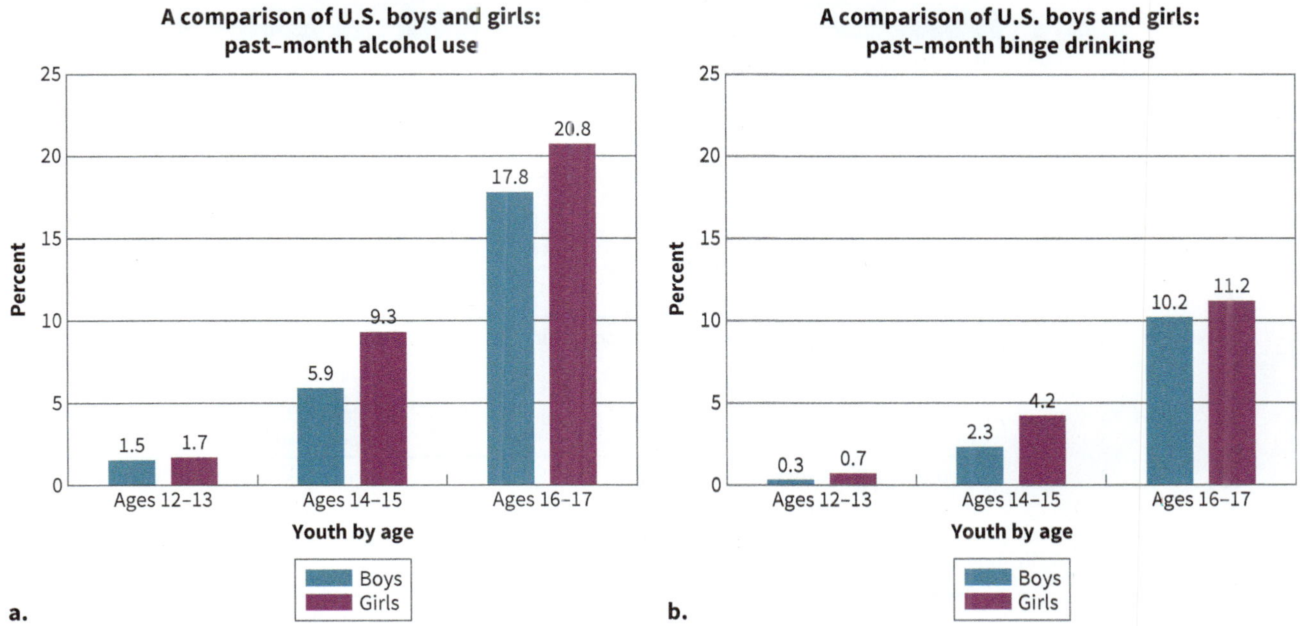

NIAAA, 2021 / U.S. Department of Health and Human Services / Public Domain

stomach lining. Alcohol has calories but no nutrient value, so your body breaks the molecule down to speed its exit from the body through excretion.

After absorption into the bloodstream, alcohol is distributed throughout the body and passes through the blood–brain barrier. About 5% of the alcohol in the bloodstream is excreted through the lungs, and this is the basis for breath tests of alcohol. The remainder is excreted by the kidneys via urine. As a general rule, 1 oz. of alcohol (30 ml) is metabolized each hour, so a person will remain sober as long as they do not ingest alcohol faster than they can metabolize it. **Blood alcohol content (BAC)**, also often referred to as BAL, or blood alcohol level) is determined by grams of alcohol per 100 mL of blood. In the United States, you are considered to be legally intoxicated at 0.08% BAC. BAC of higher than 0.40% puts you at risk of coma or death through alcohol poisoning. Alcohol poisoning is a medical emergency, so if you suspect someone has alcohol poisoning, seek help immediately. The warning signs of alcohol poisoning are presented in Table 9.4.

Behavioral effects of alcohol Although alcohol is often thought of as a stimulant, alcohol is actually a CNS depressant. The reason for this misconception is that at low doses, alcohol depresses the inhibitory functions of the cerebral cortex, often leading to silly

TABLE 9.4 Warning Signs of Alcohol Poisoning

- Vomiting
- Loss of gag reflex
- Clammy skin
- Slow heart rate
- Mental confusion
- Inability to remain awake
- Seizures
- Respiration rate slower than 8 breaths per minute
- Irregular breathing (10 seconds or more between breaths)
- Low body temperature

TABLE 9.5 Behavioral Signs and Impairments as a Function of BAC

BAC (%)	Behavior	Impairment
0.03–0.059	Relaxation Talkativeness Some euphoria Decreased inhibition	Concentration
0.06–0.099	Euphoria Disinhibition Blunted feelings	Reasoning Glare vision Peripheral vision Depth perception
0.1–0.199	Socially inappropriate behavior	Reflexes Motor control (staggering) Speech (slurring) Temporary erectile dysfunction in men
0.2–0.299	Nausea and vomiting Mood swings, including anger or sadness Decreased libido Impaired sensory perception	Loss of consciousness Memory blackouts Severe motor impairment
0.3–0.399	CNS depression Oscillating in and out of consciousness Some possibility of death	Loss of bladder function Shallow respiration Inability to stand Decreased heart rate
0.4–0.499	Severe CNS depression Coma Death possible	Respiratory paralysis Irregular heart rate
>0.5	High risk of death	

behavior. As BAC increases, the depressive effects extend from the cortex down toward the brain stem, with serious consequences (**Table 9.5**).

One critical effect of alcohol on behavior is a decrease in your ability to attend to important details in tasks such as driving. You can investigate the impact the alcohol has on the visual demands of driving in this **Check It Out: The Effects of Alcohol on Visual Search**.

Check It Out | The Effects of Alcohol on Visual Search

In this experiment, participants who either had no alcohol, a low level of alcohol, or a moderate level of alcohol viewed traffic images, and as they did so, a computer recorded information about where they were looking and for how long. Watch the video **Effects of Alcohol on Visual Search**, and then answer these questions.

1. In the first task, participants were asked to search an image for targets in a specific order. What was the effect of alcohol on performance on this task?
2. In the second task, participants were asked to pretend they were driving and to scan objects that were important to monitor while driving. What was the effect of alcohol on performance on this task?
3. Taking the results of both tasks as a whole, summarize the effect of alcohol on one's ability to operate a vehicle or a complex piece of equipment.

The issue of the use and abuse of alcohol becomes especially important when one considers the impact of alcohol use in student life. According to the NIH, in 2021, 53% of full-time U.S. college students drank alcohol in the last month and 33% binge drank in in the same period. Alcohol not only represents a direct danger to health, it also leads to other dangerous behaviors. Roughly 700,000 students in the same age range will be assaulted by another student who has been drinking, and around 100,000 students in that same age range will be the victim of a sexual assault where either or both parties had consumed alcohol. About 5,000 individuals under the age of 21 will die this year from alcohol-related causes, including 1,900 that involved vehicular accidents.

Opioids

The term **opioid** refers to a class of alkaloids that chemically resemble morphine, a derivative of opium. Other common opioids are heroin, codeine, hydrocodone, oxycodone, fentanyl, and carfentanil. Opioids may be used legally or illegally and come in pill, powder, or liquid form.

Opioids have many therapeutic uses; primary among these is relief from acute pain such as pain from a traumatic wound or surgery. In addition, they are often used for pain management for patients requiring long-term palliative care such as individuals diagnosed with terminal illnesses. In 2019, health care providers wrote 153 million prescriptions for opioid pain medication. Side effects include CNS depression, decreased respiration, and a feeling of euphoria; it is that last effect that leads to addiction.

Endogenous versus exogenous opioids In the 1950s, scientists began to investigate the mechanism by which opioids such as heroin, morphine, codeine, and opium worked. At Johns Hopkins University in the 1970s, Solomon Snyder and his graduate student Candace Pert explored this problem by using a morphine analog called naloxone that had a radioactive tracer attached; when the animal was injected with naloxone, any tissue that took up the tracer would show up on a scanner.

To their surprise, they found that there were receptors in the brains of their animal for opioids. It wasn't long before it was shown that the human brain was packed with opioid receptors (Figure 9.12). Snyder and Pert were puzzled; why would you and I have a receptor in our brain for a compound that was not encountered in nature until a few thousand years ago and was not widely available until morphine was first synthesized in 1804?

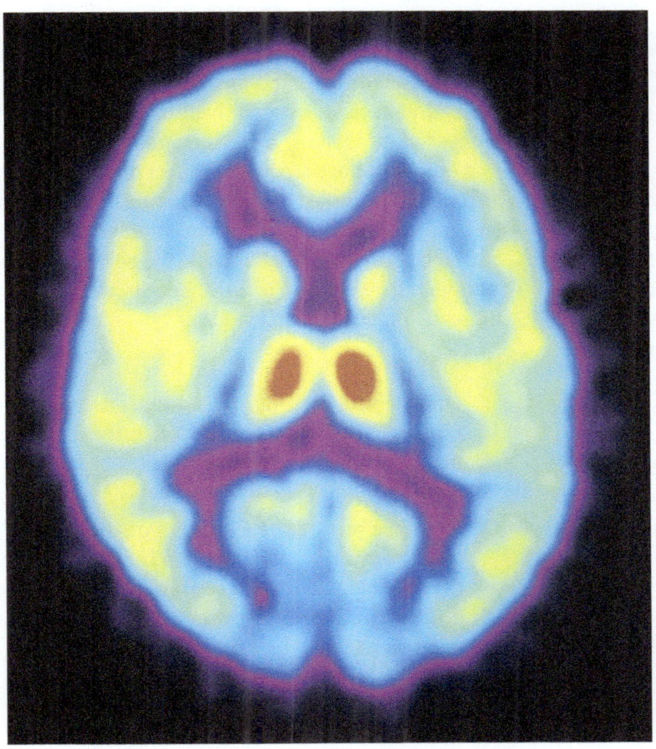

Philippe Psaila / Science Source

FIGURE 9.12 Endogenous opioid receptors The presence of endogenous opioid receptors can be located by linking a radioactive molecule to an opioid and then administering the opioid. These receptors are illuminated when the brain is scanned through positron emission tomography (PET), which detects the radioactive emissions.

Snyder and Pert solved the puzzle by turning it on its head; opioids work as they do because they mimic substances that our brain naturally creates. When you earn an A on an assignment, master a difficult skill, thrill to a lover's kiss, or see an old friend after a long absence, that pleasurable sensation is generated by your brain secreting natural opioids. When opioids are created internally, they are known as **endogenous opioids**. Opioids generated outside the body such as morphine, heroin, and OxyContin, are **exogenous opioids**. Given that exogenous opioids mimic the natural effects of endogenous opioids that are manufactured by your own brain, you can now readily see why they are so addictive. Pert's work on endogenous opioids was the start of an illustrious career, as you can see in this **Advances in Behavioral Neuroscience: Candace Pert and Opioid Receptors**.

Subsequent work with nonhuman species suggests that the opioid receptor has a long evolutionary history and one of its important functions is to enhance the development and maintenance of strong social bonds, which support pair bonding and family ties. Given the importance of social bonds among primates in general and humans in particular, it should not be surprising that exogenous opioids are very reinforcing and highly addictive.

Synthetic opioids Morphine and heroin are compounds created from opium and as such are considered natural opioids. However, chemists have also devised synthetic opioids in an effort to create an analgesic that relieves pain but is not addictive. So far that goal has not been reached, but the effort has led to synthetic opioids that are more powerful than any natural opioid. Some examples of synthetic opioids are oxycodone, methadone (often used as a substitute for heroin in treatment), fentanyl, and carfentanil. These last two are extremely potent; fentanyl is thought to be 100 times more potent than morphine, and carfentanil is considered to be 10,000 more potent than morphine. Using the potency of morphine as a reference point, **Table 9.6** provides a glimpse into the relative potency of common synthetic opioids.

Advances in Behavioral Neuroscience | Candace Pert and Opioid Receptors

Everett Collection Inc.

Sometimes in life, you have to listen to your own inner voice and do what you think is right, even when others disagree. Candace Pert was an internationally renowned neuroscientist whose curiosity and intuition about nature led her to rewrite our understanding of the link between our brain and our health.

Dr. Pert was born in 1946 in New York City; she graduated from Bryn Mawr College in 1970 and then entered the Ph.D. program at Johns Hopkins, where she worked in the laboratory of Solomon Snyder, one of the world's premier pharmacologists. Pert earned her doctorate in pharmacology from the Johns Hopkins School of Medicine in 1974.

In 1972, while still a graduate student in Snyder's lab, Pert began to think seriously about a vexing question: what is the neural mechanism that allows us to feel pleasure? Pert believed there must be an endogenous or innate opioid receptor in the human brain. Although she was told that the search for this receptor was a dead end and she should go back to her other work, she came into the lab on a weekend and conducted the experiment that made her world famous. Other discoveries would soon follow.

In a career that spanned four decades, Pert published over 250 research articles and changed the way we think about how the brain affects our overall health. In her later work, she was a pioneer of mind–body medicine and a proponent of the concept that emotions have a powerful effect on physical health. She was affectionately known as the Mother of Psychoneuroimmunology and the Goddess of Neuroscience and encouraged her students and collaborators to think outside the box of accepted knowledge. Although Dr. Pert died in 2013, her discoveries and insights continue to influence the field today.

Questions

1. **One way to end opioid addiction would be to permanently blockade opioid receptors. Assuming for a moment that this is even possible, what unpleasant consequences might this cure create?**
2. **Pert was one of the founders of the field of psychoneuroendocrinology. Search the internet for the term, and (a) summarize in one sentence what the field is about, and (b) note one disorder that the field may provide a treatment for.**

TABLE 9.6 Common Opioids and Their Relative Potency

Drug	Relative potency
Morphine	1
Codeine	~0.1
Pethidine (Demerol)	0.1
Tramadol (Ultram)	~0.1
Hydrocodone (Vicodin)	1
Oxycodone (OxyContin, Percocet)	2
Heroin	2–4
Hydromorphone (Dilaudid)	10
Buprenorphine (Subtex)	25–100
Fentanyl	~100
Sufentanil	~1000
Carfentanil	10,000

Of course, along with greater potency comes a greater risk of fatal overdose. Figure 9.13 illustrates the relative amounts of heroin, fentanyl, and carfentanil needed for a toxic dose.

Pharmacology of opioids Opioids are highly addictive because they stimulate the mesocorticolimbic pathway to both provide a reward and increase the expression of delta FosB, This in turn increases cravings while making withdrawal more severe. Opioid receptors are found throughout the body, including the GI tract and the spinal cord, but they are also found in the CNS and especially in the NAcc. When these receptors are stimulated, the effect is to blunt the perception of pain and also create euphoria. Opioids work by binding to one of three primary classes of opioid receptors known as μ, κ, and δ (mu, kappa, and delta), which differ somewhat in their function and distribution. It is beyond the scope of this chapter to delve deeply into the differences between mu, kappa, and delta receptors and the many receptor subtypes that have been discovered, but it is useful to know that each receptor acts in subtly different ways and so has different side effects and therapeutic profiles. With this knowledge in hand, you now understand why the various medical opioids differ in the extent to which they relieve pain, cause nausea or constipation, or present a risk for respiratory distress.

Opioid overdose leads to severe apnea and respiratory depression so that the individual may simply stop breathing and die of hypoxia (oxygen deprivation). Opioid overdoses can occur in first-time IV users as well as chronic users who have developed a substantial tolerance. With the increasing availability of more potent opioids, deaths from opioid overdose have risen dramatically over the past two decades (Figure 9.14).

Knowing how the opioid receptor works informs your understanding of opioid overdose treatments such as naloxone. Naloxone is a competitive antagonist with a very high affinity for the mu receptor. It replaces the opioid at the receptor site and blockades receptors not yet activated. When administered intravenously, its effects can occur in less than one minute and in 2–5 minutes via nasal spray, so it is very effective at saving the life of individuals who are in danger of dying from an opioid overdose. Naloxone is marketed under several brand names, most commonly as Narcan.

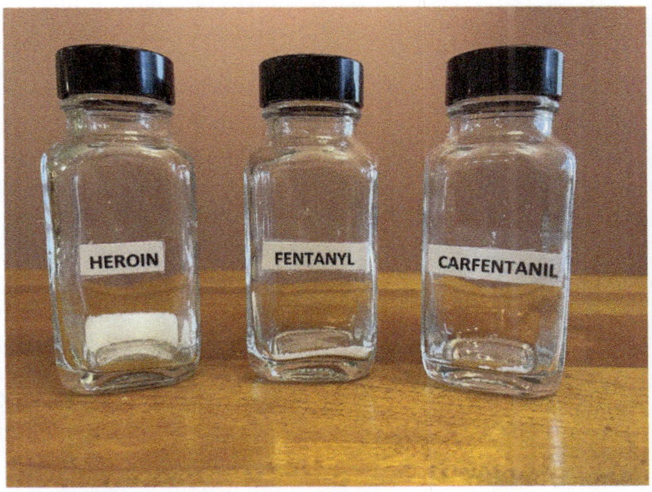

FIGURE 9.13 **Relative toxic doses of three opioids** Three vials, each containing a simulation of the amount of a common opioid that would be a toxic dose.

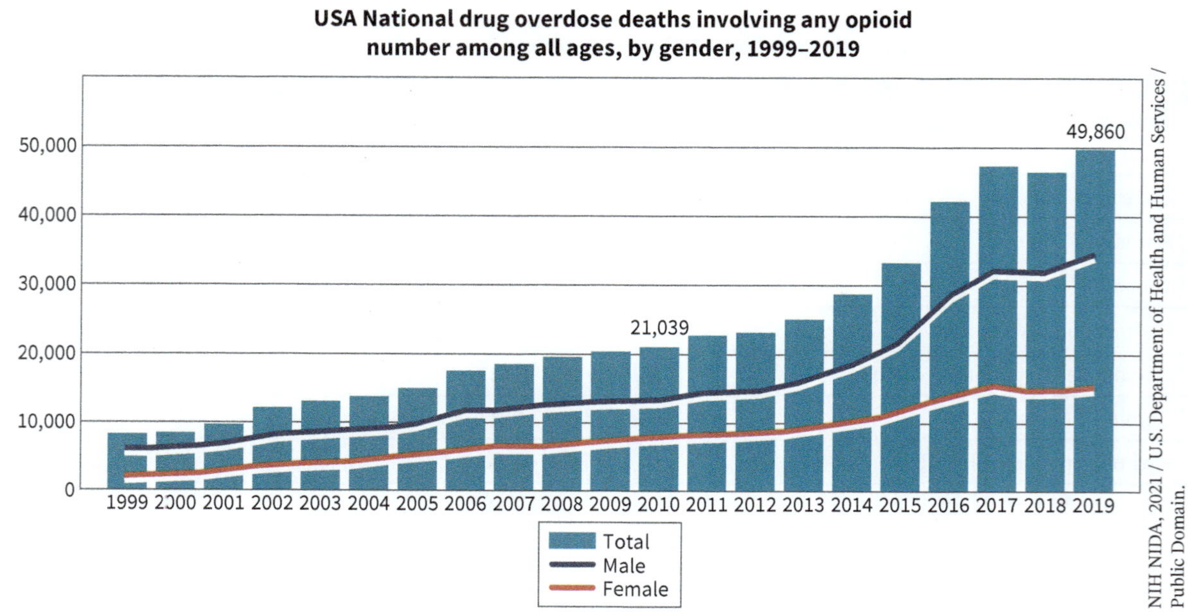

FIGURE 9.14 **Annual number of opioid overdose deaths** Over a two-decade period, the annual number of deaths due to opioid overdose increased by more than a factor of four.

Marijuana

Marijuana refers to the plant *Cannabis sativa*, which is indigenous to Asia. Marijuana is typically smoked but can also be ingested as cakes, brownies, or candy (**Figure 9.15**). Although cannabis is not particularly toxic, and in fact there has never been a report of a fatal overdose, chronic users do experience a deficit in short-term memory and attention while under the influence. Smokers incur significant risks to their lungs as a result of the inhalation of hot gases and particulates. Some evidence exists that cannabis use increases the risk of cardiovascular damage, but that evidence is contested because users often also use tobacco or other drugs. There is significant evidence that regular use of THC in any form during adolescence has a negative effect on neurodevelopment.

Society's view of marijuana has evolved from a gateway to hard drugs to an analgesic for chronic pain and an aid to individuals who struggle with nausea from chemotherapy. Ideally, some form of cannabinoid that conveys the therapeutic benefits without the mood intoxication could be found. One possibility is anandamide, an endogenous cannabinoid that is being investigated in animal studies as a therapeutic replacement for THC that would be less objectionable to society.

Pharmacology of THC There are several hundred metabolically active molecules in marijuana, including a class of compounds known as cannabinoids, but the major psychoactive ingredient in cannabis is THC (delta-9 tetrahydrocannabinol). A complete understanding of how THC influences behavior is not yet clear, but at this point we have identified two cannabinoid receptors in the brain: the CB1 is mostly found in the brain, and the CB2 is mostly found in peripheral tissue. THC exerts its influence on mood and cognition by acting as a CB1 agonist; CB1 activity then indirectly increases dopamine release, which leads to the behavioral effects associated with marijuana use. Because THC is highly soluble in lipids, it persists in the body and can be detected for weeks or longer after use.

Behavioral effects of cannabinoids THC and other cannabinoids are psychoactive drugs that have both medicinal and recreational uses. THC is an intoxicant, but it also has hallucinogenic properties at medium to higher doses. Cannabinoids create feelings of euphoria, relaxation, increased appetite, and altered perception of time and other sensory modalities. Cannabinoids typically show a biphasic effect; at low doses, most people experience positive effects, but at large doses, some experience agitation and paranoia, hallucinations, and psychotic thoughts. Most cannabis users do not develop tolerance or show withdrawal effects, although among heavy chronic users, a significant percent do report withdrawal

a. rgbspace / Getty Images

b. Oxik / Shutterstock

c. Elsa Olofsson, CBD Oracle

FIGURE 9.15 **Forms of THC** THC is available ins many forms other than the traditional joint. **a.** THC wax is smoked or vaped. **b.** THC oil is applied topically to the skin. **c.** THC edibles include gummy candies.

effects. Dependence is rare and is most likely to occur among individuals with poor self-esteem and social skills. It is important to note that although THC has euphoric properties, not all cannabinoids do. Cannabidiol (CBD) has been reported to provide relief for depression, PTSD, and anxiety without the euphoric effects of THC.

9.4 Before You Go On

Section Summary

Drugs can be used for medicinal or recreational purposes. Intoxicants such as alcohol and opioids create euphoria but are highly addictive, and their misuse creates significant social problems. One of the most important advances in our understanding of the neuroscience of behavior was the discovery of endogenous opioids, which are essential to the functioning of the brain's reward circuit. THC is an intoxicant that may have potential use as a treatment for pain and nausea.

Comprehension Check

1. Alcohol is considered to be what?
   - **a.** a stimulant that causes euphoria and relaxation
   - **b.** a nervous system depressant
   - **c.** a drug with minimal potential for addiction
   - **d.** a drug that influences neurotransmitter function in the hypothalamus
2. Alcohol intoxication is associated with what?
   - **a.** greater probability of an accident while driving
   - **b.** increased likelihood of assault
   - **c.** damage to the liver
   - **d.** all of these
3. What did Candace Pert and Solomon Snyder discover about opioids?
   - **a.** Opioids are highly addictive.
   - **b.** The mesolimbic circuit is stimulated by opioids.
   - **c.** We generate opioids internally.
   - **d.** The opioid receptor responds to opioids.
4. Although THC is an intoxicant, _____.
   - **a.** some cannabinoids have no intoxicating properties
   - **b.** it acts like an opioid at opioid receptors
   - **c.** it leaves the system in less than 4 hours
   - **d.** it has severe withdrawal potential

9.5 Stimulants and Other Commonly Abused Drugs

LO 9.5 Describe the pharmacology and behavioral effects of common stimulants, depressants, hallucinogens, and dissociatives.

Stimulants enhance attention, feelings of well-being, and energy. Common stimulants are nicotine, caffeine, and amphetamines. Stimulants provide the perception of increased alertness and energy. **Depressants** such as sedatives and tranquilizers do the opposite. **Hallucinogens** like LSD alter the perception of reality.

Nicotine

Nicotine is a stimulant that occurs naturally in tobacco as well as other plants, although not in the concentration that occurs in many species of tobacco. Nicotine is a neurotoxin that repels insects and many animals that might feed on the plant's leaves. Although nicotine can also be administered by vaping or a patch, it is most commonly associated with tobacco products. Tobacco comes in several forms, including cigars, cigarettes, snuff, and dip, and it is estimated that upward of 1.1 billion people use tobacco today.

As a result of inhaled chemicals, smoking causes a variety of diseases including cancer, heart disease, stroke, and lung diseases. According to the World Health Organization, tobacco use is the number one cause of preventable death across the world. As Figure 9.16 illustrates, nicotine itself affects systems throughout the body, whether smoked or ingested in other ways. That tobacco use persists despite being demonstrably harmful suggests that it has a powerful effect upon the brain and so is worthy of our interest.

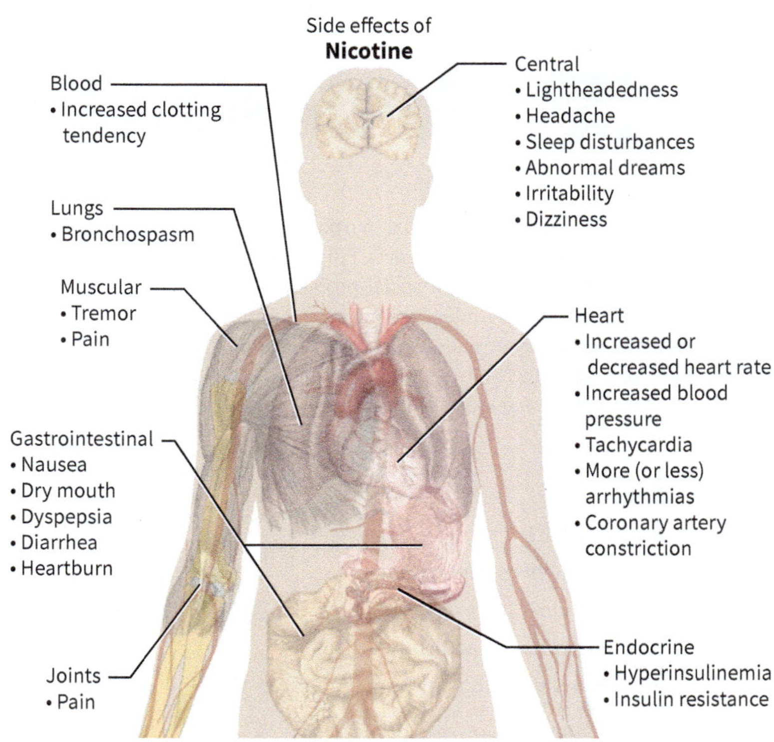

FIGURE 9.16 Systemic effects of nicotine Nicotinic receptors are found throughout the CNS, PNS, and the body, so its effects are widespread.

Pharmacology of nicotine Nicotine is a cholinomimetic stimulant; that is, it stimulates the parasympathetic nervous system in the same way as the neurotransmitter acetylcholine. Its action as a stimulant likely plays an important role in its addictive properties.

Nicotine absorbed through the lungs by smoking enters the brain within 7 seconds and stimulates the release of many chemical messengers, especially acetylcholine but also epinephrine, serotonin, dopamine, and others. It is purported to enhance concentration, create the sensation of calmness, and increase alertness, as well as improve motor coordination, although this claim of cognitive enhancement has been disputed.

As much as we understand about nicotine, there is much we still do not understand. For example, nicotine has an interesting biphasic action in that at low doses it acts as a stimulant by enhancing the action of dopamine and norepinephrine, whereas at higher doses, it enhances the effect of serotonin and has a tranquilizing and analgesic effect.

Addictive properties of nicotine The pharmacology of nicotine makes it powerfully addictive. According to the CDC, although more than half of adult smokers in the United States say they want to quit, fewer than 1 in 10 succeed in quitting each year. Why is nicotine so addictive? Nicotine affects dopamine levels in the mesolimbic pathway and increases long-term delta FosB expression, thereby leading to significant and sustained craving for the drug.

How smoking administers nicotine to the brain is significant in two ways. The first is that smoking delivers nicotine to the brain in a matter of seconds, and the second is that each individual puff delivers nicotine in a separate wave of stimulation. This pattern is known as pulsatile stimulation because each puff delivers a pulse of the drug. Rapid and pulsatile stimulation of the mesolimbic pathway is a uniquely powerful mechanism that encourages the smoker to take another puff. Because the addictive properties of nicotine are strong, withdrawal from nicotine is also strong and is characterized by irritation, loss of attention, sleep disturbances, depression, craving, and possibly increased appetite.

Research funded by the National Institute of Drug Addiction using animals concluded that acetaldehyde, a chemical found in tobacco smoke, increases the addictive properties of nicotine differentially for adolescent brains. This finding, if it were to be replicated in humans, would explain why beginning to smoke early in life leads to a much greater addiction potential.

Caffeine

Caffeine is the most widely used psychoactive drug in the world. The global popularity of caffeine is likely due to the stimulating effect it has on the CNS. Caffeine improves concentration, reduces fatigue, and improves athletic performance. College students, graduate students, physicians, professors, and shift workers often rely on caffeine to get them through their day in a productive fashion. Caffeine is typically ingested via a beverage such as coffee, tea, soft drinks, or caffeine-laden energy drinks (**Figure 9.17**). It also occurs in chocolate and other foods.

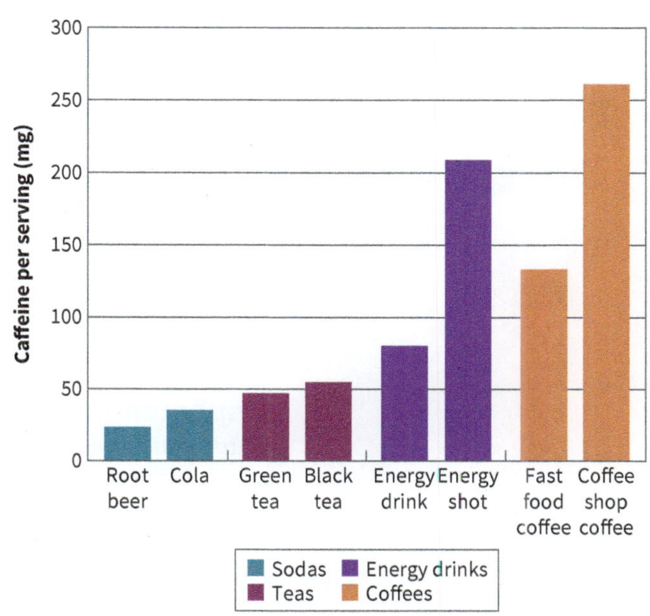

FIGURE 9.17 Caffeine levels in common beverages Caffeine levels vary widely between types of beverages and also within a beverage type.

Pharmacology of caffeine Caffeine is an adenosine receptor antagonist. Adenosine is a naturally occurring compound that has an inhibitory role in the CNS and in regulating blood flow to the heart. When you are well rested and alert, you have little adenosine in the CNS, but as adenosine builds up over time, its inhibitory effects place a greater burden on the CNS, leading to fatigue. Caffeine blocks the adenosine and decreases its inhibitory effects. As a result, you feel more alert, with increased heart rate and respiration.

Behavioral effects of caffeine Caffeine absorbed from coffee or other beverages begins to have an effect within 45 minutes and reaches a peak within an hour or two. The half-life of caffeine is several hours, but its effect varies widely among

Think About It | Do We Abuse Caffeine?

Hakan Dere / iStockphoto / Getty Images

Caffeine is the most widely used psychoactive drug. Caffeine is consumed in a number of beverages, but the most common vehicle for caffeine consumption is coffee. In 2020–2021, 165.5 million bags of coffee beans, each bag weighing 132 lbs. (60 kg), were consumed worldwide. In the United States, over 80% of adults drink coffee regularly, but Americans are not the most faithful coffee drinkers. In fact, at an average of 9.26 pounds (4.2 kg) of coffee consumed per person per year, Americans barely make it into the top 25 of coffee consumers, with the Finns leading the world at an average of 26.5 lbs. (12 kg) of coffee per year per adult. Caffeine is also present in most teas and many soft drinks and is an essential part of energy drinks. The average cup of coffee contains 95 mg of caffeine, teas average 47 mg of caffeine, and soft drinks average about 30 mg, but these numbers can be deceiving; some brands of these beverages can deliver close to 400 mg of caffeine in a serving.

Caffeine is a CNS stimulant, and its primary and most desired effect is to provide a temporary boost in alertness and energy. Clearly, people like the effect that caffeine has on their mood and energy, but how much caffeine is too much? The healthy limit appears to be about 400 mg of caffeine a day, which corresponds to four cups of coffee, ten soft drinks, or two energy drinks. Signs that you are drinking too much coffee are nervousness, headache, irritability, rapid heartbeat, muscle tremors, nausea, and insomnia.

Questions

1. **Do you think that professors drink more coffee than students, or is the opposite true? Do you see any relationship between mean daily reported caffeine dose and personality or academic achievement?**
2. **Do you use caffeinated beverages of any sort to enhance your performance or maintain an energy level? If so, how often?**
3. **Have you or anyone you know ever experienced withdrawal symptoms from caffeine? If so, describe the symptoms.**

individuals. With chronic use, tolerance reduces the physical effects of caffeine such as jittery nerves and trembling, but tolerance to the cognitive effects of increased alertness is less likely to occur. Many individuals develop a high tolerance and consume large amounts of coffee on a daily basis, more than a first-time user could tolerate. Consider why this might be so in this **Think About It: Do We Abuse Caffeine?**

Other Commonly Abused Drugs

Nicotine and caffeine are legal and ubiquitous, but many other drugs are used legally or illegally to increase energy, calm nerves, or alter one's perception of the world. In this section we will cover amphetamines, depressants, classic hallucinogens, and dissociative hallucinogens.

Amphetamines

Amphetamines are a class of drugs that stimulate the CNS. They sharpen attention, improve memory, and provide the user with an increased sense of well-being and energy. Amphetamines can be useful in a number of circumstances, and a common use is to improve the attentional performance of individuals diagnosed with attention-deficit-hyperactivity disorder (ADHD). Amphetamines are believed to block the reuptake of dopamine and norepinephrine in the presynaptic neuron. Increasing the amount of dopamine and norepinephrine in the brain enhances the ability of the CNS to inhibit impulsivity and thereby focus more on the task at hand.

Amphetamines are also prescribed to treat obesity because increasing a person's metabolism may lead to weight loss. Amphetamines may be prescribed for the treatment of narcolepsy, a condition in which the individual may suddenly fall asleep at any moment. Common examples of amphetamines are Adderall, Evekeo, Dexedrine, and Vyvanse. Individuals in extreme environments such as battlefields may be given amphetamines to maintain vigilance over extended periods of time. Because amphetamines increase heart rate, respiration, and muscle strength; speed recovery from exertion; and reduce fatigue, their use is banned at all levels of competitive athletics.

Amphetamines can also be abused. At typical low doses used for therapeutic purposes, amphetamines have a low potential for physiological addiction. At higher doses used by recreational abusers, amphetamines create euphoria and delirium, and their continued use is highly addictive. Amphetamines stimulate the mesolimbic pathway between the VTA and the NAcc. In chronic users, they raise the level of delta FosB in the NAcc, leading to an addiction that is very resistant to treatment.

Methamphetamine (meth) is an amphetamine that is mainly used recreationally, although in low doses it may be prescribed as a treatment for ADHD. Methamphetamine is highly addictive, and withdrawal effects are quite severe and persistent; as a result, meth addicts are resistant to treatment. Methamphetamine abusers often demonstrate the unusual behavior of punding, which is a compulsion to engage in repetitive and purposeless acts such as arranging objects in a line or assembling and disassembling watches or other complex objects.

Depressants In contrast to stimulants, depressants, including sedatives and tranquilizers, inhibit the CNS. **Barbiturates** are used as sedatives in the treatment of insomnia, to prevent seizures, and as an anesthetic. They work by enhancing the action of GABA (an inhibitory neurotransmitter) and blocking the action of glutamate (an excitatory neurotransmitter) and thereby reduce neural activity throughout the nervous system. Common examples of barbiturates are Nembutal, Luminal, and Seconal, which are frequently prescribed for the treatment of insomnia. Insomnia is defined as a continuing difficulty to fall asleep or stay asleep, and it afflicts roughly 40% of the adult population at some point (Figure 9.18). Chronic insomnia causes mental impairment and tiredness and is a contributing factor in accidents of every type.

Sarah Waverka / Shutterstock

FIGURE 9.18 **Sleep disorders** Your brain and body require a period of rest to clear away metabolic waste and refresh resources. Without adequate sleep, a person struggles to meet their daily challenges.

Recreational use of barbiturates creates a feeling of euphoria and relaxed contentment. Barbiturates have a high potential for addiction and overdose, and chronic users experience significant withdrawal. Because of these concerns, barbiturates have been replaced by benzodiazepines for most legal uses.

Benzodiazepines are used medically to relieve the symptoms of stress and anxiety. Common benzodiazepines are Valium, Librium, and Xanax. They work by increasing the levels of GABA, the major inhibitory neurotransmitter. In doing so, they reduce neural activity in the brain's reward circuit and thereby create a feeling of calmness but also drowsiness, lack of energy, and loss of attentional focus. Because of the mild euphoria that often accompanies their use, they are sometimes used recreationally. Benzodiazepines have a very high potential for addiction, and severe cravings and withdrawal symptoms can appear after only a few weeks of repeated use.

Classic hallucinogens Hallucinogens are drugs that alter perception and consciousness. When you consider the complexity of human cognition, it is perhaps not a surprise that some drugs create such an alteration in consciousness that the individual experiences hallucinations (Figure 9.19). Hallucinogens are divided into classic and dissociative drugs. In this first section we will briefly discuss the most common hallucinogens: LSD, psilocybin, mescaline, and MDMA.

ysuel / Adobe Stock

FIGURE 9.19 **Sensory distortions created by hallucinogens** Hallucinogens create visual, auditory, and other sensory disruptions in processing.

LSD (lysergic acid diethylamide) was first synthesized by Swiss chemist Albert Hofmann in 1938. Although the effects of LSD differ from person to person, the most common effects are visual hallucinations that include vivid colors and geometric shapes as well as distortions of time. LSD can also trigger panic attacks.

The mechanism by which LSD creates its effect is not well understood. Our current understanding is that it enhances the effect of glutamate in the cerebral cortex, particularly in layers IV and V. We now know that schizophrenia is associated with a thinner cortex and cellular disarray in cortical layers IV and V. Given that LSD creates hallucinations and psychiatric disorders such as schizophrenia also share that characteristic, it was hoped that understanding the mechanism by which LSD affects consciousness would provide better understanding of schizophrenia and thereby help develop treatments for psychoses. However, LSD was made illegal in the 1960s, so there has been very little research on its mechanism.

Psilocybin is a defensive compound produced by many varieties of mushrooms to deter predators. Psilocybin is metabolized into psilocin, a serotonergic agonist that mimics the effects of serotonin. Psilocin also indirectly enhances dopamine activity. Humans who ingest mushrooms containing psilocybin experience visual hallucinations and euphoria, although a significant percentage report paranoia and anxiety.

Mescaline is a naturally occurring compound whose chemical structure closely resembles dopamine. Like LSD and psilocybin, mescaline enhances serotonergic and dopaminergic activity.

Also known as ecstasy, X, E, XTC, or mollie, MDMA is a psychoactive drug that combines the properties of an amphetamine with those of a hallucinogen. MDMA creates feelings of euphoria, energy, and empathy (that is a feeling of being a part of the group), and it enhances perception of sound and colors. For these reasons, MDMA has become popular with individuals who attend raves, which feature loud music, a light show, and crowds of people dancing.

Because MDMA stimulates the release of serotonin, dopamine, and norepinephrine, it is known as a triple-releasing agent (TRA). The enhanced release of serotonin is likely responsible for the mood-enhancing properties of MDMA. However, the brain compensates for this excessive release by reducing serotonin synthesis, thereby causing a crash a few days after use. In this state, MMDA users often report feelings of anxiety, agitation, hallucinations, and depersonalization (a feeling of detachment from the real world). MDMA users are prone to hyperthermia and dehydration, which has led to some fatalities. MDMA has also been shown to create lesions in the serotonergic pathways as well as other consequences such as changes in white and gray matter. Cognitive impairment has also been noted in MDMA users correlated with level and duration of use.

Are hallucinogens addictive? The answer is complex. LSD and psilocybin do not appear to create a physical dependency, and repeated use does not create tolerance; however, an individual may become psychologically dependent by using hallucinogens to escape life's pressures. MMDA has been reported to physically change the DA and SE systems in the brain, and for this reason there are divided opinions on its addictive potential.

Dissociative drugs

Another type of hallucinogen is **dissociative drugs**. These also create hallucinations but differ from classic hallucinogens in that they also create a perception of detachment from one's body. Common dissociatives are PCP, ketamine, and DXM.

PCP (phencyclidine, or angel dust) was developed in the 1950s as an anesthetic for surgery. It blocks the reuptake of DA, SE, and NE in the brain and in doing so potentiates their action. At the same time, it acts as an antagonist of glutamate receptors in the brain; the net result of this action is to block the perception of pain and emotions. This combination of increased neural activity coupled with blocking of pain and emotional response may explain the common anecdotal report that individuals using PCP may behave without regard for the consequences of their actions. PCP users exhibit withdrawal symptoms such as confusion, depression, and craving, and these symptoms can continue for months after cessation of use.

Ketamine (also called Special K, vitamin K, or Kit Kat) is a fast-acting anesthetic that is commonly used to induce unconsciousness in patients about to undergo surgery. The mode of its action is not well understood, but it is known to interfere with glutamic receptors that mediate the transmission of signals from the spinal cord. When used recreationally, it creates a dissociative state that may be accompanied by visual and auditory hallucinations. Ketamine is currently considered to be an effective adjuvant (that is, a substance that enhances the action of another) to antidepressants and psychotherapy in the treatment of clinical depression. Ketamine is thought to have the potential to become addictive.

DMT (also known as N,N, dimethyltryptamine, Dimitri, or Spirit) is a drug that is illegal to manufacture, buy, possess, or use in the United States. Users report an experience characterized as very intense and brief in duration that involves hallucinations and a dissociative state. DMT is a naturally occurring psychoactive agent in several species of plants and as such has been incorporated into the religious ceremonies of some groups. Its potential for addiction is not well understood. DMT binds to certain SE and DA receptors throughout the CNS.

You learned in a prior chapter that although we commonly refer to receptors as SE-or 5-HT, DA-, or NE-receptors (for example) to indicate the neurotransmitter that is associated with that receptor. However, the reality is that within a type of receptor, there are varieties that have unique characteristics. In that light, it is worth noting that hallucinogens share the trait of being agonists of the 5-HT_{2a} receptor. For that reason, researchers consider this receptor to be an agent of interest when investigating the neurological basis of psychiatric disorders such as schizophrenia and paranoia.

9.5 Before You Go On

Section Summary

Nicotine is highly addictive and in the form of tobacco leads to serious health effects. Caffeine is a stimulant that is in common use. Stimulants and sedatives alter the general activity of the brain, and their therapeutic potential is weighed against their addictive properties; for instance, methamphetamine is highly addictive. Hallucinogens alter the perception of reality in ways that may help us to understand psychiatric illnesses, whereas dissociative drugs distort perception while creating a feeling of detachment from a person's environment and even their own self.

Comprehension Check

1. Which statement about the addictive properties of nicotine is false?
   - **a.** Nicotine is an extremely addictive drug.
   - **b.** Once addicted, breaking the nicotine addiction is extremely difficult.
   - **c.** Smoking delivers nicotine to the brain in a matter of seconds.
   - **d.** Nicotine delivered by dip, snuff, or snus is less addictive than nicotine delivered by smoking.
2. Which statement is true of caffeine?
   - **a.** It is the most commonly used stimulant in the world.
   - **b.** It is a psychoactive drug that does not generate tolerance.
   - **c.** It works by depressing the action of norepinephrine.
   - **d.** Its beneficial effects appear only at higher doses.
3. One fact about methamphetamine is that it ______.
   - **a.** is widely used as a treatment for pain
   - **b.** stimulates the adrenal glands to produce cortisol
   - **c.** is highly addictive with severe withdrawal
   - **d.** was widely used in World War I to sedate soldiers
4. When LSD was discovered, what was one use considered for it?
   - **a.** understanding the perceptual process
   - **b.** determining the cause of hypertension
   - **c.** treating psychotic behavior
   - **d.** determining the biological cause of schizophrenia
5. Which of these is a defining characteristic of dissociative drugs?
   - **a.** Their hallucinations always involve vision.
   - **b.** They create a feeling of disconnect from reality.
   - **c.** They are known to be highly addictive.
   - **d.** They are psychologically addictive.

Conclusion

We began our discussion of the biological basis of reward with the story of Roy, whose gambling addiction led to the loss of most of what he had spent his life achieving. Humans are able to work toward goals that are in the distant future and involve mastering many intermediate steps because they have a reward circuit that delivers satisfaction at each level. When this circuit becomes attached to an unhealthy goal such as gambling or drugs, addiction ensues. Common drugs of use and abuse include intoxicants, stimulants and depressants, hallucinogens, and dissociatives. Understanding how drugs affect the body and how in turn the body alters drug action is important to your understanding of how the brain creates behavior.

Critical Thinking Questions

1. How often do you crave a salty or sweet snack? If you satisfy the craving by binging on the snack, how would you describe your mood afterwards? Can you be addicted to salty or sweet snacks? Explain your argument for or against.
2. How would understanding the link between habits and maintenance of behavior help you to break an undesired habit? Provide an example from your own life.
3. It is possible that one day soon, we will be able to test people for their tendency to develop delta FosB in response to reinforcing stimuli, which is a risk factor for addiction. Would it be ethical to determine if a financial officer of a corporation or a pharmacist were overexpressing delta FosB as a factor in a hiring decision? Would you want to know if a potential spouse was prone to delta FosB overexpression? Briefly explain your answer.
4. Do you think you or your friends consume too much, too little, or just the right number of caffeinated beverages? Do you consider caffeine to be a drug or a food item? Defend your answer.
5. The mean half-life of naloxone is roughly 45 minutes, whereas the mean half-life of most opioids is longer. What are the implications of this mismatch in terms of treating an individual who is overdosing on an opioid such as heroin?
6. Hallucinogens and dissociatives may help us understand the neurological basis of psychiatric illnesses, but their use brings risks. At what level should these drugs be available? Some options are carefully controlled medical research studies, by prescription only, and freely available. Support your argument.

Dig Deeper into Behavioral Neuroscience

9.1 The Biological Basis for Reward

Berridge and Kringelbach provide an overview of the brain's reward circuit.

Berridge, K. C., & Kringelbach. M. L. (2015, May). Pleasure systems in the brain. *Neuron*, 86(3), 646–664. https://doi.org/10.1016/j.neuron.2015.02.018. PMC: 4425246. PMID: 25950633.

Deep brain stimulation has the potential to revolutionize treatment for many neurological and psychiatric disorders.

Hariz, M., & Amadio, J. P. (2015). The new era of neuromodulation. *AMA Journal of Ethics, 17*(1), 74–81.

Holtzheimer, P. E., & Mayberg, H. S. (2011). Deep brain stimulation for psychiatric disorders. *Annual Review of Neuroscience, 34*, 289–307. https://doi.org/10.1146/annurev-neuro-061010-113638

Breiter, H. C., Aharon, I., Kahneman, D., Dale, A., & Shizgal, P. (2001). Functional imaging of neural responses to expectancy and experience of monetary gains and losses. *Neuron, 30*(2), 619–639. https://doi.org/10.1016/s0896-6273(01)00303-8

9.2 Addiction as Dysfunction of the Reward Circuit

The National Institute of Drug Addiction (NIDA) is a division of the NIH that is focused on understanding the biological and behavioral basis of addictions with an eye toward developing effective treatments. The Surgeon General's Office provides regular reports on the science behind addiction and the current state of addiction in the United States.

Office of the Surgeon General. (2016, November).

Facing addiction in America: The Surgeon General's report on alcohol, drugs, and health (PDF), pp. 35–37, 45, 63, 155, 317, 338. U.S. Department of Health and Human Services..

NIH is a clearinghouse for the latest science of addiction.

https://www.drugabuse.gov/drug-topics/addiction-science

Delta FosB plays a central role in developing and maintaining addictions.

Nestler, E. J., Barrot, M., & Self, D. W. (2001). DeltaFosB: A molecular switch for addiction. *Proceedings of the National Academy of Science, 98*(20), 11042–10046. http://www.ncbi.nlm.nih.gov/pubmed/11572966

9.3 The Essentials of Psychopharmacology

Psychopharmacology can be confusing, especially for those without a strong chemistry background. Preston and Johnson provide a simple but accurate presentation, whereas Muse and Moore go into greater depth.

Preston, J., & Johnson, J. (2019). *Clinical psychopharmacology made ridiculously simple* (9th ed). Medmaster. ISBN-13: 978-1935660408

Muse, M., & Moore, B. A. (2021). *Clinical psychopharmacology for psychologists*. Wiley. ISBN-13: 978-0470907573

9.4 Intoxicants

Alcohol is widely used across cultures and is considered to have benefits, but there are significant health risks to the regular consumption of alcohol. The National Institutes of Health is a clearinghouse for information about alcohol use and abuse.

https://www.niaaa.nih.gov/publications/brochures-and-fact-sheets/alcohol-facts-and-statistics

https://www.niaaa.nih.gov/alcohols-effects-health/alcohols-effects-body

Brecher's *Licit and Illicit Drugs* is considered to be a readable overview of commonly used drugs, even though it was written about a half century ago.

Brecher, E. M. (1973). *Licit and illicit drugs. The Consumers Union Report on Narcotics, Stimulants, Depressants, Inhalants, Hallucinogens, and Marijuana—Including Caffeine.* Little Brown & Co. ISBN-13: 978-0316107174

Kalant, H. (2001). The pharmacology and toxicology of "ecstasy" (MDMA) and related drugs. *CMAJ: Canadian Medical Association Journal/Journal de l'Association medicale canadienne, 165*(7), 917–928.

9.5 Stimulants and Other Commonly Abused Drugs

Stimulants and sedatives are frequently prescribed for a wide range of medical and psychiatric disorders, and knowledge of their properties is useful to weigh their benefits with their risks. The National Institutes of Health are a reliable source of information about them.

https://www.drugabuse.gov/publications/drugfacts/prescription-stimulants

Sedatives are used to treat various disorders but are repurposed for recreational use.

https://www.drugabuse.gov/drug-topics/prescription-medicines

Hallucinogens are often used for recreation but may have important medical and psychiatric value.

https://www.drugabuse.gov/drug-topics/hallucinogens
https://www.drugabuse.gov/publications/research-reports/hallucinogens-dissociative-drugs/director

Chapter 9 Before You Go On Answers

9.1 Before You Go On Answers

1. d. all of the these
2. b. the MFB, the VTA, and the nucleus accumbens
3. a. a link between emotion and behavior; planning for reward

9.2 Before You Go On Answers

1. c. It maintains addictions in response to repeated exposure.
2. a. a disorder on a continuum from mild to severe
3. c. supported by psychological, environmental, and genetic factors

9.3 Before You Go On Answers

1. d. veins, lungs, mouth, skin
2. d. The release of the drug is spread out over time.
3. c. 2 p.m., 50; 4 p.m., 25
4. b. tolerance
5. a. has different effects at low and high doses

9.4 Before You Go On Answers

1. b. a nervous system depressant
2. d. all of these
3. c. We generate opioids internally.
4. a. some cannabinoids have no intoxicating properties

9.5 Before You Go On Answers

1. d. Nicotine delivered by dip, snuff, or snus is less addictive than nicotine delivered by smoking.
2. a. It is the most commonly used stimulant in the world.
3. c. is highly addictive with severe withdrawal
4. d. determining the biological cause of schizophrenia
5. b. They create a feeling of disconnect from reality.

Chapter 9 Glossary

abuse In the context of addiction, a pattern of compulsive drug use or other behaviors marked by recurring adverse social, occupational, economic, legal, or interpersonal consequences.

addiction A treatable, chronic medical disease with genetic, environmental, and biological causes that leads to compulsive behaviors with negative consequences.

amphetamine One of a class of drugs that stimulate the CNS, sharpening attention, improving memory, and providing the user with an increased sense of well-being and energy.

barbiturate One of a class of sedative drugs that are prescribed to induce sleep and reduce anxiety.

behavioral tolerance A drug's decreased potency acquired in conjunction with a specific context, often environmental.

binge drinking A pattern of engaging in an activity over a short period of time to excess.

biphasic effect When used to describe drugs, having opposite effects at low and high doses or at simple and complex levels of task demands.

blood alcohol content (BAC) A measure of the amount of alcohol in the blood stream; also known as blood alcohol level (BAL).

delta FosB gene A variant of the FosB gene that is a molecular switch for addictive behavior.

dependence A state in which an individual is dependent on a substance or behavior due to withdrawal when access to the item is removed.

depressant A drug that reduces the function of the nervous system.

dissociative drug A type of hallucinogen that also creates a perception of detachment from one's body.

endogenous opioid Opioids that are created within the body.

exogenous opioid Opioids that are generated outside the body.

FosB gene A gene whose variant delta FosB is a factor in addictions.

half-life The time it takes for the amount of the drug in your body to be reduced by half.

hallucinogen A drug that causes visual, auditory, or other perceptual hallucinations.

intoxicant Any drug that causes feelings of happiness or pleasure.

medial forebrain bundle (MFB) A white matter tract that connects the septal nuclei and the ventral tegmental area with the hypothalamus.

mesocortical pathway A pathway in the reward circuit that involves planning and strategy in obtaining rewards.

mesolimbic pathway A pathway in the reward circuit that involves emotions and pleasure in obtaining rewards.

nucleus accumbens (NAcc) A structure found in each cerebral hemisphere that is considered to be a part of the basal ganglia and plays a role in the brain's reward circuit.

opioid A class of drugs resembling opium in their addictive and sedative properties.

pharmacodynamics The study of reactions between a drug and the body.

pharmacokinetics The study of how the body affects drug action, including absorption and metabolism.

potentiation An interaction between two or more drugs that results in at least one of the drugs exerting a more powerful effect.

psychopharmacology The study of the effects of drugs on the brain and body that influence behavior.

sensitivity In psychopharmacology, the state in which a given amount of a drug produces a greater effect due to receptor changes.

stimulant A drug that amplifies the function of the nervous system.

tolerance When, over repeated use, higher and higher doses of drug are required to obtain the same effect.

ventral tegmental area (VTA) A structure that constitutes the floor of the midbrain and has reciprocal relationships with many subcortical regions, particularly the amygdala, the NAcc, and the prefrontal cortex.

withdrawal A group of aversive symptoms that occur when a person is unable to access an addictive substance or object.

CHAPTER 10

LStockStudio / Adobe Stock

Sex and reproduction are essential human behaviors.

Sexual Reproduction and Behavior

CHAPTER OUTLINE

10.1 An Overview of Sex and Reproduction

- Asexual and Sexual Reproduction
- Sexual Dimorphism
- Male and Female Reproductive Anatomy
- Sex versus Gender

Try It Out: Sex Roles and Gender Roles

10.2 Prenatal Sexual Organization

- Genetic Influences on Biological Sex
- Sexual Differentiation *in Utero*
- Hormonal Influences on Prenatal Sexual Development

Try It Out: The D2:D4 Ratio

10.3 The HPG Axis and Life-Span Sexual Development

- Puberty

Think About It: Environmental Effects on Puberty Onset

- Hormonal Control of the Menstrual Cycle
- Hormonal Activation of Parental Behavior
- Sexuality and Aging

10.4 Sexual and Reproductive Behaviors

- The Human Sexual Response
- Sexual Orientation

Advances in Behavioral Neuroscience: Simon LeVay and the Neural Basis of Sexual Orientation

- The Biological Basis of Attraction

Check It Out: Are Average Faces More Attractive?

LEARNING OBJECTIVES

LO 10.1 **Describe** the evolutionary advantages of sexual reproduction as well as the male and female reproductive systems.

LO 10.2 **Identify** genetic and hormonal influences on prenatal sexual organization.

LO 10.3 **Identify** hormonal influences on sexual development over the life span.

LO 10.4 **Describe** the diversity of human sexual response, including sexual orientation and attraction.

Stories of the Brain

In 1967, twin boys Bruce and Brian Reimer were born to a Canadian family. At seven months of age, a botched circumcision left Bruce's penis burned beyond repair, and physicians felt that creating a vagina for Bruce would be easier than creating a replacement penis. Psychologist Dr. John Money, a world-famous expert in sexual development and gender identity, advised the parents that, with appropriate education and hormonal treatment, Bruce could become a well-adjusted female. Money was a proponent of the idea that gender identity is gained though experience and social learning rather than being innate, and having Bruce sexually reassigned to be a female while his identical twin was raised a male was a perfect test of his theory. Based on this advice, the Reimers had Bruce surgically reassigned to be a female, renaming him Brenda.

In the years after the surgery, Money met with the Reimers annually and coached Brenda on how to engage in traditionally female sexual behaviors. For a while, this case was hailed as a great success and influenced the medical practice of sexual reassignment as well as gender theory. However, Brenda's parents had not been entirely truthful with Money about how Brenda was doing at home, and by the time she was an adolescent, Brenda was experiencing significant depression and contemplating suicide.

When Brenda was 15, her parents told her the truth about her past and the reassignment, a revelation that confirmed her inner belief that she was not a "real girl." Brenda went through reassignment surgery again, this time to build an artificial penis, and he began testosterone treatments to replace female secondary sexual characteristics such as breasts with corresponding male secondary characteristics such as facial hair. As David Reimer, he married and adopted his wife's three children. Sadly, David Reimer did not have the adult life he deserved. Marital problems, bad financial decisions, and the death of his brother led to him taking his own life at age 38. This tragic true story changed how scientists view gender identity, and today intersex infants are typically not assigned a gender until they are old enough to give informed consent with the help of counseling.

In this chapter, we will discuss sex and reproduction from a number of angles. With a grasp of how genetics, hormones, and environment interact to create biological males and females, we will ask complex questions about sexual behavior. What is sexual orientation, and what forces drive it? Is it useful or even accurate to think of sex as a binary choice? The course of any developmental process always moves from simple to complex, and as you learn about the mechanisms that regulate sexual development, identity, and behavior, you may come to think more deeply about what it means to be a male or female.

10.1 An Overview of Sex and Reproduction

LO 10.1 Describe the evolutionary advantages of sexual reproduction as well as the male and female reproductive systems.

At this point in your exploration of behavioral neuroscience, you have read about the neural processes that regulate essential functions such as hunger, thirst, temperature, blood pressure, respiration, and alertness. Without these mechanisms, we could not survive. Reproduction is a behavior as essential to the survival of humanity as maintaining homeostasis. Unlike glucoregulation or thermoregulation, however, our sexual behavior is something that is very personal. For many of us, it is at the very center of who we are.

We begin our discussion of sexual reproduction and behavior with an overview of the evolutionary context of sexual reproduction and dimorphism, a review of the primary sex characteristics, and a clarification of what we mean by biological sex.

Asexual and Sexual Reproduction MCAT

All living organisms share a biological drive for their species to survive. For that to happen, an organism must produce offspring that can carry on the traits of the parent. You may recall from a prior chapter that all living organisms contain DNA in the genes of each cell and that this DNA contains the information necessary to create a copy of the parent. Progeny can be created through asexual or sexual reproduction, and each method has advantages and disadvantages.

Asexual reproduction is common among single-celled organisms, plants, and fungi. Through asexual reproduction, such organisms need no partner to reproduce. With only one set of genetic information involved, these offspring are identical to the parent except for changes due to random mutation. By contrast, **sexual reproduction** requires two parents, who each contribute genetic material to the offspring. Sexual reproduction is common among animals, including all mammals.

There are many advantages to asexual reproduction. First, asexual reproduction is more convenient than sexual reproduction. It may be difficult to find a partner for sexual reproduction, requiring significant travel and possibly competition for the opportunity to reproduce. In contrast, an organism needs only itself to reproduce asexually. Second, asexual reproduction is more reliable than sexual reproduction. Roughly half of all ova that are fertilized fail to implant successfully in the uterus. Of that number, roughly half spontaneously abort due to a genetic defect or an environmental stressor. Finally, some pathogens use sex as a vehicle to propagate their own species, and sexually transmitted diseases (STDs) can sterilize or even kill the participants. Clearly, the odds of successfully passing on genetic material to offspring are better for an organism that can simply make a copy of itself.

Despite these compelling arguments in favor of asexual reproduction, sexual reproduction is the rule for mammals. This suggests that sexual reproduction enjoys powerful evolutionary advantages that have tipped the scales in its favor, at least for mammals. Two advantages to sexual reproduction are explained by Muller's ratchet and the Red Queen dilemma.

Muller's ratchet One disadvantage of asexual reproduction is the propensity for genetic errors to quickly accumulate in generations of offspring produced by asexual reproduction, a problem known as **Muller's ratchet**. Rather than engage in a complex discussion of genetics, a simple thought experiment might suffice to explain this phenomenon. Imagine you have a page completely filled with the letters A, G, T, and C (representing the four bases in DNA). You scan and print that page, then discard the original, and make a scan of the second generation. Now repeat that process of scanning a copy, discarding the original, and making a scan of the new original for 50 generations. The final sheet will have lost its definition due to the accumulation of errors during the reproductive process (**Figure 10.1**). In the same way, genetic errors accumulate rapidly in the form of mutations during asexual reproduction.

The Red Queen dilemma Another disadvantage of asexual reproduction is that it limits the ability of organisms to adapt to changing environmental conditions, a challenge known as the **Red Queen dilemma**. This name refers to the character of the queen in Lewis Carroll's *Through the Looking-Glass*, who said to Alice that she has to run as fast as she can just to stay in one place. In the natural world, conditions are always changing; the climate changes, food sources move or disappear, and new predators suddenly emerge. Confronted with these environmental changes, species must continually adapt just to survive.

Because sexual reproduction involves the recombination of DNA from two parents, it confers a simple but profound advantage by providing offspring with a wider mix of genetic traits than asexual reproduction. When you consider that traits that are advantageous today could be a burden in the near future, a reproductive mechanism that creates individuals with a wide range of traits is likely to contribute to the survival of the species.

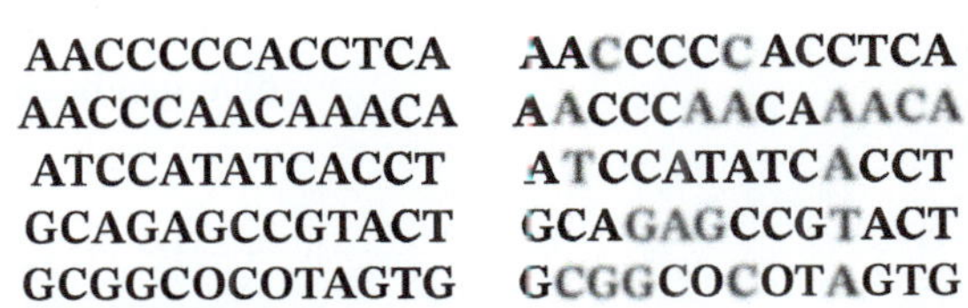

FIGURE 10.1 **Errors introduced into the genetic code from repeated duplication** **a.** Organisms that reproduce asexually create offspring whose genetic code is identical except for random mutations. **b.** Over generations, errors accumulate that may be harmful.

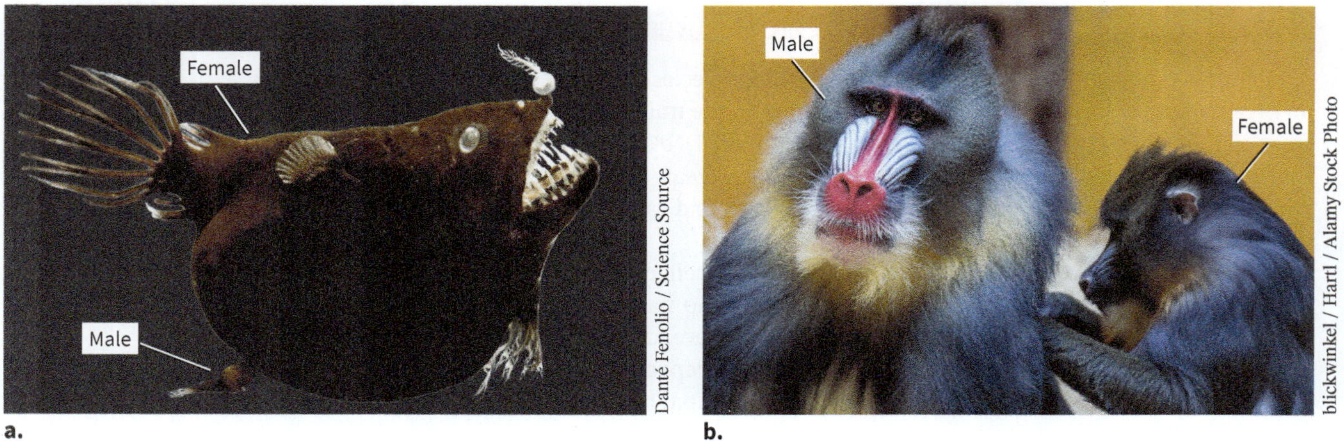

FIGURE 10.2 **Two examples of sexual dimorphism** **a.** In some species of angler fish, the female is considerably larger than the male. **b.** Adult male mandrills are three times the size of their female counterparts and sport a brightly colored snout.

The Red Queen dilemma shows that genetic diversity confers an evolutionary advantage, and one way to introduce diversity is if male and female parents differ structurally and behaviorally so that each sex contributes unique capabilities. This differentiation is known as sexual dimorphism.

Now that you see some of the advantages and disadvantages of sexual reproduction, we can begin to think for a moment about the physical differences between males and females, which are termed **sexual dimorphism**. Most species that reproduce sexually are dimorphic, that is, males and females differ noticeably in their physical characteristics. Perhaps the most startling example of sexual dimorphism is the angler fish; in some species, the male anglerfish is less than one-hundredth the size of the female, and the male mates with the female by attaching its body permanently to the female. Among primates, the mandrill may be the most dimorphic. The male mandrill sports a vibrantly colored snout and is several times larger than the plainer, more diminutive female (**Figure 10.2**).

It can be difficult to step away from the human perspective; most people probably see human males and females as being very different, particularly because we often adopt clothing and hair styles that accentuate our differences. In fact, compared to angler fish and mandrills, male and female humans are quite similar. Nonetheless, human males and females have numerous physical differences including distribution of body hair and fat, size, and strength.

What is the evolutionary advantage of sexual dimorphism for humans? Evolutionary theory suggests that dimorphic traits evolved because they conferred advantages tailored to the unique challenges faced by that sex. Human sexual dimorphism may have evolved out of different environmental pressures for males and females in our evolutionary past. The separation of male and female roles may stem from the unique challenges posed by birthing offspring that require long-term care before they are able to survive on their own. Hunting requires the hunter to travel great distances in search of game and the strength to carry the animal back to camp, whereas gathering can occur close to camp and allows for simultaneously tending a suckling infant. Over time, the optimal physical characteristics for each sex would have diverged based on which traits were more useful to each role.

Male and Female Reproductive Anatomy MCAT

Sexually dimorphic organisms typically express their sexual characteristics in two forms: primary and secondary. **Primary sexual characteristics** are those present at birth, and these consist of the genitalia. **Secondary sexual characteristics** help distinguish the sexes of the species, but they are not directly involved in reproduction. These include facial hair for males and breasts for females. We will begin with an overview of the primary sexual characteristics of humans. Secondary sexual characteristics develop during puberty, discussed later in this chapter.

Female genitalia The external structures of the female genitalia are known collectively as the **vulva**, which consists of the mons pubis; the **labia** majora and minora, which serve to protect internal tissues; and the **clitoris**, which is the sex organ primarily responsible for sexual pleasure in females. Internal structures of the female genitalia are critical to reproduction. The female gametes are eggs, or ova, which are produced in the **ovaries**, the female gonads. The two **Fallopian tubes** allow the transport of ova from the ovaries to the **uterus**, the hollow, pear-shaped organ in which the fertilized ovum implants. During the menstrual cycle, the uterine lining flourishes and then sloughs off, creating the flow known as the menses, or period. The **cervix** is a plug of tissue at the entry to the uterus, and the **vagina** is an elastic canal consisting of fibrous tissue and muscle that connects the uterus to the vulva. These structures are depicted in Figure 10.3.

Male genitalia The male genitalia are also classified as external or internal with the penis and scrotum being external. The **penis** contains the urethra and two chambers that fill with blood to cause an erection, called the corpus cavernosa and corpus spongiosum. Beneath the penis is the **scrotum**, a sac of skin that contains the male gonads, the **testes**. The testes are responsible for producing the male gametes, sperm, which are then transported by the vas deferens to the urethral opening. The prostate, along with other glands, secretes the seminal fluid that carries sperm and creates a chemical environment favorable to conception. These structures are depicted in Figure 10.4.

As we discuss the differences between male and female sexual characteristics, it is important to recognize that sexual differences are not always clear cut, and we will discuss some of those nuances later in the chapter. Before we go further, however, we must establish what we mean by the term sex and how that is different from gender.

Sex versus Gender MCAT

Discussions of the biology behind human reproduction and sexual behavior involve both the terms sex and gender. Although these terms are sometimes used interchangeably, they are not identical concepts. **Biological sex** refers to anatomical structures created by genes and

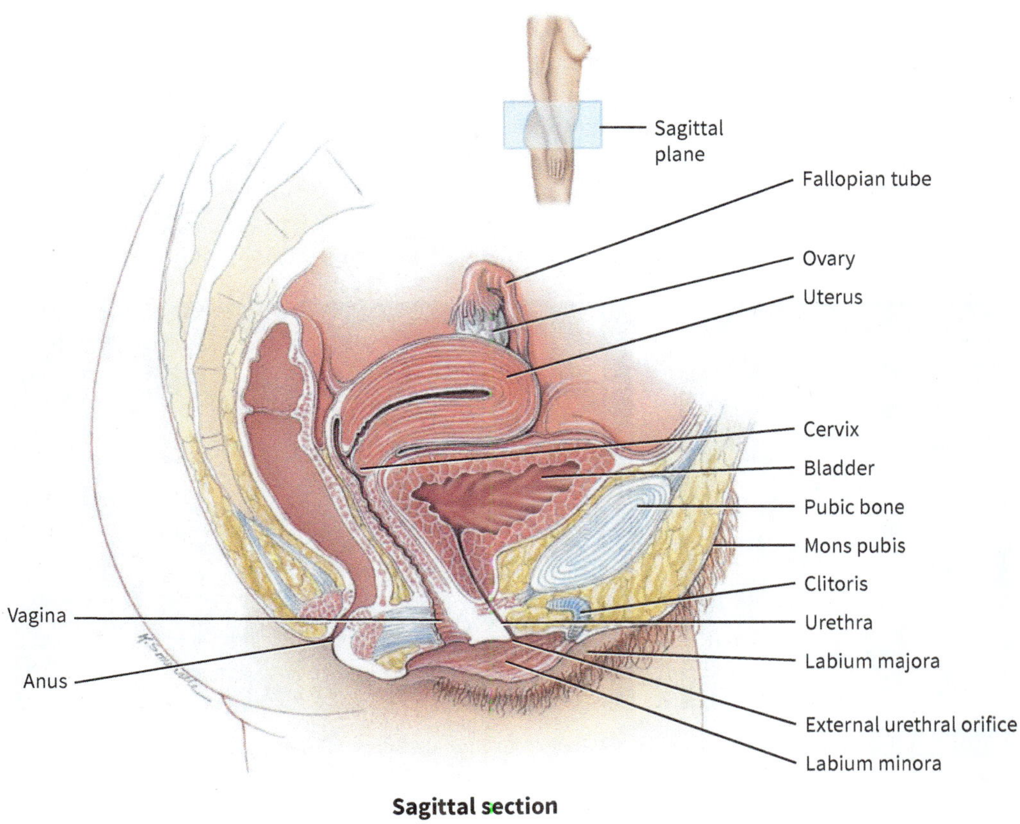

FIGURE 10.3 Female genital system The female genital system consists of external and internal structures. External structures are the mons pubis, the major and minor labia, and the clitoris. Internal structures include the vagina, cervix, uterus, Fallopian tubes, and ovaries.

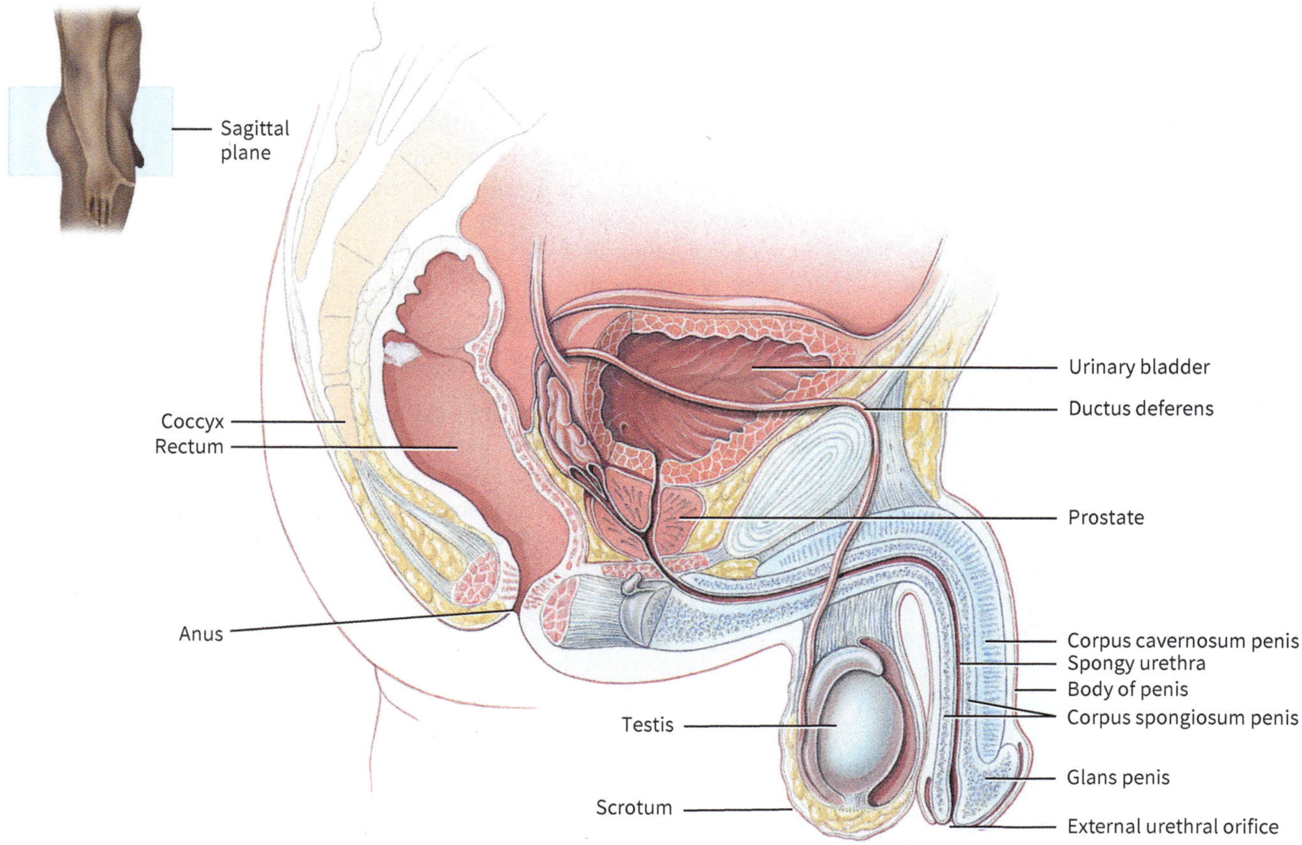

FIGURE 10.4 The male genital system The male genital system consists of external and internal structures. The external structures are the penis and scrotum. Internal structures include the testes, corpus cavernosa, corpus spongiosum, prostate gland, and vas deferens.

hormones that result in male or female genitalia and associated secondary sexual characteristics. In contrast, **gender** refers to behaviors and a sense of identity that has several determinants, which include biological sex but also social and cultural forces.

Figure 10.5 presents you with two examples of the difference between biological sex and the social construction we call gender. Roosevelt (Rosie) Grier was a 6′6″ (1.96 m), 280 lb. (129 kg) professional football player for the New York Giants who was a part of an NFL champion team; his hobby is needlepoint. Rhonda Rousey is a former Olympic champion at judo and a world-class mixed martial arts fighter. Both Grier and Rousey express gender-role behaviors that differ from typical cultural expectations.

Society and culture place expectations about gender roles that have no basis in biology. For example, in some cultures, males were traditionally thought to be strong, confident, assertive,

FIGURE 10.5 Sex versus gender Rosie Grier and Rhonda Rousey illustrate that one's biological sex is determined by genes and hormones, but the gender roles we take on are guided by our own interests and by our culture.

Try It Out | Sex Roles and Gender Roles

Sudowoodo / Adobe Stock

In this chapter, you will consider the different effects that genetics and hormones on the one hand and society and culture on the other have on one's view about men and women. In this activity, you can think about how you personally view these constructs. Complete the worksheet **Sex Roles and Gender Roles** and evaluate personality traits, activities, and careers in light of your perception of the link between sex and gender.

good at mathematics, competitive, better leaders, and to prefer the color blue. Females, on the other hand, were considered to be weaker, doubtful or tentative, poor at math, noncompetitive, empathetic and nurturing, better team players, and to prefer the color pink.

Consider for a moment what it means to decouple biological sex from the concept of gender. It may be convenient to classify people into two biological states that reflect their genitalia, but the concept of gender is more amorphous. Gender involves expectations that are created by society and culture, and a person's gender identity may not map directly onto their biological sex. A transsexual individual is someone whose bodily characteristics are being or have been altered through surgery or hormones to bring them into alignment with their gender identity. You can consider your own perspective on gender roles in **Try It Out: Sex Roles and Gender Roles**.

Having clarified the difference between biological sex and gender, we now turn to an in-depth discussion of the biological aspects of sexual development, starting with prenatal development. We will return to the consideration of the social and behavioral aspects of sex and gender at the end of this chapter.

10.1 Before You Go On

Section Summary

Asexual reproduction involves the replication of the genetic material of a single parent, whereas sexual reproduction involves the recombination of genetic material contributed by two parents, which allows a species to adapt more rapidly to a changing environment. Sexual dimorphism in humans may be the result of different environmental pressures on males and females in our ancestral past. The male and female genital systems provide us with reproductive capabilities. For humans, biological sex and gender are not the same.

Comprehension Check

1. One advantage of sexual reproduction over asexual reproduction is that sexual reproduction _____.

- **a.** creates a greater number of offspring
- **b.** involves less competition for mates
- **c.** is more pleasurable
- **d.** creates greater genetic diversity

2. What is sexual dimorphism?

- **a.** Physical traits that are different between males and females of a species
- **b.** When an individual's biological sex is inconsistent with their gender
- **c.** The development of secondary sexual characteristics during puberty
- **d.** When an individual has elements of both male and female genitalia

3. The female gonad is the _____, and the male gonad is the _____.

- **a.** vagina; penis
- **b.** egg; sperm
- **c.** ovary; testis
- **d.** gamete; seminal fluid

4. Which of these is an important difference between gender and sex?

- **a.** Gender is biologically determined, and sex is a social construct.
- **b.** Gender roles may vary considerably due to cultural differences.
- **c.** There are three stages of gender development but one of sex development.
- **d.** We know a lot more about gender than we do about sex.

10.2 Prenatal Sexual Organization

LO 10.2 Identify genetic and hormonal influences on prenatal sexual organization.

Sex has both an organizing function and an activating function. By organization we mean that hormones secreted during pre- and postnatal development change our anatomy in profound ways, and by activation we mean that this organization activates a range of sexual behaviors. In order for you to understand adult sexual organization and activation, it is essential that you see the developmental path that leads to the adult genitalia. We therefore continue our discussion with a consideration of the development of sex differences, which begins at conception.

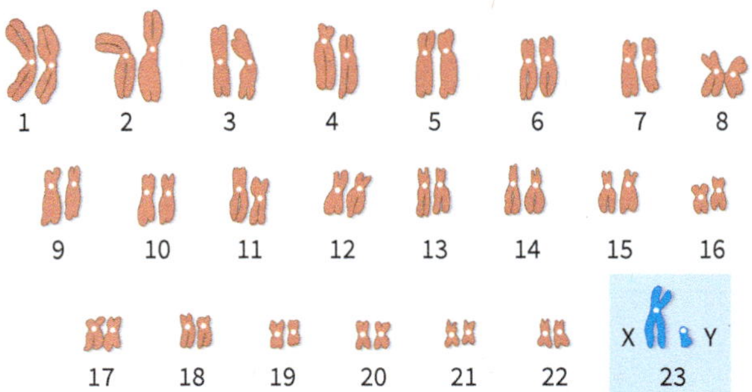

FIGURE 10.6 Male chromosomes with X and Y allosomes Humans have 23 chromosomes pairs, of which the allosomes contain the genetic information for biological sex.

Genetic Influences on Biological Sex MCAT

Humans have 46 chromosomes divided into 23 pairs, one pair of which are known as allosomes. The two **allosomes**, or sex chromosomes, transmit information related to the sex of the individual, whereas the 22 autosomes contain the genetic information for the rest of the body. Allosomes themselves come in two varieties: X and Y. In humans, females have two X allosomes, and males have one X and one Y allosome. This means that in sexual reproduction, a female always contributes an X allosome, whereas the male can contribute either an X or a Y allosome. Therefore, the genetic contribution of the male sperm decides the sex of the offspring. The XY allosomes of a typical male are illustrated in Figure 10.6.

The vast majority of us sort out genetically into either XX or XY, but nature always provides variation, so we would expect to see some individuals who do not fit neatly into this binary classification. These include X, XXY, and XYY, among others. As is often the case, exploring what sometimes goes wrong with a biological system can help us understand how processes normally work. Table 10.1 gives an overview of disorders related to sex chromosomes.

From the moment that conception occurs, the single fertilized cell that will become a person has been set onto the XX or XY path that will affect their entire lives. That is not the end of the story, and as the embryo begins to form, genetic and hormonal actors begin to differentiate males from females, the topic of our next section.

TABLE 10.1 Disorders Related to Sex Chromosomes

Genotype	Disorder	Description
X	Turner syndrome	Short stature and a short, webbed neck, infertile
XXY	Klinefelter syndrome (KS)	Taller than average and poorly coordinated; small, poorly functioning testes; infertile; one of the more common chromosomal disorders, occurring perhaps 1 in every 1,000 male births.
XYY	XYY syndrome	Faster growth than XY males and slightly taller than average; more likely to develop acne in adolescence and more likely to have a learning disability; often undiagnosed.
XX and XY in same individual	XO/XY mosaicism	Leads to wide range of sexual developmental paths, depending on the resulting proportion of XX to XY cells, ranging from ambiguous genitalia to completely developed male or female gonads, with male development being by far the most common condition.

Sexual Differentiation *in Utero* MCAT

Throughout your study of the biological basis of behavior, you have encountered mechanisms that weave excitatory and inhibitory processes to create the complexity that is a hallmark of humanity, and sexual differentiation is no different. Sexual differentiation begins with a common framework that genes and hormones elegantly sculpt to create adult females and males capable of reproduction.

Müllerian and Wolffian systems Male and female embryos initially share a common urogenital system. This primitive system includes two sets of ducts, one of which leads to female reproductive organs and the other of which leads to their male counterpart (**Figure 10.7a**). The **Müllerian ducts**, also known as the paramesonephric ducts, lead to the female reproductive organs, including ovaries, Fallopian tubes, and uterus (**Figure 10.7b**). The **Wolffian ducts**, also known as mesonephric ducts, give rise to the male reproductive organs including testes and vas deferens (**Figure 10.7c**). One way to remember which system is female and which is male is to remember that that Müllerian and mother both begin with the letter m.

All human embryos develop similarly for the first few weeks, but around the sixth week after conception, sexual differentiation begins to distinguish males from females. The genetic trigger for this differentiation comes from a gene on the Y chromosome known as the ***SRY* gene** (named for the sex-determining region on the Y gene), which encodes a protein known

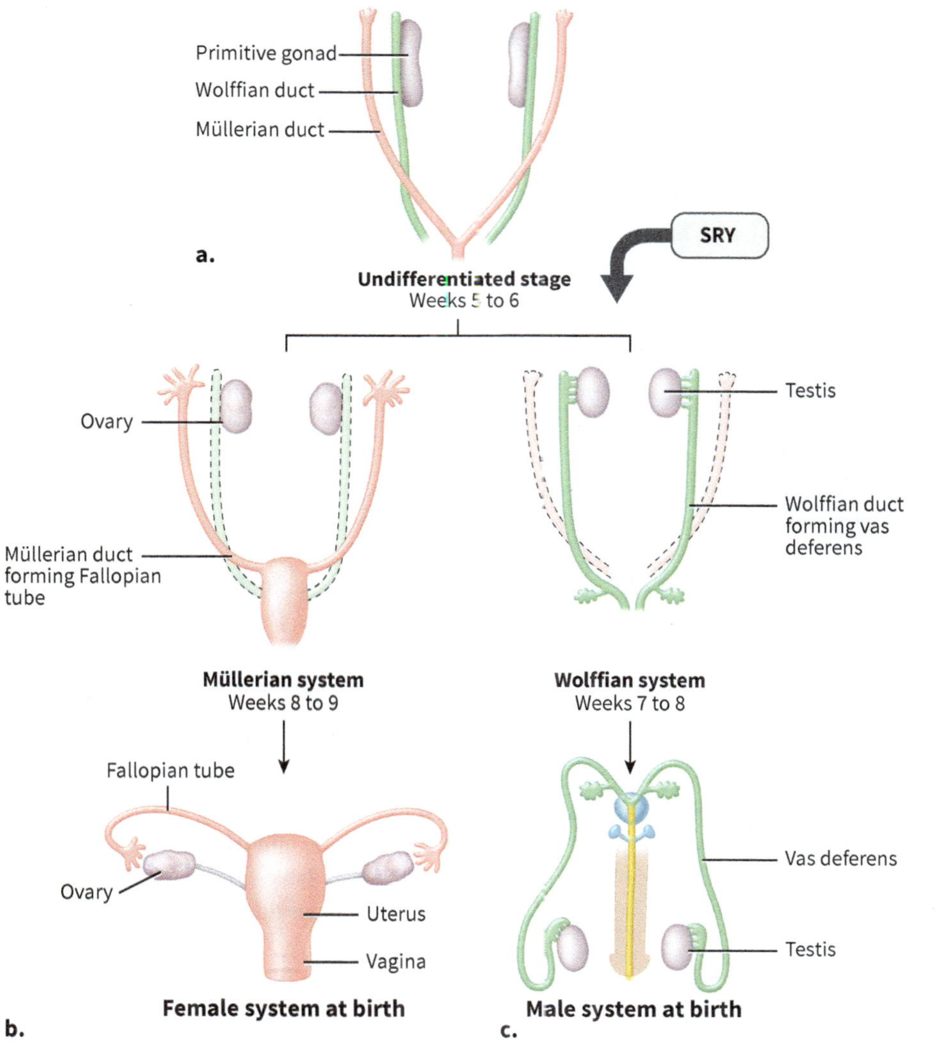

FIGURE 10.7 The development of the Müllerian and Wolffian urogenital systems **a.** The male and female urogenital systems are the same until around week 6. **b.** Without the influence of androgens, the system will develop into the female urogenital system. **c.** The presence of the *SRY* gene, which is contained on the Y chromosome, triggers the differentiation of the male system from the default female system.

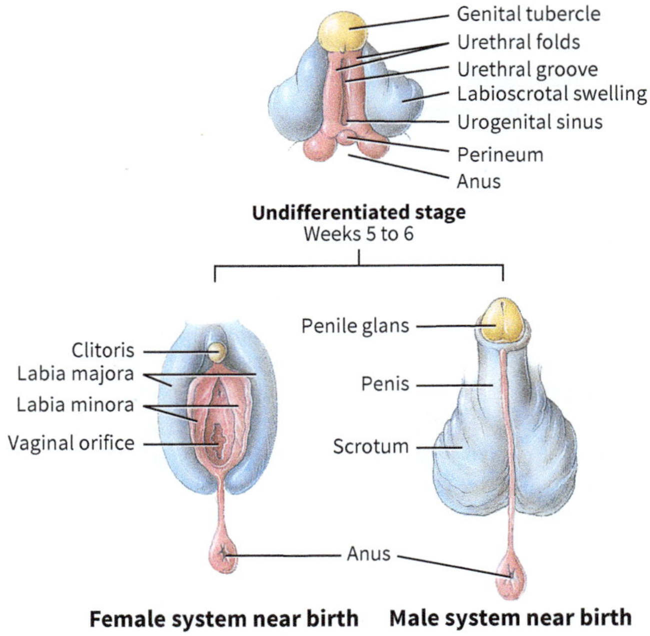

FIGURE 10.8 Human sexual divergence *in utero* For every female structure, such as the clitoris, there is a corresponding structure from the male, such as the glans, that shared the same tissue bed.

as the SRY protein. In the absence of a Y chromosome containing the *SRY* gene, the default developmental path will lead to a female. When the Y chromosome is present, the SRY protein changes the developmental path from Müllerian to Wolffian, resulting in the development of a male embryo with testes.

As early as the eighth week of gestation, these rudimentary testes begin to secrete testosterone, and uterine testosterone levels reach their peak between the tenth and fifteenth weeks. In addition, the SRY protein stimulates the testes to secrete Müllerian-inhibiting hormone (MIH), also known as anti-Müllerian hormone, which has a defeminizing effect on the embryo. The paired positive action of testosterone and the inhibitory action of MIH creates the typical male of our species.

For those embryos destined to become male, the Müllerian ducts atrophy. In the absence of the action of the *SRY* gene, a corresponding atrophy of the Wolffian ducts occurs for embryos that will become female. One would assume that fetal ovaries secrete estrogens just as the testes of the fetal male secrete testosterone. The evidence is that they do, but disentangling the effect of fetal estrogen from that produced by the mother's own ovaries is difficult. One implication of the organizational effects of genes and hormones is that biological sex is decided fairly early *in utero*.

Homologous genital structures We said at the start of this section that sexual differentiation was the result of genetic and hormonal processes. By examining the divergence of genital structures during development, one can readily see the organizing effects of genetics and hormones that lead to different outcomes. As prenatal development continues, structures begin to diverge under the influence of hormones, with homologous (corresponding) structures on the male and female models (Figure 10.8). For example, a ridge of tissue naturally develops into ovaries unless the *SRY* gene codes for the presence of the SRY protein, in which case that structure develops instead into testes. The genital tubercle becomes the clitoris in females but becomes the penile glans in males. The urethral groove folds into what will become the penile spongy tissue in males or the labia minora in females. The labioscrotal swelling develops into the labia majora in females, whereas that same tissue becomes the scrotum in males.

Hormonal Influences on Prenatal Sexual Development MCAT

Sex hormones drive the physical development of male or female urogenital differentiation *in utero*. They increase in volume during puberty and decrease in late adulthood, as we will see in the next section. In this section, we will discuss the organizing effect of androgens and estrogens on prenatal sexual differentiation.

Try It Out | The D2:D4 Ratio

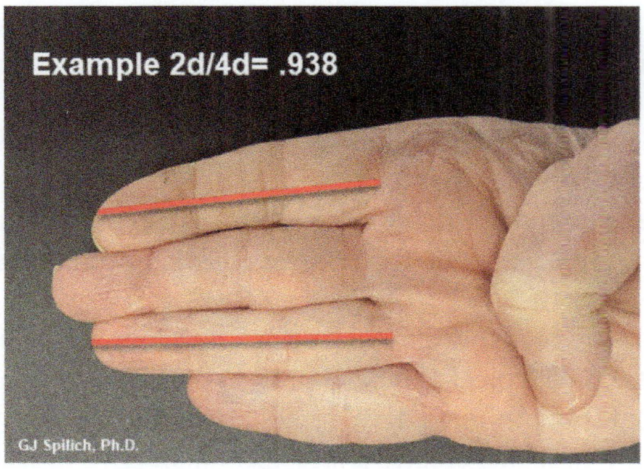

Exposure *in utero* to androgens changes the ratio of the second and fourth fingers, known as the D2:D4 ratio. The D2:D4 ratio is considered to be a rough measure of prenatal androgen exposure, with greater digit disparity correlating with greater uterine androgen exposure. Epidemiological evidence also suggests that the D2:D4 ratio correlates with some health and behavioral traits later in life. Complete the worksheet **The D2:D4 Ratio** to explore this for yourself.

Androgens Named for the Greek *andro* for male and *genos* for creator, an **androgen** is any natural or synthetic steroidal hormone whose action creates both primary and secondary male sexual characteristics. Both males and females secrete androgens although in different amounts, and their source is the testes, ovaries, and adrenal glands. Androgens operate by stimulating androgen receptors throughout the body. The primary androgens are testosterone, dihydrotestosterone, and androstenedione. In this **Try It Out: The D2:D4 Ratio**, you can explore an easily measurable trait that is related to prenatal androgen exposure.

Testosterone is the main male hormone. It is essential for the development of male genitalia and secondary characteristics such as facial hair, increased body muscle mass, deepening voice, and higher bone density. In both males and females, it plays an activating role in libido and also in preventing osteoporosis, a disorder that causes bones to become weak and brittle. Dihydrotestosterone (DHT) is created in the body from testosterone and is in fact more powerful than testosterone. DHT is essential in the development of the male genitalia. Androstenedione is an androgen created by the testes, adrenal cortex, and ovaries. It is converted to testosterone and also estrone, an estrogen. In the past, it has been used as a supplement to build muscles, although there is no evidence that small doses have any effect on muscle enhancement. Its use among professional athletes is illegal, as is its recreational use.

Estrogens Although estrogen is often considered as a single hormone, in fact, **estrogens** are a category of hormones that are responsible for the development of the female reproductive system and secondary sex characteristics. The major estrogens are estrone, estradiol, and estriol, with estradiol being the most powerful. We now know that estrogens do much more than guide female sexual development. In males, estrogens modulate sperm production. In both males and females, estrogens affect mood and bone density as well as the ability to learn new concepts and recall what has already been learned.

Hormonal disruptions of sexual differentiation You saw earlier in our discussion that genes drive sexual differentiation by stimulating and controlling the expression of hormones that then act on the brain and other bodily structures. In the prior section we considered what happens when the individual deviates from the common XX/XY allosomal dichotomy. We now turn to the other side of the mechanism to investigate the course of sexual development when hormonal disruptions occur.

In some individuals, a genetic anomaly on the X chromosome leads to the formation of insensitive androgen receptors, a condition called androgen insensitivity syndrome (AIS). The extent of the androgen receptor insensitivity varies from partial to complete. In XY individuals, the result is a male *in utero* whose testes secrete both androgens and anti-Müllerian hormone, but the androgens have no effect on sexual development. Because of the lack of the masculinizing effect of androgens, testes do not fully develop, and the vas deferens, seminal vesicles, and prostate typically do not appear. However, due to the defeminizing influence of anti-Müllerian

Kirstin Sinclair / Getty Images Entertainment / Getty Images

FIGURE 10.9 Hanne Gaby Odiele Hanne Odiele is a Belgian-born intersex model whose image has graced the covers of fashion magazines and who has been the face of many fashion industry advertising campaigns.

hormone, normal female reproductive structures are also affected, so it is common for the individual to have no uterus or Fallopian tubes and a shallow vagina.

Individuals with AIS often consider themselves to be neither male nor female or both male and female, and as such they are sometimes referred to as **intersex** individuals. Although they cannot reproduce, individuals with AIS do experience normal sexual function. Hanne Gaby Odiele is an international model, a representative for many fashion product lines, a spouse, and, as an individual with AIS, a spokesperson for intersex individuals (**Figure 10.9**).

The counterpoint to AIS is estrogen insensitivity syndrome, an extremely rare condition. In fact, less than a half-dozen cases have been identified in males and females.

Another hormone-associated disorder of prenatal sexual development is persistent Müllerian duct syndrome. In a typical male, the presence of androgens and anti-Müllerian hormone prevent the continued development of the Müllerian system in favor of the Wolffian system, which creates a biological male. In extremely rare cases, the Müllerian structures persist so that the male embryo develops some aspect of Fallopian tubes, a uterus, and possibly even a part of a vagina along with male genitalia.

You can now see that although many people fit neatly into the traditional male–female dichotomy, not all people do. We can take a moment to consider what it means to be intersex. Earlier in our discussion of sexuality, you learned that biological sex and gender identification are related concepts but not synonyms. Expectations of male and female behavior change between cultures and over time. Along with this mix of biology and culture is the question of sexual orientation, which we will discuss shortly. It is easy to see that drawing clear lines based on biological sex alone is not possible and may not even be helpful.

10.2 Before You Go On

Section Summary

Humans have 23 chromosome pairs including two sex-determining allosomes, XX for females and XY for male. The default model for sexual organization in humans is female, but if acted upon by the *SRY* gene, the primordial ovarian tissue develops instead into testes, which secrete testosterone and anti-Müllerian hormone, inhibiting the growth of female structures and driving development toward the male model. These changes are triggered by the action of androgens and estrogens. Most humans develop into either the male or the female variant, but as with any physical mechanism, variations occur that may lead to developmental paths that are neither clearly male nor female.

Comprehension Check

1. Of our 46 chromosomes, each human has _____ that determine our biological sex.

- **a.** two allosomes
- **b.** four allosomes
- **c.** two autosomes
- **d.** one autosome in males and two autosomes in females

2. The chromosomal arrangement _____ codes for female, and _____ codes for male.

- **a.** XY; XX
- **b.** XX; XY
- **c.** XX; YY
- **d.** X; Y

3. There are two urogenital developmental pathways for humans, the _____ system for women and the _____ system for men.

- **a.** Müllerian; Wolffian
- **b.** Wolffian; Müllerian
- **c.** Red Queen; Müllerian
- **d.** XX; XY

4. Why are the penile glans/clitoris and the scrotum/labia majora considered to be homologous?

- **a.** They represent a male-centric perspective on sexual anatomy and function.
- **b.** They represent anatomical understanding (from *homo*/human and *logia*/knowledge).
- **c.** They share a common antecedent, or precursor structure.
- **d.** Both begin as a primitive male structure that hormones alter to create a female.

5. How do individuals with androgen insensitivity syndrome differ from typical males?

- **a.** They are really females that have been changed by androgens.
- **b.** They have fully developed male and female genitalia.
- **c.** Their secondary sexual characteristics such as beards and body hair are missing.
- **d.** Without the effect of androgens, only anti-Müllerian hormone has an effect.

10.3 The HPG Axis and Life-Span Sexual Development

LO 10.3 Identify hormonal influences on sexual development over the life span.

At this point, you have a clear grasp of the organizational effects of genetics and prenatal hormones on the sexual dimorphism of humans. These effects continue through adolescence, adulthood, and beyond, when they lead to activation of behaviors.

In previous chapters where we discussed homeostasis and the endocrine system, you saw the importance of the link between the hypothalamus and the pituitary in controlling bodily functions through the adrenal glands. We now look at another example of hypothalamic–pituitary regulation as we discuss the hormonal regulation of puberty, sexual maturation, and adult sexual behavior through the hypothalamic–pituitary–gonadal (HPG) axis. We now turn to an overview of life-span sexual development, starting with puberty.

Puberty MCAT

Around ages 10 to 11 for females and 11 to 12 for males, the onset of **puberty** brings structural changes that have major behavioral implications. Puberty begins when hormonal and physical changes mature a child's body to the point where they are capable of sexual reproduction. Interestingly, the average age of puberty is younger now than in the 19th century, when it was 15 for girls and 16 for boys. This change is likely due to improvements in nutrition, more adipose tissue deposits related to a more sedentary lifestyle, and also possibly environmental exposure to chemicals that mimic the action of estrogens.

In a prior section, you learned about the role of hormones in organizing biological sexual assignment. At puberty, hormones again begin to organize structural changes and activate behavior.

The HPG and puberty

The mechanism that initiates puberty is moderately complicated and involves both structures and hormones. As always, we see both excitatory and inhibitory mechanisms that operate in a negative feedback mechanism to regulate the activation of puberty. The evidence suggests that puberty is initiated by the activity of the *Kiss1* gene, which stimulates cells in the hypothalamus to code for proteins known as kisspeptins (named not in reference to romance but in honor of a laboratory in Hershey, Pennsylvania). Kisspeptins in turn stimulate the hypothalamus to release **gonadotropin-releasing hormone (GnRH)** in pulses. This pulsatile stimulation causes the anterior pituitary to release **luteinizing hormone (LH)** and **follicle-stimulating hormone (FSH)**, which then stimulate the gonads (ovaries and testes) to release estrogens and testosterone. These hormones, in turn, begin to organize the reproductive system so that sexual behavior can begin.

In order for the HPG axis to properly initiate and maintain the organizational effects of puberty, there must be inhibitory processes. The release of estrogens, progesterone, and testosterone from the gonads and the release of prolactin from the anterior pituitary also act to inhibit the release of GnRh from the hypothalamus by inhibiting the activity of kisspeptin. In addition, specialized cells in the gonads (granulosa cells in the ovaries and Sertoli cells in the testes) secrete the aptly named protein inhibin, which inhibits LH and FSH production in the anterior hypothalamus.

A short summary of the complex feedback loop that initiates and maintains the emergence of sexual maturity is in order. In response to a genetic signal, hormones released from the hypothalamus initiate a cascade of other hormones from the anterior pituitary that in turn activate the gonads to release estrogens and testosterone. These processes (illustrated in Figure 10.10) have several inhibitory mechanisms to keep the entire system in check. The result is that our bodies develop primary and secondary sexual characteristics that are the hallmark of a sexually mature human.

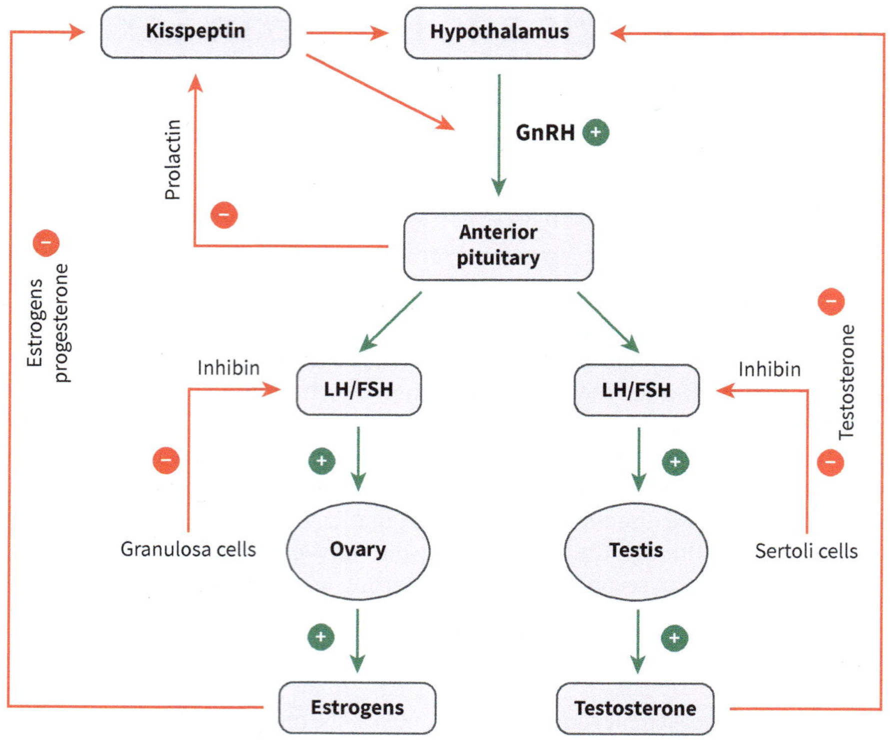

FIGURE 10.10 Hormonal regulation of puberty Kisspeptin initiates the release of GnRH from the hypothalamus, which stimulates the anterior pituitary to release LH and FSH. These, in turn, stimulate the gonads to release sex hormones that are responsible for developing mature primary sex characteristics as well as secondary sex characteristics. Several inhibitory mechanisms provide negative feedback to keep the system in check.

You can see that there is a complex web of hormones that regulate the initiation and organization of puberty, and therefore disruptions of the mechanism can occur at several points. Consider the implications of outside influences on puberty in **Think About It: Environmental Effects on Puberty Onset**.

Think About It | Environmental Effects on Puberty Onset

Phovoir / Shutterstock

Puberty can be a confusing time for young adolescents; your body is changing dramatically, and the changes can be disturbing. People may look at you differently than they did when you were a child, and taking on more adult responsibilities can be daunting. Emotions may swing wildly, and relationships that once were simple may now become complicated by romantic or sexual feelings.

From an evolutionary perspective, puberty should not occur until an individual has the cognitive maturity to deal with these issues. The evidence suggests that in our distant past, many adolescents did not reach the age of reproduction until age 15 or even as late as 17. Over time, the greater availability of food that came with civilization led to earlier puberty. Yet, there is another factor that may be driving the age of puberty down, and that is the presence of environmental agents that affect hormones involved in the HPG axis that regulates puberty.

Researchers investigating the advancing age of puberty have identified several environmental agents that act to disrupt normal endocrine function by binding to receptor sites or suppressing hormone synthesis. These chemicals, natural and synthetic, are called endocrine disruptors.

Examples are industrial chemicals known as phthalates, pesticides, and hormones in birth control products that are excreted and end up in wells and municipal water systems. In the past few years, municipal water systems have begun to filter estrogens from the water they deliver to customers. Phthalates are more difficult to remove.

QUESTIONS

1. **We are now seeing some girls reaching puberty as young as age 8 and boys as young as 9. In your opinion, are these appropriate ages to deal with the stresses of puberty? What age do you think would be most appropriate for puberty? Explain your answer.**
2. **Search online using the terms "endocrine disrupters and puberty" and describe one source of endocrine disrupters that people in your area might be exposed to.**
3. **What affect might earlier age of puberty have on society, including in schools and families?**

During puberty, hormones of the HPG axis stimulate the maturation of the gonads. The most visible changes are the development of secondary sex characteristics, which signal to others that the individual is approaching maturity.

Secondary sex characteristics We previously identified primary sex characteristics as those an individual is born with that are related to reproduction. In contrast, secondary sex characteristics are primarily physical traits that emerge with maturity in sexually dimorphic species. Examples in other species are the luxurious plumage of the male peacock, the mane of the lion, and the antlers of a stag.

Evolutionary biologists theorize that secondary sex characteristics are the result of evolutionary pressures that amplify differences between the sexes in an effort to restrict breeding to the fittest members of the species. In the case of the peacock or lion, spectacular plumage or a lush mane might signal vitality and a surplus of resources to a receptive female, and in the case of the stag, the horns may signify disease resistance. Females who select males with a high level of vitality or fitness will be more likely to have offspring who survive to reproduce. This creates a cycle that amplifies sexual dimorphism over generations.

In humans, secondary sex characteristics emerge during puberty and serve to both signal sexual maturity and differentiate males from females. Secondary sex characteristics common to both males and females include pubic hair and armpit hair. Table 10.2 lists other secondary sex characteristics.

Pubertal behavior changes The flood of hormones that accompanies puberty also drives behavioral changes. Puberty initiates a restructuring of social centering; that is, the adolescent's reference group often changes from family to social cohort. When compared with their prepubertal selves, adolescents strive for contact with their peers and seek out potential romantic partners. These changes are often accompanied by a desire for sensation-seeking and a search for new experiences and feelings. This natural striving for identity may lead to conflict between the adolescent and the family, especially the parents.

TABLE 10.2 Male and Female Primary and Secondary Sex Characteristics

	Male	Female
Primary	Penis, scrotum, testes	Vagina, ovaries
Secondary	Typically larger body	Typically smaller body
	Facial and chest hair	Less body hair
	Adam's apple	Breast enlargement
	Faster muscle development	Slower muscle development
	Broader chest	Broader hips
	Possible balding	Thicker hair on head
		More body fat

Of all of the hormonal changes associated with the development of sexual maturity, likely none is as dramatic or profound as menarche.

Hormonal Control of the Menstrual Cycle MCAT

For female adolescents, the start of menstruation, or **menarche**, represents the beginning of fertility. As such, it is considered a rite of passage in many cultures. Developing a firm grasp of the menstrual cycle can be challenging, so it is helpful to remember that there are two parts: the **ovarian cycle**, which involves the maturation and then release of the egg; and the **uterine cycle**, which involves the preparation of the uterine lining for the implantation of the egg. The two cycles run simultaneously, experiencing different effects of the same hormones.

The ovarian cycle Within the ovarian cycle there are also two distinct phases. During the follicular phase, the follicle holding the egg (ovum) matures, and during the luteal phase, the egg is released from the follicle to travel down the Fallopian tube and into the uterus. In Figure 10.11 you see that tiny, immature follicular cells are exposed to FSH and estrogens during the follicular stage, which causes some follicles to mature into follicular cells holding an egg ready to be released during ovulation. Changing concentrations of FSH, estrogens, LH, and progesterone act as triggers for the transition from phase to phase. For example, estrogens rapidly peak and then drop just prior to ovulation; simultaneous with the sudden drop of estrogens, levels of FSH and LH peak. Note also that body temperature fluctuates very slightly but reliably in concert with ovulation, a signal that some women use when trying to conceive.

Interestingly, a female fetus has as many as 6 to 7 million eggs *in utero*. By the time she is born, however, her ovaries hold as few as 1 million eggs, and those are all the eggs she will ever have. By puberty, only around 300,000 eggs remain, and of these, only around 300 to 500 will be ovulated during her lifetime.

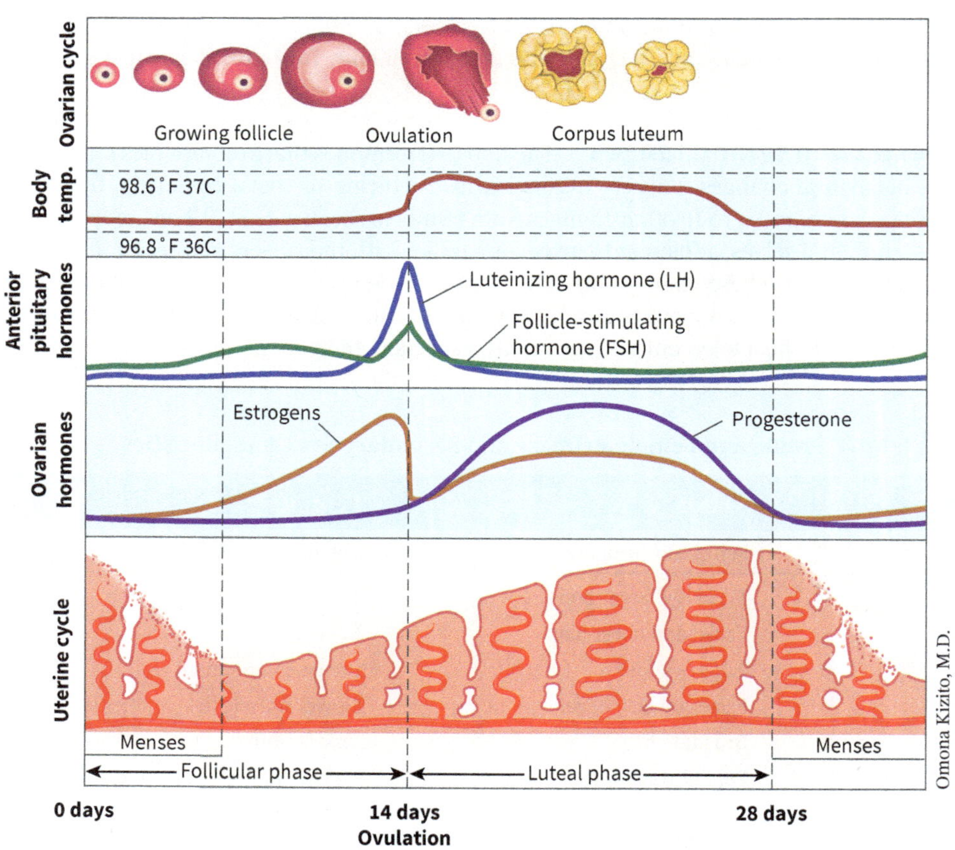

FIGURE 10.11 **The menstrual cycle** In response to hormones released by the pituitary, the ovum is prepared for release for fertilization and the endometrium is prepared to accept the fertilized ovum for implantation and growth.

The uterine cycle The second and equally important piece of the menstrual cycle is the uterine cycle, which consists of three parts: menstruation, the follicular phase, and the luteal phase. The cycle begins with the first day of a woman's period; this typically lasts four to five days as the uterus sheds its lining.

During the follicular phase, the uterine lining (also known as the endometrium) develops or proliferates to a point where it can support the implantation of a fertilized egg. Ovulation occurs as the uterine lining is approaching its optimal state; ideally, the fertilized egg traveling down the Fallopian tube reaches the uterus at the point in time when the endometrial lining is at its most developed.

The luteal phase starts at or around the same time as ovulation, when the corpus luteum (yellow substance) is formed in ovaries. This substance secretes progesterone and estrogens, which further stimulate the growth of the endometrial lining. All of this activity is designed to support the successful implementation of a fertilized egg in the endometrial lining of the uterine wall. In months when no egg is implanted, the uterine lining begins to disintegrate around day 27, the corpus luteum dissolves, and the cycle is ready to begin again.

Occasionally, an ectopic pregnancy occurs, in which a fertilized ovum implants in the wall of a Fallopian tube, a location that is incapable of providing sufficient blood supply to the developing embryo. Between the sixth and sixteenth week after conception, the structure holding the developing embryo ruptures. Ectopic pregnancies are fatal to the fetus, and they can be fatal to the mother due to internal bleeding if not treated quickly.

Before we move on, take a moment to consider the implications of the complexity of the human reproductive system. Interactions among the hypothalamus, pituitary, and ovaries on the one hand and the precise timing of their release of hormones on the other hand are all essential to successful conception. Roughly 10% of all women who are trying to become pregnant experience difficulties conceiving and you can now understand how even a small disruption in the system can result in infertility.

For many mammals, especially humans, the end of pregnancy does not spell the end of parents' interaction with their offspring. How do neurohormonal mechanisms activate parental behavior? That is the focus of our next section.

Hormonal Activation of Parental Behavior MCAT

You have seen in our discussion of sex and reproduction that hormones have both organizational effects on physiology and activating effects on behavior. One purpose of sexual behavior is to create offspring, but the story does not end there. For many species, especially mammals, their offspring are not capable of surviving independently for some time after birth, and therefore some parental behavior is required for their care. You will not be surprised to learn that hormones play a central role in activating human parental behaviors (**Figure 10.12**).

During pregnancy, the mother's neuroendocrine system is bathed in elevated levels of progesterone secreted by the adrenal cortex and the ovaries, but these levels drop as part of the neuroendocrine signal to prepare the mother for birth. In response, the endocrine glands increase their secretion of estrogen and also oxytocin. Oxytocin has been found to be associated with feelings of empathy and personal warmth. Research reports that mothers who displayed a high level of physical and emotional contact with their newborn had high levels of oxytocin postcontact, whereas mothers showing less affection had the opposite. The evidence is that physical contact between mother and child stimulates the release of oxytocin in the mother, which then creates a positive feedback loop that leads to a strong maternal bond.

Monkey Business / Adobe Stock

FIGURE 10.12 Infant with parents Hormonal mechanisms activate parental behavior in both males and females.

The hormonal activation of maternal behaviors seems obvious because mothers are the ones who give birth and lactate, but what about fathers? In some species, the paternal role ends immediately or soon after conception, but that is not true of humans. In nonhuman biparental species (that is, species where both male and female play significant roles in parenting postpartum) an increase in oxytocin has been observed in the males. In humans, research investigating parental behavior in males who are judged to be emotionally attached to the child has found high levels of oxytocin; a drop in testosterone, which is associated with libido; an increase in prolactin, which is normally expressed postorgasm; and an increase in vasopressin, which has been seen to stimulate bonding behavior. One interpretation of these observations is that postpartum, the male neuroendocrine system tends to activate behaviors that build a bond between the father and the mother–child dyad. Evolutionarily, continued paternal support of the mother and offspring would improve the survivability of that offspring.

stevecoleimages / Getty Images

FIGURE 10.13 **Renewed sexuality with age** Once past the age of reproduction and free of daily family responsibilities, many couples rekindle their romance.

Sexuality and Aging

Every beginning has an end, and that is true of fertility as well. With age, hormone production changes, and the ability to reproduce declines for both males and females, although sexual behaviors continue. In fact, many older couples report that with their children out of the house, either retired or with careers well established, and pregnancy not an issue, romance is enhanced (**Figure 10.13**).

For females, the end of fertility occurs at **menopause**, when the menses cease. Menopause is defined as 12 consecutive months without menses. This typically occurs around age 51, although there is considerable variation in that time. Because hormones play such an important role in metabolic and emotional functions, menopause is often accompanied by physical symptoms such as hot flashes, vaginal dryness, and sleep disturbances as well as mood changes.

Male fertility and sexual behavior are also affected by age, although in different ways. Erectile dysfunction (ED) is defined as the inability to sustain an erection sufficient for sexual intercourse. ED is the result of both declining testosterone levels and reductions in blood flow due to loss of cardiac fitness. Individuals can experience ED at any age, but it is found in 5% of males between the ages of 40 and 50 and in over 50% of all men older than 70.

Furthermore, the quantity and quality of male sperm begins to decline after age 50, and there is some evidence that the offspring of men over the age of 50 are at greater risk of schizophrenia and autism spectrum disorder, even when the mother is much younger. A causal relationship between advanced paternal age and a greater risk of developmental problems has not been established.

We can end our discussion with the realization that sexual behavior is a part of life that does not end at middle age but can continue on into one's later years.

10.3 Before You Go On

Section Summary

During puberty, hormones activate the development of secondary sex characteristics and the start of reproductive capabilities. The human menstrual cycle is controlled by hormones released by the pituitary, and the signaling of events occurs through relative contributions of FSH, LH, estrogen, and progesterone. Over the 28 days of a typical cycle, the endometrium is prepared for a fertilized egg to arrive at an optimal time, and when that does not happen, this tissue layer is lost, and the process begins again. Postpartum neuroendocrine processes build and maintain emotional bonds among mother, father, and child. Declining hormone levels with age affects the fertility of males and females.

Comprehension Check

1. The control of the menstrual cycle is accomplished mainly by what?

- **a.** fluctuating levels of estrogen, the female hormone
- **b.** fluctuating levels of FSH and LH
- **c.** interactions among estrogens and androgens
- **d.** fluctuations in FSH, LH, estrogens, and progesterone

2. The ovum is released from the ______, which is located in the ______.
   a. follicle; ovary
   b. follicle; Fallopian tube
   c. corpus luteum; uterus
   d. ovary; uterus

3. When parents interact physically and emotionally with their offspring, ______ is released in ______, which creates a bond.
   a. progesterone; the infant
   b. oxytocin; the mother only
   c. oxytocin; both parents
   d. progesterone; both parents

4. What is a frequently observed change in sexuality for those past middle adulthood?
   a. an increase in sexual desire but only for men
   b. an increase in sexual desire but only for women
   c. vaginal dryness for women and erectile difficulties for men
   d. the loss of interest and desire starting age 55

10.4 Sexual and Reproductive Behaviors

LO 10.4 **Describe** the diversity of human sexual response, including sexual orientation and attraction.

At this point in our discussion, you are familiar with human genital anatomy and understand the pre- and postnatal course of sexual development through puberty and adulthood. Now we turn to the behaviors that support and accompany sexuality and reproduction, beginning with an exploration of the human sexual response.

The Human Sexual Response MCAT

No discussion of sex and reproduction would be complete without considering the behavior we refer to as the human sexual response. For both females and males, there are four stages to this response: excitement, plateau, orgasm, and resolution. The response varies somewhat, however, according to sex. We will begin by looking at the female sexual response and then compare the male response.

The female sexual response The four stages of female sexual response are shown in Figure 10.14. During the excitement phase, stimulation causes glands in the vulva and vagina to secrete lubricants, and the labia and clitoris engorge with blood (**step 1**). During plateau, the outer third of the vagina constricts and is engorged with blood. The inner third expands, and the clitoris retracts (**step 2**). During orgasm, the vagina, uterus, and anus all rhythmically contract (**step 3**). Finally, during resolution, blood leaves the labia and orgasmic platform, and the clitoris fully retracts (**step 4**).

The male sexual response In males, the four stages of excitement, plateau, orgasm, and resolution follow a similar course (Figure 10.15). During the excitement phase, the corpus cavernosa and the corpus spongiosum fill with blood and the penis erects (**step 1**). During plateau, the penis is fully erect, and a small amount of fluid from the bulbourethral glands may be released (**step 2**). During orgasm, sperm mixed with fluid from the bulbourethral and prostate glands is rhythmically expelled from the penis (**step 3**). Finally, during resolution, blood leaves the corpus cavernosa, and the penis returns to a flaccid state (**step 4**).

Although the human sexual response cycle is similar for males and females, it is not identical, as Figure 10.16 illustrates. Males pass through the phases of excitement, plateau, and orgasm followed by a refractory period during which they are unable to achieve an orgasm and then a resolution phase during which a return to excitement is more difficult to achieve. In contrast, many females can achieve more than one orgasm with appropriate stimulation, and do not show a physiological refractory period postorgasm.

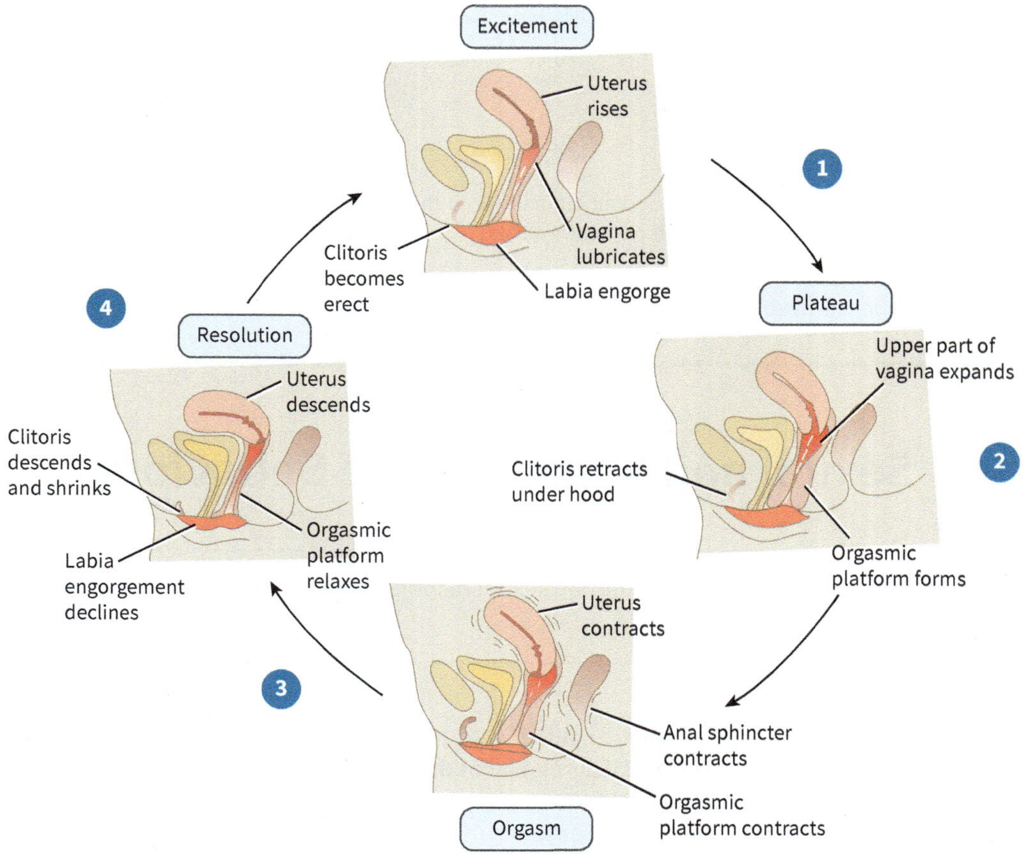

FIGURE 10.14 The female sexual response The female sexual response consists of four stages.

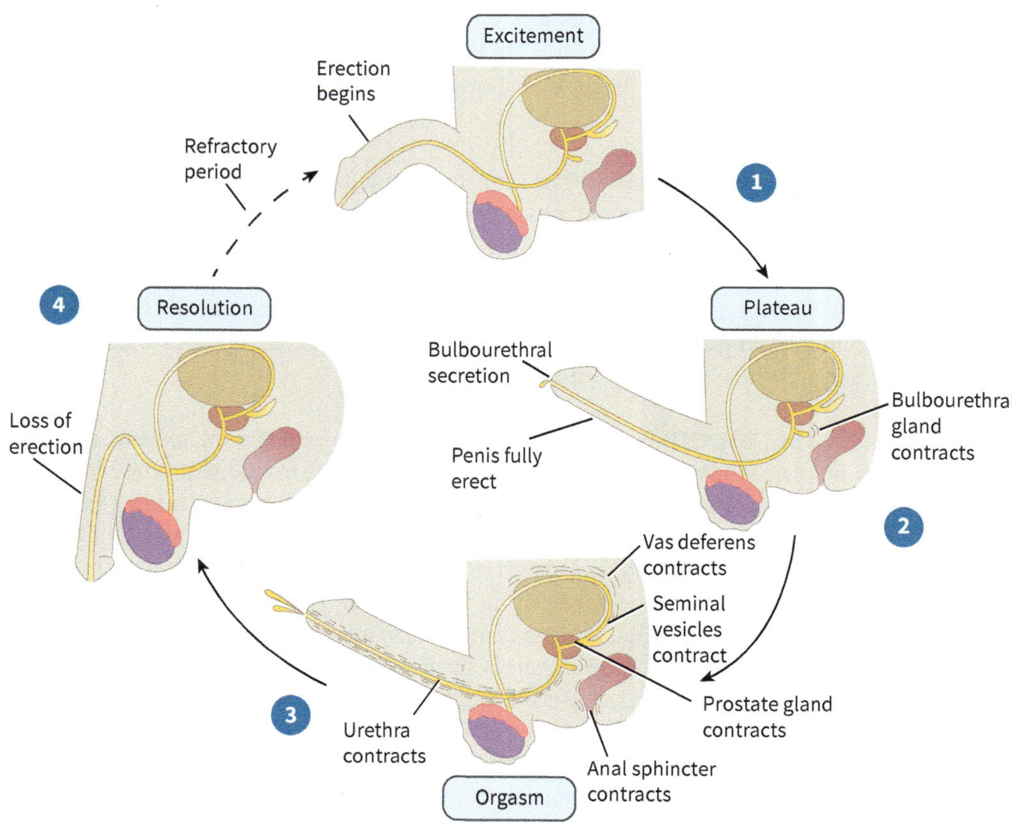

FIGURE 10.15 The male sexual response The male sexual response consists of the same four stages as the female response.

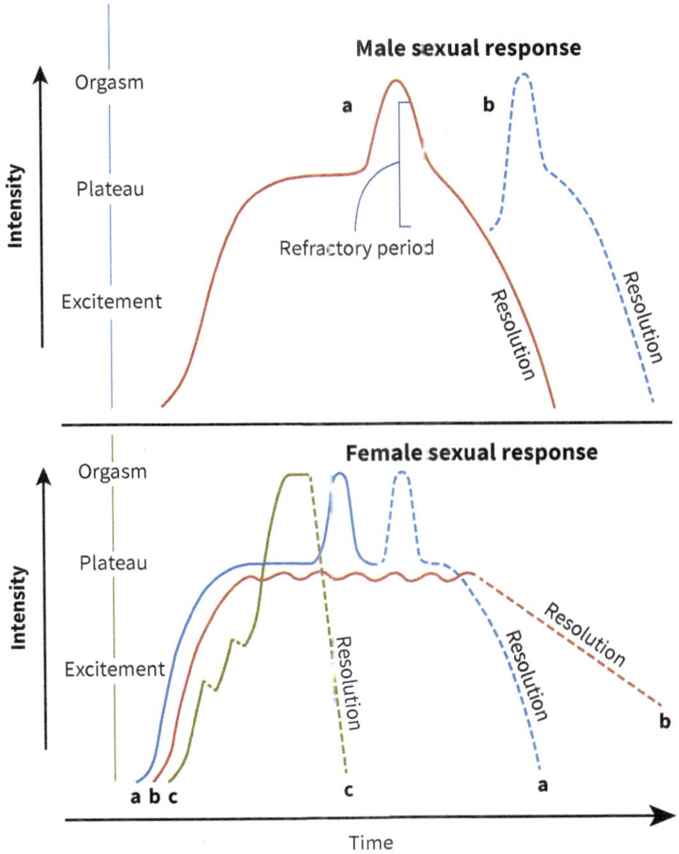

FIGURE 10.16 Male and female sexual response cycles The male sexual response cycle consists of excitement, plateau, orgasm, and resolution. After an orgasm (a), a refractory period occurs before a male can have second orgasm (b). In contrast, there are several possible response cycles for females. Person a reaches plateau and experiences two orgasms. Person b reaches plateau but does not orgasm. Person c reaches plateau and has one orgasm.

Evolutionary explanations for the human orgasm Throughout our study of behavioral neuroscience, we always ask the question, "How?" but we often also ask, "Why?" What is the evolutionary purpose of the human orgasm? The answer is not clear, but there are some facts to consider.

Clearly, species that reproduce sexually face a biological imperative to mate, and pleasure may support that essential process. We might start by asking whether other animals experience pleasure during sexual reproduction in addition to being compelled by their nervous systems to mate. That is likely to be an unanswerable question, but we can say with some assurance that more complex animals such as dogs, cats, and other primates act as if they do feel pleasure during sex. We also note that for many mammals, the female is only receptive when she is ovulating. That is not true of humans and many other primates, who are receptive to sex and experience pleasure even when pregnancy is not possible. This suggests that there may be something more at work than the biological drive to reproduce.

Another possibility is that in addition to its reproductive function, the orgasm may improve the survivability of the offspring postpartum. Some primates, and particularly humans, appear to use sex to build a social bond between parents. The human endocrine system releases oxytocin during orgasm, which is known as the cuddle hormone because of the loving feelings it creates. A strong bond between parents may be evolutionarily important because, unlike many mammals whose offspring have a short period of dependency before they can live on their own, newborn humans require years of care before they can feed and maintain themselves. The female of our species also requires assistance and care after birthing, which many other species do not. A bonded male will be more invested in the care and survival of the offspring.

So far we have been talking about the evolutionary and physiological aspects of sex, but none of these are particularly personal. By contrast, sexual orientation is an intensely personal matter. How is sexual orientation organized?

TABLE 10.3 Six Categories of Sexual Orientation

Heterosexual	Attracted to people of the opposite sex
Homosexual	Attracted to people of the same sex
Bisexual	Attracted to both men and women
Pansexual	Attractions are not limited by biological sex or gender identity
Asexual	No sexual attraction to anyone. Asexuality differs from celibacy in that celibacy is a choice and asexuality is not.

Sexual Orientation MCAT

In our discussion of sex and reproduction, you saw that within the first few weeks of a pregnancy, genetics and hormones organize the course of subsequent sexual development. However, humans are complex organisms, and as we pointed out in our earlier discussion, there is a difference between the concepts of sex and gender. Now we add the additional complicating factor of **sexual orientation**, which describes the gender(s) to which a person is sexually attracted.

Earlier in our discussion of sexuality, we pointed out that biological sex and gender orientation are different and more complex than simple binary categories, and the same is true of sexual orientation. There is still much to learn about the mechanisms that organize sexual orientation, but Table 10.3 provides one simple organization.

What determines sexual orientation? People can be subject to harassment, discrimination, or hate crimes because of their sexual orientation, so it is worthwhile to consider what the contributing factors are. Some possibilities under investigation by researchers are brain differences, genetics, and birth order.

Brain differences It is reasonable to ask in what ways male and female brains differ and if any of those differences are relevant to understanding sexual orientation. Researchers investigating sexual dimorphism discovered that several neural structures appear to organize sexual behavior.

In animal research, electrical stimulation of the medial preoptic area (MPA) in rats has been shown to be related to sexual behavior. For example, damage to the MPA in male rats led to the elimination of male sexual behavior. Other animal research has focused on the volume of a region of the hypothalamus known as the sexually dimorphic nucleus of the medial preoptic area (SDN-POA). The SDN-POA is as much as seven times larger in male than in female rats, and this dimorphism is thought to be common in mammals. Animals whose SDN-POA was damaged or destroyed exhibited unusual sexual behavior. In other studies, the volume of the SDN-POA seemed to be related to sexual attraction. Taken as a whole, these results suggest that certain neural centers are intimately involved in sexual behavior in animals, that these brain areas are sexually dimorphic, and that they are involved in the orientation of sexual behavior.

Research has also shown dimorphism related to sexual orientation in human brains. Comparisons of the third interstitial nucleus of the anterior hypothalamus (INAH-3) show that the INAH-3 is 2 to 2.5 times larger in heterosexual males than in heterosexual females. More pertinent to our discussion is that the INAH-3 in homosexual males has been reported to be much smaller than that of heterosexual males and well within the range of heterosexual females (Figure 10.17).

That heterosexual males and females show a dimorphism in the INAH-3 has been interpreted as evidence that sexual orientation has a biological basis. Evidence that homosexual males present the same INAH-3 pattern as heterosexual females is additional evidence of a biological basis for sexual orientation. You can read more about the person who discovered this dimorphism in this **Advances in Behavioral Neuroscience: Simon LeVay and the Neural Basis of Sexual Orientation**.

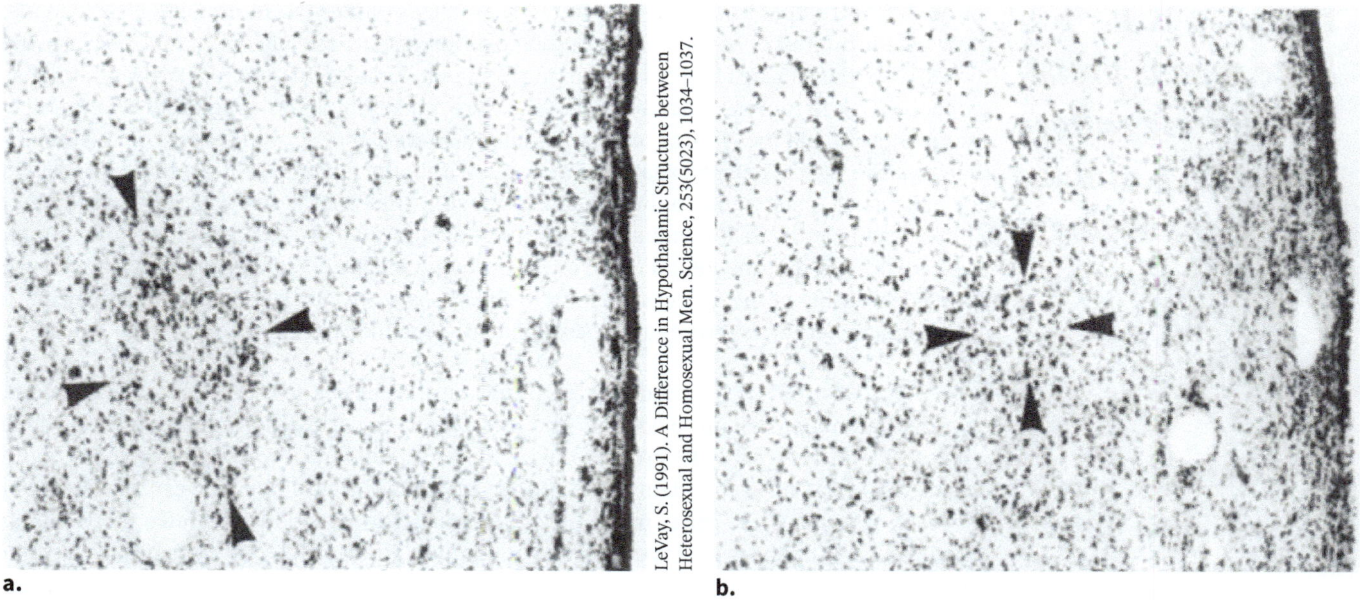

LeVay, S. (1991). A Difference in Hypothalamic Structure between Heterosexual and Homosexual Men. Science, 253(5023), 1034–1037.

LeVay, S. (1991). A Difference in Hypothalamic Structure between Heterosexual and Homosexual Men. Science, 253(5023), 1034–1037.

FIGURE 10.17 **The INAH-3 in heterosexual males and homosexual males** The INAH-3 of a heterosexual male (a) is more than twice as large on average as that of a homosexual male (b).

Advances in Behavioral Neuroscience | Simon LeVay and the Neural Basis of Sexual Orientation

Simon LeVay

Simon LeVay is a neuroscientist whose research on the neurological basis of sexual orientation is considered to be groundbreaking. LeVay was born and educated first at Cambridge in the United Kingdom and then pursued a Ph.D. in neuroanatomy at Göttingen in Germany and did postdoctoral research at Harvard. Although he has published widely on a number of topics in neuroscience, he is best known for his work on the neural basis of sexual orientation.

By the 1990s, it was well known that the INAH-3 was significantly larger in males than in females, and LeVay wondered if the size of the INAH-3 in gay males reflected their biological sex or their sexual orientation. To answer this question, LeVay autopsied the brains of 41 adult individuals whose sexual orientation was known and found that the INAH-3 in gay males was reliably smaller than in heterosexual males and in fact was within the range typically seen for heterosexual females. LeVay's work led many in the field to consider the INAH-3 as the human analog of the SDN-POA in mammals.

LeVay's work was widely covered in the press as suggesting that there was a gay center in the brain, a claim that LeVay himself expressly rejected. LeVay proposed that his research suggested several possibilities. One is that this dimorphism exists prenatally or early in life and exerts some effect on sexual orientation. A second interpretation is that sexual feelings or behavior early in life change the structure of the INAH-3, and a third is that some other event is responsible for both the changes in the INAH-3 and sexual orientation.

It should be noted that one criticism that was leveled against LeVay's work was that all the brains in his homosexual sample came from individuals who had died of AIDS, so it was possible that HIV infection had somehow diminished the size of their INAH-3. Subsequent work proved that criticism to be meritless.

Questions

1. **Based on what you have read so far, do you consider sexual orientation to be entirely neurological, entirely a matter of personal preference, or somewhere in between? Support your answer.**
2. **All of the work on the INAH-3 up to this point has compared the brains of adult heterosexual individuals with that of adult homosexuals. The difficulty is that by the time of data collection, other developmental variables may have influenced the INAH-3. Design an experiment that could in theory address this problem.**
3. **LeVay used anecdotal information about brain donors to assign them into either heterosexual or homosexual categories, but today we know that sexual orientation is more a spectrum than a binary choice. Assume that you had access to information about the size of the INAH-3 of many individuals and also information about where they are on a 5-point scale of sexual orientation (highly hetero-, mostly hetero-, mixed, mostly homo-, and highly homo-). What might you predict about the relationship between the INAH-3 and sexual orientation?**

A brain structure known as the central bed nucleus of the stria terminalis (BSTc) has also attracted the interest of behavioral neuroscientists interested in sexual orientation because, like the INAH-3, the BTSc is reliably larger in males than in females. In one study, researchers compared the BTSc in transsexual individuals who were transitioning from one sex to the other. The results showed that the BTSc of biological males who were transitioning to female was smaller than that of a typical heterosexual male and well within the range of heterosexual females. The one biological female transitioning to male in the study had a BTSc size congruent with that of a biological male. These individuals had not been using sex hormones up to this point, so there was no evidence that the neural change was hormone related. This suggests that transsexuals have at least some of the neural structures of the opposite sex, further supporting the idea that differences in neural structures affect sexual identity and behavior.

Genetic influences Studies attempting to unravel a possible genetic influence on sexual orientation are difficult to conduct and interpret. Comparing the genomes of individuals who self-identify as highly heterosexual with those of individuals who self-identify as highly homosexual has led to the identification of several genes that differ between these groups. However, whether these genes influence sexual orientation or influence personality characteristics such as openness to experiences is not known. Perhaps the best evidence at this time is a massive study out of the University of Queensland in Australia that mined databases containing genomes associated with responses to questions about sexual behavior. Based on this data, researchers concluded that genetics alone explains perhaps 1% of sexual orientation. You can read more about this study and opposing views in *Dig Deeper into Behavioral Neuroscience* at the end of the chapter.

Birth order effects It is well established that the probability of being a male homosexual is related to birth order; the more older brothers a male has, the more likely it is that he will have a sexual preference for males. Although several mechanisms have been proposed to explain this phenomenon, including the age of parents and a richer male social environment, the prevailing thought at this time is that bearing multiple male embryos somehow progressively immunizes the mother against male antigens, leading to a greater probability of an embryo organizing as a gay male.

You now know quite a bit about the biological factors that influence what sex a human is attracted to, but what factors determines whether one individual is more attractive than another? We conclude our discussion of sexual behavior with a brief foray into the biology of attraction.

The Biological Basis of Attraction

We have already discussed the evolutionary basis for how other species either audition or select a mate. Birds may build a nest and offer it to a prospective mate, lions may grow luxurious manes as evidence of their vitality, or stags may fight all comers for sexual access to a doe. What biological forces determine who is an attractive human mate?

It seems obvious that some people are more attractive than others, but we might ask on what basis we make the judgment about attractiveness. Behavioral neuroscience could provide a framework for understanding how we make this judgment. If you have ever played poker, you may be familiar with the concept of a tell, which is an involuntary behavior such as a fleeting smile or a wrinkled brow that tells others at the table if your hand is strong or weak. Humans have a wide range of views on what makes a person physically attractive, but an evolutionary perspective would propose that the perceptual experience we call attractiveness is a tell for health and fertility. Individuals who are most likely to be able to reproduce and have the vitality to bring their offspring to maturity should have an evolutionary advantage. One such tell is thought to be facial symmetry. Take a moment to view the faces of celebrities presented in **Figure 10.18**. Which of the four would you rate as more attractive? Keep in mind that all of these individuals are actors who are considered appealing for many reasons.

Most people rate individuals with highly symmetrical faces as being more attractive, and now we ask why. Symmetry is a tell for a healthy prenatal environment, so it is thought that when we see a person whose face is highly symmetrical, our nervous system is interpreting that data as signifying the individual is healthy and therefore would make a good mate.

Steve Granitz / WireImage / Getty Images
Taylor Hill / FilmMagic / Getty Images
Alberto E. Rodriguez / Getty Images Entertainment / Getty Images
Georges Biard / Wikimedia Commons / CC BY-SA 3.0

a. b.

FIGURE 10.18 **Symmetrical and asymmetrical faces** **a.** George Clooney and Kim Kardashian have highly symmetrical faces. **b.** Tyler Posey and Ellen Barkin have asymmetrical faces.

Check It Out | Are Average Faces More Attractive?

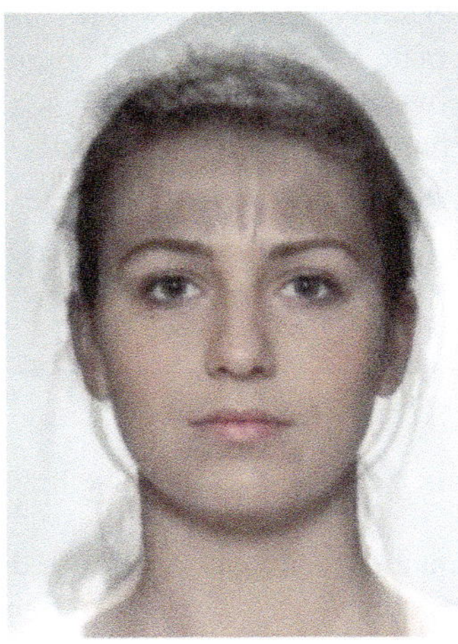

Researchers interested in the biological basis of attraction have examined a wide range of physical traits such as faces. What makes one face more attractive than another? Lisa DeBrune and Ben Jones are researchers at the University of Glasgow in Scotland whose work on the neuroscience behind attractiveness is internationally renowned. Watch the short video **Are Averaged Faces More Attractive?** and then answer these questions.

1. In your opinion, were the averaged faces more attractive than the individual faces used to create the composites?
2. Go to faceresearch.org and try your hand at creating composite faces. Were the composite faces you created more attractive than the originals? Were composites made from five or more faces more attractive than composites made from only two or three? Does your judgment on this last question support the idea that averaged faces are more attractive?

It might seem paradoxical that average faces are also considered more attractive than the individual faces that were averaged. One might think that averaging several faces into a composite face would reduce the attractive aspects of an individual to some pedestrian level, but the opposite is true. You can see this for yourself in **Check It Out: Are Average Faces More Attractive?**

10.4 Before You Go On

Section Summary

Male and female sexual responses both follow the same four-part cycle of excitement, plateau, orgasm, and resolution, but the details differ due to differences in anatomy. The factors that determine sexual orientation are complex, and evidence suggests the determination is made prenatally. People vary in attractiveness, and these differences may provide an insight into evolutionary processes.

Comprehension Check

1. What are the four phases of the sexual response cycle in order?
   - **a.** attraction, excitement, orgasm, resolution
   - **b.** excitement, plateau, orgasm, resolution
   - **c.** stimulation, orgasm, resolution, plateau
   - **d.** excitement, resolution, orgasm, refractory period

2. Along with BSTc, what neural center appears to be related to human sexual orientation?

a. SDN-POA

b. VMH and LH

c. pituitary

d. INAH-3

3. What are two factors that appear to have some influence on sexual orientation?

a. birth order and genetics

b. genetics and social environment

c. birth order and early experience

d. parental sexual orientation and environment

4. From an evolutionary perspective, physical traits such as facial symmetry are considered to be signals that an individual is what?

a. sexually receptive

b. sexually mature

c. healthy and fertile

d. attractive to others

Conclusion

Behavioral neuroscience explores the biological mechanisms that underlie behavior. At times, the topics may seem distant from your everyday life, much as understanding how your refrigerator works is not essential to knowing how to use it to store your food. However, in this chapter, we delved into a topic that is intensely personal: sexuality. The *Stories of the Brain* about David/Brenda Reimer suggests that one's identity as a male or female is based on much more than just genitals and hormones. How is the path between male and female chosen? What are the mechanisms that organize reproductive mechanisms throughout our life? Is there a biological basis for sexual attraction? We still have much to learn.

Critical Thinking Questions

1. Biological sex has not changed through history, but gender roles have. How do you think assumptions of appropriate male and female roles have changed in the last 100 years? In the last 20 years? How might they change in the next 20 years?
2. In this chapter, you saw that we know a fair amount of how biological sex is decided *in utero*. What would be the consequences if parents could choose with 100% accuracy the sex of their child? In your opinion, would this capability be an advantage or a disadvantage to society? Explain.
3. In our discussion of hormonal influences on parental behavior you learned that a positive feedback loop exists between parent and child that involves hormones such as oxytocin. Based on your knowledge of behavioral neuroscience, write two or three sentences of advice for first-time parents on how to interact with their child.
4. There is evidence that at least some of what our mind perceives as attraction is a tell to our brain of reproductive fitness. Consider the physical attributes that most people use to judge physical attractiveness for males or females. In your opinion, do these attributes signify health and/or reproductive fitness? Support your answer with examples.

Dig Deeper into Behavioral Neuroscience

10.1 An Overview of Sex and Reproduction

We now know that one's biological sex and one's gender identity are not necessarily the same. Read what the American Medical Association has to say in "Sex, Gender and Why the Differences Matter."https://journalofethics.ama-assn.org/article/sex-gender-and-why-differences-matter/2008-07

Rubinow, D. R., & Schmidt, P. J. (1996, August). Androgens, brain and behavior. *American Journal of Psychiatry, 153*(8), 974.

10.2 Prenatal Sexual Organization

This source provides an overview of the organizing role of prenatal hormones on sexual differentiation.

Sobel, V., Zhu, Y-S., & Imperato-McGlinty, J. (2004, December). Fetal hormones and sexual differentiation. *Obstetrics and Gynecology Clinics of North America, 31*(4), 837–856. https://doi.org/10.1016/j.ogc.2004.08.005

10.3 The HPG Axis and Life-Span Sexual Development

This source provides a broader understanding of the hormonal and structural effects of puberty.

Kulin, H. E., & Müller, J. (1996). The biological aspects of puberty. *Pediatrics in Review, 17*(3), 75–86.

10.4 Sexual and Reproductive Behaviors

Any student interested in human sexual response should read the seminal work by Masters and Johnson.

Masters, W. H., & Johnson, V. E. (1966). *Human Sexual Response.* Toronto and New York: Bantam Books. ISBN: 978-0-553-20429-2

Hormones play a central role in creating and maintaining sexual desire, as these sources show.

Gades, N. M., Jacobson, D. J., McGree, M. E., St. Sauver, J. L., Lieber, M. M., Nehra, A., et al. (2008, September). The associations between serum sex hormones, erective function, and sex drive. *Journal of Sexual Medicine, 5*(9), 2209–2220. https://doi.org/10.1111/j.1743-6109.2008.00924

Wåhlin-Jaconsen, S., Pedersen, A. T., Kristensen, E., Laessøe, N. C., Lundquist, M., Cohen, A. S., et al. (2015). Is there a correlation between androgens and sexual desire in women? *Journal of Sexual Medicine, 12*(2), 358–373.

The story behind David Reimer is one of the most interesting in behavioral neuroscience. Here are two readable and interesting accounts; the first is an account of the events surrounding David Reimer, and the next two are essential readings for anyone interested in the implications of sexual reassignment.

Colapinta, J. (2006). *As Nature Made Him: The Boy Who Was Raised as a Girl.* New York: Harper Collins.

Diamond, M., & Sigmundson, H. K. (1997). Management of intersexuality: Guidelines for dealing with individuals with ambiguous genitalia. *Archives of Pediatrics and Adolescent Medicine, 151*(10), 1046–1050.

Diamond, M., & Signundson, H. K. (1997). Sex reassignment at birth: Long-term review and clinical implications. *Archives of Pediatrics and Adolescent Medicine, 151,* 298–304.

What is the role of genetics in defining sexual orientation? These sources provide a broad overview of the issue.

Zietsch, B. P., Sidari, M. J., Abdellaoui, A., et al. (2021). Genomic evidence consistent with antagonistic pleiotropy may help explain the evolutionary maintenance of same-sex sexual behaviour in humans. *Nature Human Behavior, 5,* 1251–1258. https://doi.org/10.1038/s41562-021-01168-8

Anyone interested in the neurobiology of sexual orientation should read work by Simon LeVay and Dean Hamer.

LeVay, S. (1991). A difference in hypothalamic structure between homosexual and heterosexual men. *Science, 253*(5023), 1034–1937. https://doi.org/10.1126/science.1887219. PMID: 1887219. S2CID: 1674111.

LeVay, S. (1993). *The Sexual Brain.* Cambridge: MIT Press. ISBN:0-262-62093-6

Hamer, D., & Copeland, P. (1994). *The science of desire: The search for the gay gene and the biology of behavior.* New York: Simon and Schuster.

Is facial symmetry a sign of reproductive fitness? You can read more with these sources.

Jones, B. C., DeBruine, L. M., & Little, A. C. (2007, November). The role of symmetry in attraction to average faces. *Perceptual Psychophysics, 69*(8), 1273–1277. https://doi.org/10.3758/bf03192944

Grammer, K., & Thornhill, R. (October 1994). Human (*Homo sapiens*) facial attractiveness and sexual selection: The role of symmetry and averageness. *Journal of Comparative Psychology, 108*(3), 233–242. https://doi.org/10.1037/0735-7036.108.3.233. PMID: 7924253.

Chapter 10 Before You Go On Answers

10.1 Before You Go On Answers

1. d. creates greater genetic diversity
2. a. Physical traits that are different between males and females of a species
3. c. ovary; testis
4. b. Gender roles may vary considerably due to cultural differences.

10.2 Before You Go On Answers

1. a. two allosomes
2. b. XX; XY
3. a. Müllerian; Wolffian
4. c. They share a common antecedent, or precursor structure.
5. d. Without the effect of androgens, only anti-Müllerian hormone has an effect.

10.3 Before You Go On Answers

1. d. fluctuations in FSH, LH, estrogens, and progesterone
2. a. follicle; ovary
3. c. oxytocin; both parents
4. c. vaginal dryness for women and erectile difficulties for men

10.4 Before You Go On Answers

1. b. excitement, plateau, orgasm, resolution
2. d. INAH-3
3. a. birth order and genetics
4. c. healthy and fertile

Chapter 10 Glossary

allosome The chromosome that determines the sex of an individual; also known as the sex chromosome.

androgen A type of hormone that regulates the development and maintenance of male characteristics in males and contributes to growth and reproduction in males and females.

asexual reproduction A form of reproduction that does not involve the union of sex cells or gametes from another member of the species.

biological sex A biological trait that determines one's reproductive function as either male or female.

cervix A cylinder of tissue that separates the vagina from the uterus.

clitoris A highly sensitive external organ of the female reproductive system.

estrogen A type of hormone that regulates the development and maintenance of female characteristics in females and contributes to growth and development in males and females.

Fallopian tube One of two tubes on either side of the uterus down which ova travel from the ovary to the uterus.

follicle-stimulating hormone (FSH) A hormone secreted by the anterior pituitary that is essential for sexual development and reproduction in females and males.

gender A range of characteristics pertaining to the concepts of masculinity and femininity.

gonadotropin-releasing hormone (GnRH) A hormone secreted by the hypothalamus that is responsible for regulating many pituitary functions.

intersex Having the sexual or reproductive structures of both sexes.

labia Folds of skin that protect the vaginal opening.

luteinizing hormone (LH) A hormone secreted by the anterior pituitary that is essential for sexual development and reproduction in males and females.

menarche The first occurrence of menstruation.

menopause The point in a woman's life when her menstrual cycles stop for more than 12 months.

Müllerian duct One of the two embryonic ducts that form the default urogenital system for humans, which, in the absence of masculinizing hormones, will result in a female urogenital system; also known as the Müllerian system.

Muller's ratchet A process by which harmful mutations accumulate in the absence of diversity of genetic information (especially in asexual reproduction).

ovarian cycle The portion of the menstrual cycle that involves the ovaries and release of the ova.

ovary The female gonad, where ova are produced.

penis The male reproductive organ.

primary sexual characteristic One of the physical structures present at birth that consist of the external and internal genitalia.

puberty The developmental period in which hormonal and physical changes mature a child's body to the point where they are capable of sexual reproduction.

Red Queen dilemma A theory in evolutionary biology stating that a species must constantly adapt and evolve just to survive in an ever-changing environment.

scrotum A sac of skin hanging beneath the penis that contains the testes.

secondary sexual characteristic One of the features that appear during puberty in humans and at sexual maturity in other animals.

sexual dimorphism Distinct differences in physical traits of the male and female members of a species.

sexual orientation A trait based on the gender or genders to which one is sexually attracted.

sexual reproduction A form of reproduction involving the combination of genetic information from two individuals of different sexes.

***SRY* gene** A gene on the Y chromosome that contains the instructions to create the SRY protein, which organizes male development *in utero*.

testis The male gonad, where sperm are produced.

uterine cycle The portion of the menstrual cycle that involves the uterus.

uterus A pear-shaped organ in which the fertilized ova develop during pregnancy.

vagina A muscular part of the female genital tract that extends outward from the uterus to the vulva.

vulva The outer or external part of the female genital system.

Wolffian duct One of the two embryonic ducts that form the male urogenital system due to the influence of testosterone; also known as the Wolffian system.

CHAPTER 11

Anuchit Singkham / EyeEm / Getty Images

Just as a small and simple seed will grow into a large and complex tree, the human nervous system develops in size as well as complexity from conception forward.

Neuro-development Across the Life Span

CHAPTER OUTLINE

11.1 Prenatal Development of Brain Structures
- Embryonic Neurodevelopment
  Think About It: Developing Brains at Risk
- Fetal Neurodevelopment

11.2 Prenatal Neurodevelopment at the Cellular Level
- Neurodevelopment as an Organizational Process
  Check It Out: Visualizing Neuromigration
- Guidance for Neuromigration

11.3 Postnatal Neurodevelopment
- Environmental Influences on Postnatal Neurodevelopment
  Advances in Behavioral Neuroscience: Marian Diamond and Neuroplasticity
- The Adolescent Brain
  Try It Out: Are You a Risk Taker?
- The Aging Brain

11.4 Neurodevelopmental Disorders
- Defects in Neurulation
  Check It Out: Spina Bifida
- Fetal Alcohol Syndrome
- Attention-Deficit/Hyperactivity Disorder (ADHD)
- Williams Syndrome
- Schizophrenia

LEARNING OBJECTIVES

LO 11.1 **Trace** the development of brain structures through the embryonic and fetal periods.

LO 11.2 **Describe** the course of prenatal neurodevelopment at the cellular level.

LO 11.3 **Evaluate** aspects of postnatal neurodevelopment.

LO 11.4 **Relate** concepts of neurodevelopment to disorders.

Stories of the Brain

Jamal was described by his parents as a child who never slowed down, always moving from activity to activity. Although he never had a serious behavior problem in school, his teachers described him as fidgety and easily distracted. Although he was bright and usually understood the material presented in class, he often struggled to complete homework assignments in a timely manner. Jamal's parents helped him to remember deadlines and obligations and in general maintained a highly structured environment with daily routines to keep him on track.

After high school, Jamal enrolled in college without a clear idea about a major, thinking that he would find his place by taking a wide range of courses. As time went by, however, he fell into a pattern of becoming enthusiastic about a class and then losing interest as the work piled up. By the end of his second year, he had declared five different majors but was not making progress on any of them. He often daydreamed in class, sometimes not even taking notes. Jamal frequently acted impulsively, and this caused trouble in his personal life; he would party the night before an exam and miss appointments with friends because he "just got distracted and forgot."

When Jamal finally landed on academic probation, his faculty advisor convinced him to make an appointment with the study skills center. There he was evaluated and diagnosed with attention-deficit/hyperactivity disorder (ADHD). Jamal was prescribed a medication to improve his focus and attended a class that provided him with strategies for maximizing his potential. With these tools in hand, Jamal earned his way onto the dean's list and graduated with distinction.

What was happening in Jamal's brain that made it hard for him to focus, and how did medication change that? In this chapter, we will discuss the core topic of neurodevelopment and how you began your journey on the path to who you are now, and who you will become later in life. In our discussion we will consider four aspects of neurodevelopment. The first is to describe at the macroscopic level the development of the nervous system from its beginnings to the moment of birth. In an earlier chapter you saw that the neural tracts in your brain are unbelievably complicated, so our second topic is to understand how a simple neuron knows where to go to create the organization that every human shares. The third question asks if neurodevelopment ends in childhood or continues into adulthood. Our final topic applies what we have learned in this chapter to real-world issues of neurodevelopmental disorders such as ADHD. With that organization in hand, we first turn to a discussion of the physical development of the human brain.

11.1 Prenatal Development of Brain Structures

LO 11.1 Trace the development of brain structures through the embryonic and fetal periods.

Just as we cannot fully understand current events without knowing about the past, we cannot fully understand how the brain creates behavior without understanding the course of neurodevelopment. We can all agree that your brain is immensely complicated, but how did this complex device come to be?

Imagine for a moment that you are in charge of the construction project that will build your house. A house consists of many components—the foundation, electrical and plumbing systems, walls and ceilings, a roof, and so on—that must be linked together into a whole in order for the house to be habitable. As the project progresses from the foundation on up, workers must move around the work site to be where they are needed at each moment, and damaged and unused materials must be removed from the work site. Years later, you may decide based on experience that an additional bathroom and a guest room are needed and make further alterations. If you can visualize all these steps in building a house starting from scratch, then you have a schema, or mental framework, for understanding neurodevelopment across the span of your life.

We begin our exploration of the development of the nervous system just 25 days postconception with the appearance of the neural tube.

Embryonic Neurodevelopment MCAT

As the fertilized human egg divides and grows, it begins to differentiate into clearly distinct structures known as the ectoderm, the endoderm, and the mesoderm (Figure 11.1a). The **endoderm** becomes our digestive and respiratory systems, the **mesoderm** becomes our bones and muscles, and the **ectoderm** becomes our skin and neural tissue. Embryonic development progresses along a longitudinal axis defined by the notochord, a thin rod of cartilaginous tissue that acts as a scaffold for embryonic development and eventually becomes the spinal vertebrae. We turn our attention first to the process by which this ectodermal tissue, known as the **neural plate**, becomes the first step in our nervous system.

As the neural plate thickens and flattens during the third or fourth week, it begins to fold onto itself in a process called **neurulation** (Figure 11.1b). During this period, a structure forms at the top of the neural tube known as the **neural crest**. Around 20 to 30 days after conception, the folding tissue closes up to form the **neural tube** (Figure 11.1c). This simple neural tube is only 3 mm long (0.1 inch, or about as long as this line _ at three weeks. The neural tube and the neural crest are the precursors to the brain and spinal cord, respectively.

Around the third to fourth week postconception, the neural tube begins to develop three vesicles, or bulges, called the **prosencephalon**, the **mesencephalon**, and the **rhombencephalon**. With the addition of a rudimentary spinal cord, these primitive structures constitute the early nervous system (Figure 11.2a). Each of these three vesicles is a seat of future development. The prosencephalon, or forebrain, is the rostral or most anterior section; the rhombencephalon, or hindbrain, is at the rear; and the mesencephalon connects the two. Damage or disruption to the developmental process at this key point will likely lead to significant if not fatal physiological defects.

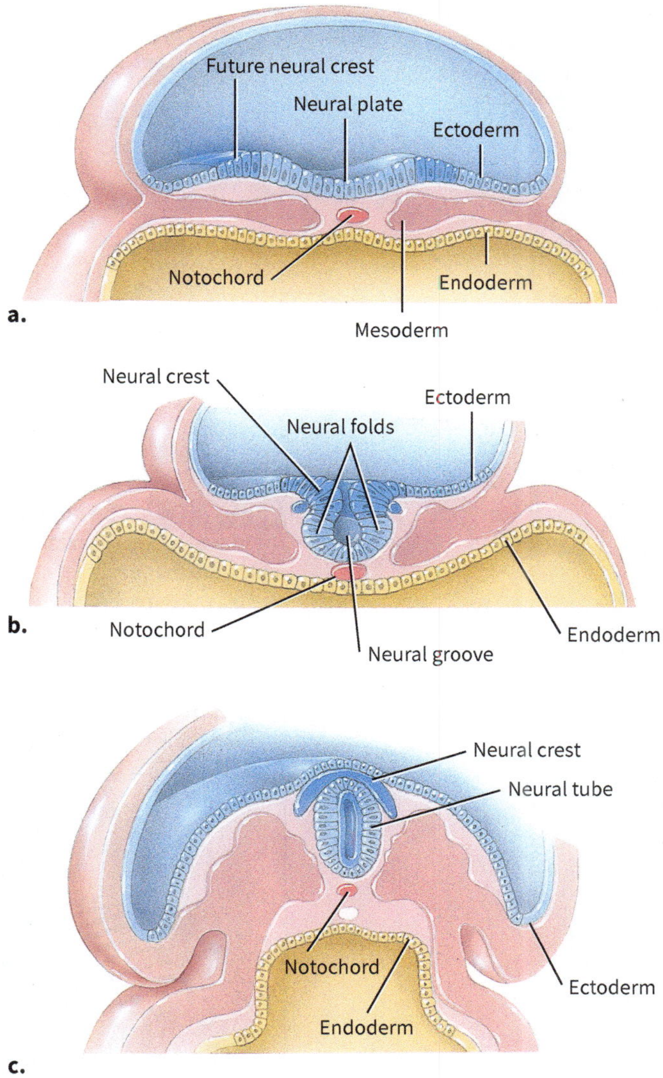

FIGURE 11.1 The beginnings of neural development
a. Early in the developmental process, the mass of cells begins to show a complex structure, first dividing into three cellular layers. **b.** During the third or fourth week, the neural groove develops, which begins to fold into itself through neurulation. The neural crest begins to form. **c.** Around 20 to 30 days after conception, the folding tissues close up to form the neural tube.

By around the seventh week post-conception, we begin for the first time to see structures that slightly resemble the adult nervous system (Figure 11.2b). The embryonic brain further differentiates into four clear divisions as the prosencephalon divides into the telencephalon and the diencephalon. The **telencephalon** becomes the cerebral cortex and basal nuclei, whereas the **diencephalon** develops into the retina, thalamus, and hypothalamus. The mesencephalon develops into the adult midbrain as well as the superior and inferior colliculi. Finally, the rhombencephalon is the seat of the pons, cerebellum, and medulla.

Just as large changes to a house plan early in the construction process will make the final product very different from its original conception, environmental factors early in development can significantly alter the natural course with catastrophic results. This is because basic structures such as the ectoderm, neural tube, and major divisions of the adult nervous system appear quite early. Agents that change the course of development are known as **teratogens**, and these include drugs, disease, and environmental events such as genetic disruptions or injuries. Alcohol is a common teratogen, and excessive exposure to alcohol *in utero* can lead to fetal alcohol syndrome. Nicotine has also been demonstrated to have teratogenic capabilities. Common diseases such as rubella (German measles), herpes simplex, and herpes zoster have been shown to have teratogenic effects. You can learn more about teratogens in **Think About It: Developing Brains at Risk**.

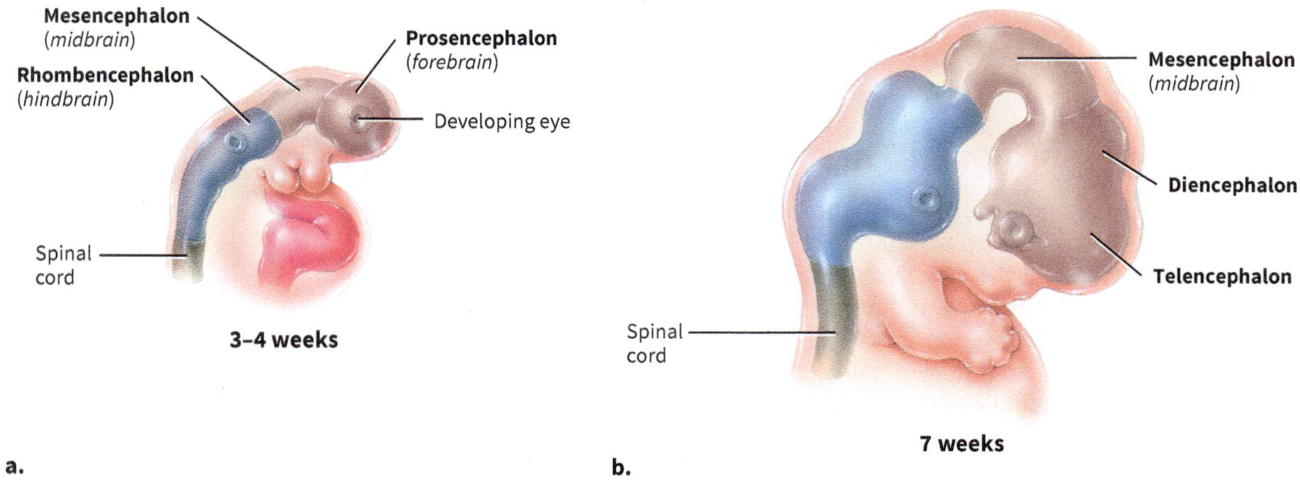

FIGURE 11.2 The embryonic central nervous system **a.** During the fourth week postconception, the neural tube begins to differentiate into the prosencephalon, mesencephalon, and rhombencephalon, which are the precursors of adult brain structures. **b.** By the seventh week, further differentiation creates the telencephalon and the diencephalon.

Think About It | Developing Brains at Risk

SpeedKingz/Shutterstock

Imagine for a moment that you are on a bicycle, poised on the top of a hill and ready to travel to the bottom. You can see that changing your direction only slightly at the beginning of your journey will mean that you arrive at a very different destination at the end of your travels. Just as changing your path a few degrees as you begin your descent will take you to a different place, environmental forces that appear early in neurodevelopment may have significant and even fatal implications for the developing individual. This is why women who are pregnant or planning to become pregnant must be aware of the teratogenic capabilities of any drugs, licit or illicit, that they take.

One tragic example occurred in the late 1950s, when a German pharmaceutical firm created a sedative known as thalidomide to reduce anxiety and treat nausea during pregnancy. The U.S. Food and Drug Administration (FDA) refused to approve the drug because of insufficient testing, but it was made widely available in countries across Europe. It was soon found that the use of thalidomide early in a pregnancy was linked to perhaps as many as 4,000 infant deaths at birth and many more children born with serious birth defects including brain damage and malformed limbs.

Thalidomide was found to create birth defects by damaging the developing embryo, and its effects varied due to when in the pregnancy it was taken. Embryos exposed in the first 20 days suffered massive damage to the nervous system, whereas later exposure selectively affected eyes, ears, or limbs. Exposure after day 40 had no effect. The thalidomide scandal led to many countries imposing more rigorous testing requirements for new drugs.

Questions

1. **Briefly explain why thalidomide's effects differed depending on when during the pregnancy the drug was taken.**
2. **Women are advised to avoid teratogens when they might become pregnant, not only when they already are pregnant. Why does this recommendation make sense?**
3. **Should avoiding teratogens be solely the choice of a pregnant mother, or does society bear some responsibility mitigating these risks? For example, should it be illegal to serve a pregnant woman alcohol? Search online for a list of common teratogens, and make a case for or against government involvement in their regulation.**

Fetal Neurodevelopment MCAT

As the embryo continues to develop, additional structures appear, especially after the eighth week, when the developing brain begins to resemble the adult brain in its appearance. At this point the embryo is called a fetus. By week 11, the major structures of the adult brain are clearly discernable in rudimentary form, including the diencephalon, midbrain, cerebellum, pons, and medulla (**Figure 11.3a**). The diencephalon will develop into the thalamus and hypothalamus. The midbrain will develop into the tectum and tegmentum, which process sensory stimuli. The cerebellum coordinates movement, and the pons and medulla provide essential functions such as respiration and heart rate. By week 40, which is typically the end of a pregnancy, the newborn brain has clearly taken on the appearance of an adult human brain (**Figure 11.3b**).

In a previous chapter, you learned that the cortex of the brain consists of gray matter and that this gray matter is where much of the information processing occurs that creates consciousness. You also learned that the cortical plate is quite large, in fact, too large to fit into the skull without being folded. One of the more dramatic developmental changes that occurs in the last 20 weeks of fetal development is the proliferation of the cortex, which not only grows in size but also folds into sulci and gyri. During the last two trimesters of the pregnancy, the total volume of the brain increases linearly, while the surface area of the cortex increases geometrically (**Figure 11.4**). This extreme folding of a large cortical surface to fit into a relatively small skull is a hallmark of the human brain.

Before we move on, let us return for a moment to our home construction metaphor for prenatal neurodevelopment. Although an electrician may install outlets in the walls, these provide no power until the home is connected to the local power company. In the same vein, the mere presence of neural structures in the fetus does not indicate that they are connected and communicating.

Investigators of postnatal nervous systems such as yours enjoy an advantage over those interested in prenatal neurodevelopment in that researchers can ask an individual what they are experiencing. Because fetal humans are not gifted with speech, we must infer as best we can the capabilities of the developing nervous system, but our understanding of neurobiology does provide some guidance. In order for a fetus to experience stimuli such as temperature, touch, or pain, certain components of the circuit must be in place. Before a developing brain could experience sensations from the body, there must be functioning sensory receptors in the skin, an ascending peripheral nerve tract, a functional thalamus, and perhaps most importantly, a cortex with connections to all of these.

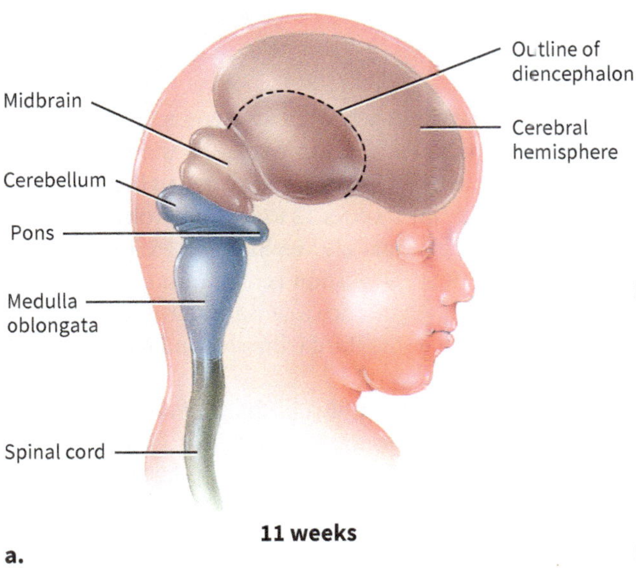

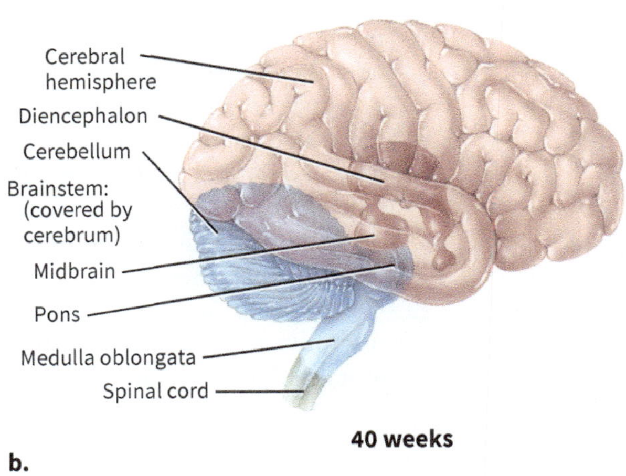

FIGURE 11.3 The human brain at 11 and 40 weeks **a.** At 11 weeks, the rudiments of the adult brain are visible. **b.** At birth, a newborn brain has all the major features of the adult brain.

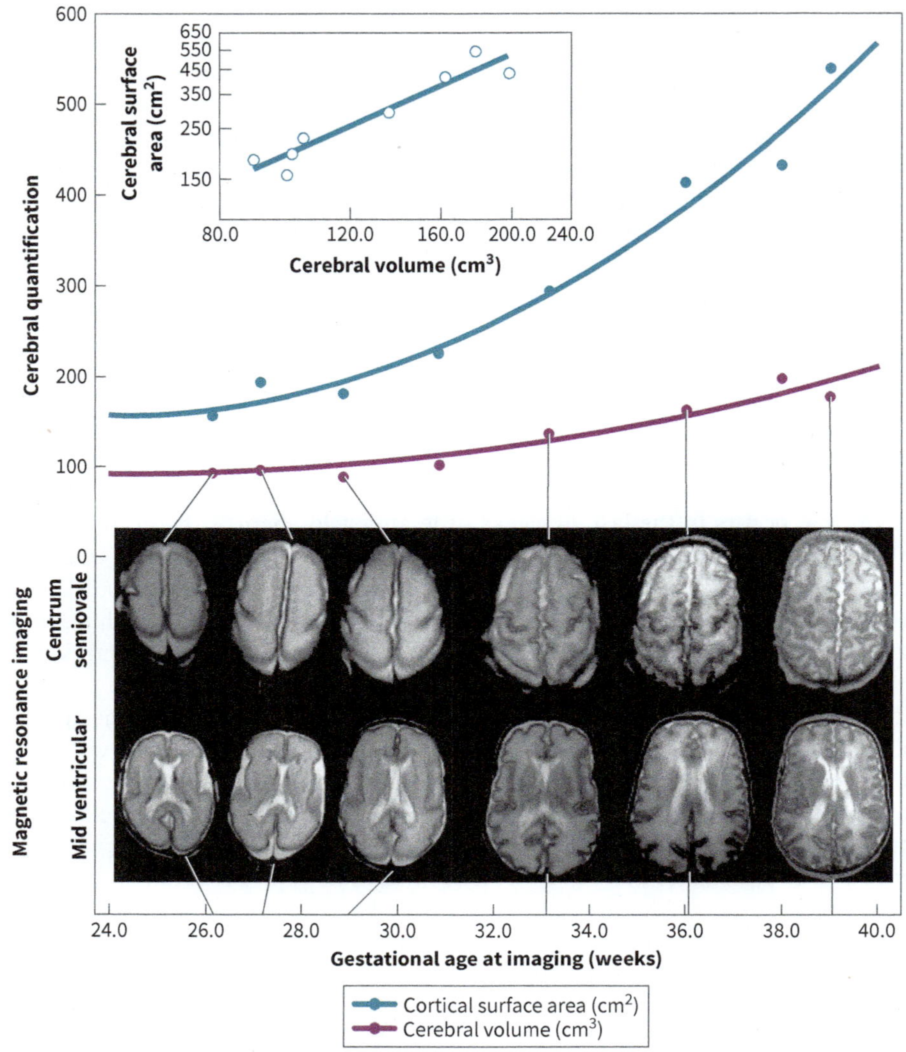

FIGURE 11.4 **Fetal development of brain volume and cortical surface area** The fetal brain grows steadily throughout the third trimester; however, the surface area of the cortex grows much faster during this period than does the rest of the brain due to the development of cortical folds.

Given that understanding, what do we know about when an awareness of sensation is possible? Free nerve endings begin to appear around the seventh week of gestation, and it is only after this point that a mechanism exists for a sensory signal to be created in the periphery. The first projections to the thalamus do not appear until weeks 12 to 16, and projections from the thalamus to the cortex do not appear until approximately weeks 25 to 26. Determining how developed a fetal cortex needs to be to experience sensation is a daunting task, but the best evidence is that a fetal brain cannot consciously experience a sensation from a peripheral nerve until at the very earliest around the beginning of the third trimester and possibly much later into that trimester.

By this point in our discussion, you have a pretty clear idea of the general course of the macroscopic development of the human nervous system; the neural tube grows in size and develops into an organ recognizable as a human brain. What is missing so far from our discussion is the answer to a fundamental question in neurodevelopment: how does a neuron make its way from its point of genesis to its eventual destination? We will begin to answer that question in our next section.

11.1 Before You Go On

Section Summary

The human nervous system begins as a plate of cells that first folds into a tube, which then differentiates by the sixth week into a structure with the rudiments of the adult brain. During the last trimester, the developing brain increases in size while the cortical surface experiences exceptional growth. By the end of the last trimester, the fetal brain begins to resemble the adult brain and connects to sensory neurons in the body.

Comprehension Check

1. The rudimentary neurodevelopmental structure that will become the CNS is the neural _____.

- **a.** crest
- **b.** epithelium
- **c.** tube
- **d.** genesis

2. By six weeks after conception, the embryo develops the _____, a structure that will eventually become the cerebral cortex.

- **a.** telencephalon
- **b.** corticoencephalon
- **c.** rhombencephalon
- **d.** medialencephalon

3. What is an important and highly visible developmental trend during trimesters two and three?

- **a.** the expansion of the prosencephalon into the rhombencephalon
- **b.** the appearance and proliferation of sulci and gyri on the cortex
- **c.** the appearance of the cerebellum at the rostral position
- **d.** the disappearance of the mesencephalon and its replacement by the thalamus

4. At what point does the fetal nervous system acquire connections between the periphery and the developing nervous system?

- **a.** the sixth week
- **b.** the second trimester
- **c.** the third trimester
- **d.** immediately after birth

11.2 Prenatal Neurodevelopment at the Cellular Level

LO 11.2 Describe the course of prenatal neurodevelopment at the cellular level.

Neural development begins as a simple collection of cells known as the neural plate, and in a relatively short period of time, it develops the highly organized structure of processing centers and information highways that you are using at this moment to read this chapter. To help you appreciate this feat, consider a modern airliner consisting of approximately six million individual parts (**Figure 11.5**). Now imagine that you toss these six million parts into an open field and, nine months later, you return to find that they have spontaneously organized themselves in the proper order to create a fully operational airliner ready to fly. As unbelievable as that would be, how much more astonishing is it that in nine months a simple neural tube is transformed into roughly 100 billion neurons organized in just the right way to form the brain you had at birth? How that happens is the subject we turn to next.

AP Images / KYDPL KYODO

FIGURE 11.5 A modern airliner in construction and in flight Modern airliners are complicated machines consisting of several million parts that must be organized in exactly the right order to create a safe and reliable airplane.

Neurodevelopment as an Organizational Process MCAT

The fact that we can experience our world and ruminate on the meaning of life or the complexity of the human brain should strike us as simply incredible. Developing neurons somehow

TABLE 11.1 Stages of Neurodevelopment

Stage	Significant development
Neurogenesis	Neurons are created.
Neuromigration	Neurons travel to their destination.
Corticogenesis	The cortex is created.
Synaptogenesis	Synapses are created.
Apoptosis	Unnecessary synapses are removed.
Synaptic reorganization	Experience fine-tunes the brain.

migrate to their proper destination in the brain, orient their dendrites and terminal buttons in the proper direction (often in conjunction with numerous others), and then create synapses with possibly tens of thousands of other neurons. How do cells born in the nursery of the neural tube become the adult human brain? Table 11.1 presents six distinct stages in neural development that lead to the development of the human central nervous system in its highly organized fashion: neurogenesis, neuromigration, corticogenesis, synaptogenesis, apoptosis, and synaptic reorganization.

Stage 1: Neurogenesis

The path from a single cell to neurons and glia is a process termed **neurogenesis**. Your nervous system is comprised of billions of cells that can be classified into neurons and neuroglia (often simply called glia). We have perhaps 100 billion neurons and 50 to 100 times more glia than neurons. How do we start with a single fertilized cell and end up with such complexity? The major driving force behind prenatal neurodevelopment is our genes; they control cell division and what type of cell develops in a particular region. We begin with **stem cells**, which have the unique ability to become any type of mature cell. In the first critical step of neurodevelopment, these cells become the neural plate and then the neural tube.

The neural tube is lined with neural stem cells that have the capacity to regenerate or replace themselves at the end of their life cycle, but also to produce **progenitor cells**, or precursor cells. Progenitor cells create the neuroblasts and glioblasts that mature into the cells that you see in brain tissue. **Neuroblasts** mature into the many specialized types of neurons seen in the nervous system, whereas **glioblasts** mature into glia. Neurogenesis appears to be complete by roughly five months of gestation. This process is illustrated in Figure 11.6.

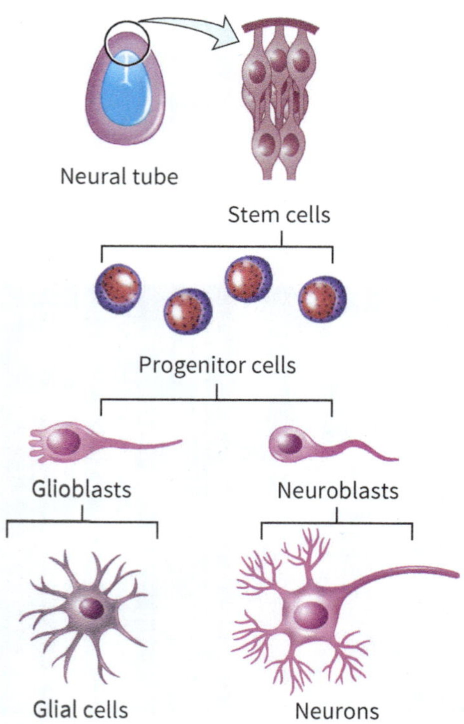

FIGURE 11.6 Neurogenesis The course of neurodevelopment begins with stem cells in the neural plate that transition to progenitor cells for the glial and neuron lines. These develop into a range of cells that become our nervous system.

Stage 2: Neuromigration

It might surprise you to discover that neurons in the mature vertebrate nervous system are typically born in a different location from their eventual destination. Therefore, the next step in neurodevelopment is the migration of neurons from the site of their genesis to their final location in the brain. During neural migration, neurons move forward, switch direction, and even change their mode of migration during development.

Neurons migrate along two dimensions. **Radial migration** occurs when neurons extend their axons deeper into the brain, and **tangential migration** is when they extend their axons across the brain (Figure 11.7). Radial migration is responsible for perhaps 90% of neural migration in the human brain. You may recall that a radius is a line extending from the edge of a circle into the center, so neurons migrating radially do so by moving either inward from the periphery or outward from the center. Migration can also occur along the length of the developing nervous system, or tangentially. In this fashion, nature weaves an elaborate tapestry containing billions of threads that interconnect hundreds if not thousands of trillions of synapses; this enchanted loom supports life and creates our conscious experience.

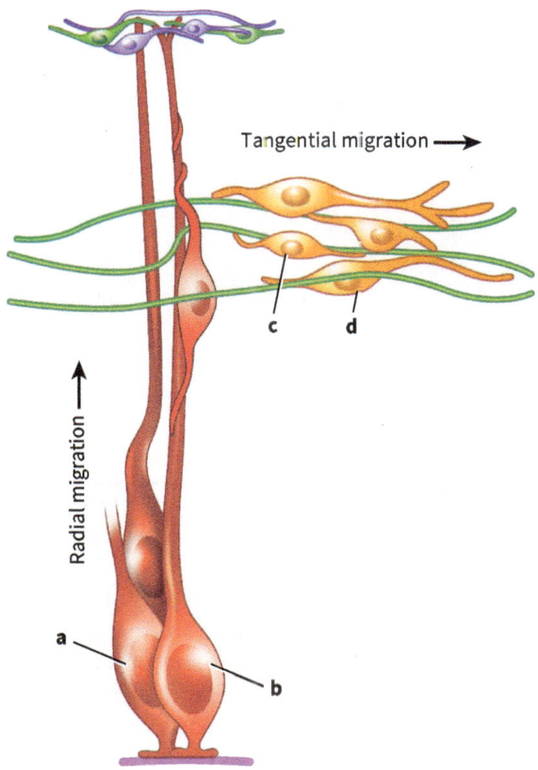

FIGURE 11.7 Radial and tangential neural migration Neurons a and b are migrating along a glia and, in doing so, are extending their axons radially, much as the spokes of a wheel point from the perimeter into the hub. Neurons c and d extend their tendrils tangentially toward surrounding structures.

You can observe the process of neural migration in this **Check It Out: Visualizing Neuromigration**.

Stage 3: Corticogenesis The formation of the cortex, or **corticogenesis**, is a complex process with several moving parts. Corticogenesis begins around week seven during the embryonic phase and continues until around week 18 of the fetal stage. **Figure 11.8** illustrates the process beginning in week 10 and continuing to birth.

Check It Out | Visualizing Neuromigration

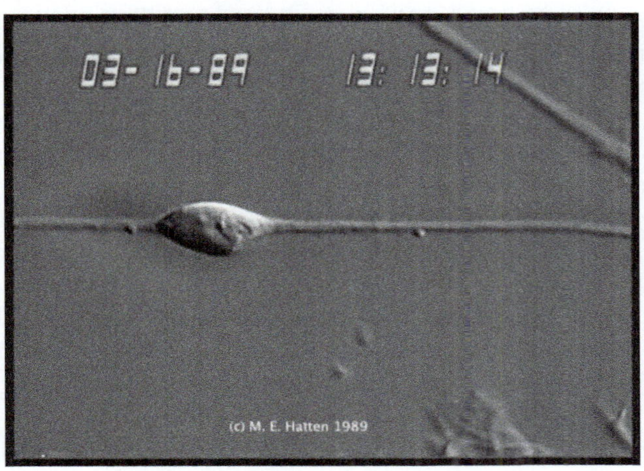

The video **Neural Migration** shows a neuron migrating along a glial scaffolding and illustrates the startling reality that your developing nervous system comes with self-assembly instructions. Watch the video, and then answer these questions.

1. If the signaling system that guides neuronal migration and turns it on and off at the right points were to go awry, do you think the results would be beneficial or harmful to the developing nervous system? Briefly explain your answer.
2. How might understanding the factors that control neural growth and migration help develop treatments for traumatic brain damage that occurs through sports, war injuries, and tumors?

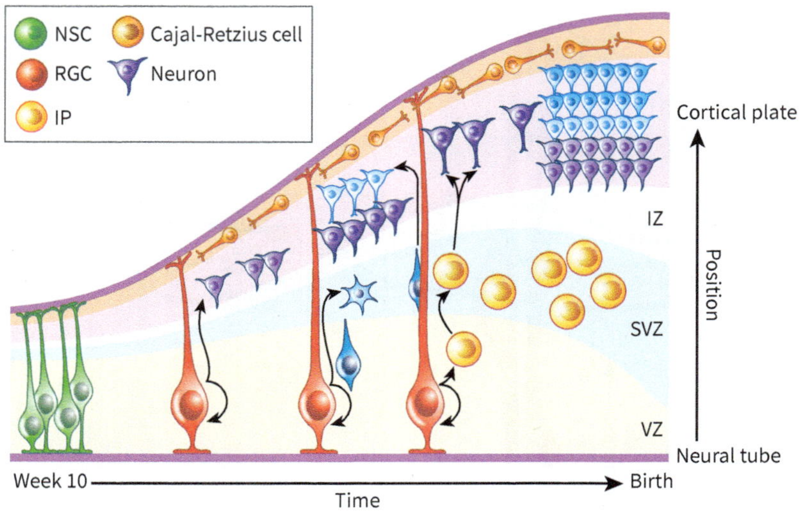

FIGURE 11.8 Corticogenesis from week 10 to birth Beginning after week eight, progenitor cells (NSCs, RGCs, IPCs, and Cajal–Retzius cells) grow, migrate, differentiate, and give rise to the glia and neurons that create the cortex. Cells migrate radially from the ventricular zone outward. Other cells migrate tangentially, creating several layers of densely packed and interconnected neurons upon which experience will create a human mind.

Corticogenesis proceeds in an inside-out fashion as progenitor cells in the epithelial layer of the ventricular zone (VZ) begin to divide, differentiate, and migrate. These primitive cells begin to expand radially from the neural tube, and as they do so, they differentiate into radial glia cells (RGCs) and intermediate progenitor cells (IPCs). There are two progenitor cell lines, one for glia that will become oligodendrocytes and astrocytes and the other for neurons. As neurodevelopment progresses, these cells divide, differentiate, and migrate radially and tangentially as they extend outward through the subventricular zone (SVZ) and the intermediate zone (IZ) and into the cortical plate (CP). As they do, they take on the characteristics we recognize as neurons.

Cajal–Retzius cells, like RGCs and IPCs, are transient cells in that they appear, perform a function, and then are absorbed. These cells secrete a chemical messenger that guides cortical cells as they migrate radially. The result is the development of six distinct cortical layers that together constitute the human cortex.

The key to understanding corticogenesis is to remember that your nervous system began as a tube, and in response to genetic direction, chemical messengers, and environmental influences, it grew from the core outward, much like a building that begins with a foundation and then adds floors one atop the other by bringing building materials up through the structure until the roof is finally put in place.

Given that the cortex is the seat of so much of what makes us human, there is a great deal of research interest in understanding the normal course of its development and what it might mean when that course goes awry. In subsequent discussions of topics such as the biological basis of intelligence and also the neurology underlying psychiatric illnesses, we will discuss what differences in cortical organization might tell us about these subjects.

Stage 4: Synaptogenesis

We know that the synapse that is the basic mechanism of information transmission in the nervous system, so the creation and proliferation of synaptic connections, called **synaptogenesis**, is at the heart of understanding behavior. As these neural processes extend and grow, they also grow synaptic connections that are essential to creating both simple and complex behaviors. A chemical messenger known as a guiding molecule guides the development of a connection between two neurons. This process of creating connections is so explosive during the prenatal period that it is called *exuberant synaptogenesis* to differentiate it from the synaptic growth that occurs throughout your life.

Much of what we know about synaptogenesis in the CNS is inferred from our understanding of the growth of synapses at the neuromuscular junction, that is, at the more easily observable site where a neuron controls a muscle fiber. What we do know is that there are chemical messengers that guide the formation of synapses. Guided by genetics, environment, and the organism's experience, this stage yields billions of potential neural connections. Picture a vast crowd of people, all shaking hands and introducing themselves, and you have an excellent metaphor for synaptogenesis.

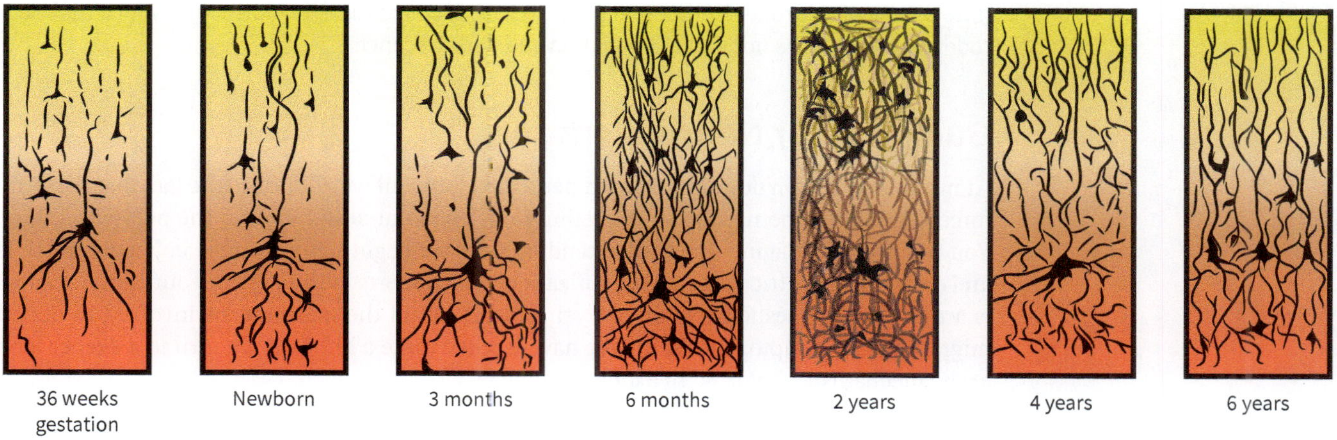

FIGURE 11.9 Synaptogenesis and apoptosis During the first few years of life, synaptogenesis occurs at a furious rate, building new connections. By age 4, apoptosis has begun to prune away unnecessary synapses.

Stage 5: Apoptosis

We all have friends at school, at home, and on social media, but some are friends for life, some are friends for now that we may lose contact with as we leave school and enter our careers, and some are individuals we would prefer to never see again. Similarly, your nervous system responds to the exuberant proliferation of neurons in the prenatal period by selectively removing those neurons that have not created useful connections in a process known as **apoptosis**, or programmed cell death (Figure 11.9).

You may have engaged an in apoptotic process if you have ever pruned your phone contacts or holiday greeting card list of individuals that have permanently left your circle of family and friends. Just as removing such individuals makes a search through your contact list faster, removing neurons that have not made useful connections enhances the efficiency of your brain.

Stage 6: Synaptic reorganization

While apoptosis removes the neural deadwood, an additional process of synaptic reorganization takes over that will continue for the rest of your life. Synaptic connections that are useful and frequently activated are strengthened, whereas connections that are unused are dissolved. If apoptosis can be viewed as removing individuals from your contact list who you no longer associate with, synaptic organization is similar to putting frequent contacts on a favorites list.

Neurotrophins are chemical messengers that regulate the creation and function of neural pathways in the brain. There is much we do not understand about this process, but it appears that the presence of neurotrophins during the pruning stage suppresses cell death among frequently used neural pathways. Development is a life-span process, so as your career and life require you to exercise some cognitive skills and abandon others, both synaptic growth and synaptic pruning occur.

Most people think that the brain starts out with few neurons at birth and development adds more neurons. In fact, the opposite is true. The brain of a newborn is luxuriously provided with neurons, and experience with the world eliminates neurons and connections that are unused. It is thought that as much as half of the neurons present at birth are pruned away in infancy, and some regions may lose much more. Do not consider the developing brain to be an unfolding flower; it is more accurate to visualize it as a block of marble that a sculptor chips away at until it assumes its final shape. You can summarize synaptic organization with these two golden rules: "Neurons that fire together, wire together," and, "If you don't use it, you lose it."

As we conclude our discussion of the stages of prenatal development, take a moment to look at Figure 11.10, which portrays the

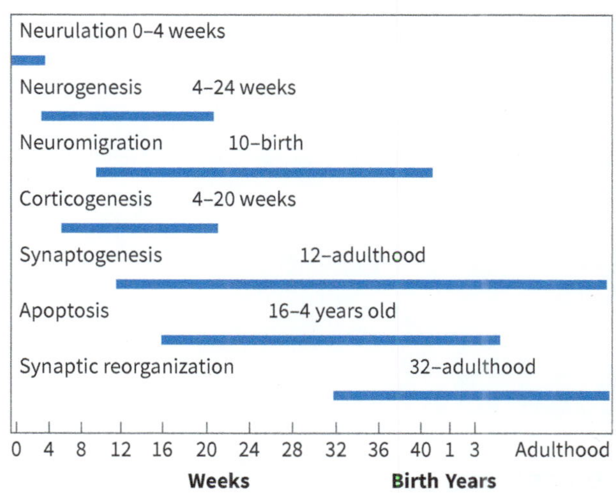

FIGURE 11.10 A timeline of human neurodevelopment Some neurodevelopmental processes begin early and end early, some begin early but end before birth, and some begin before birth and continue on into adulthood.

time course of neurodevelopmental processes from conception to adulthood. Notice that neurodevelopment does not end at birth or even in adolescence.

Guidance for Neuromigration

At multiple points in our discussion of neurodevelopment we alluded to the fact that there are forces that guide the neuron in migrating to its eventual destination in the nervous system. You might well imagine that understanding how we begin with a simple collection of cells and end with the intricate neural organization that supports life and creates our consciousness is an important question. We do not yet understand all the mechanisms involved in neural migration and synaptogenesis, but we have learned quite a bit. Next we turn to a deeper look at the mechanisms involved in axonal guidance.

Growing axons develop **growth cones**, at the end of which are finger-like extensions known as **filopodia** (from the Greek for thread-like feet) that sense signals in the cellular environment (Figure 11.11). Differential growth of the cytoskeleton on one side of the growth cone causes the tip to point in the direction of the signaling gradient; in this way, signaling messengers guide the development of neural connections. There are several chemical signaling systems that are responsible for attracting as well as repulsing the filopodia that are extending their reach into new regions. Neurotrophins are secreted by the target tissues, and they control the development of the CNS.

Pioneer neurons often use immature glia as a framework to guide their growth. In fact, pioneer neurons and glia most likely coordinate their growth through some signaling system. Finally, **guidepost neurons** are cells that appear at intermediate points along a pathway and serve as way stations that break up the gap between a neuron and its eventual target. Your rudimentary nervous system first creates a scaffold of guide cells; then neurons and glia migrate along this framework in a highly organized fashion. Chemical messengers guide one neuron to make a physical connection to another, and over the course of nine months (and continuing into adulthood) these processes create the complex structure we call the human brain. Researchers are keenly interested in understanding more how these messengers work, as they may hold the key to unlocking ways to regenerate brain tissue that was damaged by disease or injury.

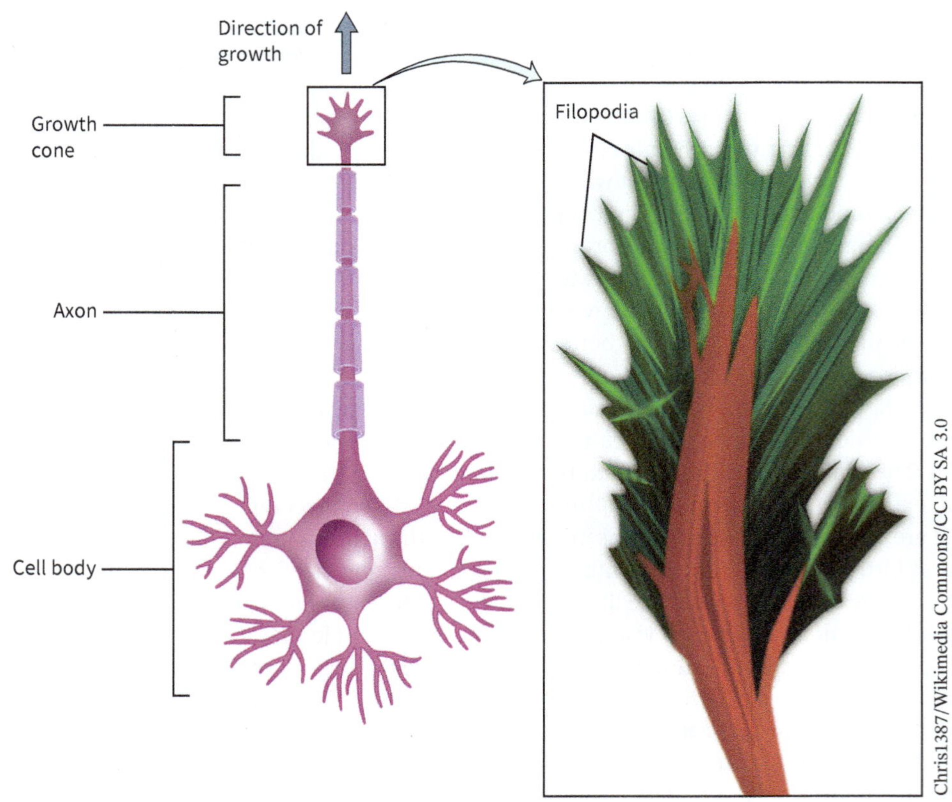

FIGURE 11.11 Growth cone with filopodia extended Growth cones develop filopodia from their ends that seek out connections with adjacent neurons.

11.2 Before You Go On

Section Summary

Neurodevelopment involves the birth of neurons, the migration of their cell bodies and axons to their eventual locations, the differentiation of cells into their eventual cell type, and the building of connections followed by the pruning of unnecessary connections. Radial growth and tangential growth create a neural structure that is highly interconnected. Neurons extend filopodia from growth cones to make new connections, and chemical messengers known as neurotrophins guide those filopodia. All of these steps create an efficient nervous system.

Comprehension Check

1. Select the correct order of neurodevelopmental stages.

- **a.** neurulation, neuromigration, synaptogenesis, apoptosis, reorganization
- **b.** neurulation, synaptogenesis, neuromigration, apoptosis, reorganization
- **c.** synaptogenesis, neurulation, neuromigration, apoptosis, reorganization
- **d.** synaptogenesis, neuromigration, neurulation, apoptosis, reorganization

2. Radial neural migration is like _____, whereas tangential migration is like _____.

- **a.** a spider web; a pinwheel
- **b.** the spokes of a wheel; the strings on a tennis racket
- **c.** the legs of a chair; the spokes of a wheel
- **d.** a pinwheel; the blades of a fan

3. What is the structure that extends toward another neuron to build a synapse?

- **a.** presynaptic extension
- **b.** neural finger
- **c.** growth cone
- **d.** synaptic exchange

4. Chemical messengers that stimulate the filopodia of a growth cone to extend toward another neuron are called what?

- **a.** growth factor
- **b.** neurotransporter factor
- **c.** defenestration factor
- **d.** neurotrophins

11.3 Postnatal Neurodevelopment

LO 11.3 Evaluate aspects of postnatal neurodevelopment.

Prenatal neurodevelopment sets the stage for much of human behavior, but it is a mistake to think that the nervous system is complete at birth or even during childhood. Neurodevelopment continues into adulthood, and the implications are important for understanding behavior across the life span. Therefore, we now turn to two interesting topics in life-span neurodevelopment: the adolescent brain and the aging brain.

Up to this point, we have been discussing neurodevelopmental processes that are genetically directed. However, we know that such mechanisms interact with the environment and the combination of genetics and environmental input led to the person we are today. Let's take a moment to touch on the effects of a richly stimulating environment as well as poverty of stimulation and ask if there are critical periods when input from the environment is essential to normal development.

Environmental Influences on Postnatal Neurodevelopment MCAT

We ended our conversation about the course of prenatal neurodevelopment with a discussion of synaptic reorganization. Does the reorganization of synapses end at birth? That topic was on the mind of Marian Diamond and Mark Rosenzweig, neuroscientists who challenged the prevailing belief that the brain's structure was fixed by the end of childhood. They argued that synaptic connections in the brain could change over the life span in response to learning and the environment, a trait called **neuroplasticity** (see **Advances in Behavioral Neuroscience: Marian Diamond and Neuroplasticity**). To test this theory, Diamond and Rosenzweig raised rats in either a sparsely furnished cage or in an enriched environment full of toys, ladders, and tunnels (**Figure 11.12**).

Advances in Behavioral Neuroscience | Marian Diamond and Neuroplasticity

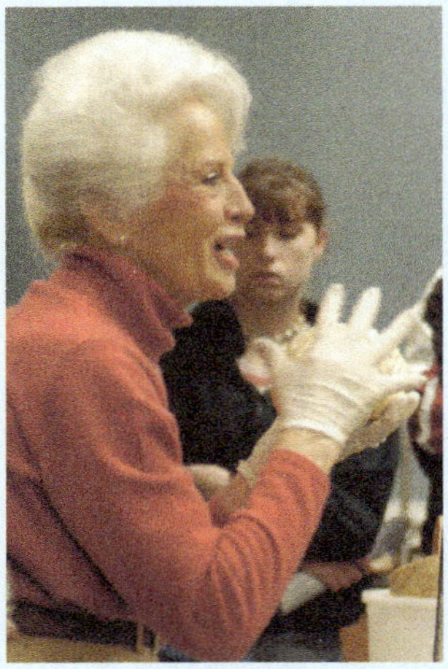

Image courtesy of Peg Skorpinski

Dr. Marian Diamond was a pioneer in the field of neurodevelopment. Aside from being one of the first women in the field, her research led to a total reversal of one of the oldest beliefs about the brain.

Marian Cleeves Diamond was born in California to an English physician and a Latin teacher. After her undergraduate studies at the University of California–Berkeley, she was the first woman to enroll in the graduate program in anatomy and also the first to graduate with a doctorate in anatomy from UC–Berkeley. After working as a research assistant at Harvard, she was the first female on the faculty at Cornell University, where she taught human biology and comparative anatomy. She was invited to return to UC–Berkeley first as an instructor and then as a tenure-track assistant professor, a rank she maintained until she retired as an emeritus professor. Although she was a productive researcher, Dr. Diamond was also known to be a highly popular teacher.

Her scientific career spanned over five decades and led to many publications, but among her many accomplishments, perhaps two stand out. One of the most basic doctrines in neuroscience was that brain structure was determined by genetics and was fixed and unchangeable. Diamond, along with colleagues Mark Rozenzweig and David Krech, completely overturned that theory by showing that environmental influences could make significant changes in brain structure. Her work was the basis of our current interest in neuroplasticity, and if you think environmental influences such as nutrition, stress, and social experience guide neurodevelopment, you can thank Marian Diamond. A second area where her creativity was at work came about when she was given four segments of Albert Einstein's brain to study. She realized that his brain differed slightly but significantly from typical brains in that the left hemisphere had significantly more glial cells than is normal.

Marian Diamond was a pioneer, not only in breaking a path for women in neuroscience, but also a scientific pioneer whose work showed the importance of environmental enrichment as a factor in neurodevelopment.

Questions

1. **Diamond and Rosenzweig's research has implications for all of us. Provide two examples of environmental enrichments that a parent can provide an infant that are inexpensive yet stimulating.**
2. **If it is true that the brain can reorganize, think of a neurological condition that might benefit from technology that assists the brain in reorganizing. Explain how this might work.**

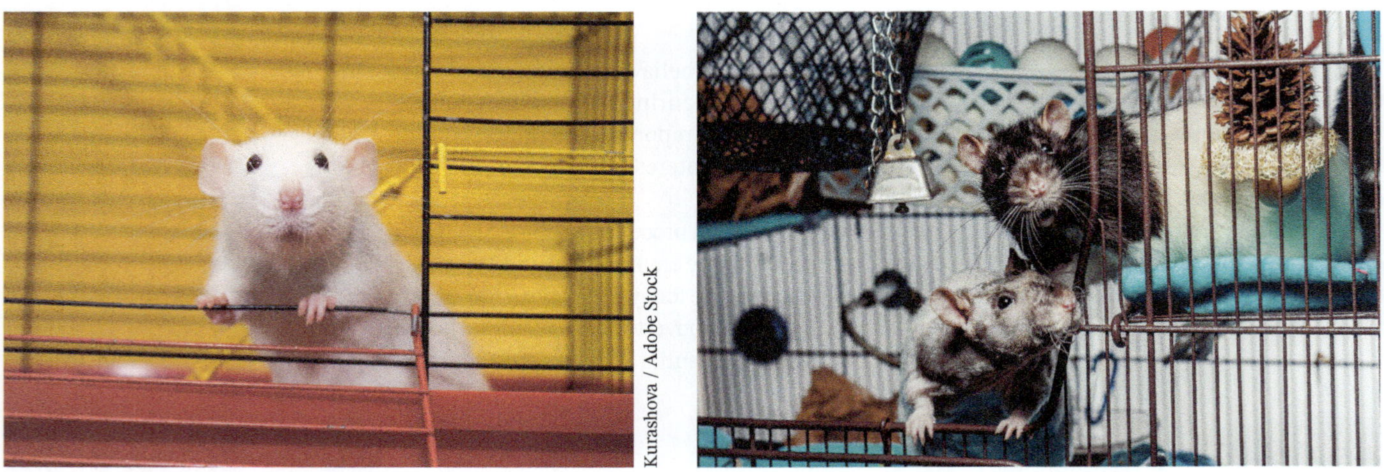

Kurashova / Adobe Stock

a.

Ezume Images / Adobe Stock

b.

FIGURE 11.12 **A common rat cage and an enriched environment cage** **a.** This rat lives in a simple cage that provides water and food but is lacking in stimulation. **b.** In addition to basic amenities, this rat is also provided with engaging sensory stimulation.

Rats raised in the enriched environment had thicker cortices with 25% more synapses and glia than the controls, and their dendritic arbor (such as the length of dendrites and the number of dendritic spines) was superior. In addition, the enriched environment rats had more acetylcholine receptors in the cortex. This finding led scientists to reconsider the question of brain plasticity in new ways; for example, will an enriched environment across the life span lead to neurological changes that enhance cognitive performance and greater cognitive reserve in old age?

Diamond, Rosenzweig, and others who investigated the impact of stimulation on neurodevelopment proved that exposure to an enriched environment will have a substantial and positive effect upon the infant's subsequent development. What of the opposite? Are there discrete periods where the brain is expecting to receive input and without that input, subsequent neurodevelopment takes a turn for the worse?

I came face to face with this issue early in my career when I met Clarence, who had been raised in a barn and only interacted with his parents and farm animals until adulthood. When a caseworker found Clarence and he was moved to our facility, he was pleasant and gentle but considerably intellectually challenged. Knowing Clarence led me to think about the role of sensory and social experience in neurodevelopment and especially the question of critical periods.

Diamond and others demonstrated that a stimulating environment facilitated neurodevelopment, but are there points in neurodevelopment where such stimulation is essential? That is, are there times where environmental enrichment exerts an especially powerful effect? Nobel Laureates David Hubel and Torsten Wiesel addressed this question in the 1960s by sewing shut one eye of a newborn cat for the first three months of life. What was the effect of monocular deprivation on the visual cortex of these cats? The answer is that a lack of stimulation from the occluded eye created impoverished ocular dominance columns in the primary visual cortex for that eye. Furthermore, those regions of the primary visual cortex (PVC) that were serviced by the deprived eye showed less neural development. Finally, input from the intact eye took over part of the PVC that is normally dedicated to the eye that was occluded. These results were interpreted to mean that normal neurodevelopment requires appropriate sensory input.

In addition, researchers found that there were periods where this sensory input must occur; providing it later did not resolve the deficit. We now know that the developing nervous system expects certain input at a specific time, known as a **critical period**. If stimulation does not occur during that critical period, permanent deficit will occur. Infants born with cataracts in one or both eyes will not develop normal vision unless those cataracts are surgically repaired in the first six months.

A critical period has been demonstrated similarly with language development. Research with linguistically deprived infants as well as normal infants suggest that the first six months of life constitute a critical period in which the developing brain is sensitive to the speech sounds known as phonemes that they hear. Examples of common phonemes in English are /ph/, /ta/, and /ba/ sounds. The ability to automatically process these phonemes and discriminate one from another appears to be hardwired by age six months, after which time it becomes increasingly more difficult to distinguish those sounds. The evidence suggests that by age 12, the ability to easily process the unique phonemes of a different language is gone, and this makes learning a second language more difficult. For example, Japanese does not have phonemes that correspond to our /r/ and /l/; instead, there is a single phoneme that is a blend of the two. As a result, native Japanese speakers who were not exposed to English as young children struggle to differentiate the two phonemes when speaking English. If you have ever tried to learn a new language as an adult, you may have experienced the difficulty of learning to form novel phonemes.

We know quite a bit about neurodevelopment and critical periods, but we are still learning. Science progresses slowly but steadily, and over time, our understanding of how one's environment affects postnatal neurodevelopment has changed. Originally, it was thought that almost any environment that was rich in stimuli could result in greater synaptic proliferation, but subsequent research has provided the more nuanced perspective about the effects of different kinds of stimuli.

Researchers such as Allison Mackey hypothesized that adverse childhood events (ACEs) might lead to a neurodevelopmental path that has negative consequences. Examples of ACEs are extreme poverty, food insecurity, and exposure to violence. Mackey and her colleagues reasoned that the developing nervous system might adapt in some way to maintain homeostasis in an ACE environment. Remember that homeostasis, the drive to maintain balance in the face of a changing environment, is a core concept in behavioral neuroscience and biology in general.

What you just learned about synaptic proliferation, apoptosis, and cortical organization might provide a clue as to what Mackey found in her experimentation. We know that in childhood and early adolescence, the developing brain is plastic, that is, easily modifiable. This extended period of plasticity is unique to human neurodevelopment in how it enhances our ability to acquire new information. By collecting measures of cortical thickness and organization, Mackey found that children in ACE environments appeared to end that period of extreme plasticity much earlier than children raised in a more supportive environment. One interpretation of her findings is that in a challenging environment, our nervous system responds by maturing as early as possible, but potentially to the detriment of subsequent neural organization.

The Adolescent Brain MCAT

The human brain reaches 90% of its eventual adult weight by around age six, and significant structural changes continue until around age 25. During this period, white matter increases at the expense of gray matter and a significant amount of neural pruning occurs to strengthen the connections between the prefrontal cortex and the rest of the cortex.

Recall that gray matter serves more as computational hardware, whereas white matter is responsible for information transmission throughout the cortex. MRI scans of adolescent brains ages 5 to 20 show that during this period, gray matter is slowly replaced with white matter as the brain prunes away inefficient connections and replaces them with myelinated tracts that speed the transfer of data throughout the brain. The thickness of the cortical gray matter decreases, but a corresponding increase in gray matter density occurs that favors the frontal lobes where the traits of abstract thinking and judgment are located (**Figure 11.13**).

To summarize this neurodevelopmental trend, during childhood, through adolescence, and into early adulthood, white matter tracts that act as communication channels increase at the expense of gray matter, which is computational. At the same time, the distribution of gray matter favors the frontal lobe where our more abstract thinking is based.

During early adolescence, there is at first a proliferation of synaptic connections followed by a pruning of those connections that are not frequently used, thereby leading to a cortex that is organized to be maximally efficient. If the paradoxical nature of this process confuses you, this metaphor might be helpful. Imagine you had to decide where to put walkways across a busy grassy lawn. One possibility would be to allow people to walk across the space as they wish for a time, choosing for themselves the most efficient routes to the places they visit most often. Over time, the grass would wear away on the most used pathways, and the contractor can use this as a plan for where to pave. In much the same way, the adolescent brain preserves those synaptic pathways that are most useful and efficient, pruning away the rest.

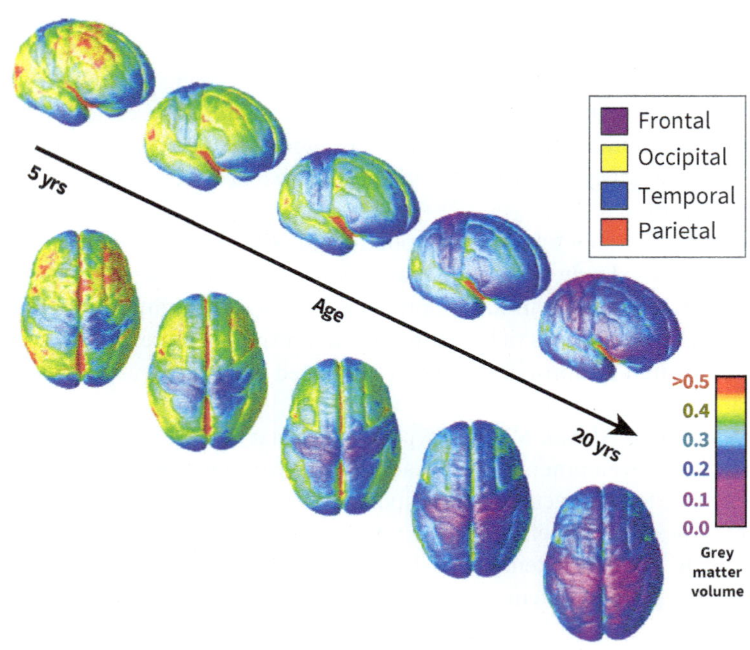

FIGURE 11.13 Changes in cortical tissue from childhood to adolescence From childhood into early adulthood, the neurodevelopmental trend favors the frontal lobe where abstract thinking and judgment reside.

Try It Out | Are You a Risk Taker?

Greg Epperson/Adobe Stock

Psychologists have devised questionnaires for assessing risk-taking among adolescents. Such instruments are tested on a large sample of individuals and validated against driving records, police incidents, and self-reported behaviors. Complete the **Risky Behavior Worksheet** anonymously, and then answer these questions.

1. As you look over your own answers to the questionnaire, do you see yourself as a risk taker or a risk avoider?
2. If a brain scan were available that would tell you if you are prone to risky behavior, would you want to know? Would the knowledge that you are a risk taker change your behavior? Explain your answer.
3. If you are older than age 25, would your answers have been different earlier in your life? Provide an example.

An equally important feature of adolescent neurodevelopment is the pattern of this synaptic pruning. The above-mentioned maturation occurs first in primary sensory areas and the motor areas. Only in the early 20s does this maturation occur in the frontal and prefrontal areas, regions known to be associated with impulse control, complex planning, and risk evaluation.

Because the prefrontal cortex is associated with the evaluation of risks and rewards as well as impulse control, you would expect that adolescents are more likely to engage in risky behavior until their brains fully mature, at some time around age 25 or 26 years. In fact, injuries and fatalities involving adolescents are more likely to involve vehicles, alcohol, or unprotected sex, and the evidence suggests that at this age they are less able to evaluate the relative risks and benefits of their behaviors. You can test your own tolerance for risky behavior in this **Try It Out: Are You a Risk Taker?**

The Aging Brain MCAT

As we age, it is normal to lose some speed and efficiency in cognitive functions. There is a great deal of variability in normal cognitive aging; some elderly individuals begin to show noticeable loss of cognitive function in their 70s, whereas others live independently into their 90s. Still others endure the abnormal brain changes associated with dementia or Alzheimer's disease.

Normal brain changes with age

As we progress into our 70s and beyond, it is normal to see changes on a number of levels: neurons, structures, and even vascular function. Starting around age 50, brain volume begins to shrink roughly 5% per decade (**Figure 11.14**).

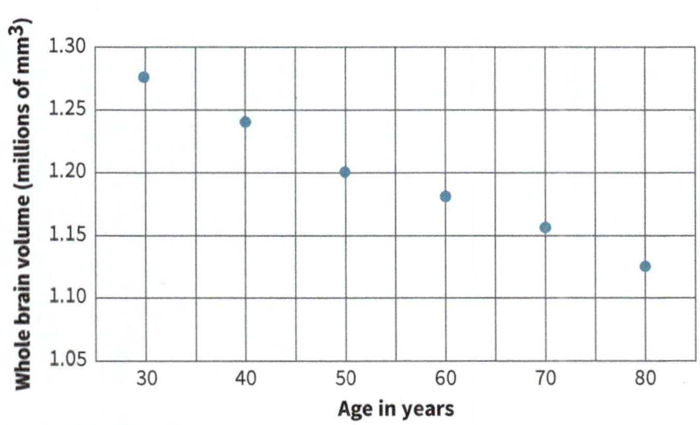

FIGURE 11.14 Changes in brain volume in adulthood Between ages 30 and 80, the volume of the brain as measured by MRI decreases each decade.

There are suggestions that changes in both gray and white matter begin to occur in one's 70s, and there is clearer evidence that some areas are more affected than others, with the prefrontal area and also the hippocampus showing the greatest loss of volume. Given that the prefrontal area and the hippocampus are responsible for inhibition, evaluation of complex messages, memory, and also wayfinding, it is perhaps not surprising that the elderly are at greater risk for financial and identity scams, forgetfulness, and losing track of objects and even their location.

Other changes associated with advanced age are diminishing neurotransmitter levels and changes in vascular function. Dopamine and serotonin levels drop by as much as 10% per decade among the elderly, and changes in vascular function often lead to hypertension and a corresponding increase in transient ischemic strokes and small vessel disease.

In summary, there are regular changes in function and structure among those 70 and beyond that are associated with diminishing cognitive ability. However, because the elderly are a diverse group, it is risky to make global conclusions about individuals. It is clear that how you live your life has an effect on how your brain ages.

David Snowden, an epidemiologist and neurologist, followed almost 700 nuns ranging in age from 72 to 102 to investigate how we maintain our intellectual faculties into our last decades. Snowden found that dealing well with stress and a lifelong appetite for learning both predict good cognitive health into your later years. Some elderly people exercise, avoid excessive use of alcohol and tobacco, stay mentally active, and are fortunate to have a genetic endowment that, to an extent, protects them from the ravages of time. There are numerous examples of individuals over 70 graduating with honors from college, successfully running a business, and producing literature, art, and music. The most effective neuroprotective medicine may be simply to stay mentally active and engaged.

Neurodevelopmental changes associated with dementia So far we have been discussing normal brain changes as a result of aging, but brain disorders also become more common with age. **Dementia**, a group of brain disorders defined by a loss of memory and diminished cognition, affects tens of millions of individuals worldwide. The worldwide cost of patient management of individuals with dementia approaches a trillion dollars per year. The most common form of dementia is **Alzheimer's disease (AD)**, which accounts for approximately 70% of all dementias.

The brain of an individual with AD is characterized by a loss of neurons in the cortex and atrophy of the cortex, especially regions such as the frontal and parietal lobes. Figure 11.15 contrasts a normal brain with that of an individual with AD, which shows cortical atrophy, ventricular enlargement, and damage to subcortical structures such as the hippocampus that are involved in memory and attention. The speed of progression of AD varies greatly, but the typical life expectancy of an individual diagnosed with AD is 3 to 7 years.

At this time, there is no treatment for AD because we do not yet understand the mechanism that creates the damage. Some theories suggest that AD is an inflammatory disease, possibly caused by a low-level infection, and that environmental factors such as tobacco use, pollution, and stress may play a part in its development and progression. It is known that repeated head injuries earlier in life are a risk factor for the development of dementia later in life. Perhaps the answer is that there are several pathways for the development of AD. It is difficult to treat a disease where several causes may interact in different ways in many individuals.

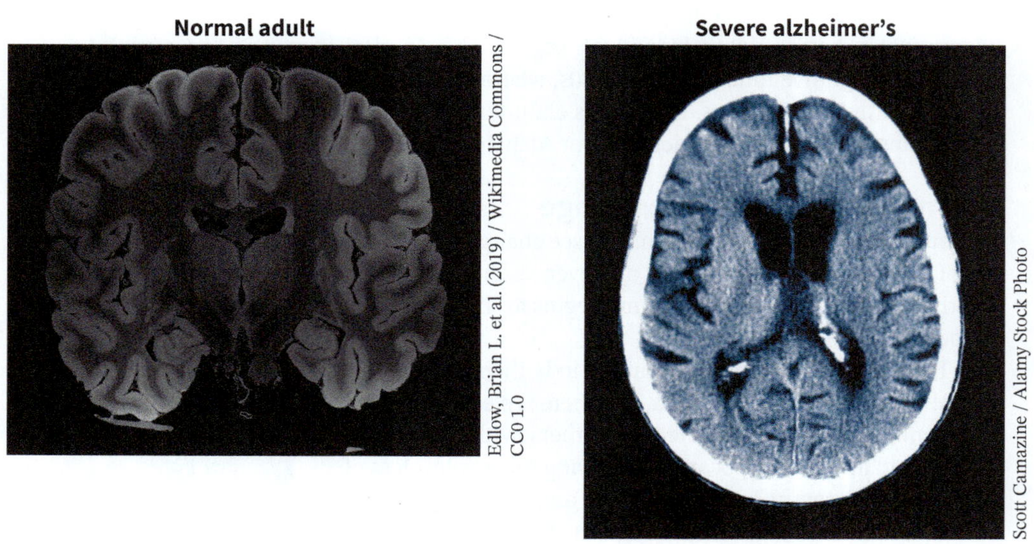

FIGURE 11.15 Comparison of normal and Alzheimer's brain In these two coronal views, you see a comparison between a normal brain and a brain damaged by Alzheimer's disease. **a.** The normal brain fills the skull, ventricles are small, and the cortical surface is highly folded. **b.** In a brain with advanced AD, the white and gray matter have atrophied, the ventricles have dramatically expanded, and the surface of the cortex is severely diminished.

As individuals live longer, their probability of their developing AD increases, as do the associated costs of their care; recent estimates put the cost of caring for AD individuals in the United States at $277 billion annually. There is no way to put a dollar value on the heartache and family disruption associated with caring for an AD individual. Understanding the causes of AD and developing an effective treatment is a scientific problem whose solution would greatly benefit society.

11.3 Before You Go On

Section Summary

Experience interacts with and guides the course of neurodevelopment, and critical periods exist in which stimulation is required for proper neurodevelopment. The adolescent brain is still pruning gray matter away to achieve maximum efficiency, and cortical structures that are involved in planning and risk evaluation are not yet fully connected. Your brain is a physical device, and just as any device wears out over time, your brain will show the ravages of decades. Alzheimer's disease is different from normal aging, and although there are many possible causes, at this time there is no effective treatment.

Comprehension Check

1. What is the importance of critical periods?

- **a.** They are the source of critical neurodevelopmental changes.
- **b.** When they are disrupted, the influence of guidance neurons becomes apparent.
- **c.** They occur in different orders for different people.
- **d.** They demonstrate that genes open a receptive period for a limited time.

2. What happens to your brain as you age into your 20s?

- **a.** The gray matter layer in the cortex thins.
- **b.** The gray matter layer in the cortex gets thicker.
- **c.** White matter tracts in the left hemisphere become thicker.
- **d.** White matter tracts in the right hemisphere become thinner.

3. What is one reason why adolescents are more likely than others to engage in risky behavior?

- **a.** The connection between their gray and white matter is not complete.
- **b.** Their amygdala is not yet mature, so they feel less fear.
- **c.** Apoptosis has not yet pruned away all the unnecessary connections.
- **d.** Their frontal lobes are still maturing.

4. What is a typical brain change that comes with advanced age?

- **a.** The limbic system is activated.
- **b.** Brain volume decreases.
- **c.** There is damage to the brainstem.
- **d.** There is a loss of cerebrospinal fluid.

11.4 Neurodevelopmental Disorders

LO 11.4 Relate concepts of neurodevelopment to disorders.

In this chapter, you have followed the development of the human nervous system from a tiny neural tube to the adult brain and considered the mystery of how neurons organize themselves into the complex structure that creates the human mind. In most cases, the processes you learned about come to fruition without an error, but errors do occur, and the knowledge you acquired while reading this chapter provides a basis for understanding some of these disorders.

Defects in Neurulation MCAT

As you might imagine, the proper development of the neural tube, called neurulation, is essential to normal development and in rare cases, the tube does not close properly.

Spina bifida One defect associated with neurulation is **spina bifida**, or split spine, which occurs when the tube does not completely close at a point closer to the caudal (tail end) of the tube. In these cases, a portion of the spinal cord may protrude through an opening in the spinal vertebrae. This congenital defect leads to a number of physical difficulties. Spina bifida is one of the most common birth defects, occurring in roughly 1 in 1,000 births worldwide.

Check It Out | Spina Bifida

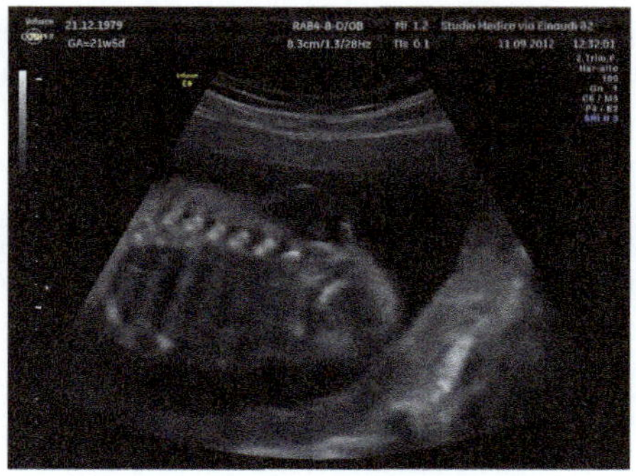

Spina bifida is one of the most common birth defects. In some cases, defects are easily repaired by surgery, but more severe cases can lead to life-long disabilities. Watch the short video **Spina Bifida**, and answer these questions.

1. Diets low in fruits, vegetables, and grains such as beans, peas, and lentils place a developing fetus at risk for neural tube dysfunctions such as spina bifida. Do you think that some education about the importance of proper diet is worth making a part of the high school curriculum? Explain your answer.
2. Search online using the phrase "Famous people with spina bifida" and mention one person whose name you recognize that you did not know had spina bifida.

When the neural tube fails to close properly, a segment of the spinal cord protrudes through the gap between two vertebrae. Surgery can repair the gap, and in mild cases, the individual should have little or no residual effects. Moderate to severe cases may leave the individual with a lifetime of mobility problems, bowel and urinary control issues, and other difficulties. You can see a sonogram of a fetus with spina bifida and consider the implications of this disorder in **Check It Out: Spina Bifida**.

The causes of birth defects such as spina bifida are complex, often a combination of genetics, environmental influences, and so on. For example, white people and Hispanic people have a greater risk for spina bifida, as do females. One factor that we know affects the odds of developing spina bifida is folate deficiency in the mother. Folate, the natural form of vitamin B_9, helps protect the embryonic nervous system from developing any breaks in the neural tube and is included in prenatal vitamins.

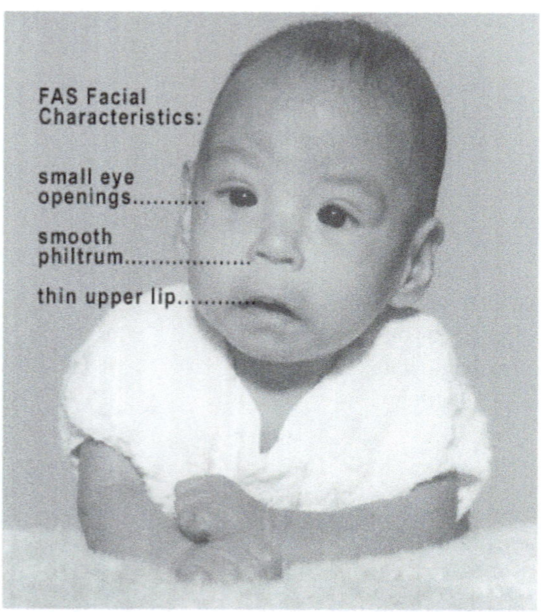

FIGURE 11.16 Physical characteristics of fetal alcohol syndrome A child who is exposed to significant amounts of alcohol *in utero* may develop FAS, which is characterized by a constellation of facial signs as well as intellectual impairment.

Anencephaly

Other defects are caused when the neural tube fails to close at the rostral (head) end. This type of defect leads to **anencephaly**, from the Greek *an* for *not* and *en kephal* for *in the head*. In these cases, the brain stem is typically present, but because the telencephalon did not develop, there are no cerebral hemispheres. As a result, even the lowest level of awareness is not possible. Such infants are generally blind, deaf, and insensitive to pain; they rarely survive beyond birth.

Fetal Alcohol Syndrome

Pre- and postnatal exposure to toxins such as lead, pesticides, and alcohol can lead to subsequent developmental problems. Alcohol easily passes the blood–brain barrier; it just as easily passes the blood–placental barrier. The consequences of exposure to significant alcohol in the development of an infant can be devastating and lifelong.

Prenatal exposure to significant amounts of alcohol can lead to **fetal alcohol syndrome (FAS)**. Individuals with FAS have a constellation of physical aberrations that are diagnostic of the condition. Conspicuous among these are a thin upper lip, indistinct philtrum, and small midface as illustrated in **Figure 11.16**. FAS individuals have significant neurological issues that are likely due to a reduction in the proper migration of neurons during neurogenesis. This leads to improper and incomplete development of the corpus callosum, hippocampus, cerebellum, and other structures critical to cognitive performance.

TABLE 11.2 ADHD Traits and Associated Behaviors

Trait	Associated behaviors
Impulsivity	Continuous talking, interrupting while others are talking, blurting out answers, inability to wait one's turn, risk taking
Hyperactivity	Forgetfulness, fidgeting with hands or legs, inability to stay seated for extended periods of time, always moving
Inattention	Organizational difficulties, short attention span, distractibility, trouble focusing, poor attention to detail

Attention-Deficit/Hyperactivity Disorder

In this chapter's *Stories of the Brain*, you met Jamal, a young man who struggled with attention and impulsivity. In fact, you probably know someone who has been diagnosed with **attention-deficit/hyperactivity disorder (ADHD)**, given that it is the most commonly diagnosed neurodevelopmental disorder. ADHD behavior is characterized by three traits: impulsivity, or the inability to inhibit inappropriate behaviors; hyperactivity, or an inordinate level of physical activity; and inattention, or an inability to sustain attention as well as being easily distractable. Typical behaviors associated with ADHD traits are presented in Table 11.2.

The incidence of ADHD is between 5 and 7% depending on the diagnostic criteria used, and a male is three times as likely to develop ADHD as a female. Most children diagnosed with ADHD grow out of the disorder, but it is thought that at least a third continue to show clear ADHD behaviors into adulthood. What do we know about ADHD?

There appears to be a clear genetic component to ADHD as the concordance rate between identical twins approaches 75%. That being the case, some argue that the presence of ADHD confers an evolutionary advantage to a group as a whole because an individual who is restless and compelled to move about and explore their environment would benefit the group even if the behavior is injurious to the individual. We do not yet have a complete understanding of ADHD, but much of the general picture regarding differences between ADHD and neurotypical brains is clear. There are subtle but meaningful differences in brain structure that distinguish the ADHD brain.

Given that impulsivity and risk-taking are hallmarks of ADHD, you should not be surprised that the prefrontal cortex of ADHD individuals is smaller as this region is thought to be associated with executive function. In addition to volumetric differences, the ADHD brain matures at a slower rate. You can see in Figure 11.17 that the neurotypical cortex depicted on the bottom has achieved its maximal thickness almost entirely by the start of the study at age 7, whereas a similar level of completion is not reached in the ADHD brain until ages 11 or 12 or beyond.

In addition to anatomical differences in the ADHH brain, there are significant physiological differences. Figure 11.18 illustrates one important physiological difference between neurotypical and ADHD brains, and that is hypoactivity, or a lower-than-normal level of metabolic activity. In one study, glucose was tagged with a radioactive molecule and then metabolic activity in the brain was measured by the amount of glucose taken up by neurons. Detectors similar to Geiger counters positioned in an array over the cortex detect the radiation emitted by the glucose and use that data to create an image of activity in the cortex. The cortical metabolic activity of the ADHD brain was found to be significantly lower than that of controls, particularly in the frontal region, which controls executive functions such as judgment, impulse control, and emotional regulation.

There is also strong evidence that neurotransmitter dysfunction is a part of the ADHD picture. Both dopamine (DA) and norepinephrine (NE) are implicated in ADHD. The mesolimbic pathway for DA services the nucleus accumbens, and the noradrenergic pathway for NE services the locus coeruleus. Both these regions are involved in controlling impulsivity and also regulating sustained attention. It is likely that other neurotransmitter systems are affected by ADHD.

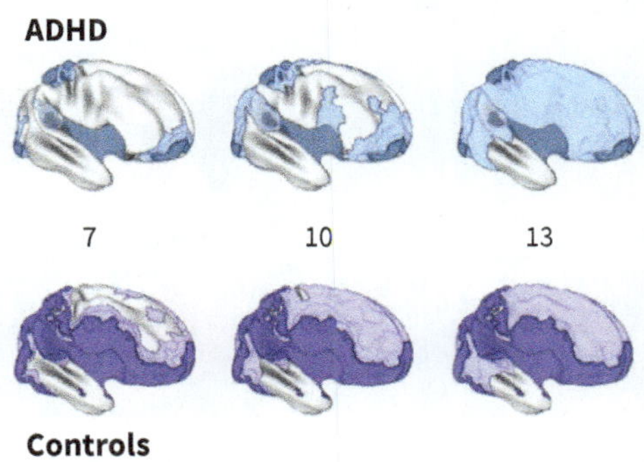

FIGURE 11.17 Brain maturation in ADHD and neurotypical controls These brains illustrate the development of the cortex for composite ADHD and control brains at ages 7 through 13. Darker colors represent a region of the cortex that has achieved its maximum thickness.

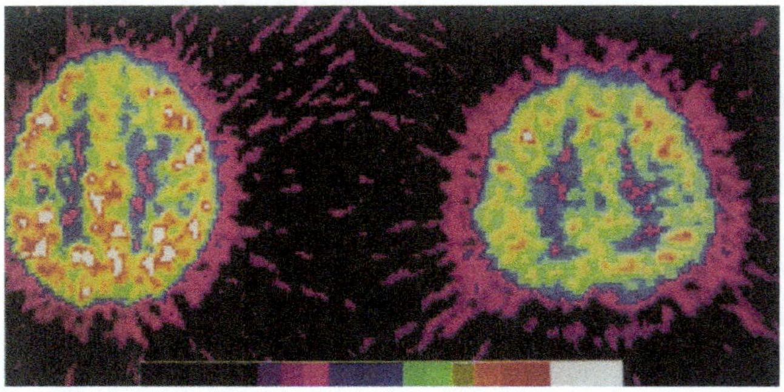

Alan J. et.al, 1990 / Massachusetts Medical Society

FIGURE 11.18 Cortical metabolic activity in neurotypical and ADHD brains More neural activity is reflected in red or white, and lower-than-average metabolic activity is portrayed in green and blue. The cortical metabolic activity of the ADHD brain (right) is significantly lower than that of controls (left).

Williams Syndrome

You might be surprised to find that one neurodevelopmental disorder confers unusual talents. Williams is caused by a deletion error on chromosome 7, and this deletion leads to deficits in vision, a narrowing of the aorta that leads to health problems, low muscle tone, and lower-than-average IQ. Williams individuals have a smaller brain, less white matter, lower density in gray matter, and relatively less volume in the parietal and occipital lobes as well as the thalamus. What is most interesting is the accompanying cognitive abilities. Williams individuals typically have social skills and verbal abilities above those of IQ-matched individuals, and their speech is described as emotional, rich in descriptive words, and lyrical. Many Williams individuals express unusual musical talents and quickly learn to sing or play instruments, while negative behaviors associated with Williams are impulsivity and temper tantrums.

Schizophrenia

Early in this chapter we discussed the process of neuromigration, and likely you were amazed at the forces that direct neurons and their associated structures to their proper places. At that point, you may also have wondered what happens if the process goes awry. If so, you are not the only person with that creative thought, and one theory is that schizophrenia may be caused by disruptions in neuromigration.

Schizophrenia is a mental disorder characterized by psychotic behavior such as hallucinations, delusions, and disordered thinking; loss of motivation and feelings of pleasure; and problems with attention and processing information in a normal fashion. Schizophrenia typically appears in late adolescence or early adulthood and affects roughly 1% of the population, but it is a serious disorder that is difficult to treat. We will discuss psychiatric illnesses in great detail in a subsequent chapter, but here we will consider the possible neurodevelopmental aspects of schizophrenia.

Researchers interested in schizophrenia have wondered if changes to the brain in adolescence might provide a clue to the causes of schizophrenia. Because schizophrenia is a disorder of thought, many studies have looked for abnormalities in the cortex, the seat of cognition. Comparisons of schizophrenic and control cortices have indicated that the neurons in schizophrenic cortices have a somewhat disorganized appearance, which is evidence that the process of corticogenesis is faulty in schizophrenics.

You might imagine that this neuronal disorganization is in some way related to the cognitive symptoms of schizophrenia, and you would be correct, although it is likely that many factors are responsible for schizophrenic behavior. For example, one study compared the slight but normal loss of cortical tissue that accompanies the pruning and reorganization of the cortex through adolescence with cortical loss in schizophrenic people (**Figure 11.19**). Compared with their normal controls, schizophrenic individuals lost a significant amount of cortical tissue in the dorsofrontal region and the temporal and parietal lobes, all regions associated with higher cognitive function, language, and motor coordination. This loss was greater for males than females.

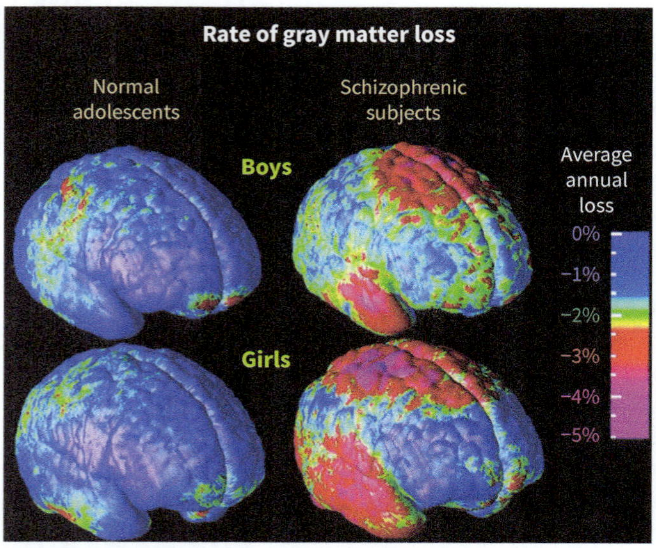

Thompson, P.M., et. al. (2001) / Proceedings of the National Academy of Sciences

FIGURE 11.19 Annual cortical tissue loss in normal and schizophrenic adolescents During adolescence, the cortex becomes highly organized, and in nonschizophrenic adolescents, cortical tissue loss is negligible (left side of the image). Schizophrenic adolescents, on the other hand, lose cortex in regions associated with higher-order cognitive processes (right side of the image).

11.4 Before You Go On

Section Summary

Defects in neurodevelopment due to genetic or environmental forces can lead to serious consequences. When the neural tube is not fully formed either at the back or at the front, it can lead to spina bifida or anencephaly, respectively. Deviations from the normal developmental path may lead to a wide range of disorders such as ADHD, Williams syndrome, and schizophrenia.

Comprehension Check

1. When the neural tube fails to close completely at the caudal end, one consequence may be what?
   - a. anencephaly
   - b. spinus interruptus
   - c. spina bifida
   - d. spina inconsequential
2. Fetal alcohol syndrome (FAS) is characterized by what?
   - a. intellectual difficulties, poor impulse control, and distinguishing facial signs
   - b. low birthweight, lethargy, and dental problems
   - c. adult depression, long spidery limbs, and a tall forehead
   - d. increased incidence of dementia later in life
3. Which of the following statements about ADHD is false?
   - a. ADHD has a significant genetic component.
   - b. Most people with ADHD grow out of it by adulthood.
   - c. ADHD may be related to developmental differences in brain maturation.
   - d. ADHD is caused by prenatal vitamin deficiencies.
4. Individuals with William's syndrome demonstrate what?
   - a. small stature, large philtrum, and large eyes
   - b. tall stature, quiet personality, and agoraphobia
   - c. diminished IQ, long fingers, and mathematical ability
   - d. small stature, cardiac problems, and musical abilities
5. One possible cause of schizophrenia is what?
   - a. disruption in cortical migration
   - b. thicker cortex
   - c. thinner cortex
   - d. disruption in the neural crest

Conclusion

We all share a common neurodevelopmental plan, but subtle differences in the execution of that plan lead to variation. We began our discussion of neurodevelopment with an anecdote about an individual whose variation in neurodevelopment, in the form of ADHD, initially led to difficulties.

Neurodevelopment begins *in utero*. From a simple plate of cells that forms days after conception, a genetic plan creates an unfolding of increasing complexity over nine months, creating a brain that has all the structures necessary to make a human. Neurodevelopment continues throughout one's life in an interaction with experience and environmental influences. When the normal neurodevelopmental process goes awry, the consequences can be serious. Understanding the neurodevelopmental process and the implications of disruptions prepares you to understand topics such as psychiatric and neurological illness, the focus of a later chapter.

Critical Thinking Questions

1. Search online for "risk factors for neurodevelopment." Pick one risk factor, describe it in a sentence, and suggest a way to lessen its impact.
2. Strokes, cancers, and traumatic brain injuries take a heavy toll on individuals, their families, and society. If we could learn how to turn axonal guidance signaling back on in adults, we might be able to treat these disorders. Search online using the phrase "latest advances in axonal guidance" and summarize one thing you learned.
3. Enriched sensory environments appear to have a significant positive effect on brain development. List three things a parent can do to create an enriched environment that are inexpensive and likely to be efficacious.
4. The incidence of spina bifida is greatly reduced by taking vitamin B_9 during pregnancy. The cost of nine months of B_9 is approximately \$12, whereas the cost of surgery for spina bifida is approximately \$30,000, and this does not include the cost of a lifetime of disability for the most severe cases. Do you think that would be worthwhile to provide free B_9 supplements for women contemplating pregnancy if they test for low levels of folate? Explain your answer.

Dig Deeper into Behavioral Neuroscience

11.1 Prenatal Development of Brain Structures

For a general discussion of neurodevelopment:

Stiles, J., & Jernigan, T. L. (2010). The basis of brain development. *Neuropsychology Review*, 20(4), 327–348.

To read about risk factors for neurodevelopmental disorders:

Carlson, T., Molander, F., Taylor, M., & Bolte, S. (2020, June 24). Early environmental risk factors for neurodevelopmental disorders—a systematic review of twin and sibling studies. *Developmental Psychopathology*, 1–48. https://doi.org/10.1017/S0954579420000620

11.2 Prenatal Neurodevelopment at the Cellular Level

Axonal guidance is complicated and not well understood.

Stoeckli, E. T. (2018, May 14). Understanding axon guidance: Are we there yet? *Development, 145*(10), dev151415. https://doi.org/10.1242/dev.151415

11.3 Postnatal Neurodevelopment

Understanding the effects of environmental influences on neurodevelopment is an important research topic. To read about environmental enrichment:

Rosenzweig, M. R., Krech, D., Bennett, E. L., & Diamond, M. C. (1962, August). Effects of environmental complexity and training on brain chemistry and anatomy: A replication and extension. *Journal of Comparative and Physiological Psychology, 55*(4), 429–437.

The effect of congenital cataracts upon adult visual skills:

Kurson, R. (2007). *Crashing through: A true story of risk, adventure, and the man who dared to see.* Random House.

Adolescents are prone to risky behavior, and their neurology might explain why.

Mariam, A., Haque, M., Johal, L., Mathur, P., Nel, W., Rais, A., et al. (2013). Maturation of the adolescent brain. *Neuropsychiatric Disease and Treatment, 9*, 449–461. https://doi.org/10.2147/NDT.S39776

Kann, L., Kinchen, S., Shankin, S. L., Flint, K. H., Hawkins, J., Harris, W. A., et al. (2014). Youth risk behavior surveillance: United States 2013. *Center for Disease Control and Prevention Morbidity and Mortality Weekly Report, 63*, 1–168.

Holmes, A. J., Hollinshead, M. O., Roffman, J. L., Smoller, J. W., & Buckner, R. L. (2016). Individual differences in cognitive control circuit anatomy link sensation seeking, impulsivity, and substance use. *Journal of Neuroscience, 36*(14), 4038–4049.

The aging brain:

Peters, R. (2006, February). Ageing and the brain. *Postgraduate Medical Journal, 82*(964), 84–88. https://doi.org/10.1136/pgmj.2005.036665

Aging and glymphatic flow:

Aalling, N., Munk, A. S., Lundgaard, I., & Nedergaard, M. (2015, December). The glymphatic system: A beginner's guide. *Neurochemical Research, 40*(12), 2583–2589. https://doi.org/10.1007/s11064-015-1581-6

Read more about Snowden's study on how a group of nuns maintained intellectual performance into their later years.

Snowden, D. (2001). Aging with grace: What the nun study teaches us about leading longer, healthier, and more meaningful lives. Bantam Books.

11.4 Neurodevelopmental Disorders

These sources provide additional information about common neurodevelopmental disorders. Fetal alcohol syndrome:

The National Institutes of Health maintains a comprehensive site describing research and treatment of FAS. https://www.cdc.gov/ncbddd/fasd/index.html

ADHD:

Faraone, S., Asherson, P., & Banaschewski, T. et al. (2015). Attention-deficit/hyperactivity disorder. *Nature Reviews Disease Primers 1*, 15020. https://doi.org/10.1038/nrdp.2015.20

Schizophrenia:

Owen, M. J., O'Donovan, M. C., Thapar, A., & Craddock, N. (2011, March). Neurodevelopmental hypothesis of schizophrenia. *British Journal of Psychiatry, 198*(3), 173–175. https://doi.org/10.1192/bjp.bp.110.084384

Chapter 11 Before You Go On Answers

11.1 Before You Go On Answers

1. c. tube

2. a. telencephalon

3. b. the appearance and proliferation of sulci and gyri on the cortex

4. c. the third trimester

11.2 Before You Go On Answers

1. a. neurulation, neuromigration, synaptogenesis, apoptosis, reorganization

2. c. the legs of a chair; the spokes of a wheel

3. b. neural finger

4. d. neurotrophins

11.3 Before You Go On Answers

1. d. They demonstrate that genes open a receptive period for a limited time.

2. a. The gray matter layer in the cortex thins.

3. d. Their frontal lobes are still maturing.

4. b. Brain volume decreases.

11.4 Before You Go On Answers

1. c. spina bifida

2. a. intellectual difficulties, poor impulse control, and distinguishing facial signs

3. d. ADHD is caused by prenatal vitamin deficiencies.

4. d. small stature, cardiac problems, and musical abilities

5. a. disruption in cortical migration

Chapter 11 Glossary

Alzheimer's disease (AD) A progressive neurological disorder that damages memory and essential neural control of bodily functions.

anencephaly A birth defect that results in a child being born without parts of the brain and skull.

apoptosis Cell death that occurs as a normal course of events.

attention-deficit/hyperactivity disorder (ADHD) A common neurodevelopmental disorder characterized by impulsivity, hyperactivity, and inattention.

corticogenesis The process by which the cortex is formed.

critical period A period in development during which a skill or trait is most able to be acquired.

dementia A disorder that involves the loss of memory.

diencephalon The caudal part of the forebrain, which contains the hypothalamus, thalamus, and third ventricle.

ectoderm The germ layer of the embryo that is the source of the nervous system and skin.

endoderm The innermost layer of cells in the germ layer of the embryo, which is the source of the gut and associated structures.

fetal alcohol syndrome (FAS) A spectrum of disorders that result from exposure to alcohol *in utero*.

filopodia Finger-like extensions on a growth cone that respond to signals by growing toward positive stimuli.

glioblast A type of cell that has the ability to differentiate into several types of cells.

growth cone A structure on a developing axon that is seeking a synaptic target.

guidepost neuron A cell that assists in the organization of the nervous system.

mesencephalon Another term for the midbrain.

mesoderm The germ layer of the embryo that is the source of bones, muscles, and the circulatory system.

neural crest A temporary structure atop the neural tube that becomes the PNS.

neural plate The first structure to form in the development of the vertebrate nervous system.

neural tube A simple structure that emerges early in development that forms the basis for the brain and spine.

neuroblast An embryonic cell that develops into nerve fibers.

neurogenesis The process by which neurons are formed.

neuroplasticity The ability of the brain to modify itself throughout life.

neurotrophin A protein that stimulates and controls neurogenesis.

neurulation The process by which the neural plate folds in on itself and becomes the neural tube.

pioneer neuron A type of cell originating in the neural plate that are precursors to cortical cells.

progenitor cell A descendent of a stem cell that can differentiate to become one of many types of cells.

prosencephalon Another term for the forebrain.

radial migration The most common form of neural migration, which begins at the neural tube and extends outward.

rhombencephalon Another term for the hindbrain.

schizophrenia A mental disorder characterized by psychotic behavior such as hallucinations, delusions, and disordered thinking; loss of motivation and feelings of pleasure; and problems with attention and processing information in a normal fashion.

spina bifida A birth defect in which the membrane surrounding the spinal cord does not fully close.

stem cell A cell with the unique ability to develop into many different types of cells.

synaptogenesis The stage of neurodevelopment that involves the creation and proliferation of synaptic connections.

tangential migration Neural migration that is tangential to the ventricular surface.

telencephalon The part of the forebrain that contains the cerebral hemispheres.

teratogen An agent that changes the course of development, including drugs, disease, and environmental events such as genetic disruptions or injuries.

CHAPTER 12

Learning and Conditioning

Ursula Alter / Getty Images

Someone who enjoys eating hot peppers may salivate at the sight of them, whereas someone who once had a bad experience with peppers might feel ill. These responses are both instances of conditioning.

CHAPTER OUTLINE

12.1 Classical Conditioning

- Innate Versus Learned Behavior
  Check It Out: Experience Conditioning
- Pavlov and the Discovery of Classical Conditioning
  Try It Out: Classical Conditioning in Everyday Life
- Watson and the Little Albert Experiment
  Think About It: Ethics in Psychology Research
- Conditioned Taste Aversion

12.2 Operant Conditioning

- The Discovery of Operant Conditioning
- Reinforcement and Punishment
  Try It Out: Turning Vices into Virtues
- Conditioning Phenomena
- Reinforcement Schedules
  Try It Out: Operant Conditioning in Everyday Life

12.3 Learning and the Brain

- Where Memories are Stored
- The Neuronal Basis of Learning
- Long-Term Potentiation
  Check It Out: Long-Term Potentiation
- NMDA and AMPA Receptors
  Advances in Behavioral Neuroscience: Karl Deisseroth and Optogenetics

LEARNING OBJECTIVES

LO 12.1 **Describe** the processes of classical conditioning.

LO 12.2 **Describe** how operant conditioning occurs.

LO 12.3 **Explain** where and how information is stored in the brain.

Stories of the Brain

After a few years in her first job, Veronica felt she was ready for a change, so she found a new position halfway across the county. Her coworkers were happy for her good fortune but sad to see her go, so the week before she left town, a friend threw a going-away party to send her off with their good wishes.

As a special treat, the host provided the group with a dinner of steamed and seasoned crabs, quite an unusual delicacy for individuals who lived in the Midwest. Veronica had never tasted crab served this way before, so it made an impression on her. Unbeknownst to the host, the food was tainted, and all the guests developed food poisoning that night and were violently ill all the next day.

When Veronica arrived at her new job on the east coast, her colleagues invited her over for a meet and greet and, in a welcoming gesture of hospitality, served the food that the region was known for: highly seasoned steamed crabs. As she gazed upon the main course, Veronica felt a wave of nausea come over her.

Take a moment to consider Veronica's story. She was sick after eating crabs, so she now associates crabs with nausea. Her association of steamed crabs (a stimulus) with illness (a bodily state) happened over one exposure and will likely continue for some time, despite the fact that she enjoys seafood and is unlikely to encounter tainted crabs ever again.

Understanding how the nervous system adapts to its environment is one of the most basic questions in behavioral neuroscience. How do living organisms modify their behavior? How do they learn? Organisms that are locked into the behaviors they were born with cannot profit from experience. They are at a great disadvantage compared to organisms that can modify their behavior and adapt to changing circumstances.

In this chapter, we will focus on the simplest forms of learning that humans share with the panoply of living organisms that make up our world. Many human behaviors are built on the foundation of simple learning, and animal models of potentially disruptive behavior such as gambling can help us understand compulsive gambling in humans and design more effective treatment plans. We will also consider the biological mechanism for simple learning and how learning changes the brain. We begin with a discussion of classical conditioning.

12.1 Classical Conditioning

LO 12.1 Describe the processes of classical conditioning.

As you become more familiar with neuroscience research, you will realize that it often involves using animal models to investigate some aspect of simple learning such as time to acquire a behavior, learn a maze, or associate a stimulus with a response. Therefore, it is worth our time to delve into the mechanism involved in simple learning. Higher-order mental processes such as problem solving, reading, and language are considered cognitive processes, and these will be discussed in a subsequent chapter.

Innate Versus Learned Behavior

All living organisms come equipped with some hard-wired behaviors known as reflexes that are essential for their survival. Reflexes have an innate, or inherited, basis and are seen in all members of a species. For example, bacteria may move toward light or away from oxygen, depending on their needs; flowers open their petals and turn their faces toward the sun in the morning and then close their petals at dusk; all animals naturally retreat in the face of fire and shiver in response to deep cold. These are all innate behaviors, which means that they are automatic and unthinking responses to a stimulus. Reflexive reactions to potentially dangerous stimuli help us survive because when we encounter a threat, we react swiftly without the need to consider a course of action.

Humans have their own innate responses; for example, an object unexpectedly looming in our face will cause us to blink, and a loud noise will create a defensive reaction. One example of an innate response to a stimulus is the Moro reflex seen in infants. A loud noise or a sensation of falling will cause an infant to arch their back, then fling their arms out and open their hands, and finally bring their knees and hands in close to their body (Figure 12.1). It is thought that the evolutionary basis of the Moro reflex is to clutch the mother and bring her close in case of danger.

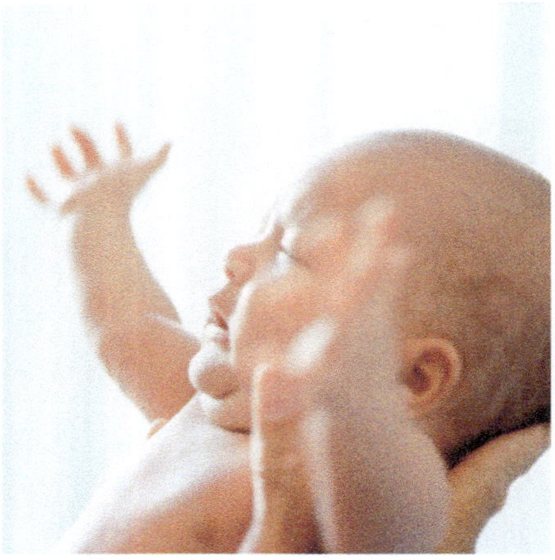

Steve Prezant / Getty Images

FIGURE 12.1 The Moro reflex When startled, an infant responds reflexively with outstretched arms and open hands followed by a clutching response.

As useful as reflexes are, imagine for a moment what an advantage it would be for any organism to be able to create new reflexes. They could then adapt to changing conditions by associating a new and more useful behavior with a situation they regularly encounter. Such an adaptation would be a powerful tool in the competition of life. This ability to learn from experience is one important contributor to the evolutionary success of humans.

Behavioral scientists often use the word learning differently than the average person, so we begin with a definition. In behavioral science, **learning** is the permanent or semipermanent acquisition of new behavior or knowledge, or the modification of prior behavior or knowledge. Learned behaviors differ from reflexes in that learned behaviors are not innate and so do not happen automatically in response to a stimulus.

Psychologists and behavioral neuroscientists refer to this simple form of learning as **conditioning**. Conditioning can occur unconsciously through repeated everyday experience, or it can be implemented purposefully through training, as is commonly seen with pets who are housebroken and trained to play fetch through conditioning (Figure 12.2).

Petr Jilek / Shutterstock

FIGURE 12.2 A dog demonstrating an acquired behavior Dogs have an innate desire to chase a thrown object and through conditioning can be taught to bring the object back.

Consider the differences between innate and learned behavior in this example. The fact that you salivate in response to the presence of food is a reflexive behavior that you were born with and that you share with all other humans and many other animals. The production of saliva is an essential first step in digestion, so it makes sense that it would be reflexive. However, salivation can also occur as a learned behavior. The fact that you begin to salivate when grandma says, "Come to the table," is an example of a conditioned behavior because past experience has shown that you will eat after coming to the table.

The human nervous system is designed to condition, regardless of our conscious awareness of the process. You experience conditioning in your own life, especially in situations where a setting is paired with an emotional experience. For example, have you ever been in a crowd when you heard a phone ring tone, and then you and many others reflexively reach for your phones? Perhaps as you walked through a mall, you began to salivate as the aroma of freshly baked cookies wafted into your nostrils. For many students, simply watching an instructor hand out exams can cause a rapid heartbeat and anxiety. All of these are examples of classical conditioning in our daily lives. You can experience the ease with which conditioning occurs in Check It Out: Experience Conditioning.

If you are a fan of spicy foods, take a moment to visualize yourself sitting down to eat a spicy Indian curry, a pepper-based Thai dish, or nachos with jalapeno peppers. You might find that you are starting to salivate *at this moment*. How does this happen? The answer to this question began to emerge more than a century ago in a laboratory in Russia.

Pavlov and the Discovery of Classical Conditioning MCAT

One of the first questions that psychology investigated was how we acquire new behaviors. An early systematic exploration into learning was conducted by Ivan Pavlov (1849–1936), a physiologist who was the head of the Institute of Experimental Medicine in St. Petersburg, Russia, and whose research led to the Nobel Prize in Medicine or Physiology. Pavlov was interested in the physiology of digestion and used canine models to understand the process. To

Check It Out | Experience Conditioning

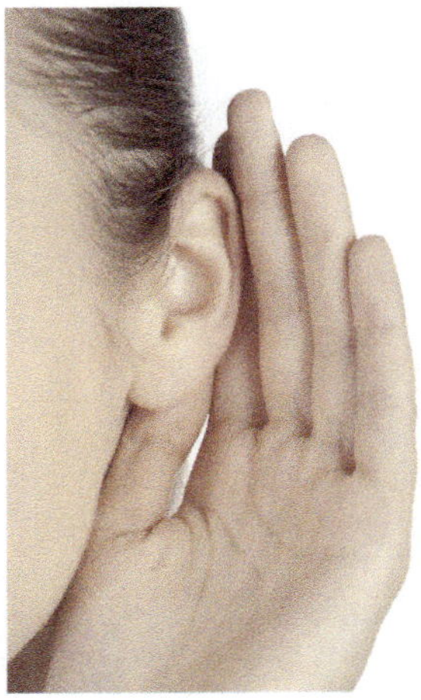

DenisZbukarev / Getty Images

In this activity, you will be presented with a series of tones that generate a UCR in the form of a slight physiological response. Before you begin, adjust the volume on your speakers or headphone to a normal level. Play the video **Experience Conditioning**, and then answer these questions.

1. When an individual watches this video while attached to sensors, a small rise in heart rate and blood pressure can be observed after the last tone. Is this response a UCS, UCR, CS, or CR? Explain.
2. Did you experience a slight physical or emotional response at the end? If so, briefly describe it.

this end, he surgically implanted a cannula, or small tube, in the mouth of each of his canine subjects in order to collect their saliva when they salivated in response to meat powder that he administered.

Pavlov noticed that the dogs in his kennel became excited even before they were fed when a caretaker appeared in the kennel. This made him wonder if he could connect an unrelated event such as the ringing of a bell to their salivary response. He chose a bell because it had no intrinsic relationship to feeding. When the bell was rung without being paired with meat powder, the animal did not respond by salivating. However, after he repeatedly paired meat powder with the sound of a bell ringing, the dogs began to salivate as soon as the bell was rung, even if no meat powder immediately appeared. Pavlov realized that the dogs had learned a novel behavior in response to the pairing of meat powder with the sound of a bell. Pavlov had discovered **classical conditioning**, in which a new stimulus is associated with an innate response.

We can use the example of Pavlov's dogs to explore the standard components of classical conditioning. Prior to conditioning, the meat powder is an **unconditioned stimulus (UCS)** that naturally causes the **unconditioned response (UCR)** of salivation in the hungry dogs. Pavlov showed that he could link an unrelated stimulus (ringing a bell) to the existing salivary reflex by consistently pairing the bell with the appearance of food. In this example, the bell is the **conditioned stimulus (CS)**, and salivation to the sound of the bell is a **conditioned response (CR)**. It is important to note that as a result of conditioning, the original UCR (salivation) became a conditioned response. This relationship among stimuli and responses is summarized in **Figure 12.3**.

Pavlov demonstrated how conditioning progresses over repeated trials by measuring when the dogs began to salivate during each trial. On the first trial of the experiment, he rang the bell 15 seconds prior to the delivery of the meat powder (UCS), and no salivation (UCR) occurred until the appearance of the UCS. By the tenth trial, the dogs began to salivate in response to the bell slightly in advance of the delivery of the meat powder. By the twentieth trial, the ringing of the bell (now the CS) had become associated with the appearance of the meat powder, and the bell immediately elicited a conditioned salivary response (now the CR). **Figure 12.4** shows how the response changed over time.

Researchers soon realized that conditioning is more or less successful depending on the time that elapses between the CS and the UCS. Conditioning occurs most rapidly when the CS immediately precedes the UCS. Conditioning is slower or fails altogether if too much time elapses between the CS and the UCS or if the CS follows the UCS rather than preceding it. This is because as the time interval between the CS and the UCS increases, the likelihood of the subject associating the two stimuli is diminished. An example of this is the bell (CS) that signals the arrival of an elevator (UCS). When you hear the bell, you have been conditioned to expect the elevator to arrive immediately, so you orient toward the door before it even opens,

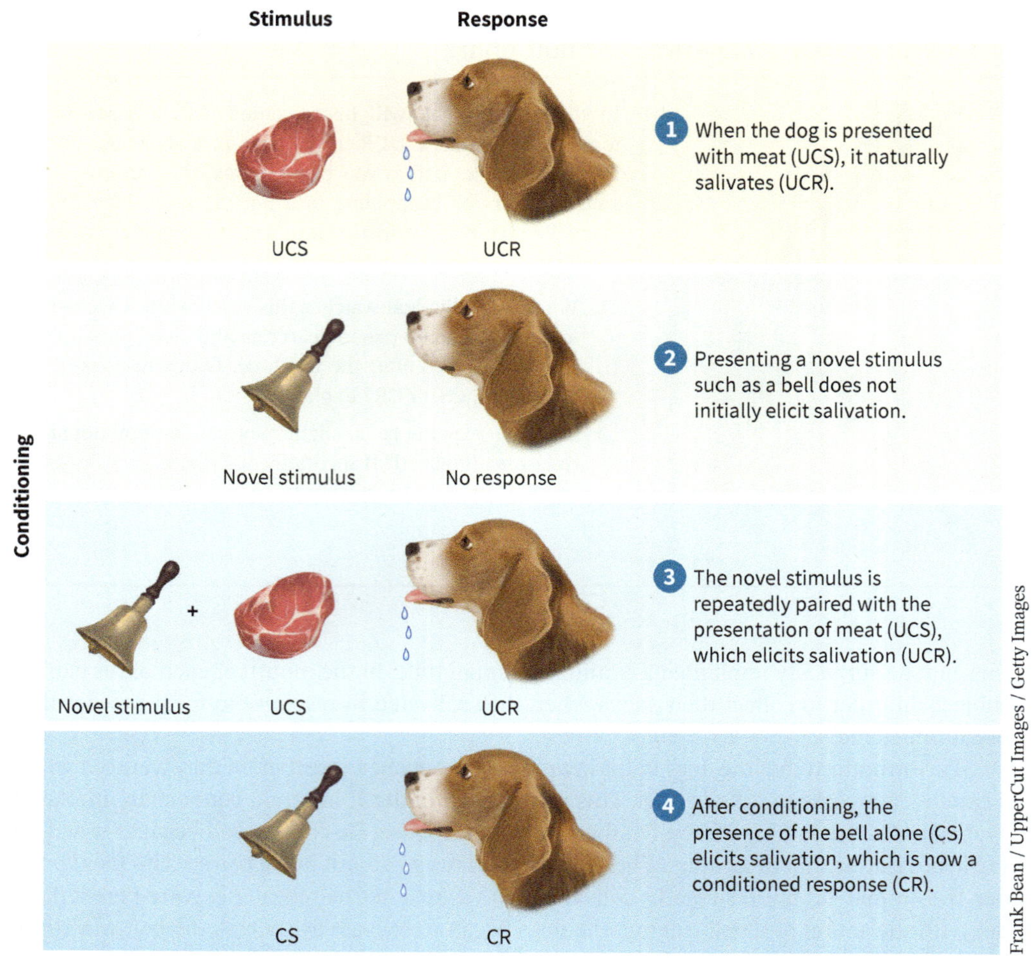

FIGURE 12.3 Classical conditioning as demonstrated by Pavlov Classical conditioning associates a stimulus (UCS) that elicits an innate response (UCR) with a neutral stimulus repeatedly until the neutral stimulus alone (CS) elicits the response (CR).

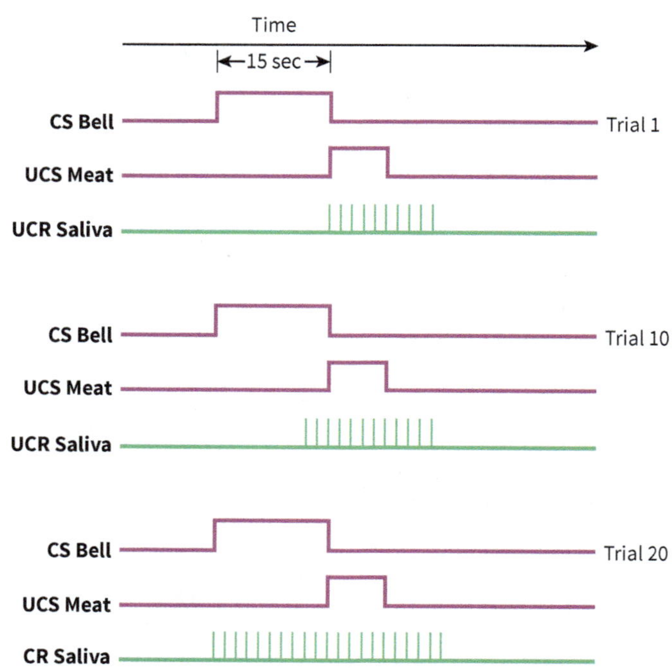

FIGURE 12.4 Conditioning When the bell was first paired with the meat powder, it had no effect on saliva production. By the tenth pairing, the bell had increased saliva production. By the twentieth pairing, the bell alone could initiate salivation.

and this makes loading the elevator more efficient. If the bell preceded the elevator's arrival by 2 minutes, or rang when you were already inside, you can see why the conditioning would not be effective.

Now that you understand the relationship between the UCS, UCR, CS, and CR, you can consider how classical conditioning works in everyday life in **Try It Out: Classical Conditioning in Everyday Life**.

Try It Out | **Classical Conditioning in Everyday Life**

We tend to equate classical conditioning with training pets, but it can also explain many human behaviors. In this activity, you will consider several everyday situations from the perspective of classical conditioning. Complete the worksheet **Classical Conditioning** to identify components of the situation with the UCS, UCR, CS, and CR.

Watson and the Little Albert Experiment

Researchers soon tested Pavlov's methods of classical conditioning on humans to understand its implications for understanding human behavior and psychiatric illnesses. One famous experiment was conducted by behavioral psychologist John B. Watson a century ago. Watson obtained parental permission to conduct an experiment using a nine-month-old infant known in the literature as Little Albert as his participant.

Watson first demonstrated that the infant showed no fear response to a wide range of objects such as a white rat, a rabbit, and a monkey. Then he repeatedly struck a metal bar with a hammer near Little Albert, making a loud sound (UCS) that frightened him and caused him to cry (UCR). Having established the UCS–UCR association, Watson then paired the appearance of a white rat (CS) with the striking of the metal bar. In short order, Little Albert responded to just the presence of the rat and visually similar animals with wailing and tears (**Figure 12.5**).

The Little Albert experiment raised the possibility that psychiatric behaviors such as phobias might be created by classical conditioning to aversive events. If this experiment that conditioned fear in an infant makes you uncomfortable (and it should), you should know that the Little Albert study would never be permitted today. Learn more about the regulation of human research in **Think About It: Ethics in Psychology Research**.

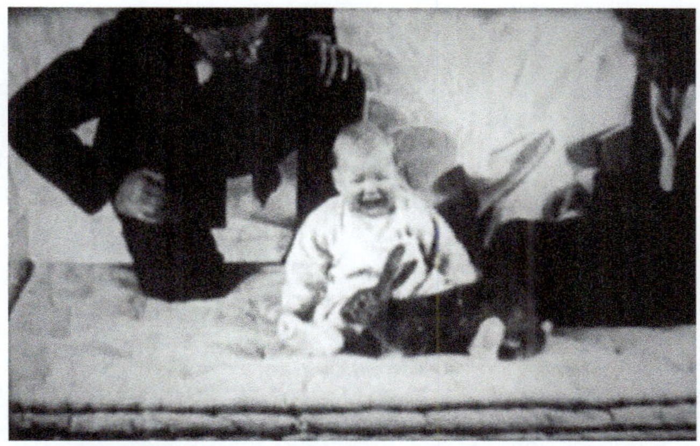

John B. Watson / Wikimedia Commons / Public domain

FIGURE 12.5 **The Little Albert experiment** Little Albert showed no fear of the rabbit prior to conditioning, but after the pairing of the aversive sound with the rabbit, he responded with distress to the rabbit's presence.

Think About It | **Ethics in Psychology Research**

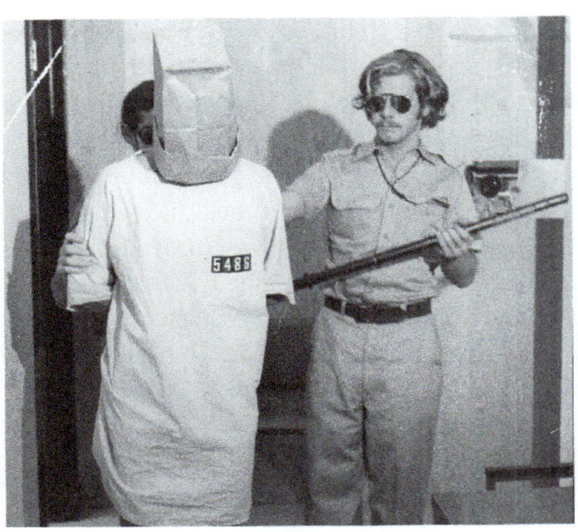

San Francisco Chronicle / Hearst Newspapers via Getty Images / Hearst Newspapers / Getty Images

When we read about significant discoveries in the history of science, we sometimes find that the experiments that led to the discoveries had a negative impact on their human subjects that would be unacceptable to us today. The Little Albert study is one of these.

A more recent problematic psychological study took place at Stanford University in the early 1970s. The object was to determine if personality or environment was the key to understanding why people in control might lose empathy for those under their power. Researchers assigned student participants to the roles of either prisoners or prison guards and housed them in a simulated prison where they would live for 2 weeks. The "prisoners" were subject to so much humiliation and abuse from the "guards" that the study was shut down after only 6 days.

Although the study investigated questions with real social value, the science behind the experiment has been criticized, and its ethics condemned.

Today psychological studies like the Little Albert experiment or the Stanford prison experiment would not be allowed. Beginning in 1974, academic and other research institutions in the United States are required to create an Institutional Research Board (IRB) to review research proposals involving human subjects in order to ensure that the rights of participants are protected. A similar review board to oversee research using animals in the United States was mandated in 1985.

Beyond these federal regulations, the American Psychological Association (APA) code of ethics requires that researchers using human participants adhere to certain rules when designing their studies, which are paraphrased here:

Informed consent: Participants must be informed about the purpose and duration of the study and what their participation will involve.

Right of refusal: Participants can end their participation at any time without penalty.

Risks and benefits: Researchers must inform participants of all potential risks of participation.

Respect for confidentiality and data protection: Personal information about study participants must be kept confidential with safeguards for the preservation of anonymity.

Questions after participation: Individuals have the right to ask questions about their study after their participation ends.

Treatment options: For research investigating disorders or aversive experiences, participants must be informed about available treatments.

As you continue in your study of behavioral neuroscience, notice that some discoveries we discuss were made using experimental techniques that would violate today's ethical codes.

Questions

1. **Describe how you would adapt the Little Albert experiment so that it would meet the requirements in the APA code of ethics.**
2. **A researcher is rewarding study participants with a small credit that they can use to boost their grade. One participant decides that they want to withdraw partway through and is warned that if they don't finish the experiment, they will not get the extra credit. Using the APA code of ethics, argue for or against this researcher's methods.**
3. **A researcher posts a notice on the campus public bulletin board asking any gay or lesbian student who wishes to participate in a study to come to room 210 in the Student Union at 7 PM this Saturday. Using the APA code of ethics, argue for or against this researcher's methods.**

We now know that psychiatric disorders such as phobias and panic disorder may in part be a result of conditioning. Therefore, understanding how a behavior can be conditioned can provide insights into how to end the association. In fact, many forms of therapy involve the use of conditioning phenomena to modify unhealthy behaviors into healthier versions. One example of this is systematic desensitization, in which an individual is exposed to a fear stimulus in gradually increasing levels until the fear response is extinguished. A therapist treating a person with a phobia of flying might have the individual first drive past the airport without stopping, then park but not leave the car, then drive to the airport and have a meal before leaving, and so on until they feel comfortable boarding the plane.

Now that you understand how conditioning usually occurs, we will turn to one special case of conditioning.

Conditioned Taste Aversion

The rules of conditioning typically require that the UCS quickly follow the CS and that the association be repeated multiple times. There are exceptions to every rule, however, and one such exception is a special case of conditioning known as **conditioned taste aversion** (or the Garcia effect) that was identified by psychologist John Garcia (1917–2012). Conditioned taste aversion refers to the fact that a strong aversion to a specific food can appear after only a single exposure that results in an aversive experience such as nausea or allergic reaction, even when that reaction occurs hours after the original food stimulus. This was exactly the situation portrayed in the *Stories of the Brain* that opened this chapter, in which Veronica developed a lasting aversion to crab after one incident of food poisoning by crab.

If you think about it for a moment, it makes sense that we have an innate mechanism designed to protect us from poisonous foods. After all, you may not have a second opportunity to learn that eating a particular mushroom is a deadly mistake. For this reason, it is not surprising that a single negative response to a food often results in a long-term food aversion, even if there is reason to believe that the illness was not the result of ingesting the food. In fact, this evolutionary protection may explain why children are naturally wary of novel foods (**Figure 12.6**).

FIGURE 12.6 A child expressing disgust Children are naturally wary of novel foods, a behavior thought to be protective in our evolutionary past.

Food neophobia (a persistent reluctance to try new foods) is not uncommon in childhood and may have an evolutionary basis as a way to ensure that our ancestors avoided potentially dangerous foods. This innate caution about new foods can usually be overcome through conditioning. Rather than press the child to try the food and possibly amplify the response, parents are advised to continue to make the novel food occasionally available during meals. After a month or so, the child will probably become habituated to its appearance and aroma and be willing to try it.

The important thing to remember is that food aversions are a special case of learning in that the aversive stimulus can follow the UCS of food by many hours, and conditioning can occur even after a single exposure.

12.1 Before You Go On

Section Summary

We are born with innate reflexes that help us survive, but learning allows us to adapt to changing situations by creating new behaviors based on experience. A stimulus can be associated with a response through classical conditioning when the conditioned stimulus slightly precedes the unconditioned stimulus and is repeated multiple times. Conditioned responses can contribute to phobias or other disorders, and therefore conditioning can also be used in their treatment. Taste aversion is an exception to the standard rules of conditioning in that a negative association with a certain food may occur even when the negative event occurred only once and many hours later.

Comprehension Check

1. Which of the following is an innate behavior?

- **a.** being annoyed when you find you missed your flight
- **b.** feeling happiness as you see the sunset
- **c.** being startled in response to a sudden loud sound
- **d.** feeling excited on your birthday

2. Which of these is an example of classical conditioning?

- **a.** learning to read and evaluate risks and benefits
- **b.** linking a behavior such as making a joke with being in a comedy club
- **c.** associating an existing innate behavior such as blinking with a tone
- **d.** drawing a connection between being stuck with a needle and then feeling pain

3. When a nurse enters the room to draw your blood, you begin to feel faint. In this example, the UCS is ______, the UCR is ______, the CS is ______, and the CR is ______.

- **a.** the needle; fear; faintness; the phlebotomist
- **b.** the needle; fear; the phlebotomist; faintness
- **c.** the phlebotomist; faintness; the needle; fear
- **d.** fear; the needle; faintness; the phlebotomist

4. Which of these is an example of conditioned taste aversion?

- **a.** feeling nauseated at the thought of shellfish after shellfish made you sick
- **b.** a child showing disgust at a new food and refusing to try it
- **c.** finding unfamiliar food from other cultures disgusting
- **d.** salivating at the image of a cheeseburger in an advertisement

12.2 Operant Conditioning

LO 12.2 Describe how operant conditioning occurs.

Classical conditioning explains how an association can be created between a neutral stimulus such as a bell and an existing unconscious behavior such as salivation or heart rate. In classical conditioning, the association is created when the stimulus *precedes* the behavior. Behavior can also be modified by the consequences that *follow* behavior through **operant conditioning**, sometimes also called instrumental conditioning.

In operant conditioning, the consequences of a behavior modify the frequency of that behavior in the future. Another distinguishing aspect is that classical conditioning builds on involuntary and reflexive behaviors, whereas operant conditioning operates on voluntary behaviors. The key to understanding operant conditioning is its relationship with consequences of behavior, as opposed to its antecedents.

The Discovery of Operant Conditioning

Operant conditioning was first studied by Edward Thorndike (1874–1949), who proposed what is known as the **law of effect**. This law simply states that actions followed by pleasurable consequences are likely to increase in frequency, whereas actions followed by unpleasant consequences are more likely to decrease in frequency or even disappear.

A better-known pioneer of operant learning is B. F. Skinner (1904–1990), who spent his career investigating the applications of operant conditioning in animals and humans. Skinner believed that free will is an illusion and all our actions are shaped by the consequences of our behaviors. Skinner's proposal during WWII to create a ship-to-ship missile controlled by a pigeon was not well received. Perhaps the most strident proponent of operant conditioning in everyday life, Skinner was the inventor of an experimental device called the **operant chamber**, also known as the Skinner box, shown in Figure 12.7.

In a typical operant conditioning scenario, a rat is placed in the operant chamber that includes red and green lights, a loudspeaker, and a response lever that, when pressed, will deliver a food pellet to the hungry rat. The task to be learned might be to press the response bar to deliver a pellet when the green light is lit but not the red light, or to press the bar when a high-pitched tone is presented but not a low-pitched tone. The bottom of the chamber consists of metal bars that can be used to deliver an electrical shock for incorrect responses.

The operant chamber allowed researchers to explore the link between actions and consequences methodically. Such experiments created a body of work known as **behaviorism**. Behaviorists such as Skinner thought that human and animal behavior could be explained entirely through conditioning, without reference to higher-level behaviors such as thoughts, intentions, or feelings. Today we understand that some kinds of behaviors can be explained by conditioning, but many others cannot. Nonetheless, behaviorism did uncover several phenomena that humans and other organisms exhibit regularly. Primary among these are the roles of reinforcement and punishment in modifying behavior.

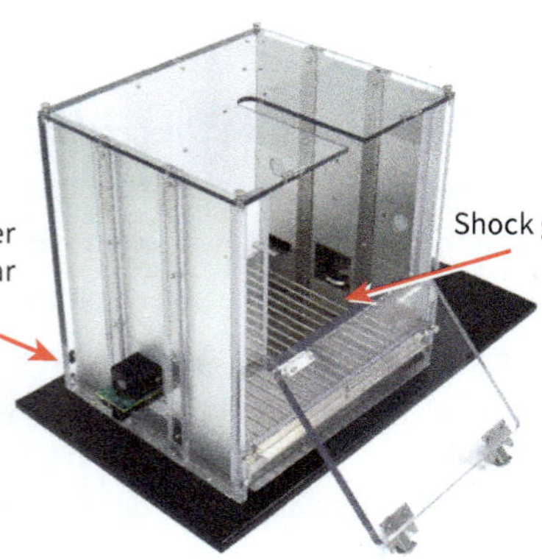

Image Courtesy Lafayette Instrument company, Inc.

FIGURE 12.7 **A typical operant chamber** An operant chamber is commonly used in experiments investigating learning using rodents as participants. It typically consists of a mechanism to deliver a food pellet in response to the rat's behavior and an electrified floor to deliver an aversive but not dangerous shock.

Reinforcement and Punishment MCAT

The mechanism that alters behavior in operant conditioning is the relationship between a behavior and its consequences. A behavior that is followed by reward, or reinforcement, will be modified to occur more frequently, whereas a behavior followed by punishment will occur less frequently.

Reinforcement The mechanism that causes a behavior to be expressed with greater frequency is **reinforcement**. Reinforcement can be positive or negative. Positive reinforcement occurs when the behavior is followed by a positive stimulus. Rewarding a rat with a food pellet for pressing a bar is one example; giving a discount to regular shoppers is another instance of positive reinforcement.

Negative reinforcement occurs when the target behavior is followed by the removal of an unpleasant stimulus. An example of negative reinforcement is a high-school student whose parents limit the student's privileges until their grades are raised. It is important to notice that negative reinforcement is not the same as punishment. Restoring the privileges is actually a reward for the student increasing studying behavior.

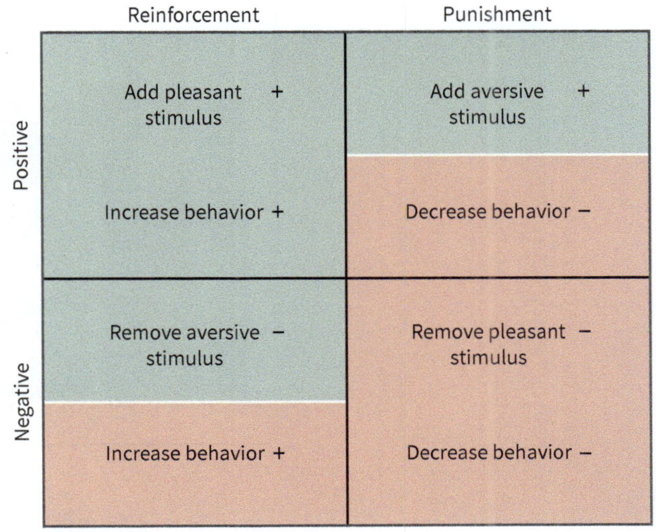

FIGURE 12.8 Reinforcement and punishment Both punishment and reinforcement can be either positive or negative.

Punishment In contrast to reinforcement, where the goal is to increase the frequency of the behavior, the goal of **punishment** is to decrease the frequency of the behavior. Punishment can also be either positive or negative. Positive punishment consists of presenting a negative stimulus after the target behavior in order to decrease or eliminate the behavior. An example of positive punishment is the shock that follows brushing up against an electric fence. The shock motivates the individual to avoid brushing up against the fence a second time.

By contrast, negative punishment is the removal of something pleasant to decrease or eliminate behavior. An example of negative punishment is taking away a child's phone after they misbehave. The parent hopes that losing the phone will encourage the child to decrease bad behavior in the future. Most people confuse negative reinforcement with punishment, but they are not the same; negative reinforcement takes away something unpleasant, whereas punishment takes away something pleasant. Figure 12.8 summarizes the relationship between reinforcement and punishment.

According to legend, students of a famous behaviorist decided to condition him into a novel behavior as a prank. Whenever the instructor lectured from the right side of the classroom, the students all looked puzzled and confused, but when he moved to the left side, they all smiled and nodded. By the end of the semester, the professor had been conditioned to go to the left side of the room whenever he wanted to make a point. While this story is likely untrue, instructors do look to their students' faces to gauge their comprehension of a lecture and adapt their teaching in response, which is a kind of conditioning.

Examples of operant conditioning in everyday life are numerous. If you decide that you must study until 10 PM to earn a treat, you are using operant conditioning to motivate your studying. Advertisers rely on operant conditioning to motivate you to buy their products. They use celebrities to promote their products with the expectation that you will want to buy the product because you associate it with your admiration for the celebrity. Many people have a step counter on their phone or watch that they set to alert them when they have reached a certain number of steps; the message of reaching the goal is reinforcing. You can also use your knowledge of conditioning to eliminate habits that you find undesirable. In **Try It Out: Turning Vices into Virtues**, you can see how to use operant conditioning to your advantage.

Conditioning Phenomena MCAT

There are several phenomena associated with all types of conditioning that merit a moment of our time. Habituation, extinction, sensitization, generalization, shaping, and recovery from extinction are all are conditioning phenomena that apply to everyday life.

Habituation occurs when repeated presentation of a stimulus such as a UCS no longer elicits the CR, which must be an involuntary or unconscious behavior. An example of habituation is when individuals who live near an airport or busy city street become accustomed to the

Try It Out | Turning Vices into Virtues

Syda Productions / Adobe Stock

Habits are decisions about how we act in specific situations that occur repeatedly. We encounter hundreds of choices in our daily lives, and habits help us speed through many of them as we act in a way that has worked for us in the past. But habits can be good or bad, and bad habits are difficult to change. The worksheet **Turning Vices into Virtues** can guide you as you identify the stimuli and reinforcements that maintain an undesirable habit and replace that habit with a desirable one.

noise (Figure 12.9). The first time a jumbo jet roars overhead during takeoff, the individual will react to the noise, but after some time, the UCS of the noise no longer elicits any response. Doctors in training also often show habituation to the sight of blood; they may faint the first time they see an operation, but after repeated exposures, they are able to view such stimuli without anxiety.

If involuntary behaviors habituate, what of voluntary behaviors that are no longer reinforced? When a behavior that has been acquired through operant conditioning is no longer reinforced, that behavior will eventually fade away through **extinction**. When the elevator you take to work has been out of order for some time, you will eventually walk past the doors toward the stairs without even looking to see if the lights are on; the behavior of stopping at the elevator has been extinguished.

Sensitization is the opposite of habituation and describes the phenomenon in which repeated application of a stimulus leads to a lower threshold for response. An example of sensitization is that substance abusers can experience a heightened response to intoxicating substances; this may play a role in relapse behavior for some addicts. Sensitization may also explain the paradoxical perception of increased pain that some patients experience while

Tyrannosaurus Rex / Shutterstock

mura / Getty Images

FIGURE 12.9 Examples of habituation If you live near a significant source of noise like an airport or busy intersection, you will find that over time you no longer hear the noise. This is an example of becoming habituated to a stimulus.

suffering from chronic pain; continuous or repeated pain can lead to a decrease in threshold and an increase in perceived pain. The mechanism by which the receptor site becomes increasingly sensitive to a stimulus is not well understood.

Generalization is the principle that a novel stimulus similar to an already conditioned stimulus may also be conditioned, even though that object or event was never actually paired with a UCS. In the Little Albert experiment, Albert was conditioned to express fear when exposed to a white rat. He was then shown a white rabbit. Due to generalization, Albert expressed distress at the sight of a second white, furry animal even thought it had never been paired with the loud noise.

Recovery from extinction refers to the rapid reestablishment of a learned behavior that had been extinct. Recovery from extinction is a phenomenon that is applicable to many real-world situations such as parenting. Imagine a child who whines incessantly for a treat whenever their parent approaches the supermarket checkout. The parent typically refuses to reinforce this behavior by buying a treat, so the child eventually gives up whining and the behavior is extinguished. However, if in the near future the child whines and the parent gives in "just this one time," then it is likely that the whining will recover at full strength (**Figure 12.10**). The key point about recovery from extinction is that a behavior that took several or even many trials to acquire, once placed in extinction, can recover quickly and durably if reinforcement is delivered.

SpeedKingz / Shutterstock

FIGURE 12.10 Recovery from extinction When an undesirable behavior has been established and then placed on extinction, a single reinforcing event can result in the return of the previously extinguished behavior.

When you are trying to condition a response to an especially complex behavior, **shaping** is a useful technique. In shaping, you reinforce for any approximation to the desired behavior. Once a connection is created, you continuously tighten the criteria for reinforcement until the target behavior is reached. For example, a parent trying to get their child to straighten up their room might at first reward for just putting dirty clothes in a hamper. After that behavior is established, reinforcement can then be withheld as additional behaviors such as vacuuming and making the bed are established.

Reinforcement Schedules

Research into operant conditioning has investigated how changing the timing of the reinforcement, or the reinforcement schedule, affects conditioning. Schedules of reinforcement can be classified as either continuous or intermittent, and they apply to both reinforcement and punishment.

Continuous reinforcement (also known as CR, not to be confused with a conditioned response) is when reinforcement is delivered after every time the behavior is emitted. For example, a vending machine delivers a candy bar every time you put in the correct amount of money. Operant conditioning of a behavior using continuous reinforcement is the quickest way to teach a new behavior. A disadvantage of continuous reinforcement is that when the reinforcement stops, the conditioned behavior quickly ceases. How long would you continue to put money into a vending machine once it failed to produce candy? Also, continuous reinforcement does not reflect most real-world scenarios, so it is not an ideal model for understanding human behavior.

Intermittent reinforcement (IR), sometimes called partial reinforcement, is when reinforcement does not follow the behavior every time. Intermittent reinforcement schedules are frequently used in research as a better approximation of real-world situations in which reinforcement rarely comes immediately after every instance of behavior. Intermittent reinforcement can occur after a specific interval of time following the behavior or following the last reinforcement. Alternatively, it may depend on the number or ratio of responses since the last reinforcement.

Fixed-interval (FI) reinforcement only delivers reinforcement after a specific interval of time. For example, an FI20 schedule would delay additional reinforcement until 20 seconds after the last reinforced response, regardless of how many correct responses occurred in the

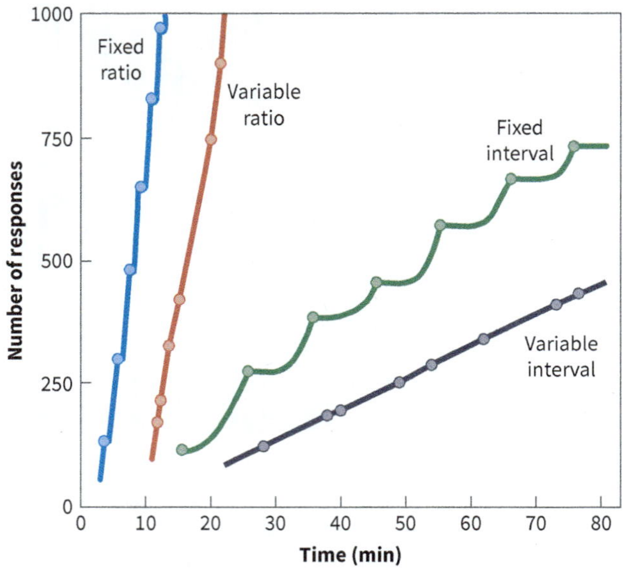

FIGURE 12.11 Common intermittent reinforcement schedules The time it takes to condition a behavior depends on the type of reinforcement schedule used. Ratio schedules lead to rapid responding, and interval schedules lead to slower rates of response.

interval. If the additional reinforcement occurs around 20 seconds post-reinforcement as the mean but sometimes at 22 seconds and sometimes 17 seconds, that is known as **variable-interval (VI) reinforcement**.

Reinforcement can also occur as a function of the number or ratio of responses. Unlike an interval schedule, in which reinforcement is delayed until the interval ends, **fixed-ratio (FR) reinforcement** and **variable-ratio (VR) reinforcement** deliver reinforcement based on frequency of response. An FR15 schedule delays the next reinforcement until 15 responses are made, and a VR12 schedule delays reinforcement around a mean of 12 responses, for example, sometimes after 10 responses and sometimes after 14.

VR schedules have been shown to be especially resistant to extinction. Card games, slot machines, and dice all pay off in a VR schedule, so this explains why gambling addiction is particularly hard to overcome. It is estimated that approximately 2 million U.S. citizens are heavily addicted to gambling and perhaps 20 million gamble to the extent that it is seriously disruptive to their life. Unlike the vending machine example, where you would stop inserting money almost immediately when you saw it was not producing candy, you are conditioned to expect slot machines to pay out some times and not others. The possibility of a reward on the next try motivates you to put money in again and again.

Different reinforcement schedules are more or less effective for different goals, varying in the speed at which conditioning is accomplished as well as in the number of responses generated. Both types of ratio schedules condition faster than interval schedules. In a FR or VR schedule, the faster you respond correctly, the more reinforcement you receive. This is not true for interval schedules, in which a rewarded response will always be followed by a time in which additional correct response are not rewarded.

As Figure 12.11 shows, among all intermittent reinforcement schedules, fixed-ratio reinforcement typically generates the greatest number of conditioned responses in the shortest time, whereas variable-interval reinforcement is slowest at conditioning. Fixed-interval reinforcement is unique in that it demonstrates periodic dips in responding, as indicated by scalloped shapes in the FI line on the graph. If you have ever been in a course where the instructor gave a quiz every Friday and you only studied on Thursday nights, you have experienced the limitations of an FI-reinforcement schedule. Because you know when to expect the assessment, the conditioned behavior (studying) falls off during the period when you know there will be no reinforcement.

Now that you understand the principles of operant conditioning, you can apply what you have learned to the real world in Try It Out: Operant Conditioning in Everyday Life.

Try It Out | Operant Conditioning in Everyday Life

JuSun / Getty Images

People use operant conditioning to train their pets, but we also acquire new behaviors through conditioning to people, events, and even our own behavior. You can consider just how common conditioning or simple learning is by completing the worksheet **Operant Conditioning in Everyday Life**.

Now that we have covered how simple conditioning occurs, we turn to a related and equally important question in the next section: how do we store that newly acquired adaptive behavior?

12.2 Before You Go On

Section Summary

Unlike classical conditioning, which involves associating an involuntary behavior with stimuli that come before the behavior, operant conditioning involves associating a consequence that follows a voluntary behavior. Reinforcement and punishment increase or decrease the frequency of the behavior, respectively, and conditioning phenomena such as habituation and extinction play an important role in all forms of conditioning. The schedule used to present consequences to a behavior greatly affects conditioning.

Comprehension Check

1. A key difference between classical and operant conditioning is that classical conditioning _____, and operant conditioning _____.

a. applies to humans; applies to rats

b. takes longer; is faster

c. involves events that precede the behavior; involves events that follow it

d. involves voluntary behaviors; involves involuntary behaviors

2. The difference between reinforcement and punishment is that reinforcement _____, and punishment _____.

a. is good; is bad

b. is more effective; is less effective

c. removes an existing behavior; adds a new behavior

d. increases the frequency of a behavior; reduces its frequency

3. Your dog has developed the unpleasant habit of greeting you by jumping up on you. You don't care for this behavior, and whenever it occurs, you turn away, fold your arms close to your chest, and avoid eye contact with the dog. This is an example of what?

a. habituation

b. generalization

c. extinction

d. shaping

4. The schedule of reinforcement that most rapidly leads to conditioning is _____, and the schedule that creates a behavior most resistant to extinction is _____.

a. CR; VI

b. VR; VI

c. FI; CR

d. CR; FR

5. To get you to gamble as much money as possible, slot machines are designed to deliver a reward (winning) on what type of schedule?

a. FR

b. VR

c. FI

d. continuous

12.3 Learning and the Brain

LO 12.3 Explain where and how information is stored in the brain.

Acquiring a new behavior is meaningless if we cannot store it and then retrieve it when needed in the future. Imagine if your computer had no mechanism for storage, and each day all your prior work was lost. Behavioral neuroscientists focus on the biological mechanism for storing these changes in behavior and the physical location where a memory is stored. How exactly is information stored in our brain? Classical and operant conditioning provide mechanisms for studying the brain–behavior relationship, and researchers have slowly uncovered where and how information is permanently stored.

Where Memories Are Stored MCAT

Consciousness is a property of the mind, but the storage of information is a property of the brain, and in some way, learning must involve a physical change in the brain's structure. The **engram** (sometimes referred to as the memory trace) refers to the physical change in the brain that is involved in storing a memory or an association. One question that scientists have struggled with is understanding where the engram is stored in the brain.

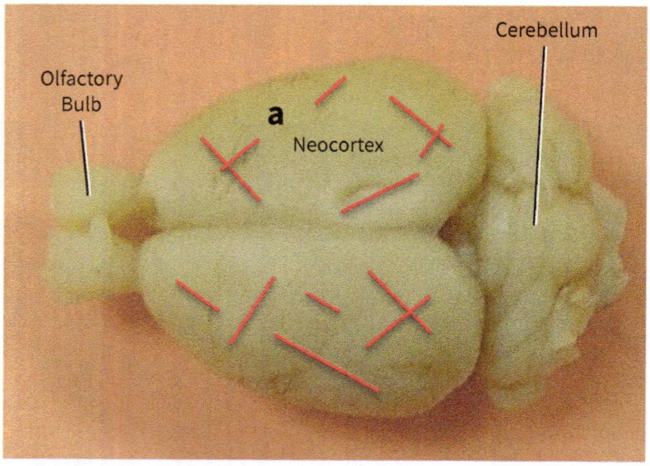

FIGURE 12.12 **Lashley's test of localization** Lashley conditioned a rat to learn a maze and then cut sections of the rat's cortex to interrupt communication between point a, where the engram might be stored, and other brain areas.

When a rat learns to make a response to a stimulus such as pressing a bar to receive a food pellet, a connection is being created between the rat's cortex and the motor neurons that control its movements. Where exactly in the rat's brain is that association stored? When conditioning was first being understood, some scientists felt that an association learned through conditioning must be stored in a specific physical location in the cortex, a theory called **localization**. Others thought that such associations must be stored in a distributed or equal fashion across the entire cortex.

One way to understand the difference between these two theories is to compare the functioning of the pituitary gland to that of the liver. Your pituitary consists of the anterior and posterior lobes, and damage to one lobe results in the loss of some hormones but not others, which means that pituitary function is localized. In contrast, your liver secretes its chemical messengers as a single unit, so damage to 30% of your liver results in losing a proportion of all of its functions, not just certain secretions.

About a century ago, the neurophysiologist Karl Lashley proposed a study that he hoped would address the question of whether the engram was stored one physical location or throughout the entire brain. He first taught the rat a simple visual task (learning the path through a maze) through operant conditioning. Once that conditioned behavior was established, Lashley made cuts in the cortex of the rat, as illustrated in Figure 12.12, and then tested the rat on the same maze to see how the cuts had affected the conditioned behavior.

Lashley reasoned that if the engram was stored in a single physical location, such as point a in the photo, then severing the connection between the engram and the rest of the brain would result in the total loss of the learned behavior. However, if the engram was stored throughout the cortex, then the learned behavior would be at least partially retained elsewhere. What Lashley found when he cut the cortex was that all skills declined roughly in proportion to the number of cuts that he made. Based on his observation that the rat did not entirely lose its ability to recall the maze, Lashley proposed that learning is distributed across the entire brain, a theory known as the **mass action principle**. Lashley also proposed the principle of **equipotentiality**, which states that all areas of the brain are equally able to perform a task.

The equipotentiality hypothesis was put to the test by behavioral neuroscientist Richard Thompson (1930–2014), who used classical conditioning of rabbits to pair a tone (CS) with a puff of air in the eye (UCS) and a blink (UCR) until the CS elicited an eyeblink (CR). With that CS–CR engram established, Thompson chemically deactivated certain cerebral regions temporarily. His thinking was that if the CS–CR engram disappeared when he deactivated a region and reappeared when that region was reactivated, then that must be the physical location of the memory.

Thompson found that the CS–CR association disappeared when a region in the rabbit cerebellum was deactivated. He therefore concluded that at least part of the engram was stored in the cerebellum specifically and nowhere else, refuting Lashley's mass action principle. Subsequent research suggests that the cerebellum might be where engrams for classical conditioning of simple reflexive behaviors are stored, but other types of memories are likely stored elsewhere.

The question of where and how information is stored and processed in the brain is one of the most important topics in all of behavioral neuroscience, so a brief summary is in order of where Lashley was right and where he was not. Lashley proposed that learning is distributed across the entire brain and every part of the brain is equally capable as another at processing information. Although Lashley was correct in assuming that the whole brain is involved in learning and remembering, he was wrong in assuming that every function involved in creating the engram is available throughout the brain. In fact, the reason the rat was able to remember some of the previously learned ability to navigate through the maze was because that behavior involved many different skills that were located in different areas of the brain.

Subsequent work showed that the brain is comprised of highly specialized processing centers located throughout the brain and that these many centers, acting together, constitute a complex web of processing, as depicted in **Figure 12.13**.

You might think of the brain in Lashley's model like a craftsperson who takes raw materials and works alone at one workbench to create a chair entirely by themselves. The reality of neural organization is more like a modern factory in which raw materials pass through the hands of many individuals located at different stations in the factory, each worker adding to the process to create the end product of a functional chair. The fact that our brain is comprised of a number of specialized processing centers that share information is essential for understanding how our brain creates higher-order cognitive processes such as language comprehension and also for understanding the consequences of brain damage, topics that are explored in other chapters.

Understanding that the engram is distributed throughout the cortex and likely other structures raises the question of how the engram is stored, the topic of our next section.

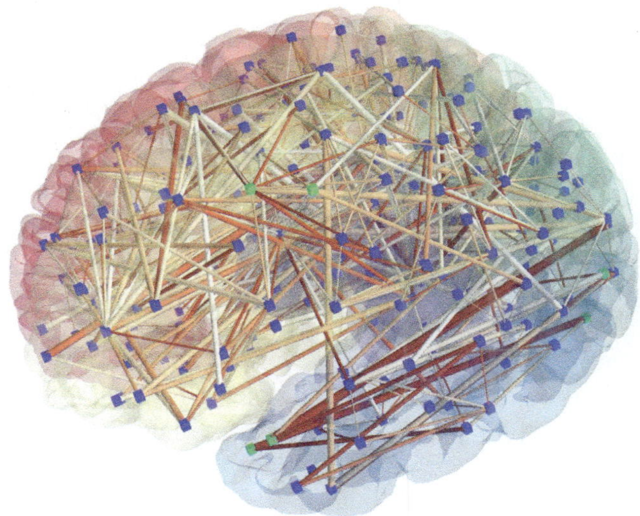

Patric Hagmann

FIGURE 12.13 The brain as a processing network The human brain consists of a collection of specialized processing centers that continuously communicate with each other.

The Neuronal Basis of Learning MCAT

We have touched on the location of memory storage, but how is the engram stored in the brain? One strategy for solving an impossibly complicated problem is to look for the simplest case, understand it, and then apply what you have learned to more complex systems. Neuroscientist Eric Kandel reasoned that to understand how data is stored in the human brain, he needed to find a simple animal model and begin there to unravel the mysteries of learning. He chose the California sea slug, or *Aplysia* as it is known in scientific circles, as his model (**Figure 12.14**). *Aplysia* has several advantages as a research platform for the study of the neurobiology of learning. There are roughly 100 billion neurons in the human brain, but *Aplysia* has only 20,000. In contrast to our neurons, many of *Aplysia*'s neurons are 1 mm in size, large enough to be seen by the naked eye.

Aplysia reflexively withdraws its gill in response to a jet of water, so pairing a CS with the gill-withdrawal reflex provided Kandel and his colleagues with a platform for investigating neuronal changes that accompany learning. Kandel employed the phenomena of habituation and sensitization, which were discussed earlier in this chapter, to shed light on neuronal changes that accompany the simple learning that classical conditioning creates.

Kandel was able to measure the electrical activity of a single neuron, so he compared the activity of the sensory and motor neurons before and after the behavior of the gill was changed by learning. Observing that neither the sensory neuron receiving the water jet nor the motor neuron that moves the gill were changed through learning, Kandel reasoned that habituation and sensitization must involve a physical change in the synapse. His research showed that when a neuron habituates in response to repeated stimulation, there is a decrease in the amount of neurotransmitter at the synapse. In contrast, sensitization occurs when reuptake of the neurotransmitter is blocked at the synapse, which leads to greater amounts of the neurotransmitter being available at the receptor site. Learning was actually changing how signals were transmitted between neurons.

Image Courtesy Dr. Genny Anderson

FIGURE 12.14 ***Aplysia*** Although many students associate the rat with neuroscience research, the sea slug *Aplysia* is important because its large neurons make research easier to conduct.

Exploring the behavioral eccentricities of a sea slug may seem trivial, but revealing how synaptic processes create behavioral phenomenon such as habituation and sensitization is an essential step in understanding human maladies like depression and post-traumatic stress disorder as well as learned helplessness and addictive behaviors. Kandel's work formed the foundation for much of our understanding of basic neural process and was awarded the Nobel Prize in Medicine or Physiology.

Now that we have identified the synapse as the site of brain changes related to memory storage, we turn to the question of how the storage mechanism works.

Long-Term Potentiation MCAT

Kandel demonstrated that in order for an engram to be stored, some physical change at the synapse must occur. The evidence clearly suggests that repeated firing across the synapse from neuron A to neuron B creates a physical change, most likely in both the pre- and postsynaptic neurons. The fact that this connection has been repeatedly accessed in the past translates into easier access in the future. Think of a restaurant that you have patronized so frequently in the past that when you show up at their door, you are immediately shown to your favorite table. This connection is known as a Hebbian synapse, and the process of learning this association is called Hebbian learning after Donald Hebb who proposed the mechanism in the 1940s.

How could researchers investigate a process that operates at the level of the synapse? Bliss and Lømø (1973) reasoned that if learning consisted of a physical change at the synaptic level that subsequently facilitated the connection between neurons, they should be able to see changes that reflect a physical alteration at the synapse by measuring the postsynaptic response before and after learning.

Bliss and Lømø were skilled at recording from cells in the rabbit hippocampus and were able to observe the current pulse through neurons known as the **excitatory postsynaptic potential (EPSP)** that accompanies the excitatory process. Bliss and Lømø stimulated a pathway entering the hippocampus and measured the EPSP that emerged. They were not surprised to see that a post-output EPSP followed excitation of the presynaptic pathway. Next, they simulated learning by increasing the frequency of the stimulation so that the same neural pathway was stimulated more often. They were surprised to see that when the presynaptic pathway was stimulated at a high frequency (say, 100 Hz), the amplitude of the postsynaptic EPSP increased (**Figure 12.15**). That result seemed to indicate that more frequent repetition of the stimulus changed something at the level of the synapse that increased the amplitude of the response. This change is called **long-term potentiation (LTP)**.

One way to visualize their results is to imagine a dam with a river at its feet. When Bliss and Lømø opened the flood gates and let 1,000 gallons past, a wave was created, and its crest traveled down the river. Now imagine that after repeatedly and quickly opening the flood gates several times, the same 1,000 gallons now creates a wave twice as large as it did previously. That change would surprise you, just as the increased amplitude of the postsynaptic EPSP surprised Bliss and Lømø.

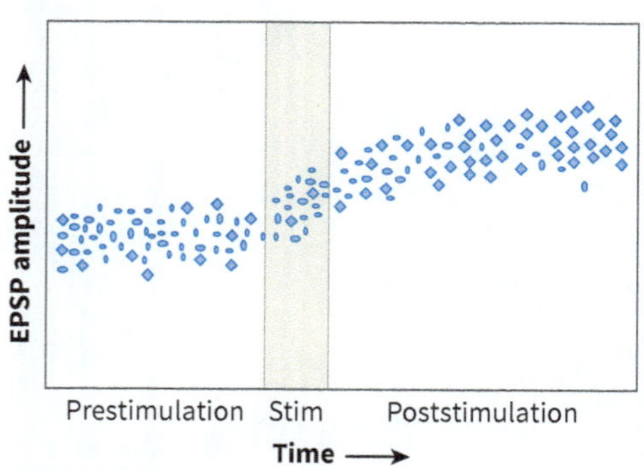

Adapted from Le Ray, D., and Cattaert, D. 1999.

FIGURE 12.15 Amplitude of response pre- and poststimulation Repeated stimulation of the postsynaptic receptor leads to greater amplitude of the receptor response.

What is the mechanism responsible for LTP? There are two likely scenarios. High-frequency stimulation either creates increased neurotransmitter release from the presynaptic neuron or higher receptor sensitivity at the postsynaptic neuron (**Figure 12.16**). At this point in time, it is not clear which alternative is correct, and there is reason to believe that both pre- and postsynaptic changes are occurring.

The possibility exists that nitric oxide (NO) acts as a retrograde neurotransmitter; that is, it may be released by the postsynaptic dendrite and travel back across the synapse to bind to the axon terminal of the presynaptic neuron. In this hypothesis, postsynaptic NO release acts in a feedback loop to build and maintain presynaptic neurotransmitter release and sustain the physical changes associated with LTP. The details of this process are not clearly known at this time.

To summarize, it appears that repeated activation of a pre- and postsynaptic connection leads to a physical change in the synaptic environment. It is likely that this change involves both an increase in neurotransmitter release

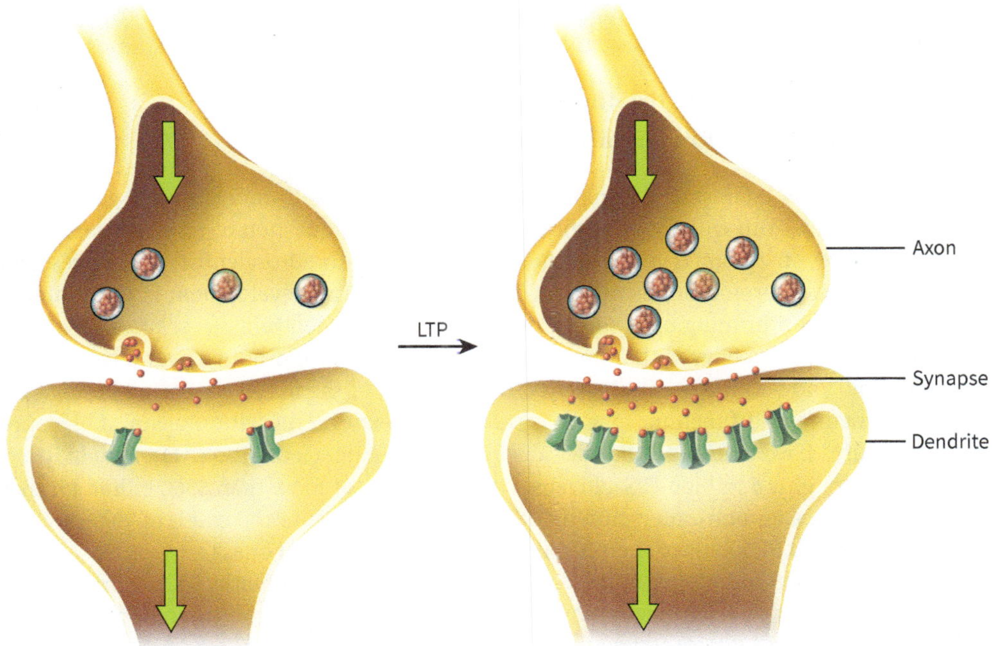

FIGURE 12.16 Long-term potentiation With repeated activation, a physical change occurs that likely involves both increased release of neurotransmitter by the presynaptic neuron and higher receptor sensitivity on the postsynaptic neuron.

and increased sensitivity on the part of the receptors themselves. You can see a simulation of this process in **Check It Out: Long-Term Potentiation**.

Now that you know that high-frequency stimulation of a synaptic junction leads to a change in the synapse, you might ask what happens to the EPSP amplitude when stimulation is less frequent. The answer is that low-frequency stimulation (10 Hz) leads to decreased levels of EPSP amplitude. This phenomenon is known as **long-term depression (LTD)**. LTD is just as important at LTP; if you imagine that LTP is the mechanism for creating new connections,

Check It Out | Long-Term Potentiation

© dononeg / iStockphoto

Long-term potentiation is the basis of all learning. Without LTP, our brains would be like computers without a storage device; once we shifted our attention to another topic or fell asleep, we would be right back where we started. Watch the video **Long-Term Potentiation**, and then answer the following questions.

1. In an earlier chapter you learned about upregulation at the synapse. Based on what you learned in the video, what effect would blocking upregulation have on the development of LTP?
2. We are all different in many ways, and it is likely that some of us create LTP connections faster than others. If there was a drug that enhanced your ability to create and then store memories, would you want to take it? Why or why not?
3. Think of a situation where someone might want to selectively reduce LTP for a memory and thereby make the recollection of that memory more difficult. Briefly explain why you chose that situation and what you think the outcome would be.

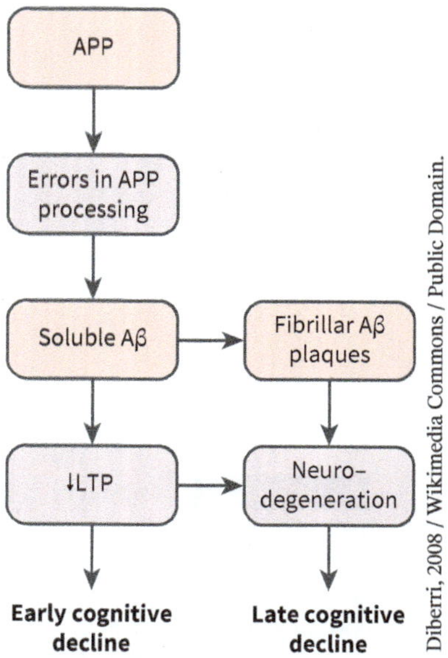

FIGURE 12.17 A role of LTP in Alzheimer's disease Changes in the processes that underlie LTP may be responsible for the memory loss in dementias such as Alzheimer's disease.

then there would need to be a way to prune connections that are no longer useful or necessary. The take-home message from LTP and LTD can be summarized in these two lines:

Neurons that fire together, wire together.

Neurons that fire out of sync, lose their link.

LTD is of interest to researchers studying neurological disorders as well as addiction mechanisms. You will not be surprised to hear that the question of how neurons rewire themselves in response to experience is of interest to neuroscientists whose research focuses on neurological disorders. A hallmark of neural development is the proliferation and then pruning of synapses in response to experience. Until relatively recently, however, it was believed that mature neural structures could not recover after damage. We now know that **neural plasticity**, or the ability of the nervous system to reorganize its structure and function, may be a process that can be turned on again after neural structures have matured. One such disorder is Alzheimer's disease (AD), a malady that afflicts roughly 6 million U.S. citizens with a total cost of more than $300 billion for their healthcare. Can our understanding of LTP provide a clue to the cause of this disorder, which robs individuals of their life and families of their loved ones?

We know that hippocampal damage is one of many hallmarks of AD. You learned elsewhere that proteins play an important role in synaptic regulation and the amyloid precursor protein (APP) is thought to play a part in synaptic formation and repair. Therefore, an alteration in this protein would lead to abnormal functioning, and the result of this functional disruption would be a decrease in the efficiency of LTP. A second and more severe result of sustained errors in APP processing would be a disruption of normal amyloid clearance from the cell. Instead of normal clearance, there would be a buildup in the cell of soluble fragments known as amyloid β, which has been linked to the development of neurofibrillary tangles. This buildup of Aβ damages cells, impairs LTP, and may be a causal factor in AD. That model is depicted in Figure 12.17.

NMDA and AMPA Receptors

We began our discussion of the neuronal basis of learning by discussing changes at the synapse. A neuron is 10 microns long, and it would take 100,000 neurons to stretch 1 meter. Yet, to understand LTP, we need to focus on the phenomenon at a level that is even more minuscule. Is it possible to discern what is happening at the receptor site during LTP? The answer is yes, and it involves two glutamate receptors discovered in the 1980s, NMDAR and AMPAR.

You will recall from a previous discussion of neurotransmitters that glutamate is the most abundant excitatory neurotransmitter in the nervous system. There are several types of glutamate receptors, and one is known as the **NMDA receptor (NMDAR)** because it has a special affinity to a molecule known as NMDA. Another type of receptor is the **AMPA receptor (AMPAR)**, or the quisqualate receptor, which is similarly defined by its affinity for the artificial glutamate AMPA. NMDARs and AMPARs are common receptors in the CNS. You know from your understanding of how neurons function that in order for the neurotransmitter to signal the postsynaptic neuron, a receptor in the postsynaptic dendrite must activate. The evidence suggests that NMDA and AMPA receptors are at the heart of LTP.

Our current understanding of LTP at the receptor level is depicted in Figure 12.18. First, glutamate binds to NMDARs and AMPARs **(step 1)**. When the AMPAR channel opens, Na^{+} flows into the postsynaptic dendrite, and it depolarizes **(step 2)**. This depolarization leads to the ejection of a magnesium ion (Mg^{2+}) into the extracellular space, which allows the NMDAR channel to open **(step 3)**. NMDARs are permeable to both Na^{+} and Ca^{2+}. Ca^{2+} enters the cell and triggers upregulation (an increase) in the AMPARs, which leads to greater EPSPs **(step 4)**. The negative feedback mechanism that keeps NMDA and AMPA receptors in check is the availability of glutamate, which decreases when this system becomes overactive.

We have already seen that an increase in EPSP amplitude is a hallmark of LTP, so we now have at least an outline of how learning occurs at the level of the receptor site. It will not surprise you that AMPAR dysfunction may hold the key to some neurological disorders. Both overactivation and underactivation of AMPARs have been implicated in seizure disorders and Alzheimer's, respectively. Given this, they are promising targets for therapeutic drug development.

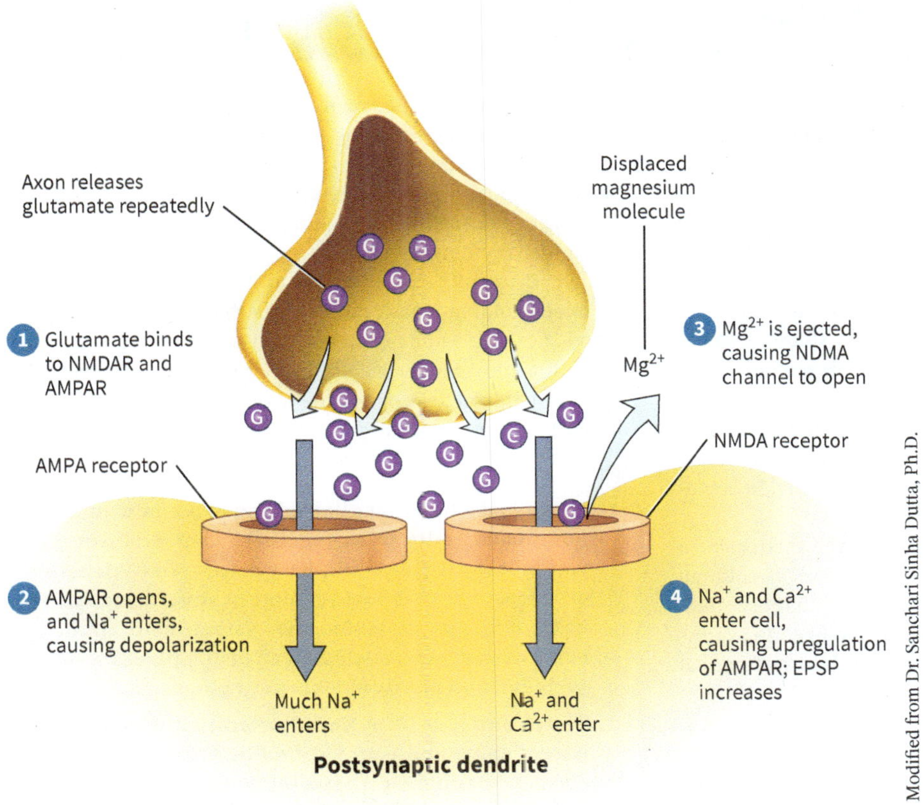

FIGURE 12.18 NMDA and AMPA receptors and LTP NMDA and AMPA receptors cooperate to create LTP.

As we leave this topic, we can take a moment to mention the relationship between AMPAR activity and dendritic spine plasticity. Dendritic spines are dynamic; they may appear and disappear over the course of minutes in response to the neural activity of the organism. AMPAR levels are inversely related to spine growth, and NMDAR activity is positively related to dendritic spine growth. We now know that maintenance and increase in dendritic spines is related to LTP. When an organism acquires a new behavior, dendritic spines flourish between neurons. Repeated activation of the neural connections leads to a strengthening of the connection as illustrated in **Figure 12.19**.

Take a moment to consider the effect of tools on our understanding of the natural world. When astronomers first turned a telescope to the heavens, their eyes were opened to a limitless universe; similarly, the invention of the microscope opened the door to our understanding of microbes and disease. Behavioral neuroscience also benefits from technological advances; for example, imagine if you could actually see neurons connecting. Optogenetics uses gene splicing to insert chemicals into neurons that light up as the neurons operate in specific ways. Read more about optogenetics in **Advances in Behavioral Neuroscience: Karl Deisseroth and Optogenetics**.

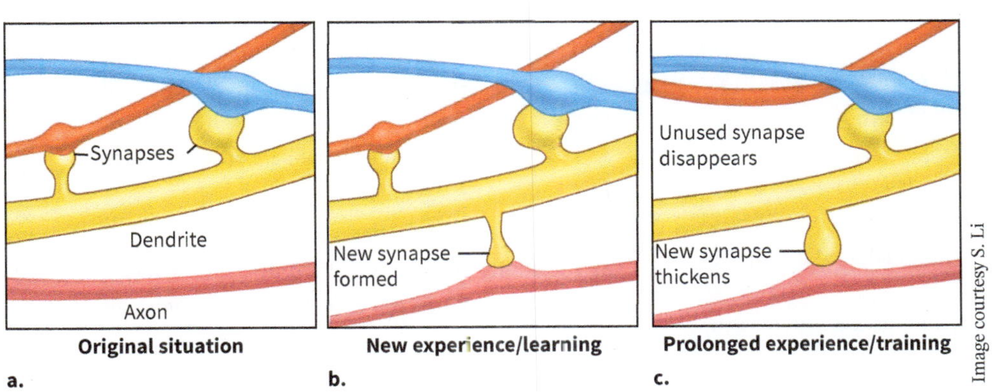

FIGURE 12.19 Dendritic spine plasticity in response to learning **a.** Two neurons are in close proximity but not physically connected. **b.** Learning has occurred, and a synaptic connection now exists. **c.** Repeated activation of the connection leads to a stronger connection, whereas a second, unused synaptic connection has disappeared.

Advances in Behavioral Neuroscience | Karl Deisseroth and Optogenetics

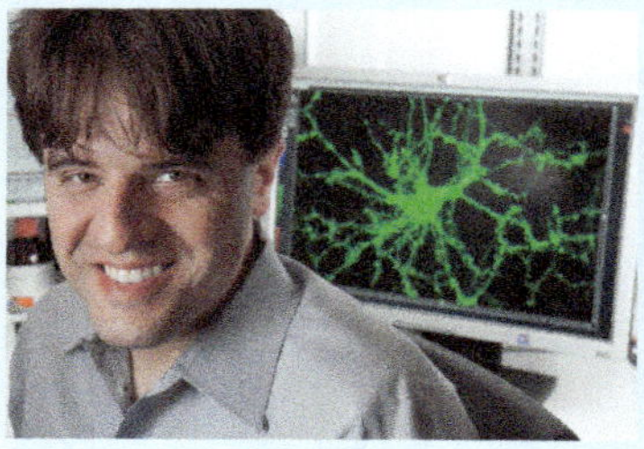

Linda A. Cicero / Stanford News Service

Exploring the inner workings of neurons is a difficult task to be sure; they are miniscule in scale, and the act of removing them from an organism for study changes them profoundly. In addition, many of the methods for investigating the workings of neurons do not easily lend themselves to focusing on the processes involved in a single neuron. Until recently, it was a complicated problem to turn on and off the mechanism of an individual neuron, possibly beyond solution.

Dr. Karl Deisseroth and his colleagues solved that problem in 2005 by demonstrating an optogenetic response in cultured mammalian neurons. Optogenetics is a technique that uses light to control neurons that have been genetically modified to express light-sensitive ion channels. By inserting genetic information that controls chemicals known as opsins into the genetic code of a mouse, researchers could use light to alter the workings of a neuron. It is even possible to specify which type of neuron or where in the mouse brain the adapted neuron will end up. Optogenetics can be used to see how individual neurons connect or how whole structures work in concert. Optogenetics has already helped us better understand what happens during a stroke, and it may hold promise for novel treatments for diseases such as Parkinson's. Optogenetics has been used to map circuits in the amygdala that contribute to fear, explore the workings of the olfactory system, and to gain a deeper understanding of the mechanisms that control cardiac rhythms.

Dr. Deisseroth's contribution to neuroscience is advancing our understanding of basic mechanisms that translate into important practical applications. In 2010, optogenetics was chosen as the Method of the Year across all fields of science and engineering and was highlighted as one of the scientific breakthroughs of the decade by the prestigious journal *Science*.

Questions

1. **If you had a tool that allowed you to see how individual neurons actually work together, what problem would you investigate? It could be a disease like Alzheimer's or stroke; it could be a regulatory system such as your heart, kidneys, or endocrine system; or it could be a behavioral circuit such as fear or happiness. Choose a problem, and briefly explain your choice.**
2. **Search online for "Dr. Karl Deisseroth" and also "karldeisseroth.org" and you will find he is both scientist and author. Keeping Dr. Deisseroth's situation in mind, do you think it is better to focus your career on one thing or to spread yourself more widely? Explain.**

12.3 Before You Go On

Section Summary

New information is stored in an engram and long-term potentiation is the process by which a physical change is created at the level of the neuron to store the engram. Understanding LTP may be the key to developing treatments for neurological disorders such as dementia. NMDA and AMPA receptors appear to control synaptic plasticity by facilitating the proliferation of dendritic spines and thereby play a central role in LTP.

Comprehension Check

1. What is an engram?

- **a.** a measure of weight equal to one half gram
- **b.** a measure of the strength of a memory
- **c.** the location in the brain of a stored memory
- **d.** the location of a processing center in the brain

2. Equipotentiality theory assumes that _____, and localization theory assumes that _____.

- **a.** every part of the brain can perform all functions; brains centers are specialized
- **b.** brain centers are specialized; every part of the brain can perform all functions
- **c.** only the cortex is specialized; other regions are not
- **d.** damage is difficult to recover from; damage is relatively easy to recover from

3. Why did Kandel choose to work with *Aplysia*?

- **a.** They are easy to find and train.
- **b.** Their brain, although quite small, is similar to a human brain.
- **c.** They have a special receptor that responds to several neurotransmitters.
- **d.** Their neurons are large and easy to work with.

4. From the perspective of behavioral neuroscience, what does learning involve?

- **a.** being able to remember what happened in the past
- **b.** a physical change in the brain
- **c.** being able to think creatively
- **d.** something in humans that is different from rats

5. What appears to be the role of NMDA and AMPA receptors in LTP?

- **a.** stimulating receptor downregulation
- **b.** stimulating inhibition of glutamate antagonists
- **c.** stimulating the hypothalamus to increase metabolic activity
- **d.** regulating the growth of dendritic spines

Conclusion

One of the most basic questions in behavioral neuroscience is how the nervous system modifies itself to benefit from experience in the process known as learning. The simplest form of learning is conditioning, and conditioning occurs in two ways: the linking of a novel stimulus to an already existing involuntary behavior, termed classical conditioning; and the linking of a behavior-modifying consequence following a behavior, termed operant conditioning. At the neural level, learning must involve an alteration to the physical structure of the brain, and this process is known as long-term potentiation (LTP). LTP is still not completely understood, but we do know it involves several glutamate receptors that regulate the development of dendritic spines.

We began this chapter with an example of how one experience modified a behavior through conditioning, and we concluded our discussion by exploring how that experience might be stored in the brain. Understanding conditioning and LTP is essential to our ability to treat psychiatric maladies such as PTSD and neurological disorders such as dementia and traumatic brain damage.

Critical Thinking Questions

1. Some organisms are born with a wide range of reflexes to deal with life's challenges, whereas others have lesser or greater abilities to learn new behaviors. Would you consider this flexibility to adapt a hallmark of intelligence among humans? Are less adaptable individuals apt to be less successful in life? Explain why or why not.
2. Many individuals that we rely on, such as police, firefighters, emergency medical personnel, and the military, have to act in a calm and professional fashion in situations that may cause involuntary behaviors such as heart rate, blood pressure, and cortisol response to skyrocket. Describe a conditioning program that would help one of these groups keep their involuntary responses at a manageable level.
3. Advertisers use operant conditioning to get us to attend to and then buy their products. Give one example that you have observed of how an advertiser has employed conditioning to attempt to influence you. Alternatively, if you have ever been inside a casino, describe how casinos use conditioning to encourage and then maintain gambling behavior.
4. One day we might have a drug that enhances LTP and a drug that disrupts LTP. Choose one of these drugs and briefly describe how it might be useful and/or how it might be dangerous.

Dig Deeper into Behavioral Neuroscience

12.1 Classical Conditioning

You can learn more about Ivan Pavlov and classical conditioning in these sources.

Windholz, G. (1997). Ivan P. Pavlov: An overview of his life and psychological work. *American Psychologist*, *52*(9), 941–946. https://doi.org/10.1037/0003-066X.52.9.941

McSweeney, F. K., & Murphy, E. S. (2014). *The Wiley Blackwell Handbook of Operant and Classical Conditioning*. Malden, MA: John Wiley & Sons. p. 3. ISBN: 9781118468180

The Little Albert experiment, although unethical by today's standards, is still a classic.

Beck, H. P., Levinson, S., & Irons, G. (2009). Finding Little Albert: A journey to James B. Watson's infant laboratory. *American Psychologist*, *64*(7), 605–614. http://libres.uncg.edu/ir/asu/f/Beck_Hall_2009_Finding_Little_Albert.pdf

Conditioned taste aversions are of great interest to therapists, nutritionists, and parents.

Garcia, J., Kimeldorf, D. J., & Koelling, R. A. (1955). Conditioned aversion to saccharin resulting from exposure to gamma radiation. *Science*, *122*(3160), 157–158.

Seligman, M. E. P., & Hager, J. L. (1972, August). Biological boundaries of learning: The sauce-bearnaise syndrome. *Psychology Today*, *V6*, 59–61, 84–87.

12.2 Operant Conditioning

Operant conditioning is used in marketing, therapy, and everyday life.

Skinner, B. F. (1938).*The Behavior of Organisms: An Experimental Analysis*. New York: Appleton-Century-Crofts.

Staddon, J. E. R., & Cerutti, D. T. (February 2003). Operant Conditioning. *Annual Review of Psychology, 54* (1), 115–144. doi:10.1146/annurev.psych.54.101601.145124. PMC 1473025. PMID 12415075.

B. F. Skinner was and still is a controversial figure in the history of behavioral science.

Bjork, D. W. (1997). *B. F. Skinner: A Life*. American Psychological Association.

An article by Skinner's daughter dispels many of the falsehoods about her life:

Buzan, D. S. (2004). I was not a lab rat. *The Guardian*. https://www.theguardian.com/education/2004/mar/12/highereducation.uk

Both casinos and therapists who specialize in gambling have a keen interest in the operant conditioning of gambling behaviors.

Zentall, T. R., & Stagner, J. (2011). Maladaptive choice behaviour by pigeons: An animal analogue and possible mechanism for gambling (sub-optimal human decision-making behaviour). *Proceedings of the Royal Society B: Biological Sciences, 278*, 1203–1208.

Zentall, T. R., & Laude, J. R. (2013). Do pigeons gamble? I wouldn't bet against it. *Current Directions in Psychological Science, 22*(4), 271–277.

12.3 Learning and the Brain

Although we no longer consider Lashley's conclusions about the engram to be accurate, the search for the engram goes on.

Eichenbaum, H. (2016). Still searching for the engram. *Learning & Behavior, 44*(3), 209–222. https://doi.org/10.3758/s13420-016-0218-1

Eric Kandel is an important figure in behavioral neuroscience, and his 2013 book still stands the test of time.

Kandel, E. R. (2013). The new science of mind and the future of knowledge. *Neuron*, 80, 546–556.

Here is additional reading on long-term potentiation.

Bliss, T., & Lømo, T. (1973). Long-lasting potentiation of synaptic transmission in the dentate area of the anaesthetized rabbit following stimulation of the perforant path. *The Journal of Physiology, 232*(2), 331–356. PMC 1350458. PMID 4727084.

Lømo, T. (2003). The discovery of long-term potentiation. *Philosophical Transactions of the Royal Society of London. Series B, Biological sciences, 358*(1432), 617–620. https://doi.org/10.1098/rstb.2002.1226http://www.ncbi.nlm.nih.gov/pmc/articles/PMC1693150/

Researchers interested in Alzheimer's disease have an interest in LTP.

Rowan, M. J., Klyubin, I., Cullen, W. K., Anwyl, R. (April 2003). Synaptic plasticity in animal models of early Alzheimer's disease. *Philosophical Transactions of the Royal Society of London. Series B, Biological Sciences, 358*(1432), 821–828. doi:10.1098/rstb.2002.1240. PMID 12740129. PMC: 1693153, http://journals.royalsociety.org/openurl.asp?genre=article&issn=0962-8436&volume=358&issue=1432&spage=821

You can learn more about optogenetics here.

Editorial. (2011). Method of the Year 2010. *Nature Methods, 8*(1), 1. doi.org/10.1038%2Fnmeth.f.321

Deisseroth, K. (2010). Optogenetics: Controlling the brain with light. *Scientific American*, October 20, 2010.

Chapter 12 Before You Go On Answers

12.1 Before You Go On Answers

1. c. being startled in response to a sudden loud sound
2. c. associating an existing innate behavior such as blinking with a tone
3. b. the needle; fear; the phlebotomist; faintness
4. a. feeling nauseated at the thought of shellfish after shellfish made you sick

12.2 Before You Go On Answers

1. c. involves events that precede the behavior; involves events that follow it
2. d. increases the frequency of a behavior; reduces its frequency
3. c. extinction
4. a. CR; VI
5. b. VR

12.3 Before You Go On Answers

1. c. the location in the brain of a stored memory
2. a. every part of the brain can perform all functions; brains centers are specialized
3. b. Their brain, although quite small, is similar to a human brain.
4. a. being able to remember what happened in the past
5. d. regulating the growth of dendritic spines

Chapter 12 Glossary

AMPA receptor (AMPAR) A glutamate receptor that is critical to synaptic plasticity.

behaviorism A theory of human and animal learning, which proposed that all behavior can be explained in terms of conditioning without reference to higher-order concepts such as thoughts or feelings.

classical conditioning A type of learning that involves associating stimuli with involuntary or unconscious behaviors.

conditioned response (CR) An automatic response created by conditioning.

conditioned stimulus (CS) A previously neutral stimulus that elicits a conditioned response as a result of conditioning.

conditioned taste aversion An aversive response to a food that is associated with a prior illness, often after only one negative occurrence; also known as the Garcia effect.

conditioning A form of simple learning in which a stimulus or a consequence is associated with a behavior.

continuous reinforcement A reinforcement schedule in which reinforcement follows every instance of a behavior.

engram The physical change in the nervous system that accompanies learning.

equipotentiality A view of the brain that assumes all parts are involved in every mental action.

excitatory postsynaptic potential (EPSP) A postsynaptic potential that enhances the probability that the postsynaptic neuron will fire in the future.

extinction When a response that was conditioned operantly ceases after it is no longer reinforced.

fixed-interval (FI) reinforcement A reinforcement schedule in which reinforcement occurs after a specific time interval.

fixed-ratio (FR) reinforcement A reinforcement schedule in which reinforcement occurs after a set number of responses.

generalization A conditioning phenomenon in which a conditioned response transfers to stimuli that are similar to the original CS.

habituation When an innate response diminishes in frequency or strength after repeated pairings with a stimulus.

intermittent reinforcement Reinforcement that only occasionally follows the behavior.

law of effect The principle that behaviors that lead to pleasant outcomes are likely to be repeated, whereas behaviors that lead to adverse outcomes are likely to be reduced or eliminated.

learning A change in behavior in response to experience.

localization A perspective that mental functions are located in specific brain structures.

long-term depression (LTD) A long-lasting or permanent decrease in efficiency of a synaptic connection.

long-term potentiation (LTP) A long-lasting or permanent increase in efficiency of a synaptic connection.

mass action principle A theory in neuroscience that the extent of behavioral damage is proportional to the amount of damaged brain tissue and not the specific tissue damaged.

neural plasticity The ability of the nervous system to modify its function or structure in response to experience or damage.

NMDA receptor (NMDAR) A glutamic receptor involved in establishing LTP.

operant chamber A common laboratory device used to operantly condition an animal; also known as a Skinner box.

operant conditioning A type of learning in which the frequency of a behavior is modified by reward or punishment; also known as instrumental conditioning.

punishment A consequence to a behavior that decreases its frequency.

recovery from extinction A phenomena in which a behavior thought to be extinct suddenly reappears following minimal reinforcement.

reinforcement In operant conditioning, anything that increases the frequency of a behavior.

sensitization A conditioning phenomenon in which repeated exposure to a stimulus leads to an amplified response.

shaping A technique in operant conditioning in which initially any simple variant of a desired behavior is reinforced and then the reinforcement criteria are slowly tightened until only the target behavior is reinforced.

unconditioned response (UCR) An unlearned or innate response to an unconditioned stimulus.

unconditioned stimulus (UCS) A stimulus that leads to an innate response.

variable-interval (VI) reinforcement A reinforcement schedule in which reinforcement is delivered after a varying amount of time.

variable-ratio (VR) reinforcement A reinforcement schedule in which reinforcement is delivered after a varying number of responses.

CHAPTER **13**

Sleep, Dreaming, and Circadian Rhythms

Shine Nucha / Shutterstock

Sleep is essential for life. When we cannot sleep, the consequences are serious.

CHAPTER OUTLINE

13.1 The Mystery of Sleep
- Evolutionary Patterns of Sleep
  Think About It: Has Civilization Changed How We Sleep?
- The Purpose of Sleep

13.2 Sleep Stages and Dreaming
- The Study of Sleep
- The Stages of Sleep
  Try It Out: What's Your Sleep IQ?
- Dreaming
  Try It Out: Are You a Lucid Dreamer?

13.3 The Biological Basis of Sleep
- Sensory Stimulation and Sleep
- The Reticular Activating System and Sleep
- The Brainstem, Hypothalamus, and Sleep
- Neurotransmitters and Sleep
  Advances in Behavioral Neuroscience: Gina Poe and the Role of Sleep in Memory

13.4 Sleep Disruptions
- Drugs and Sleep
- Sleep Disorders
  Think About It: Sleep Insufficiency

13.5 Circadian Rhythms
- Melatonin and the Sleep–Wake Cycle
- The Suprachiasmatic Nucleus
- When Diurnal and Circadian Rhythms Are Mismatched
  Try It Out: The Dangers of Microsleep

LEARNING OBJECTIVES

LO 13.1 **Describe** sleep, its evolution, and its purpose.

LO 13.2 **Describe** the stages of sleep and their functions, including dreaming.

LO 13.3 **Identify** the brain mechanisms that create sleep.

LO 13.4 **Describe** common sleep disorders.

LO 13.5 **Relate** circadian rhythms to the larger phenomenon of sleep.

Stories of the Brain

On a crisp and clear December morning in 2013, a commuter train was approaching its eventual destination in New York City. The train carried people on their way to work, shop, visit museums, and sight-see. The scenery flew by in a blur as the train was moving at a rapid pace.

As the train neared its destination, it entered a rather sharp curve near the Henry Hudson Bridge in the Bronx, and a few of the regular passengers who were familiar with the route looked up from their newspapers when they realized the train was not slowing down as it usually did. Moments later, their surprise turned to terror as the train jumped the tracks and tumbled down the slope toward the water. Passenger cars overturned, riders were tumbled about, and some unfortunate passengers were ejected through the windows and then crushed to death as the cars rolled over them. When the train finally came to a halt at the river's edge, four passengers were dead and more than 50 were seriously injured. How could this accident have occurred? Was a signal incorrectly set? Was there a brake failure?

Subsequent investigation by the National Transportation Safety Board (NTSB) found that the brakes were working properly and there was no signal failure. Although the posted speed limit on the curve was 30 mph (48 kph), the train entered the curve at just over 80 mph (130 kph). The engineer who was driving the train was neither intoxicated nor under the influence of prescription or illicit drugs, and he was familiar with the route. How then did this accident occur? The NTSB concluded that the engineer had been working many different shifts and also was suffering from an undiagnosed disorder known as sleep apnea that made it difficult for him to get a good night's sleep. Although his eyes were wide open as he drove the train into the curve, his brain was in a state like that of a person fast asleep.

In this chapter, we will begin by defining sleep, consider why sleep is universal throughout the animal kingdom, and then explore the various stages of sleep and consider the benefits of sleep. With this background in hand, we will turn our attention to clinical aspects of sleep disorders. No discussion of sleep would be complete without a consideration of dreaming, so we will take a moment to consider that fascinating topic. We will also step back to see sleep as part of a larger system that alters our behavior to be in sync with daily and seasonal changes. By the end of this discussion, one explanation for the tragic train accident that occurred that day may become apparent.

13.1 The Mystery of Sleep

LO 13.1 Describe sleep, its evolution, and its purpose.

A general definition for **sleep** is a regular period of significantly decreased muscle activity and a lessened awareness of the environment coupled with a heightened threshold to rouse the individual back to their more typical alert state. Being asleep differs from being unconscious in that a sleeping person can be aroused, whereas an unconscious individual does not respond to external stimuli. Mysteries remain about the purpose of sleep, but in this first section, we will begin to explore the evolutionary basis for sleep and some theories about the purpose of sleep.

Evolutionary Patterns of Sleep MCAT

Consider for a moment this puzzle: the typical human closes their eyes and becomes senseless and defenseless for a good portion of every day. There must be a very good reason why you and I spend as much as a third of our lives asleep, so we turn our attention to understanding the evolutionary basis for this phenomenon.

Sleep appears to be a universal phenomenon; mammals, birds, fish, reptiles, and even insects all appear to sleep or at least regularly enter a sleeplike state. Given that every member

TABLE 13.1 Sleep Periods for Various Mammals

Species	Average sleep period (hrs.)
Armadillo	18.1
Lion	13.5
Cat	12.5
Mouse	12.0
Dog	10.1
Chimpanzee	9.7
Rabbit	8.5
Human	8.0
Sheep	5.0
Deer	4.5
Cow	4.0
Horse	2.9

of the animal kingdom sleeps, it seems reasonable to ask what evolutionary advantages sleep provides. Perhaps looking at sleep patterns across species will offer a clue.

Sleep patterns across species It might be an easy task to determine if your roommate or sibling is fast asleep, but it's not always so easy to identify the sleep state in other organisms. Not surprisingly, the way different species sleep has evolved to meet their unique environments and needs, so there is a great deal of variance in the amount and manner of sleep across the animal kingdom.

Horses and elephants sleep standing up by locking their knees. Dolphins need to return to the water's surface regularly to breathe, so they have the unique capability of having one hemisphere sleep while the other attends to respiration and other duties; after an hour or so, the hemispheres switch roles. Fish and insects go through periods of quiescence when they do not move but are still alert and responsive to stimuli. Reptiles appear to sleep but do so in short time segments as opposed to a single, long stretch. All mammals sleep, but their sleep patterns vary greatly, as illustrated in Table 13.1. As you look at the large differences in sleep times for these animals, do you see a pattern that might teach us something about the evolution of sleep?

One interpretation is that predators sleep more than prey, although prey such as the armadillo are an exception due to their protective scales. Rabbits and mice are considered prey, but they live in secure burrows in the ground and spend much of the day safe asleep. In contrast, vulnerable animals such as sheep, cows, and deer who spend their lives above ground sleep very little.

In addition to differences in length of sleep, predators and prey often differ in the pattern of their slumber. Organisms that sleep in one uninterrupted stretch experience **monophasic sleep**, whereas those who break their sleep into two segments experience **biphasic sleep**. Predators are more likely to sleep in one extended segment, whereas prey who do not have the luxury of a secure den are likely to sleep less and in an interrupted fashion. Deer doze off for 30 seconds to a few minutes at a time and then rouse to check their surroundings for a brief moment, often moving to a different location between sleep episodes (Figure 13.1).

Human sleep patterns If you consider the nature of sleep across the animal kingdom as an adaptive response to the environment, you might ask yourself if the human pattern of sleep also serves a purpose. In our distant history as a species, humans were at risk for predators, but as we developed technology such as fire and weapons, we became more predator than prey. Today, most humans sleep in one extended segment of 6–8 hours, but some have asked if monophasic sleep was always the case for us. You can consider whether monophasic sleep is our natural state in **Think About It: Has Civilization Changed How We Sleep?**

JHVEPhoto / Adobe Stock

KenCanning / Getty Images

a. b.

FIGURE 13.1 **Predator and prey** **a.** Predators such as the lion may sleep longer because they are only at risk from larger predators. **b.** Prey such as the deer sleep less because they must be constantly on guard against a wide range of threats.

Think About It | Has Civilization Changed How We Sleep?

Gorodenkoff / Adobe Stock

Humans are a remarkably diverse species, but if there is one trait that binds us all together, it is sleep. Everyone sleeps, regardless of sex, culture, or personal beliefs. Given the universality of sleep, you might think that how we sleep has not changed over time, but there is considerable evidence to suggest that our current sleep habits are a recent development in our history as a species.

In modern industrialized nations, people often wake after dawn, go to work or school, come home and stay awake several hours after sunset, and then sleep through the night in one continuous stretch. This sleep pattern is known as monophasic sleep because it consists of a single continuous event. But what if monophasic sleep is not our natural sleep pattern? Before the invention of electric lights, candles and lanterns were an expensive and not very effective source of illumination, and before they were invented, our only source of illumination after dark was a fire.

Historians have recognized in reading diaries and other personal accounts of life in the preindustrial age that many people engaged in biphasic sleep. They went to bed as night fell, slept 2–4 hours, awoke around midnight and looked after their children or perhaps tended the fire, and then went back to sleep for 4–5 more hours before awakening for their day. The Industrial Age and its appetite for workers who could be present for a long and continuous shift plus the easy availability of electric lights may have dramatically changed a basic human behavior.

Questions

1. **Search the internet using the terms "biphasic sleep or second sleep in history." Do you think the evidence for second sleep in the preindustrial age is convincing? Why or why not?**
2. **As a general rule, predators sleep more than prey. Humans today sleep like predators, yet some argue that our natural sleep pattern is biphasic sleep. Would biphasic sleep be more adaptive for our distant ancestors whose existence in the African savanna was more tenuous?**
3. **It is assumed that the Industrial Age created an environment in which it was advantageous to replace biphasic sleep with monophasic sleep. In today's world, people may be part of a team that is distributed across several time zones. Would biphasic sleep be an advantage for a person whose career requires them to interact part of the day with people on the other side of the globe?**

Holger Leue / Getty image

FIGURE 13.2 **Metabolic waste removal during sleep** Just as the by-products of modern life must be removed, the mental activity required to support life and the creation of complex behaviors requires a great deal of glucose to be converted into energy, leaving metabolic waste to be removed during sleep.

The Purpose of Sleep MCAT

Given that all animals sleep, it would seem reasonable to consider what benefits sleep provides you and me. There are several possible answers to this, including for refreshment, clearing of metabolic waste, processing memories, and cortical development.

Refreshment Perhaps the most obvious reason we sleep is to restore and refresh the body. Just as your cell phone or laptop requires you to regularly recharge its battery if you expect it to work, we need to recharge our own physical battery. We expend calories every moment that we are alive, and we must consume enough calories to cover the energy demands of our daily metabolism. You burn fewer calories while sleeping, so sleep gives you a chance to catch up and replenish your energy stores.

We now know that regular sleep is an important contributor to health. Our immune system is less efficient when we go without sleep for an extended period. White blood cell count (an indication of the strength of our immune system) goes down with extended sleep deprivation, and there is evidence that chronic sleep deprivation is linked to a greater risk of obesity and developing type 2 diabetes. Going without sleep can be dangerous to your health.

Clearance of metabolic waste from the CNS A second perspective can be easily understood by anyone who has lived in a large city. Late at night after most people have gone to bed, a whole host of workers start their day. They collect trash and make repairs to roads, bridges, computer networks, and other infrastructure, and then as the sun comes up, they retreat into the background. This model suggests another reason for sleep: the brain uses the downtime to cleanse itself of the by-products of the activity created by its heavy responsibilities during the day (Figure 13.2).

Oxygen is essential for life, but oxygen is a very corrosive gas, as anyone who has seen rust can attest. You know from a prior chapter that our cellular mitochondria create ATP (adenosine triphosphate) as an energy source, and a by-product of metabolism is the creation of highly reactive and unstable atoms or molecules known as free radicals, of which the most important in this context are oxidants. Oxidants damage cells, so our body creates antioxidants to neutralize their effects. However, obesity, diets high in sugar and fat, smoking, excessive alcohol consumption, exposure to pollutants, and personal stress reduce the body's ability to detoxify free radicals and cellular damage ensues, as illustrated in Figure 13.3.

In an earlier chapter, you learned about the GAS (general adaptational syndrome) and that repeated stresses without ample opportunity to recover leads to disorders. We respond to stress by activating our autonomic nervous system, which increases our metabolism. This, in turn, results in an increase in free radicals in the body. Reducing stress and adding the powerful antioxidant vitamin E along with foods such as blueberries, dark chocolate, strawberries, and grapes can increase your supply of antioxidants.

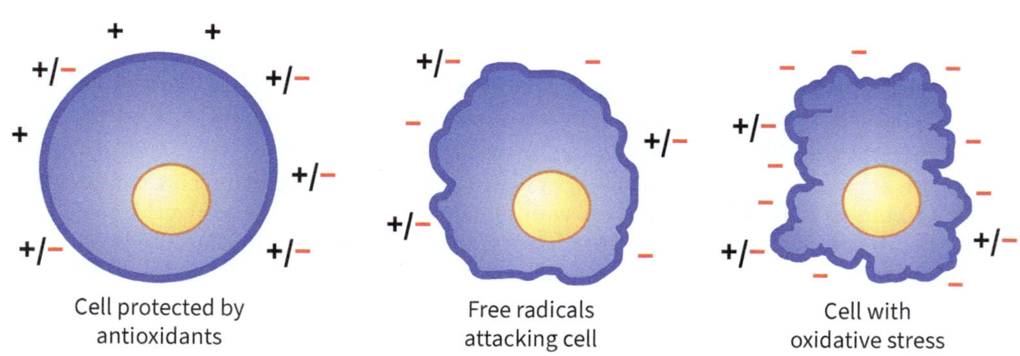

FIGURE 13.3 **Antioxidants and free radicals** Normal cellular metabolism creates free radicals (indicated by a –) and antioxidants (indicated by a +) that keep them in check. When the ability to neutralize free radicals declines, cellular damage results.

It has been theorized that during sleep, our glymphatic system clears free radicals and other waste products from the CNS. By comparing the efficiency of the glymphatic system during sleep and wake cycles, it appears that extracellular spaces expand during sleep to facilitate the clearance of waste products such as free radicals that accumulate during the wake cycle. Just as waste removal and highway and bridge repairs are more easily accomplished at night during times of low use, it makes sense that our brain can more easily cleanse itself while the pace of mental activity is slow.

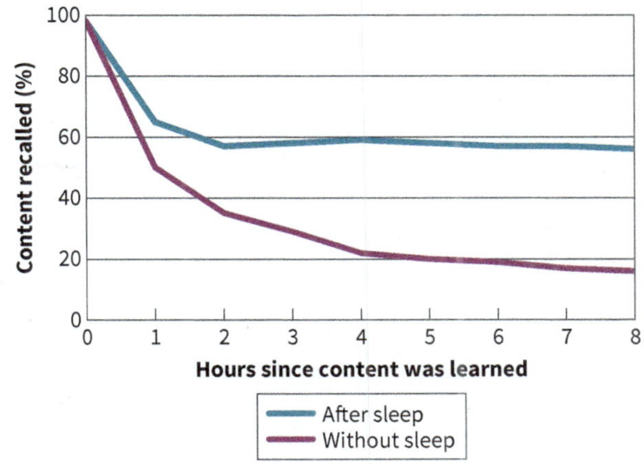

FIGURE 13.4 The effect of sleep on recall Jenkins and Dallenbach (1924) had participants learn a series of syllables and then recalled them at 1-hour intervals over 8 hours. Half the participants slept while the others engaged in normal activities. Those who slept recalled more of the syllables throughout the 8-hour period.

Sleep and memory There is substantial evidence that sleep is critical to the formation of durable memories. The ability to create a memory of an event that is stable and long lasting is a useful adaptation for any organism. As far back as 1924, Jenkins and Dallenbach reported that participants in a simple memory study performed better after a period of sleep than after a similar period of time awake. In their famous study, participants learned a series of syllables either in the morning or in the evening just before retiring, and then they were tested for their memory of the series 1, 2, 4, or 8 hours later. Individuals who learned the material after a night's sleep recalled more syllables than individuals who recalled later in the day (Figure 13.4).

As you inspect Figure 13.4, you might consider the benefits of preparing for an exam over some days and getting a good night's sleep the night before rather than staying up late to study. In the century since their study was published, considerable research into the processes underlying memory has demonstrated that memories of events that occurred during the day are consolidated or made more permanent during sleep.

Ontogenesis: Stimulation of the developing brain Sleep might play a very important role in cortical development early in life. In our discussion of the synapse in a previous chapter, you learned that repeated activation of neurons leads to the creation of a strong and durable synaptic connection, and when we dream, there is significant cortical activity. It is believed that dreaming early in life facilitates the development of useful synaptic connections. You may already have encountered the term *ontogeny*, which refers to the development of the individual from birth through adulthood. The hypothesis that dreaming early in life may lead to the proliferation of synaptic connections is known as the ontogenetic hypothesis because it addresses organizational and developmental issues that begin early in life but have an impact later in life.

Now that we have explored the evolutionary basis for sleep and the possible purposes of sleep, we take a closer look at what happens while we are sleeping.

13.1 Before You Go On

Section Summary

Sleep consists of a period of inactivity and lowered threshold to stimuli. Although every animal sleeps, the nature of their sleep differs based on each organism's environmental needs. The reasons for sleep among humans are varied; biological refreshment, metabolic waste removal, memory consolidation, and neurodevelopment have all been noted as benefits to sleep.

Comprehension Check

1. What is the most accurate definition for sleep?
   - **a.** a dreamlike state in which we rehearse the day's events
   - **b.** a period of immobility and lowered reactivity to stimuli
   - **c.** a time when we are safe from predators
   - **d.** a period when we are not dreaming
2. In what way do animals differ in the amount and type of sleep they need?
   - **a.** Aquatic animals do not need to sleep.
   - **b.** Larger animals need to sleep more because of greater metabolic needs.
   - **c.** Predators sleep the least because they need to hunt to survive.
   - **d.** Prey typically sleep less than predators.

3. Which of these is *not* considered a reason for sleep in humans?
   a. refreshing metabolic resources
   b. eliminating waste products
   c. consolidating memories
   d. resting our eyes

4. Our understanding of sleep suggests that sleep _____.
   a. is a luxury we can do without
   b. is different for males and females
   c. has several purposes or functions
   d. will become unnecessary as we evolve

13.2 Sleep Stages and Dreaming

LO 13.2 Describe the stages of sleep and their functions, including dreaming.

Given that humans have been sleeping throughout their existence, you might think we have always had a deep understanding of sleep. In fact, it was only in the 1920s that we began t sleep, and there are still many unsolved mysteries about why we sleep and dream. We begin by establishing how scientists study sleep.

The Study of Sleep

Dr. Nathaniel Kleitman set up the first modern sleep laboratory at the University of Chicago in the 1920s. Kleitman would pay participants to sleep overnight in his laboratory while a graduate student monitored various physiological measures. Such a study is known as **polysomnography**, from the Greek for *many sleep written measurements*.

Polysomnograms record the physiological state of a sleeper, including heart rate, blood pressure, respiration, blood oxygen level, galvanic skin response (a measure of physiological arousal), muscle movements, and EEG. **EEG (electroencephalography)** uses sensors on the scalp to record electrical activity in the brain that corresponds to mental activity. Polysomnography, also known as a sleep study, is helpful in diagnosing sleep disorders (**Figure 13.5**).

To understand what is happening while we sleep, sleep scientists use EEG to record brain activity. You already know that activity in your nervous system involves the transmission of tiny electrical potentials down an axon. By using EEG, those electrical potentials can be detected, amplified, and then recorded and analyzed. By the 1920s, researchers had discovered that EEG frequency bands are associated with different states of mental activity (**Figure 13.6**). High-amplitude and low-frequency waves are associated with lower levels of conscious awareness, such as drowsy relaxation. These levels of lower awareness are delta, theta, and alpha frequency bands, with delta being the lowest level of awareness. By contrast, low-amplitude and high-frequency waves are associated with higher levels of awareness such as alertness and mental focus. These levels of higher awareness are beta and gamma frequency bands, with gamma being the most alert.

When an individual transitions from wakefulness to slumber, their EEG progresses from the gamma and beta of active alertness through alpha as they relax, into theta as they become drowsy, and finally arriving at delta as they fall asleep. Prior to Kleitman's research, it was assumed that a sleeping individual would remain in delta sleep until they awoke, at which time the sequence of EEG stages that correspond falling asleep would reverse. However, what happened in Kleitman's lab changed our view of sleep forever, as you will read in our next section.

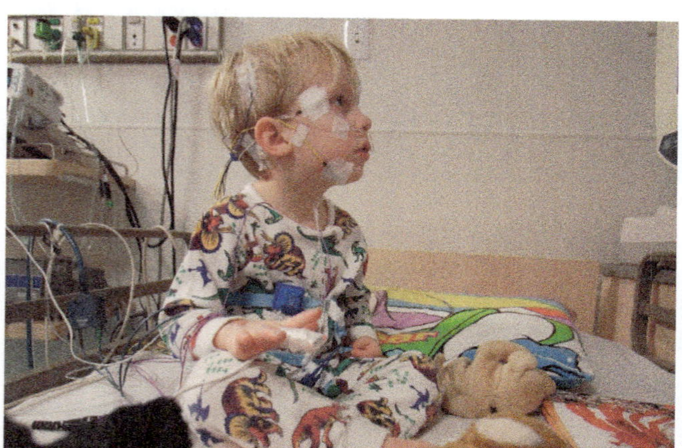

Rklawton

FIGURE 13.5 A child prepared for a polysomnography study Physicians request polysomnography to collect EEG, blood oxygen levels, heart rate, respiration, and even eye and leg movements in order to diagnose sleep disorders.

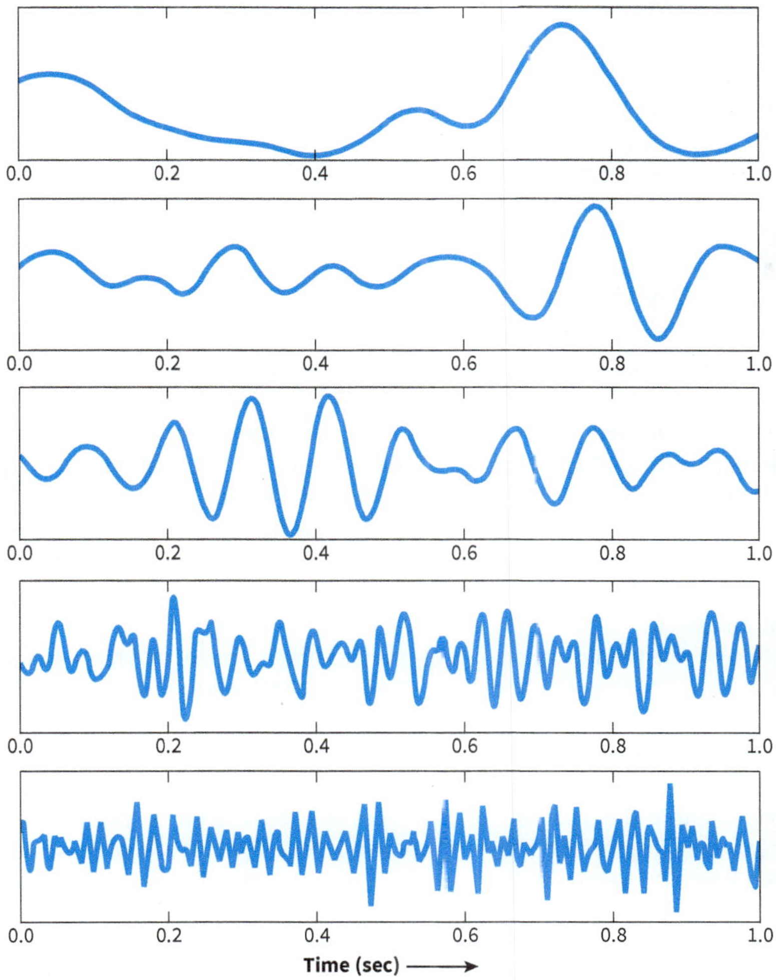

FIGURE 13.6 EEG frequency bands and associated mental states Low-frequency, high-amplitude EEG is associated with lowered level of conscious awareness, and higher-frequency and lower-amplitude EEG is associated with greater levels of conscious awareness and alert, focused cognition.

The Stages of Sleep

Although sleep appears to be a time of mental inactivity, sleep studies in Kleitman's lab showed that there were periods every night in which a person's brain was as actively engaged as if they were awake.

Kleitman had an observant graduate student by the name of Eugene Aserinsky who spent hours upon hours collecting data from sleeping participants. (As a side note, Aserinsky's personal life also demonstrates a serious lesson about the importance of sleep: he died in a car crash in 1998 after he likely fell asleep at the wheel.) Aserinsky recruited his son, Armond, to be a participant in a sleep study that included EEG and also measured the activity of muscles that move the eyes. To the older Aserinsky's astonishment, Armond's EEG during the night indicated he was awake and his eyes were moving, yet when Aserinsky checked on his son, he found him sound asleep.

Aserinsky began to wake his participants when the polysomnogram indicated the sleeper's eyes were rapidly moving, and when he asked them what they were experiencing, they almost always responded that they were dreaming. Kleitman and Aserinsky realized that **rapid eye movement sleep (REM sleep)** was qualitatively different from sleep when the person was not in REM, called non-REM or **NREM sleep**. During REM sleep, the individual is asleep and may be difficult to rouse, yet their EEG shows that their cortex is wide awake. It might interest you to know that humans are not the only organisms that show REM. Pet owners can attest that dogs and cats experience REM, and the evidence suggests that all mammals experience REM.

We spend roughly 25% of our sleep time in REM, and most people REM within the first 90 minutes of falling asleep and have four to six REM episodes each night. But what about the other 75% of our slumber time? NREM sleep is divided into several stages.

NREM Stage 1: Stage 1 (NREM1) is the border between wakefulness and sleep. The eyes open and close and skeletal muscles are still active. People awakened from this stage will often report that they were not actually sleeping. At times, a person in NREM1 will experience the feeling of falling, which may cause an abrupt muscle contraction known as hypnic myoclonus, or the myoclonic jerk. NREM1 lasts 5–10 minutes.

NREM Stage 2: During Stage 2 (NREM2), the eyes are closing and there is an accompanying increase in theta activity. The body is preparing to enter deep sleep. NREM2 typically lasts 10–25 minutes.

NREM Stage 3: This stage is characterized by relaxed muscles, slow deep breathing, and a drop in blood pressure. NREM3 is also known as slow-wave sleep or deep sleep; in this stage, the sleeper is exhibiting delta activity in their EEG and is relatively unresponsive to their environment. An individual awakened from slow-wave sleep may need several minutes to come to full alertness. NREM3 lasts 20–40 minutes and alternates with REM sleep.

NREM Stage 4: This is the deepest level of sleep. Many sleep researchers today feel that the distinction between NREM3 and 4 is arbitrary. All sleep beyond NREM2 is defined as physical immobility coupled with slow-wave EEG and relative insensitivity to external stimulation.

Once the sleeper reaches NREM3/4, it is typical that they ascend briefly to NREM2, enter a period of REM sleep, and then return to NREM3/4 to complete an NREM–REM cycle. These NREM–REM cycles take about 70–100 minutes at first, but the cycles may lengthen to 90–120 minutes as the night progresses.

The cyclical nature of sleep is illustrated by a specialized graph known as a **hypnogram**. Figure 13.7 shows a typical hypnogram with five different REM episodes spread across 8 hours of sleep.

Another of Kleitman's graduate assistants was a young medical student by the name of William Dement. Dement also explored the connection between REM and dreaming, and in fact, he was instrumental in the discovery of the stages of NREM sleep. After earning his degree, Dement began his sleep research by investigating the effects of sleep deprivation and opened the Sleep Research Center at Stanford University, one of the first sleep research centers. Dement and his students and colleagues over the years performed groundbreaking research on narcolepsy, sleep apnea, and the effects of sleep deprivation on driving performance. When Dr. Dement was asked why people sleep, his answer after studying sleep and sleep disorders for 50 years was that "the only reason we need to sleep that is really, really solid is because we get sleepy."

You may wonder if sleep patterns change over the course of your life, and the answer is that they do. Infants spend most of their life sleeping and in REM, which is believed to facilitate neuronal proliferation. Adults sleep 7–8 hours a day and spend less time in REM than children. Elderly individuals sleep only 6 hours per night on average and show a marked decrease in REM time compared with their younger counterparts. Individuals in their 60s and

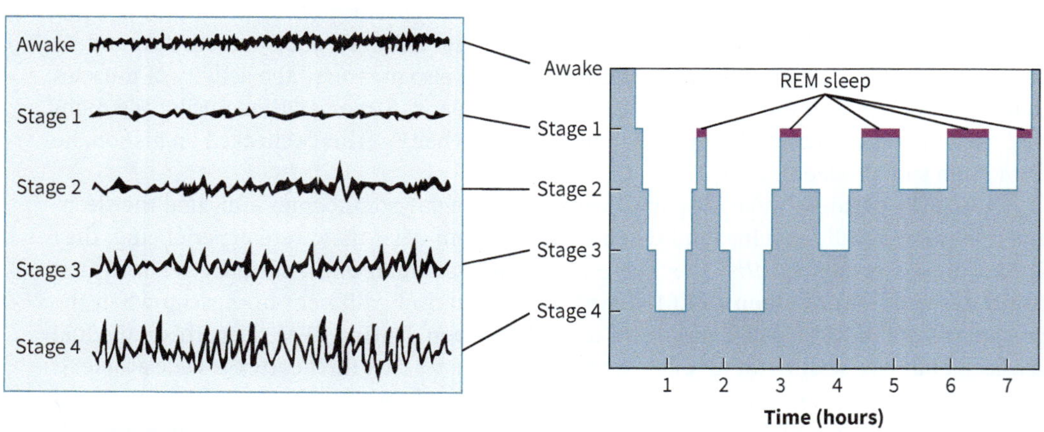

FIGURE 13.7 **A hypnogram illustrating typical sleep cycles** This hypnogram shows that the sleeper quickly went to NREM4 within the first hour of sleep, had five REM episodes with the longest beginning around the 6-hour mark, and spent most of the second half of their sleep in NREM2.

Try It Out | What's Your Sleep IQ?

Ninell / Shutterstock.com

By 21 years of age, the average person has slept in excess of 61,000 hours. You would think that we would all be experts in a subject that we have been engaged in for a third of our life, but that is not the case. In fact, most people have several misconceptions about the factors that affect their sleep. Take the **Sleep IQ test** and see for yourself if you are knowledgeable about sleep.

beyond who report getting less than 6 hours of sleep each night are significantly more likely to develop dementia. You know from earlier in the chapter that our glymphatic system cleanses our brain of harmful metabolic waste products as we sleep, so adequate sleep may help to preserve cognitive function in our later years.

Although you might think you know a lot about sleep, most people know less about sleep than they realize, as you may discover in **Try It Out: What's Your Sleep IQ?**

Dreaming MCAT

Once thought to be messages from the gods or our ancestors, we now know that dreams are the product of our nervous system. **Dreaming** can be defined as mental activity that occurs while asleep, characterized by vivid imagery of possible or impossible events. The content of dreams may include strong emotions sufficient to cause the sleeper to wake. Yet, despite the potency of these images and emotions, the content of dreams is rarely remembered even a few minutes after awakening. Dreaming generally occurs during REM sleep but can occur during any sleep stage. Often the longest dream occurs in the very early morning, not long before you wake. Most of us can remember some of what we dream, especially if the last REM episode ends right before awakening.

Sleep paralysis While you are dreaming, cortical regions are active that would be active during your waking hours, such as your visual cortex as if you were watching a movie or your motor strip as if you were fighting a velociraptor. You may wonder why a vivid dream in which you were fighting the forces of darkness does not cause you to thrash about in bed. The answer is that while dreaming, your pons and ventromedial medulla (VMM) suppress skeletal muscle activity by the action of the inhibitory neurotransmitters GABA and glycine, which in turn inhibit motor neurons in the CNS. If you have ever awakened suddenly and found that for a brief moment you were unable to move a muscle, you have experienced the **sleep paralysis** that normally accompanies REM sleep. Your brainstem was just a bit a tardy at reversing the inhibitory action of the pons and VMM as you came out of a REM episode.

Why we dream We all dream every night, regardless of whether we remember our dreams. If dreaming is universal among humans, and evidence suggests that other mammals also dream, then there must be a purpose for dreaming. Several theories have attempted to address this question.

One of the earliest theories about why we dream is the psychodynamic orientation of Sigmund Freud, Carl Jung, and Fritz Perls, who viewed dreams as a window into the unconscious and a way to fulfill forbidden wishes, suppressed conflicts, and ignored desires. These perspectives have not been supported by empirical evidence.

A second theory that has garnered significant support is that while we dream, we consolidate memories of the day's events into a more stable form for storage. This theory states that during REM sleep, our hippocampus and cortex communicate to create a more compact

Try It Out | Are You a Lucid Dreamer?

fran_kie / Adobe Stock

The content of our everyday dreams is often quite fanciful; we may dream experiences that are remotely possible such as being a swashbuckling pirate, or we may dream experiences that are impossible such as being able to fly. One characteristic of most dreams is that we are unaware that we are dreaming, but lucid dreams are the exception. Take the survey **Try It Out: Are You a Lucid Dreamer?** to see if you might be a lucid dreamer.

but accessible version of events and may also prune away unnecessary synaptic connections. Evidence for this theory is contradictory. Some believe that anxiety-related dreams are a way for the person to deal with impending events that are on that person's mind.

Yet another theory, the **activation-synthesis theory**, proposes that while we sleep, fragmentary and unconnected data is activated, and a dream is our brain's attempt to make a coherent story out of the mix. It may be that we dream for several reasons: to consolidate the memories of the day, to deal with ongoing anxieties, and simply to make sense of random neural events. Typical themes of anxiety dreams are of being chased, having your teeth fall out, being lost or unable to accomplish some task, falling, or appearing naked in public. Professors may dream the night before the first day of classes at the start of the school year of finding themselves standing in front of the class in their underwear and any moment, the students will look up from their notes. Anxiety dreams are normal and not a concern unless they excessively disrupt sleep.

Lucid dreaming Dreams are typically defined as perceptual experiences in which the dreamer is insensitive to the outside world, experiences events that may not have a connection to reality, and perceives the dream to be real. However, some individuals experience **lucid dreaming** in which they experience a dreamlike state and REM but are aware that they are in a dream.

Lucid dreamers report that they can alter the content of the dream; for example, if the dreamer was being chased by a tiger, they could choose to alter the dream by equipping themselves with a gun for defense. EEG studies of lucid dreamers have reported that one difference between lucid dreams and normal dreams is what part of the brain is active. A region of the anterior prefrontal cortex known as the dorsolateral prefrontal region that is associated with executive function such as planning and attention is active during lucid dreaming, but less active during normal dreaming. To find out if you are one of these individuals, see **Try It Out: Are You a Lucid Dreamer?**

We may perceive sleep as a time in which nothing much happens besides a pleasant dream; the reality is that we cycle through several distinct stages. In the next section we will consider the mechanisms that create this complex behavior.

13.2 Before You Go On

Section Summary

We progress through sleep cycles consisting of several different phases every night, and this progression is portrayed in a hypnogram. During REM sleep, when most dreams occur, our brain activates similar to how it activates during real events. Dreaming may be part of the brain's way of cleaning up the loose ends of the day's events, more permanently fixing the memories of the day, or dealing with impending events that are on the dreamer's mind. Lucid dreaming is when the dreamer is aware they are dreaming and can influence the course of the dream.

Comprehension Check

1. What did the discovery of REM reveal about sleep?
   a. Most people sleep in two phases.
   b. Our brain can be active and alert while we are sleeping.
   c. Most people do not get enough sleep.
   d. During NREM sleep we are generally dreaming.
2. What happens during a typical night's sleep?
   a. We spend most of our time in REM sleep.
   b. We initially spend most of our time in NREM2 but progress to NREM3/4.
   c. We experience three to five REM episodes with the longest early in the night.
   d. We quickly go into NREM3/4 and then have three to five REM episodes alternating with NREM2.
3. If you were dreaming about swimming across the English Channel, you would predict increased activity in what part of the brain?
   a. hippocampus
   b. cortical motor strip
   c. cerebellum
   d. temporal lobe
4. Lucid dreams differ from normal dreams in that lucid dreams _____.
   a. involve impossible actions
   b. always involve people we know
   c. reflect current anxieties and challenges
   d. are when the dreamer is aware they are dreaming

13.3 The Biological Basis of Sleep

LO 13.3 Identify the brain mechanisms that create sleep.

In science, the first step is always to describe, and the second step is to uncover the underlying mechanism that creates the phenomenon. So far in our discussion, we have attempted to define sleep and then describe it. There must be some mechanism that rouses our brain from its sleeping state and keeps us alert to our surroundings and responsive to stimulation from without and from within. The search for this mechanism has taken many turns, and tracing the path of discovery will help us understand the entire mechanism. What are the underlying neurological processes that regulate sleep?

Sensory Stimulation and Sleep MCAT

It was first thought that sleep is a baseline or resting state of our brain, and when sufficient sensory stimulation enters our nervous system, we become active and alert. In the 1930s, Belgian neurophysiologist Frédéric Bremer tested the theory that sensory stimulation would cause an animal to awake from slumber. He first cut the brainstem of a cat at a point between the diencephalon and the hindbrain, or roughly between the superior and inferior colliculi in humans. This procedure eliminated sensory input to the CNS and is known as *cerveau isolé* (CI; pronounced *ser-voh e-so-lay*), which is French for isolated brain. This procedure eliminated all sensory input to the brain and resulted in a cat that was permanently in a deep sleep.

Bremer then sectioned a second cat's brainstem between the bottom of the medulla and the top of the spinal cord, a procedure he labeled *encéphale isolé* (EI; pronounced *en-sef-al e-so-lay*), or isolated head. This procedure eliminated sensory input from the body but preserved input from the olfactory and optic nerves as they enter the brain at a point higher than the cut created by the EI. In contrast to the CI cats, who were denied all sensory input and as a result were permanently asleep, the EI cats maintained a sleep–wake cycle. From this experiment, Bremer concluded that a lack of sensory input led to sleep. The approximate location of the two cuts as they would appear on a human brain with the medulla highlighted in red are depicted in Figure 13.8.

An example may help to make Bremer's experiment and conclusions easier to understand. Many of us have several electronic devices at work or in our home such as computers

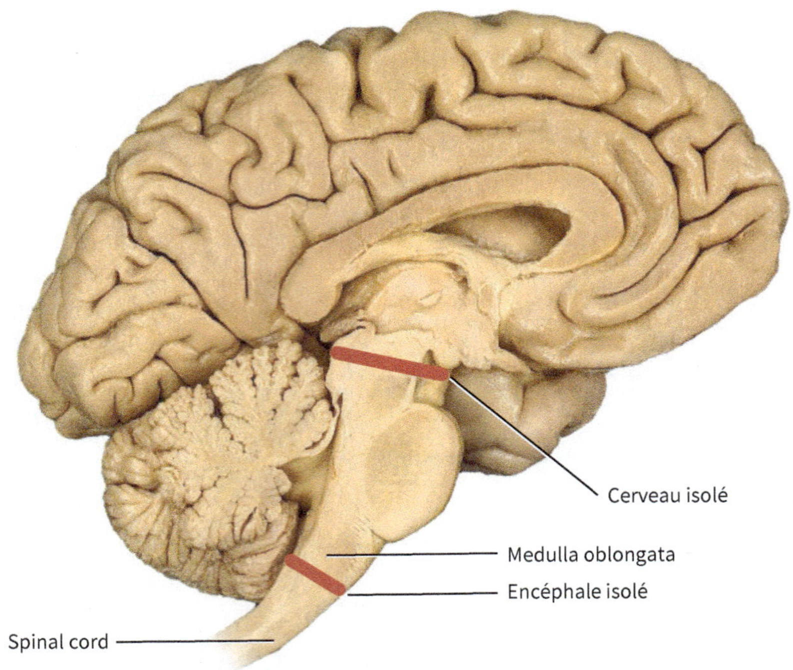

Dissection Shawn Miller
photograph Mark Nielsen

Photograph Mark Nielsen

FIGURE 13.8 CI and EI procedures as they would appear in the human brain Bremer tested the theory that sensory input creates wakefulness by separating the cerebrum from sensory input at two levels: at the CI level, which eliminated all sensory input; and at the EI level, which preserved some input. Only cuts at the CI level led to permanent sleep.

or printers that communicate with each other through a router and also communicate with the larger world through a modem (**Figure 13.9**).

If you were to disconnect the modem, your devices could still communicate with each other, but you would lose internet service to the outside world. However, if you were to disconnect the router, you would lose the ability to connect to the internet, and in addition your devices would not be able to communicate with each other. Disconnecting the modem is similar to an EI, and unplugging the router is like a CI.

This simple model of the sleep–wake mechanism was attractive, but it did not take long for people to question it. You can probably think of some criticisms yourself; for example, if stimulation prevents a person from sleeping, then individuals in a highly stimulating environment such as driving a car, operating a commuter train, serving in combat, or listening to your instructor lecture on the biological basis of sleep should never fall asleep. Yet, we know that even in circumstances of extended wakefulness such as guard duty in wartime when the penalty for falling asleep is potentially death, people will eventually fall asleep. Continued sensory input does not maintain a wakeful state indefinitely, so the search for an explanation for sleep was not concluded. From here, our story gets even more interesting.

pictafolio / Getty Images

FIGURE 13.9 Sensory pathways in the brain The EI procedure is like disabling the modem in a network, in that local devices can still communicate with each other. In contrast, the CI procedure is like disabling the router, in that it prevents both internal and external communication.

The Reticular Activating System and Sleep MCAT

In 1949, two neurophysiologists, Giuseppe Moruzzi and Horace Magoun, were investigating the neural basis of sleep and found that stimulation of certain brain regions in the brainstem created a state of wakefulness in a cat. Their search for the biological basis of sleep and arousal then led to intense study of a structure known as the **reticular activating system (RAS)**. The word reticulum means netlike, an apt description of the RAS, which is a collection of almost 100 structures that communicate with each other and also with the entire brain (**Figure 13.10**).

The RAS consists of a constellation of nuclei in the brainstem that take in sensory input from the body and in response communicate with the entire cerebral cortex. There are two components of the RAS: the ascending reticular activating system (ARAS) and a complementary descending pathway known as the DRAS. The ARAS plays an important role in maintaining wakefulness and even consciousness. Damage to the ARAS may result in a permanent loss of consciousness or coma. The DRAS leads from the brainstem to the cerebellum and plays a part in integrating posture and equilibrium.

In your study of behavioral neuroscience, you have seen time and again that regulation of essential functions such as homeostasis is accomplished by interconnected systems that exist at the level of large circuits, small regulatory centers, and finally, neurotransmitter or hormonal messengers. Sleep is no different. The RAS is an essential circuit that maintains alertness and consciousness, but there must be smaller structures and even chemical messengers also involved. We now turn to these important parts of the puzzle.

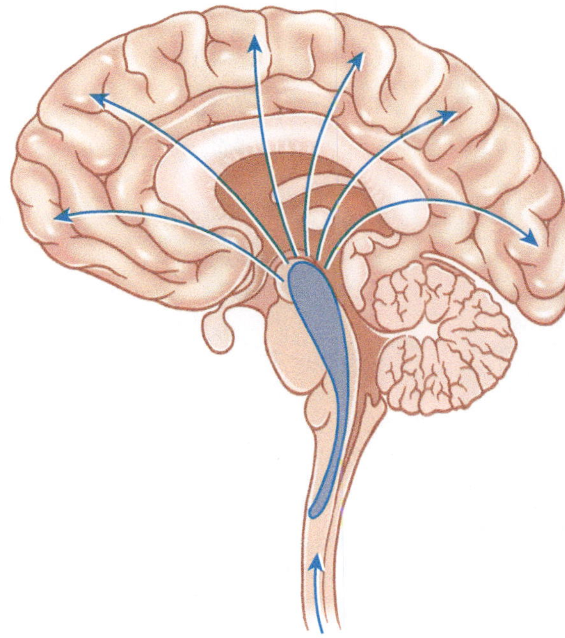

FIGURE 13.10 The reticular activating system The RAS is a weblike network of nuclei in the brainstem whose projections to the hypothalamus, thalamus, and cortex sustain alertness and consciousness.

The Brainstem, Hypothalamus, and Sleep MCAT

At the end of World War I, the Great Pandemic of 1918 (erroneously known as the Spanish flu despite there being no evidence that it originated in Spain) raged across the globe. Soldiers moving between the United States and Europe as the war ended played a major role in spreading the disease, as this image from a military emergency hospital in 1918 clearly shows (**Figure 13.11**).

Between 1918 and 1920, half a billion people were infected and around 50 million people died; by comparison, 20 million died in World War I, 60 million in World War II, and about 50 million died from the bubonic plague. Of the survivors of the Great Pandemic of 1918, most recovered without any residual effects, but for a small number of patients, their lives were forever changed for the worse. Some of these unlucky patients fell into a state of lethargy or coma and died without ever recovering consciousness. Others found that they could not sleep at all, and they also died after a few days.

Constantin von Economo, a Viennese neurologist, autopsied some of the survivors and found brain differences that corresponded to these different outcomes. Those patients who slipped into an irreversible coma had damage to the posterior hypothalamus, and those who died unable to sleep had damage to the preoptic nucleus of the anterior hypothalamus. These observations led von Economo and others to hypothesize that our brain contains a sleep center and a wakefulness center.

We now believe that the ventrolateral preoptic nucleus is responsible for initiating NREM sleep, and the posterior hypothalamus initiates wakefulness by stimulating the RAS in the brainstem. After von Economo's discovery, it became apparent that damage to the brainstem often resulted in an irreversible coma and that damage or disease could create either pathological sleep or pathological wakefulness. Researchers realized that a complex system for controlling sleep and wakefulness existed at least partially in the brainstem. The search for understanding was on.

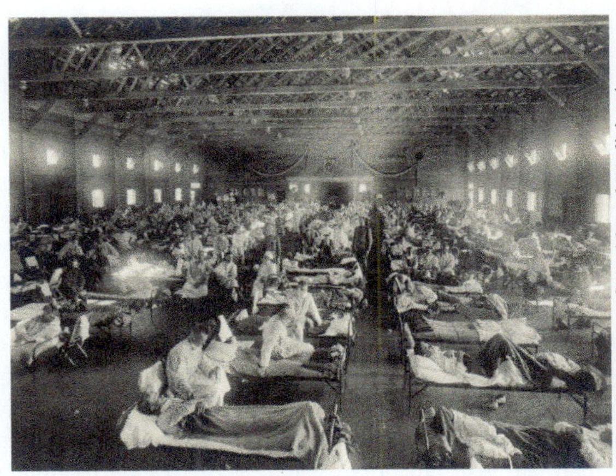

FIGURE 13.11 An influenza hospital ward in Camp Funston, Kansas So many soldiers fell ill during the 1918 epidemic that overflow patients from hospital wards were housed in enormous tents.

Neurotransmitters and Sleep

Up to this point we have examined a number of factors that may be responsible for being awake and then falling into sleep. Stimulation plays a part, but we know that stimulation alone is not the answer. The RAS keeps us alert. Add to this the discovery that damage to centers in the hypothalamus results in an animal that sleeps constantly or stays awake until it dies. But we do not live in a world where we are either fully awake and alert or deeply asleep; there are as many fine gradations of consciousness and slumber as there are colors in a sunset. This suggests the presence of a neurochemical messenger system capable of making subtle adjustments to the sleep–wake cycle.

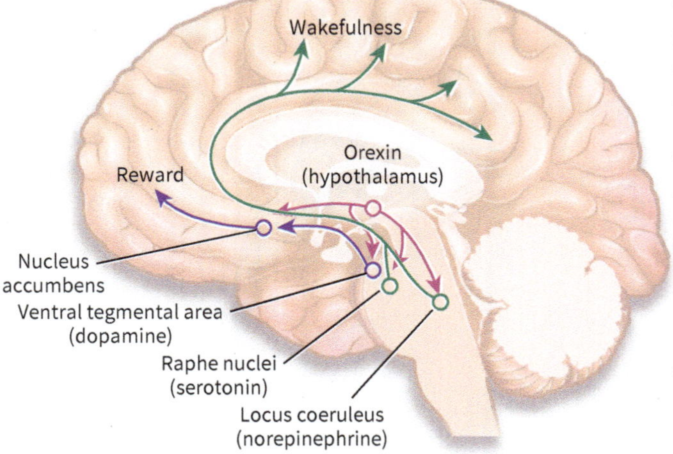

Scammell, T. E., & Saper, C. B. (2007). Orexins: looking forward to sleep, back at addiction. Nature Medicine. Reproduced with permission of Springer Nature.

FIGURE 13.12 **The orexigenic pathway** Orexigenic neurons in the hypothalamus interact with a wide range of subcortical structures, thereby creating wakefulness.

There are at least two such messengers that earn our attention, and the first is **orexin**, a neuropeptide that you may already have encountered as you learned about hunger and satiety. Orexin regulates arousal and wakefulness through a small collection of orexigenic neurons, perhaps as few as 20,000. These neurons are located in the lateral hypothalamus, and their axons project throughout the CNS. Orexigenic neurons can make subtle adjustments to our level of arousal and alertness by activating the RAS and several other regions of the brain: the locus coeruleus, which secretes norepinephrine; the Raphe nuclei, which secrete serotonin; and the ventral tegmental area, which secretes dopamine. These orexigenic neurons also stimulate the nucleus accumbens, which creates motivation and reward for being alert (**Figure 13.12**).

You know that every mechanism has an opposite and equal mechanism, so if there is a sleep center and a wakefulness center, there must be a soporific, or sleep-inducing, companion to wakefulness-inducing orexin. The agent that causes a person to experience sleepiness is **adenosine**. You may recall from earlier discussions of neural processes that adenosine triphosphate (ATP) is a source of energy that powers our cells. When ATP releases its energy in a chemical reaction, one by-product of that reaction is adenosine. There are adenosine receptors in the striatum and the hypothalamus that, when stimulated, act to inhibit orexin production, thereby inducing sleepiness.

As we make our way through our daily routine, expending energy and consuming ATP, our levels of adenosine rise in areas such as the RAS that control wakefulness. With rising levels of adenosine, we feel increasingly fatigued. As you sleep, adenosine levels slowly decline, until you become more likely to wake. This elegant negative feedback loop is one reason why you feel tired after a strenuous day and energetic after a good night's sleep. You can read about one researcher who is exploring the relationship between neurotransmitter events that occur during sleep and memory in **Advances in Behavioral Neuroscience: Gina Poe and the Role of Sleep in Memory**.

Advances in Behavioral Neuroscience | Gina Poe and the Role of Sleep in Memory

The Regents of the University of California

Dr. Gina Poe is a professor in the Department of Integrative Biology and Physiology at UCLA, where she studies the role of the hippocampus and other brain structures in memory consolidation during sleep. One area of her work investigates how and when hippocampal processes activate and consolidate memory traces. For instance, when you file a document on your computer, the computer writes the file to your hard drive and clears temporary files from its memory. Dr. Poe reasoned that our hippocampus might act in a similar fashion. She and her colleagues and students examine the role that dysfunctional sleep mechanisms may play in disorders as disparate as PTSD and Alzheimer's disease.

Although Dr. Poe is now considered to be an expert on sleep and memory, her path toward prominence did not start so auspiciously. Her original intention was a career in public health, but a job in a research lab led her to an interest in the neural processes that occur during sleep. During her graduate studies at UCLA, a fellow student told her that one of her professors had said that a woman of color could not be successful in the field. Determined to prove him wrong, she worked even harder than before and earned her doctorate.

In addition to being a world-class researcher, Dr. Poe uses her position to facilitate the careers of underrepresented groups in neuroscience, including by directing outreach and education programs at UCLA and co-directing the Society for Neuroscience's Scholars program, which provides support and mentoring.

Questions

1. **Search online for "Dr. Gina Poe" and read her faculty biography. Identify one area of her research that you find interesting and briefly state why you selected it.**
2. **Poe's research indicates that fluctuations in neurotransmitters that are unique to sleep are essential to the formation of stable, long-term memories. If you could adjust these neurotransmitter levels during your sleep to maximize your ability to remember tomorrow what you learned today, would you do so? Why or why not?**

The gradual buildup of adenosine levels in the brain creates the perception of drowsiness and is likely responsible for episodes of **microsleep**, which is defined as a temporary episode of sleep that lasts from a few seconds to as long as 30 seconds. Individuals who are fatigued may experience microsleep without being aware of it. Microsleep is the result of fatigue, sleep deprivation, sleep apnea, or shift work, and it is most likely to occur when the individual is engaged in a monotonous task such as driving a vehicle on a highway. The consequences of even a short microsleep episode can be serious if not fatal.

13.3 Before You Go On

Section Summary

There are several neural mechanisms that interact to keep us alert. The RAS serves as a global mechanism for maintaining an alert state, and centers in the brainstem and hypothalamus contribute. The neurotransmitter orexin maintains wakefulness, but as adenosine levels rise throughout the day, they inhibit the activity of orexigenic neurons, thereby creating the need to sleep.

Comprehension Check

1. The experiment in which Bremer made cuts in the brains of cats was designed to test whether sleep was the result of _____.
   - **a.** a lack of energy
   - **b.** disinhibition of the dopaminergic system
   - **c.** a lack of sensory input
   - **d.** overstimulation of the senses, leading to receptor fatigue
2. The reticular activating system is so named because it _____.
   - **a.** is located in the reticular nucleus of the hypothalamus
   - **b.** is a netlike structure with connections throughout the brain
   - **c.** activates the pons and medulla
   - **d.** directly controls the release of serotonin
3. When von Economo autopsied victims of the Great Pandemic of 1918, he discovered that damage to the _____ and damage to the _____.
   - **a.** medulla led to coma; pons led to wakefulness
   - **b.** pons led to coma; thalamus led to wakefulness
   - **c.** hypothalamus led to wakefulness; thalamus led to coma
   - **d.** posterior hypothalamus led to coma; anterior hypothalamus led to wakefulness
4. Orexin and adenosine are _____.
   - **a.** excitatory and inhibitory, respectively, with regard to sleep
   - **b.** hormones that stimulate appetite and satiety
   - **c.** neurotransmitters that regulate the RAS
   - **d.** neurotransmitters that regulate cortical alertness via the thalamus

13.4 Sleep Disruptions

LO 13.4 Describe common sleep disorders.

One goal of this textbook is to present a behavior that has a neurological mechanism, describe it, explain how it works, and then pivot to discuss how the process can go awry. We now turn our attention to a few disorders of sleep. Before we put the topic of sleep to bed, let us take a moment to provide an overview of agents that affect sleep quality and quantity.

Insomnia comes from the Latin meaning no sleep and is defined as a difficulty in either falling asleep or obtaining sufficient duration of sleep. Each year, 30%–40% of adults experience an episode of insomnia, and 10%–15% report chronic insomnia. Some causes of sleep difficulties lie outside the focus of this chapter, such as stress from work, finances, or relationships, and also chronic pain. Here we will briefly touch on the causes of insomnia that are relevant to the topics currently under consideration, including drugs, alcohol, and sleep disorders.

Drugs and Sleep MCAT

Many common substances, including but not limited to prescription drugs, have a negative effect on sleep, both in making it harder to fall asleep and in inducing drowsiness when we want to stay awake.

Three examples of legal or licit drugs that affect sleep are alcohol, caffeine, and cold remedies. Alcohol significantly changes the normal REM–NREM sleep cycle, as illustrated in

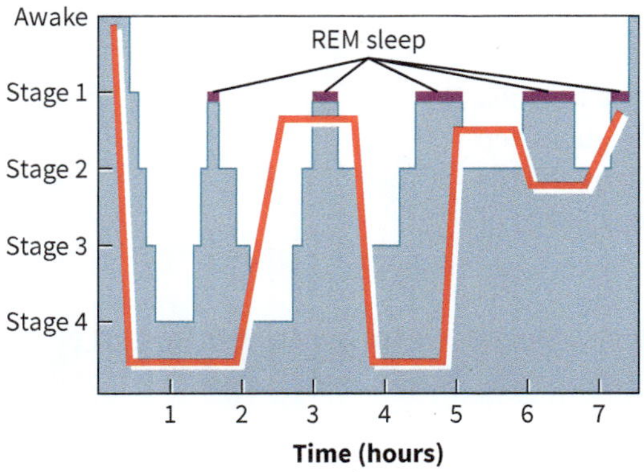

FIGURE 13.13 The effect of alcohol on the sleep cycle The red line indicates the sleep cycle of an individual who has consumed alcohol prior to retiring, overlaid on the normal pattern of sleep cycles from Figure 13.7. Alcohol causes a rapid drop into NREM4 deep sleep but disrupts the REM cycle, leading to fewer REM episodes and more NREM2 sleep than is normal.

Figure 13.13. In contrast to the normal pattern of several REM cycles spread throughout the night, alcohol decreases the time to fall asleep but then suppresses REM in the first half of the night. This suppression is followed by a REM rebound in the remainder of the night. This disruption to the normal sleep cycle explains why a person who drinks heavily will experience a sleep that is of normal length yet not refreshing.

Students and instructors lead demanding lives, and many rely on caffeine to help them stay alert for extended periods. As noted in a previous chapter, caffeine is typically consumed in coffee or caffeinated beverages such as soft drinks and energy drinks. Caffeine is a stimulant, and the evidence suggests that caffeine increases sleep latency (the time it takes to fall asleep), shortens total sleep time, and decreases sleep quality but does not affect REM sleep. Because the half-life of caffeine in your system is roughly six hours, these negative effects can be reduced by consuming the day's caffeine before noon so that its effects are wearing off by nightfall.

Before we move on to prescription drugs, we can take just a moment to consider the effects of cold and flu remedies on the sleep–wake cycle. You likely have noticed that some are marketed to make you drowsy and help you get a good night's sleep, whereas others are advertised for daytime use. Understanding how they work is quite simple. Many of us have allergic reactions to pollen, bee venom, and certain foods that act as allergens or allergic-response generators. When we encounter an allergen, our immune system responds by releasing histamines, chemicals that activate a response to repel the allergen, and that response includes a runny nose and coughing. There are histamine receptors in the CNS and throughout the body. When histamine receptors in the hypothalamus are stimulated, they have an excitatory effect on the sleep–wake cycle. Cold and flu remedies contain antihistamines that block this effect, leading to drowsiness. Some over-the-counter cold and flu remedies contain diphenhydramine (Benadryl) or doxylamine succinate, two agents that block histamines throughout the body and brain, leading to both drowsiness and a decrease in allergic reaction. Other antihistamines are designed to affect bodily histamine receptors but not CNS receptors, ideally leading to a decrease in allergic response in the body without creating drowsiness.

Amphetamines such as Ritalin and methylphenidates such as Concerta, which are prescribed for ADHD, have a detrimental effect on sleep. These drugs block dopamine and norepinephrine transporters, leading to an increase in DA and NE at the receptor site. A consequence of this increase is heightened alertness, but this comes at the cost of sleep difficulties and possibly a decrease in REM sleep. It is estimated that perhaps 6% of children in middle and high school regularly take some form of medicine for ADHD and that among college students, perhaps one in six has used one of these pharmaceuticals illegally as a study aid.

Other drugs induce drowsiness, including **benzodiazepines** such as Ativan, Valium, Halcion, Klonopin, and Xanax. These drugs cause drowsiness but typically reduce the amount of deep sleep or REM sleep. NREM3 and REM sleep are essential for a good night's sleep, so you can see that an individual using these drugs might not feel refreshed even after 8 hours of sleep. Melatonin is sometimes thought of as a sleep aid, but it affects seasonal or circadian rhythm mechanisms, so we will discuss its role when we consider that topic.

Finally, you can imagine that the inability to achieve a good night's rest leads to all manner of unpleasant outcomes. Poor health as well as job and family stress are common consequences of chronic insomnia. For this reason, researchers have been motivated to find treatments that can cure or at least ameliorate incidents of insomnia for patients who do not respond well to current treatment modes. Common examples of this class of drugs are the so-called Z-drugs (zolpidem, eszopiclone, and zaleplon), which are marketed under the trade names Ambien, Lunesta, and Sonata, respectively.

These drugs are classified as **nonbenzodiazepines**, which means that they act as benzodiazepines but have a very different structure at the molecular level. Nonbenzodiazepines have the advantage that their action is primarily limited to the brain because they have an affinity for GABA receptors in the brain but less so for GABA receptors in the spine. Their inhibitory effect creates a sensation of drowsiness and may help the user to stay asleep. Although benzodiazepines and nonbenzodiazepines have been useful in treating insomnia in many patients, in rare cases, individuals using them have been found to enter a fugue state in which they might get out of bed and walk around their home, cook a meal, leave their house, or drive a car, and later have no recollection of their actions. The mechanism that is responsible for such behavior is currently not well understood.

Sleep Disorders MCAT

It is a sad fact that disorders of normal processes can teach us about the mechanisms involved in that process. Some sleep disorders are narcolepsy, night terrors, seasonal affective disorder, and sleep apnea.

Narcolepsy One malady that helps us to clarify our understanding of the mechanisms that create sleep and wakefulness is **narcolepsy**. Narcolepsy is quite rare, affecting perhaps 0.05% of the population. Narcolepsy should not be confused with dozing off; we all have felt sleepy in class or found ourselves dozing off in the evening while watching a show. Narcoleptics will suddenly fall asleep for a few minutes to a half-hour mid-conversation, while operating heavy machinery such as a chain saw, or while sitting at the dinner table, despite having enjoyed a full night's sleep.

We know two things about narcolepsy that should make sense to you now; the REM sleep of narcoleptics is atypical, and even more to the point, there is a loss of orexin-producing neurons in the lateral hypothalamus. The causes of narcolepsy are unclear at this time, but genetics plays a part, and there is evidence that for some individuals, their immune system or a *Streptococcus* infection may have damaged orexigenic neurons.

Night terrors It can be disconcerting when a person sits bolt upright in bed in the middle of the night, eyes wide open with a panicked look on their face and the perception that their heart is racing out of control. They may thrash about with their limbs in what appears to be an attempt to protect themselves from harm. They may appear to be responsive, and when asked if something is wrong, they may even respond in a simple way. When my brother-in-law was about 8 years of age, he would occasionally wake in the middle of the night with eyes wide open and begin to howl in terror. This would go on for 2–3 minutes, and then he would fall back asleep and not remember a bit of the event in the morning. He was experiencing a sleep disorder called **night terrors**, or sleep terrors, which most often occurs among children. Night terrors differ from nightmares in that when a person is having a nightmare, they can be awakened and may remember the details of the dream for a while. Individuals experiencing night terrors may have open eyes and may even respond to simple questions, but they are not actually coherent and usually do not remember the details of their dream.

Night terrors are rare (perhaps 5%–6% of children experience them), and they typically go away during adolescence. They usually occur during NREM3, when the individual is exhibiting slow-wave delta sleep. Their cause is unknown, but night terrors do run in families, and concordance is highest in identical twins. Children who experience night terrors generally grow out of them with no increased risk of psychiatric illness. (My brother-in-law is now a highly respected family physician.) Adults who experience repeated episodes of night terrors should see a physician because several serious disorders such as epilepsy, hypoglycemia, and PTSD are linked to night terrors in adults.

Seasonal affective disorder As fall gives way to winter, and the days become shorter and the nights longer, some people find themselves feeling depressed and lacking energy. They may experience a decreased libido, have trouble waking up, and overeat at mealtimes and in between, often craving carbohydrates. The likely diagnosis is **seasonal affective disorder (SAD)**. SAD is considered to be a type of depression that is seasonal, is serious, and includes both trouble getting to sleep and trouble waking up from sleep. As our current focus is on sleep and not depression, we will simply say that seasonal changes in daylight adversely affect hypothalamic processes in some individuals, leading to SAD.

Women are more likely to suffer from SAD than men, and, as exposure to daylight is involved in SAD, individuals living farther from the equator are more likely to experience SAD than those who live closer to the equator, where days are generally longer and also more uniform across seasons. Residents of Japan and Iceland are noteworthy exceptions to the rule that people living in higher latitudes are prone to SAD, and there is some thought that their steady consumption of fish high in vitamin D may protect them from the effects of extended seasonal periods of darkness.

Courtesy of The SunBox Company

FIGURE 13.14 **An individual with SAD using a light box** The symptoms of seasonal affective disorder can be reduced by regular exposure to a light box that simulates normal sunlight.

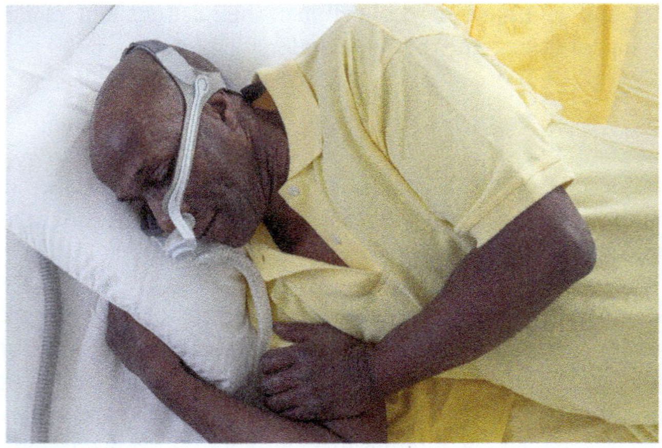

JPC-PROD / Adobe Stock

FIGURE 13.15 **A CPAP nasal mask** Individuals who suffer from obstructive sleep apnea can benefit from sleeping with a CPAP mask that forces air into the airway, thereby relieving the symptoms of hypoxia.

You might be able to predict one treatment for SAD. If you guessed that light is the answer, you are on the right track. To effectively combat SAD, a light source must mimic certain characteristics of natural light, which is why sitting in front of a computer screen or being in a room with traditional lights is ineffective. Light boxes that mimic sunlight have been found to be effective in the treatment of SAD; sitting in front of a light box with eyes open for an hour each day provides many people relief from their symptoms (Figure 13.14).

Sleep apnea One sleep disorder that is a serious medical condition is **sleep apnea**. This condition is characterized by interruptions or even cessation in breathing during sleep, often followed by gasping for breath as the individual experiences hypoxia, or a lack of blood oxygen. There are two forms of sleep apnea: obstructive sleep apnea and central sleep apnea.

Obstructive sleep apnea is caused by an obstruction in the airway. Risk factors are obesity and possibly a defect in the palate of the throat that blocks the passage of air. Perhaps one in fifteen adults experience obstructive sleep apnea at some point in their life. In middle age, the incidence of obstructive apnea climbs for individuals who smoke, do not exercise, or are overweight. A common form of treatment for obstructive sleep apnea is to force air into the airway during sleep through a continuous positive airway pressure (CPAP, pronounced *see-pap*) mask (Figure 13.15).

Central sleep apnea, which is rarely seen, is caused by a dysfunction of structures in the medulla. When functioning normally, these structures monitor blood levels of carbon dioxide (CO_2) and react to rising levels of CO_2 by stimulating breathing.

All of the sleep disruptions mentioned here can lead to insufficient sleep. Some medical professionals believe that many people do not get an adequate night's sleep on a regular basis. In Think About It: Sleep Insufficiency, you can learn more about sleep adequacy in the United States.

Think About It | Sleep Insufficiency

Elnur / Adobe Stock

Humans use technology to adapt the world to their needs; we can have daylight in our homes even after the sun sets, and we can create a climate inside our homes to suit us, regardless of the season. These advances make our lives comfortable, but there are times when our attempts to bend the natural world to our needs works against us. For instance, many countries have adopted some variant of daylight saving time (DST), in which clocks are advanced an hour in the spring and are set back an hour in the fall; the idea being to better align clocks with daylight Statistics on vehicle crashes show that the switch back from DST to standard time, which lessens that night's sleep by one hour, is accompanied the following week by a 6% increase in fatal car crashes. There is much we still do not know about sleep patterns in the modern world.

The Centers for Disease Control (CDC) serves as a national clearinghouse for information about sleep disorders and treatments. You might be surprised to learn that in the United States there are reliable geographic differences in sleep adequacy, as defined by the percent of the adult population that average seven or more hours of sleep per night. There are also reliable differences in sleep adequacy that are related to age, sex, and ethnicity or race. Search on "CDC Sleep Disorders Statistics" and review the data before considering these questions.

Questions

1. **Search the internet using the terms "workplace accidents following DST." What did you discover? If you managed a factory where your employees worked in a hazardous environment, can you think of anything you could do to minimize the risk for the day after the switch?**
2. **The CDC presents data on sleep adequacy that are categorized by criteria such as geography, age, sex, and ethnicity. Pick one such criteria and examine the data. Why do you think this pattern exists?**
3. **In the same page, the CDC lists diseases that are associated with inadequate sleep. Pick two that you think are important and propose a way for society to reduce the risk to sleep adequacy that these two disorders present.**

13.4 Before You Go On

Section Summary

Modern life places many demands on us, and a good night's sleep prepares us to meet the demands of school, work, and our personal life. Sleep disorders interfere with our ability to do our best and can seriously affect our health. Some sleep disorders are caused by life-style choices or seasonal effects, and others may have a neurolcgical basis.

Comprehension Check

1. One fact about sleep disorders is that they _____.
   - **a.** are typically only seen in children
   - **b.** affect at least 15% of adults at some time during their lives
   - **c.** can be easily treated with caffeine
   - **d.** are always the result of lifestyle choices
2. Alcohol affects sleep by making it _____ to fall asleep and then creating _____.
   - **a.** harder; nightmares
   - **b.** harder; more restful sleep
   - **c.** easier; more restful sleep
   - **d.** easier; less restful sleep
3. One fact about seasonal affective disorder is that it _____.
   - **a.** appears to be related to changes in sunlight
   - **b.** can be treated by caffeine and nicotine
   - **c.** is mostly seen in people living near the equator
   - **d.** tends to disappear with advanced age
4. Which statement is true of sleep apnea?
   - **a.** It affects mostly males.
   - **b.** It is often treated with nonbenzodiazepines.
   - **c.** It is not a serious health risk.
   - **d.** Obesity is a risk factor for sleep apnea.

13.5 Circadian Rhythms

LO 13.5 Relate circadian rhythms to the larger phenomenon of sleep.

Life on Earth is described by many cycles rather than by a straight line. Each year, salmon and wildebeest migrate hundreds of miles, and the Arctic tern lives in the Arctic but migrates to the Antarctic to breed and nest, a round trip of 44,100 miles (71,000 km). Plants sprout,

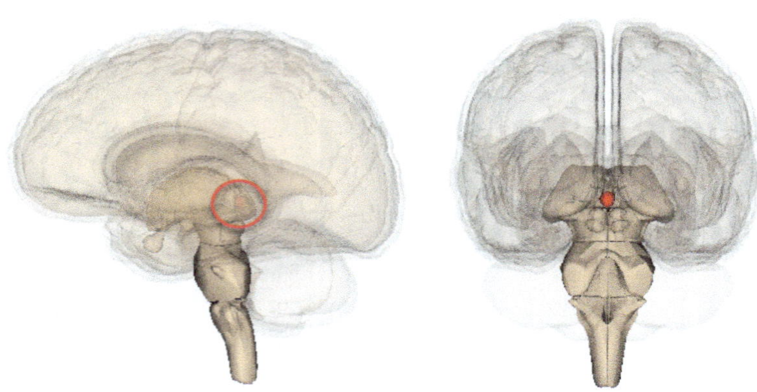

FIGURE 13.16 The pineal gland *in situ* The pineal gland is located on the brain's midline and at the rear of the thalamus. It secretes melatonin in response to the daily cycle of light and dark.

flourish, and then go dormant or die. These and other organisms behave in a predictable way as the seasons go through their endless cycle.

You and I also have cycles in that our brains are sensitive to the daily (diurnal) transition from light to dark and also to seasonal changes in the length of the day. Our distant ancestors woke at sunrise and slept when it was dark; that light–dark, wake–sleep rhythm became a part of our natural cycle. These mechanisms create what are known as circadian rhythms or cycles from the Latin *circa* meaning around and *diem* meaning day. A **circadian rhythm** refers to a cycle of around 24 hours in length, whereas the **diurnal rhythm** or cycle refers to the typical day–night cycle. For the majority of us who regularly go to sleep sometime after dark and arise when it is light, these two cycles are roughly synchronized. As you are about to see, when the diurnal cycle and the circadian cycle become desynchronized, the consequences can be serious. We begin with a discussion of how circadian rhythms relate to sleep.

Melatonin and the Sleep–Wake Cycle MCAT

We are all familiar with light-sensitive security lights with photocells that constantly measure the level of ambient light. As the day ends and dusk approaches, the photocell signals the device that it is time to turn on the light. At dawn, the photocell detects the increase in illumination and turns off the light. We have a circuit in our brain that works in a similar way. A part of the endocrine system known as the **pineal gland** (Figure 13.16) is the source of the hormone **melatonin**, which is part of the circuit that creates the feeling of drowsiness.

Light stimulates the retina when you are awake during daylight hours, and the retina communicates with the pineal gland to inhibit the release of melatonin. When it gets dark or when your eyes are closed during sleep, the inhibition ends, causing the pineal gland to secrete melatonin. In this fashion, melatonin levels rise and fall in a daily cycle, with blood concentrations greatest between 2 AM and 4 AM as illustrated in Figure 13.17.

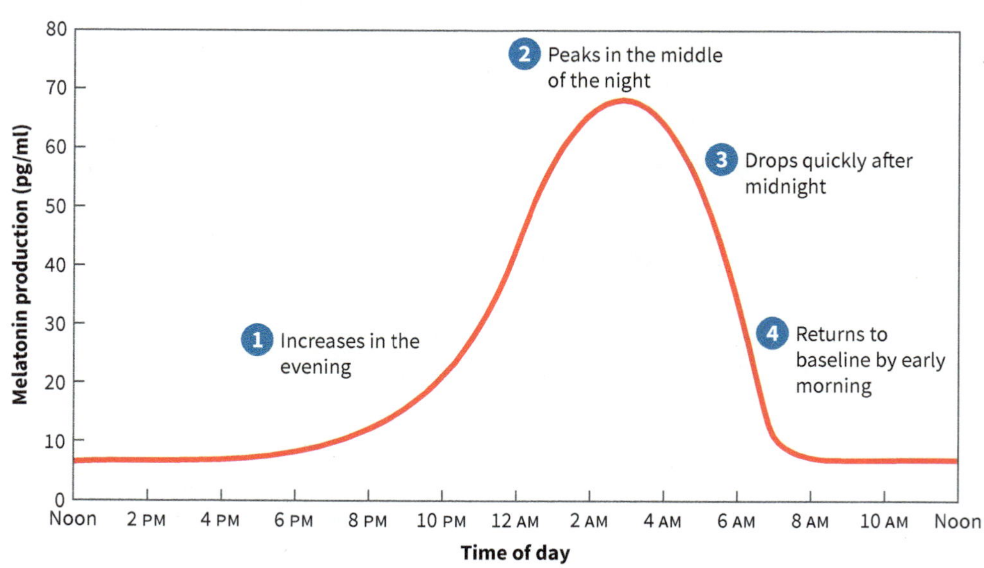

FIGURE 13.17 Melatonin diurnal cycle Melatonin production follows a diurnal cycle, increasing in the hours before bedtime and decreasing in the hours before waking.

For this reason, melatonin supplements are often one of the first remedies suggested to those who have difficulty getting to sleep or staying asleep. However, clinical evidence for the therapeutic effect of oral melatonin as a treatment for sleep difficulties is mixed.

The Suprachiasmatic Nucleus MCAT

As you have often seen in your study of the brain, there is rarely a simple answer to our most basic questions, and that rule also holds true as we consider circadian rhythms. The secretion of melatonin into the bloodstream is an important part of the sleep–wake cycle, but it is not the only part, and it may not even be the most important part. We now understand that a small structure known as the **suprachiasmatic nucleus (SCN)** of the hypothalamus (named for its position atop, or superior to, the optic chiasm) is an important regulator of the sleep–wake cycle.

Like the pineal gland, the SCN, which consists of a mere 20,000 or so neurons, receives information from the retina about levels of light. In response to signals that indicate darkness, the SCN and the pineal gland stimulate the sensation of drowsiness and related behaviors such as lowered body temperature and metabolic slowing that are associated with sleep. **Figure 13.18** clearly shows that circadian rhythms affect a wide range of metabolic processes.

When Diurnal and Circadian Rhythms Are Mismatched

We end our consideration of sleep, dreaming, and circadian rhythms with a topic that may bring the complexity and the importance of these topics into a clear focus. Sleep researchers categorize the various mechanisms involved in daily or diurnal rhythms as **Process S** (for sleep) and the mechanisms that regulate circadian rhythms as **Process C** (for circadian). Process S responds to the daily rhythm of day and night and is regulated by chemical messengers such as adenosine and melatonin. Process C is a 24-hour rhythm of metabolic functions that is regulated by the suprachiasmatic nucleus (SCN) of the hypothalamus and is independent of the day–night cycle.

If you have ever stayed up all night and then experienced a wave of energy mid-morning, you can thank Process C for providing you with that second wind as your metabolism was energized by your SCN in an attempt to bring you back into accord with your normal daily rhythm. In general, these two cycles coincide nicely; the circadian rhythm of Process C is in synch with the reticular, brainstem, hypothalamic, and neurotransmitter components of Process S that are driven by being awake during the day. What happens when these two processes become desynchronized?

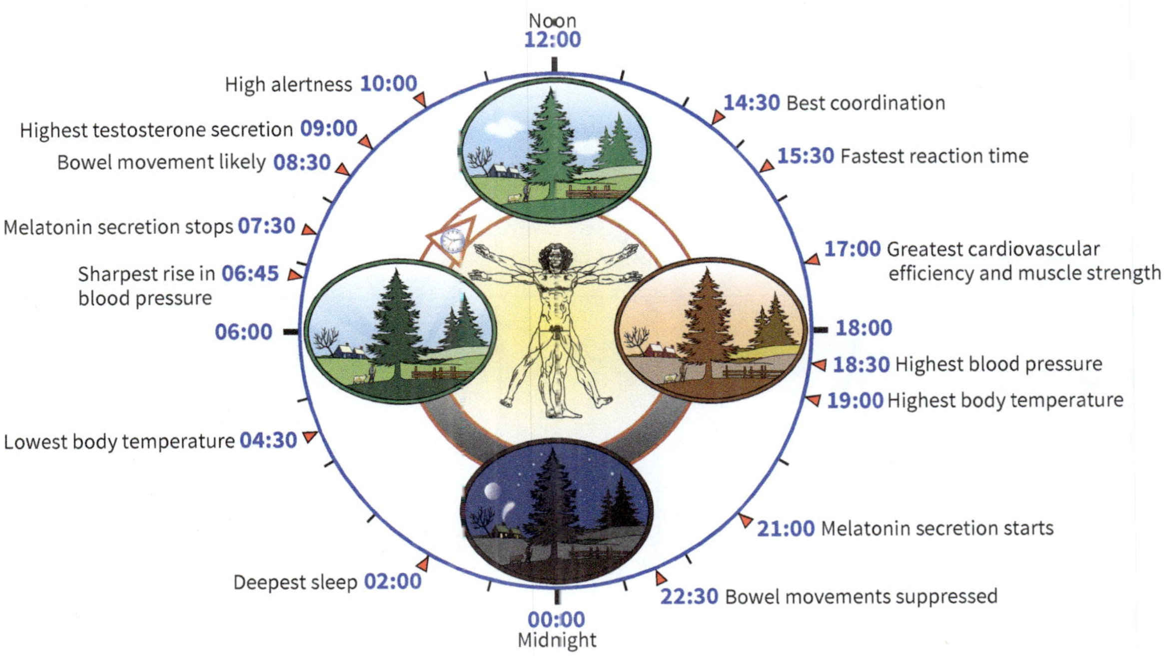

FIGURE 13.18 Features of a typical circadian rhythm During each day, physiological mechanisms such as mental alertness, blood pressure, and muscle strength ebb and flow.

You may be familiar with the concept of jet lag, where a person traveling over more than two time zones in a single day will experience sleep disruption and even difficulties concentrating. This occurs because the normal circadian rhythm is out of synch with the diurnal cycle until your brain is able to resynchronize the two cycles. Eastward travel takes longer to adjust to than westward travel because advancing the circadian rhythm seems to be more difficult than delaying it. The rule of thumb is for each time zone crossed going east, two-thirds of a day to a full day is needed to adjust, and for traveling west, one-half day to two-thirds is sufficient.

Not everyone flies across the country regularly, but some of us must be awake in the middle of the night to provide the essential functions that modern civilization requires. We expect that medical staff such as physicians, nurses, and associated health care professionals will be available 24 hours a day should we need them, and firefighters, police, and those who maintain our infrastructure also must be available at all hours. Manufactured goods and groceries must be transported from their point of origin to distant markets, and truckers are the people who provide that service, sometimes driving all night.

When an individual arises in the morning, goes to work, and then retires at night, their Process S and C cycles align with the natural light–dark cycle, and they become drowsy at a point where sleep is appropriate as shown in Figure 13.19a. However, when an individual's work schedule consists of shifts that constantly change, where one month they might work 11 PM to 7 AM, the next month from 7 AM to 3 PM, and the month after that from 3 PM to 11 PM, their Process S and C cycles become disconnected from their daily rhythm, which leads to drowsiness and potentially microsleep at an inappropriate time of day, as shown in Figure 13.19b.

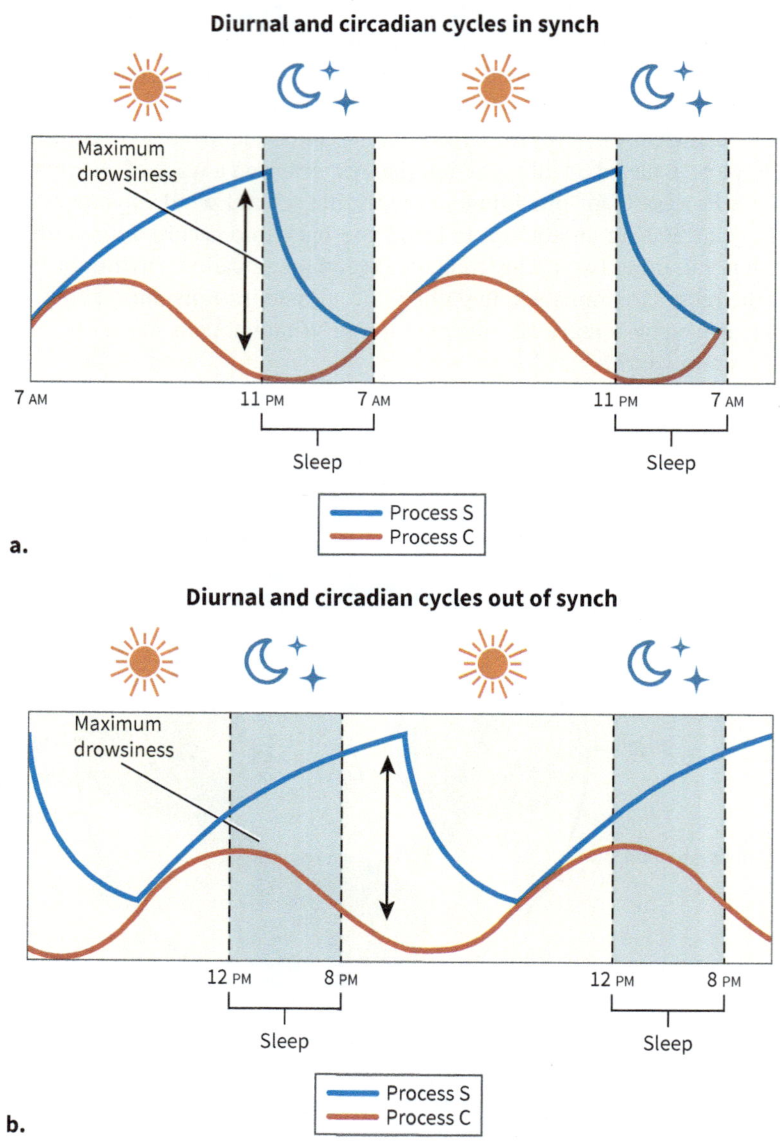

FIGURE 13.19 Diurnal and circadian cycles **a.** When diurnal and circadian cycles are synchronized, drowsiness peaks at 11 PM as bedtime approaches. **b.** When diurnal and circadian cycles are out of synchrony with their daily routine, maximum drowsiness occurs early in the morning while they may still be at work, driving a car, or operating machinery.

Try It Out | The Dangers of Microsleep

National Transportation Safety Board / Wikimedia Commons / Public domain

We began this chapter with a story about a train engineer who, although familiar with the route he was assigned that day, entered a curve at about 50 mph faster than was safe and the derailment resulted in several deaths and many injuries. In the worksheet **The Dangers of Microsleep**, you will be presented with the facts of the case that were discovered in the review by the National Transportation Safety Board (NTSB), and then you will consider the event from the perspective of behavioral neuroscience.

We will conclude this chapter by returning to the *Stories of the Brain* that you read at the beginning of this chapter. This tale described an incident when an experienced and professional train engineer did not see the danger that was directly in front of him. In this **Try It Out: The Dangers of Microsleep**, you may see the incident in a different light.

13.5 Before You Go On

Section Summary

Diurnal organisms have evolved a mechanism to link sleep and wakefulness to both the daily solar rhythms and seasonal changes in the length of a day. When light stimulates the retina, one output channel goes to the pineal gland and the suprachiasmatic nucleus of the hypothalamus, and this circuit stimulates the secretion of melatonin, which creates the perception of drowsiness. The diurnal and circadian cycles keep us in congruence with seasonal changes, but disruptions to that congruence can lead to difficulties.

Comprehension Check

1. A circadian rhythm is defined as a biological rhythm that ______.

a. follows a 24-hour cycle

b. repeats every lunar cycle

c. keeps us awake in the winter

d. solely regulates the sleep–wake cycle

2. When light enters the retina, it sends a signal to the ______ to regulate circadian rhythms.

a. lateral hypothalamic nucleus and optic chiasm

b. medial hypothalamus and optic chiasm

c. suprachiasmatic nucleus and pineal gland

d. pineal gland and circadian nucleus of the hypothalamus

3. Which of these is an important hormonal messenger that regulates our sleep–wake cycle?

a. leptin

b. valerian

c. theanine

d. melatonin

4. According to Figure 13.18, the time when you are probably at your very strongest is ______, and the time when you would be most alert would be ______.

a. noon; 6 PM

b. 7:30 AM; 2:30 PM

c. 5 PM; 10 AM

d. 9 AM; 7 PM

Conclusion

Before you read this chapter, you may have considered sleep to be a simple phenomenon; we close our eyes for several hours in which nothing much happens except perhaps a dream. But the story at the beginning of the chapter raised puzzling questions about the very nature of sleep and the complex web of mechanisms that regulate it. You have seen in your study of behavioral neuroscience that essential behaviors such as homeostasis often have an evolutionary basis and involve mechanisms interacting at several levels to carefully regulate the behavior, and sleep is no different. We still do not fully understand all the details of these several mechanisms and how they interact, but as we gain a clearer understanding of sleep, we may be better able to translate that knowledge into effective treatments for sleep disorders.

Critical Thinking Questions

1. Why do you think that animals have such different sleep requirements? Note two animals with different sleep schedules and explain how their sleep schedules match their particular needs.
2. Drugs such as benzodiazepines lengthen Stage 3/4 sleep at the expense of REM. Imagine that an individual who is experiencing insomnia is prescribed a benzodiazepine. What is your prediction about the effect this will have on their sleep satisfaction?
3. Many dreams contain fanciful content, but occasionally the content reflects tensions or anxieties in our lives. Briefly describe a dream you have had in which impossible events occurred or which you expressed anxiety over an impending event. Did the dream affect you in any way?
4. We all have occasionally said, "I just need more hours in the day." Suppose there was a pill that reduced your nightly sleep to 3 hours. Would you think this is a good or bad idea? Explain your answer.
5. If physical and mental activity expends glucose and one by-product of that activity is adenosine, what should be the relationship between level of physical activity and the need for sleep?
6. Medical professionals, firefighters, and police traditionally require individuals to work in rotating shifts. Do you see this as a problem? How might we use knowledge of sleep to make their lives and ours safer?

Dig Deeper into Behavioral Neuroscience

13.1 The Mystery of Sleep

The National Institutes of Health (NIH) is a source for a wide range of information about sleep.

NIH, National Institute of Neurological Disorders and Stroke. (2017). Brain basics: Understanding sleep. https://www.ninds.nih.gov/Disorders/Patient-Caregiver-Education/Understanding-Sleep

For anyone interested in the role of sleep in memory consolidation, this article by Susan Sara is a must read.

Sara, S. J. (2017, January 18). Sleep to remember. *Journal of Neuroscience, 37*(3), 457–463. https://doi.org/10.1523/JNEUROSCI.0297-16.2017

Jenkins and Dallenbach provided the seminal view into the role of sleep in memory consolidation.

Jenkins, J., & Dallenbach, K. (1924). Obliviscence during sleep and waking. *The American Journal of Psychology, 35*, 605–612. https://doi.org/10.2307/1414040

13.2 Stages of Sleep and Dreaming

Dr. William Dement's story of his career as a sleep researcher is a classic worth reading.

Dement, W. C. (1974). *Some Must Watch While Some Must Sleep*. W. H. Freeman. ISBN: 0-7167-0769-1

A readable overview of sleep and dreaming from a neurological perspective is provided in this textbook by McNamara.

McNamara, P. (2019). *The Neuroscience of Sleep and Dreaming*. Cambridge University Press. ISBN-13: 978-1316629741

Research on the neurological basis of dreaming continues to this day.

Hobson, J. A., Pace-Shott, E. F., & Stickgold, R. (2000). Dreaming and the brain: Toward a cognitive neuroscience of conscious states. *Behavioral and Brain Sciences, 23*, 793–1121.

13.3 The Biological Basis of Sleep

Hypothalamic control of sleep is an important topic for researchers and clinicians.

Saper, C. B., Chou, T. C., & Scammell, T. E. (2001). The sleep switch: Hypothalamic control of sleep and wakefulness. *Trends in Neuroscience, 24*(12), 726–731.

This article discusses Aserinsky and the discovery of REM.

Brown, C. (2003). The stubborn scientist who unraveled a mystery of the night. *Smithsonian Magazine*. https://www.smithsonianmag.com/science-nature/the-stubborn-scientist-who-unraveled-a-mystery-of-the-night-91514538

Understanding the mechanism by which orexin and adenosine influence the sleep–wake cycle is important to understanding sleep.

Inutsuka, A., & Yamanaka, A. (2013). The physiological role of orexin/hypocretin neurons in the regulation of sleep/wakefulness and neuroendocrine functions. *Frontiers in Endocrinology (Lausanne), 4*, 18. https://doi.org/10.3389/fendo.2013.00018. PMC: 3589707. PMID: 23508038.

13.4 Sleep Disruptions

Understanding the neurology and the neurochemical basis of sleep may provide the treatments for disorders such as narcolepsy.

De la Herrán-Arita, A. K., Guerra-Crespo, M., & Drucker-Colin, R. (2011, April 18). Narcolepsy and orexins: An example of progress in sleep research. *Frontiers of Neurology.* https://doi.org/10.3389/fneur.2011.00026

13.5 Circadian Rhythms

The effects of sleep deprivation on driving performance and risk of vehicular accident are a very hot topic.

Williamson, A. M., & Feyer, A. M. (2000). Moderate sleep deprivation produces impairments in cognitive and motor performance equivalent to legally prescribed levels of alcohol intoxication. *Occupational and Environmental Medicine, 57*(10), 649–655. https://doi.org/10.1136/oem.57.10.649. PMC: 1739867. PMID: 10984335.

Martiniuk, A. L. C., Senserrick, T., & Lo, S., et al. (2013). Sleep-deprived young drivers and the risk for crash. *JAMA Pediatrics, 167*(7), 647–655. https://doi.org/10.1001/jamapediatrics.2013.1429

Chapter 13 Before You Go On Answers

13.1 Before You Go On Answers

1. b. a period of immobility and lowered reactivity to stimuli
2. d. Prey typically sleep less than predators.
3. d. resting our eyes
4. c. has several purposes or functions

13.2 Before You Go On Answers

1. b. Our brain can be active and alert while we are sleeping.
2. d. We quickly go into NREM3/4 and then have three to five REM episodes alternating with NREM2.
3. b. cortical motor strip
4. d. are when the dreamer is aware they are dreaming

13.3 Before You Go On Answers

1. c. a lack of sensory input
2. b. is a netlike structure with connections throughout the brain
3. d. posterior hypothalamus led to coma; anterior hypothalamus led to wakefulness
4. a. excitatory and inhibitory, respectively, with regard to sleep

13.4 Before You Go On Answers

1. b. affect at least 15% of adults at some time during their lives
2. d. easier; less restful sleep
3. a. appears to be related to changes in sunlight
4. d. Obesity is a risk factor for sleep apnea.

13.5 Before You Go On Answers

1. a. follows a 24-hour cycle
2. c. suprachiasmatic nucleus and pineal gland
3. d. melatonin
4. c. 5 PM; 10 AM

Chapter 13 Glossary

activation-synthesis theory A theory of dreaming that assumes dreams are the brain's effort to make sense of random neural activity.

adenosine An inhibitory neurotransmitter that stimulates drowsiness.

benzodiazepine A sedative whose core chemical structure is a benzene and diazepine ring.

biphasic sleep Sleep that occurs in two discontinuous segments.

circadian rhythm Biological and behavioral changes that follow a 24-hour cycle that may or may not be synchronized with the day–night cycle.

diurnal rhythm A biological cycle that is synchronized with the typical day–night cycle and may or may not be synchronized with the circadian cycle.

dreaming The act of experiencing images, thoughts, or emotions while sleeping.

EEG (electroencephalography) Recording of electrical activity from the scalp that represents mental activity.

hypnogram A graph that represents a person's sleep stages.

insomnia A sleep disorder in which the individual cannot fall asleep and/or stay asleep.

lucid dreaming A type of dreaming in which the dreamer is aware that they are dreaming and may influence the dream.

melatonin A hormone primarily secreted by the pineal gland that regulates sleep.

microsleep Brief, unintentional episodes in which attentional loss occurs, typically while fatigued and engaged in a monotonous task.

monophasic sleep Sleep that occurs on one continuous segment.

narcolepsy A long-term neurological condition in which the individual can no longer regulate sleep–wake cycles.

night terrors A sleep disorder in which the person has feelings of panic and may scream or flail about with open eyes, and they do not remember the event in the morning,

nonbenzodiazepine One of a class of psychoactive drugs with effects similar to those of benzodiazepines.

NREM sleep Non-rapid eye movement sleep characterized by slow breathing and little movement.

orexin A neurotransmitter secreted by the hypothalamus that stimulates wakefulness.

pineal gland A small gland in the brain that is essential in regulating sleep and circadian rhythms.

polysomnography A test used in sleep studies in which measures of heart rate, blood pressure, respiration, and EEG are collected.

Process C The 24-hour rhythm of metabolic functions including sleep that is regulated by the suprachiasmatic nucleus of the hypothalamus and is independent of the day–night cycle.

Process S The system of sleep mechanisms that respond to the daily rhythm of day and night and are regulated by chemical messengers such as adenosine and melatonin.

REM sleep (rapid eye movement sleep) Sleep in which the eyes move rapidly back and forth; associated with dreaming.

reticular activating system (RAS) A network of neurons in the brain that activate the cortex.

seasonal affective disorder (SAD) A type of depression that typically occurs in the winter months and especially at higher latitudes and is thought to be prompted by shorter daylight hours.

sleep An altered state of consciousness characterized by reduced responsiveness and lowered metabolism.

sleep apnea A sleep disorder in which the individual repeatedly stops breathing.

sleep paralysis Temporary paralysis when asleep or waking up.

suprachiasmatic nucleus (SCN) One of a pair of nuclei above the optic chiasm that is associated with circadian rhythms.

CHAPTER 14

Sergey Nivens / Adobe Stock

Our two hemispheres collaborate to comprehend language and music.

Language, Music, and the Brain

CHAPTER OUTLINE

14.1 The Evolution of Human Language
- Communication in the Animal Kingdom
  Check It Out: The McGurk Effect
- The *FOXP2* Gene
- Evidence for Speech in Our Ancestry

14.2 The Neural Basis of Language
- The Discovery of Broca's and Wernicke's Areas
- The Geschwind–Wernicke Model of Language
- The Two-Circuit Model of Language
- Contributions of the Right and Left Hemispheres
  Try It Out: The Dichotic Listening Task

14.3 The Neural Basis of Reading and Writing
- The Invention of Reading and Writing
- Reading Processes
- Eye Movements and Reading
- Reading and Writing Disorders
  Check It Out: Reading Through the Eyes of a Person with Dyslexia

14.4 The Neural Basis of Music
- The Evolutionary Basis of Music
  Think About It: Williams Syndrome
- Characteristics of Music
- Music Processing in the Brain
  Advances in Behavioral Neuroscience: Isabelle Peretz and Neural Processing of Music
  Try It Out: Test Your Own Tone Discrimination

LEARNING OBJECTIVES

LO 14.1 **Explain** the evidence for the evolution of language in humans.

LO 14.2 **Discuss** how different brain structures process language.

LO 14.3 **Discuss** the importance of reading and writing and its neural basis.

LO 14.4 **Summarize** how the brain analyzes music.

Stories of the Brain

One day in the 1830s, Louis Victor Leborgne lost the ability to speak other than to utter the syllable *tan* twice in succession, accompanied with expressive hand gestures that helped to convey his meaning. After several months with no improvement, his family took him to the Bicêtre, a Parisian hospital that specialized in mental illnesses. Medical understanding at the time equated verbal fluency with intelligence, so doctors judged Leborgne to be mentally defective and confined him to the hospital, where he was renamed Tan and remained for the rest of his life.

In 1851, Leborgne came to the attention of a physician at the hospital, Paul Broca. Broca examined Leborgne and realized that although he was unable to create useful speech, he understood language perfectly. When Leborgne died, Broca performed the autopsy and found a damaged area on Leborgne's left frontal lobe, above the temporal lobe and near the lateral sulcus. Broca encountered a second patient with similar symptoms, an elderly groundskeeper named Lazare Lelong who could only utter five simple words but, like Leborgne, was mentally alert. Upon autopsy, Broca found damage to that same area in Lelong.

Based on his observations, Broca concluded that speech production was controlled by a single area on the left hemisphere of the brain, an observation that called into question the principle of equipotentiality of function that was dominant at that time. Scientists and neurologists had assumed that each hemisphere was a mirror image of the other in structure and function, but Broca showed that the brain was both more complex and more interesting than originally thought.

We will begin our discussion by first considering the evidence for the emergence of complex language in humans. We will then consider the neural basis of spoken and then written language. The invention of writing completely changed the course of human history, and so understanding how our brain comprehends written language is an important topic. Finally, we will consider how the brain processes music, a behavior that is unique to humans. In all these discussions, we will see how evidence from the clinic and the laboratory combine to create a more complete picture of how we communicate through language and music. We begin by thinking about the purpose of language and its evolution.

14.1 The Evolution of Human Language

LO 14.1 Explain the evidence for the evolution of language in humans.

The goal of behavioral neuroscience is to understand how a brain creates behavior, and in the course of your study, you have seen that humans share much with other organisms. Basic neural functions, the use of our senses, the regulation of our internal environment, and the hormonal control of sexual maturation and behavior are all behaviors whose underlying processes are shared across the animal world. Yet humans are different in many ways. We have developed the capacity for language, culture, and technology; we have created great literature and art, we have sent spacecraft beyond the solar system, and we have pondered the meaning of life. All these accomplishments rely on the uniquely human capacity to communicate through language.

Communication in the Animal Kingdom

The ability of one organism to communicate its intentions to another organism provides an advantage in the struggle to survive. Dogs snarl or wag their tails, cats hiss or purr, and birds sing different songs to indicate danger or an interest in mating. Many organisms use physical posture or coloring to signal danger; for example, the rattlesnake's rattle warns intruders to keep their distance, and the strawberry poison-dart frog is conspicuously colored to ensure that predators recognize it as a fatal meal. Animals convey information through sound, posture, gestures, coloring, and other means of communication (**Figure 14.1**).

Bill Gorum / Alamy Images

a.

Alfredo Maiquez / Shutterstock

b.

FIGURE 14.1 **Nonverbal forms of communication** The rattlesnake uses sound to communicate its intentions, and the strawberry poison-dart frog uses coloring to signal to predators that consumption may be fatal.

Humans also communicate through postures and clothing to convey information about ourselves, but humans have developed oral language to a degree not seen elsewhere in the animal kingdom. Speech that enters our auditory processing stream is combined with information gleaned from watching the speaker's lips, and together they create the complex signal we refer to as language. In normal conversations, the two information streams are in accord, but violations of this congruence can be surprising, as you will see in **Check It Out: The McGurk Effect**.

Humans have developed the ability to express complex thoughts that can then be communicated immediately or saved for the future through oral traditions or writing. How did humans develop language? Human language likely began as grunts and gestures, but then something happened that enabled our ancestors to create a remarkable new tool of communication. The evidence is that we inherited the gift of language due to a simple genetic mutation that had profound consequences.

Check It Out | The McGurk Effect

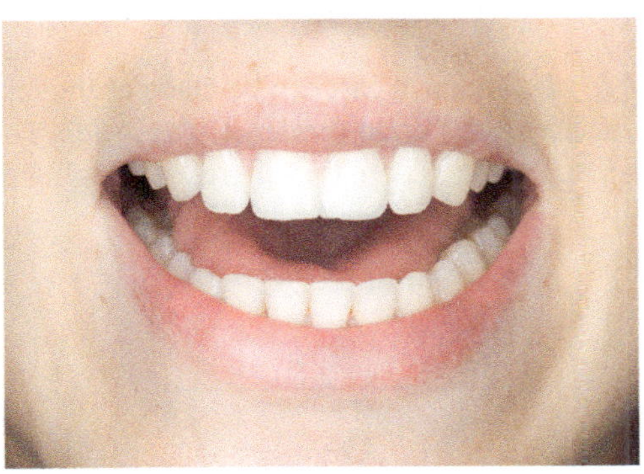

4FR / iStock / Getty Images

When you are speaking with another person, where do you look? Most people will say they look at the other person's eyes, but the evidence is that we unconsciously spend much of the time lip-reading and adding that information to what we hear. But what if those two data streams differ? Watch the video **The McGurk Effect**, and then answer this question.

1. The McGurk effect demonstrates that proper oral language comprehension involves integrating multiple data streams from hearing and vision into one complex signal. What are the implications for language comprehension if mental capacity is diminished, perhaps when the listener is highly stressed or is impaired in some way?
2. In many countries it is considered appropriate to wear a mask if you must be out in public with a cold or flu. Given the McGurk effect, might obscuring the mouth with a mask have a slight but negative effect on spoken language comprehension?

The *FOXP2* Gene MCAT

Our most distant ancestors left no records of the first time a member of our species uttered a meaningful word, but there are some clues. Most vertebrates carry a gene known as the ***FOXP2* gene** that is essential for learning the patterns of vocalizations. Two animals for whom the neural basis of vocalizations has been extensively studied are mice and birds. Mice constantly vocalize to express anxiety, territoriality, and mating intention, but they do so in the frequency range of 30 kHz–110 kHz, which is considerably above the 20 kHz upper limit of human hearing. Individual species of birds have unique songs, and different geographic groups have their own dialects just as human languages differ across a country. In mice, birds, and other organisms, damage to the *FOXP2* gene leads to a change in their ability to vocalize. Does this gene hold the key to understanding the origins of human speech?

All mammals have the *FOXP2* gene, but humans have a unique variant, and when the human version is damaged, a language disorder known as **developmental verbal dyspraxia** occurs in which the individual struggles to speak clearly and distinctly. Interest in the *FOXP2* gene was sparked by the discovery of the "KE family" in the United Kingdom who exhibit this disorder over several generations due to a mutation in their *FOXP2* gene (**Figure 14.2**).

As a consequence of the mutation in their *FOXP2* gene, affected individuals in the KE family have impaired articulation for some speech sounds due to facial muscular dysfunction, but in addition, they have intellectual and linguistic deficits. Perhaps the most interesting aspect of their disability is a deficit in understanding grammatical relations, a behavior that is essential to human communication and normally develops between the ages of 2 and 3.

If human speech depends on the presence of the human variant of the *FOXP2* gene, then those ancestors who expressed it may also have had language. Paleogenetic research using DNA extracted from fossils indicates the *FOXP2* gene was present at least 240,000 years ago in the genes of both our direct ancestors and our Neanderthal relatives. But there is additional evidence to consider in the search for the origins of human language.

Evidence for Speech in Our Ancestry

At the beginning of this chapter, you read about Broca's discovery that a region in the left hemisphere of Leborgne's brain was involved in language production. When during human evolution did this cortical region appear? The brains of our distant ancestors do not survive as fossils, but skulls do, and these ancient containers for a brain may provide a clue. Your brain exerts a slight amount of pressure on your developing skull, leaving a faint but discernable pattern on the inside of the skull, much as a shoe that you wear daily slightly adapts to the unique contours of your foot. This pattern gives hints about the original underlying structures of the brain. In modern humans, there is a slight depression in the left side of the skull over this language area reflecting the slightly larger cortical area on the left hemisphere, which is dedicated to language.

Although ancient brains are not available for dissection, a cast can be made of the inside of a skull to provide a model of the external surface of the brain that once was inside. These **endocasts** tell us about the brains of our distant ancestors. There is evidence of a slight bulge

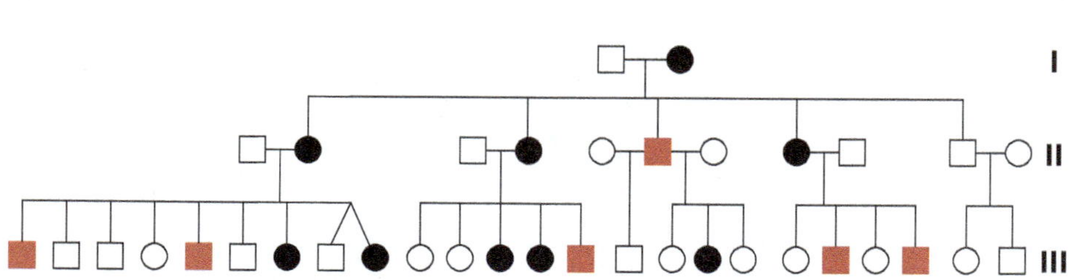

Adapted from Watkins et al, (2002) Brain; 125(3).

FIGURE 14.2 Genetic pedigree of verbal dyspraxia The chart shows three generations of the KE family (I, II, and III), with males represented by squares and females by circles. Individuals affected with developmental verbal dyspraxia are indicated with a red, filled shape and unaffected with an open shape. From an original pairing where the mother was affected, four of five children were affected, and out of 24 grandchildren, 10 were affected.

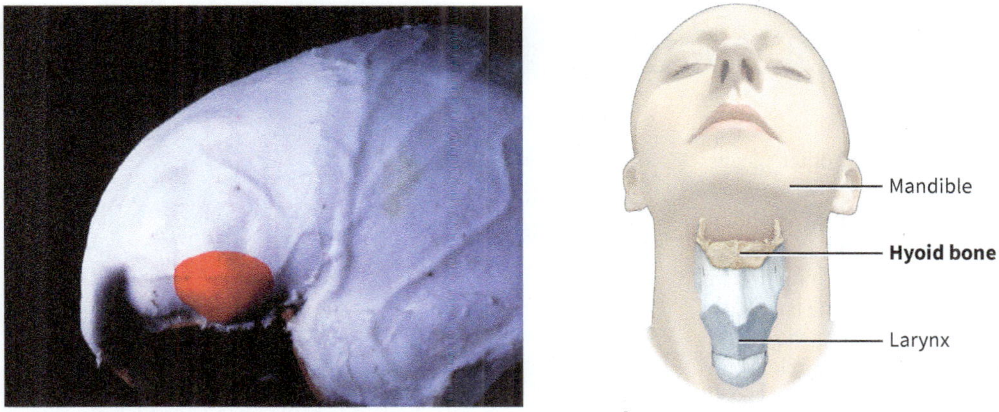

a. b.

FIGURE 14.3 **Evidence for early language** **a.** Endocasts of the left hemisphere of *Homo erectus* suggest that the ability to create language may have been present in our distant ancestors. **b.** The human variant of the hyoid bone is essential for the production of human speech as we know it.

over the location of the language area in ancestors as distant as *Homo erectus*, which dates to 2 million years BCE (**Figure 14.3a**).

The second piece of evidence for dating the evolution of speech in our ancestry is the presence of the **hyoid bone** (**Figure 14.3b**), which anchors the tongue in the throat. Many animals including fish, mammals, and primates have a hyoid, but the variant found only in humans and Neanderthals enables our vocal apparatus to create the wide range of sounds that make up human speech. The oldest surviving human hyoid is dated at 60,000 years BCE.

One final piece of evidence to consider is the development of culture. The motivation to create art and music implies individuals capable of abstract thought and suggests the existence of communities that share information. If human language is the expression of complex thought, then evidence of symbolic representations through art or the desire to create meaning through music could be evidence of language. The earliest evidence of art comes from a cave in Borneo and dates from 40,000 years BCE (**Figure 14.4a**). This rudimentary flute recovered from a cave in Slovenia is dated at 43,000 years BCE (**Figure 14.4b**).

The evidence from paleogenetics (the study of the past from preserved ancient genetic material), paleontology (the study of ancient fossils), and archaeology (the study of ancient human artifacts) suggest that humans had language at least 40,000 to 50,000 years BCE and possibly much longer ago than that. The evolution of the ability to express complex thoughts through language is a large part of the story of our species, so we will continue our discussion by turning to an exploration of the structures that create our ability to understand and produce language.

a. b.

FIGURE 14.4 **Paleolithic art and musical instrument** **a.** This image of a bull or cow was found in the Lubang Jeriji Salah cave in Borneo and dates from 40,000 years BCE. **b.** This fragment of a bear's femur recovered from a cave in Slovenia is dated at 43,000 years BCE. The arrangement and size of the holes suggests it was used as a flute.

14.1 Before You Go On

Section Summary

Animals communicate through sounds, gestures, and other means to attract, warn, or otherwise change the behavior of other animals around them, but humans are unique in the extent of their ability to communicate their internal states, intentions, and knowledge through language. The evidence suggests that a random mutation in the *FOXP2* gene long ago provided the possibility to create language. The presence of the hyoid bone and an enlargement over the region in the left hemisphere associated with language comprehension suggests that our ancestors were capable of language at least 60,000 years ago.

Comprehension Check

1. Which of these are forms of communication used by the animal kingdom?

- **a.** vocalizations
- **b.** posture
- **c.** gestures
- **d.** all of these

2. Where is the *FOXP2* gene found?

- **a.** in most vertebrates
- **b.** only in mammals
- **c.** only in primates
- **d.** only in humans and their direct ancestors

3. One reason why the KE family interests neuroscientists is that _____.

- **a.** they only speak in simple sentences
- **b.** their disability is limited to being unable to speak clearly
- **c.** they provide evidence that human language has a genetic basis
- **d.** only the males exhibit the disorder

4. Skull endocasts of our ancestors from 200,000 years BCE appear to show what?

- **a.** the progress of evolution of the hyoid bone
- **b.** that the right hemisphere evolved language capabilities
- **c.** that language can occur in the absence of a specialized region
- **d.** that a specialized language area may have present

14.2 The Neural Basis of Language

LO 14.2 Discuss how different brain structures process language.

In the prior section, you considered the genetic and evolutionary evidence that humans are equipped at birth to create and comprehend language. We now turn to an exploration of the neural basis of language, and the story that began with Broca and Leborgne becomes more complicated and illustrates how knowledge gained in the clinic and the research laboratory combine for a deeper understanding. With those ideas in mind, we turn our attention back to Leborgne and Broca.

The Discovery of Broca's and Wernicke's Areas MCAT

In this chapter's *Stories of the Brain*, you learned about Broca and his discovery that a lesion in an area of the posterior frontal lobe just superior to the temporal lobe led to a disturbance in language comprehension. Broca was not the first to note this connection; 25 years prior to Broca, French physician Marc Dax had also observed that a lesion in this location resulted in a language disturbance, but because Dax's observations did not come to light until later, Broca is generally credited with the discovery that a region in the left hemisphere was directly involved in speech comprehension. Accordingly, this part of the brain is now known as **Broca's area**.

A language disturbance resulting from damage to the brain from an interruption of the blood supply (known as a stroke) or physical trauma is known as aphasia. When the disturbance results specifically from damage to Broca's area, it is called **Broca's aphasia**, also referred to as expressive aphasia. Broca's aphasia is characterized by speech that is halting, with incorrect word order and sparse grammar. An important characteristic of individuals with Broca's aphasia is that they are frustrated by their struggle to understand and be understood. These characteristics of halting, telegraphic speech coupled with the awareness of the problem initially led to the belief that Broca's aphasia was a deficit of language production alone rather than comprehension as well, an assumption that was later disproven.

Broca's discovery inspired further research into the left hemisphere, and in 1874, the German physician Carl Wernicke (pronounced *Ver-nick-ee*) reported that a second cortical region posterior to Broca's area was responsible for a very different language disturbance. We now refer to this cortical region as **Wernicke's area**, and damage to this region causes **Wernicke's aphasia**. In contrast to Broca's aphasia, Wernicke's aphasia is characterized by speech that is fluid, fluent, and grammatically correct but devoid of meaning. In addition, individuals with damage to Wernicke's area are prone to linguistic errors such as neologisms and paraphrastic errors.

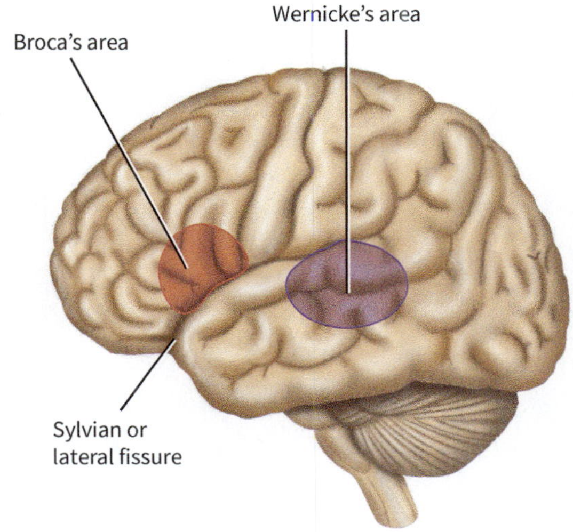

FIGURE 14.5 **Broca's area and Wernicke's area** Broca's area lies at the juncture of the frontal and temporal lobes and is associated with the production of speech, whereas Wernicke's area is at the posterior of the temporal lobe and is associated with speech comprehension.

Neologisms are words that are possible given the rules of a language but do not actually exist. For instance, words in English must have at least one vowel, so *blim* is a nonsense word that is possible in English, whereas *zykvx* is not. A second common linguistic error in Wernicke's aphasia is **paraphrastic errors**, in which the individual substitutes a syllable or a word for the intended word. For instance, the word *branch* might become *nanch*, the word *leaf* become *lele*, and *coffee* become *lofee*. Paraphrastic errors include the substitution of a related word for the intended word, for example *sunbulb* for *lightbulb*.

Wernicke's aphasia is also known as receptive aphasia because expression is preserved but meaning is lost. In contrast to individuals with Broca's aphasia, who are aware of their difficulty, those with Wernicke's aphasia are generally unaware of their deficit. The occurrence of neologisms and paraphrastic errors along with fluid production and lack of awareness are diagnostic criteria for Wernicke's aphasia. These two regions are depicted in **Figure 14.5**.

Examples of the speech of these types of aphasia can help to clarify the differences between expressive aphasia and receptive aphasia. In a typical aphasia screening battery, an individual is shown a series of simple line drawings, such as an image from the Kentucky Aphasia Battery (**Figure 14.6**), and asked to describe the scene. In this example, a boy balancing on an unsteady chair to replace a light bulb drops the bulb while a girl reads a book, and a dog carries something away in its mouth.

The description of the scene in Figure 14.6 that is typical of Broca's aphasia is telegraphic and sparse in nature, often only consisting of only nouns and verbs and produced with some effort. In contrast, the description typical of Wernicke's aphasia is characterized by fluidity, a lack of meaning, and the insertion of neologisms and paraphrastic errors.

You might predict that these two brain regions involved in language comprehension and production must have a physical connection, and in fact they do. A large band of white matter tracts known as the **arcuate fasciculus** connects Broca's and Wernicke's areas (**Figure 14.7**). Damage to this structure results in **conduction aphasia**, a rare condition in which comprehension of speech is preserved, production is paraphrastic, and repetition is severely impaired. Despite the relative preservation of language comprehension and production, individuals with conduction aphasia struggle to repeat a word or a phrase.

Broca's (expressive) aphasia: *House.....chair....chair....dog .....girl.....book.. reading.. reading....* (with the individual clearly showing frustration).

Wernicke's (receptive) aphasia: *Oh, what a nice family. The girl is having such a time; her langtin is right there and I can see the tale of the tarry and its so nice. The boy is having a plaser and its all fun. The boy lopped the lobe, how silly. When the tilly takes away the tarn, everything will become so very blem.*

Marshall & Wright (2007) / American Journal of Speech-language pathology

FIGURE 14.6 **An image from an aphasia battery** Having an individual verbally describe a scene provides essential information in the diagnosis of aphasia. Descriptions of this scene by individuals with Broca's aphasia differ markedly from those with Wernicke's aphasia.

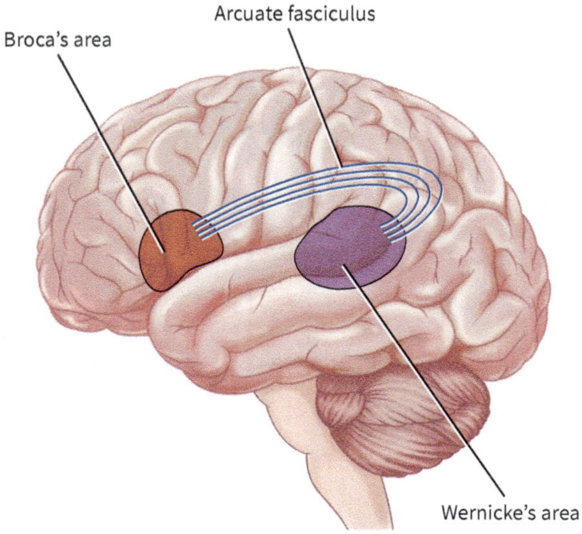

FIGURE 14.7 The arcuate fasciculus Damage to specific regions in the left hemisphere leads to language disturbances. Broca's area is associated with expression, Wernicke's area with comprehension, and the arcuate fasciculus with the coordination of these two areas.

FIGURE 14.8 Grammatical comprehension in Broca's aphasia Although *The girl is chasing the dog* and *The girl is being chased by the dog* are almost identical sentences, only one is a correct description of this image. Individuals with Broca's aphasia often judge incorrectly.

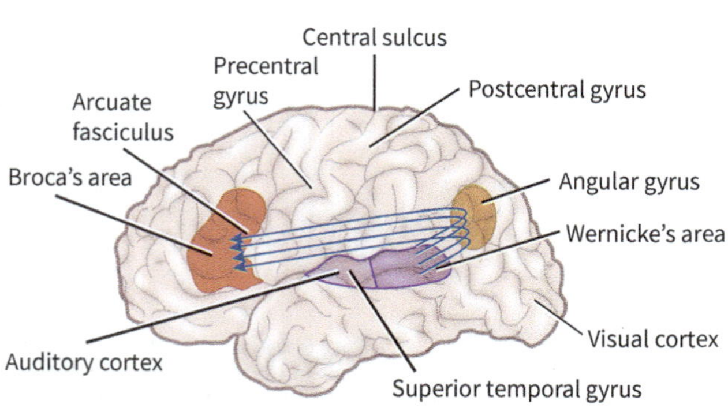

FIGURE 14.9 The Geschwind-Wernicke model of language comprehension and production The Geschwind-Wernicke model of language comprehension and production proposes that comprehension occurs in Broca's area, whereas production occurs in a larger circuit involving Broca's and Wernicke's areas as well as the angular gyrus and the primary motor cortex which is located on the Precentral gyrus.

The Geschwind–Wernicke Model of Language MCAT

As researchers and clinicians encountered additional examples of aphasia, they realized that the simple division of aphasia into three categories, one for each of the three anatomical structures involved in the Broca–arcuate fasciculus–Wernicke circuit, was inadequate to explain the variety of symptoms they were seeing in the clinic. Recall for a moment that Broca's aphasia is characterized by speech that is sparse, telegraphic, and consists mostly of nouns with few verbs and generally no articles such as *the* or *an* or prepositions such as *on*, *over*, or *between*. Researchers and clinicians assumed from the nature of Broca's speech that comprehension was relatively intact in these individuals, but were they actually understanding what they heard?

Language constructs meaning not just from the meaning of individual words but also from grammatical relations among those words. For example, when asked if the picture in **Figure 14.8** is described by the sentence *The dog is chasing the girl*, an individual with Broca's aphasia is likely to say yes. However, if given the same task but now the sentence is *The girl is chasing the dog*, the same individual is just as likely to say yes. It became apparent that for many individuals with Broca's aphasia, they were only comprehending the nouns and verbs separately rather than their grammatical relationships. In essence, they only comprehended *girl*, *dog*, and *chase* and made a reasonable judgment about what might have been said.

Based on observations like this one, it became clear that the Broca–arcuate fasciculus–Wernicke model of language comprehension and production was inadequate to explain what was being seen in clinical settings. For example, some individuals are able to comprehend and produce language but unable to read, whereas others are unable to write but can read. Other patients present with symptoms that are not clearly a pure production or pure comprehension deficit but somewhere in between. Some individuals can understand speech very well, repeat words, and read moderately well, but they struggle with finding words when speaking or writing and so substitute ambiguous words or descriptions for nouns and verbs. An individual with **anomic aphasia** (from the Greek for without words) might describe the image portrayed in Figure 14.6 as, "The boy is up on the ... sitting thing and working ummm with the flashlight thing while the girl is looking at the word thing."

These cases were problematic to fit within the simple Broca–Wernicke model, as were other examples of aphasia. A more complex model known as the Geschwind–Wernicke model added a region in the angular gyrus to the original Broca–arcuate fasciculus–Wernicke circuit (**Figure 14.9**).

This model proposed that language comprehension occurs as speech sounds enter the auditory processing stream via the ear, then are transmitted by the auditory nerve to the primary auditory cortex on the temporal lobe, and finally arrive at Broca's area where meaning is extracted. Speech begins in Wernicke's area where the meanings of words are assembled in the angular gyrus; then that signal is transmitted via the arcuate fasciculus to Broca's area, and from there to the primary motor cortex, which activates the facial muscles. Over time, the continued accumulation of data from clinical cases led to a more detailed classification of aphasias defined by three behaviors: fluency, comprehension, and the ability to repeat what was said (**Figure 14.10**).

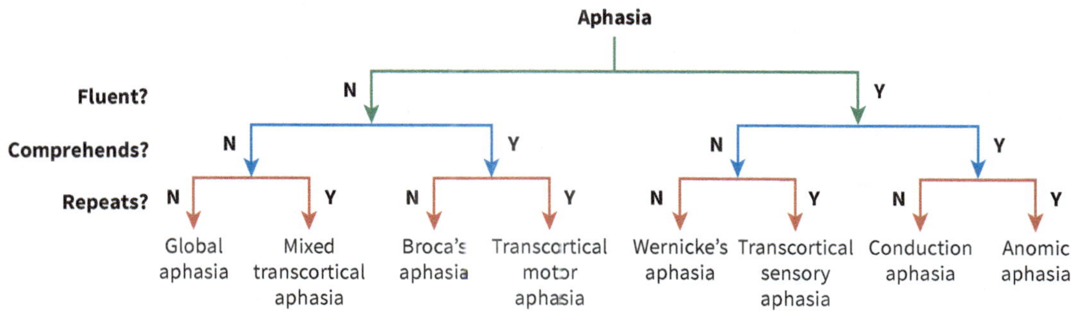

FIGURE 14.10 Types of aphasia A useful clinical categorization of aphasia is based on the individual's verbal fluency, ability to comprehend, and ability to repeat what they hear.

The different symptoms for these forms of aphasia are summarized in **Table 14.1**.

About 1 million people in the United States experience aphasia, and 180,000 new cases are diagnosed each year with a majority in individuals middle-aged or older as the incidence of stroke increases with age. Broca's, Wernicke's, and global aphasia are the aphasias most frequently seen, and woman are more likely to experience Wernicke's or global aphasia, whereas men are more likely to experience Broca's aphasia.

Before we move on, you might wonder if individuals who express themselves through American Sign Language (ASL) can also experience some form of aphasia, and the surprising answer is that they do. Deaf individuals who experience damage to Broca's area have trouble creating signs but maintain reasonable comprehension skills, and those individuals with damage to Wernicke's area struggle to comprehend signs but can produce them. The fact that the brains of congenitally and profoundly deaf ASL speakers who have never heard spoken language use the same regions for sign language comprehension and production as hearing individuals speaks to the innate organization of critical skills such as language in the human brain. In the absence of auditory input, the left hemisphere of ASL speakers simply takes over the task of comprehending and producing visual sign language in the absence of spoken language. This plasticity of structure to accomplish function is a hallmark of our brain.

The Two-Circuit Model of Language Comprehension and Production MCAT

It would be convenient if language comprehension and production could be entirely explained by the Geschwind–Wernicke model, but that is not the case. When we discussed homeostasis, the hormonal regulation of sexual development and various metabolic processes, and the regulation of sleep–wake cycles, we saw that there was no simple mechanism consisting of an on/off switch. Rather, there were a number of mechanisms involved, and the way the brain handles language is no different.

The original Broca–arcuate fasciculus–Wernicke model and the later Geschwind–Wernicke model of language comprehension and production were developed based on what neurologists were seeing in the clinic, but later technological advances in neuroimaging showed that both models were inadequate. Noninvasive imaging techniques such as PET (positron emission tomography) and fMRI (functional magnetic resonance imaging) now

TABLE 14.1 Types of Aphasia and Their Symptoms

Aphasia	Description
Global aphasia	Can produce or comprehend few words; most severe form of aphasia; results from widespread damage to multiple speech areas.
Mixed transcortical	Similar to global, but repetition is preserved.
Broca's	Speech production is damaged but can understand speech fairly well.
Transcortical motor	Speech production is damaged, but comprehension preserved.
Wernicke's	Fluent but nonsensical speech and impaired thought.
Transcortical sensory	Poor comprehension, intact repetition, and fluent speech with paraphasias or word substitutions.
Conduction	Expression and comprehension relatively intact, but repetition impaired.
Anomic	Speech is fluid, but individual struggles to name objects.

allow clinicians and researchers to peer into the active brain by measuring metabolic activity or changes in blood oxygenation and flow that occur in response to changes in neural activity. In addition to advances in how we understand functional systems, new imaging techniques provide a much clearer view of structures. More recently, DTI (diffusion tensor imaging) permits us to visualize white matter tracts in the living brain, which has changed how we see structures such as the arcuate fasciculus. What once was considered a simple band of tissue connecting Broca's and Wernicke's areas is now seen as a two-way highway whose tendrils extend from deep in the temporal lobe and into the frontal lobe (**Figure 14.11**).

With such tools at their disposal, researchers were able to investigate the neural basis of language comprehension in a way that would have been previously impossible, and an example may help make that point clear. In one experiment, an individual is placed in an fMRI scanner and then asked to consider the meaning of concrete nouns such as *desk* and abstract nouns such as *justice*, or they may be asked to process sentences in the active voice ("The boy was eating a sandwich") and also sentences in the passive voice ("The sandwich was eaten by the boy"). Changes in blood flow or metabolic activity are measured to identify those regions that activate to process concrete nouns but not abstract nouns, or sentences in active but not passive voice. In this way, researchers construct a model of how the brain processes the granular levels of language.

We now realize that language comprehension and production is created by two diffuse networks that engage different areas across the left hemisphere depending on the grammatical complexity and content of the communication. This explains why aphasias do not always fit into neat categories, because damage from lesions, strokes, or trauma rarely occupies the exact regions that anatomists identify in their textbooks, and it is common for individuals with aphasia to express a mix of symptoms.

Figure 14.12 provides are more realistic picture of the two large circuits involved in language comprehension and production. One area, a language-production system, is highlighted in red. Damage to this area creates expressive difficulties, especially as the damage is closer to Broca's area. The second circuit is a language-understanding and -reception system, highlighted in blue. Damage to this area is more likely to create a comprehension problem, again with increasing proximity to Wernicke's area leading to more severe symptoms.

Contributions of the Right and Left Hemispheres

With the left hemisphere so involved in language, can we conclude that the right hemisphere plays no part in language? In truth, the right hemisphere is heavily involved in spatial and other

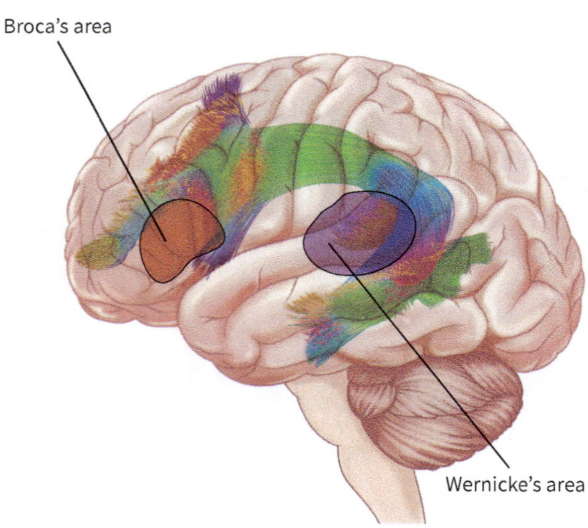

Adapted from Yeh FC, et al. Population-averaged atlas of the macroscale human structural connectome and its network topology. Neuroimage. 2018.

FIGURE 14.11 The arcuate fasciculus illuminated by DTI The arcuate fasciculus was originally seen as a simple connection between Broca's and Wernicke's areas, but modern neuroimaging tools have shown it to be a structure that connects the frontal, temporal, and parietal lobes into a complex processing circuit capable of creating the diversity of human language.

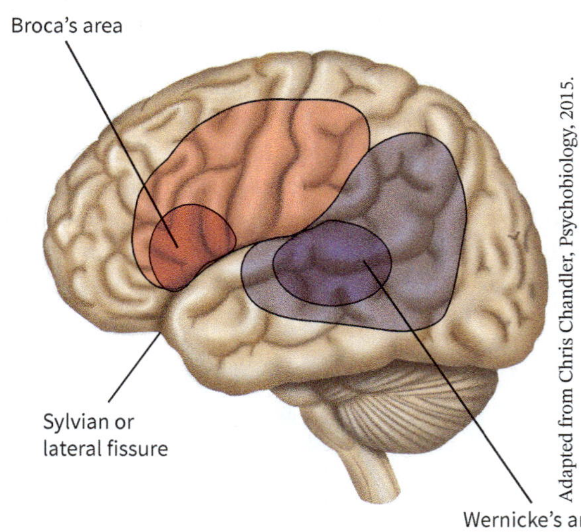

Adapted from Chris Chandler, Psychobiology, 2015.

FIGURE 14.12 The language understanding and production circuits The evidence suggests the existence of a language-production circuit that is centered around Broca's area but includes cortical regions nearby as well as part of the motor cortex (shown in red). In addition, a language-comprehension circuit is centered around Wernicke's area and includes part of the temporal and parietal lobes (shown in blue). Researchers have identified small centers in each of these circuits that activate very specific aspects of language.

skills, a topic that will be discussed in detail in subsequent chapters. If you have come to believe that the left hemisphere is the sole seat of language, you are mistaken. The right hemisphere makes important contributions to language comprehension and production, including prosody and analysis of the shapes of letters and words during reading. **Prosody** refers to the varying pitch, rate, and rhythm of speech. If you have ever heard a talk in which the speaker delivered their remarks in a monotone, you understand the value of prosody. Damage to the posterior inferior portion of the right frontal cortex leads to a loss of prosody, a region associated with language production in the left hemisphere.

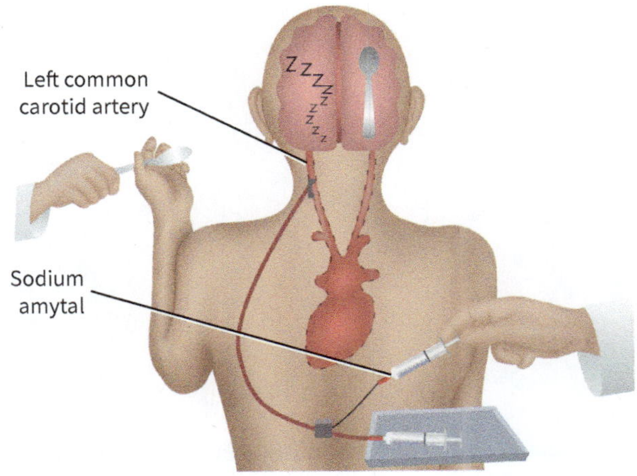

FIGURE 14.13 The Wada test When one hemisphere is temporarily anesthetized, loss of motor control and interruption of those cognitive functions that are controlled by the contralateral hemisphere ensues. In the situation depicted above, sodium amytal is injected into the left hemisphere, and the individual's right arm goes limp and they are unable to speak the word *spoon* when asked to identify the object.

The Wada test

It should be clear by now that the left hemisphere controls the majority of language processing for most people, but there are exceptions. For this reason, neurosurgeons intending to operate and remove damaged tissue around the language and memory areas may wish to be certain which hemisphere is dominant for these functions prior to operating in order to assess the risk of serious postoperative loss of function. One technique for identifying the dominant hemisphere for language is the **Wada test**, devised by neurologist Juan Wada (Figure 14.13). Sodium amytal is introduced into the carotid artery of one hemisphere, and for a brief period, that hemisphere is anesthetized so that functions controlled by that hemisphere are interrupted. The procedure is then repeated with the other hemisphere. During the test, the individual is asked to hold their hand up while counting from one to one hundred. When the left hemisphere is anesthetized, the right arm falls limp and, in most individuals, counting ceases for a short time until the anesthesia is cleared from that hemisphere. By comparing the response of the two hemispheres, the neurosurgeon can estimate the extent to which cognitive functions will be damaged by removing tissue in that region

We cannot provide you with the materials for conducting a Wada test, but there is another task that identifies the hemisphere that is specialized for language comprehension called the **dichotic listening task**. The logic behind this task is as simple as catching a ball; when someone throws a ball to you, you instinctively catch it with your favored hand. In the same way, when each hemisphere is simultaneously presented with a word to comprehend, the hemisphere that is specialized for language takes the lead. You can discover which of your hemispheres is dominant for language in **Try It Out: The Dichotic Listening Task**.

Try it Out | The Dichotic Listening Task

Alexander Demyanenko / Adobe Stock

Sounds entering an ear are preferentially processed by the contralateral hemisphere, so speech entering the right ear is analyzed by the left hemisphere that is specialized for language. As a result, that signal has an advantage over speech entering the left ear at the same moment. This fact led researchers to devise what are known as dichotic listening tasks, in which words are simultaneously presented via headphones to both hemispheres and the listener records what they heard. In this activity, you can participate in a dichotic task and determine which hemisphere is dominant for language. Download the document **Dichotic Listening Worksheet** and read the directions. When you are ready, play the video **The Dichotic Listening Task** and record your responses.

1. Did you show an advantage for one ear over the other? If so, which ear and associated hemisphere showed an advantage for language processing?
2. Some languages such as Japanese or Chinese use both symbols that represent sounds, as do Western languages, and also, ideograms or images that represent whole words. For those whose native language is conveyed through ideograms, do you think they will show a right ear, left hemisphere advantage? Explain.

Monkey Business / Adobe Stock

FIGURE 14.14 Public speaking and stuttering Some people enjoy public speaking, but others find that the anxiety created by speaking in front of a group can intensify stuttering.

Stuttering Roughly 1% of the world's population experiences stuttering, the involuntary repetition of a syllable. The severity of stuttering varies from mild to severe, and stressful situations such as giving a public talk, asking questions in class, or being in a large group can stimulate or exacerbate stuttering (**Figure 14.14**). You might assume that stuttering prevents an individual from achieving success in careers that require verbal fluency, yet actors James Earl Jones, Samuel L. Jackson, Sam Neil, Harvey Keitel, and musicians Kendrick Lamar, B.B. King, Ed Sheeran, Marc Anthony, and John Lee Hooker are all individuals whose stuttering has not prevented them from being successful in their careers.

There are many theories about the cause of stuttering that involve genetic and anatomical differences between stutterers and nonstutterers, but in the context of our current discussion about the different contributions of the left and right hemispheres to language production, one theory is noteworthy. PET scans that demonstrate cortical activation show that in nonstutterers, the left hemisphere is more active than the right, whereas in stutterers, the right hemisphere is more active than is typical. A simple metaphor may help to make the issue clearer. In a factory that makes automobiles, different components must be installed in a specific order; for example, the wheels must not be attached until the axels have been mounted. You will see as you read this chapter that when we speak or read, a number of processes occur in hundredths of a second and they need to occur in a specific order for language comprehension and production to be optimal. In stutterers, the right hemisphere may be out of synch with the processes of the left hemisphere and thereby interfere with speech production. We will briefly return to stuttering at the end of this chapter when we discuss the neural basis of music.

14.2 Before You Go On

Section Summary

Damage to Broca's area results in receptive aphasia, and damage to Wernicke's area results in expressive aphasia, each with very different symptoms. Damage to the arcuate fasciculus leads to an inability to repeat what was heard. However, the simple model of language comprehension and production proposed by Broca and Wernicke was insufficient to explain the many varieties of aphasia that are seen in the clinic. A more nuanced classification of aphasias is possible by categorizing aphasia symptoms according to the three symptoms of fluency, comprehension, and repetition. Advances in neuroimaging have shown that rather than assigning speech comprehension and production to two distinct cortical regions, these tasks involve two large networks of processing centers arranged around Broca's and Wernicke's areas. Although much of language processing occurs in the left hemisphere, the right hemisphere is involved in some aspects.

Comprehension Check

1. Which of these symptoms are characteristic of Broca's aphasia?
   - **a.** the inability to repeat words
   - **b.** halting, telegraphic speech and frustration
   - **c.** fluid speech, neologisms, and paraphrastic errors
   - **d.** the inability to write
2. Which of these symptoms are characteristic of Wernicke's aphasia?
   - **a.** the inability to repeat words
   - **b.** halting, telegraphic speech and frustration
   - **c.** fluid speech, neologisms, and paraphrastic errors
   - **d.** the inability to write
3. Which of these symptoms are characteristic of conduction aphasia?
   - **a.** the inability to repeat words
   - **b.** halting, telegraphic speech and frustration
   - **c.** fluid speech, neologisms, and paraphrastic errors
   - **d.** the inability to write
4. The simple Broca–Wernicke model of language comprehension and production was abandoned because it failed to _____.
   - **a.** recognize the importance of the arcuate fasciculus
   - **b.** recognize the role of the angular gyrus in language production
   - **c.** categorize the aphasia symptoms that were seen in the clinic
   - **d.** provide a model for treatment
5. The Wada test is used prior to neurosurgery to determine which hemisphere _____.
   - **a.** is essential to life
   - **b.** is most in control of language
   - **c.** controls homeostatic functions
   - **d.** is most in control of vision

14.3 The Neural Basis of Reading and Writing

LO 14.3 Discuss the importance of reading and writing and its neural basis.

How did writing and reading develop? Just as it seems logical to assume that the earliest forms of spoken language were simple vocal signs and gestures, written language might have evolved from marks on trees and rocks to indicate a favored hunting or fishing spot, a hazard to be avoided, or the path back to camp. Over eons, these primitive symbols evolved into something capable of conveying more complex ideas. Consider for a moment the power of being able to record a thought or experience in such a way that others far away from you in time and distance can still hear your voice. Writing forever changed humans.

José-Manuel Benito / Wikimedia Commons / Public Domain

FIGURE 14.15 The Kish tablet The Kish tablet is the earliest surviving example of written language, dating from 3300 BCE. It was written by Sumerian authorities in the city of Kish and is thought to contain instructions of some sort.

The Invention of Reading and Writing

The earliest examples of writing were found in Mesopotamia and are dated to roughly 5300 BCE, although other civilizations such as the Chinese and Mayans independently created writing on their own. These early examples of writing began as a means of recording the number of animals in one's herd for tax and other purposes, and over time, symbols were invented to represent a wide range of ideas. These early written languages are logographic systems, which means that each image represents a single word. The Kish tablet, circa 3300 BCE, is the oldest written document in the world (Figure 14.15). It contains images of heads, threshing boards, numbers, feet, and hands and is thought to represent the use of a logographic alphabet, an advance over older examples of simple counting marks.

Around 3500 BCE, the Egyptians and Sumerians invented alphabets, a system of writing in which symbols represented each speech sound. The basic speech sounds of a language are referred to as **phonemes**. Languages differ in the number of their phonemes; English has 40, French 36, Cantonese 36, Spanish 25, and Southern Khoisan 161. The invention of alphabetic writing accelerated the development of writing and reading because instead of having to memorize a large set of symbols, one for each idea, a person only has to memorize a small set of symbols that represent the speech sounds of a language. Examples of alphabetic languages are Spanish, English, French, Ukrainian, Balinese, Korean, and Italian. Chinese and Japanese kanji are logographic, whereas Japanese kana is alphabetic. Table 14.2 presents the word *brain* in several languages, all equally effective at communicating the idea.

TABLE 14.2 The Word *Brain* in Various Languages

Logographic languages	
Chinese	腦
Japanese	脳
Alphabetic languages	
English	brain
Greek	Εγκέφαλος
Arabic	خ
Korean	뇌

FIGURE 14.16 Reading as a uniquely human behavior When we read, we have access to the thoughts and experiences of people we may never meet. In this way, people who are long gone may still influence our behavior.

Reading Processes

Given the considerable importance of reading in society, a great deal of research has been focused on the processes involved in reading as well as on reading disorders that have a neurological basis. You may recall from our earlier discussion of the human visual system that an image entering the visual processing stream first stimulates line and edge detectors that then communicate with each other to form a more complex representation of the world. This framework will help you understand what happens in your brain when you read. The ability to share ideas and experiences with a wide audience through reading the writing of another person is one of the greatest achievements of our species (**Figure 14.16**).

Feature detectors in the eye and downstream in the visual cortex identify the lines and curves of letters, and this information is then fed to higher-level analyzers that match the incoming perceptual image to our mental library of letter shapes, thereby identifying a letter. That letter is added to the previously analyzed letters and a decision is made as to what word it might be. The meaning of that word is then activated in the brain and added to the meanings of recently decoded words. When that meaning makes sense in the context of previously decoded words, the process continues. When it does not make sense, the reader stops and goes back into the text to search for more clues.

Bottom-up processes such as edge and letter detection occur simultaneously with top-down processes such as context to continuously improve comprehension, and the two work seamlessly to speed comprehension as illustrated by this everyday example of that cooperation. If an inkblot partially obscures part of the letter *b* in the word *brain*, your bottom-up feature detectors may be unable to analyze the lines and curves to create the representation of the letter, but your top-down processes note that the sentence being comprehended is "The part of your body responsible for controlling behavior is your ..." and fills in the needed information. These processes are illustrated in **Figure 14.17**.

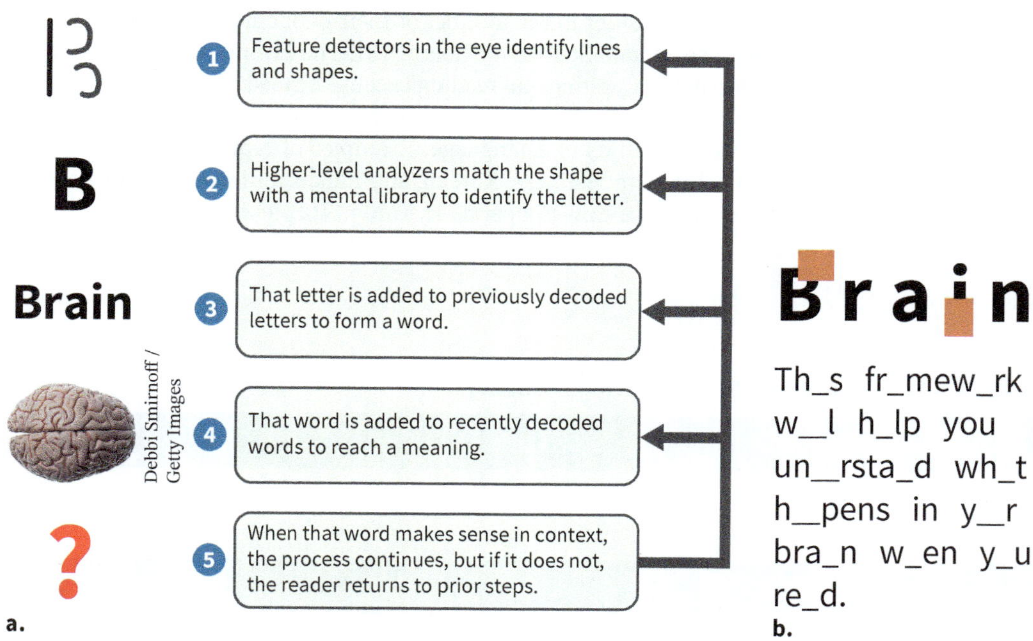

FIGURE 14.17 Processing levels of reading When we read, feature detection, letter recognition, and word identification processes all interact with contextual processes to create comprehension. **a.** Bottom-up processing begins by identifying line, then letter, then word. **b.** Top-down processing occurs when higher-level processes adjust bottom-up processes so that comprehension still occurs even if the letter or word is incomplete.

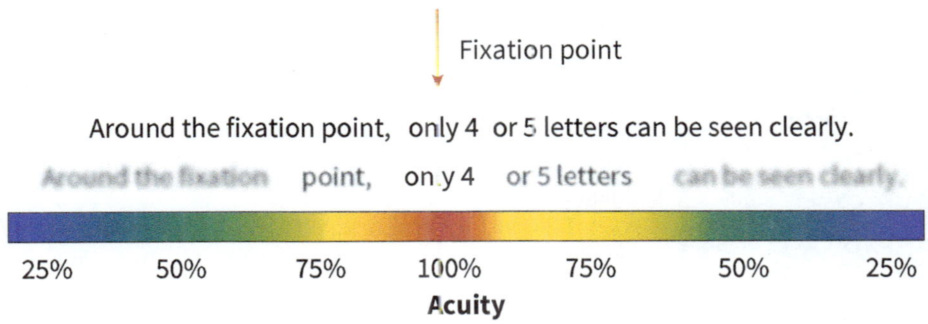

Source: Hans-Werner34 at English Wikipedia

FIGURE 14.18 The information taken in during a single fixation while reading Only a small amount of information is taken in during an eye fixation. The reading span is slightly asymmetric, favoring the right for languages like English that read from left to right.

Decoding individual letters into phonemes and then these sounds into words is one way to convert writing into meaning, but most readers process written language by whole words because their experience has allowed them to recognize the shape of words, given the context of what they are reading.

Eye Movements and Reading

Our eyes do not move smoothly across the page when we read, taking in words one at a time. In reality, our eyes focus on a section of the text, where we stop and process what the fovea can take in, and then our gaze jumps to a point farther on in the text. Each of these points of focus is called a **fixation**, and the flight of the eye to the next fixation is a **saccade**. During normal reading, the fovea takes in only four or five letters in a single fixation, which lasts between 200 and 250 ms. This window, known as the **reading span**, is slightly asymmetrical, and in languages that read from left to right across the page, the window is slightly longer on the right or leading side, allowing letters and words that are just ahead to contextually influence the analysis occurring at the fixation point (Figure 14.18). In languages that read in a different direction, the asymmetry of the span is appropriate to each language.

When we read, our eyes jump across the page in a series of fixations. With each fixation, we take in information and add the contents of that processing cycle to the previous cycles. We continue in this way unless we are confused, at which point we regress and go back in the text to a prior point to resolve the issue. When the complexity of the text is low, the duration of fixation and the distance between saccades is longer. In contrast, when the text is very dense with information or is written in a confusing fashion, saccades are more frequent, and fixations are longer in duration. Similarly, less skilled readers have longer fixations and more frequent saccades than more skilled readers.

A **scanpath** records the movement of the eye during reading. The relative duration of a fixation is shown by the size of the circle, and the saccadic movement from fixation to fixation through the text is indicated by a straight line that links fixation points. Figure 14.19 shows

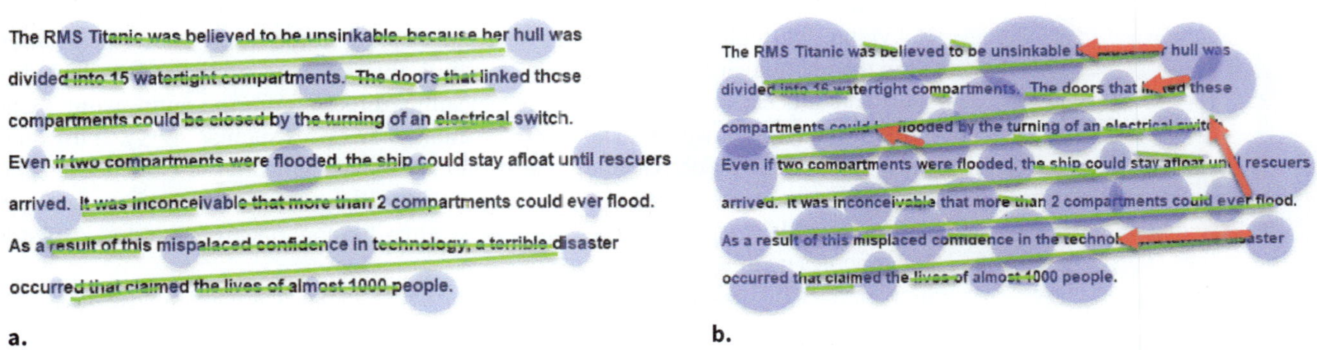

FIGURE 14.19 Scanpaths of a skilled and a slow reader A scanpath tracks eye movements as the subject reads. The circles indicate fixations, with the size of the circle indicating duration, and the lines show where the eyes moved in a saccade. **a.** Skilled readers require fewer and briefer fixations to comprehend text. **b.** Less skilled readers require more and longer fixations and may occasionally regress back into previously processed text (shown in red) to achieve comprehension.

FrankBoston / Adobe Stock

FIGURE 14.20 Reading disorders create a struggle When the normal reading processes are disrupted, reading can become virtually impossible for those with alexia and difficult and frustrating for those with dyslexia.

the eye movements of two readers as they read the same passage. Figure 14.19a depicts the scanpath of a skilled reader whose fixations are briefer and generally progress from left to right, line by line. Contrast that with the scanpath of the less skilled reader in Figure 14.19b. This reader required more and longer fixations, and there were regressions back to previously processed text, most likely due to some confusion about meaning.

Take just a moment to consider your amazing ability to read. As you read this sentence, your eyes fixate on a phrase in the sentence. Your visual processing stream considers edge, lines, and curves to make judgments about letters while other processes operating simultaneously are making judgments about words. These processes then feed into a system where the meaning of the words is activated, and that meaning is added to what you comprehended in the last fixation and guides what you extract from the next fixation. All of this is happening in processing cycles of less than a quarter of a second on average. Reading is one of humankind's greatest achievements.

Reading and Writing Disorders MCAT

There are many reasons an individual might struggle to read, including intellectual deficits, visual disorders, and, sadly, because they never had the opportunity to learn; however, our interest is the neural basis of behaviors, so we will conclude our discussion of reading with two disorders that have a neural basis: alexia and dyslexia.

There are two different but complementary routes that your brain uses to create meaning from written symbols. The first is the **phonological route** (sometimes called the surface route) in which the sounds associated with individual letters are activated as you read. This is likely the route you used when you were learning to read; you sounded out the letters C, A, and T and then merged the sounds to read the word *cat*. The second route is known as the **whole-word route** (sometimes called the direct route), in which the reader recognizes the word based on its visual characteristics and meaning is assigned from there. The direct route explains how we can read irregular words with silent letters such as *knife*, *bridge*, or *colonel* for which the sum of the phonemes does not correspond to the word as it is spoken in English. In everyday life, skilled adult readers rely on both routes with the whole-word route being dominant, but what happens when these processes break down and letters, words, and sentences become a confusing jumble (Figure 14.20)?

Alexia

The inability to read in the absence of a visual or intellectual deficit is known as **alexia**. Alexia provides us with neuropsychological evidence for the existence of separate reading comprehension pathways, because one route is disrupted while the other is spared. Alexia is quite rare and is typically the result of cerebral damage caused by a stroke.

Individuals with phonological alexia have a disruption in the whole-word (direct) route but not in the phonological (surface) route. Therefore, they can read familiar words, including irregular words, because they are extracting meaning directly from the entire word shape. They fail on reading nonwords such as *toin* or unfamiliar words such as *syzygy*. Individuals with surface alexia have a corresponding disruption in the whole-word route but not the phonological route. As a result, they misread irregular words because the sounds have no correspondence to words that are familiar. They also confuse homophones, or sound-alike words, so *know* becomes *no* or *won* becomes *one*.

Individuals with deep alexia show many of the characteristics of phonological alexia, but in addition, they make semantic paraphrastic errors; that is, they misread a word as another word, but one with a related meaning. Someone with deep alexia might misread *lunch* as *dinner* or *ocean* as *lake*. They also struggle with abstract words and function words such as pronouns and conjunctions.

An individual with pure alexia cannot read unless they spell the letters of a word one by one and thereby identify the word. The tedious nature of their reading makes the activity very difficult. Those with pure alexia can comprehend and produce spoken language and can write, but they cannot read their own writing. Pure alexia is almost always a result of a stroke or blockage of the left posterior cerebral artery (left PCA), which directly services the left occipital lobe and indirectly services the left temporal lobe via the left temporal lobe artery. Damage to the PCA generally damages the splenium, the thickest and most posterior region of the corpus callosum. The splenium connects occipital, temporal, and posterior regions of the two hemispheres, so it is not surprising that disruption of blood flow through the left PCA leads to disorders in visual processing and language comprehension.

Dyslexia In light of the importance of reading in the modern world, reading-comprehension problems can be a serious impediment to success. **Dyslexia** refers to a difficulty in reading that is not due to intelligence or access to education. Dyslexia is not uncommon; perhaps 5% of the U.S. population are clearly dyslexic, and 15% are dyslexic to some extent. You saw in our discussion of reading that several processes operating in concert are responsible for our reading ability, so you should not be surprised that disorders of varying degrees in any of the several mechanisms would result in different symptoms. You may have heard that dyslexia causes letters to be seen backward; in reality, reversing letters such as d, b, p, and q is common in all children up to age 8. Some individuals with dyslexia report that letters seem to move about in the word or appear indistinct; others report that letters and words appear to move about the page. Males are more likely to be diagnosed with dyslexia, but some argue that females are equally at risk but are better at compensating for their disorder. Dyslexia is found across all ethnic groups and economic levels. There is a genetic component to dyslexia; a child with a dyslexic parent is thought to have a 40%–60% increased risk of being dyslexic themselves.

The ability to read comes easily to many students, so it may be fruitful to experience what one severely dyslexic individual sees when they read in **Check It Out: Reading Through the Eyes of a Person with Dyslexia.**

Dyslexia makes it difficult to read, and what we know about the visual-processing stream and language-processing stream provides a model for understanding what might be happening to cause this difficulty. You now know that the left hemisphere is heavily involved in language, and that when we read, two different processes combine to create comprehension: one relying on the identification of the speech sounds associated with language and another more direct route in which the word as a whole directly accesses meaning. That framework suggests there might be different forms of dyslexia, one that results from a disruption in the letter-to-sound path and the other in the whole-word path, and that appears to be the case.

Phonological dyslexia involves substituting or *intruding* incorrect phonemes into a word such as reading the word *prescribe* as *describe*. Some believe that phonological dyslexia is merely a mild case of deep alexia. Individuals with surface dyslexia have the opposite problem; their letter-to-sound path is intact, but the direct word-to-meaning path is damaged. As a

Check It Out | Reading Through the Eyes of a Person with Dyslexia

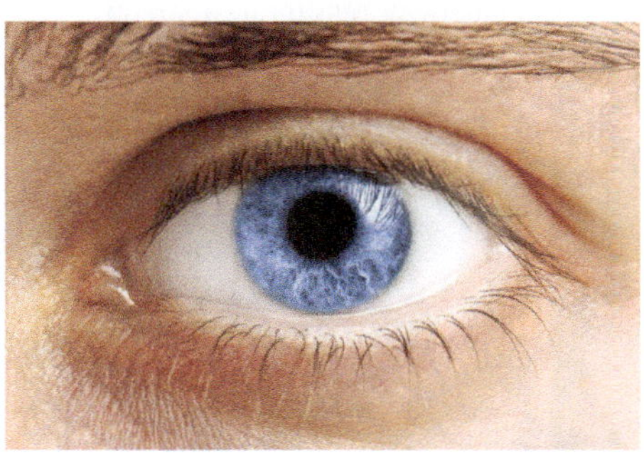

Anton / Adobe Stock

You have learned in this chapter about how an eye-gaze tracker captures where and for how long an individual looks at something and creates a track showing how their gaze point moves across the page. You can see the scanpaths of a neurotypical and a very dyslexic adult by watching the video **Normal and Dyslexic Readers**. After doing so, answer these questions.

1. In a sentence or two, compare the scanpaths of the two individuals. If your normal reading process was similar to that of the dyslexic reader whose scanpath you just saw, how might that affect your academic success and potential career choices?
2. Search online using the terms "current research in dyslexia" and briefly summarize one advance in our understanding that you found.

result, common words are recognized, but infrequent words are a problem. Finally, there are individuals with deep dyslexia who experience damage to both the phonological and whole-word paths. Deep dyslexics substitute semantically related words for the actual word; *cow* may be read as *bull*, and *state* may be read as *region*.

Advances in neuroimaging are beginning to shed light on whether a dyslexic brain differs from a neurotypical brain and if so, whether those differences are related to dyslexic processing. Small but regular differences in the size of the corpus callosum have been observed in individuals with dyslexia. These differences may slow the transfer of information between the two hemispheres, affecting how processing centers in both hemispheres participate in the decoding of letters and words as well as access to their meanings. In addition, there is some evidence that dyslexia is linked to a slightly thinner layer of gray matter in the cortical area associated with language in the left hemisphere.

One intriguing theory regarding dyslexia proposes that a dysfunction in the magnocellular layers of the lateral geniculate nucleus can play a role. You recall that M-cells of the LGN are responsible for many functions of visual processing including contrast sensitivity. Some argue that M-cell dysfunction as evidenced by decreased contrast sensitivity in dyslexia could make it harder to analyze the shape of letters, thereby leading to problems in converting letters into speech sounds. Not all studies of contrast sensitivity in dyslexia support this position.

By now you realize that reading is complicated, and a dysfunction in one or more of several mechanisms can create a reading disorder whose symptoms vary widely and therefore is not easily treated. Because dyslexia may have many causes, one type of remediation may be effective for some and not for others. As we learn more about the neural basis of dyslexia, we may be able to devise treatments that directly address each variant. One example of the potential for remediation may be instructive. A group of 20 dyslexic children and 12 neurotypical controls were asked to perform a rhyming task while activation of cortical areas was being measured. The dyslexic children showed less activity in regions of the left hemisphere when compared to the controls. Following the initial brain scan, the dyslexic children were engaged in a computerized rhyming practice task for 100 minutes a day 5 days a week for 8 weeks. Following this intervention, the cortical activation of the dyslexic group increased in those regions that had previously been seen as less active, and this change was accompanied by an improvement in reading skills (**Figure 14.21**).

As we learn more about the neural basis of reading, our ability to tailor remediation to these disorders should improve. Being dyslexic is not a measure of one's intelligence, nor does it need to be an insurmountable obstacle to success. Two of my former students who are extremely dyslexic illustrate how a person can rely on their strengths; one is the owner of a construction company, and the other is a tenured professor of biostatistics at an Ivy League school. Individuals you may be familiar with but not know that they are dyslexic include Albert Einstein, Leonardo Da Vinci, actors Tom Cruise and Kiera Knightley; musicians John Lennon and Guy Richie; filmmaker Steven Spielberg; and in the business world, Walt Disney and Richard Branson.

Agraphia

You will not be surprised to discover that there is a parallel between reading disorders and the writing disorder **agraphia** given that they may share some of the same underlying mechanisms. You saw in our discussion of reading that there is a phonological pathway and a whole-word pathway. Disruption to each pathway results in a specific type of

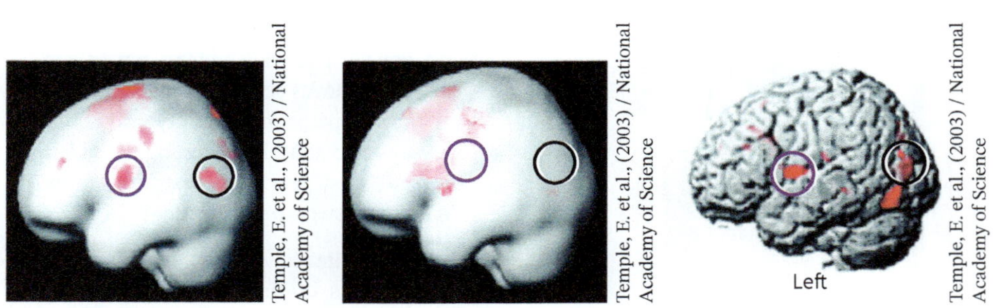

a. b. c.

FIGURE 14.21 Cortical activation in neurotypical and dyslexic children **a.** Neurotypical children show activation while rhyming in the indicated regions of the left hemisphere. **b.** Dyslexic children show less activation in these same areas. **c.** After intensive practice in rhyming, the dyslexic children showed an increase in activation in these same areas.

agraphia. Individuals with phonological agraphia can spell regular and irregular words but struggle with nonwords. Damage to the parietal lobe above the Sylvian fissure is found with phonological agraphia. In contrast, individuals with lexical agraphia can spell words and nonwords but struggle with irregular words. Damage to the posterior portion of the parietal lobe closer to the visual processing areas is a hallmark of lexical agraphia.

We might conclude our discussion of language with the thought that our ability to communicate with other members of our species is one of the most remarkable of all the achievements of humankind.

14.3 Before You Go On

Section Summary

We read through two complementary processes: one analyzes phonemes, or sounds assigned to letters, and the other goes directly from the image of the whole word to its meaning. When we read, our eyes jump from point to point in movements called saccades, and during fixations, we take short segments of text into our visual processing stream. Reading disorders such as alexia and dyslexia and writing disorders such as agraphia reflect breakdowns in these processes.

Comprehension Check

1. There is evidence for writing as far back as _____ years ago.
   - **a.** 50,000
   - **b.** 20,000
   - **c.** 3,500
   - **d.** 1,000
2. Two mechanisms involved in converting written letters and words into meaning are the _____ pathways.
   - **a.** phonological and surface
   - **b.** deep and whole-word
   - **c.** whole-word and phonological
   - **d.** articulatory and phonological
3. What is it called when the eyes jump from one segment of text to another while reading?
   - **a.** fixation
   - **b.** visual span
   - **c.** visual leap
   - **d.** saccade
4. An individual with pure alexia can _____ but struggles to _____.
   - **a.** comprehend speech and write; read even their own writing
   - **b.** read common words; read infrequent words
   - **c.** read regular words; read irregular words
   - **d.** read words in very large type; read smaller type

14.4 The Neural Basis of Music

LO 14.4 Summarize how the brain analyzes music.

What is the purpose of music? Every culture creates music, and musical instruments dated to roughly 40,000 years BCE suggest that music has always been a part of the human experience. The fact that humans have always created music suggests that there is something intrinsically important about creating a rhythmic sequence of tones and also that this ability must be hardwired into the brain.

The Evolutionary Basis of Music

Why humans create music is a puzzle with no obvious answer. The human brain has an innate ability to detect patterns, so creating rhythms may simply come naturally to humans. Humans are social organisms, and the rhythm of music has long been a tool to coordinate the work of groups who are engaged in physical tasks. Finally, humans evolved emotions to motivate adaptive behaviors such as parenting and clan bonding; listening to music triggers an emotional response in many people. It may simply be that music is an evolutionary accident; as evolutionary psychologist Steven Pinker has opined, music may be auditory cheesecake for the brain because it pleasantly tickles multiple cognitive faculties. Yet, just as you learned that the *FOXP2* gene may provide the basis for language in humans, there is intriguing evidence for a strong genetic influence on musical ability.

You learned in a prior chapter that when chromosomes duplicate, errors can occur, and individuals with one specific deletion are born with **Williams syndrome**. These individuals have an IQ of around 55 but are gifted with an outgoing personality and in many cases, innate musical ability. Imaging studies comparing the brains of neurotypical individuals with those of individuals with Williams indicates that their planum temporale is both larger than expected and bilaterally symmetrical, which is notable. In addition, neurons in the primary auditory cortex show greater connectivity. You can learn more about Williams syndrome in **Think About It: Williams Syndrome**.

Characteristics of Music

Before we begin to consider the neurological basis of music, let us take a moment to define some of its characteristics. Music has both tonal and rhythmic properties. Tone refers to sound characteristics such as pitch and timbre. Pitch corresponds to the musical concept of a note as

Think About It | Williams Syndrome

AB Forces News Collection / Alamy Stock Photo

When we think about genetic disorders, we generally assume that only deficits can result from genetic mutations, but that is not necessarily the case as individuals with Williams syndrome illustrate. Williams syndrome is caused by a deletion of roughly 25 genes from chromosome 7 (specifically, 7q11.23) during conception and can occur at either the sperm or the egg. Williams is relatively rare, occurring in perhaps 1 in 10,000 births. Individuals with Williams are born with cardiovascular problems, difficulties with their intestinal tract, and visual defects. They perform well at face recognition and language tasks, but struggle with drawing and any task that involves spatial relations, such as assembling blocks to create a specific pattern. Despite these challenges, they are gifted with an outgoing personality and a propensity for music. They respond emotionally to music and often express an innate sense of rhythm and an ability to remember lyrics and melodies of songs that are incongruent with their intellectual abilities. Not all individuals with Williams are musically adept, but a great many are, and the incidence of perfect pitch (the ability to identify a single note by name) is much higher than that found in the general population. To better understand the complexity of Williams syndrome, take a moment to learn about Gloria Lenhoff.

Ms. Lenhoff is a woman in her 60s with an IQ of about 55. She struggles with many everyday tasks that involve numbers; for example, she cannot make change for a dollar, and she cannot subtract seven from ten. What she can do is sing more than 2,500 songs from memory in more than 25 different languages, including Japanese, Korean, and Hebrew. She knows hundreds of songs on the accordion, despite being unable to read a single musical note. She has been a featured performer at several opera houses across the world.

Questions

1. **Search online using the terms "Gloria Lenhoff and Williams syndrome" and view one of the many videos of her performing. Describe what you saw in terms of her musical ability.**
2. **Williams syndrome involves behaviors that many would find appealing (musical enjoyment and skill as well as a sunny personality). If you could change your genetic code to enhance just one intellectual or personality trait, what would you choose, and why?**

well as the physical concept of frequency. Hitting the middle B (B^4) on a piano corresponds to a sound of 493.8 hertz (Hz). Timbre corresponds to the harmonics accompanying the note that create the differences in musical experience that allow us to distinguish between the same note played on different instruments, as demonstrated in **Figure 14.22**.

In addition to tonal characteristics, music also has temporal characteristics such as rhythm and meter. Rhythm refers to the duration of notes. For example, when we sing the alphabet song, A, B, C, D, E, and F all get one beat but then G gets twice as many beats. Meter refers to the rhythm but in a larger sequence; if you have ever danced the waltz, you have experienced a 1-2-3 meter.

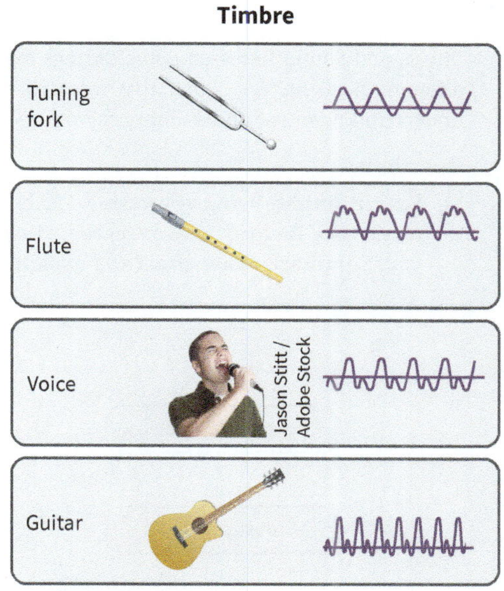

FIGURE 14.22 Differences in timbre Differences in the harmonics of sounds create very different perceptual experiences.

Music Processing in the Brain

We saw with language that most, but not all, of the processing occurs in the one hemisphere, and the same is true for music. Most, but not all, processing of musical tones occurs in the right hemisphere, but as we saw with language perception, there are centers or regions where specific aspects of music are processed. For example, simple noise activates a region in the primary auditory cortex (PAC), whereas focusing on pitch activates the inferior frontal gyrus, and attending to melody activates the PAC and associated temporal region. In contrast, processing rhythm activates the supplementary motor area (SMA) on the cortex and its connection to the basal ganglia within the cerebrum (**Figure 14.23**). These findings suggest that the human brain has a circuit dedicated to processing the tones and rhythms of music.

Music is an important part of almost everyone's life; we may listen to music while we drive, do chores, or exercise, or hum to ourselves while we work. Yet, until recently, investigating the neural mechanism that underlie musical perception and associated disorders has not reflected the value of music. You can learn about a person whose work has been instrumental in providing an understanding of how we process music in this **Advances in Behavioral Neuroscience: Isabelle Peretz and Neural Processing of Music**.

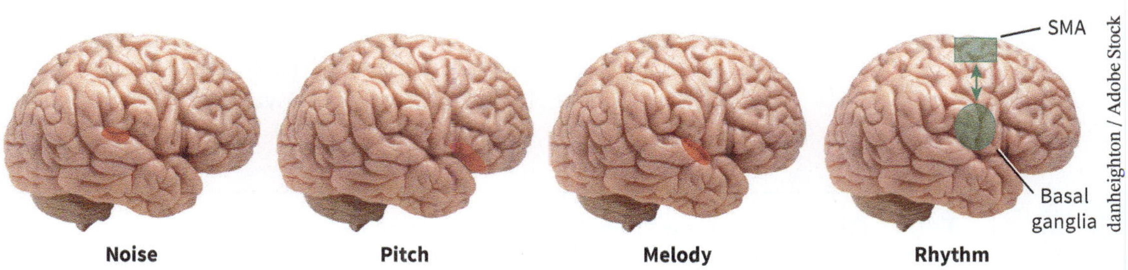

FIGURE 14.23 Activation of cortical regions by tone and rhythm The pattern of cortical activation by noise, pitch, and melody leads to different cortical responses in the right hemisphere. Rhythm activates the SMA at the dorsal surface of the cortex as well as the basal ganglia within the brain.

Advances in Behavioral Neuroscience | Isabelle Peretz and Neural Processing of Music

We all enjoy music as both an accompaniment to daily activities or as the very focus of our attention, but few have asked how the brain takes a series of tones and creates the experience we call music. Luckily for us all, Dr. Isabelle Peretz has asked that question throughout her career as the Casavant Research Chair in Neurocognition of Music at the University of Montreal in Canada.

Peretz is one of the world's leading experts on the neuroscience of musical perception, and her work in this area informs her clinical research into amusia, a disorder of musical perception. She has published roughly 400 papers in scientific journals that have been cited over 25,000 times by her peers, and she is a frequent keynote speaker at professional conferences. In 2020, she was awarded the Prix du Québec for her exceptional contributions to science.

Beyond her own stature as a world class scientist, her laboratory at the University of Montreal is a training ground for the next generation of researchers in the neurocognition of music. In 2005, she founded BRAMS, the International Laboratory for BRAin,

Music, and Sound Research, which brings together experts from different fields in a collaborative research effort dedicated to understanding music and auditory cognition.

Questions

1. **Search online using the terms "Isabelle Peretz laboratory," and identify one research effort that interested you. Summarize that effort and explain why you chose it.**
2. **Peretz's work extends beyond uncovering the neuroscience of musical cognition to applying that understand to amusia. Search online using the term "amusia." How common is congenital amusia? Finally, consider the consequences of amusia. How would having this disorder affect you personally?**

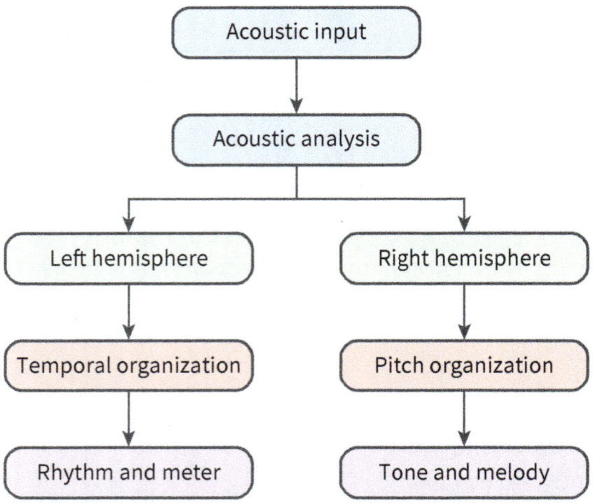

Adapted from Peretz, I. & Coltheart, M. (2003). Modularity of music processing. Nature Neuroscience.

FIGURE 14.24 A model of musical processing When music enters our acoustic processing stream, it is first analyzed by the auditory system as is every sound and then split into two processing streams: one to process pitch aspects and the other to process the rhythmic properties of the music.

Like language, music is processed in a network of connected regions, each contributing to the overall processing stream. Perhaps the best model of musical processing has been proposed by Peretz and Coltheart. Acoustic input in the form of sounds enters the brain through the auditory system, which, in the case of music, then analyzes tone and melody separately from rhythm to create the experience of music. In adult humans, rhythm is mostly processed in the left hemisphere, whereas the right hemisphere is more sensitive to melody and tone (**Figure 14.24**).

The Peretz and Coltheart model proposes separate streams for tone and rhythm processing, and one implication of that assumption is that some individuals could have a deficit in one processing mode and not the other. You have learned by now that damage to neural structures is often informative, and that is true for musical perception just as it was for language. Some individuals experience **amusia**, or tone deafness, a deficit in musical perception in which the individual cannot reliably distinguish tones. Congenital amusia is rare, and even rarer is amusia resulting from physical trauma or a cerebral lesion. The term "tone deaf" is often but inaccurately assigned to individuals with a mild inability to distinguish tones, and it is thought that 10%–20% of us whose tone discrimination is somewhat lacking struggle to sing in tune. You can learn about a famous individual who is thought to have had amusia in this **Try It Out: Test Your Own Tone Discrimination.**

Try It Out | Test Your Own Tone Discrimination

Bain News Service/Wikimedia Commons / Public Domain

In the New York City high society of the 1930s and 1940s, there was a phenomenon that captured the interest of the musical world. Socialite Florence Foster Jenkins (1868–1944) was an opera singer whose musical performance has led many to believe that she suffered from amusia.

Play the **Tone Discrimination Task** in which you will hear a short sample of her operatic abilities and then test your own musical skills at a task designed to identify individuals who may be experiencing tone deafness, a mild form of amusia.

1. Do you believe that your performance on the tone deafness test is an accurate measure of your own musical perception? Why or why not?

2. In this chapter, you learned that although language and music processing occur mostly in different structures and in different hemispheres, they do share some important connections. Tone deafness may impair a person's ability to recognize the emotional component of the speech of others, leading to misunderstandings. Search using the terms "tone deafness and emotions" and summarize your view on this question.

Earlier in our discussion of language, we briefly mentioned that for some stutterers, a miscommunication between the left and right hemispheres may be the cause of the problem. You may know that most stutterers can sing without stuttering. Because of this, one technique for speaking without a stutter is to sing your speech in such a way that others perceive you to have a nicely lyrical speaking voice. Now consider what you have just learned. If stuttering is the result of the right hemisphere taking an oversized role in language production, and melody is normally the province of the right hemisphere, then allowing the right hemisphere to take a leading role in speech by activating the melody circuit may explain why stutterers can sing without stuttering.

14.4 Before You Go On

Section Summary

Although there is no obvious survival value to creating music, humans have been making music for at least 70,000 years. Individuals with Williams syndrome, a genetic disorder, often express unusual musical ability, which suggests that there may be some genetic basis for humans to make music. Music consists of tones and rhythms, and there are separate mechanisms for the processing of each with the left hemisphere more sensitive to rhythm and the right more sensitive to tone and melody. Damage to these areas can result in amusia, or tone deafness, a condition in which the individual struggles to distinguish tones.

Comprehension Check

1. Musical instruments have been dated to be as old as _____ years BCE.

- **a.** 10,000
- **b.** 40,000
- **c.** 70,000
- **d.** 120,000

2. An individual with Williams syndrome is likely to have an interest in what?

- **a.** reading
- **b.** video games
- **c.** dance
- **d.** music

3. Tones and pitch are mostly processed in the _____, and rhythm is mostly processed in the _____.

- **a.** left hemisphere; right hemisphere
- **b.** thalamus; hypothalamus
- **c.** right hemisphere; left hemisphere
- **d.** thalamus; arcuate fasciculus

4. Amusia is defined as an inability to _____.

- **a.** detect tones above about 4,000 Hz
- **b.** distinguish tones
- **c.** follow a rhythm
- **d.** play a musical instrument

Conclusion

The complexity of human language is remarkable. Consider the nuanced shades of meaning that one can create by choosing among the words glance, gaze, and stare in a sentence; they all refer to a visual fixation, and yet they convey very different meanings. Because reading and writing are woven into our culture, it is easy to take them both for granted, and yet the invention of a system of symbolic representation that allows a person to make transient human thoughts permanent and transmittable across time and space is an accomplishment that is greater than or equal to any other in the history of our species. Consider for a moment that today you can read the words of Merer, an Egyptian official who more than 4,500 years ago was tasked with organizing the delivery of materials to the construction site of the Great Pyramid at Giza. You might ask yourself what documents written today might still be read four millennia from now. Why we also invented music is a puzzle, but given that music exists in every culture, it is likely that it serves a purpose beyond the simple pleasure of listening. When you next listen to music, you are engaging in a behavior that links you to your distant ancestors.

Critical Thinking Questions

1. A major theme in this chapter is the richness of human language, but other organisms do communicate through vocalizations, gestures, posture, and even chemical messengers. They do so to establish a territory, attract a mate, coordinate behavior with others, and manage their young. Search the internet to learn more about communication among organisms other than humans. Identify and describe at least three species that enjoy a complex and sophisticated language.
2. In this chapter, you learned about aphasia. Search the internet to find out answers to the following questions. How common is aphasia? What are the risk factors for aphasia? Search using the terms "new treatments for aphasia" and mention one.
3. Before writing was invented, humans used oral language to convey important information from person to person and from generation to generation. Epic poems such as the *Iliad* and the *Odyssey* were created that could be memorized and then presented to groups of people to convey social concepts such as honor and fidelity. Such works were immense; the *Iliad* consists of 15,693 lines, and other works from antiquity were even longer. If living in a preliterate society where important information must be memorized led to the development of memory, what effect might the current availability of online resources have on the development of memory skills? What would be the consequences of this for humans?
4. Music is not essential to our survival, yet every society has developed some form of music. One theory about the evolution of music is that humans have evolved to look for patterns and music has clear patterns in the chord progressions and scales of each culture. We know that when we listen to music, we release dopamine (DA) in the reward circuit, and this leads to a pleasurable feeling. Given this, what effect do you think that DA blockers have on one's pleasure from music?

Dig Deeper into Behavioral Neuroscience

14.1 The Evolution of Human Language

An interesting and comprehensive overview of the evidence for the appearance of language in humans is available in this book.

Fitch, W. T. (2010). *The Evolution of Language.* Cambridge Press. ISBN-13 978-0521677363

You can learn more about the KE family and the evidence for the genetic basis for language in humans.

Vargha-Khadem, F., Watkins, K., Alcock, K., Fletcher, P., & Passingham, R. (1995). Praxic and nonverbal cognitive deficits in a large family with a genetically transmitted speech and language disorder. *Proceedings of the National Academy of Sciences, 92*(3), 930–933. Bibcode, 1995PNAS...92..930V

Many of us have access to a wide range of films and videos through our school and home subscriptions. Two members of the KE family are interviewed in the National Geographic documentary *Human Ape*.

14.2 The Neural Basis of Language

Paul Broca and the discovery of Broca's area was a starting point for much of our current understanding of specialization of function in the brain. You can read more about his discovery and see images of Leborgne's and Lelong's brains, which have been preserved for over 150 years.

Dronker, N. F., Plaidant, O., Iba-Zizne, M. T., & Cabanis, E. A. (2007, May). Paul Broca's historic cases: High resolution MR imaging of the brains of Leborgne and Lelong. *Brain, 130*(5),1432–1441. https://doi.org/10.1093/brain/awm042

14.3 The Neural Basis of Reading and Writing

Given the importance of reading to academic and personal success, you will not be surprised to see that a considerable amount of research has been devoted to understanding the process, how best to teach reading, and how best to deal with reading disabilities. This source is a current overview of all these topics.

Snowling, M. J., & Hulme, C. (Eds.). (2022). *The Science of Reading: A Handbook.* Wiley. ISBN-13: 978-1405168113

Dyslexia is more common than most people know. The National Institute of Neurological Disorders and Stroke at NIH maintains an informative site on dyslexia research. https://www.ninds.nih.gov/Disorders/All-Disorders/Dyslexia-Information-Page

The history of the invention of writing is a fascinating topic. This site provides a comprehensive and interesting overview of the evidence for the evolution of writing from simple marks to record goods to modern writing.

Schmandt-Besserat, D. (2014). The evolution of writing. In J. Wright (Ed.), *International Encyclopedia of Social and Behavioral Sciences.* Elsevier. (This article is available online).

14.4 The Neural Basis of Music

Understanding the neural basis of music has been the life's work of Isabelle Peretz and her colleagues.

Peretz, I., & Coltheart, M. (2003). Modularity of music processing. *Nature Neuroscience, 6*, 688–691. https://doi.org/10.1038/nn1083

Peretz, I., & Zatorre, R. J. (Eds.). (2005). *The Cognitive Neuroscience of Music.* Oxford Press. ISBN-13: 978-0198525202

A very readable and yet academic coverage of music and the brain has been provided by Daniel Levitin, who holds the unique position of having played guitar in a rock band, worked as a recording engineer for Carlos Santana and Blue Oyster Cult, and earned a Ph.D. in neuroscience from the University of Oregon.

Levitin, D. (2007). *This Is Your Brain on Music: The Science of a Human Obsession*. Penguin. ISBN-13: 978-0452288522

Oliver Sacks puts his hand on every neurological topic in an entertaining and informative fashion, and music is no exception.

Sacks, O. (2008). *Musicophilia: Tales of Music and the Brain*. Vintage. ISBN-13: 978-1400033539

Chapter 14 Before You Go On Answers

14.1 Before You Go On Answers

1. d. all of these

2. a. in most vertebrates

3. c. they provide evidence that human language has a genetic basis

4. d. that a specialized language area may have present

14.2 Before You Go On Answers

1. b. halting, telegraphic speech and frustration

2. c. the inability to write

3. a. the inability to repeat words

4. c. categorize the aphasia symptoms that were seen in the clinic

5. b. is most in control of language

14.3 Before You Go On Answers

1. c. 3,500

2. c. whole-word and phonological

3. d. saccade

4. a. comprehend speech and write; read even their own writing

14.4 Before You Go On Answers

1. b. 40,000

2. d. music

3. c. right hemisphere; left hemisphere

4. b. distinguish tones

Chapter 14 Glossary

agraphia An impairment or loss of the ability to write.

alexia An impairment or loss of the ability to read or comprehend written language.

amusia An impairment or loss of the ability to process pitch or rhythm; commonly known as tone deafness.

anomic aphasia A type of aphasia in which the individual cannot recall the names of common objects.

aphasia A neurological disorder defined by an inability to comprehend or produce speech.

arcuate fasciculus A band of white matter tissue connecting the temporal cortex and inferior parietal cortex with the frontal lobe.

Broca's aphasia A neurological language disturbance characterized by a deficit in speech production; also known as expressive aphasia.

Broca's area A region in the frontal lobe of the left hemisphere that is concerned with the production of speech.

conduction aphasia A neurological language disorder characterized by an inability to repeat words.

developmental verbal dyspraxia A neurological disorder characterized by the inability to plan, organize or coordinate the movements of the tongue, lips and other body parts involved in speech.

dichotic listening task A task in which words are simultaneously presented to each ear in order to determine which hemisphere is dominant for language.

dyslexia A disorder involving difficulty in reading due to problems identifying sounds of whole words that is independent from intellectual or visual problems.

endocast A three-dimensional representation of the space within a cavity such as a skull that can give an indication of brain structures.

fixation In eye-gaze research, the act of maintaining gaze upon a single point.

***FOXP2* gene** A gene whose function is associated with speech and language.

hyoid bone A bone in the throat attached to the muscles that move the tongue and is essential to speech production.

logographic system A writing system in which each symbol represents an entire word.

neologism A pseudoword that is grammatically possible but does not exist.

paraphrastic error An error in reading or speaking where a sound or syllable is substituted for the correct item.

phoneme A basic unit of sound in a language.

phonological route A process by which letters are transformed into sounds to create words; also known as the surface route.

prosody Patterns of rhythm, stress, and intonation in a language.

reading span The amount of letters or words that can be taken in during one fixation.

saccade The rapid movement of the eye between fixations.

scanpath The path the eye creates as it moves between fixations.

Wada test A test for hemispheric dominance for language that involves successively anesthetizing each hemisphere and recording brain activity while the subject is speaking.

Wernicke's aphasia A neurological disorder in which the individual has lost the ability to comprehend speech but can produce fluid speech lacking in meaning; also known as receptive aphasia.

Wernicke's area A region in the posterior portion of the left temporal lobe that is associated with language comprehension.

whole-word route A process of language processing in which whole words are directly translated into their meaning; also known as the direct route.

Williams syndrome A developmental disorder characterized by somatic deficits, an outgoing personality, and often unusual musical abilities.

CHAPTER 15

Cognitive Neuroscience

freshidea / Adobe Stock

The machinery that creates consciousness is complex.

CHAPTER OUTLINE

15.1 Memory
- The Three-Store Model of Memory
- Sensory Memory
  Check It Out: Experience Echoic Memory
- Working Memory
  Try It Out: Measure Your Working Memory
- Long-Term Memory
- Amnesia
  Think About It: A Life Lived Moment by Moment

15.2 Attention
- Aspects of Attention
- The Neural Basis of Attention
  Check It Out: Change Blindness
- Hemineglect
  Think About It: Why Do Actors Enter from Stage Left?

15.3 Object Recognition
- Visual Object Recognition
- Challenges to Object Recognition
- Visual Agnosia
- Prosopagnosia
  Check It Out: What Is It Like to Be Face blind?

15.4 Executive Function
- Aspects of Executive Function
- The Neural Basis of Executive Function
  Advances in Behavioral Neuroscience: Marie Banich and Executive Function
- Evaluating Executive Function
  Try It Out: The Stroop Task
- Psychosurgery as a Treatment for Executive Dysfunction
  Think About It: Moniz, Psychosurgery, and the Lobotomy

15.5 Hemispheric Specialization
- Anatomical Differences Between the Hemispheres
- Sperry and the Split-Brain Procedure
  Check It Out: Do We Have Two Brains?
- The Mystery of Left-Handedness
  Try It Out: Are You Right- or Left-Handed?

LEARNING OBJECTIVES

LO 15.1 **Describe** the different types of human memory and amnesia.

LO 15.2 **Summarize** the functions of attention.

LO 15.3 **Explain** how the brain creates a mental representation of objects.

LO 15.4 **Explain** the role of executive function in cognitive behaviors.

LO 15.5 **Describe** the different functions of each hemisphere.

Stories of the Brain

Henry Molaison was a healthy and active child who was normal in every way until age 7, when a bicycle accident caused a serious head injury. Due to that injury, Henry suffered from epileptic seizures that increased in frequency and severity as he grew older. One treatment for epileptic seizures that do not respond to drugs is the removal of the cortical tissue that is the source of the seizures, and surgeon Henry Beecher Scoville felt that the hippocampus was the cause of Henry's epilepsy. Scoville had previously removed *either* the right or left hippocampus of other patients without apparent serious consequences, so he decided he could take out *both* Henry's left and right hippocampus. When Molaison was 27 years old, Scoville removed a portion of the medial temporal lobe including the left and right hippocampus from both of Molaison's hemispheres.

The procedure was successful in that it eliminated the seizures, but to Scoville's surprise, Henry Molaison was unable to create new memories from that day forward. He spoke perfectly, understood others with ease, and knew the purpose of everyday objects. However, his recall of events from his past was deficient, and he was unable to form new memories about people or events, although over considerable exposure he could learn new motor skills. This outcome highlights the mystery and complexity of the human ability to form memories.

In your study of behavioral neuroscience, you have learned that the nervous system controls a wide range of physiological functions to maintain our internal environment, allows us to sense the external world, and conditions us to respond adaptively to stimuli. Yet humans are more than simple reflexive organisms. We create great works of literature and art, we build bustling cities, and we have created technology that allows us to feed millions of people, travel from one continent to another through the air, and communicate instantly with friends or colleagues on the other side of the globe. The basis of these and many other human accomplishments is our remarkable higher-level cognitive processes.

In this chapter, we will consider cognitive neuroscience; the neurological basis of higher-order mental process such as memory, attention, recognizing objects, and judgment; as well as the advantages of having each hemisphere perform different cognitive tasks.

15.1 Memory

LO 15.1 Describe the different types of human memory and amnesia.

How our nervous system analyzes, stores, retrieves, and actively manipulates information is a compelling question. More than 2,000 years ago, the Greek philosopher Archimedes said, "If you give me a lever and a place to stand, I can move the world." In science, theoretical models provide us with the lever and the ground on which we can pose testable hypotheses, so we begin our consideration of human memory with a simple but powerful model.

The Three-Store Model of Memory

One way to parse the complex phenomenon of memory is to divide it into three stages, or stores: **sensory memory**, **working memory** (or short-term memory), and **long-term memory** (Figure 15.1).

Sensory information enters the visual or auditory systems and then passes through three distinct processing stores. In each of the three stores, specific processing operations occur, and information is lost in specific ways.

According to the three-store model of memory, unanalyzed sensory input enters the nervous system and moves through three different processing stages that take in the signal, extract information from the signal (a process known as **encoding**), and then either pass the result on to the next stage or discard the information from the nervous system. These

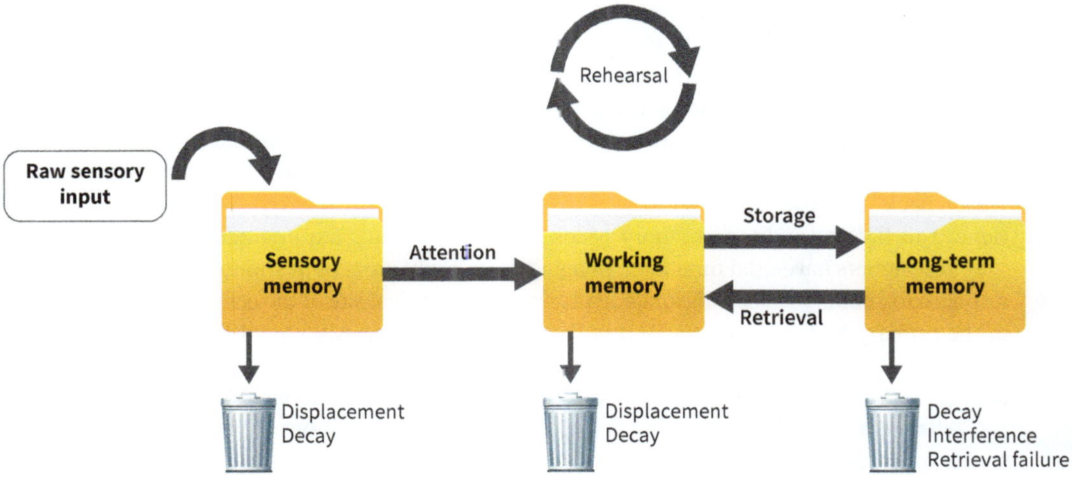

FIGURE 15.1 **The three-store model of memory**

three stages differ in the length of time that information resides there, the type of analysis that occurs, the durability of the memory created, and the processes by which information is lost.

For sensory memory and working memory, information decays (is forgotten) over time or is displaced by incoming information. In long-term memory, information is lost through decay when the memory is not activated over time, through interference when competing memories interfere with retrieval, or through retrieval failure when the memory, or engram, is simply unavailable due to a lack of retrieval cues. An example of interference is giving someone your old phone number that you had for years instead of the new number. If you've ever struggled to recall a word that was on the tip of your tongue, you've experienced retrieval failure.

The characteristics of these three stores are summarized in **Table 15.1**.

A metaphor about a factory that makes wooden furniture may help you grasp the characteristics of the three memory stores. In the first stage, raw materials are sorted into piles depending on their eventual use. In a second stage, the wood is shaped and manipulated by many tools into a useable piece of furniture. In the final stage, the finished product is stored in a warehouse for future delivery to a customer when needed. Each of these three stages has its parallel in your memory system: input to the system of raw materials (sensory memory), manipulation and elaborate processing (short-term memory), and storage until needed (long-term memory). We begin with the first store, sensory memory.

Sensory Memory MCAT

When raw or unanalyzed information enters your sensory system, the first stage of information processing involves extracting physical information from the stimulus. The sensory memory system ingests raw, unprocessed data, holds it briefly while extracting information, and then either passes the output on to the next stage or discards it. The acoustic version of this system is known as **echoic memory**, and the visual version is called **iconic memory**. We turn our attention first to the auditory system because it is easier to conceptualize.

TABLE 15.1 **The Three-Store Model of Human Memory**

Characteristic	Sensory memory	Working memory	Long-term memory
Duration	Iconic: 200–250 ms Echoic: 500 ms	15–30 seconds	Potentially a lifetime
Encoding	Physical	Acoustic	Semantic (meaning)
Capacity	Huge	Around seven items	Unlimited
Forgetting	Decay or displacement	Decay or displacement	Decay, interference, or retrieval failure

Echoic memory When sounds enter the auditory processing stream, they are stored in echoic memory, which extracts simple physical information from the raw signal. Either new sounds enter to displace the current contents or the information fades within 3–4 seconds. You may have experienced echoic memory during a conversation when your attention wanders for a moment, and suddenly you realize that your companion has just asked you a question, but you are not sure what the question was. You have the perception at that moment that you "reach back" into your brain to retrieve the question you were just asked. You can directly experience this fleeting auditory memory system in **Check It Out: Experience Echoic Memory**.

Researchers have used functional imaging to highlight areas of the brain that are activated in response to auditory tasks. This imaging places the anatomical site of the echoic store for each ear in the primary auditory cortex (PAC) in the contralateral (opposite) hemisphere. Children with damage to the PAC typically show developmental problems with language acquisition; in adults, damage to the PAC that affects the echoic store leads to problems comprehending spoken language.

Although our sense of hearing is important, humans rely more on vision than any other sense, so we can assume there is a visual counterpart to echoic memory. That system is your iconic memory.

Iconic memory Visual information is first processed by iconic memory. Information in iconic memory is held briefly, typically for a few hundred milliseconds (ms). In most cases, the data is then transferred to working memory for processing; if not selected, the data decays or is displaced by new data entering iconic memory.

You know from the discussion of reading that we take information into our visual processing stream in a series of still images called fixations that are linked together by flights of the eye known as saccades. Visual data collected during fixation is then sent to the iconic memory system for processing. The typical duration of information in iconic memory is roughly 200 ms. This information is decaying just as the next fixation brings in a new image for analysis roughly 250 ms later. If a set of data from a second fixation enters iconic memory before the first set is transferred to working memory, then the first set is ejected from iconic memory without a trace, a phenomenon known as visual suppression, or masking.

You can experience the suppression of information in the iconic store by standing in front of a mirror and looking at one eye and then shifting your gaze to the other eye. You will not see any movement of your eyes because during the saccade your visual system suppresses the flow of information. Consider for a moment why this is essential. If the information in the iconic store did not fade by the time the next fixation occurred, our visual world would be a blurry jumble.

Iconic processing occurs in the occipital lobe, most likely in the ventral, or "what," processing stream. There is some evidence that individuals developing various forms of dementia such as Alzheimer's have a deficit in iconic memory in which information fades very rapidly so that incomplete encoding may be passed on to the next processing stage.

Check It Out | Experience Echoic Memory

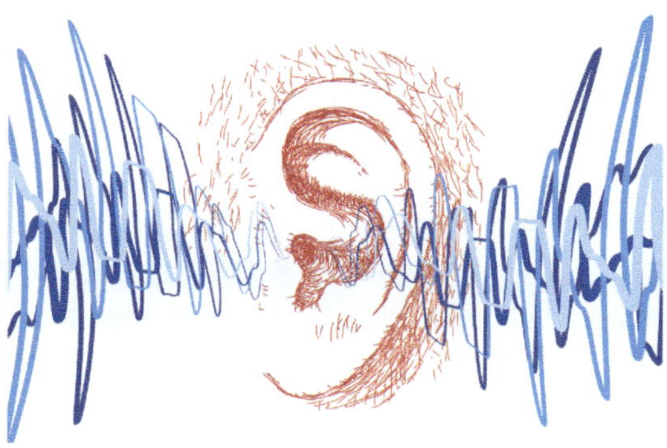

robodread / Adobe Stock

In this activity, you will experience echoic memory and see how researchers can use the characteristics of our nervous system to illuminate processes that might otherwise be unknown. Watch the video **Echoic Memory**, and then answer the following questions.

1. There is evidence that children's echoic memory capacity increases as they age. Why might this be?
2. It appears that individuals with Alzheimer's disease have little or no capacity in their echoic memory. How might this affect their comprehension of spoken language? Knowing this, how might you adapt your speech to communicate with such an individual?

Working Memory MCAT

We may depend on sensory memory to begin the analysis of information from our visual and auditory systems, but we rarely experience the workings of sensory memory consciously. In contrast, working memory is where consciousness exists and our thoughts are organized from moment to moment. The experience of writing a term paper for a course is a good metaphor for your working memory. You might sit at a table or desk with several sources: your textbook, perhaps a few other books or articles that are pertinent, various browser tabs to sources that you have identified as relevant to the assignment, a list of points you need to cover to satisfy the instructor, and so on. As you progress through the assignment and cover specific points that the assignment requires, you move some books and articles off the table or close those browser tabs and replace them with new sources. The concept of making items available and then manipulating them in your consciousness is at the very heart of working memory (**Figure 15.2**).

Working memory is thought to consist of several processes, and the most common model is presented in **Figure 15.3**. Your conscious intentions are represented by the central executive. Verbal information is maintained in the phonological loop where it is encoded acoustically—that is, represented by speech sounds— and maintained by rehearsal. Information will be lost by decay or displacement within 30 seconds if not passed to long-term memory as the familiar example of trying to remember a phone number illustrates. You repeat the number over and over, trying to hold it in memory, and if you are distracted or stop rehearsing, the number is lost. It is not by accident that phone numbers are seven items long, as that is the approximate capacity of working memory.

Visuospatial information is similarly maintained in the visuospatial sketchpad. Data from these two sources (sometimes referred to as buffers) are combined in the episodic buffer, the seat of consciousness. Should the central executive decide that information from long-term memory is needed, it can access that information and add it to the contents of the episodic buffer.

The central executive is maintained in the prefrontal cortex, information in the phonological loop is stored in the temporal lobe, and the visuospatial sketchpad is maintained in the parietal-occipital region, particularly in the right hemisphere. Although this model of working memory may at first appear confusing, a familiar activity will make it clear and easy to remember. Consider the experience of assembling a piece of furniture that arrives in a box in pieces. You open the box and spread out all the pieces where you can see them and check the contents against the list of parts (central executive) because you remember that in the past, you were delivered an item where one piece was missing (long-term memory). The

valentinrussanov/Getty Images

FIGURE 15.2 Information organization and working memory Writing a term paper involves acquiring access to multiple sources and then arranging and organizing them to create the best outcome. In working memory, we combine information from our senses and from memory to create the conscious experience.

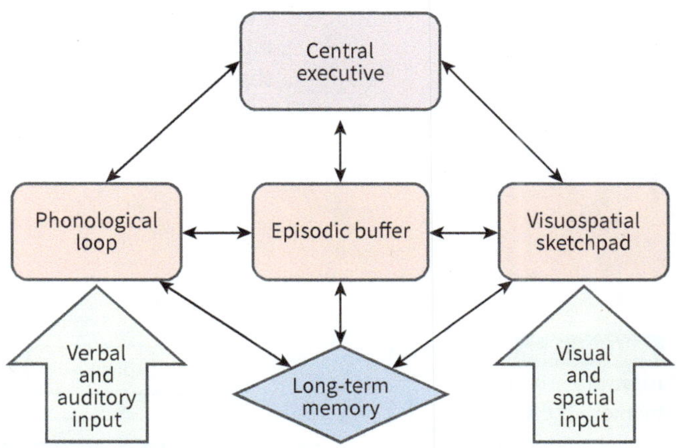

FIGURE 15.3 A model of working memory Information that is in conscious awareness is held in the episodic buffer. The phonological loop and the visuospatial sketchpad act as supporting systems to manipulate verbal and visuospatial information, respectively. The central executive coordinates the interaction among these systems, drawing information from long-term memory when needed.

Try It Out | Measure Your Working Memory

Gajus/Adobe Stock

Working memory involves holding information in consciousness and manipulating it much as you might arrange blocks into a design. Neuropsychological tests often include some measure of working memory because it is essential to many cognitive processes. In this activity, you can test your working memory. Download the **working memory worksheet**, view the video **Measure Your Working Memory**, and then answer the questions.

1. Performance on the first of these two tasks is more resistant to the effects of aging and neurological disorders, whereas performance on the second is more sensitive to these effects. Why might that be?
2. Think of two careers where being able to hold information in the episodic buffer and manipulate it might be essential to success. Briefly justify your choices.

instructions tell you that the first step is to make sure that the seven connectors are oriented to the left, so you repeat several times to yourself "connectors must face left" (phonological loop) as you complete that step. As you assemble the item, you regularly refer to an image of the finished product for clues about how to assemble the item while manipulating the pieces to see how they might fit together (visuospatial sketchpad).

You now see that our working memory is both the seat of consciousness and an active component of cognition. Disease or disorders that damage any of the components of working memory seriously disrupt conscious behavior. You can explore the characteristics of your own working memory system in **Try It Out: Measure Your Working Memory**.

Long-Term Memory MCAT

We began our discussion of human memory by considering sensory memory, where raw sensory information is analyzed, and then working memory, which acts as the desktop of our mind. We continue by exploring long-term memory, the enormous repository of experiences and facts. Unlike sensory and working memory, whose contents are lost rapidly, information that is stored in long-term memory may never be completely lost, although it may be inaccessible after many years without activation. Long-term memory is divided into explicit and implicit memory and then into additional subcategories, presented in **Figure 15.4**.

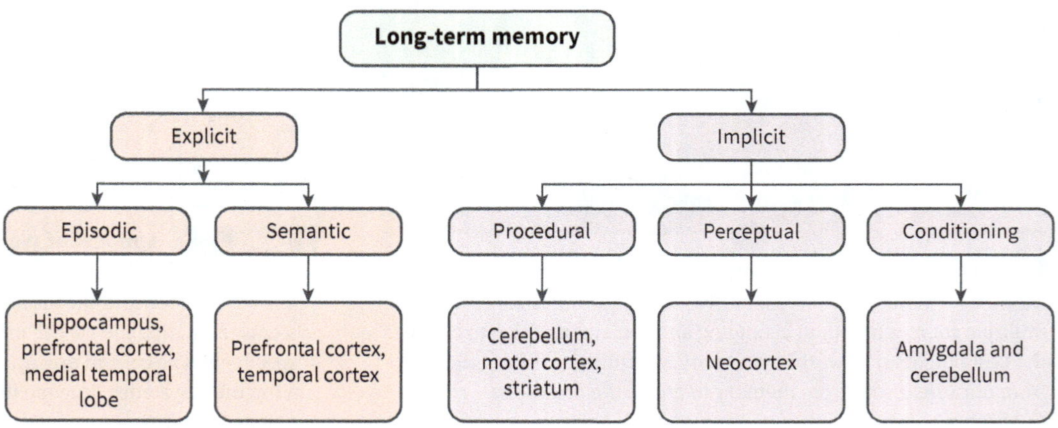

FIGURE 15.4 Long-term memory Long-term memory is divided into explicit and implicit domains. Explicit, or declarative, memory consists of episodic and semantic memories, whereas implicit memory contains procedural, perceptual, and conditioning memories.

Explicit memory

Your conscious or intentional memory is called **explicit memory** (also known as declarative memory) and is divided into episodic and semantic memory. **Episodic memory** is your memory for events that you witnessed. Memories of where you live, what classes you are taking, or what you had for breakfast are all examples of episodic memory. In contrast, **semantic memory** is your memory for words but also concepts and general knowledge about the world. Knowing that the word *exuberant* means "enthusiastic or high spirited" and knowing that the Eiffel Tower is in Paris are both examples of semantic memory. Evidence that episodic and semantic memory are distinct comes from individuals with Alzheimer's and other neurological disorders, who typically lose access to episodic memory long before their semantic memory is gone.

Explicit memories are stored in the prefrontal cortex and the medial temporal lobe. The hippocampus has an integral role in storing and retrieving explicit memories, as was apparent in the case of Henry Molaison in this chapter's *Stories of the Brain*. Molaison's surgery involved the removal of the left and right hippocampus, parts of both temporal lobes, as well as the entorhinal cortex, which provides sensory input to the hippocampus (Figure 15.5). Without a hippocampus, Henry Molaison could no longer create new explicit memories and was severely hampered at retrieving existing explicit memories.

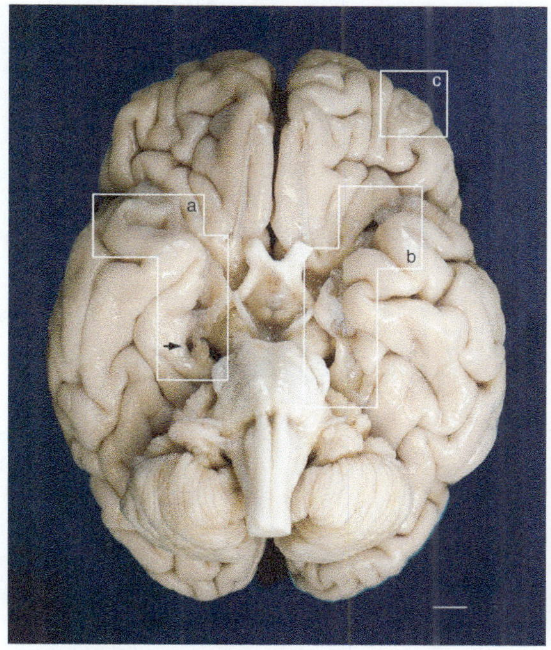

Annese, J., Schenker-Ahmed, N., Bartsch, H. et al. Postmortem examination of patient H.M.'s brain based on histological sectioning and digital 3D reconstruction. Nat Commun 5, 3122 (2014)

FIGURE 15.5 Henry Molaison's brain In this ventral view of Molaison's brain, the areas of the temporal lobe including the hippocampi that were surgically removed are indicated in areas **a** and **b**. The black arrow indicates a stain left by a surgical clip. Area **c** indicates an area of damage to the orbitofrontal gyrus.

Implicit memory

We have a second memory system that complements our explicit memory system. Explicit memory contains conscious recollections of events or word meanings, whereas **implicit memory** contains information that is not in the conscious realm yet influences our behavior. Implicit memories appear to be stored in areas associated with motor functions and emotional processing. The exception is when priming occurs; primed concepts are activated in the neocortex but without conscious awareness. For example, if I ask you to think of a famous landmark and tourist attraction in a European nation, what comes to mind? It is likely that the first image that came to mind was the Eiffel Tower, despite there being many possibilities across the continent. The concept *Eiffel Tower* was primed in your implicit memory by its mention in a prior paragraph and so is likely to be activated before any other possibility.

Implicit memory consists of procedural memory, perceptual memory, and conditioning. We have discussed conditioning in some detail in an earlier chapter, so we will devote our attention here to procedural and perceptual memory. **Procedural memory** refers to skilled behaviors that you have acquired through practice that have little or no verbal component to them. If you can ride a bicycle or shift gears on a stick-shift transmission without thinking, you are using your procedural memory. It is worth noting that procedural memory is very resistant to forgetting; if you learned how to ride a bicycle in your youth, you are unlikely to ever lose that skill.

Perceptual memory is the acquisition of a sensory skill that develops due to experience and knowledge. Textile experts perceive differences in fabrics that are not obvious to ordinary people, and gourmets may experience subtle shades of flavor in foods that are not apparent to the rest of us.

Amnesia MCAT

As the unfortunate case of Henry Molaison demonstrated, neurological damage can result in **amnesia**, or loss of the ability to recall people, places, facts, or events. Amnesia is typically the result of damage to the medial temporal lobe and its associated structures such as the hippocampus, mammillary bodies, and fornix.

There are two different types of amnesia. **Retrograde amnesia** is loss of the ability to recall past events; an individual who suffers a traumatic brain injury from an automobile accident may no longer be able to remember anything that happened before the accident, although distant memories are often preserved. Whereas retrograde amnesia is amnesia for events occurring *prior* to the injury of disease, **anterograde amnesia** is a loss of memory for events that *follow* the precipitating event (Figure 15.6).

FIGURE 15.6 **Retrograde and anterograde amnesia** **a.** An individual experiencing retrograde amnesia cannot recall memories formed before the damaging event, but new memories can still be created. **b.** An individual experiencing anterograde amnesia cannot recall events occurring after the point where damage occurred, but memories prior to the event may still be available.

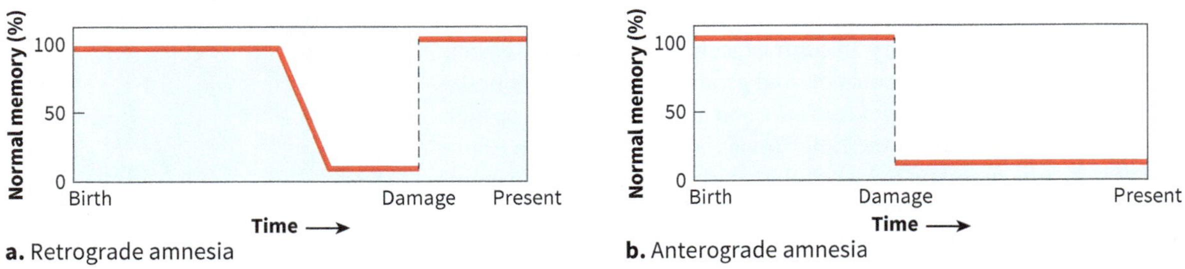

Henry Molaison expressed aspects of both types of amnesia; he recalled the events of his childhood up to about 3 years prior to the surgery but could not consciously remember anything he experienced after the surgery. Although Henry Molaison may be the most famous amnesiac, you can meet another interesting individual in this **Think About It: A Life Lived Moment by Moment.**

Think About It | A Life Lived Moment by Moment

John Dee/Shutterstock

Clive Wearing was a musician, composer, and conductor whose talents had brought him into the limelight of classical music. He had a particular interest in music from the 17th, 18th, and 19th centuries and founded the Europa Singers to present the music of that period to appreciative audiences. The BBC engaged him to recreate with authentic instruments and scores the music from a Bavarian royal wedding of 1568; this music was presented during the wedding of Prince Charles and Diana Spencer in 1981. He was at the top of his career, respected, and in demand.

In 1985, Wearing contracted encephalitis (an inflammation of the brain). His infection destroyed both hippocampi and left him with the most profound case of retrograde and anterograde amnesia recorded. Wearing literally lives in a bubble of less than 30 seconds of consciousness; to this day, at every moment he believes that he has just become conscious for the first time. Objects such as a piece of chocolate appear newly arrived within seconds of having taken a bite; he forgets what he is eating before the food can move from plate to mouth. He keeps a daily diary in an effort to understand what is happening, and he may write, "I have just awakened for the very first time," and moments later, cross out that line and write underneath it, "Who is writing rubbish?" Other than his wife Deborah, Clive recognizes no one, including the individuals who have been responsible for his care for decades.

It is interesting to note that although Wearing's amnesia is profound, some new memories have been created in the more than a quarter century since his illness. After watching certain movies hundreds of times over the years, he appears to anticipate events although he is not consciously aware of the plot. Some information appears to be stored in his implicit system after years of repetition. Search for online video content using the terms "Clive Wearing living without memory," and then answer these questions.

Questions

1. **Wearing cannot remember events from 30 seconds ago, yet he can speak and understand language. Does that argue for or against the idea that episodic and semantic memory share the same neurological circuit?**
2. **Wearing cannot remember his past or present since his illness more than 35 years ago, but he can read music and play the piano at a professional level. How is this possible?**

15.1 Before You Go On

Section Summary

Human memory can be thought of as a processing stream of three parts: sensory memory, which extracts simple physical information; working memory, which permits manipulation of verbal and spatial information; and long-term memory, where information about events and the meaning of words is stored permanently and available to the working memory system. Explicit memory is information in conscious awareness, and implicit memory is not in consciousness but can affect behavior. Procedural memory occurs through repetition of action. The PFC, medial temporal lobe, and particularly the hippocampus are essential for explicit memory, and damage to these areas results in amnesia.

Comprehension Check

1. What are the three major divisions of human memory?

- **a.** long-term memory, declarative memory, and sensory memory
- **b.** sensory memory, iconic memory, and short-term memory
- **c.** long-term memory, sensory memory, and working memory
- **d.** long-term memory, declarative memory, and implicit memory

2. Explicit memories are those memories that _____.

- **a.** you are consciously aware of
- **b.** go through the hippocampus and then the cerebellum
- **c.** are stored mostly in the temporal lobe
- **d.** are strong and explicit in detail

3. You walk into a friend's kitchen and see an unusual detail in the wood trim around the door. You think that you have seen that detail before, although you have no memory of having noticed that detail in the past. That memory was most likely stored where?

- **a.** iconic memory
- **b.** episodic memory
- **c.** procedural memory
- **d.** implicit memory

4. What is the cerebral region that is involved in both storage and retrieval of long-term memories?

- **a.** occipital lobe
- **b.** parietal lobe
- **c.** hippocampus
- **d.** basal ganglia

5. Karen cannot remember events that came after she suffered a blow to the head. This is an example of what type of amnesia?

- **a.** anterograde
- **b.** declarative
- **c.** retrograde
- **d.** implicit

15.2 Attention

LO 15.2 Summarize the functions of attention.

Every moment that you are conscious, your nervous system is dealing with a torrent of information. One of the reasons that humans are so successful is that our brains are skilled at choosing to attend to some events and not others. We avoid obstacles as we walk or drive, we manage to contribute to several conversations at the same time, and we stop hearing the radio while driving when traffic becomes heavy. In these and a thousand other events that occur daily, we rely on a mechanism that detects salient or important objects in our environment, focuses our attention on them, maintains that focus until it is appropriate to focus elsewhere, and can even juggle that focus among more than one task. Sometimes due to task demands, environmental factors, or disorders, this elaborate mechanism breaks down. How does our brain know what to attend to, when to attend, and when not to attend?

Aspects of Attention MCAT

Our brain can take in an almost unlimited amount of information but has a limited ability to process that information. Consequently, our nervous system must select the most important information for additional processing and discard the rest. This process of selecting and discarding is the basis of **attention**. There are several ways in which our attentional mechanism can be conceived, and they are alertness, vigilance, selective attention, and divided attention.

FIGURE 15.7 Multitasking The demands of modern life often require us to juggle multiple tasks.

Alertness is a basic function of the nervous system. We are alert when awake, less so when asleep, and not at all when in a coma. Vigilance, or sustained attention, is our ability to stay on task for an extended period. When a task is repetitive or an individual is fatigued, maintaining focus can be challenging. Selective attention is our ability to choose which information to attend to. Some information that we select is more "bottom-up," and some is more "top-down"; for instance, while walking to class, your sense of balance alerts you that the stairs you are climbing are slippery (bottom-up information), so you grasp the handrail. At the same time, you might be rehearsing in your mind a question you will pose to your instructor (top-down information). In both cases, you are choosing to prioritize certain information. Finally, most of us are capable of multitasking or attending to more than one task at a time (Figure 15.7).

Our ability to simultaneously attend to multiple tasks is related to the general level of our physiological arousal, but that relationship is not linear. More than a century ago, it was discovered that this relationship is best represented by a bell curve. When an individual is lethargic, their ability to attend is poor, and with increasing arousal, performance reaches a peak. At even higher levels of arousal, performance declines until a point is reached where the attentional capacity is almost zero. In other words, being somewhat aroused is good for your performance, like when you are alert and ready to take a test, but being overaroused is bad for your performance, like if your heart is racing with panic. This relationship, known as the **Yerkes–Dodson curve**, is depicted in Figure 15.8.

You may have read about events at large stadiums where the crowd, panicked by some incident, overwhelmed the exits and the resulting crush led to injuries and fatalities. Individuals whose mental resources are limited by overarousal may be incapable of holding more than one thought in working memory, and that is to flee, even if the exit leads to a dead end. With this overview of attention in hand, we now turn to the neural basis of attention and what can happen when the mechanism is damaged.

The Neural Basis of Attention MCAT

Our description of attention demonstrated the breadth of the mechanism, which ranges from processes that maintain consciousness to those that allow some individuals to juggle lit torches while holding a conversation. You should not be surprised then that there are several mechanisms at different levels of the nervous system that maintain our attention. At the most basic level, our reticular activating system (RAS) maintains states of consciousness, and damage to the RAS may result in an irreversible coma. As we look about the world, we must orient our eyes to objects that appear to be of importance, and the superior colliculi in the midbrain pay a critical role in shifting attention from one object to another. Interestingly, individuals with massive cortical damage that makes it impossible for them to have conscious thought may still visually track a moving object if the superior colliculi are intact.

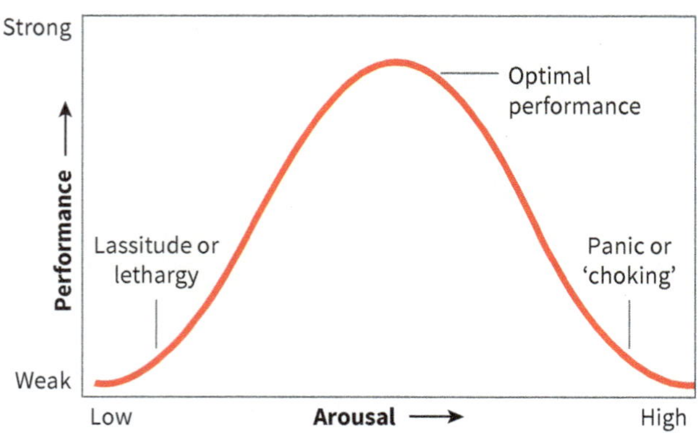

FIGURE 15.8 The Yerkes–Dodson curve When arousal is low, one's ability to attend to multiple concepts is poor. With increased arousal, the ability to multitask increases to an optimal level, but at higher levels of arousal performance degrades.

Moving up from the midbrain farther into the CNS, we see that some thalamic nuclei share with the RAS the task of maintaining alertness while the pulvinar nucleus has a hand in selective attention. Evidence from animal studies and humans with pulvinar lesions suggest that when there are competing objects in our visual field, the pulvinar nucleus acts to direct attention to some objects while suppressing attention to distractors.

As we move from the thalamus to the cortex, neural mechanisms involved in attention become more sophisticated. You may remember from an earlier discussion of the visual system that we have two streams in the visual system: a "where," or dorsal, stream and a "what," or ventral, stream. The parietal lobe is part of the

Check It Out | Change Blindness

wavebreakmedia/Getty Images

By its very nature, attention involves selecting some parts of our perceptual world while ignoring others. In this activity, you will see for yourself that we can be blind to what is happening directly in front of us. Watch the video **Change Blindness**, and then answer these questions.

1. Change blindness can be a disadvantage for drivers negotiating busy intersections or air traffic controllers. How might we reduce the probability of inattentional blindness interfering with performance in such roles?
2. Driving or operating heavy machinery places heavy demands on our ability to process information, and distraction can have serious repercussions. What distractions might an operator of a vehicle eliminate when driving in heavy traffic or treacherous conditions?

"where" pathway as it directs attention to objects in our visual field. Finally, the prefrontal cortex enjoys a supervisory role in attention; we can consciously decide to attend to some aspects of our environment and ignore others.

Attention allows us to select some information for additional processing and to either attenuate or discard other information. It may surprise you to discover that discarding information is one of the most important functions of our information processing system because our attentional mechanism has limitations in the number of items it can track simultaneously. As a result of our ability to focus, we may become blind to elements of a scene outside that focus. In **Check It Out: Change Blindness**, you can see for yourself that attending to some aspects of a scene results in being unaware of others.

Hemineglect MCAT

Behavioral neuroscientists devise theories to explain the human behavior they see every day, but sometimes that behavior is so far from expectations that it opens a door into new perspectives. One such phenomenon is **hemineglect**, or as it is commonly called, left-side neglect.

Individuals with damage to the right parietal lobe often fail to attend to objects on the left side of their midline. Perhaps the most startling example is an individual who sits down to a meal, dines upon everything on the right side of the plate, and appears to be unaware that there is another half meal on the left side (**Figure 15.9**).

Take a moment to consider what you just read. An individual who has lost one eye or has a portion of their retina damaged is generally aware of the loss of visual function and compensates by moving their head back and forth to be sure that they see their surroundings. In contrast, individuals with left-side neglect are simply unaware of objects in their left visual field. **Figure 15.10** shows how left-side neglect affects performance on a variety of tasks.

It is important to recognize that left-side neglect is not a sensory impairment but occurs farther along in the visual processing stream. This is illustrated clearly in the research on left-side neglect by the Italian neuropsychologists Bisiach and Luzzati. The researchers asked patients with left-side neglect to imagine they were standing at one end of the famous Piazza del Duomo in their fair city of Milan and describe what they saw (**Figure 15.11**).

These individuals described landmarks on the right side of their mental image but not landmarks on the left. Bisiach and Luzzati then

FIGURE 15.9 Hemineglect dinner An individual with right-hemisphere parietal lobe damage may show hemifield neglect and only perceive the portion of their meal to the right of the midline.

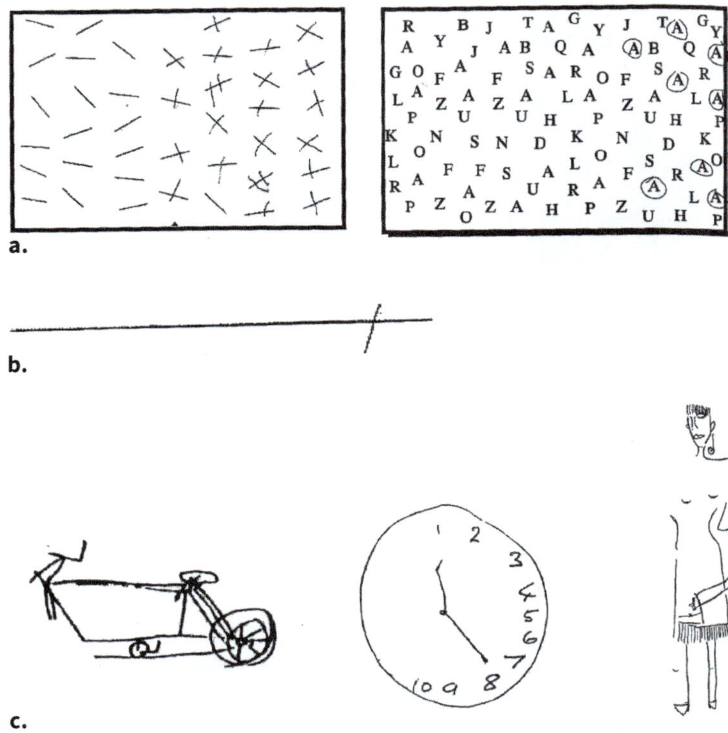

FIGURE 15.10 Examples of left-side neglect Individuals with left-side neglect are unable to attend to objects in the left side of their visual field. **a.** When asked to check off lines or circle the letter A in a field of letters, only those items on the right are completed. **b.** When asked to bisect a line, their choice is skewed to the right. **c.** When asked to draw a familiar object, they often leave off a portion that falls in the left side of their visual field.

FIGURE 15.11 Piazza del Duomo The Piazza del Duomo in Milan is a famous public space in Milan that is well known to local residents.

asked the same individuals to imagine they were now standing at the exact opposite end of the Piazza and to again describe the square; these same individuals now described the landmarks they had previously ignored, which were now on their right, and failed to mention the landmarks they had just described in the first task, now on their left. These findings suggest that in some way we create an internal mental representation of the space around us, much like an architectural model of a building or a city, which is disrupted by damage to the parietal lobe.

The importance of the parietal lobe in visual attention is hard to overemphasize. Right hemisphere parietal damage almost certainly leads to some disability in attending to objects in one's visual world, although not necessarily as overt as left-side neglect. You can get a feel for the importance of the right parietal lobe in **Think About It: Why Do Actors Enter from Stage Left?**

Think About It | Why Do Actors Enter from Stage Left?

The connection between the parietal lobe and visual attention is important to your understanding of how we allocate attention in our complex world. Damage to the right parietal lobe often results in some form of neglect, which illustrates that the right parietal lobe has a dominant position in the creation of our visual world. You do not have to damage your right parietal to see the that the right parietal lobe has an advantage in guiding visual attention, as you are about to see for yourself.

Consider this image; your task is to decide which of the two bands is the darker.

Most people will judge the bottom band to be darker, but in fact, the two bands are identical except for one being flipped on its long axis. This task, known as the grayscale task, demonstrates the superiority of visual attention in the right hemisphere as its contralateral connection favors the left visual field. This phenomenon is sometimes referred to as pseudoneglect because it describes a slight but normal preference for attending to objects in a person's left visual field.

This effect is small but pervasive in humans. Artists wishing to use light as a central part of the scene typically have illumination entering from the left, and in complex scenes, the most important items are more likely to be positioned to the left of the picture's midline.

Questions

1. **Actors who are hoping to enter a scene without being obvious may prefer to do so from stage right, whereas an actor who is hoping to make a dramatic entrance may prefer to enter the scene from stage left. Why might that make sense?**
2. **Students often wish to be noticed by the instructor in some classes and, conversely, to avoid eye contact with the instructor in another class. How might an understanding of pseudoneglect help students achieve these goals?**

15.2 Before You Go On

Section Summary

Attention is an essential part of dealing with a complex world. Alertness, the ability to select salient features while ignoring other features, as well as managing multiple tasks are all examples of different aspects of attention. The RAS, midbrain, thalamus, and frontal lobes all are involved in attention, but the right parietal lobe plays a central role in visual attention. Damage to the right parietal may lead to left-side neglect, or the inability to attend to objects in that visual field.

Comprehension Check

1. What is the basic task of our attentional circuit?

- **a.** reduce the amount of information by discarding that which is less important
- **b.** quickly switch between different objects of thoughts in working memory
- **c.** ignore half of what we see
- **d.** perform several tasks at the same time

2. When your instructor is telling you about the final exam while your neighbor is telling you about their plans for the weekend, which aspect of attention would lead to academic success?

- **a.** vigilance
- **b.** selective attention
- **c.** multitasking
- **d.** arousal

3. In a crowded stadium, something happens that causes the crowd to panic and rush for the exits despite the obvious fact that the exits cannot possibly handle the entire crowd at once, and injuries result. This is an example of what?

- **a.** selective attention
- **b.** multitasking
- **c.** vigilance
- **d.** resource limited attention

4. You see an individual whose hair is combed on only one side of their head and whose clothes are not properly arranged on the same side of their body. You might assume that this person has experienced damage to what part of the brain?

- **a.** RAS
- **b.** thalamus
- **c.** superior colliculus
- **d.** right parietal lobe

15.3 Object Recognition

LO 15.3 Explain how the brain creates a mental representation of objects.

Take a moment to consider the human visual system. Every waking moment, we are immersed in an everchanging sea of visual stimuli that we process into manageable information. In a prior chapter, we considered how the visual system extracts information about brightness, color, and location from this sea of information. We now turn to an even more complex topic: how do we recognize objects in the visual word?

You understand that our visual processing stream is comprised of a series of analyzers that begin by detecting edges and lines. At each successive step, they extract more complex information from the visual world. You may recall that we have two processing streams: a dorsal stream (known as the "where" pathway) and a ventral stream (known as the "what" pathway). How do simple analyzers in the ventral pathway create the perception of objects, and what happens when there is a disorder in that processing stream? We now turn to an overview of how our visual system creates the perception of objects.

Visual Object Recognition MCAT

Visual processing begins with what neuroscientist David Marr termed the creation of the **primal sketch**. Our visual system takes in the scene and uses edge detectors to create a simple image that is the first step in judging what an object is. Complex analysis converts subtle changes in illumination and texture to identify edges and then joins edges together to create objects. Figure 15.12 illustrates what a primal sketch might look like for a scene.

FIGURE 15.12 Creation of the primal sketch Feature analyzers identify edges in a scene and then attempt to link them into the outlines of objects.

Vision scientist Irving Biederman has proposed that our visual system takes the primal sketch and uses it to create simple three-dimensional forms known as **geons**. If you have ever taken a beginning drawing class, you were probably taught to capture complicated objects such as faces, fruit, or landscapes by decomposing them into simple geometrical shapes that can then be tuned to create realistic images of the word. In like fashion, our visual system combines geons into the familiar objects of our visual world, as demonstrated in Figure 15.13.

Geons have the property of being comprised of only a few attributes such as straight lines, curved lines, and convex or concave surfaces. This combination of simplicity and high differentiability means that conversion from primal sketch to complex object is almost instantaneous. In addition, the simplicity of recognizing objects by decomposing them into simple geometric shapes makes it easier to recognize objects regardless of their orientation. Figure 15.14 shows two images of a common object that you likely recognized instantly, even though one of the views was not one you often see. Vision researchers define the two views in Figure 15.14 as canonical and noncanonical views.

You may be familiar with the term *canon* as it is used in literature, where it refers to a body of books or other sources that are considered most influential for that genre. In the context of visual neuroscience and the artificial intelligence of vision, a **canonical view** is a view that is highly representative and might be the first view that comes to mind when you mentally image the item. In contrast, a noncanonical view is very different from what you typically imagine.

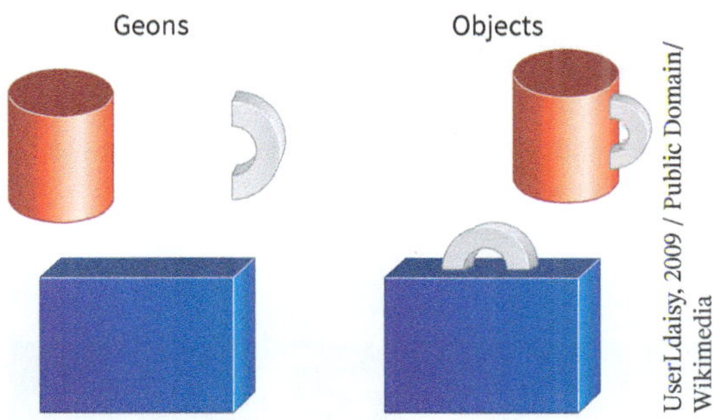

FIGURE 15.13 **Geons** By combining simple forms, our visual system creates complex three-dimensional objects such as a mug and a suitcase.

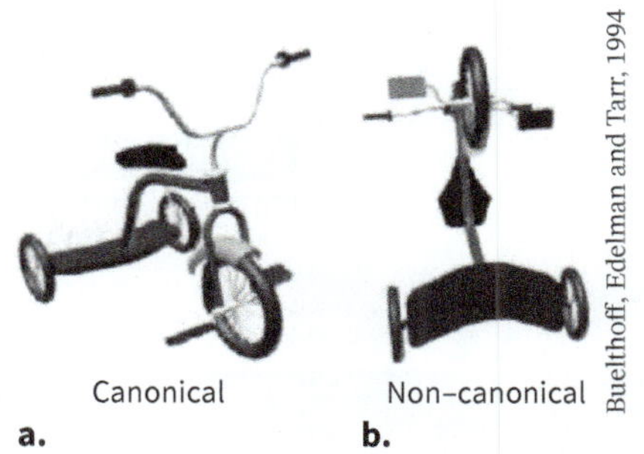

FIGURE 15.14 **Canonical and noncanonical views** **a.** A canonical view is one that is highly representative of the object. **b.** A noncanonical view must be mentally manipulated in some way before it is recognized.

The final step in object recognition is to find a match in semantic memory for the canonical view that the visual system has created. An individual seeing the noncanonical view of the tricycle in Figure 15.14b would have to mentally rotate the image into the canonical view and then match it with their internal representation of a tricycle to recognize the object; this additional processing would increase the time to recognize the item.

Challenges to Object Recognition

Because our visual processing stream is so efficient at its task, it is easy to take all that it does for granted. Imagining what it would involve to construct such a system is one way to understand its complexity. Consider for a moment the obstacles that computer scientists face when creating an artificially intelligent system capable of object recognition. Any system that might be tasked with driving a vehicle must be able to process information almost instantly and recognize objects such as pedestrians, other vehicles, and objects in or near the side of the road, regardless of their orientation.

Just how difficult is the task of recognizing objects regardless of their orientation? You may be familiar with the **captcha**, an image that presents a simple perceptual problem as a gateway to entry to a website; these are used to allow humans to access a site while preventing software applications known as bots from gaining unauthorized access. Figure 15.15 presents a captcha in which you must identify the squares where a traffic light is present. You probably easily identified the six squares where a traffic light is depicted; however, because of the difficulty of identifying this object in several noncanonical views, a bot would almost certainly fail.

A second type of captcha is one where a line is superimposed over letters and the individual must read or otherwise interpret the words; humans can easily ignore the line, but it would interfere with the letter-identification software in an automated system (Figure 15.16).

When lines are interrupted, feature detection may fail to recognize the outlines of an object. Given that the visual system of all animals at the most primitive level uses lines and edges in the first stages of detecting an object, you would predict that some organisms would use this fact to their advantage. This explains why camouflage can be so successful. By breaking down the outlines of predator or prey in a way that disrupts object recognition, an animal or object can become almost invisible, as illustrated in Figure 15.17.

Captchas and camouflage illustrate challenges to object recognition, but brain damage can cause a total failure of object recognition, the topic of our next section.

FIGURE 15.15 **An object recognition captcha** Captchas present an object-recognition task that a human can easily solve but a computer cannot. In this case, the user is asked to choose the squares that include a traffic light.

FIGURE 15.16 **A word-recognition captcha** Because the line interferes with every letter in the phrase, it interrupts automated letter- and word-recognition software.

FIGURE 15.17 **Examples of camouflage** Camouflage involves blending into the background by color and creating lines that interrupt the natural edges that define the object or person wishing to remain unobserved.

Visual Agnosia MCAT

Consider for a moment the power and elegance of your visual system. Millions of photons enter the eye each second and are channeled through a series of increasingly sophisticated analyzers that create a complex internal representation that allows us to identify objects in our environment. But object recognition, like any other neural process, can break down, as demonstrated in this true story.

Dr. P. was a renowned musician whose behavior had become odd, including failing to recognize students and mistaking fire hydrants for small children. The neurologist Oliver Sacks examined Dr. P. and found him quite witty in conversation, but as Dr. P was preparing to leave, he reached out and grasped his wife's head, trying to put it on his own head. Dr. P. had mistaken his wife's head for a hat. How could anyone make such a mistake?

Visual agnosia (from the Greek for *without knowledge*) is the inability to recognize objects due to a breakdown in the visual processing stream that is not attributable to a simple eye defect or memory disorder. An individual with visual agnosia who is shown a cat might describe it as a small object moving about with four tubes projecting from underneath. Yet the moment the cat rubbed against the person's leg and purred or the individual touched the cat, the identification would be immediate. How can a person whose eyes are functional and who knows what a cat is be unable to recognize what they are looking at?

Before you read on, take a moment to review what you know about object recognition. Lines and edges are combined into more complex forms; those forms are assembled into a three-dimensional representation, which is then matched against the contents of semantic memory to create recognition. Where in the processing stream might dysfunction lead to visual agnosia? Clinical evidence supports our understanding of object recognition as a two-phase process: the first involves a cascading series of ever more complex perceptual analyses that results in the creation of a canonical representation of an object; the second involves matching that representation with elements in semantic memory. Damage to different parts of the brain can disrupt either phase of object recognition.

Apperceptive agnosia Some visual agnosics are able to make simple judgments about lines, color, and brightness of objects but are unable to combine these basic elements into the more complex arrangement that constitutes normal object recognition. Individuals with apperceptive agnosia cannot copy a simple line drawing of a familiar object, even when they are looking at an example (**Figure 15.18**).

You likely recall from previous discussions that information from the retina ends up in area V1 of the occipital lobe and from there is directed through several additional processing areas in the ventral processing stream. Although visual agnosia is rare, the evidence from available cases is that diffuse damage in the occipital lobe is the common basis for apperceptive agnosia.

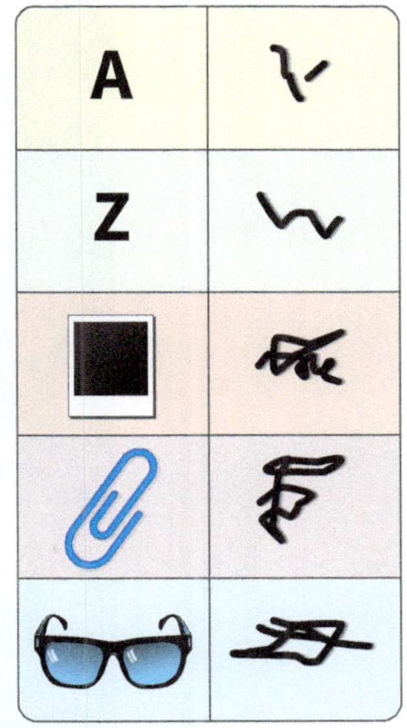

FIGURE 15.18 **Apperceptive agnosia** Individuals with apperceptive agnosia fail to copy simple objects because of their inability to combine primitive perceptual information into a cohesive object.

Associative visual agnosia Unlike apperceptive agnosia, which results from dysfunction in the earlier stages of visual perception, associative agnosia is the result of an inability to associate the product of the perceptual process with a representation in semantic memory. Associative agnosics can copy figures, often in great detail, but then are unable to name the object that they so faithfully copied (**Figure 15.19**).

Unlike apperceptive visual agnosia, which results from diffuse damage in the occipital lobe, apperceptive agnosia is characterized by damage to the occipitotemporal region. Given that the occipital lobe analyzes visual information, and the temporal lobe is associated with language production and comprehension, it is interesting to consider that the match between the canonical representation and its semantic representation occurs through the coordination of these two regions.

To summarize, apperceptive agnosics cannot create the primal sketch or the geons that create the canonical representation, so they have only primitive features such as color, texture, or luminosity to inform a guess about an object's identity. In contrast, associative agnosics can create a more holistic representation but fail to connect it to an object held in memory.

In an earlier chapter, we referred to **blindsight**, a phenomenon which individuals who have lost their sense of vision as a consequence of damage to the primary visual cortex still demonstrate some aspects of visual perception without conscious awareness. These individuals report they are functionally blind, but when asked to point to an object in their visual field, they do so at a level better than chance. One individual known in the literature as DB, when asked to guess about objects that he claimed not to be able to see, was above chance at reporting the object's color, shape, or movement. How could a person detect information about objects in their visual field when clear evidence of global damage to the PVC exists? The most likely answer is that there exists a circuit that bypasses V1 and terminates in the extra-striate cortex (areas V2–V5/MT) that perform complex visual processes that normally follow object recognition and so the output of these areas is created in the absence of analysis of the identity of the object itself.

You have seen many times in our discussions that complex behaviors involve coordination of many processing structures in the brain, and some of these structures have surprisingly specific tasks. A breakdown in these systems can help us understand how they normally function, as in a specific case of object recognition: human faces.

FIGURE 15.19 **Associative visual agnosia** Someone with associative visual agnosia is able to reproduce a drawing but cannot identify the objects depicted.

Prosopagnosia MCAT

Take a moment to inspect **Figure 15.20**. How would you describe what you see to another person? Most of us would describe a fanciful or even grotesque representation of a person, but an individual with severe prosopagnosia would instead provide you with a list of fruits and vegetables without reference to a person.

Up to this point, we have only considered visual agnosia as a disorder in which individuals struggle to recognize objects in general. There exists a very specific form of visual agnosia known as **prosopagnosia**, or as it is more commonly known, face blindness. Prosopagnosia results from damage to a very specific brain structure, the **fusiform face area (FFA)**, a small region of the fusiform gyrus located in the inferior portion of the temporal lobe (**Figure 15.21**).

As with any neurological malady, the extent to which an individual experiences prosopagnosia varies from mild to severe. In the most severe cases, all faces appear to be expressionless masks

Giuseppe Arcimboldo/Wikimedia Commons/Public domain

FIGURE 15.20 **Vertumnus** Giuseppe Arcimboldo (1526–1590) was an Italian artist known for his paintings that incorporated fruits, vegetables, flowers, and other food objects into whimsical portraits. This portrait was of Rudolf II the Holy Roman Emperor, depicted as Vertumnus, the Roman god of the seasons.

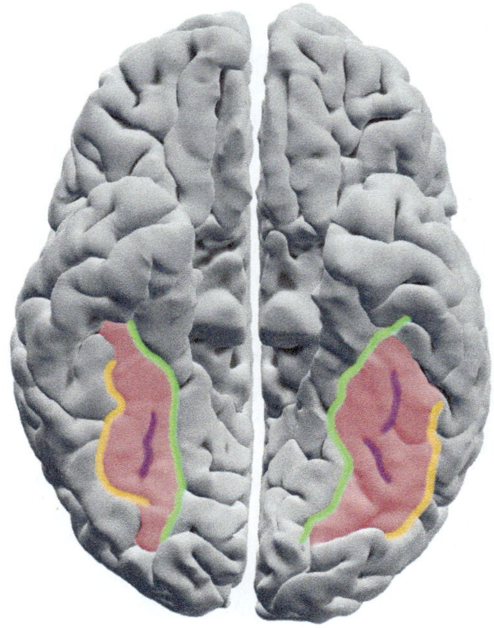

FIGURE 15.21 **The fusiform face area** The fusiform face area (FFA) is a processing center specialized for facial recognition. Damage to the FFA results in face blindness.

that only differ in their skin tone or the characteristics of hair. In **Check It Out: What Is It Like to Be Face blind?** you can meet Kristen and learn firsthand what it is like to have prosopagnosia.

You might be surprised to find that Apple co-founder Steve Wozniak, primatologist Jane Goodall, actor Brad Pitt, and Colorado governor John Hickenlooper are all believed to experience some aspects of prosopagnosia.

Check It Out | What Is It Like to Be Face blind?

wavebreakmedia/Adobe Stock

It can be hard to imagine the everyday life of an individual with prosopagnosia given that recognizing the faces of others seems to be at the heart of our social experience. How might face blindness change your behavior? Watch the video **What is it like to be face blind?** and then answer these questions.

1. Was Kristen a child or a young adult when she learned she had prosopagnosia? What does that tell you about the adaptability of humans?
2. Kristen spoke about how her face blindness was an impediment and how it might be an asset. How might you have accommodated being face blind in a similar situation?
3. In behavioral neuroscience we often see that behaviors or functions that are essential for our survival are hardwired into their own dedicated structure. What does the existence of the FFA suggest about the importance of the ability to recognize who is in our clan and who is not in the evolution of humans?

15.3 Before You Go On

Section Summary

In a series of increasingly complex stages, our visual system builds on the perception of simple lines and edges to create a three-dimensional mental representation of objects in our world. Visual agnosia is one result of damage to this system. The fact that a person can have an agnosia that is specific for faces suggests that humans have an innate propensity to pay particular attention to the face

Comprehension Check

1. The first step in object recognition is to identify the _____.
   - **a.** texture and color of the object
   - **b.** edges of the object
   - **c.** geons that make up the primal sketch
   - **d.** purpose of the object
2. Why do objects in a noncanonical view take longer to recognize?
   - **a.** They are unfamiliar to us.
   - **b.** Their edges and geons are not apparent.
   - **c.** They must be mentally rotated into a canonical view.
   - **d.** They confuse the lateral geniculate nucleus.
3. What is the basic principle of camouflage?
   - **a.** breaking up the typical edges of the object
   - **b.** using color to confuse the viewer
   - **c.** making the object appear to be of a different size
   - **d.** making the object appear to be a different and more familiar object
4. **a.** An individual with visual agnosia _____.could not recognize or describe a banana
   - **b.** can recognize only familiar objects
   - **c.** cannot identify a banana even when holding it in their hand
   - **d.** struggles to identify objects by vision alone
5. Prosopagnosia is characterized by an inability to recognize _____.
   - **a.** faces although other objects are recognizable
   - **b.** faces that are uncommon
   - **c.** the faces of famous people in the media
   - **d.** new faces only

15.4 Executive Function

LO 15.4 Explain the role executive function in cognitive behaviors.

A large corporation has many workers with a wide variety of responsibilities, but there is also an executive whose responsibilities are more global. The executive sets goals for the company and judges which actions likely to be successful and which are too risky. Similarly, our nervous system consists of many mechanisms that regulate our internal state and maintain consciousness, but we also have an executive system that provides the ability to plan, weigh alternatives, and be aware of the effect that our behavior has on others.

To understand the uniqueness of human behavior, one must understand executive function. In this discussion, we define the behaviors involved in executive function, review the neurological basis of these functions, and then provide an overview of the behavioral consequences of damage to or dysfunction of those same areas.

Aspects of Executive Function MCAT

Executive functions are behaviors involved in planning, goal setting, and execution of the plan to reach the goal. They are summarized in Table 15.2.

How important are executive functions to the success of humanity? All living organisms regulate their internal environment. Squirrels bury up to 10,000 nuts or seeds each year and are adept at remembering where their treasure trove of food is stored when needed, birds communicate through song, and whales appear to have a form of language with syntax. Many organisms create clans whose social behavior in hunting and defending their turf helps them survive, but no other organism matches human in their expression of executive functions. As a student you undoubtedly employ all the varieties of executive function to achieve success, including planning for a future career, completing long-term assignments, or turning down a tempting invitation because you need to prepare for an exam.

TABLE 15.2 Executive Functions

Executive function	Example
Mental flexibility	Adapting in response to changing conditions
Judgment	Weighing the consequences of actions and acting accordingly
Abstract thought	Thinking in terms of principles
Planning	Breaking down a complex problem into smaller steps to reach the goal
Multitasking	Performing more than one task at a time
Self-monitoring	Being aware of one's strengths and weaknesses and how one's actions affect others
Inhibition	Controlling impulsive behavior that would divert from attaining a goal

The Neural Basis of Executive Function MCAT

Identifying cortical and subcortical structures that support executive function is more complicated than identifying the neural basis of memory or object recognition because the behaviors associated with executive function are expressed in many ways. For instance, a person with a deficit in executive functioning might be capable of changing their plans, but totally incapable of seeing a situation from another person's perspective. The evolution of the frontal lobe provides a clue to the emergence of executive function in humans. Figure 15.22 compares the human frontal lobe with those of other animals.

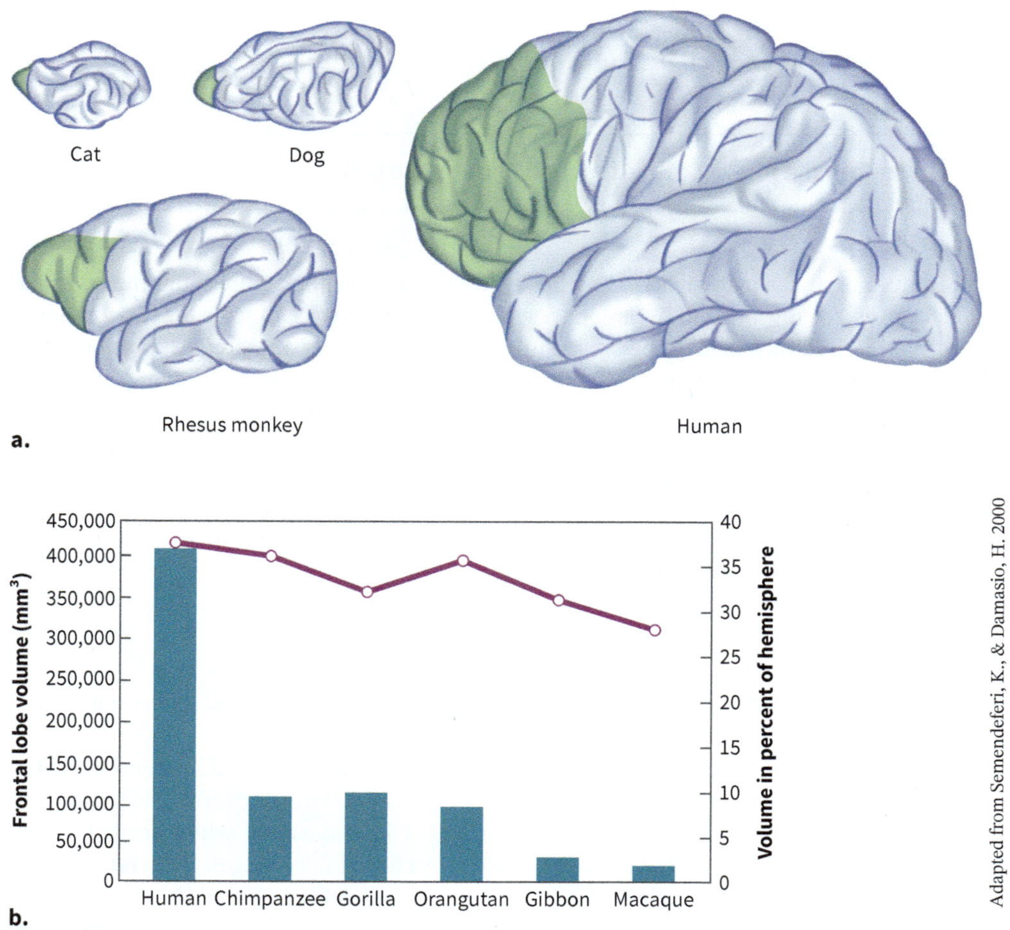

FIGURE 15.22 Comparison of frontal lobes across several species **a.** Humans have a much larger frontal lobe compared with these common mammals. **b.** Humans clearly have a large frontal lobe, but the proportion of cortex devoted to the frontal lobe is similar to that in nonhuman primates.

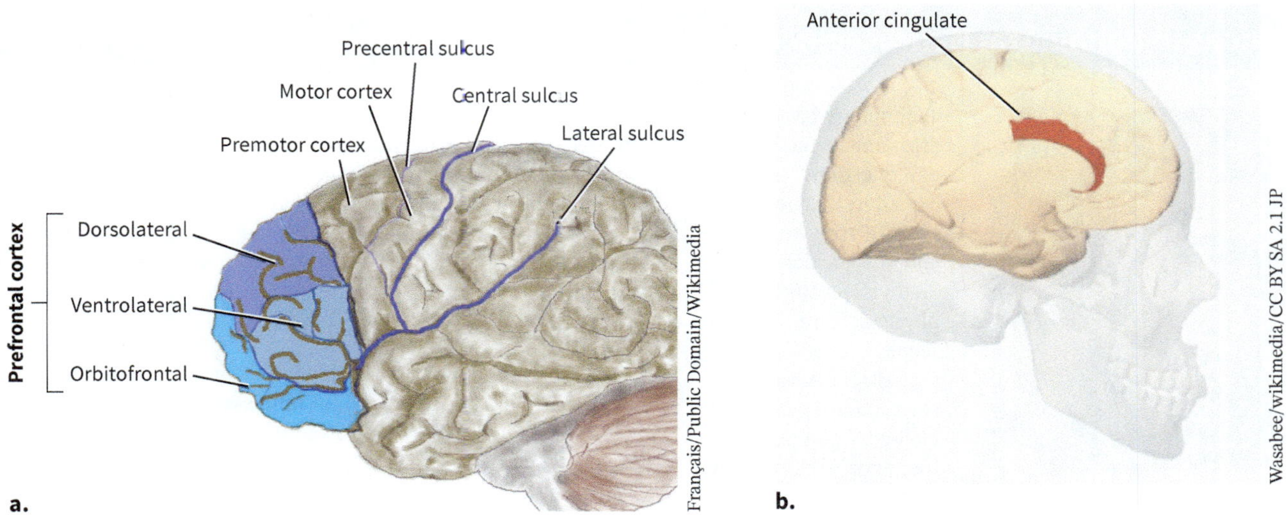

FIGURE 15.23 Regions involved in executive function **a.** The dorsolateral, ventrolateral, and orbitofrontal regions of the prefrontal cortex are involved in executive function. **b.** The anterior portion of the cingulate cortex is also a part of the executive function circuit.

Figure 15.22 tells an interesting tale that might not be obvious at first glance. Humans have the luxury of a large frontal lobe, but the proportion of frontal lobe to cortex in an orangutan or chimpanzee is only slightly less than ours; nonetheless, these nonhuman primates do not carry day planners or show evidence of having considered the meaning of life. This suggests that the *amount* of frontal lobe is more important to executive function than its *proportion* to the brain as a whole, and a metaphor may make that point memorable. Having more memory, or RAM, in a computer means that more applications can be open at once and that individual applications will run faster. Perhaps a nervous system requires an enormous number of neurons and synaptic connections to create the complex cognitive processes involved in executive function.

Neuroscientists, clinical psychologists, and neurologists often refer colloquially to executive function as a frontal lobe process; in reality, executive function is primarily the result of processes in the prefrontal area of the frontal lobe. Several areas of the **prefrontal cortex (PFC)** have been implicated in executive function, and among these are the dorsolateral, ventrolateral, and orbitofrontal regions, as well as a subcortical structure at the anterior tip of the corpus callosum called the anterior cingulate (**Figure 15.23**). **Table 15.3** presents the properties of these four regions.

Our understanding of executive function and the mechanisms by which the PFC creates and coordinates these functions has been dramatically enhanced by neuroimaging tools that tie brain activity to cognitive processes. You can learn more about research into the role of the PFC in **Advances in Behavioral Neuroscience: Marie Banich and Executive Function**.

TABLE 15.3 Areas of the Brain Involved in Executive Function

Region	Left-hemisphere functions	Right-hemisphere functions
Dorsolateral	Response selection and inhibition of inappropriate response	Monitoring of response and sustained attention to task
Ventromedial	Encoding of meaning	Maintenance of visual and spatial information
Orbitofrontal	Emotional processing and evaluation of rewards and risks	
Anterior cingulate	Evaluating response conflict and error detection	

Advances in Behavioral Neuroscience | Marie Banich and Executive Function

Marie Banich university of Colorado

Technology has a powerful impact on the type of questions science can address. The invention of the telescope and the particle accelerator revolutionized physics, and microscopes and genetic sequencers have had a corresponding effect on biology. Cognitive neuroscience was equally revolutionized by the marriage of modern neuroimaging tools to the study of higher cognitive processes.

Marie Banich has used those tools to develop a world-class research program dedicated to understanding executive function. Using neuroimaging tools that measure changes in metabolic activation or increases in neurotransmitter activity that accompany executive tasks, Banich has brought a greater understanding of these functions in neurotypical individuals as well as those with developmental challenges. Her work has led to numerous awards and accolades. She is a professor of psychology and neuroscience at the University of Colorado, where her lab investigates executive function from numerous perspectives.

Questions

1. **Search online using the terms "Marie Banich Laboratory." Pick one topic that interests you and briefly describe it and why you picked it.**
2. **One of the interests of Banich's lab is to examine executive function from a developmental perspective. Understanding when an adolescent is neurologically capable of making informed judgments that rely on executive processes has interesting legal and psychiatric implications. Consider one such application and briefly describe it.**

Evaluating Executive Function MCAT

Executive functions such as goal setting, cognitive flexibility, and inhibition of inappropriate responses are critical to our success. In fact, measures of executive function correlate with performance on standardized IQ tests. Patients with lesions of the prefrontal cortex generally show difficulties with planning, goal setting, and dealing with conflicting information, so several tests have been devised to measure an individual's performance on these functions, including the Wisconsin Card Sorting Test, the Tower of Hanoi, and the Stroop task.

The Wisconsin Card Sorting Test (WCST) An important aspect of executive functioning is the ability to create an abstract representation of a problem and then alter one's behavior in response to changing circumstances. In the WCST, an individual is shown four cards with symbols that differ in color, shape, and number (Figure 15.24).

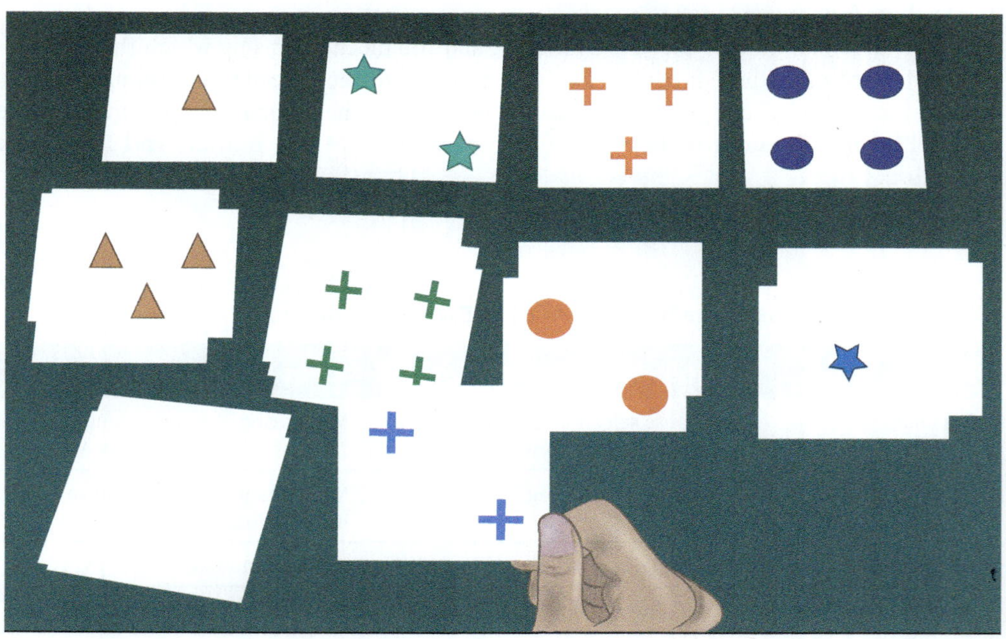

FIGURE 15.24 The Wisconsin Card Sorting Test The individual is asked to match cards from a pile to the test cards on the top row according to a rule deduced from feedback. In this example, the individual is considering whether to match the card based on color (blue), number (two), or shape (cross).

Participants are told they need to match each of the four cards according to a rule but are not told what the rule is. The individual attempts to sort the cards one at a time and is only told whether that attempt was correct. After the person finally deduces the rule, the experimenter changes the rule. The two measures of interest are number of trials to deduce the original rule and number of trials in which the individual continues to use the original rule after it has been changed. Those individuals with executive dysfunction struggle to deduce the rule. The WSCT also identifies **perseveration**, the inability to shift responses appropriately when circumstances call for it.

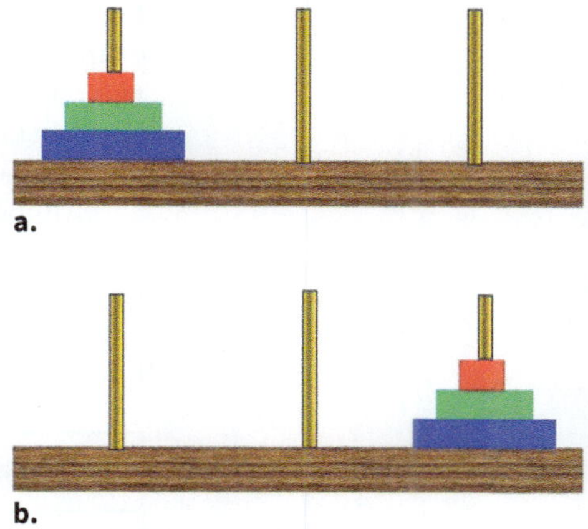

FIGURE 15.25 **The Tower of Hanoi** **a.** At the beginning of the task, all the disks are on one post. **b.** The task is completed when all the disks have been moved to the far post by moving one disk at a time and never placing a disk on one that is smaller.

The Tower of Hanoi One of the most popular executive function tasks, the Tower of Hanoi measures both working memory and planning. To successfully complete the task, one must move all the disks from the far-left post to the far-right post while moving one disk at a time and never placing a larger disk on a smaller one. Figure 15.25 depicts a simple Tower of Hanoi with three disks to move. The more disks to move, the harder the task becomes.

Individuals with damage to the prefrontal area, particularly the dorsolateral cortex, require many more moves to solve the problem than do intact individuals, suggesting that they are solving the problem by trial and error instead of creating a set of rules for moving the plates that would lead to a quick solution.

The Stroop task In a world of complex stimuli, data entering the nervous system may contain conflicting information as well as errors. An important role of the dorsolateral and cingulate cortices is to resolve such issues and thereby maintain focus. These processes are generally automatic and occur without the conscious awareness of the individual. Sometimes these automatic tendencies must be suppressed to achieve a new goal, and one task that reflects this aspect of executive function is the Stroop task.

This task presents an individual with a word that is the name of a color (for example, RED or BLUE) but printed in ink of another color. For example, the word RED is presented in blue letters. The executive system is responsible for overcoming the conflict between the automatic recognition of the word meaning and the color identification. You can experience this task yourself in Try It Out: The Stroop Task.

Psychosurgery as a Treatment for Executive Dysfunction

MCAT

Because abstract thought, goal-driven behavior, inhibition of impulses, and self-monitoring are all behaviors associated with the frontal lobe, damage to this area leads to behavioral issues that can be quite serious. Frontal lobe damage may be a consequence of an auto accident, fall, or physical trauma resulting from a blow to the head. Neurological disorders such as Alzheimer's, encephalitis, and supranuclear palsy typically have a frontal component as does

Try It Out | The Stroop Task

RED
GREEN
BLUE

A common test of executive function is the Stroop task, which asks participants to name the color of a word when it conflicts with the name of the word. Play the video The Stroop Task, and then answer these questions.

1. Did you find it easier to read the words in the first list than in the second? One explanation is that word meaning and color identification are both automatic processes that the Stroop task places in competition. Why might automatic processes be an important part of executive function?
2. Many maladies affect the functioning of the prefrontal area, including depression and obsessive-compulsive disorder (OCD). Consider the many functions of the executive system and propose how dysfunction of the PFC could play a part in a psychiatric disorder.

frontotemporal disorder; these and other neurological disorders are discussed in a separate chapter. However, no discussion of executive function and the frontal lobe would be complete without mentioning a dark episode in neurology and neuroscience, and that is the intentional severing of the connections between the PFC and the rest of the brain (a lobotomy) as a treatment for behavioral disorders.

Although the procedure was touted as an effective treatment for schizophrenia and emotional disorders, it was not efficacious for those conditions. In fact, it was often used to treat females who were judged as behaving in a sexually adventurous fashion, a treatment not applied to males expressing the same behaviors. You can learn about the development of the lobotomy in **Think About It: Moniz, Psychosurgery, and the Lobotomy**.

Think About It | Moniz, Psychosurgery, and the Lobotomy

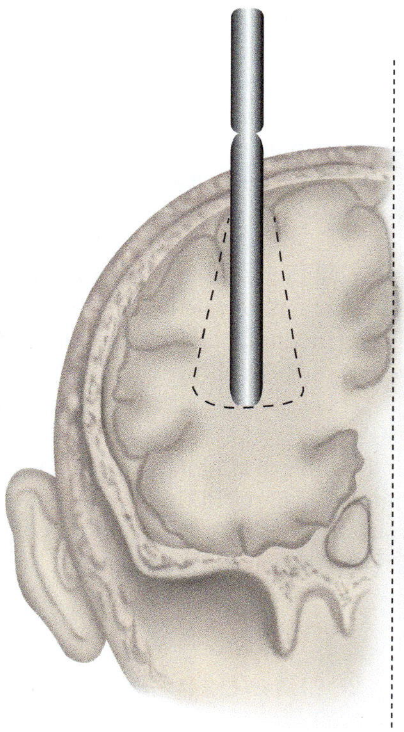

Before the discovery of psychotropic drugs to treat psychiatric disorders such as schizophrenia, depression, anxiety, and mania, physicians had very little in the way of treatments to offer a patient other than institutionalization. With the development of antiseptics and more modern surgical techniques, neurosurgeons turned their attention to developing surgical treatments for such disorders. In 1935, neurologist Egas Moniz proposed that individuals who were exhibiting psychotic or overly aggressive behaviors could be treated by separating the connection between the frontal lobes and the rest of the brain. Moniz pioneered a procedure in which he drilled several small holes in the skull and, inserting a tool, rotated it to sever white matter tracks connecting the frontal lobe to the rest of the brain.

Hearing of Moniz's work, American surgeons Walter Freeman and James Watts simplified the procedure considerably so that it could be done in an office and with only local anesthesia. Instead of drilling a hole in the skull, the Freeman–Watt procedure involved inserting an icepick-like instrument underneath the upper eyelid and through the eye holes in the skull. One sharp tap of a hammer broke though the thin bone and into the brain. Once the tool was inside the brain, it was rotated back and forth to cut white matter fibers, and then the same procedure was performed on the other side.

In the absence of effective pharmacological treatments for mental illness, the lobotomy was hailed by Moniz, Freeman, and Watts as an effective treatment for mental disorders, an opinion that was not supported by subsequent independent research. In fact, lobotomized individuals were often worse off after the procedure.

Questions

1. **Search online for the answer to the question, "How many lobotomies were performed in the United States?" What was the answer? Was this more or less than you would have thought?**
2. **We no longer perform lobotomies; instead, we use psychoactive drugs to target basic neural processes that underlie psychiatric disorders and often combine that with psychotherapy. Why might this approach be better than surgery?**
3. **Early in my career, I worked in a residential facility with an individual called Walter, who had been lobotomized in his youth. One morning I found that Walter had broken into my workspace and consumed more than a gallon of coffee. Given what you know about lobotomies and executive function, why might coffee have been particularly appealing to Walter?**

Take a moment to reflect on how the evolution of executive functions has helped humans succeed. When you imagine what your future might be and how you can behave today to turn that dream into a reality, you are demonstrating the power of the human executive system. We continually adjust and refine our behavior in a strategic fashion, and we are capable of stepping back and viewing our actions in the light of abstract principles. Tasks that measure the ability to think abstractly, flexibly, and with a view to understanding our own motives are valuable clinical tools. Because these abilities are at the heart of human behavior, damage to the executive system can cause serious behavioral consequences.

15.4 Before You Go On

Section Summary

Our executive system permits us to set goals and create a plan to reach those goals. It helps us consider the consequences of our behavior and make inferences from incomplete data. Psychologists have devised tests that measure aspects of executive function, and we know that the prefrontal cortex has a central role in these functions. The past use of lobotomies to eliminate the action of the PFC shows the dangers of surgical intervention without understanding the executive system.

Comprehension Check

1. Which of these behaviors is an example of executive function?
   - **a.** recognizing a friend while walking to class
   - **b.** identifying a familiar song
   - **c.** being startled by nearby thunder
   - **d.** creating a course schedule that will take you to graduation
2. What is the cortical region most associated with executive function?
   - **a.** temporal lobe
   - **b.** prefrontal cortex
   - **c.** occipital lobe
   - **d.** insular cortex
3. The Stroop task measures executive function by asking the subject to ______.
   - **a.** deal with conflicting and distracting information
   - **b.** suppress one automatic process while performing a second
   - **c.** create a set of goals to solve the task
   - **d.** consider ethical issues
4. Among the reasons why lobotomies are no longer performed is the fact that ______.
   - **a.** they resulted in death for more than 50% of patients
   - **b.** they amplified the symptoms of schizophrenia
   - **c.** they were not an effective treatment
   - **d.** lobotomized individuals became hypersexual

15.5 Hemispheric Specialization

LO 15.5 Describe the different functions of each hemisphere.

During your exploration of the neural basis of behavior, you occasionally learned that the two hemispheres, despite appearing to be identical twins, are more like fraternal twins who are quite similar but differ in some important ways. We conclude our discussion of cognitive neuroscience with an overview of the topic of hemispheric specialization.

Anatomical Differences Between the Hemispheres

At first glance, the two hemispheres appear to be identical partners, but a closer inspection shows several interesting differences. The left hemisphere has a slight but discernable asymmetry known as **cerebral torque**, in which the right hemisphere is slightly warped to the left. A corresponding warp to the right exists in the left hemisphere. Additionally, the planum temporale is slightly larger in the left hemisphere and has been known to be associated with language for over two hundred years. Heschl's gyrus is larger in the right hemisphere than the corresponding region in the left hemisphere, and this structure is an important part of the

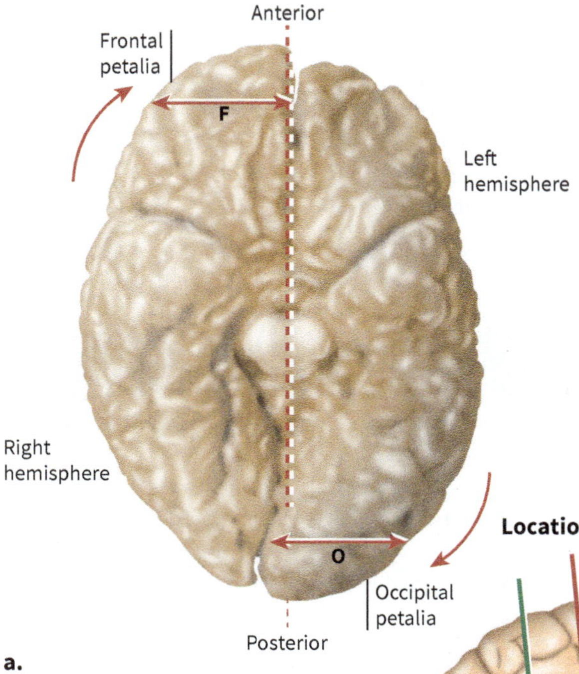

FIGURE 15.26 **Asymmetries in the human brain** **a.** Most people show a slight anatomical variation from hemispheric symmetry with the left hemisphere being slightly twisted, or torqued, across the midline of the brain. **b.** The planum temporale is larger in the left hemisphere. **c.** Hershel's gyrus is larger in the right hemisphere. **d.** The frontal operculum in the left hemisphere is larger than its counterpart in the right hemisphere. **e.** The slope of the lateral fissure that defines the temporal lobe is gentler in the left than its counterpart in the right hemisphere.

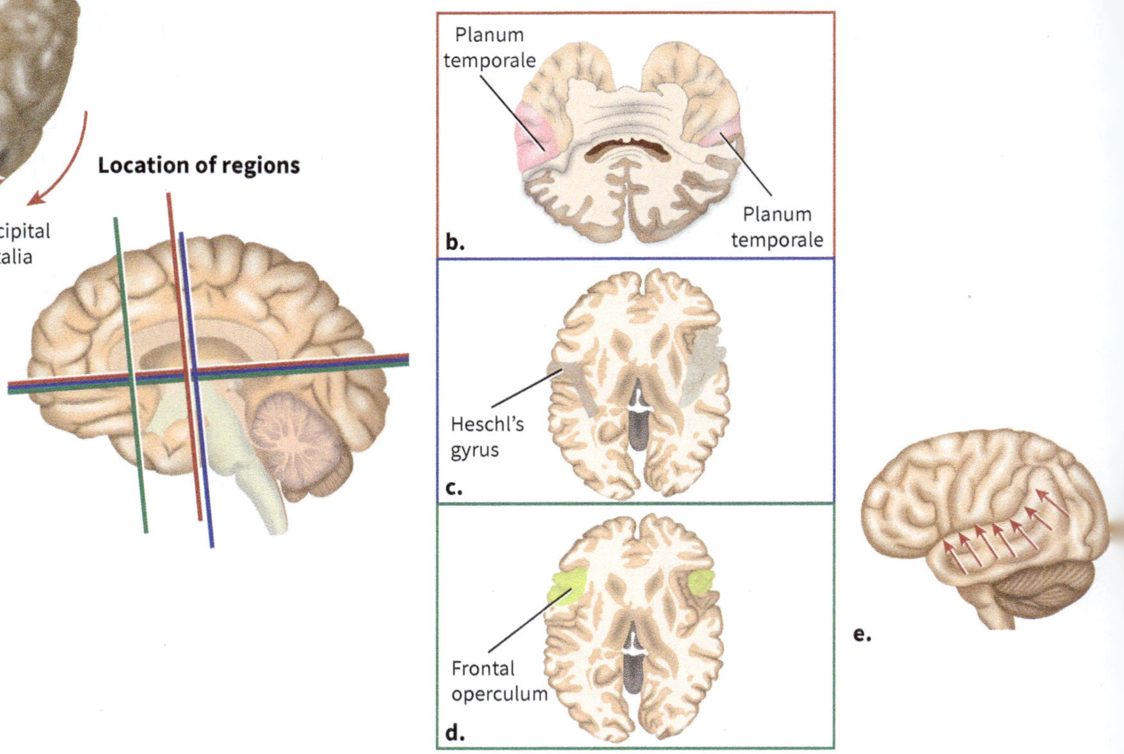

primary auditory cortex. Finally, the frontal operculum (from the Latin for *little lid*) is a region that has been shown to be closely involved in speech production. This region was damaged in Leborgne and Lelong, two individuals whose stories were important in our discussion of language and aphasia. These three regions are highlighted in Figure 15.26.

In addition to these anatomical differences, it has been noted that there is more gray matter in the left hemisphere and more white matter in the right hemisphere. There are also higher levels of dopamine in the left hemisphere and higher levels of serotonin in the right hemisphere. The full story of how these differences impact behavior is not yet known, but it is logical to assume that anatomical differences create different roles for the two hemispheres. One person's interest in hemispheric specialization led to the Nobel Prize in Medicine and Physiology, as you are about to see.

Sperry and the Split-Brain Procedure MCAT

Compelling evidence for specialization of the two hemispheres came from the work of neuropsychologist Roger Sperry. Sperry was interested in the function of the corpus callosum, which led him to work with neurosurgeons treating individuals with severe epilepsy. In the most extreme cases, it was thought that separating the two hemispheres by cutting the corpus callosum in a procedure known as a **commissurotomy** might control the spread of epilepsy from one hemisphere to the other. In the early 1960s, Sperry brought patients who had experienced a commissurotomy into his laboratory, and what happened next forever changed our view of the brain.

Because of the way the human nervous system is organized, input from both eyes crosses the midline at the optic chiasm. Sperry's participants had their corpus callosum sectioned, or cut, but the optic chiasm was spared. Sperry presented a picture of a screwdriver to the left visual field and therefore the right hemisphere; when participants were asked what they saw, their language-oriented left hemisphere responded that they saw nothing. The individual was then asked to reach out to the table where several objects were available but not visible. The left hand, connected to the right hemisphere, selected the screwdriver. By contrast, when an image of a baseball was presented to the left hemisphere, which is dedicated to language, the individual was able to name the object. Effectively, Sperry was testing individuals with two separate brains, one of which interpreted the word through an image (**Figure 15.27**).

Sperry's work transformed our understanding of hemispheric function and the role of the corpus callosum in interhemispheric communication and led to a share of the Nobel Prize in Medicine and Physiology in 1981. Subsequent research with split-brain individuals and intact undergraduate research participants by his former student and later collaborator Michael Gazzaniga led to a deeper understanding of how the two hemispheres understand the world. We now see the two hemispheres as engaging in different but complementary processing, with the verbal left hemisphere acting as an interpreter that combines the processing of both hemispheres into a coherent whole. You can see how the two hemispheres of a split-brain individual interpret the world in **Check It Out: Do We Have Two Brains?**

Through the work of Gazzaniga and many others, we now have a fuller understanding of the different processing roles of each hemisphere (**Table 15.4**). We will consider the different roles of each hemisphere in more detail in our discussions of neurological disorders.

Before we end our reflection on hemispheric specialization, we can consider one of behavioral neuroscience's most intriguing questions: why are some people left-handed?

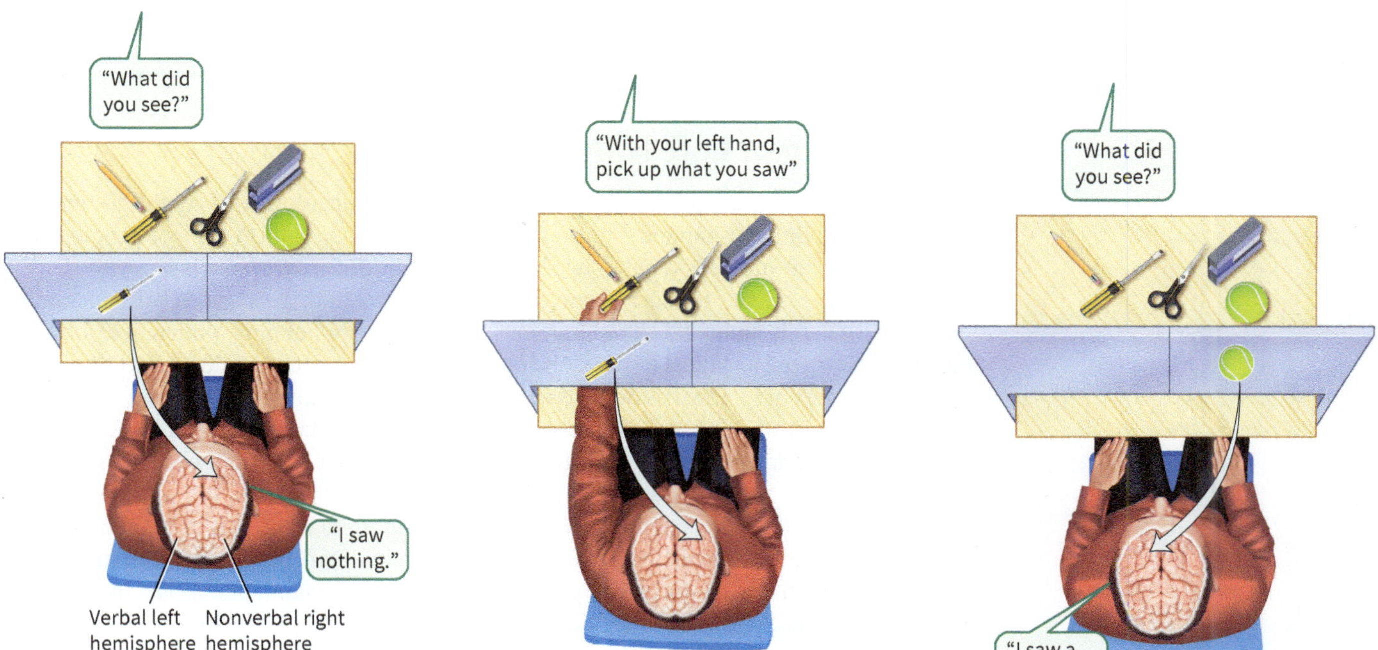

1 When a split-brain patient is asked to stare straight ahead while a photo of a screwdriver is flashed only to the right hemisphere, he will report that he "saw nothing."

2 However, when asked to pick up with his left hand what he saw, he can reach through and touch the items hidden behind the screen and easily pick up the screwdriver.

3 When the left hemisphere receives an image of a baseball, the split-brain patient can easily name it.

FIGURE 15.27 Sperry's split-brain experiment The split-brain experiment demonstrated that each hemisphere has a specialized role in object recognition.

Check It Out | Do We Have Two Brains?

freshidea/Adobe Stock

Although it is popular to think of language comprehension as the sole province of the left hemisphere, Gazzaniga's work with split-brain patients shows that both hemispheres are involved in processing a wide range of information. Search using the terms "Severed corpus callosum Gazzaniga Joe" and choose one of several short videos to watch that show Gazzaniga with "Joe," and then answer these questions.

1. In the videos of these experiments, Joe is presented with a word to his right hemisphere, which he claims not to have seen, yet he can draw the object with his left hand. Why is Joe unable to draw the object with his right hand?
2. What would be the advantage to having the brain consist of a collection of many individual processors distributed throughout the brain, as opposed to having all elements of a skill located in a one specific region of one hemisphere?

TABLE 15.4 Processing Modes of the Left and Right Hemispheres

Left hemisphere	Right hemisphere
Verbal (but prosody in right hemisphere)	Visuospatial
Rhythm in music	Melody in music
Global perspective	Local details
Positive emotions	Negative emotions

The Mystery of Left-Handedness MCAT

In light of all you have learned about hemispheric specialization, you might assume that everyone should be right-handed, but that is not the case. Roughly 10%–12% of people worldwide identify as left-handed. What do we know about handedness?

Having a preference to start a movement with one hand, paw, or foot is not unique to humans; what is unique is that humans have an overwhelming preference for right-handedness. Even the technical terms for being right-handed and left-handed reflect this dichotomy; right handers are known as dextrals from the same root as dexterity, whereas left handers are known as sinistrals from the same root as sinister. Scissors, can openers, cork screws, and school desks are designed to be used by right-handed individuals. What causes some individuals to be left-handed?

One theory is that damage to the left hemisphere *in utero* leads to a reversal of handedness. However, identical twins have a slightly higher rate of sinistrality (15%) than fraternal twins (12%), which argues for some genetic involvement in handedness. Having two left-handed parents raises one's chance of also being sinistral to 25%. There are very small differences in cortical structure that distinguish sinistrals from dextrals; in general, sinistral brains are less lateralized. That is, wherever in the brain of a dextral person there is an asymmetry in the size or composition of a structure, that asymmetry is less in sinistral people.

It is noteworthy that left-handed individuals are more likely to be schizophrenic but are also more likely to win a Nobel Prize or be members of MENSA, an organization of individuals with IQs in the top 2% of the population. Why sinistrals have an advantage in membership in such groups is unclear, but one possibility is that deviations in white and gray matter organization tend to result in either a measurable improvement in native ability or a significant cognitive disorder.

Try It Out | Are You Right- or Left-Handed?

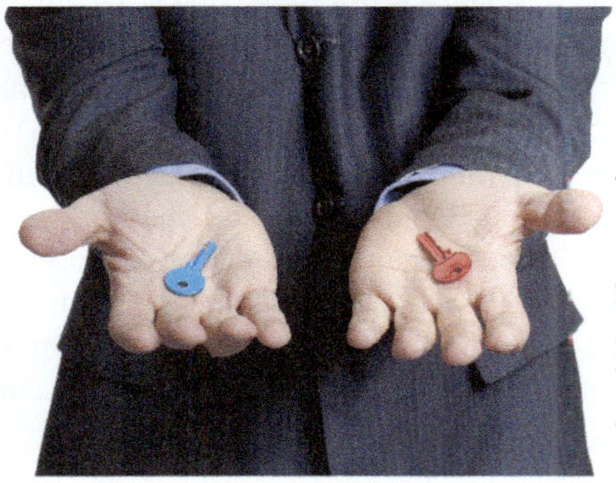
gosphotodesign/Shutterstock

Are you strongly dextral or sinistral or somewhere in between? Download the questionnaire **Are You Right- or Left-Handed?** to identify your dominant hand for most daily activities, and then consider these questions.

1. It has been said that left-handed individuals are more prone to certain disorders and have slightly shorter life spans than their right-handed counterparts. Search online using the terms "Lefthanded mortality disease" and summarize what you found.
2. Prior to 1950, many children who were naturally left-handed were forced in elementary school to write right-handed; having their hand smacked with a ruler was a common punishment for writing with the left hand. This conversion sometimes led to a slight language disturbance such as stuttering during childhood. Why might forcing a child to change their dominant hand interfere with the normal course of language production?

Handedness is not a clear dichotomy; some people may use either hand or only prefer a given hand when performing certain tasks. You can measure the extent of your own handedness in **Try It Out: Are You Right- or Left-Handed?**

The differentiation of the human brain into two hemispheres that are both complementary in function and subtly different in how they achieve these functions is a key to our success as a species. Rather than have one hemisphere simply mirror the functions of the other, there is specialization in function that gives the human brain enormous computational power.

15.5 Before You Go On

Section Summary

Our two hemispheres differ slightly anatomically, and it seems reasonable to assume that in some way their structural differences create important behavioral differences. The work of Roger Sperry, Michael Gazzaniga, and many others demonstrated the right hemisphere is not a silent partner to the left but a coprocessor whose functions are complementary and equally important. Although the majority of people are right-handed, some individuals are clearly left-handed, and the reason for this in unclear.

Comprehension Check

1. What is the primary communication path for the transfer of information between the two hemispheres?

- **a.** prefrontal cortex
- **b.** motor and supplementary motor cortices
- **c.** cerebellum
- **d.** corpus callosum

2. Roger Sperry's research investigated the effects of disconnecting the _____.

- **a.** two hemispheres on cognitive processes
- **b.** occipital lobe from the rest of the brain
- **c.** optic chiasm on visual perception
- **d.** temporal lobes from the occipital lobe

3. Michael Gazzaniga's research with split-brain individuals demonstrated that words presented to the _____ visual field that went to the _____.

- **a.** left; right hemisphere could be comprehended
- **b.** left; right hemisphere could not be comprehended
- **c.** right; right hemisphere could be comprehended
- **d.** right; right hemisphere could not be comprehended

4. Left-handed individuals are statistically more likely to _____.

- **a.** become airplane pilots
- **b.** become textbook authors
- **c.** have a high IQ or schizophrenia
- **d.** develop speech slowly in childhood

Conclusion

One of the most basic questions in philosophy and in science is this: what makes us human? Some part of that answer comes from cognitive neuroscience. Humans' exceptional memory coupled with a powerful ability to separate the important from the unimportant has enabled us to create technology that transforms our lives. We are social beings, and the ability to recognize members of our clan is a foundation for the creation of communities that have their own purpose. We can think abstractly, set goals, and make plans for events that may occur beyond our own life span. All these capabilities result from a brain that is really two processors in one, each with unique functions but also a shared mission. What you have learned about the neuroscience that creates conscious awareness will inform your thinking as we turn to a discussion in subsequent chapters of disorders of consciousness as well as the neuroscience behind social behavior.

Critical Thinking Questions

1. Our cognitive processes are the result of hundreds of thousands of years of evolutionary pressures, and features that were useful in the past may not be useful now. One example is that of the fallibility of human memory. Humans are very adept at remembering they have seen something before, but not *where* they have seen it before. Why might frequency of a memory be more important than source to us?
2. Many important tasks require sustained attention, but it is difficult to maintain attention on boring tasks where very little happens over long periods of time. Consider the job of a screener at an airport whose task is to look for weapons in carry-on luggage. Tens of thousands of bags may pass through the X-ray machine before a single unwelcome item is seen, but missing that one item can have serious consequences. How might you arrange their task to maximize their performance?
3. What you have learned about object recognition in humans should inform your understanding of the difficulties involved in creating self-driving cars and other automated systems. You may not be surprised to discover that artificial intelligence researchers are very interested in understanding how our nervous system performs object recognition as a model for their programs. Explain how one task in modern life would be improved by having an AI capable of object recognition.
4. Executive functions are perhaps the high point of human behavior; the ability to see a higher purpose, set plans that might come to fruition long after we are gone, and think in abstract terms has placed humans as the apex species in this world. As our world becomes more complex, we may need the functions of the prefrontal lobe even more. Search online using the terms "common critical thinking errors" and identify two that that you think are particularly prevalent today. Explain your choices.
5. When a behavioral or biological trait is maladaptive, it is generally selected out of the species over time. Some have made the argument that sinistrality is an example of a trait that is not adaptive, yet the evidence shows that sinistrality has not been selected out of the human genome. Even more interesting, it has maintained the same frequency of 10%–12% over tens of thousands of years. What advantages might left-handers contribute humanity so that this trait is sustained over time?

Dig Deeper into Behavioral Neuroscience

15.1 Memory

Every student of behavioral neuroscience or neurology should know about Henry Molaison. Suzanne Corkin is a neuroscientist who has studied Henry Molaison for decades and knew him as both a research participant and friend.

Corkin, S. (2013). *Permanent Present Tense: The Unforgettable Life of the Amnesic Patient.* H.M. Basic Books. ISBN 978-0465031597.

Perhaps the most powerful and personal account of Clive Wearing's illness was written by his spouse, Deborah Wearing. Her account of their life together is fascinating.

Wearing, D. (2004). *Forever Today: A True Story of Lost Memory and Never-Ending Love.* Transworld Publishing. ISBN 978-0385606264

15.2 Attention

Michael Posner is one of the most prominent attention researchers in the world. This book provides as overview of the neuroscience behind this critical cognitive function.

Posner, M. (2011). *Cognitive Neuroscience of Attention* (2nd ed.). Guilford Press. ISBN 978-1609189853.

15.3 Object Recognition

Every student with an interest in visual agnosia should read Oliver Sacks' account of a patient with a profound visual agnosia.

Sacks, O. (2021). *The Man Who Mistook His Wife for a Hat: And Other Clinical Tales.* Vintage. ISBN 978-0593466674

15.4 Executive Function

Students interested in executive function should read the appropriate chapter in Banich's textbook, which covers the research that led to our current understanding of the topic.

Banich, M., & Compton, R. J. (2018). *Cognitive Neuroscience* (4th ed.). Cambridge University Press. ISBN 978-1316507902

Your brain is a complicated device, so understanding the causes of mental illness has been a struggle. Perhaps for this reason, psychosurgical procedures such as the lobotomy were embraced uncritically. This book tells the tale of the rise and eventual fall of lobotomy as a treatment for mental illness.

Kragh, J. V. (2021). *Lobotomy Nation: The History of Psychosurgery and Psychiatry in Denmark*. Palgrave Macmillan. ISBN 978-3030653057.

15.5 Hemispheric Specialization

Pop psychology tells us that people are either left- or right-brained and that the left hemisphere is logical while the right is intuitive. You know now that this is simplistic and inaccurate. Everything you ever wanted to know about the two hemispheres can be found in this book by two neuropsychologists.

Springer, S., & Deutsch, G. (2001). *Left Brain Right Brain*. Freeman. ISBN 978-0716731115.

Chapter 15 Before You Go On Answers

15.1 Before You Go On Answers

1. d. long-term memory, declarative memory, and implicit memory
2. a. you are consciously aware of
3. d. implicit memory
4. c. hippocampus
5. a. anterograde

15.2 Before You Go On Answers

1. a. reduce the amount of information by discarding that which is less important
2. b. selective attention
3. d. resource limited attention
4. d. right parietal lobe

15.3 Before You Go On Answers

1. b. edges of the object
2. c. They must be mentally rotated into a canonical view.
3. a. breaking up the typical edges of the object
4. d. struggles to identify objects by vision alone
5. a. faces although other objects are recognizable

15.4 Before You Go On Answers

1. d. creating a course schedule that will take you to graduation
2. b. prefrontal cortex
3. b. suppress one automatic process while performing a second
4. c. they were not an effective treatment

15.5 Before You Go On Answers

1. d. corpus callosum
2. a. two hemispheres on cognitive processes
3. a. left; right hemisphere could be comprehended
4. c. have a high IQ or schizophrenia

Chapter 15 Glossary

amnesia A severe form of memory loss caused by disease or trauma.

anterograde amnesia A condition in which a person is unable to create new memories after brain damage.

attention The ability to prioritize some information while suppressing other information.

blindsight A condition in which an individual with damage to the primary visual cortex responds to visual stimuli without conscious awareness of the stimuli.

canonical view A view of an object that is most representative of the object.

captcha A task designed to prevent automated devices from accessing a web page that relies on cognitive skills such as object recognition.

cerebral torque The clockwise asymmetry of the human brain in which the right hemisphere is slightly warped clockwise.

commissurotomy The surgical separation of the two hemispheres by cutting the corpus callosum; also known as the split-brain procedure.

echoic memory The form of sensory memory that briefly holds auditory information.

encoding The act of transforming a signal into a format that holds more information than before.

episodic memory A type of long-term, explicit memory that involves events.

executive function A collection of cognitive skills that are associated with goal setting and abstract thinking.

explicit memory A type of long-term memory that involves events that can be consciously recalled.

fusiform face area (FFA) A region in the inferior temporal lobe that is dedicated to face recognition.

geon A simple three-dimensional geometric form that is believed to be an essential part of object recognition in humans.

hemineglect Unawareness of objects or people in the left visual field caused by brain injury; also known as left-side neglect.

iconic memory A form of sensory memory that briefly holds visual information.

implicit memory A type of long-term memory that involves events that are not in consciousness but affect behavior.

long-term memory The repository of knowledge of events and word meanings that the brain has stored over time.

perceptual memory A type of long-term, implicit memory that is primarily sensory rather than semantic or episodic in nature.

perseveration The repetition of a behavior or thought after it has lost its meaning.

prefrontal cortex (PFC) That region of the cortex that is at the rostral, or anterior, portion of the frontal lobe.

primal sketch A proposed first step in visual object recognition that involves identifying the edges and outlines of objects.

procedural memory A form of long-term, implicit memory that represents how we perform motor acts such as riding a bicycle.

prosopagnosia An inability to recognize people's faces due to a neurological disorder.

retrograde amnesia A condition in which a person is unable to retrieve memories that occurred before a brain injury.

semantic memory A type of long-term, explicit memory that involves the meaning of words.

sensory memory The initial memory system where sensory stimuli are briefly stored before being either passed on to working memory or discarded.

visual agnosia A condition caused by trauma or disease in which an individual can no longer visually recognize objects.

working memory A memory system in which information is briefly active while it is manipulated in consciousness to guide our behavior; also known as short-term memory.

Yerkes–Dodson curve A curved line on a graph that describes the effect of overarousal on attention.

CHAPTER 16

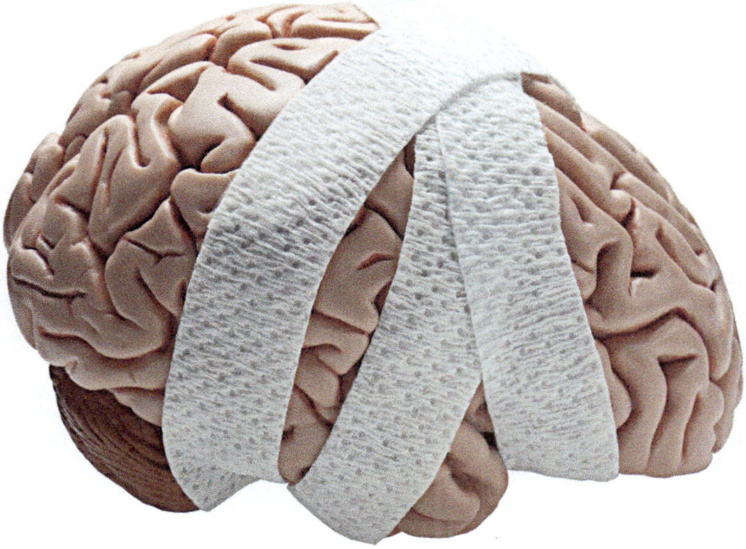

Neurological dysfunction has serious behavioral consequences.

Psychiatric and Neurological Disorders

CHAPTER OUTLINE

16.1 Mood Disorders
- An Overview of Depression
  Advances in Behavioral Neuroscience: Avshalom Caspi and the Genetic Basis of Depression
- The Neural Basis of Depression
- Treatment of Depression
- Bipolar Disorder

16.2 Anxiety Disorders
- General Anxiety Disorder
- Panic Disorder
- Obsessive–Compulsive Disorder
  Think About It: OCD, Compulsions, and Rituals

16.3 Consciousness Disorders
- Schizophrenia
  Think About It: The Cost of Schizophrenia
- Delusions of Identity

16.4 Neurological Disorders
- Bacterial and Viral Infections
- Cerebrovascular Accidents
  Check It Out: The Facts About Stroke
- Traumatic Brain Injury
  Try It Out: Concussions and Cognition
- Dementia
- Movement Disorders
  Check It Out: Understanding Tourette's Syndrome

LEARNING OBJECTIVES

LO 16.1 **Discuss** the causes and treatment of mood disorders.

LO 16.2 **Summarize** the causes and treatment of anxiety disorders.

LO 16.3 **Identify** the causes and treatments of schizophrenia and other delusions of identity.

LO 16.4 **Describe** the causes and effects of common neurological disorders.

Stories of the Brain

An intelligent and even-tempered 25-year-old, Phineas Gage was the foreman of a crew tasked with building a railroad bed through rural Vermont. Blasting through the granite hills that lay in their path was physical and dangerous work involving drilling a hole into the granite, laying an explosive charge in the bottom of the hole, covering the charge with a layer of sand that was tamped down with an iron rod, and inserting a fuse. On September 13, 1848, someone failed to place the layer of sand atop the gunpower, and as Gage attempted to tamp down the charge, the iron bar struck a spark on the granite, igniting the gunpowder. The bar, which was 43 inches long (1.1 m) and 1.25 inches (3.2 cm) in diameter and weighed 13.25 lbs (6 kg), entered Gage's head directly below his left eye, traveled through his frontal lobe, exited through the top of his skull, shot high into the air, and landed 80 feet (24.4 m) away.

Alive and conscious, Gage was taken to the home of Dr. John Harlow, where he greeted the physician with the comment, "Here is work enough for you, doctor." Gage surprised everyone by surviving the accident, although he was blinded in one eye. He returned to work after a month convalescing, but his personality had changed completely; once even tempered, he was now quick to anger; once thought of as being graced with common sense and intelligence, he was now impulsive and reckless. Soon after, he was let go by the railroad, and although he held a number of jobs in the remainder of his life, he was unable to keep any of them for very long. The damage caused by the rod passing through his brain had completely changed Gage's personality.

The mechanisms that create our consciousness and identity as humans are as complicated as they are fascinating. Just as damage to homeostatic or endocrine systems creates behavioral disorders, dysfunction in the mechanisms of consciousness creates behavioral disorders with serious consequences. You should not be surprised to discover that everything you have learned so far has a place in this discussion. When we try to understand the causes of mental illness and devise treatments, we must consider genetics, neurotransmitters, the synapse, cerebral structures, and the circuits that knit these elements together into functional units. We now turn to an overview of psychiatric disorders.

16.1 Mood Disorders

LO 16.1 Discuss the causes and treatment of mood disorders.

Psychiatric disorders are illnesses characterized by disruptive changes in mood, emotions, or thinking. **Mood disorders** are a subset of psychiatric disorders characterized by an inability to change or control one's mood. Mood disorders are the most common psychiatric disorders, affecting 10% of adults each year. Two common mood disorders are depression and bipolar disorder.

Andrew Parker/Getty Images

FIGURE 16.1 Emotional extremes in mood disorders Most of us experience highs and lows of emotion, but individuals with a mood disorder experience excessive swings or find themselves trapped on a euphoric peak or in the valley of despair.

You experience a wide range of emotions every day. You feel satisfied after a pleasant meal, happy to encounter a close friend, and sad to discover that your friend's parent is seriously ill. Mood, on the other hand, is a more persistent emotional state that influences how we view the world. External events can trigger moods, but we can also intentionally behave in ways that affect our mood; you might stroll through a park to lighten your spirits or select a horror movie to experience fright in a safe context. If you think of happiness as a hill and sadness as a valley, the typical human experience is to travel over mildly undulating terrain each day. By contrast, someone with a mood disorder is stuck on the hilltop or deep in a dark valley or is oscillating uncontrollably between the heights and depths of emotion (Figure 16.1).

An Overview of Depression MCAT

We all experience events that make us sad. We miss a friend who has moved away, lose a treasured pet, perform poorly on an exam, or have a romantic or professional relationship turn sour. **Major depressive disorder (MDD)**, often called clinical depression, is a debilitating disorder that goes far beyond sadness. A depressed individual may not get out of bed for days, may not wash or eat, and may feel that the only path out of the dark hole they are in is to end their lives. This account from an individual whose depression culminated in a suicide attempt illustrates how depression can distort reality.

> **I began to feel that life had no meaning. The small events of every day that normally created joy such as the sight and scent of a bouquet of flowers; sharing a meal with family or friends; or enjoying my hobby of creating stained glass decorations were without pleasure. I began to obsess over events that previously were unimportant but now were glaring examples of abject failures and I was unable to keep them out of my consciousness; for instance, at one point I was plagued for days with the memory that years ago, I was a few minutes late for an appointment. I could not escape the feeling that this event, along with hundreds of others, were clear evidence of my worthlessness as a human being and that everyone around me would be better off if I were dead.**

Although this individual has every reason to be happy, they are inexplicably held hostage to inconsequential events in their past to such an extent that they have considered suicide. To understand how this mismatch between one's self-image and reality can occur, we will first define depression and then consider its causes.

Diagnostic criteria Psychiatric illnesses are quite complicated and may shade one into another, so the American Psychiatric Association has created a classification system for diagnosing psychiatric illnesses known as the ***Diagnostic and Statistical Manual, 5th Edition* (*DSM-5-TR*)**. We will describe all the psychiatric maladies discussed in this chapter with reference to the *DSM-5-TR*.

The *DSM-5-TR* defines major depressive disorder as experiencing five or more of these symptoms over a 2-week period, with at least one symptom being either mood change or loss of interest or pleasure.

- Depressed mood for most of the day, nearly every day.
- Markedly diminished interest in activities and an inability to derive pleasure from participating in those abilities most of the day, nearly every day.
- Significant weight loss when not dieting, a weight gain of more than 5% of body weight in a month, or a marked decrease in appetite.
- Insomnia or hypersomnia (excess sleeping) nearly every day.
- Feeling restless or listless nearly every day.
- Feelings of worthlessness or inappropriate guilt (which may be delusional).
- Diminished ability to concentrate.
- Recurring thoughts of death or suicide.

Finally, these symptoms must not be caused by the direct action of a drug or a medical condition. They must not be the result of bereavement (the loss of a loved one) unless they persist for more than 2 months.

Demographics Major depression is the most common mental illness, afflicting as many as 17% of individuals at some point in their lives. There is a significant sex difference in diagnoses of depression, with 13% of males experiencing an episode compared to 21% of females. The reason for this sex difference is unclear. On average, females may experience more stress than males, or there may be a difference in how their HPA axis responds to stress.

In recent decades, depression appears to be diagnosed for the first time at younger ages, as illustrated in this study from Denmark, which examined the age of first diagnosis of depression for over 9 million individuals (**Figure 16.2**). Between the years 1996/1997 and 2016, there was a dramatic shift toward younger diagnosis. Before this time, the distribution could be

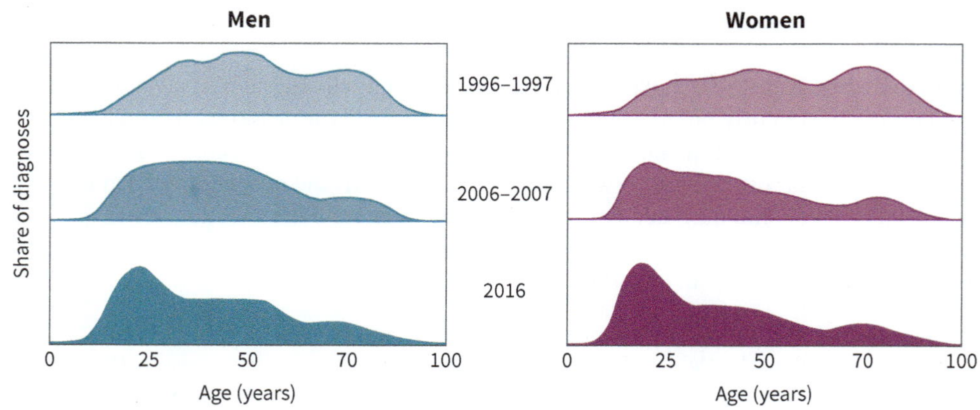

FIGURE 16.2 Changes in age of diagnosis of depression Over the period from 1996 to 2016, the age of first diagnosis for depression advanced for both men and women. That diagnosis was relatively rare for young individuals in 1996, but by 2016, first diagnosis occurred most frequently during young adulthood.

described as more varied with most individuals first experiencing depression between the ages of 30 and 55 years (with a second rise near the end of life); however, by 2016, there was a definite skew toward earlier diagnosis during late teens and early 20s. There are many reasons for this shift, including better diagnoses, greater acceptance of depression as a disorder and not a character flaw, and possibly also greater stress among the younger cohort.

U.S. president Abraham Lincoln, astronaut Buzz Aldrin, author J. K. Rowling, comedian Ellen DeGeneres, and singer Lady Gaga are all public figures who have dealt with depression.

Genetic and environmental influences Studies comparing the frequency of depression in identical and fraternal twins indicate a clear genetic component to developing depression. The concordance rate for identical twins is roughly 50%, and the corresponding rate for fraternal twins is less than 20%. It may be that a predisposition for depression results from the coordination of genes in response to environmental factors such as stress.

You learned in a prior chapter about the general adaptational syndrome (GAS), which describes the effects of chronic stress on the body without an opportunity to recover. Individuals who experience significant and chronic stress are more likely to experience depression. You can learn more about a researcher who is investigating the interaction among genetics, environment, and psychiatric illness in **Advances in Behavioral Neuroscience: Avshalom Caspi and the Genetic Basis of Depression**. We will look at these influences in greater detail in our discussion of the neural basis for depression.

Postpartum depression Having a child is a life-changing event. For the first few months, the parent is sleep deprived, and first-time mothers may be especially prone to anxiety about the challenges of parenting. Some sadness is not unusual at first, but about 15% of mothers experience severe and debilitating depression known as **postpartum depression**, or perinatal depression. The exact cause of postpartum depression is not understood, but we do know that risk factors include a family history of depression, stress due to sleep deprivation or personal circumstances, hormonal irregularities, and a history of nicotine use.

The Neural Basis of Depression MCAT

In an earlier chapter, you learned that serotonin (SE) is a neurotransmitter involved in mood regulation, so you will not be surprised to hear that serotonin depletion is closely associated with depression, although other neurotransmitters such as norepinephrine (NE) and dopamine (DA) are involved. If serotonin depletion alone were the cause of depression, then medications that increase SE at the receptor site should be accompanied by an immediate improvement in symptoms. In fact, improvement generally begins to occur a few weeks after pharmacological intervention, which suggests that the therapeutic effect has less to do with the amount of SE available at the synapse and more to do with SE receptors acting in concert with genetic, environmental, and neural circuits.

Evidence that there is more to depression than just serotonin depletion came from studies in which individuals who were diagnosed as depressed and who responded well to antidepressants were given a diet low in tryptophan, a precursor to serotonin. It was predicted that

Advances in Behavioral Neuroscience | Avshalom Caspi and the Genetic Basis of Depression

As you have learned about behavioral neuroscience, you have seen that sometimes important questions are simple to state but complicated to answer. How does our genetic endowment shape our response to the environment? We all have met individuals who have suffered unimaginable stress in their lives and somehow continue to function, whereas others struggle to overcome the usual minor stumbling blocks that life places before us. Neuroscientist Avshalom Caspi is interested in whether some of us naturally respond more unfavorably to stressful events due to preexisting genetic differences.

Avshalom Caspi is the Edward M. Arnett Professor of Psychology and Neuroscience at the Duke University Center for Genomic and Computational Biology and a professor at the Institute for Psychiatry, Psychology, and Neuroscience at King's College, United Kingdom. He was born in Israel, earned degrees from University of California–Santa Cruz and then Cornell, and was previously a member of the faculties of Harvard and University of Wisconsin.

With his colleague and spouse, Dr. Terrie Moffitt, Caspi hopes to better understand adult mental and physical health problems by uncovering the factors that influenced them early in the developmental process. Their method is to identify a cohort of babies born at a specific time and place and then follow the development of that cohort, conducting regular, interdisciplinary assessments throughout their lives. Caspi and Moffitt's work on the possible role that genes may play in response to environmental stress has been highly influential, although also controversial.

Questions

1. **Search online using the terms "Caspi and Moffitt Lab" and read more about the work of that laboratory. List one current research project that you find interesting and briefly explain why you chose it.**
2. **Assume for a moment that there is a test that determines your genetic predisposition to develop depression as a response to stress. Would you want to know your genotype? For careers that are inherently stressful, such as air traffic controller, police officer, ER physician, or airline pilot, would it be fair if applicants for these careers were tested for the genetic propensity to handle stress? Briefly explain your answer.**

they would experience a relapse in symptoms as their serotonin levels fell, and in fact this happened within a few hours. When control subjects who were not depressed were put on the same diet, those with no family history of depression showed few to no depressive symptoms. Meanwhile, among those controls who *did* have a family history of depression, a third of them showed depressed symptoms for the first time. These results suggest that depression does not necessarily develop in response to serotonin depletion but may develop when a combination of stressful life events and a genetic propensity for depression coexist.

What is the genetic basis for depression? Remember that when a neurotransmitter such as serotonin (also known as 5-HT) is released from the postsynaptic receptor back into the synaptic cleft, it is returned to the presynaptic neuron by a reuptake transporter. The creation of the serotonin reuptake transporter is controlled by the *5-HTT*, or serotonin transporter gene. *5-HTT* has two sections, a promoter region that controls how much transporter is made, and a translator region that contains instructions for actually making the transporter. It is easy to remember the roles of transporter and promotor if you think of the promotor as the order for how many cakes are to be baked and the translator as instructions for baking the cake. There appear to be two alleles (versions) of this gene that differ only in the length of the promoter region. The allele with a shorter promotor region is thought to result in slower uptake of serotonin (**Figure 16.3**).

How might this genetic difference play out in the real world? In a study involving almost 1,000 individuals in New Zealand, researchers asked subjects about stressful life events such as loss of a job or serious health issues and also if they had ever experienced depressive symptoms. Responses were then correlated with which allele of the serotonin transporter gene they carried. We get one copy of the gene from each parent, so an individual may have two long versions (L/L), one each of long and short (S/L), or two short versions (S/S). **Figure 16.4** portrays the relationship between the number of stressful life events and the probability of MDD by each of the three *5-HTT*, or serotonin transporter, gene groups.

Individuals experiencing depression report feelings of helplessness, lack of self-worth, and listlessness, so you might expect

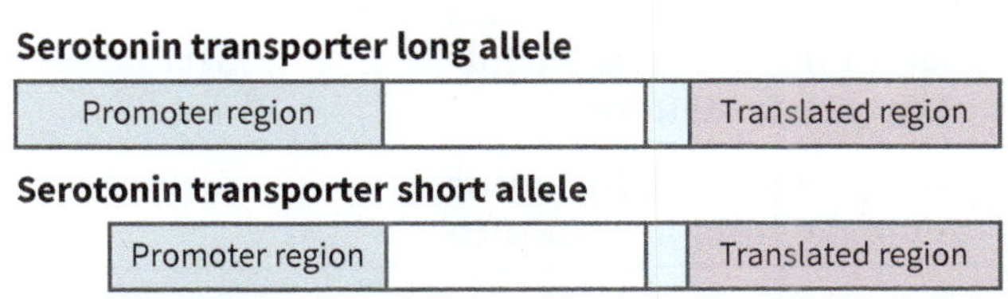

FIGURE 16.3 Serotonin transporter alleles The two alleles of the *5-HTT* gene differ in the promotor area, which regulates how much *5-HTT* is made, and not in the translator area, which contains the instructions for making *5-HTT*. These differences may affect how an individual responds to stress.

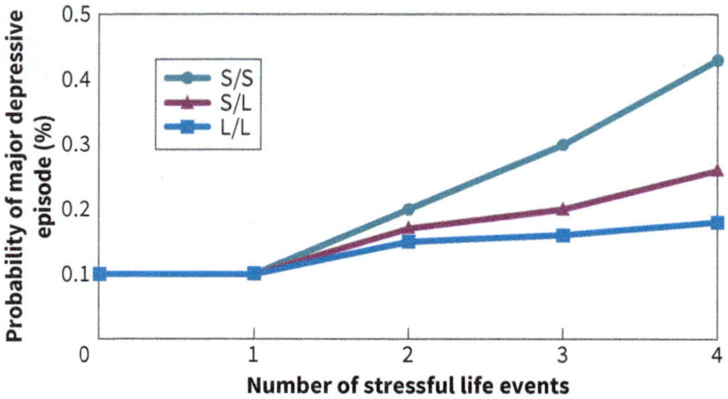

FIGURE 16.4 Life stress and probability of MDD by gene variant Individuals who were identified as having two long serotonin transporter gene alleles (L/L), one each of long and short alleles (L/S), and two short alleles (S/S) of the SE transporter gene differed in the probability of experiencing a major depressive episode. The long allele appeared to convey a protective influence, particularly when both long alleles were present.

that the prefrontal cortex (PFC), hippocampus, and amygdala would be hypoactive (less active) in the brains of depressed individuals. In reality, these individuals show greater metabolic activity in these regions. How might this make sense? Consider for a moment the roles of these three areas. The PFC is a structure that communicates with many cerebral regions, taking in information, evaluating its salience, and choosing a response. One perspective is that in depressed individuals, a dysfunctional PFC is sending misinformation to the amygdala, which in turn creates **anhedonia**, or an inability to experience joy. Your brain is always trying to make sense of the world, so to validate those feelings of sadness, your hippocampus retrieves memories of negative events, real or imagined. This theory would explain why many depressed individuals describe being overwhelmed with feelings of sadness accompanied by obsessive and intrusive thoughts of events that they view as personal failures.

Two other brain regions that have been identified as areas of interest are the cingulate cortex and the nucleus accumbens. The cingulate is part of the attentional circuit, and depressed individuals report an inability to focus and concentrate their attention on tasks due to intrusive thoughts. The nucleus accumbens is an important part of your reward system, and separating reward from behavior is part of the diagnostic criteria for depression. It is a part of our genetic heritage to create a plan to reach a goal, work toward that goal, and experience pleasure and satisfaction upon reaching that goal. When this reinforcement for goal-directed behavior is removed, life feels meaningless.

Treatment of Depression MCAT

Because depression is widespread and debilitating, many treatments have been developed, including drugs, therapy, and nonpharmacological procedures. Unfortunately, depression is a serious illness that does not always respond to treatment.

Antidepressant drugs The causes of depression are complex, so drug treatments have been developed targeting several different processes in the brain (Table 16.1). Selective serotonin reuptake inhibitors (SSRIs) are the drugs of first choice because of their proven effectiveness and manageable side effects. Patients who do not respond to SSRIs may be given drugs from other classes that target different neurotransmitters or processes. All of these drugs have side effects such as headache, nausea, diarrhea or constipation, and sexual difficulties that must be weighed against the benefits they offer.

TABLE 16.1 Classes of Antidepressants

Drug class	Action	Examples
Selective serotonin reuptake inhibitors (SSRIs)	Increase serotonin at the receptor by blocking the transporter mechanism responsible for reuptake	Lexapro, Prozac, Luvox, Paxil, and Zoloft
Serotonin–norepinephrine reuptake inhibitors	Similar to SSRIs but also affect reuptake of NE	Effexor and Cymbalta
Norepinephrine–dopamine reuptake inhibitors (NDRIs)	Inhibit NE and DA reuptake	Wellbutrin and Zyban
Ketamine	Stimulates neurogenesis	Spravato
Tricyclic antidepressants	Block SE and NE transporters, which results in elevated levels of these neurotransmitters at the synapse	Elavil
MAO inhibitors	Inhibit monoamine oxidase, an enzyme that breaks down neurotransmitters such as SE, NE, and DA	Nordil, Parnate, and Marplan

Drug treatment is indicated for depression, and there is compelling evidence that their effect is enhanced by psychotherapy and **cognitive behavioral therapy (CBT)**, which teaches the individual techniques to recognize negative thoughts and unhelpful behavior patterns and replace them with positive thoughts and behavior. Finally, exercise has been shown to be helpful in reducing the severity of depression, particularly in individuals over the age of 60.

It is common for a physician to try several different drugs before finding one that results in improvement. In addition, a drug that was initially efficacious may lose its therapeutic effect over time. You can easily imagine the many compensatory mechanisms that respond to the effects of a drug by altering the system to bring in back into the unhealthy state.

For severe depression coupled with suicidal thoughts that is resistant to standard antidepressants, ketamine has recently been considered as a possible although controversial treatment. Ketamine is an anesthetic that is commonly used to quickly induce unconsciousness during surgery. It is also a popular drug of abuse (sometimes called Special K) because it creates feelings of dissociation, and its use as a recreational drug can be fatal. Its mode of action in treating depression is unclear, but one hypothesis is that ketamine stimulates neurogenesis, which allows for new connections between episodic memories, thereby suppressing or replacing unhealthy pathways to negative thoughts.

Nonpharmacological treatments for depression Antidepressant drugs are the first course of treatment for depression, but when these are ineffective, there are several nonpharmacological treatments for depression, including electroconvulsive therapy, repetitive transcranial magnetic stimulation, and deep brain stimulation, that may be used for the most severe cases.

In the 1930s, researchers noted that individuals experiencing depression or schizophrenia experienced a temporary improvement in symptoms after an epileptic seizure. This observation suggested to neurologists that precipitating a seizure by passing a small electrical current through the brain could lead to an improvement in symptoms of depression, a procedure called **electroconvulsive therapy (ECT)**, once known as shock treatment. ECT does result in significant improvement in many individuals; best results are found with several treatments over the course of a few weeks. How ECT relieves the symptoms of MDD is not understood, but one theory is that the ECT resets the synaptic connection between the executive and unpleasant memories in long-term memory; once these memories are inaccessible, they are unable to influence behavior. ECT fell out of favor in the 1970s but has since reclaimed its reputation as an effective treatment for MDD, especially for individuals who have not responded to antidepressants, have expressed suicidal thoughts, or have attempted suicide. Side effects of ECT include temporary nausea, headache, confusion, and slight memory loss for recent events.

In addition to electrical stimulation through ECT, magnetic stimulation can be used to treat depression. Recall from the earliest chapters that neurons are electrochemical devices that create and then transmit an electrical signal down a nerve tract. Because magnetic fields disrupt electrical transmission, it was thought that rapidly pulsing a weak magnetic field over the skull would temporarily disrupt the normal function of the neurons directly underneath, a procedure called **repetitive transcranial magnetic stimulation (rTMS)**, illustrated in **Figure 16.5**. The magnetic field created by the rTMS stimulator is about 1,000 times more powerful than a refrigerator magnet and lasts less than 200 ms. Unlike ECT, rTMS does not induce seizures and is not known to have side effects other than temporary headaches. The use of rTMS over the PFC appears to increase the activity of the cingulate gyrus and amygdala, two areas implicated in mood. One potential use for rTMS would be as a therapy to stabilize a patient in the interval between when they start antidepressants and when those drugs begin to have an effect.

In **deep brain stimulation**, an electrode is implanted in a cerebral structure such as the cingulate, nucleus accumbens (NAc), or medial forebrain. The individual is then given a handheld stimulator that allows them to trigger stimulation of this region whenever they are feeling very depressed. Currently, deep brain stimulation is experimental and is reserved for the most severe cases of MDD that are resistant to other forms of therapy.

At this time, there are approximately 37 million individuals in the United States over the age of 18 who are prescribed antidepressants. Every year, around 100,000 individuals receive ECT for their depression, perhaps 65,000 are treated with rTMS, and the use of deep brain stimulation for the most severe cases is still uncommon but increasing.

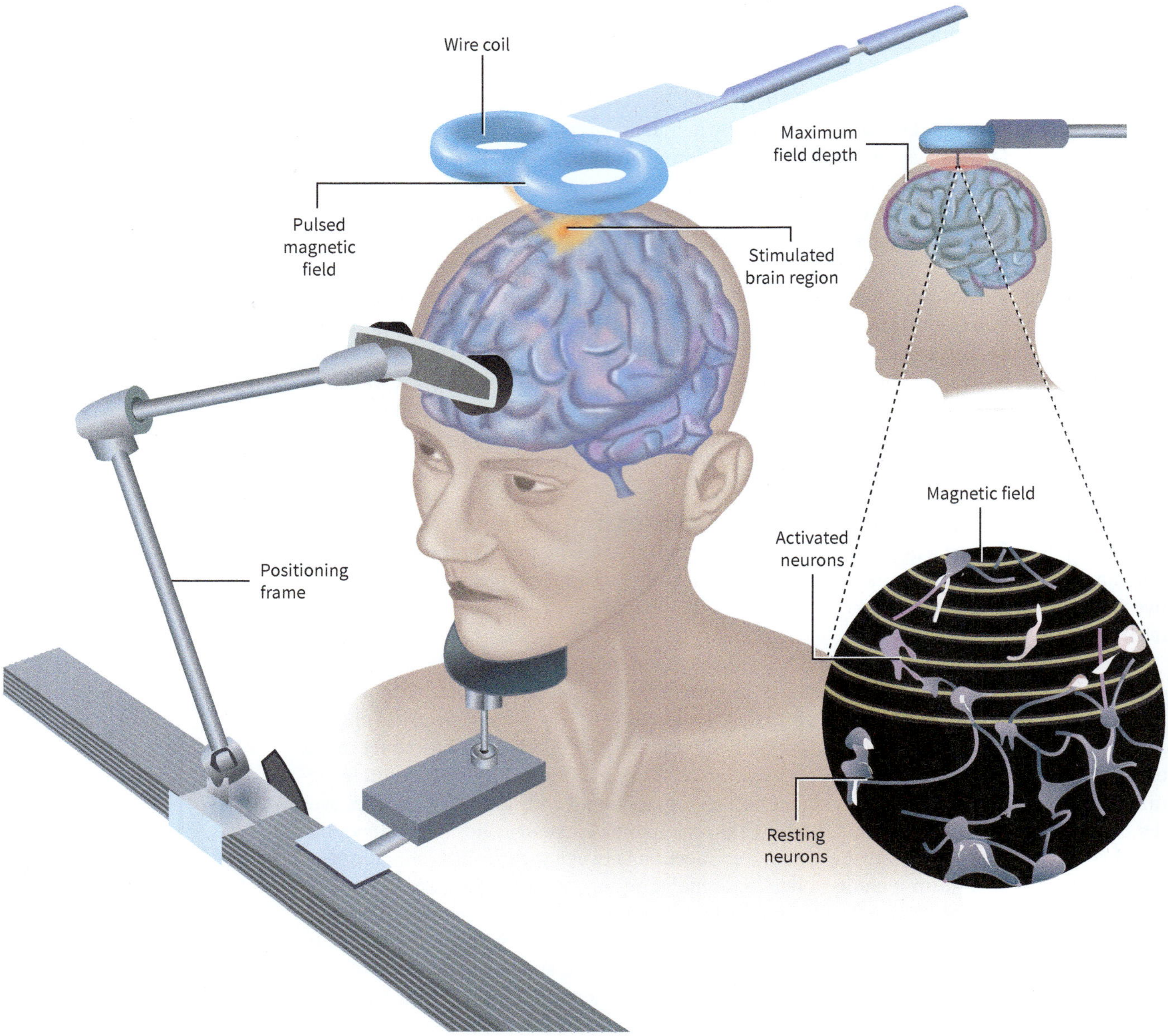

FIGURE 16.5 **Transcranial magnetic stimulation** When an electrical current is passed through a wire coil, it creates a magnetic field that passes through the brain and (depending on the strength of the field) extends 1–3.5 cm down into the brain. Repetitive transcranial magnetic stimulation (rTMS) involves multiple sessions of 30–40 minutes over several weeks. No anesthesia is needed.

Bipolar Disorder MCAT

Bipolar disorder (formerly known as manic depression) is a mental disorder characterized by mood swings that range from extreme elation to depression. These mood swings are accompanied by changes in activity levels and concentration that impair the ability to perform everyday tasks. The following account from a bipolar individual illustrates the symptoms.

> **When I was in my manic phase, I might decide at 2 am that rearranging all the closets in the house was a wonderful idea, especially since I had envisioned a way to arrange their contents that was revolutionary. As I tore the house apart, I was consumed with excitement that I could create an internet-based business in which I would sell the concept and become wealthy. I would go on spending sprees and buy all manner of things that I had no possible use for. My thoughts would race a mile a minute and I felt irritated at individuals who could not keep up. At one point, I was convinced that I was the child of a very famous individual who somehow had been separated at birth and if I could just contact them, they would welcome me back into the family.**

As this account suggests, bipolar disorder can lead to risky behavior with serious personal consequences.

Diagnostic criteria

The *DSM-5-TR* specifies that criteria for both manic and depressive disorders must be met.

- **Manic episode**: A period of abnormally elevated or irritable mood lasting at least one week with three of more of the following behaviors observed: Inflated self esteem or grandiosity, decreased need for sleep, excessive talking, racing thoughts, distractibility, or engaging in risky behavior. The manic episodes must lead to personal, social, or occupational consequences and must not be due to drug use.
- **Depressive episode**: Five or more of the following symptoms: feelings of hopelessness, sadness, or emptiness; diminished interest or pleasure in everyday activities, significant weight loss or gain of more than 5% of body weight, insomnia or hypersomnia, agitation, fatigue, diminished ability to concentrate, or recurrent suicidal ideation.

The *DSM-5-TR* classifies bipolar disorder into three categories, based on the nature and severity of the symptoms. Bipolar I is characterized by depression, a decreased need for sleep, uncontrolled speech, racing thoughts, distractibility, increased activity, and an affinity for engaging in risky behaviors. Bipolar I is the most severe form, and one defining characteristic is that the individual is unable to function normally and may be a hazard to themself and others due to irrational beliefs in their abilities. Individuals diagnosed as bipolar I must have experienced a manic episode lasting at least one week. Bipolar II is diagnosed when the symptoms are less severe, and the individual must have had a manic and a depressive episode. Finally, those individuals whose mood swings are mild are categorized as cyclothymic (**Figure 16.6**).

Demographics

Compared with depression, bipolar disorder is much less frequent, with 2%–3% of people experiencing bipolar disorder at some point in their life. Men and women are equally likely to develop bipolar disorder, and the average age of onset is the middle 20s. Individuals who have identified as bipolar include astronaut Buzz Aldrin; actors Carrie Fisher, David Harbor, and Richard Dreyfuss; singers Mariah Carey and Demi Lovato; and director Francis Ford Coppola.

Genetic and environmental influences

The causes of bipolar disorder are not well understood. There is a clear genetic component because individuals who have a parent or sibling with bipolar disorder have an elevated risk of developing the disorder. Among identical twins, the concordance rate is above 50%, whereas the rate for fraternal twins is about 5%. As with depression, there is some evidence that environmental stressors interact with genetics to place some individuals at greater risk for bipolar disorder.

Neurological basis

Dysregulation of serotonin and norepinephrine are involved with depression, so we would expect that these are also involved with the depressive phase of bipolar disorder. It is thought that the cyclic nature of bipolar disorder is partially due to excess dopamine (DA) at the receptor site, which leads to downregulation of DA receptor sensitivity, resulting in depression. In response to this, upregulation of the DA receptor leads to mania, and the cycle then repeats.

As you saw with depression, there is more to a psychiatric illness than neurotransmitter dysregulation. At this point in time, two different circuits in the prefrontal cortex have been identified that regulate cognitive control and emotional response and show less activity in bipolar individuals than in individuals not experiencing psychiatric illness. A ventral (emotional) circuit appears to be overactivated while a dorsal (executive control) circuit is underactivated.

Maria Maarbes/Shutterstock

FIGURE 16.6 Bipolar disorder Individuals with bipolar disorder cycle through episodes of mania that then transform into depression.

Treatment Therapy for bipolar disorder involves a combination of pharmacology and psychotherapy. Drugs that regulate levels of neurotransmitters such as serotonin, dopamine, and norepinephrine are most effective when coupled with psychotherapy that helps the individual recognize when their behavior is trending out of control. Lithium reduces the frequency and severity of both the manic and depressed aspects of bipolar, but how it does so is not understood, and its use may be accompanied by diarrhea, nausea, and hand tremors. Anticonvulsants such as carbamazepine (Tegretol) and valproic acid (Depakote) are also often prescribed. Drowsiness, nausea, diarrhea, tremors, and abdominal pain are common side effects of anticonvulsants.

16.1 Before You Go On

Section Summary

Depression is a serious and relatively common mood disorder in which an individual finds no joy in life, loses appetite, and struggles to engage in everyday tasks; the most severe cases are at risk for suicide. Depression results from a dysfunction of neurotransmitters, particularly serotonin, but it also involves several cerebral centers, including the PFC, hippocampus, and amygdala. Treatment involves adjusting the levels of neurotransmitters such as serotonin, typically by SE reuptake blockers. Individuals with bipolar disorder cycle through extreme elation and depression; the cause is unclear, but a disruption of a circuit that regulates emotional response seems to be at the root.

Comprehension Check

1. Depression is a serious disorder that afflicts roughly _____ of adults, but particularly _____.

- **a.** 10%; males over the age of 85
- **b.** 10%; females over the age of 80
- **c.** 17%; females and minorities
- **d.** 30%; males and females older than 80

2. Which of the following circumstances fits the *DSM-5-TR* criteria for major depressive disorder?

- **a.** an immediate response to the death of a spouse or child
- **b.** a response to chronic alcohol or other drug use
- **c.** a reaction to a life-threatening medical crisis
- **d.** feelings of sadness lasting more than 2 weeks that interfere with daily life

3. Evidence from twin studies and other sources indicate that depression is _____.

- **a.** mostly caused by environmental forces
- **b.** influenced by one's genes, but environmental stress also plays a part
- **c.** almost entirely a result of genetic predispositions
- **d.** more severe in identical twins than in fraternal twins

4. Based on our current understanding, what is the primary neural basis of depression?

- **a.** lack of serotonin only
- **b.** lack of serotonin and norepinephrine only
- **c.** dysregulation of a neural circuit involving serotonin
- **d.** hypoactivity of the PFC and the amygdala

5. What is a treatment for MDD that was developed in the past, fell out of favor, and now is often used for individuals who do not respond to antidepressants?

- **a.** tricyclics combined with SSRIs
- **b.** SSRIs combined with deep brain stimulation
- **c.** transcranial stimulation combined with ketamine
- **d.** electroconvulsive therapy over several sessions

6. Individuals experiencing bipolar disorder _____.

- **a.** are prone to engage in risky behavior during the manic phase
- **b.** experience depression but only mildly
- **c.** do not develop symptoms until their late 50s
- **d.** are treated with GABA antagonists

16.2 Anxiety Disorders

LO 16.2 Summarize the causes and treatment of anxiety disorders.

You learned in prior chapters that our nervous system has evolved to react quickly to threats. When our ancestors were faced with a mastodon or saber tooth tiger, their HPA axis signaled the sympathetic nervous system to release adrenaline and cortisol, thereby increasing heart rate and blood pressure to prepare the individual for fight or flight. We no longer find ourselves face to face with such predators, but modern life has its own threats. Seeing a person walking toward you with a weapon, hearing that your job may be eliminated, or having a physician tell you that there is something worrisome in your test results are all reasonable causes for anxiety. However, some individuals experience extreme and even debilitating anxiety when no actual threats are present, as this individual explains.

> **As long as I can remember, I've been a worrier. I worry about things that make sense like money, my job, and relationships, but I also worry about things that make no sense. When I'm out, I worry that my house might burn down, or a wild animal will suddenly appear and attack me. I worry that if I accidently copied two words from a book I've read into an assignment or exam, my professor will charge me with plagiarism, and I'll be dismissed from school and shunned by my family and friends. I can't stop worrying about everything.**

Constant worrying at a level that is not commensurate with actual threats is characteristic of an anxiety disorder.

file404/Shutterstock

FIGURE 16.7 **Anxiety** Uncontrolled anxiety can result in a debilitating sense of impending doom.

General Anxiety Disorder MCAT

Worrying about unpleasant outcomes is adaptive if the individual can respond in ways that reduce anxiety. When a response that reduces anxiety is not possible and the anxiety never decreases, that individual may be experiencing **general anxiety disorder (GAD)**. Typical topics of worry include health, crime, finances, family, relationships, and work. These are reasonable concerns for anyone, but individuals with GAD experience unremitting and irrational anxiety over these issues (Figure 16.7).

Diagnostic criteria

The *DSM-5-TR* defines GAD as follows.

- Excessive and pervasive anxiety about events or topics that lasts more than 6 months and is unwarranted by circumstances.
- The anxiety is labile; that is, if the alleged concern is removed, the anxiety shifts to another topic.
- The anxiety is accompanied by at least three of the following symptoms: fatigue, impaired concentration, restlessness, irritability, sleep difficulties (too much or too little), and muscle aches.
- The symptoms must cause significant distress or impairment to social, occupational, or other functions.

Demographics

Estimates of the lifetime incidence of GAD range between 7% and 9%. Those who develop GAD are likely to experience an anxiety disorder multiple times during their lives. Women are twice as likely to be diagnosed with GAD, and individuals of African American heritage are also more likely to experience GAD; environmental factors such as poverty and discrimination may be the cause of their overrepresentation. Individuals with GAD include singers Selena Gomez, Missy Elliott, and LeAnn Rimes; actor Kim Basinger; and comedian and television show host Stephen Colbert.

Genetic and environmental influences

Individuals who have a near relative diagnosed with GAD are more likely to develop it than those who do not, but evidence suggests that environment influences when and if that genetic predisposition leads to a disorder.

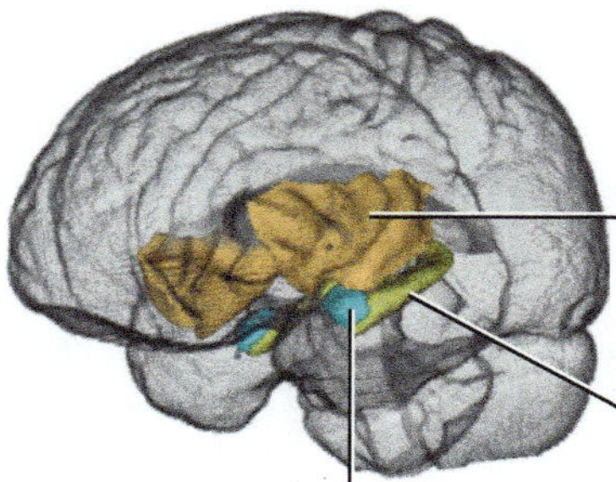

FIGURE 16.8 A circuit involved in anxiety A circuit connecting the insula, amygdala, and hippocampus may be dysfunctional in general anxiety disorder.

Neurological basis We believe that the normal connection between the amygdala and regions such as the insula and hippocampus are somehow involved in GAD (**Figure 16.8**). This circuit evaluates the importance of negative stimuli. For example, if the thought entered your mind that a tidal wave might inundate you, your amygdala would signal fear. However, the insula would note that the worry was irrational because you live in Kansas and suppress additional information that your hippocampus might retrieve about floods. If the connection between the insula, amygdala, and hippocampus is disrupted, that inhibitory mechanism is lost.

Treatment Individuals with GAD respond to a combination of cognitive behavioral therapy and SSRIs. Cognitive behavioral therapy is a form of therapy in which the individual is taught to replace negative and maladaptive thought processes with positive, adaptive ones, with the goal that behavior will follow cognition. When combined with antianxiety medication, this form of therapy can be very effective. Individuals with GAD experience anxiety in an unremitting way, but some anxiety disorders are characterized by periodic but short-lived anxiety, as is the case with panic disorder.

Panic Disorder MCAT

Whereas GAD is characterized as a generalized state of anxiety that is long lasting, **panic disorder** is characterized by brief (<10 minutes) panic attacks consisting of very intense feelings of debilitating anxiety, as described by this individual.

> **I have anxiety disorder, and panic attacks are completely different. Anxiety disorder is an irrational fear about things like money, relationships, or health that are things one can reasonably be anxious about, and you always have this low level of anxiety in the background, like a pot on simmer. Panic attacks come out of nowhere and have no relation to what you are doing right at the moment. They are like a pot boiling over, and for about 10 minutes they overwhelm you with the feeling that you are going to die.**

Panic attacks often appear at inconvenient times such as before an exam or job interview, while driving a car, or while meeting friends, leaving the individual unable to deal with the situation (**Figure 16.9**).

Diagnostic criteria The *DSM-5-TR* defines panic disorder as an unanticipated and recurring feeling of intense fear that rapidly reaches a peak and then fades after a few minutes. During the attack, four or more of these symptoms must be present.

- Pounding heart, accelerated pulse, or palpitations
- Sweating
- Trembling or shaking
- Nausea
- Shortness of breath
- Feeling light-headed, dizzy, or faint.
- Chills
- Fear of "going crazy"
- Feeling depersonalized (no longer yourself)
- Fear of dying

Photographee.eu/Shutterstock

FIGURE 16.9 Panic disorder Panic attacks can occur without warning and cause extreme and debilitating distress until they recede.

Demographics Panic disorder typically appears during late adolescence or early adulthood. Roughly 3% of adults experience panic disorder each year, and about 5% experience panic disorder sometime in their lives. Females are twice as likely (~4%) as males (2%) to experience panic disorder. Public figures who experience panic attacks include actors Johnny Depp, Kate Moss, Emma Stone, and Scarlett Johansson; and singers Adele and Fred Durst.

Genetic and environmental influences As with generalized anxiety disorder, genetics and environment interact to increase the odds of developing panic disorder. Environmental factors include experiencing chronic stress, a history of child abuse, and smoking.

Neurological basis Panic disorder is not well understood. The evidence suggests some disruption in a circuit mediated by norepinephrine that connects the amygdala to other brain regions.

Treatment As with GAD, cognitive behavioral therapy and SSRIs, SNRIs, tricyclics, and benzodiazepines are indicated in the treatment of panic disorder.

Generalized anxiety disorder and panic disorder share the characteristic that they create debilitating anxiety. The next disorder, OCD, also involves anxiety, but the focus of the anxiety is a specific compulsion such as fear of contamination, a need to count or rearrange objects, or a need to perform a behavior a specific number of times.

Obsessive–Compulsive Disorder MCAT

We have all left our home and felt the need to circle back to make sure that we locked the door or turned off the coffeemaker or the stove, but most of us are satisfied with one return to set our minds at ease. An individual experiencing **obsessive–compulsive disorder (OCD)** might be unable to put this worry aside and feel compelled to come back again and again. Consider this account from an individual with a severe case of OCD:

> **I would leave my house to drive to class, and part way there, I would become obsessed with the idea that maybe I struck someone with my car who was walking on the side of the road and they were lying in a ditch and so I would have to drive all the way back home to be sure, but then as I started again to drive to school, I'd have the same thoughts and in the end I might repeat the drive ten or twelve times. My 10-minute commute might take 90 minutes.**

Individuals with severe OCD may become hostage to their obsessions or compulsions.

Diagnostic criteria The *DSM-5-TR* defines OCD by the presence of obsessions and/or compulsions. The behaviors may not be the result of a drug or another medical condition.

Obsessions:

- Recurrent or persistent thoughts, impulses, or images that are intrusive, inappropriate, and cause anxiety; examples are fears of contamination, irrational doubts, a need for order.
- These must not be related to real life events, the individual must attempt to ignore them, and the individual must be aware that these thoughts, impulses, or images are generated in their mind.

Compulsions:

- Repetitive behaviors such as handwashing, counting steps, or repeating words that must be performed to avoid a dreaded event despite the individual's awareness that there is no reality-based relation between performing the compulsion and forestalling the event.
- Performing the compulsion must take up an hour or more each day and interfere with normal life. Some individuals with OCD will wash their hands until the skin is cracked and raw (**Figure 16.10**).

Vicki Smith/Getty Images

FIGURE 16.10 Compulsive behaviors in OCD Obsessions about contamination often lead to compulsive behaviors such as repetitive handwashing among individuals with OCD.

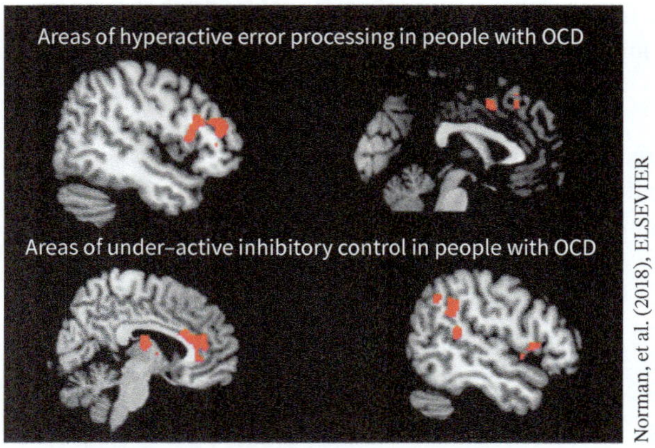

FIGURE 16.11 Cerebral circuits in OCD Behavioral initiation areas are hyperactive and inhibitory control areas are hypoactive in OCD.

Demographics OCD is relatively rare, afflicting 2%–3% of the population. Public figures who have OCD include actor Leonardo DiCaprio; musicians Justin Timberlake, Katy Perry, and Fiona Apple; and radio personality Howard Stern. Comedian Howie Mandel, who is often a judge on shows where contestants compete for prizes, cannot shake contestants' hands due to his fear of contamination.

Genetic and environmental influences There appears to be a genetic basis for OCD; concordance for identical twins is 50%–80%, and for fraternal twins, 20%–40%. If a close relative has OCD, you have a 25% chance of developing it yourself. There is some evidence that maladaptive parenting and a highly stressful environment interact with genetics to increase the odds of developing OCD.

Neurological basis Serotonin and dopamine have a role in OCD, but there also appears to be dysfunction in a circuit that links the orbitofrontal cortex (a part of the PFC), the striatum, and the thalamus. In previous discussions of homeostasis, you learned that we have feedback mechanisms that regulate our blood pressure, hunger, thirst, and satiety. An essential part of these mechanism is a controller that detects when the goal is reached and disengages the mechanism. Similarly, we have a system to disengage when we have successfully completed an action.

Evidence from imaging studies that compare cerebral activation of neurotypical and OCD individual suggest that the orbitofrontal cortex is involved in initiation of goal-directed behaviors, and the anterior cingulate cortex acts to disengage when the goal is reached (**Figure 16.11**). When the inhibitory part of the circuit is dysfunctional, the individual responds by repeating an action compulsively, for example, washing their hands more than 50 times a day, turning the locks five times, repeating certain words, or returning home over and over to make sure that the door is securely locked.

Treatment Individuals diagnosed with OCD are typically treated with a combination of therapy and medication, typically SSRIs. The therapy involves systematically desensitizing the individual to the object or situation that causes distress in a series of exposures that gradually increase the contact. In the example of an individual who fears contamination, the therapist might first have the individual imagine a soiled doorknob, then stand near a public doorknob without touching it, then touch it briefly, and so on until the individual can learn not to associate the object with revulsion or fear. You can consider how compulsions might be a maladaptive version of an adaptive behavior in **Think About It: OCD, Compulsions, and Rituals**.

Think About It | OCD, Compulsions, and Rituals

Consider for a moment the common behaviors expressed by individuals with OCD: handwashing, avoiding touching others, and checking to ensure that doors and windows are locked. These are all behaviors that spring from an innate desire to be safe. That desire to control our environment is reflected in rituals that some people adopt in an effort to control their environment, and superstitious behavior is common in athletics.

NBA star Michael Jordan always wore University of North Carolina practice shorts under his NBA team shorts; tennis star Serena Williams wore the same pair of socks throughout a tournament and would not wash them unless she lost. Major League Baseball Hall of Famer Wade Boggs had to eat chicken before every game, and college football coach Les Miles would eat a small amount of the grass on the field before each game. In each case, the athlete believes that following this ritual will enhance the odds of winning.

Although there is no connection between engaging in superstitious rituals and developing OCD, they may share an evolutionary basis. The world is a complex place where we often cannot control events, so the idea that a person can improve their odds of earning an A+ by wearing their lucky shorts provides confidence, which in itself improves performance.

Questions

1. **We are not all premier athletes, but we all have challenges in our lives where success matters. We hope to earn top grades, ace that job interview, or impress a person we find attractive. Consider your own behavior. Do you have something you do before an exam, interview, or date that you do in the belief that this will bring you success? Briefly describe the behavior, and if possible, how it developed.**
2. **Psychotherapy for OCD often involves exposing the individual to the object of their obsession, be it contamination or touching others. Why might learning that no consequences follow from violating their compulsion help to break the cycle?**

16.2 Before You Go On

Section Summary

Feeling anxious in the face of challenges or threats is appropriate, but feeling anxious in the absence of a reasonable threat is maladaptive. Individuals with general anxiety disorder experience constant, debilitating anxiety that is out of proportion to the actual threat level. Panic disorder involves incapacitating panic attacks that occur spontaneously and without a connection to a real threat. Obsessive–compulsive behaviors may have their origin in behaviors that keep us safe from contamination or predators, but the compulsion to reduce anxiety leads to maladaptive behaviors.

Comprehension Check

1. A defining characteristic of generalized anxiety disorder is anxiety _____.
   - **a.** occurring in 2- to 3-week cycles
   - **b.** over unlikely events such as developing a rare disease
   - **c.** over threats to loved ones but not to the individual themselves
   - **d.** over actual threats but out of proportion to the level of threat
2. Panic attacks differ from general anxiety disorder in that they _____.
   - **a.** never occur in individuals over the age of 60
   - **b.** are acute and transitory
   - **c.** have no hereditary basis
   - **d.** are more common among individuals with European ancestry
3. What do the compulsions and obsessions in OCD often involve?
   - **a.** wild animal attacks, contracting rare diseases, or falling into caves or sinkholes
   - **b.** the activities of people in the news
   - **c.** future career plans and hopes for relationships
   - **d.** contamination and safety of self and others

16.3 Consciousness Disorders

LO 16.3 Identify the causes and treatments of schizophrenia and other delusions of identity.

There may be nothing more astounding in nature than the ability of the human brain to create consciousness, which has allowed humans to build cities, cure diseases, talk to people on the other side of the world, and send spacecraft to explore distant planets. Sadly, sometimes the mechanism that creates a rational understanding of the world breaks down, and schizophrenia and delusions may result.

Schizophrenia MCAT

A neurotypical brain creates a representation of the world that enables a person to be successful and adaptive. For some people, however, their representation of the world does not seem to be based on reality. I believe that dogs do not speak to me about their plans for the week; that I am not Elvis, Moses, or Jesus; and that if I opened the skin on my arm, I would not see wires, gears, and electrical components, but these are all delusions I have heard from individuals I worked with early in my career.

When a person has a view of the world that is in direct contradiction with generally accepted reality and interferes with their everyday life, they may be suffering from **schizophrenia**. Schizophrenia is a term coined by German psychiatrist Eugen Bleuler from the Greek words *schizen* (to split) and *phren* (mind) to indicate a separation between perception and reality. Consider the story of this individual who was diagnosed with schizophrenia.

> **I used to be a cowboy but when the moon comes up over the hills the cows stop talking. Once I was an angel, but then I fell to earth and the doctors put me here because I lost tossed caught a fish which left the summer sunset only to see the leaves. I like leaves. I've seen the sun; it's my father.**

It is normal and even healthy to fantasize about winning the lottery, scoring the goal that brings your team the championship, or waking up to discover you are attractive and popular beyond your wildest dreams. Such fantasies can provide the motivation to pursue rational personal and professional goals. However, the account you just read by the schizophrenic individual reflects a dysfunctional perception of reality.

Diagnostic criteria The *DSM-5-TR* categorizes schizophrenic behavior as being either positive or negative, but not in the sense of good and bad. Positive symptoms are behaviors that an individual with schizophrenia *does* that neurotypical individuals do not. Negative symptoms are behaviors that someone with schizophrenia does *not* do that neurotypical individuals do. The behaviors may not be the result of a drug or another medical condition.

Positive symptoms:

- Hallucinations (auditory and/or visual)
- Delusions (typically bizarre and often persecutory)
- Disordered and disorganized thinking
- Believing that someone or something is inserting thoughts into one's head

Negative symptoms:

- Flat affect (i.e., inability to appropriately convey emotions though voice or facial expression)
- Anhedonia (being unable to experience pleasure)
- Social isolation (being unwilling to be with others)
- Lack of motivation

Perceptual disturbances in schizophrenia People with schizophrenia exhibit serious disturbances in cognitive processes. Visual disturbances that change how edges and lines are processed are thought to be a result of dysfunction of the magnocellular layers of the lateral geniculate nucleus. Perhaps the most stunning examples were produced by English artist Louis Wain (1860–1939), who was famed for his drawings of cats. Over the course of his adult life, Wain's behavior became increasingly erratic, leading him to be institutionalized several times. Figure 16.12 depicts four of the portraits of cats that he created during his life. When you compare the realistic depiction that he created early in his life to the later images, you can see the progression of visual disruption that schizophrenia sometimes causes.

Demographics of schizophrenia Schizophrenia typically develops during young adulthood (around ages 20–25) and while males and females show a similar prevalence of the disorder, female onset is typically 3 to 5 years later than males. About 20% of schizophrenics recover to lead a fruitful life. Without treatment, schizophrenia often leads to homelessness and poverty. Many self-medicate with drugs or alcohol. Individuals you may be aware of who are or who are thought to have been schizophrenic include Mary Todd Lincoln, Nobel laureate

a. Louis Wain paintings/Wikimedia Commons/Public domain

b. Louis Wain paintings/Wikimedia Commons/Public domain

c. Louis Wain paintings/Wikimedia Commons/Public domain

d. Louis Wain paintings/Wikimedia Commons/Public domain

FIGURE 16.12 **Louis Wain's art** Wain's early art was highly realistic **(a)**, but as his symptoms increased his portraits became disturbing **(b)**, then increasingly less representational **(c)**, and finally highly abstract **(d)**.

and economist John Nash, artist Vincent van Gogh, musicians Peter Green and Syd Barrett, dancer Vaslav Nijinsky, and actor/comedian Darrell Hammond.

Genetic and environmental influences Concordance rates of schizophrenia among identical twins is very high, possibly as high as 80%, whereas the rate among fraternal twins is about 40%. Because the biological basis of consciousness is quite complicated, you will not be surprised to hear that although schizophrenia runs in families, several if not many genes are involved.

Schizophrenia researchers were captivated by the case of the Genain quadruplets, four identical girls born in 1930 who all developed schizophrenia (Figure 16.13). Their story demonstrates how genetics and environment can combine to create health or illness. There was family history of mental illness and evidence that the four girls were physically abused by their father. Their care was eventually taken over by the National Institute of Mental Health (NIMH), after which they lived in obscurity while

PA Images / Alamy Stock Photo

FIGURE 16.13 **The Genain quadruplets** The Genain quadruplets (not pictured) shared a family history of mental illness and an abusive home environment. All developed schizophrenia by age 24, and all but Myra were hospitalized for schizophrenia at least once.

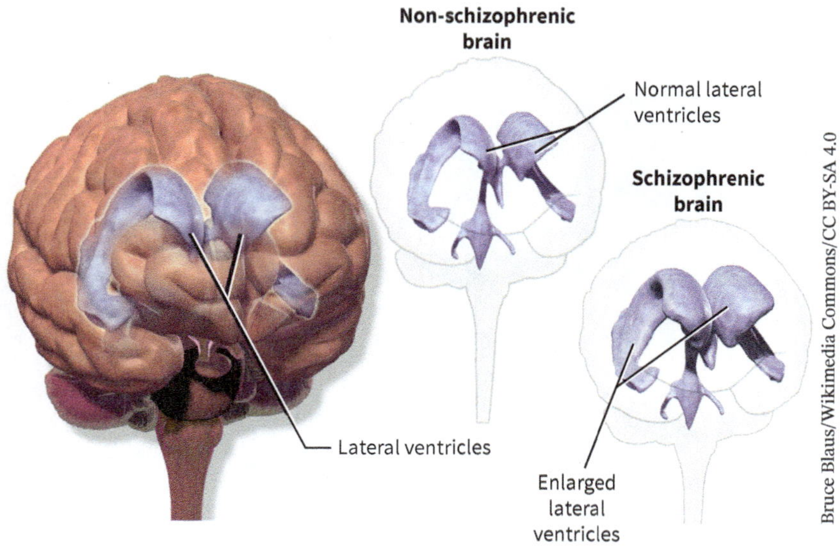

FIGURE 16.14 Ventricular enlargement in schizophrenia Schizophrenia is marked by enlargement of the ventricles at the expense of surrounding cerebral tissue.

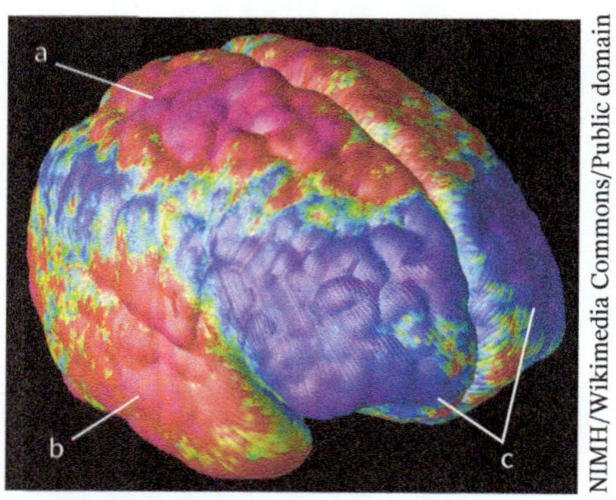

FIGURE 16.15 Gray matter loss in schizophrenia Gray matter loss in schizophrenia is most pronounced in the parietal **(a)** and temporal **(b)** lobes, as indicated in red, and to a lesser extent, the frontal lobe **(c)**, as indicated in blue.

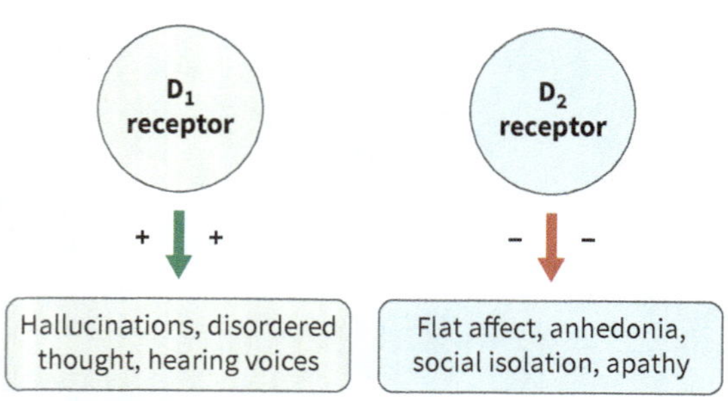

FIGURE 16.16 Dopamine receptors and schizophrenic behaviors Overactivity of D_1 receptors creates positive symptoms such as delusions, and overactivity of D_2 receptors creates negative symptoms such as flat affect and social isolation.

being studied. (Their given surname is not Genain; they were assigned this name, meaning dire birth, to protect the identity of the family, and the girls are referred to in the literature as Nora, Iris, Myra, and Hester—names beginning with the letters NIMH.) As of 2019, only Myra was still alive.

Neurological and anatomical basis of schizophrenia

To better understand the neural basis of schizophrenia, we will look at differences in structure, organization, and neurotransmitters. Perhaps the most obvious structural difference between schizophrenic and neurotypical brains is enlargement of the lateral ventricles (Figure 16.14). Ventricles are areas where there are no neurons or glia, so larger ventricles mean less cerebral tissue. It is generally believed that ventricular enlargement is an indication that cerebral abnormalities are at the root of schizophrenia, not that ventricular enlargement causes schizophrenia.

Perhaps more to the point is that schizophrenic brains show a loss of gray matter in the frontal, parietal, and temporal lobes (Figure 16.15), structures that are essential to cognitive processes such as comprehending reality, a core diagnostic feature of schizophrenia.

There are significant differences in the organization of cells in the cortex and the hippocampus. Schizophrenic brains show a disorganized structure and lower integrity of white matter connections between the frontal lobe and the rest of the brain. It appears that ventricular enlargement, less gray matter in the cortex, and lessened cerebral connectivity are all hallmarks of schizophrenia.

In addition to these structural differences, neurotransmitters are involved in schizophrenia. In the 1940s, a drug called Thorazine that was developed to treat allergies was found to have calming properties, and it was used to treat schizophrenia with encouraging results. The discovery that Thorazine blocked the action of dopamine led to the **dopamine hypothesis**, which proposed that excess dopamine causes schizophrenia. Researchers reported that patients with schizophrenia improved when given any drug or diet that decreased the action of dopamine, and psychotic behavior might result when they took drugs such as cocaine that increase dopamine in the brain.

Earlier in our discussion, you learned that schizophrenic behaviors are classified as positive (what they do but should not) and negative (what they do not do but should). Reducing the activity of the D_1 dopamine receptor leads to improvement in positive symptoms, whereas similarly reducing activity of D_2 receptors decreases negative symptoms, as illustrated in Figure 16.16.

Treatment of schizophrenia First-generation antipsychotics such as Thorazine and Haloperidol are known as typical antipsychotics. Thorazine and Haloperidol are both DA antagonists, which means they block transmission by binding to the receptor site. A second generation of antipsychotics (known as atypical antipsychotics) block DA but also 5-HT and other neurotransmitters. They are slightly more effective and generally have fewer side effects. Examples are Risperdal, Zyprexa, Abilify, and Seroquel. Although too much dopamine may lead to schizophrenia, too little has its own problems, and individuals who take antipsychotics that block DA are prone to develop movement disorders such as tardive dyskinesia, discussed later in this chapter. Other side effects of antipsychotics include dry mouth, weight gain, and fatigue. For those individuals for whom antipsychotic medication significantly reduces their schizoid thought processes, psychotherapy that provides insight into their condition may be helpful; it is not uncommon for individuals with schizophrenia to see improvement in symptoms through medication, decide to stop medication, and experience a relapse as a result.

A fascinating observation about schizophrenia is that estimates of smoking prevalence is as high as 90% among those with schizophrenia. You learned in an earlier chapter that nicotine mimics the action of acetylcholine (ACh), and you know that ACh enhances memory and attention, and so perhaps schizophrenic individuals self-medicate with nicotine in an effort to enhance their mental processes to better deal with their symptoms.

Would it be better to identify individuals at risk for schizophrenia before they develop serious symptoms? You can consider that question in this **Think About It: The Cost of Schizophrenia**.

Think About It | The Cost of Schizophrenia

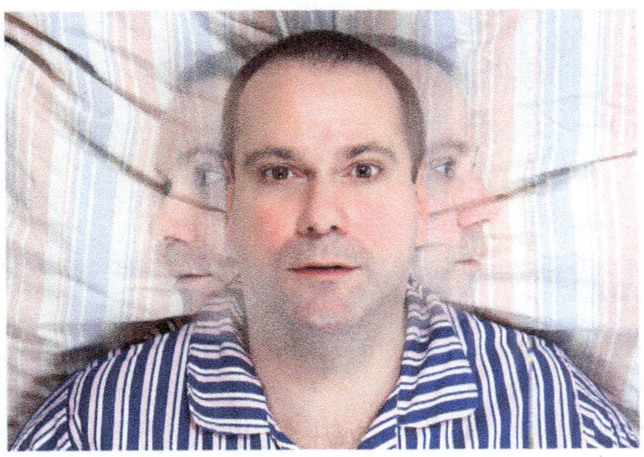
Milkovasa/Shutterstock

Schizophrenia is a disorder that appears during late adolescence or early adulthood, but without treatment it may continue for the remainder of a person's life. The majority of those with schizophrenia are not violent, but because they struggle to deal with the world rationally, they are prone to drug abuse, poverty, being the victims of crime, and homelessness.

Treating an individual with schizophrenia is very costly; in fact, the cost of their treatment is twice that of an individual with major depression and four times that of an individual with an anxiety disorder. The cost of treatment for schizophrenia in the United States is roughly $60 billion per year, and that cost does not include the intangible loss of that person's family and social life and the collateral damage to their lives.

Recently, researchers have begun to look more closely at identifying during childhood those individuals who are at risk for schizophrenia and offering treatment designed to ameliorate and even prevent the development of schizophrenia. This possibility offers hope but also raises ethical considerations.

Questions

1. **Search online using the terms "Can we predict schizophrenia?" and summarize the latest developments in this field.**
2. **On the one hand, schizophrenia is a disorder that comes with great costs to the individual, their family, and society. On the other hand, not everyone at risk for schizophrenia will develop it, so there are ethical issues in treating people who have no symptoms. Should those at risk be forcibly treated, offered treatment, or left alone? Support your argument.**

Delusions of Identity MCAT

Delusions of identity can occur with schizophrenia, but they are also associated with brain injury or dementia. Although the specific delusions discussed in this section are quite rare, they do teach us about the relationship between the brain and the mind.

The **Capgras delusion** is a disorder in which the individual believes that someone close to them such as a spouse, parent, child, or friend has been replaced by an imposter who is identical in every way to the actual individual. Capgras is found in conjunction with schizophrenia, traumatic brain injury, or dementia. Its cause is not understood, but one possibility is a disconnection between the fusiform face area (FFA), which is involved in facial recognition, and the emotional centers in the temporal lobe. Seeing a familiar face is typically accompanied by appropriate emotions; if an individual failed to have this emotional response to recognizing someone close to them, they might question whether it was really their loved one.

In contrast to the Capgras delusion, individuals with the **Fregoli delusion** believe that different people, typically strangers, are actually a single person who is constantly changing their appearance. Individuals exhibiting the Fregoli delusion often believe that this chameleon-like individual is persecuting them. Damage to the right temporal lobe and the FFA are involved. Long-term use of L-DOPA, a treatment for Parkinson's disease, has been found to be a contributing factor.

Finally, individuals with the **Cotard delusion** believe that they are dead or missing many internal organs. As with Capgras and Fregoli delusions, Cotard is comorbid with schizophrenia. The FFA is involved and possibly the amygdala. To treat all of these delusions, antidepressants and antipsychotics are prescribed.

16.3 Before You Go On

Section Summary

Schizophrenia is a relatively rare disorder of cognition, characterized by positive symptoms such as hallucinations, delusions, and disordered thought and negative symptoms such as apathy and flat emotional responses. The causes are not well understood, but heritability is high among family members. Ventricular enlargement, cortical thinning, and white matter disorganization are all characteristics of schizophrenia as is overactivity of dopamine, although other neurotransmitters are clearly involved. Reducing DA levels leads to improvement in many individuals, but sometimes at the cost of developing a movement disorder. In rare cases, bizarre delusions of identity such as Capgras, Cotard, and Fregoli delusions affect how one recognizes others and oneself.

Comprehension Check

1. The incidence of schizophrenia is roughly _____, and symptoms typically appear ______.

- **a.** 5% of men and 3% of women; after age 45
- **b.** 1% of people; before age 10
- **c.** 5% of people; around age 30
- **d.** 1% of people; between ages 18 and 25

2. An example of a positive symptom of schizophrenia is experiencing _____, and an example of a negative symptom is experiencing _____.

- **a.** elation; anger
- **b.** delusions; hallucinations
- **c.** apathy; panic
- **d.** delusions; social isolation

3. Pharmacological treatments for schizophrenia typically involve _____.

- **a.** reductions in serotonin
- **b.** dopamine antagonists
- **c.** GABA agonists coupled with serotonin antagonists
- **d.** dopamine agonists

4. One reason it is challenging to develop a pharmacological remedy for schizophrenia is that _____.

- **a.** it must be an agonist for some DA receptors and an antagonist for others
- **b.** DA receptors do not easily downregulate
- **c.** patients typically get better on their own after about 5 years
- **d.** drugs that affect dopamine also affect heart rate

5. Long-term use of antipsychotics may have the side effect of _____.

- **a.** increased heart rate that leads to arrhythmias
- **b.** liver damage that interferes with drug action
- **c.** kidney damage that causes blood poisoning
- **d.** movement disorders due to chronic dopamine depletion

6. Joaquin believes that a single person is constantly changing their appearance to follow him, whereas Mari believes that her spouse has been replaced by a very clever imposter. Joaquin has the _____ delusion, and Mari has the _____ delusion.

- **a.** Fregoli; Capgras
- **b.** Capgras; Cotard
- **c.** Cotard; Capgras
- **d.** Fregoli; Cotard

16.4 Neurological Disorders

LO 16.4 Describe the causes and effects of common neurological disorders.

Psychiatric illnesses are defined as disturbances in emotional behavior, and these often have a biological basis. However, there are other behavioral disorders that are more clearly a reflection of a disease or dysfunction of the nervous system due to infections or damage to blood vessels or the brain. We will also consider some movement disorders that have a behavioral component or reflect our focus on neurotransmitter circuits.

Bacterial and Viral Infections MCAT

Just as your heart and lungs are vulnerable to bacterial and viral infections, your brain is a target for these same agents. Although our blood–brain barrier prevents most of these pathogens from entering the brain, when they do infrequently gain entry, there can be serious consequences.

Meningitis In a prior chapter, you learned that the brain is covered by several protective layers collectively known as the meninges. This protective layer can be the target of several bacterial, viral, and fungal infections. Inflammation (identified by the suffix *-itis*) of the meninges is called **meningitis**. Symptoms of meningitis include a sore and stiff neck, drowsiness, double vision, extreme sensitivity to bright lights and loud noises, fever, nausea, and disorientation. Meningococcal meningitis (the bacterial form) is rare, with a frequency of about 1 in 100,000 individuals, but college students living in residence halls are at slightly greater risk because of living in close proximity to many others. A vaccine for meningococcal meningitis is available to prevent infection.

Inflammation causes pus to build under the meninges as well as swelling of brain tissue. Figure 16.17 illustrates the consequences of severe meningitis at autopsy; the yellow pus is clearly visible where the meninges have been retracted. Meningitis can be caused by bacterial, viral, or fungal infections, but only the viral and bacterial forms are contagious. Behavioral consequences in children may include learning difficulties. In adults, meningitis can lead to memory loss, confusion, seizures, and brain damage.

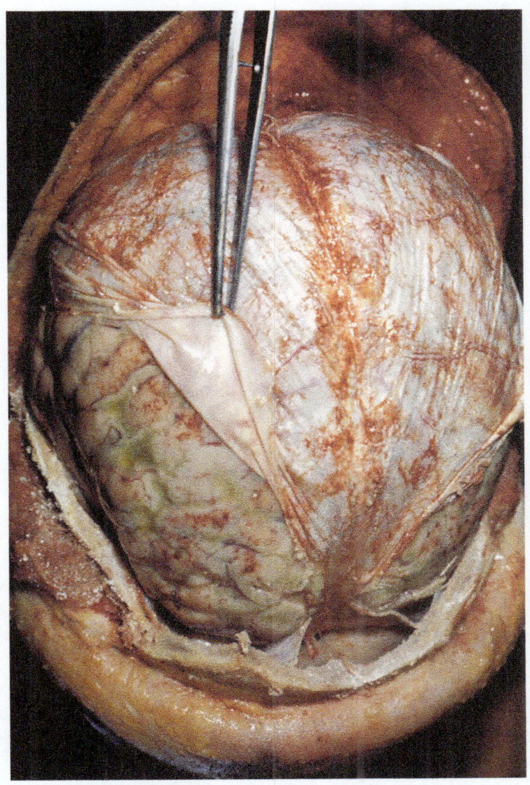

FIGURE 16.17 **Meningitis** The meninges are retracted to show yellow pus covering the cortical surface due to infection by *Streptococcus*. Infections of the meninges surrounding the brain and spinal cord can lead to serious health problems.

Encephalitis When a bacterial, viral, or other agent infects and inflames the brain itself, that condition is known as **encephalitis** (from the Greek *enkephalos* for brain and *itis* for inflammation). In a prior chapter you read about Clive Wearing, whose profound amnesia was caused by encephalitis. Encephalitis shares the same symptoms as meningitis, such as fever, confusion, drowsiness, fatigue, and even convulsions. Approximately 20,000 cases of encephalitis are diagnosed in the United States each year, but the majority are mild and quickly treated with antibiotics, antivirals, or antifungals, depending on the source of the infection. The most common cause of encephalitis is exposure to the viruses that cause rabies, herpes, and measles.

Cerebrovascular Accidents MCAT

Your brain is serviced by a complex web of blood vessels that deliver oxygen and nutrients while carrying away waste products (Figure 16.18). As a result, damage to the vascular system is a serious threat to the integrity of the brain.

If you think of your cerebrovascular system as a series of pipes or hoses, then it is easy to see that the blood supply to your brain can be interrupted by

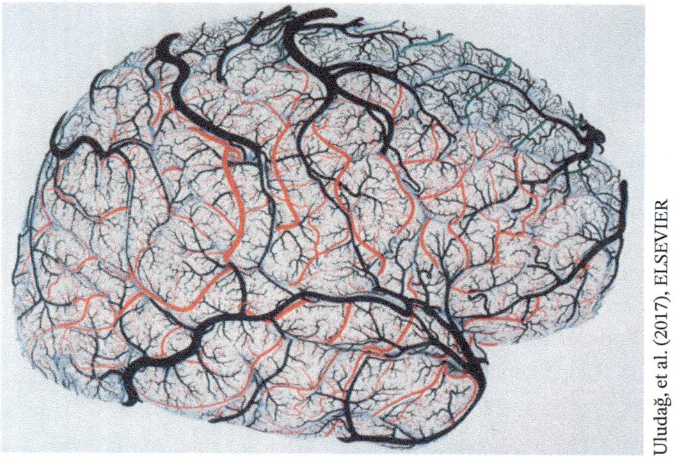

FIGURE 16.18 **Vascularization on the surface of the cortex** Your brain is supplied with blood by a dense system of large, medium, and small blood vessels that service every part of the cerebrum.

either a leak or a blockage, causing a **cerebrovascular accident (CVA)**, commonly known as a stroke. When a blood vessel wall is weak and bulges out, it is known as an **aneurysm**. A ruptured aneurysm leads to a **hemorrhagic stroke**, in which blood leaks out of the vessel, crushing surrounding brain tissue (**Figure 16.19**). Blockage of a vessel by a blood clot causes an **ischemic stroke** (pronounced *iss-key-mick*), which deprives the downstream brain tissue of blood, leading to cell death (Figure 16.19). The region of necrotic (dead) neural tissue resulting from an ischemic stroke is known as the **infarct**, and the surrounding region that is dying due to inadequate blood supply is the **penumbra**. Damage in the infarct is irreversible, so the goal of therapy in an ischemic stroke is to rescue as much of the penumbra as possible.

It has been known for some time that vascular blockage caused by an ischemic stroke leads to a decrease in the blood supply in the penumbra resulting in damage if not quickly treated; however, it was thought that the cellular damage in a hemorrhagic stroke was limited to the displaced tissue. Recent work measuring cellular activity in the region surrounding the hemorrhage suggests a decrease in activity in these areas, so the concept of a penumbra may be useful in both ischemic and hemorrhagic strokes.

The consequences of a stroke depend on the location and extent of the damage. A stroke to the left temporal hemisphere may result in language comprehension and production difficulties as well as some paralysis of the right side of the body; a right hemisphere stroke may result in left-side paralysis and difficulties with spatial tasks. A CVA in the occipital lobe will

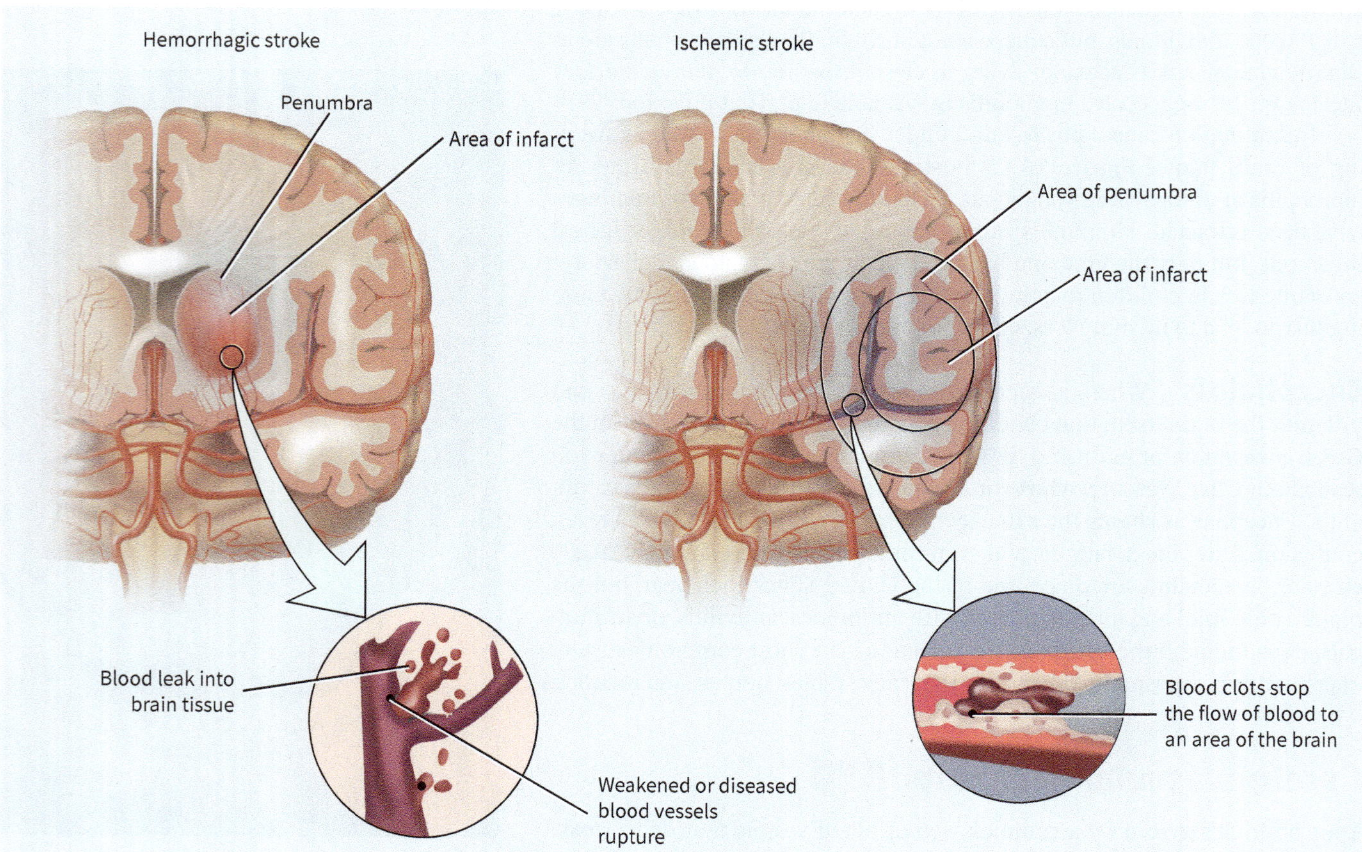

FIGURE 16.19 **Hemorrhagic and ischemic strokes** In a hemorrhagic stroke, an aneurism of a cerebral vessel fails and releases blood into the brain, which crushes nearby neural tissue. In an ischemic stroke, blockage of a blood vessel deprives the downstream regions of blood, leading to cell death.

Check It Out | The Facts About Strokes

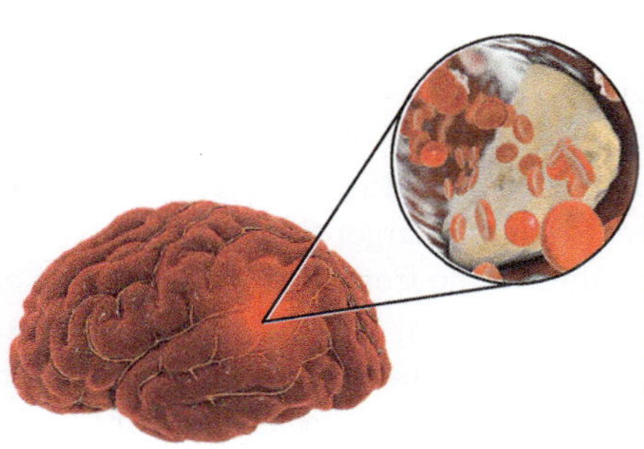

KATERYNA KON/SCIENCE PHOTO LIBRARY/Getty Images

It is very likely that you either know someone who has had a stroke or have seen a person who exhibits some of the common symptoms such as asymmetrical weakness, drooping facial muscles, or an orthotic brace that stabilizes the ankle. Watch the video **Recognize the Signs and Symptoms of a Stroke** and then answer the following questions.

1. What does the acronym F.A.S.T. stand for? Why would a drooping eye, facial muscles, slurred speech, or sudden weakness in one limb result from a stroke?
2. Now search online using the term "risk factors for stroke" and identify the biggest risk factor for a stroke and at least two others. Given this information, what behavioral changes might you suggest to someone you know who is at risk for a stroke?

likely result in blindness in some part of the visual field opposite the location of the damage, and a stroke to any portion of the brainstem may interfere with essential functions such as breathing or heartbeat.

Every year, roughly 800,000 people in the United States suffer a stroke, and of these, the majority are ischemic strokes. Given the prevalence of this disorder and its potential consequences, it is important to recognize the symptoms of a stroke and seek immediate medical help to limit brain damage. Symptoms of stroke include trouble speaking or slurring of words; weakness, numbness, or paralysis of one side of the body; blurry vision in one eye; headache; and trouble walking.

CVAs, or strokes, are a serious health issue. In **Check It Out: The Facts About Strokes** you can educate yourself about the risk factors for strokes.

Treatment The severity of a stroke is a function of the extent of the ischemia or hemorrhage, so some individuals experience only minor symptoms, whereas others experience serious and even fatal consequences. In all cases, immediate medical attention is warranted. Hemorrhagic strokes are surgically treated by placing a clip on the aneurism and draining the pooled blood, and ischemic strokes are treated by "clot-buster" drugs intended to dissolve the clot and return blood flow to the affected region. These drugs must be administered in the first few hours following the stroke to have any effect.

Traumatic Brain Injury MCAT

Until relatively recently, it was thought that any head impact that you could walk away from was not serious. Athletes were taught to shake off the confusion and get back into the game. While we have known for some time that high-contact sports such as boxing can lead to neurological damage, there is now evidence that sports once considered to be relatively safe may be more hazardous than we thought. There is also evidence that high numbers of even mild blows to the head may be as detrimental as a few severe concussions.

Traumatic brain injury (TBI) is in the news because of increased awareness of the risk to athletes, but TBI can result from vehicular accidents or falls around the home as well as wartime injuries and criminal assaults. The terms **mild traumatic brain injury (mTBI)** and concussion are often used interchangeably, and both refer to the neural and behavioral consequences of a blow to the head. Mild TBI are not rare, especially among young adults. Estimates of concussions among high school athletes run as high as 300,000 per year. Infants and toddlers are most at risk, followed by adolescents and young adults. The consequences of a TBI vary depending on the extent and location of the injury but can include persistent headaches, seizures, visual disturbances such as hemianopia, memory loss, dizziness, mood swings, aphasia, apraxia, coma, and death.

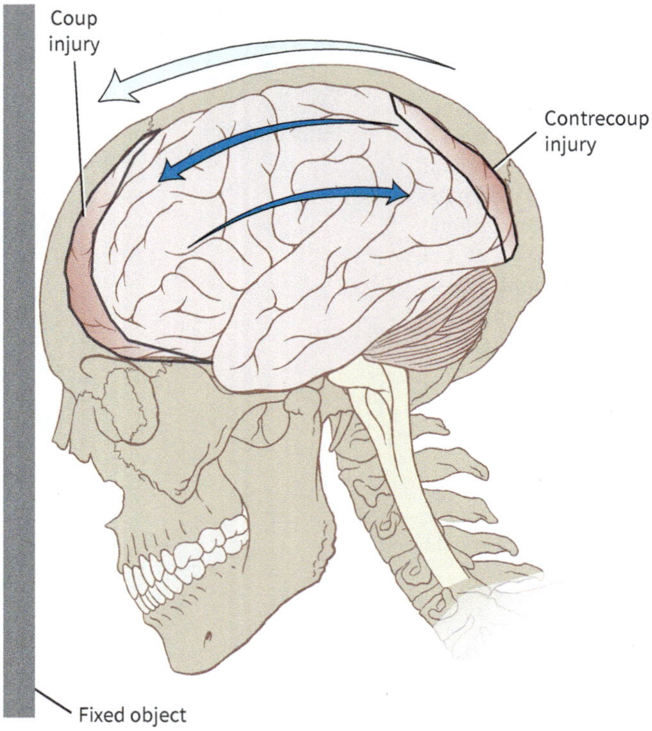

FIGURE 16.20 **Coup and contrecoup injuries** Rapid acceleration and deceleration of the brain results in two separate injuries: bruising or tissue damage at the point of impact and additional injury at a location opposite the point of impact.

Understanding what happens to the brain in a TBI is key to understanding the behavioral consequences of head injury. The most obvious cause of a TBI is the direct impact between an object and the individual's head, but that is not the only cause and, surprisingly, not always the most damaging force. Remember from a prior chapter that your brain is supported by the stalk of your brainstem and floating in cerebrospinal fluid within a container that is slightly larger than the brain. When an individual suffers a blow to the front of the head as in a fall or vehicular accident, the brain first hits the front of the skull and then rebounds backward into the rear of the skull, creating one injury in the front and a second injury at the rear (**Figure 16.20**). The initial injury is known as the coup (pronounced *coo*, from the French for hit) and the secondary injury is the contrecoup (pronounced *contra-coo*).

The famous case of Phineas Gage presented in *Stories of the Brain* is worthy of attention for several reasons besides its fame; recent critical reanalysis of the data strongly supports a reevaluation of Gage's post-injury behavior in a way that fits with our current understanding of the underlying neurology of cognition and personality. **Figure 16.21** presents Gage's skull, showing the entry and exit points for the tamping rod, the projected path of the rod through Gage's brain, and the regions disrupted by the passage of the rod.

The most important lesson from Gage's story is that damage to cerebral structures, including white matter tracts connecting the frontal region to the rest of the brain, is likely to result in intellectual and personality changes. Most of us are not at risk for a tamping rod to pass through our brain, but traumatic brain injuries are more common than you might realize.

Symptoms of TBI

Individuals experiencing a TBI show the following symptoms with signs of more severe injury at the end of the list:

- Loss of consciousness after the injury for more than a moment
- Persistent headache
- Vomiting or nausea

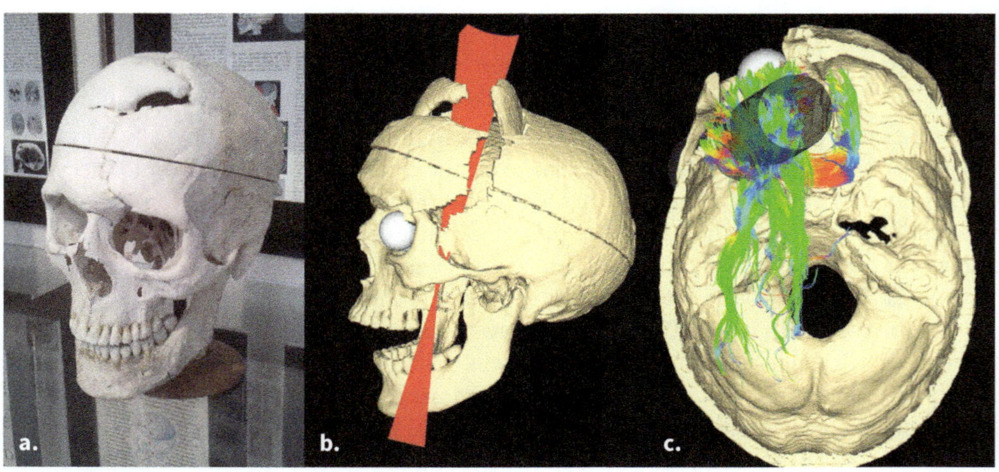

JD Van Horn, et al., (2012), PUBLIC LIBRARY OF SCIENCE (PLOS)

FIGURE 16.21 **Phineas Gage's injury** **a.** Gage's skull shows that the rod entered above his jaw and below his left eye. **b.** A digital reconstruction indicates that the rod passed through his left frontal lobe and exited through the top of his skull. **c.** There was extensive damage to white matter tracts connecting the left frontal lobe to the rest of the brain.

Try It Out | Concussions and Cognition

kdshutterman/Shutterstock

Young adults are at risk for concussions that occur on the athletic field or are the result of inattention or ingestion of alcohol or other intoxicating substances. In this activity, you can interpret data from a study investigating the effects of repeated mild TBI on the working memory of young adults. Download the worksheet **Concussions and Working Memory** and consider the topic.

- Seizures
- Dilation of one or both pupils
- Clear fluid draining from nose or ears

As you are about to discover in **Try It Out: Concussions and Cognition**, they can have serious consequences.

Dementia MCAT

Dementia literally means to lose one's memories, and because memories of the events of our lives and the meaning of words and objects around us create our reality, dementia is a disorder that first robs a person of their social life, then their ability to live independently, and finally their life.

Diagnostic criteria

The criteria for diagnosing dementia are as follows:

- Memory impairment
- At least two of these: aphasia (language disturbance), apraxia (a movement disorder), agnosia (failure to recognize object), or a disturbance of executive function
- Cognitive deficits must significantly impair social or occupational functioning when compared with the person's past.

The symptoms typically progress over 7–10 years and are subtle at first, as the following example illustrates. "Frank" was a research chemist who had earned a Ph.D. from a prestigious university and had a remarkable career in industry. In his early 60s, he became slightly forgetful, but at first no one was concerned. He retired early when he made a mistake in his laboratory that led to an explosion; luckily, no one was injured. As he passed into his 70s, he became very forgetful but also easily irritated and sometimes abusive of family members. Although a lifetime recreational athlete, he could no longer hit a golf or tennis ball or ride a bicycle, and he had to have someone with him on a neighborhood walk so he would not get lost. By his late 70s, he could no longer dress, feed, or toilet himself, and he required full-time care until his death at age 83.

The most prevalent form of dementia is **Alzheimer's disease (AD)**, which accounts for approximately 70% of all dementia diagnoses. The disorder that bears his name was identified by Alois Alzheimer, a German physician and neuropathologist. For many years, Alzheimer treated a patient named Auguste Deter who suffered from dementia. Upon her death, Alzheimer had her brain sent to him for autopsy and discovered amyloid plaques and neurofibrillary tangles that led to her dementia and eventually to her death. Public figures who developed Alzheimer's are former U.S. President Ronald Reagan; actors Charles Bronson, Charlton Heston, Rita Hayworth, and James Doohan; musicians Glenn Campbell and Aaron Copeland; and civil rights icon Rosa Parks.

Genetic and environmental factors We know that there is a genetic component to AD, but genotype confers a risk rather than a certainty of developing AD; at most, having a relative with AD raises one's risk of developing the disorder only slightly. As with many disorders, several genes have been identified as contributing to AD, with *APOE* having the greatest significance. Environmental causes such as repeated head trauma clearly increase the odds of developing dementia.

Neurological basis Alzheimer's results in the loss of semantic and episodic memory, as well as general cognitive skills, and those behavioral symptoms are reflected in the underlying neurology of AD. The hippocampus shrinks, as does the entorhinal cortex, two structures involved in declarative memory and, in the case of the entorhinal cortex, spatial cognition such as how to hold utensils or tools. Ventricular enlargement and cortical atrophy result in the loss of gray matter and white matter tracts, leading to a general loss of cognitive functions. These changes are depicted in Figure 16.22.

In addition to the damage that occurs at the structural level, there is cellular damage characteristic of AD. Two defining characteristics of AD are **beta amyloid plaques** and **neurofibrillary tangles**. Amyloid plaques are proteins that build up in the brains of individuals with AD and disrupt neural functioning. Neurofibrils are threadlike filaments in axons that provide scaffolding and are involved in transport of cellular assets along the axon. It is thought that individuals with AD develop tangles in these neurofibrils due to the inability of the brain to clear beta amyloid plaques, and these tangles then disrupt the functions of the neurofibrils (Figure 16.23).

You recall that acetylcholine (ACh) and glutamate are neurotransmitters that influence memory and attention. Individuals with AD have a loss of ACh but an excess of glutamate that may lead to overstimulation (known as excitotoxicity), which damages neurons.

Treatment At present there is no cure or even a moderately effective treatment to slow the progression of the disease. Cholinesterase inhibitors such as Donepezil (Aricept) have been employed to increase the availability of ACh at the synapse, but benefits have been small to nonexistent. Drugs such as Memantine that block glutamate receptors are often paired with a cholinesterase inhibitor, but again, their efficacy is minimal over time. Drugs that attempt to reduce the number of amyloid plaques or neurofibrillary tangles have also met with little therapeutic success, perhaps because by the time they are prescribed, processes leading to widespread damage are already in place. In the later phases of AD when the individual is often agitated, combative, or hallucinating, sedatives or antipsychotics are prescribed.

Putting aside important humanistic concerns of losing a parent, spouse, or child to Alzheimer's disease, the cost of treating Alzheimer's patients is staggering. The cost of AD in the United States alone is roughly $277 billion per year. With people living longer and health costs increasing, that number will climb. If we cannot cure AD, can we prevent it? Avoiding diseases such as hypertension and diabetes has a correlational relationship with lower incidence

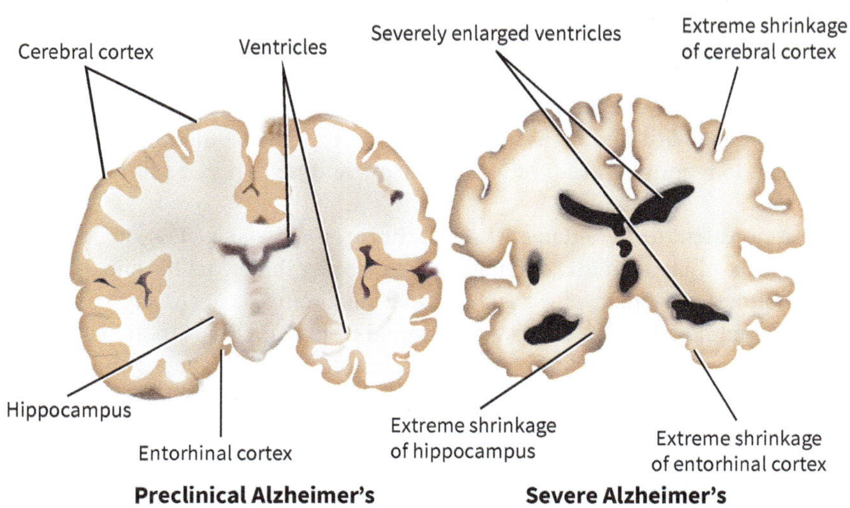

FIGURE 16.22
Neurodegeneration due to Alzheimer's disease Prior to developing AD, this individual shows normal cortical development and no loss of hippocampal and entorhinal tissue. During the late stages AD, ventricular enlargement and cortical atrophy have reduced the cerebral volume, and the hippocampus and entorhinal cortex are particularly damaged.

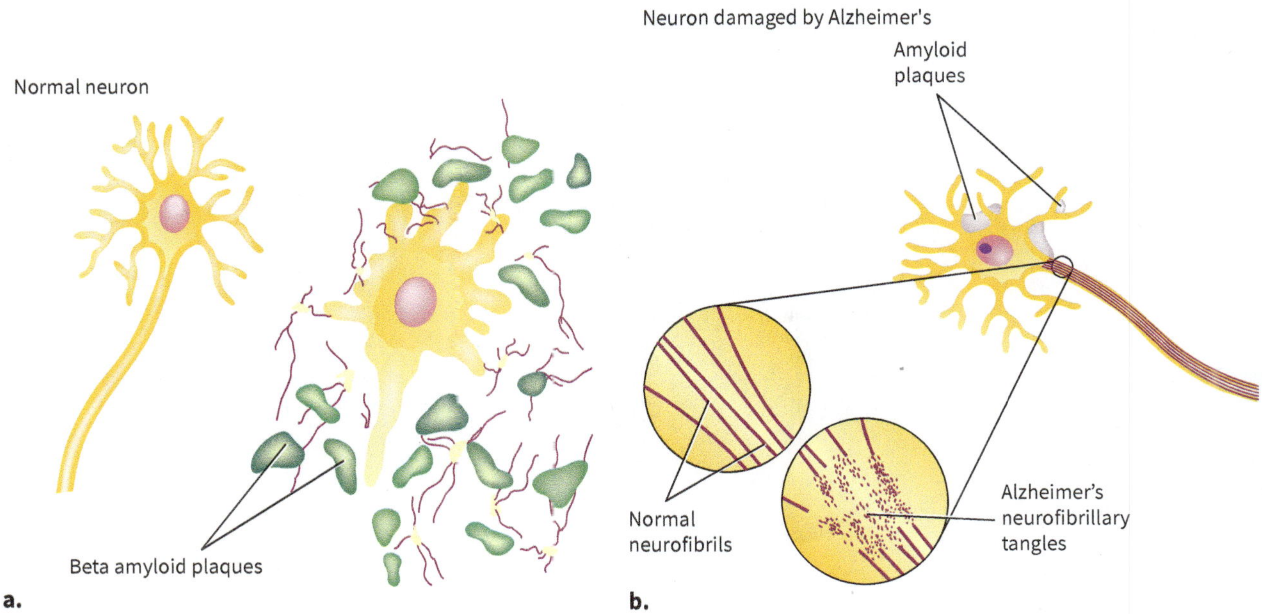

FIGURE 16.23 **Beta amyloid plaques and neurofibrillary tangles** **a.** Beta amyloid plaques build up in the brains of individuals with Alzheimer's disease. **b.** When the plaques are not cleared, neurofibril damage may result.

of AD, as does avoiding smoking and regularly engaging in moderate exercise. Finally, there is evidence that staying intellectually engaged in late adulthood through reading, taking classes, and participating in social activities all convey some protection against AD.

Movement Disorders MCAT

You should not be surprised that neurological disorders can affect every aspect of behavior, including movement. Here we briefly discuss three of the more common movement disorders: Parkinson's disease, tardive dyskinesia, and Tourette's syndrome.

Parkinson's disease You may remember from an earlier chapter that Parkinson's disease (PD) is a degenerative disorder of the CNS. PD is a significant health problem that affects 1 million people in the United States and about 7 million worldwide. PD is a progressive disease that typically takes a decade or more to advance from diagnosis to a bedridden state. In sufficiently advanced cases, death may occur from an inability to swallow.

Reference to patients with PD-like symptoms can be found as far back as several thousand years ago, but the definitive description of the constellation of symptoms is attributed to the English physician James Parkinson, whose name became associated with the disorder. People with Parkinson's include boxer Muhammad Ali, NBA star Brian Grant, actor Michael J. Fox, and musician Johnny Cash.

The primary symptoms of PD involve at least two of the following: a tremor or shaking of the limbs, bodily rigidity and gait problems, muscle weakness, and bradykinesia (slow movement). PD typically begins with a resting tremor of one limb that disappears with voluntary movement. Over time, the tremor spreads to other limbs and is joined by a characteristic shuffling gait. PD individuals often demonstrate pill rolling, where the index finger and thumb move as if they are rolling a pill back and forth. Other common symptoms include a masklike expression, a forward-leaning gait that often leads to a fall, rigidity of muscles, and cognitive deficits including memory, attention, and executive dysfunction.

There is a slight but significant genetic influence on Parkinson's as about 15% of individuals with PD have a close family member also afflicted. The cause of most cases of PD is unknown, although there is a link between PD and chronic exposure to insecticides. Head injuries are also linked to the subsequent development of PD, and in fact, there is a specific form of PD known as dementia pugilistica that may be seen in boxers and other athletes who

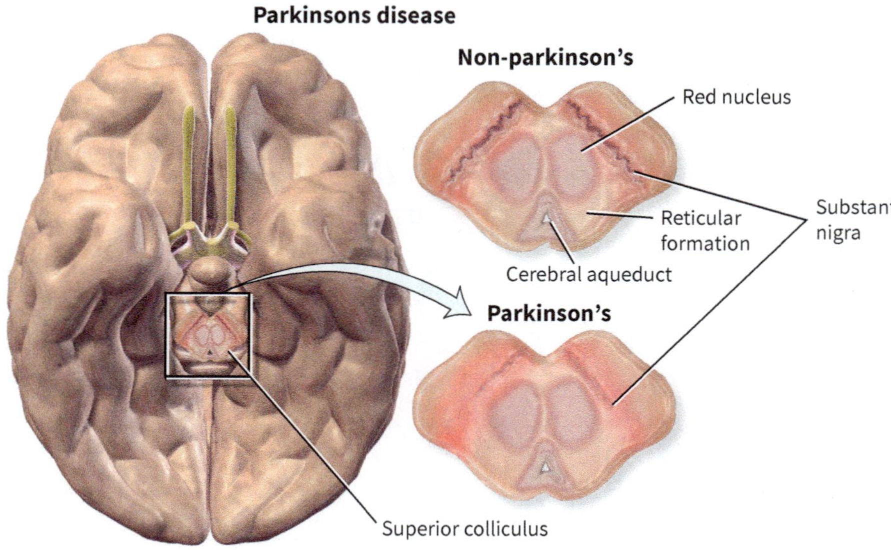

FIGURE 16.24 Parkinson's disease Comparison with a healthy brain demonstrates the destruction of DA cells in the substantia nigra due to Parkinson's disease.

suffer repeated TBI. Interestingly, there is a negative correlation between smoking and PD. Students (and especially your professors) will be glad to hear that caffeine consumption also appears to have a prophylactic (disease-preventing) relationship with PD that is dose related; that is, heavier caffeine consumers are less likely to develop PD.

The factors that lead to PD are unclear, but we do know a great deal about its neurological basis. You learned in an earlier chapter that a set of structures known as the basal ganglia control voluntary movements and play a role in cognition and emotion. The basal ganglia exert an inhibitory effect on motor movements that allows you to be at rest when it is right to do so. As you decide to reach for your pen, coffee cup, or this textbook, your dopamine (DA) system activates and the inhibitory brake is released.

The cause of Parkinson's is a loss of dopamine-producing cells in the substantia nigra, an important part of the basal ganglia (**Figure 16.24**). As DA levels drop in this structure, the effort needed to "release the brake" on voluntary movement increases to the point where initiating any movement is difficult.

You might assume that because Parkinson's symptoms are caused by a lack of DA, the disease could be treated by simply replacing the missing neurotransmitter. Sadly, this is not the answer. The DA molecule is too large to pass the blood–brain barrier, so DA cannot be provided through a pill or injection. Although the whole molecule cannot pass the barrier, however, it is possible to pass the components of DA (called precursors) into the brain with the hope they will then be assembled into DA inside the brain. Providing your brain with dopamine precursors is a bit like taking apart a large piece of furniture to get it through a doorway and into a room where it can then be reassembled. On this principle, the most effective treatment for PD currently is levodopa (L-DOPA), a precursor of DA that may allow creation of sufficient DA within the basal ganglia to ameliorate the symptoms of PD. Other effective treatments are the use of DA agonists that mimic the effects of DA and MAO inhibitors, which inhibit the breakdown of DA at the receptor site.

Why is PD so difficult to treat? The answer is a model for understanding the complexity of devising treatments for neurological and psychiatric disorders. One of the first things you learned in this text was that your brain is designed to keep you in homeostasis with the world; you react to thickening blood by experiencing thirst, you experience hunger when blood glucose levels drop, and your muscles shiver to heat them up when you are cold. Similarly, as your DA-producing cells in the substantia nigra die and L-DOPA is administered, the remaining DA receptors become ultrasensitive to DA in an effort to maintain homeostasis. This counteracts the therapeutic effects of treatment, bringing a return of motor symptoms such as tremors as well as executive function disorders.

Tardive dyskinesia A constant thread running through this text is that of homeostasis and your brain's need to keep everything in balance. Some psychiatric disorders such as psychoses result from dopamine dysregulation, so some antipsychotic drugs block DA reuptake to increase DA levels at the synaptic level. The problem arises when these drugs are used for long periods of time to manage the debilitating symptoms of psychoses. Individuals treated with typical antipsychotics may develop **tardive dyskinesia (TD)**, which causes lip smacking, grimacing, repetitive eye blinking, and involuntary movements of the torso and limbs. The cause is thought to be the development of D_2 receptor supersensitivity in the DA nigrostriatal pathway, which mediates movement. Support for the involvement of DA in tardive dyskinesia comes from studies that show a relationship between antipsychotic drug dose and TD symptoms as well as studies using D_2 agonists and antagonists.

One difficulty in treating psychoses is that the treatment often ameliorates one set of symptoms (the psychotic behavior), but the price of that treatment is another set of problems (such as TD). To avoid this, scientists and medical researchers developed atypical antipsychotics, which work in a different way and so may provide some relief to individuals experiencing psychoses without placing them at risk for movement disorders.

Tourette's syndrome One of the most interesting neurological disorders is Tourette's syndrome, which is characterized by motor and verbal tics. Once considered a rare condition that emerges in childhood, Tourette's is now thought to be more common, with many individuals incorporating the typical symptoms (coughing, eye blinking, facial expressions, and so on) into normal actions so that it may not be diagnosed. In rare cases, the individual is compelled to utter obscene or insulting comments known as **coprolalia** to those around them.

Research shows that Tourette's has a genetic component, but expressing Tourette's does not necessarily mean that your offspring will also express symptoms; its full cause is still unknown. People who have Tourette's report that the compulsion to tic is like the itch of a mosquito bite that builds until expressing the tic releases the pressure. Tourette's and OCD are likely to occur together, which suggests they may be related at a neural level. It does appear that the basal ganglia are involved in Tourette's. It is noteworthy that some athletes with Tourette's believe that their ability to respond rapidly is actually enhanced by their Tourette's. People with Tourette's include singer Billie Eilish, athletes David Beckham and Timothy Howard, and actor Dan Aykroyd.

Treatment that blocks or lessens the availability of dopamine are given in the most severe cases, but as you know, there are side effects to long-term dopamine reduction therapy. Therefore, mild cases are often treated by teaching the individual to manage their behavioral symptoms. Many children outgrow their tics as they enter late teens or early adulthood.

It is likely that you have been in contact with someone who has Tourette's and thought them antisocial or intentionally inappropriate without understanding that their tics or vocalizations were involuntary. **Check It Out: Understanding Tourette's Syndrome** will help you better understand these unfamiliar behaviors.

Check It Out | Understanding Tourette's Syndrome

Shutterstock / Rob Hyrons

Individuals with Tourette's may struggle to contain their tics. To the uninformed, they may appear to be intentionally disruptive. Recognizing Tourette's signs will help you understand the behavior of individuals with this malady. Watch the video **Understanding Tourette's**, and answer these questions.

1. Stress, anxiety, and boredom exacerbate tics. How might you help a child deal with their tics in a way that would minimize their Tourette's behavior?
2. Most people grow out of Tourette's or learn to minimize their tics, but a very small minority develop severe motor tics such as blinking, eye rolling, jumping, grimacing, touching of objects, or verbal tics such as grunting, tongue clicking, saying or repeating random words or phrases, or swearing. How might Tourette's affect a person's social or professional life?

16.4 Before You Go On

Section Summary

Neurological disorders have many causes. Bacterial infections such as meningitis or encephalitis damage neural tissue. The vascular system that supplies blood to the brain can be compromised by hemorrhage or blockage, leading to a stroke with serious consequences. Damage at the cellular level can create dementia, while damage to the structures and circuits that control movement can result in unwanted movements such as tics or an inability to initiate or control movements. Understanding the causes of such disorders will help us to create effective treatments and ideally way to prevent the occurrence of such disorders.

Comprehension Check

1. Jaxon has a headache with nausea, a stiff neck, sensitivity to bright lights, and confusion. What might he be suffering from?
   - a. a mild TBI
   - b. Tourette's syndrome
   - c. a stroke
   - d. meningitis or encephalitis
2. One reason it is so important to treat stroke quickly is that without treatment _____.
   - a. more penumbra will become infarction
   - b. infarction will become penumbra
   - c. aneurysms become penumbra
   - d. aneurysms become infarction
3. Gabriela falls backward and hits her head hard on a concrete floor. Where would you expect her to have bruising?
   - a. on the back of her brain
   - b. on the back of her brain and the temporal lobes
   - c. on the back of her brain and at a point directly opposite in the front
   - d. around the motor strip
4. Neurologically, what happens to the brain in Alzheimer's disease?
   - a. The pons atrophies.
   - b. Brain tissue swells.
   - c. Cerebrospinal fluid collects in the meninges.
   - d. Neurons die.
5. What is one characteristic common to many movement disorders?
   - a. loss of tissue in the frontal lobes
   - b. dysregulation of dopamine
   - c. cerebellar swelling
   - d. a recessive gene inherited from a parent

Conclusion

Phineas Gage's experience tells us a great deal about the human brain. He was grievously injured and yet survived. Although still capable of taking care of himself, damage to his brain led to substantial loss of his ability to regulate his mood, emotions, and mental faculties. The human brain is a remarkable device; it coordinates the action of over 600 muscles and 206 bones, regulates our internal environment, and creates the experience of consciousness. Sadly, sometimes the mechanisms we rely on are damaged, leading to disorders of mood, anxiety, consciousness, or movement that range from inconvenient to debilitating or even fatal.

Memory, emotions, and your ability to understand the world involve complex and interconnected process, so attempts to understand the basis of these disorders and then create meaningful remedies is an extremely difficult task. As researchers gain a clearer understanding of these mechanisms, our ability to repair them will improve, and this is a challenge for all behavioral neuroscience.

Critical Thinking Questions

1. Has what you have learned in this chapter about depression changed your perspective on the disorder in any way? If you could give advice to a friend who is experiencing depression, what would you say?
2. Modern life is stressful; we may struggle to deal with pressures from school, work, relationships and more. At some point, we all have felt that life is spiraling out of control, and we experience apprehension or dread about events, become irritable, and feel restless. Search online for "Things you can do to reduce anxiety" and list three that you can use in your own life.
3. Although it is sometimes said that madness and genius are two sides of the same coin, the evidence that there is a correlation between schizophrenia and genius IQ is controversial. Search online for the answer to this question and write a short statement that represents your opinion on the issue. Finally, if you could take a pill that would make you a genius but put you at risk for schizophrenia, what would you do?

4. There is a great deal of recent interest in the long-term effects of repeated head injuries that may occur during athletic activities. Search online using the terms "concussion and sports" and use the evidence you find to make an argument for or against sports participation based on the relative risks of mTBI compared to the obvious benefits of fitness and camaraderie.

5. Choose two areas that you would like to see the government invest in (for example, forgiveness of student loans, free college, national health care, infrastructure repair) and search to find an estimate of the cost of addressing those issues. How would they compare with the cost of servicing individuals with Alzheimer's disease?

Dig Deeper into Behavioral Neuroscience

16.1 Mood Disorders

Depression is a common mood disorder and has an association with suicidal thoughts and behavior. You can learn more about depression from the National Institute of Mental Health.

https://www.nimh.nih.gov/health/topics/depression

Bipolar disorder is very disruptive to life. The National Institute of Mental Health provides a clearinghouse for learning more about the disorder and current treatments.

https://www.nimh.nih.gov/health/topics/bipolar-disorder

You can read more about genetic and environmental factors in depression here.

Nabeshima, T., & Kim, H-C. (2013). Involvement of genetic and environmental factors in the onset of depression. *Experimental Neurobiology*, *22*(4), 235–243.

16.2 Anxiety Disorders

The National Institute of Mental Health provides a site for information about general anxiety disorder.

https://www.nimh.nih.gov/health/statistics/generalized-anxiety-disorder

Panic disorder can make everyday life difficult because you never know when an attack may occur. The National Institute of Mental Health is a clearinghouse for information about this disorder.

https://www.nimh.nih.gov/health/statistics/panic-disorder

16.3 Consciousness Disorders

Few diseases have the impact of schizophrenia. The National Institute of Mental Health provides information about advances and often about clinical studies that are available.

https://www.nimh.nih.gov/health/topics/schizophrenia

The American Psychiatric Association provides professionals and patients with current information about advances in schizophrenia treatment. Their website is www.psychiatry.org.

Capgras syndrome is one of the most unusual neurological disorders. Here is one source.

Jain, S. B., & Wadhwa, R. (2022). *Capgras Syndrome.*
https://www.ncbi.nlm.nih.gov/books/NBK570557/

The following article provides additional information about Cotard delusion.

Sahoo, A., & Josephs, K. A. (2017). A neuropsychiatric analysis of the Cotard delusion.
https://doi.org/10.1176/appi.neuropsych.17010018

To learn more about the Fregoli delusion, this review is helpful.

Langdon, R., Connaughton, E., & Coltheart, M. (2014). The Fregoli delusion: A disorder of person identification and tracking. *Topics in Cognitive Science*, *6*(4), 615–631.
https://doi.org/10.1111/tops.12108

The book *Hidden Valley Road* is an account of a family with 12 children, six of whom were diagnosed with schizophrenia.

Kolker, R. (2020). *Hidden Valley Road: Inside the Mind of an American Family.* Doubleday. ISBN-13: 978-0385543767

V. S. Ramachandran's highly readable book discusses several examples of neurological disorders of consciousness.

Ramachandran, V. S. & Blakeslee, S. (1998). *Phantoms of the Brain: Probing the Mysteries of the Human Mind.* Mariner. ISBN: 978-0688172176.

16.4 Neurological Disorders

The Encephalitis Society is a clearinghouse for information about research and treatment.

https://www.encephalitis.info/

The Mayo clinic provides a wide range of information and services regarding meningitis.

https://www.mayoclinic.org/diseases-conditions/meningitis/symptoms-causes/syc-20350508

The CDC provides information about the causes, symptoms, and treatment of stroke.

https://www.cdc.gov/stroke/index.htm

This readable book by a neuroscientist who suffered a stroke is worth a read.

Taylor, J. B. (2009). *My Stroke of Insight: A Brain Scientist's Personal Journey.* Penguin. ISBN: 978-0452295544.

The Alzheimer's Organization provides current information that is directed toward families of individuals with AD.

https://www.alz.org/alzheimers-dementia/what-is-alzheimers

The CDC is a clearinghouse for current information about TBI. https://www.cdc.gov/traumaticbraininjury/

Every student of behavioral neuroscience should be familiar with the case of Phineas Gage and the recent work using modern neuroimaging techniques to better understand his injury.

Harlow, J. M. (1868). Recovery from the passage of an iron bar through the head. Originally published in *Publications of the Massachusetts Medical Society* 2, 327–347. This document is available on the internet.

Damasio, H., Garbowski, T., Frank, R., Galaburda, A. M., & Damasio, A. R. (1994). The return of Phineas Gage: Clues about the brain from the skull of a famous patient. *Science 264*(5192), 1102–1105. https://doi.org/10.1126/science.8178168

Van Horn, J. D., Irimia, A., Torgerson, C. M., Chambers, M. C., Kikinis, R., & Toga, A. W. (2012). Mapping connectivity damage in the case of Phineas Gage. *PLoS ONE 7*(5), e37454. https://doi.org/10.1371/journal.pone.0037454

The Parkinson's Foundation provides information about PD and guidance toward support groups.

https://www.parkinson.org/

The NIH is a source of current information about Parkinson's.

https://www.nia.nih.gov/health/parkinsons-disease

The National Institute of Neurological Disorders and Stroke provides information about tardive dyskinesia that is appropriate for professionals, patients, and their families.

https://www.ninds.nih.gov/health-information/disorders/tardive-dyskinesia

The CDC is a clearinghouse for information about Tourette's syndrome.

https://www.cdc.gov/ncbddd/tourette/facts.html

Families of individuals with Tourette's syndrome and other tic disorders will find this organization very helpful.

https://tourette.org/

Chapter 16 Before You Go On Answers

16.1 Before You Go On Answers

1. c. 17%; females and minorities
2. d. feelings of sadness lasting more than 2 weeks that interfere with daily life
3. b. influenced by one's genes, but environmental stress also plays a part
4. c. dysregulation of a neural circuit involving serotonin
5. d. electroconvulsive therapy over several sessions
6. a. are prone to engage in risky behavior during the manic phase

16.2 Before You Go On Answers

1. d. over actual threats but out of proportion to the level of threat
2. b. are acute and transitory
3. d. contamination and safety of self and others

16.3 Before You Go On Answers

1. c. 5% of people; around age 30
2. d. delusions; social isolation
3. b. dopamine antagonists
4. a. it must be an agonist for some DA receptors and an antagonist for others
5. d. movement disorders due to chronic dopamine depletion
6. a. Fregoli; Capgras

16.4 Before You Go On Answers

1. d. meningitis or encephalitis
2. a. more penumbra will become infarction
3. c. on the back of her brain and at a point directly opposite in the front
4. d. Neurons die.
5. b. dysregulation of dopamine

Chapter 16 Glossary

Alzheimer's disease (AD) A progressive neurological disorder that damages memory and essential neural control of bodily functions.

aneurysm A bulge in a blood vessel that is the result of a weakened blood vessel wall.

anhedonia The inability to experience pleasure.

beta amyloid plaques Hard clumps of protein that clump together between neurons in Alzheimer's patients.

bipolar disorder A mood disorder characterized by periods of depression alternating with periods of abnormal elation.

Capgras delusion A disorder in which the person believes an identical duplicate has replaced someone significant to them.

cerebrovascular accident (CVA) A loss of blood supply to the brain that causes damage; also known as stroke.

cognitive behavioral therapy (CBT) A type of psychotherapy in which the individual is taught to recognize negative thought and behavior patterns and replace them with positive alternatives.

coprolalia Involuntary and repetitive use of obscene language as a symptom of a psychiatric or organic neurological disorder.

Cotard delusion A disorder in which the individual believes that they are dead.

deep brain stimulation A surgical procedure in which an electrode is implanted in a specific brain area to stimulate that area and thereby reduce psychiatric or neurological symptoms.

dementia A disorder that involves the loss of memory.

dopamine hypothesis The hypothesis that schizophrenia is caused by an excess of dopamine.

Diagnostic and Statistical Manual, 5th Edition (DSM-5-TR) A classification system for mental disorders.

Electroconvulsive therapy (ECT) Involves passing a small electrical current through the brain to cause a seizure.

encephalitis An infection of the brain.

Fregoli delusion A disorder in which the person mistakes strangers for a familiar person in disguise.

general anxiety disorder (GAD) Debilitating and persistent worrying or anxiety that is out of proportion to the risk.

hemorrhagic stroke A cerebrovascular accident caused by blood leaking into the brain and crushing nearby neural tissue.

infarct Tissue death or necrosis due to inadequate blood supply to the affected area caused by artery blockage, rupture, mechanical compression, or vasoconstriction.

ischemic stroke A cerebrovascular accident caused by inadequate blood supply to part of the brain due to blocked vessels.

major depressive disorder (MDD) A mood disorder characterized by persistent feelings of sadness and apathy.

meningitis An inflammation of the meninges.

mild traumatic brain injury (mTBI) A traumatically induced physiological disruption of brain function; also known as a concussion.

mood disorder One of a class of mental illnesses that primarily affect a person's emotional state.

neurofibrillary tangle The abnormal accumulation of the tau protein inside a neuron that is associated with Alzheimer's disease.

obsessive–compulsive disorder (OCD) A mental health disorder characterized by obsessions and compulsions.

panic disorder An anxiety disorder characterized by sudden and intense fear accompanied by chest pain, heart palpitations, and shortness of breath.

Parkinson's disease (PD) A neurological disorder characterized by uncontrollable movements, stiffness, and balance difficulties.

penumbra The hypoperfused region surrounding irreversibly damaged brain tissue where cell death slowly occurs.

postpartum depression A type of depression that may follow giving birth.

psychiatric disorder An illness characterized by disruptive changes in mood, emotions, or thinking.

repetitive transcranial magnetic stimulation (rTMS) A form of therapy in which a magnetic field is used to selectively disrupt neural functions associated with a disorder.

schizophrenia A chronic and severe mental disorder that leads to faulty sensory perception, withdrawal from reality, inappropriate behavior, and an inappropriate perception of reality.

tardive dyskinesia (TD) A neurological disorder characterized by involuntary movements of the face and jaw.

Tourette's syndrome A neurological disorder characterized by sudden, repetitive, and unintended movements or vocalizations.

traumatic brain injury (TBI) An injury that affects how the brain works.

CHAPTER 17

Social Neuroscience

Radachynskyi / iStock / Getty Images

Our brain is wired to support complex social interactions.

CHAPTER OUTLINE

17.1 Emotions

- Defining Emotion
- The Evolutionary Basis of Emotions

  Advances in Behavioral Neuroscience: Leda Cosmides and Evolutionary Psychology
- Theories of Emotions
- The Neural Basis of Emotions

17.2 Communicating Internal States

- Facial Expressions

  Try It Out: Cultural Differences in Emotional Expressions
- Social and Duchenne Smiles

  Check It Out: What's in a Smile
- Deception as an Adaptive Strategy

  Check It Out: Can You Tell the Truth from a Lie?
- Theory of Mind

17.3 Beauty and Attraction

- External Tells for Health and Reproductive Fitness
- Ovulation and Attractiveness
- Hormonal Influences on Pair Bonding

17.4 Human Social Networks

- Evolutionary Advantages of Group Membership
- Signaling Group Membership
- Ingroups and Outgroups
- Prejudice and Stereotypes

  Think About It: The Implicit Association Test

17.5 Autism Spectrum Disorder and Emotional Blindness

- Autism Spectrum Disorder
- Difficulties in Processing Emotional Cues

  Check It Out: Lindsay's Social World
- The Neurology of Asperger's Syndrome

LEARNING OBJECTIVES

LO 17.1 Describe the adaptive value of emotions.

LO 17.2 Discuss how humans use facial expressions to communicate with and influence others.

LO 17.3 Summarize the ways that humans signal health and reproductive fitness.

LO 17.4 Review the evidence for an evolutionary preference for ingroups and its negative consequences.

LO 17.5 Discuss the neural and behavioral aspects of Autism Spectrum Disorder.

Stories of the Brain

I first met Lindsay when she enrolled in my introductory psychology class. She sat in the front row, quietly taking notes. One morning while preparing for class, I found her standing in my office doorway. Lindsay told me that she had something important that she wanted to share with me, and we chatted for a few minutes until she became comfortable. I noticed her eyes seemed to sweep over every detail in the room yet somehow never met mine for more than a fleeting moment. Eventually Lindsay told me that she was mildly autistic and, as such, had considerable difficulty understanding social situations that other students her age handled effortlessly. She often misread social situations because she had trouble interpreting facial expressions and postural cues that others use to signal intent, emotions, or interest.

Despite her difficulty navigating a confusing world of social messages, Lindsay was a remarkably talented individual. She was an excellent short story writer, and her poems were carefully crafted pieces that were beautiful to behold. She tried her hand at writing a play, enrolled in drama classes, and acted in campus productions. Lindsay was meticulous in her attention to detail, rarely missed an assignment or deadline, and was always prepared for class; as a result, she was a straight-A student.

Most interesting was that she recognized her challenges and worked hard to remedy them in ways that you might not imagine. For instance, she ran for student government positions and volunteered for all manner of campus activities that would bring her into contact with others. Her considerable intellect allowed her to identify situations where she was at a disadvantage, and then she used every tool at her disposal to develop socially and lead a full life. At graduation, she was both anxious and eager to face the world.

Understanding the biological basis of social behaviors is the focus of social neuroscience, and research into the neural basis of social behavior is a fertile field of investigation that enriches our understanding of how our brain creates the complexity of the human experience. In this chapter, we will consider selected topics in social neuroscience that may shed light on how humans have become so successful as a species. We will begin by discussing emotions and their biological basis, how we use facial expressions to communicate with others, and the ways that humans use these expressions to build trust or to deceive others. We will briefly consider how faces and other physical attributes that we consider to be signs of beauty may actually be signals of reproductive fitness. We will ask why people may be prejudiced against those they do not know and have never met, and then conclude our discussion by asking what it would be like to be unable to comprehend social signals.

17.1 Emotions

LO 17.1 **Describe** the adaptive value of emotions.

As you go through your day, you experience numerous emotions. Perhaps you were happy to drink your morning coffee or tea, eager to make it to class on time, sad to see that a friend was not in their usual seat near you, and anxious to hear the results of the recent quiz in a demanding course. We navigate the paths of our lives through a sea of complex and often conflicting emotions, but what are emotions, how do we experience them, and what advantages do they confer?

Defining Emotion

Dr. Joseph LeDoux is a world-class researcher who has spent a lifetime studying the neural systems that underlie emotion. As he puts it, “One of the most significant things ever said about emotion is that everyone knows what it is until they are asked to define it” (LeDoux, 1996). We may find his words to be true, but we can begin by saying that an **emotion** is an internal or mental state created by the nervous system through the action of neural structures, neurotransmitters, and even hormones.

FIGURE 17.1 Emotion-producing situations People vary greatly in their emotional reaction to situations such as jumping out of an airplane or speaking in front of a large group of individuals.

Beyond that ambiguous description, there is no clear scientific definition of the internal state we call an emotion. Emotions guide our behavior; we gravitate toward events and people we enjoy and avoid people and places that vex us. Emotions also regulate our level of activation by exciting us through anticipation of future events and causing us to become lethargic when we are sad or depressed. Emotions are often unique to the individual; one person's reaction to jumping out of an airplane or delivering a talk in front of a large group might be delight, whereas another person would find those same activities extremely anxiety producing (**Figure 17.1**). Our emotions are a complex cocktail of internal physiological states and responses to the world as we experience it.

Why do we have emotions? Many members of the animal kingdom such as snakes and alligators apparently get along quite well without them, and dispensing with feelings of sadness, remorse, frustration, inadequacy, guilt, and loneliness might sound preferable to many. To answer this question, we must first consider the evolutionary basis of emotions and then the neurological mechanisms that support them.

The Evolutionary Basis of Emotions MCAT

To understand emotions, it is not enough to ask what they are; we must also ask how humans came to have them. The course of evolution often involves taking existing structures or behaviors and repurposing them to create new and more complex forms. Emotions may have evolved to motivate us; joy, disappointment, anger, and disgust help us survive. But humans have enhanced what we call emotions to use as a vehicle for communicating our internal state and our intentions to others.

Any discussion of the evolutionary basis of emotions begins with Charles Darwin's seminal work, *The Expression of Emotions in Man and Animals*, published in 1872. Darwin theorized that emotions evolved just as physical traits evolved, in response to environmental pressures. Emotions and the facial and other behavioral expressions of internal states evolved because the ability to signal our intentions to others while also discerning their intentions toward us is a trait associated with greater survivability. Think for a moment what an advantage it would be to know as another animal approaches whether it intends to eat you, avoid you, or mate with you. In light of our focus on social behavior, you might also determine whether the animal was a member of your clan or pack.

Darwin proposed that basic emotional expressions are adaptive and have a genetic basis. For example, when animals including humans feel threatened, they may respond with a threat display to scare away the source of the threat, including baring their teeth (**Figure 17.2**). Teeth baring is an emotional expression with a biological basis and is part of the genetic heritage we share with all vertebrates (a group including primates and mammals).

FIGURE 17.2 Examples of mammalian threat displays Threat displays such as teeth baring are similar across a wide range of mammals, which suggests that such expressions have an adaptive and genetic basis.

Today the field of evolutionary psychology focuses on tracing back our contemporary human cognition and behavior to the pressures felt by our earliest human ancestors. Leda Cosmides of the Center for Evolutionary Psychology at the University of California at Santa Barbara, along with her colleague John Tooby, has drawn some interesting connections between evolutionary theory and psychology. Cosmides and Tooby suggest that we are operating in today's space-age world with a stone-age brain, that is, a brain designed to handle the stresses of an environment tens of thousands of years in the past. You can learn more about her work in this **Advances in Behavioral Neuroscience: Leda Cosmides and Evolutionary Psychology**.

We will return to Cosmides and her study of evolutionary psychology in our discussion of deception, but next we turn to the theories of how emotions relate to our physiological state.

Theories of Emotions MCAT

Psychologists and neuroscientists have long been interested in the mind–body connection, so exploring how the body and brain create emotions and interpret emotional expressions of others has been a topic of great interest. Here we explore three of the most influential theories of emotion: the James–Lange theory, the Cannon–Bard theory, and the Schachter–Singer theory.

The James–Lange theory

One of the first attempts to explain the link between the mind and body was proposed by William James and Carl Lange in the late 19th century. The **James–Lange theory** proposed that emotions are created *in response to* physiological changes. That is, specific physiological responses to external stimuli lead to specific emotions. For example, when we see a snake, we first experience rapid breathing and sweating, and then we recognize that we are afraid.

Although this was a simple and attractive model, shortcomings were soon apparent. For one, the James–Lange theory does not easily explain how disparate emotional states can produce similar physiological responses. For example, sweating accompanied by elevated heart rate and breathing are hallmarks of such different emotions as love and fear.

Advances in Behavioral Neuroscience | Leda Cosmides and Evolutionary Psychology

Dr. Leda Cosmides began her undergraduate career at Harvard University studying biology, but her interest soon turned to psychology, specifically how evolutionary theory might explain aspects of human behavior. Cosmides earned her Ph.D. in cognitive psychology at Harvard and traveled to Stanford for a post-doc with Roger Shepard, an influential cognitive psychologist. Dr. Cosmides began her career at the University of California at Santa Barbara where she is presently a distinguished professor. Dr. Cosmides is considered, along with her husband John Tooby, to be among the founders of the field of evolutionary psychology. Her research explores how the hunter-gatherer model of human behavior that makes up our genetic heritage affects how we behave in this modern world. Among Dr. Cosmides's many honors are the AAAS prize for behavioral science research, the APA award for Early Career Contribution, and a Guggenheim Fellowship.

It is worth noting that although Dr. Cosmides is now considered to be one of most important and influential living behavioral scientists, there were many times early in her career when her ideas were not welcome. Her dissertation, now considered to be a seminal article in the field, was rejected multiple times before finally being published. Her prevailing interest in exploring the connection between evolution and human behavior was not a popular idea with mainstream psychologists at first. Nevertheless, she persisted, and today the field of evolutionary psychology is mainstream, in no small part due to her work.

Questions

1. **Search online using the terms "Leda Cosmides laboratory." Identify two of her research interests and describe each in a sentence.**
2. **The Center for Evolutionary Psychology, which Cosmides and Tooby direct, explores a wide range of topics in that discipline. Visit the website of the Center for Evolutionary Psychology at the University of Santa Barbara and, after examining their current research projects, choose two that interest you and briefly describe them.**

A second piece of evidence against the James–Lange theory was that artificially creating physiological states did not lead a person to perceive the associated emotion. For example, when injected with adrenalin, which causes the sympathetic nervous system to increase heart rate, respiration, and other fight-or-flight responses, individuals report that they feel *as if* they are scared or anxious; they do not report *being* scared or anxious. Clearly, emotions and the physiological responses associated with them have a more complex relationship than this early theory could explain.

The Cannon–Bard theory These and other criticisms of the James–Lange theory led Walter Cannon and Philip Bard in the late 1920s to propose that our brain senses the change in physiology and *then* interprets it as an emotion. The **Cannon–Bard theory** proposed that a physiological response in the thalamus and hypothalamus occurs simultaneous to, and independent of, the emotional response and that the two together are what we experience as an emotion.

A simple mnemonic can help you to remember the difference between these two theories. To remember James–Lange, recall that A comes before E just as arousal comes before emotion according to the theory. To remember Cannon–Bard, recall that there are two Ns together in Cannon just as physiology and emotion occur simultaneously according to the theory.

Just as the James–Lange theory had its critics, so did the Cannon–Bard theory. For instance, Cannon–Bard assumes that physical behaviors do not influence emotions; however, individuals who are asked to frown or smile will often report that after they do so, their mood moved in the direction of the facial expression. Because Cannon–Bard emphasized physiology over a person's mental processes that evaluate physiology, many researchers felt that the theory was insufficient to explain emotions.

The Schachter–Singer theory In the 1960s, Stanley Schachter and Jerome Singer came up with a new theory of emotion. They agreed with Cannon–Bard that emotion does not follow the physiological response, but they further proposed that we make a cognitive judgment about the context in which that response occurs, and the result is an emotion. The essence of the **Schachter–Singer theory** is that the physiological state of elevated heart rate and respiration would be evaluated differently depending on the context. The emotion associated with that physiological state might be excitement when an individual is riding a roller coaster, but fear if they are being chased by a bison.

Schachter and Singer tested their theory by injecting volunteers with a drug they were told would enhance their vision, but which was actually epinephrine, a hormone that increases respiration and heart rate. Then the subjects were placed in a room with someone they thought was another study participant but who was actually a confederate of the experimenter instructed to act either angry or euphoric, both emotions associated with the physiological response to epinephrine.

The study showed that when the confederate acted angry, the participant was more likely to report being angry themselves, but if the confederate acted euphoric, the participant was more likely to report being euphoric. In other words, the participant seemed to be labeling the same physiological state (increased respiration and heart rate) with different emotions depending on the context suggested by the behavior of the other person in the room. Efforts to replicate this study were not uniformly positive; therefore, although Schachter and Singer were correct in recognizing that physiological arousal is not the only determinant of emotions and that cognitive appraisal of the context is important; people may vary greatly in how they interpret that context.

These three theories are illustrated in Figure 17.3.

If the expression of emotions provides an organism with an advantage, then we can assume that there is a neural mechanism to create emotions, and this is our next topic.

The Neural Basis of Emotions

Our understanding of the biological basis of emotion and social organization reflects the developmental process itself in that early scientists began with a simple idea that then increased in complexity. Originally, neuroscientists searched for a single center for emotions in the same

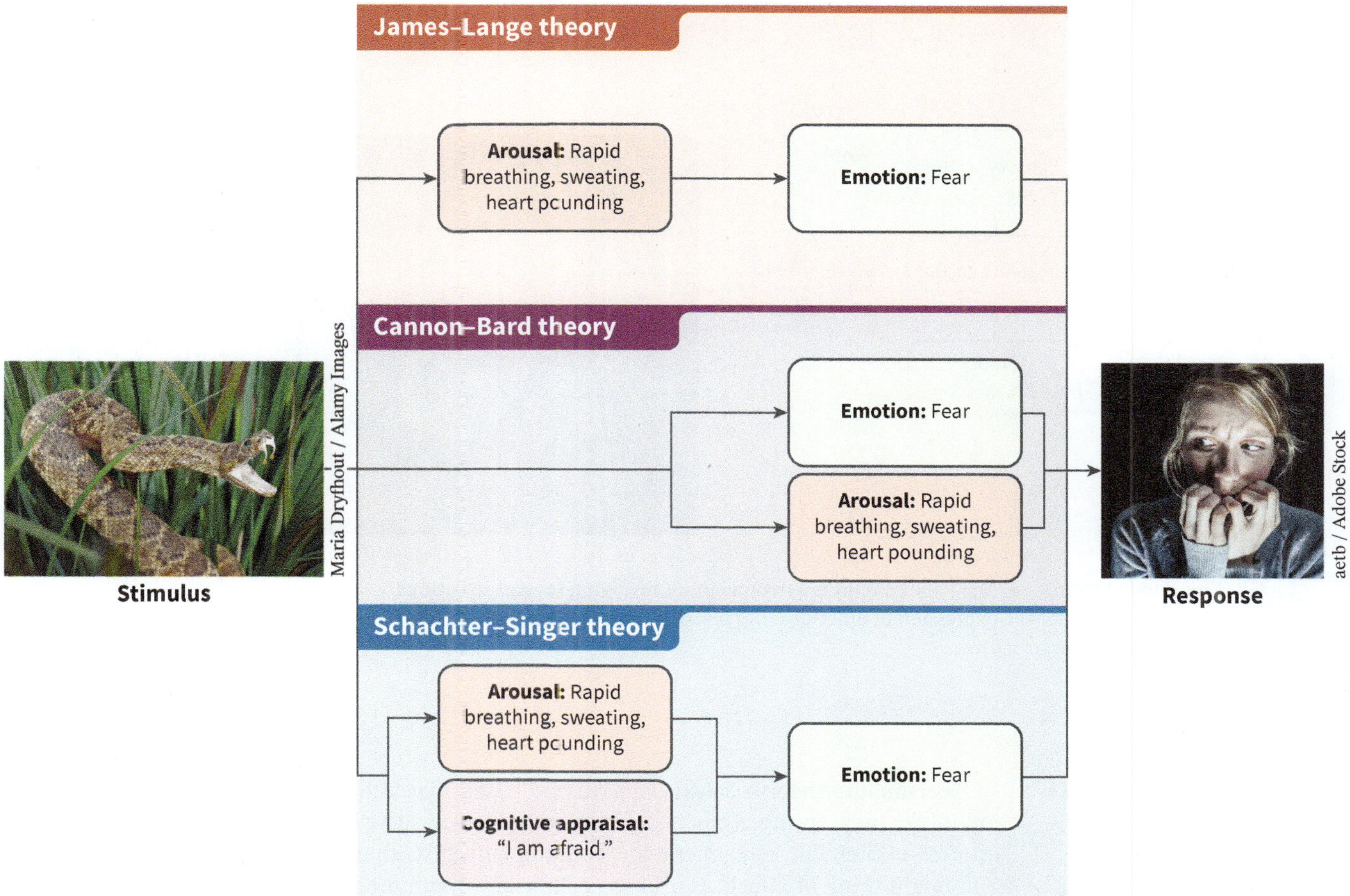

FIGURE 17.3 Three theories of emotion James–Lange proposed that physiological change precedes emotion, Cannon–Bard proposed that physiological response accompanies the sensation of emotion, and Schachter–Singer theorized that a cognitive appraisal of the physiological response leads to the creation of an appropriate emotional response.

way that Broca and Wernicke found centers responsible for speech comprehension and production. For a brief period, it was thought that the thalamus might be the source of all our emotions. As with the models of speech comprehension and production, however, research eventually uncovered more complex neural origins for emotions.

Emotions and neural circuits Extensive research by James Papez and others in the 1920s eventually demonstrated that emotions were not centered in any one place but in fact were the result of several subcortical areas working together as a circuit. This **Papez circuit** included four basic interconnected structures: the hypothalamus, the anterior thalamic nucleus, the cingulate gyrus, and the hippocampus (Figure 17.4). In the 1960s, Paul McLean and others improved on Papez's idea by adding additional structures such as the amygdala, septal nuclei, and nucleus accumbens to the original circuit; this larger circuit is the **limbic system** you are already familiar with from a prior discussion.

The take-home message in the context of social neuroscience is that emotions are created by a circuit involving multiple structures that communicate interactively. Of course, there is more to the neural basis of emotions than just circuits linking structures into sophisticated regulatory mechanisms. As with many behavioral functions, emotions are regulated not just by neural circuits but by a confluence of other agents such as neurotransmitters and hormones.

Neurotransmitters, hormones, and emotions A current theory proposes that a complex interaction among neurotransmitters is a part of the creation of basic emotions such as anger, excitement, joy, disgust, and shame.

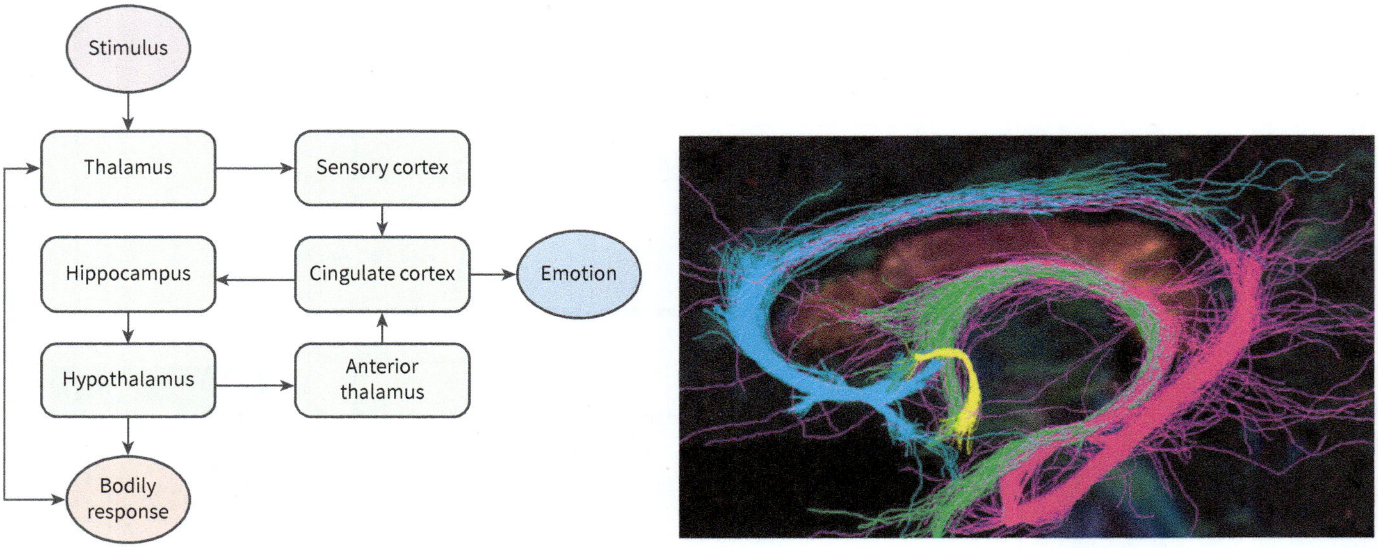

FIGURE 17.4 The Papez circuit **a**. The Papez circuit is a neural circuit involving several structures that is thought to be responsible for the expression of emotions. **b**. This MRI image clearly shows the circular connections among these structures.

Personality theorist Silvan Tomkins proposed that although emotions are complex and muddled, they spring from a discreet set of seven affects: distress–anguish, anger–rage, interest–excitement, enjoyment–joy, contempt–disgust, shame–humiliation, and fear–terror. You might think of affects as primary colors, and combining these primaries in endless ways leads to the wide palette of emotional hues we see in everyday life.

Hugo Lövheim hypothesized that the three monoamine neurotransmitters dopamine (DA), serotonin (SE), and noradrenaline (also called norepinephrine or NE) work together to create these basic affects. For example, in this model, contempt is created under conditions of low dopamine, low noradrenaline, and high serotonin. By contrast, anger is the result of high dopamine and noradrenalin coupled with low serotonin. The relationships among neurotransmitter levels and the basic affects are depicted in his model, known as the **Lövheim cube** (Figure 17.5).

Neurotransmitters are not the only chemical messengers involved in emotional regulation; hormones are also involved. You saw in earlier discussions of the endocrine and reproductive systems that estrogens and androgens exert powerful effects on mood. Estrogens increase serotonin levels, thereby indirectly influencing depression and anxiety. Androgens share with estrogens the ability to influence libido, and testosterone, the most common androgen, is associated with aggressive behavior. Furthermore, oxytocin promotes romantic-pair bonding along with feelings of **empathy** (the ability to understand another's feelings), trust, and attachment. It is easy to see that recognizing the effects of neurotransmitters and hormones is extremely useful to psychopharmacologists seeking treatments for depression, anxiety disorders, mania, and other psychiatric disorders.

Research into the biological basis of social behavior in humans, although not in its infancy, is in its adolescence. We have learned a lot, but there is still much to learn, and you will see later in this chapter that the discovery of mirror neurons and the default mode network have changed how we view the neuroscience of social behavior.

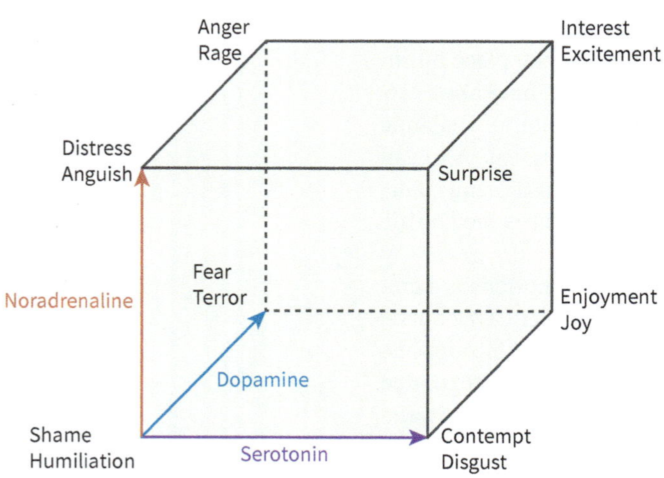

FIGURE 17.5 The Lövheim cube of emotions The Lövheim cube proposes that basic emotions are created at the neurotransmitter level by different actions of serotonin, dopamine, and noradrenaline.

17.1 Before You Go On

Section Summary

Emotions are a way to communicate internal states, and humans share this ability with a wide range of organisms. Theories of the basis of emotions disagree over whether physiological changes precede or accompany emotions, and the current thinking is that for humans, an intellectual appraisal of each situation is important. There is clear evidence that our emotions are created by a neural circuit involving several cerebral structures as well as neurotransmitters and hormones.

Comprehension Check

1. What are emotions?
   - **a.** an internal mental state created by the nervous system, hormones, and neurotransmitters
   - **b.** an appropriate response to threatening stimuli
   - **c.** a class of behaviors that only mammals experience
   - **d.** a recent evolutionary development

2. One difference between the James–Lange and Cannon–Bard theories of emotions is that Cannon–Bard claims that _____.
   - **a.** physiology precedes the emotion
   - **b.** physiology follows the emotion
   - **c.** physiology and emotion arise simultaneously
   - **d.** emotions are illusions

3. An important difference between Schachter–Singer on the one hand and James–Lange and Cannon–Bard on the other is that Schachter–Singer _____.
   - **a.** stresses that physiological responses such as heart rate and blood pressure create emotions
   - **b.** claims that a person's intellectual or cognitive appraisal of the context for physiological responses is critical
   - **c.** discounts the relevance of physiological responses in emotions
   - **d.** stresses the role that evolution plays in understanding emotions

4. The neural basis of emotions is currently thought to include which five structures?
   - **a.** cortex, hippocampus, anterior thalamus, basal ganglia, and temporal lobe
   - **b.** cortex, hippocampus, cingulate, insula, and temporal lobe
   - **c.** hippocampus, hypothalamus, cingulate, insula, and temporal lobe
   - **d.** cingulate, hippocampus, amygdala, hypothalamus, and thalamus

5. The Lövheim cube proposes that emotions are in part the result of the interaction among which neurotransmitters?
   - **a.** norepinephrine, glutamate, and GABA
   - **b.** serotonin, dopamine, and norepinephrine
   - **c.** serotonin, GABA, and glutamate
   - **d.** serotonin, dopamine, and glutamate

17.2 Communicating Internal States

LO 17.2 Discuss how humans use facial expressions to communicate with and influence others.

Now that you have a grasp of what emotions are and how they occur, we turn to the question of how we communicate our internal state and interpret the internal state of others. The human ability to understand and empathize with others gives us a distinct evolutionary advantage. We begin by investigating the language of social expression and then explore the neural mechanisms behind understanding other people.

Facial Expressions MCAT

We swim in a sea of ever-changing social signals. We are constantly evaluating the social signals sent by others as we send out our own messages in the hope we will be understood. Is my professor pleased or disappointed by my answer to a question posed in class? What did my boss's expression tell me when I asked for time off? Was my suggestion for a weekend activity taken seriously or with scorn by my friends? Our ability to infer the internal emotions and intentions of others is essential to our success. Facial expressions are an important signal of someone's internal state.

If sensing the intentions and internal states of other beings is beneficial, then we might predict that there is a common alphabet of facial expressions that signal the most basic emotions. Psychologist Paul Ekman, a pioneer in the study of emotions as they are displayed through

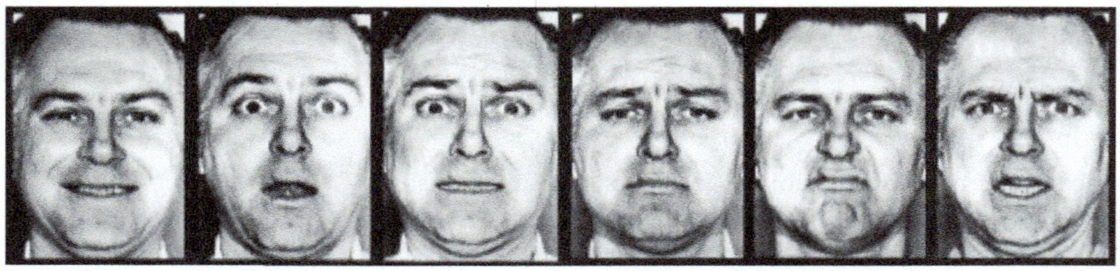

Calder, A.J (2001), Springer Nature

FIGURE 17.6 **Ekman faces** Ekman and others have proposed that all humans share a basic set of emotions and that these emotions are accompanied by specific facial expressions.

facial expressions, has proposed that all humans share the basic emotions of joy, surprise, fear, sadness, disgust, and anger; further, he suggests that the facial expressions that signal these emotions are shared across cultures (Figure 17.6).

The idea that facial expressions are rooted in biology and therefore the same across cultures has its critics. Anthropologist Margaret Mead's research on cultures supported the view that social expressions such as emotions might be learned and therefore differ from culture to culture. Mead's work with her husband Gregory Bateson supported the view that Samoan culture led to different expressions of basic emotions.

Today many theorists believe that basic emotional expressions are shared across most cultures, but culture may play a role in the expression and evaluation of facial expressions in some instances. You can explore this concept for yourself in this Try It Out: Cultural Differences in Emotional Expression.

Social and Duchenne Smiles

One well-studied example of the nuances of conveying and interpreting facial expressions is the smile. Evolutionary psychologists believe that the showing of teeth in a smile rather than a baring of fangs originated among our primate ancestors as a signal of submission and threat avoidance. Today a smile can convey a wide range of emotions: mirth, amusement, irony, and even sadness (Figure 17.7). We smile when we are about to make a wish and blow out our birthday candles, but we may also smile when we are receiving bad news but do not want anyone to know our inner sadness. Sometimes a smile may accompany a condescending retort, an insult, or even a threat. Mostly, we smile to express joy or convey agreeableness.

Distinguishing a social smile from an authentic smile is a good example of the nuances of interpreting social cues. A **social smile** is voluntary. It originates in the frontal lobe and travels through the motor cortex and the **pyramidal motor system**, which is responsible for the control of voluntary muscles and activates muscles that pull the corners of the mouth upward into a smile.

Try It Out | Cultural Differences in Emotional Expressions

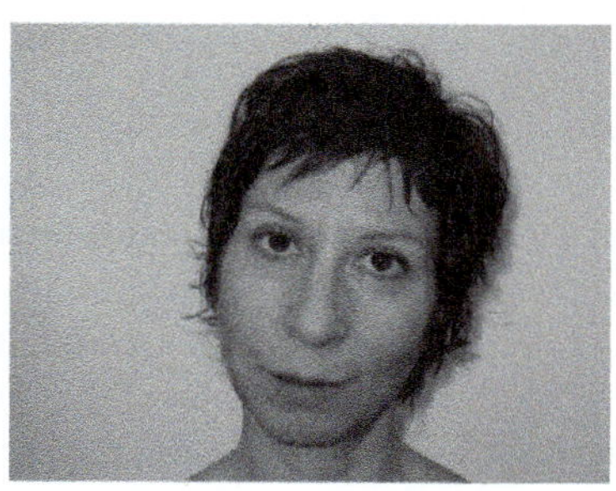

Every day we navigate a series of social interactions, and as we do so, understanding what another person is feeling is essential to our personal success. But do all cultures share the same facial expressions? In this activity, you will see several facial expressions, and your task is to decide which emotion is being depicted by that expression.

Watch the video Cross-Cultural Emotional Expressions, and then answer the questions on the worksheet Cross-Cultural Emotions.

drbimages / Getty Images

nico_blue / iStock / Getty Images

vladimirfloyd / Adobe Stock

FIGURE 17.7 **Smiles conveying different emotions** Smiles can convey more than happiness; they can convey **a**. hostility, **b**. skepticism, and even **c**. confusion.

By contrast, a **Duchenne smile** (named after the 19th-century French anatomist Guillaume Duchenne, who studied the physiology of emotions) is involuntary. It is created by a different circuit in the **extrapyramidal motor system**, which includes the brainstem and the limbic system. Because the Duchenne smile does not originate in the frontal complex as the social smile does, the individual may not even be aware that they are smiling. In addition to activating the same muscles as the social smile, a Duchenne smile activates muscles that pull up the corners of the eyes, creating a wrinkling effect. Understanding the differing roles of the pyramidal system (in which voluntary movements originate in the motor cortex) and the extrapyramidal system (in which involuntary movements originate in the brainstem) explains why a stroke patient may be unable to smile on command but can still grin when told a funny joke.

Humans use social smiles to lubricate the wheels of social exchanges, so we might expect people to be skilled at distinguishing a social smile from a Duchenne smile. You can test your own skill at distinguishing a social smile from a Duchenne smile in **Try It Out: What's in a Smile?**

The social smile is useful when conveying a specific message is more important to the situation than conveying your true emotions. In fact, the human ability to deceive is an important survival tool in our social world.

Deception as an Adaptive Strategy

Parents, television psychotherapists, and religious leaders all teach us not to lie, yet deception is commonplace. We lie about why we were late to class, what we really think of our roommate's new outfit, and whether this is really our last chance to renew our car's warranty. We

Check It Out | What's in a Smile?

MIA Studio / Adobe Stock

In this activity, you are provided with the opportunity to test your ability to interpret the intentions of others through the most universal of social signals, the smile. Download the **Social or Duchenne** worksheet, watch the video **What's in a Smile?** and then perform the associated activity on the worksheet.

deceive not only through our words but through our actions; hair dye, control undergarments, spray tans, and cosmetic surgery all tell a tale about us that is not exactly the truth. Why do we lie? The answer is simple: lying provides us with an edge in the competition of life. If we lie about our skills, our wealth, our connections, and how much we care about others, then we may gain an advantage over our more truthful peers.

Some evolutionary theorists propose that the explosion in human brain size that occurred tens of thousands of years ago was the result of increasing competition among our ancestors for resources such as food and shelter. An enhanced ability to construct social bonds and the mechanism to navigate them, even deceitfully, provided our ancestors with an advantage in the struggle to survive. An individual who successfully lies to gain an advantage over competitors would be more likely to reproduce, so this trait could be expected to become more common over generations.

If there is an evolutionary basis for lying, then there should also be an advantage to being able to detect lies. The evolutionary psychologists featured earlier in the chapter, Cosmides and Tooby, suggest that it would be to our evolutionary advantage to be able to distinguish someone trustworthy from someone who is a cheat. Individuals who have a "lie detector" would have an advantage in life because they would be able to detect individuals who will take from them but not share in return. We would expect that over time, a behavioral arms race would occur pitting increasingly sophisticated liars against increasingly sophisticated lie detectors. How good are we at telling the truth from a lie? In **Check It Out: Can You Tell the Truth from a Lie?** you can find out if you are skilled at distinguishing the truth from a lie.

Theory of Mind MCAT

You learned in an earlier chapter about the concept of **theory of mind (TOM)**, which refers to the capacity to understand others by being able to imagine their emotional state. TOM explains how we navigate complex social communication and how we experience empathy. Now we consider the neural structures that appear to be responsible for these uniquely human abilities: the default mode network and the mirror neuron system.

The default mode network You probably already know that when you are trying to remember where you parked your car, the parts of your brain involved in memory increase their activity like revving a car engine. By the same logic, it is reasonable to assume that when you are not thinking of anything in particular, activity across the brain should decline, just

Check It Out | Can You Tell the Truth from a Lie?

GlobalStock / Getty Images

Most people believe that they can tell when they are being lied to, but is that true? Are some people better at detecting lies than others? Test your own ability to tell the truth from a lie watching the video **Can You Tell the Truth from a Lie?** and then answering the questions on the associated **worksheet**.

as when you stop your car at a stop light, the car engine idles. However, neurologist Marcus Raichle and his colleagues demonstrated that the brain of a person who is mentally engaged in a demanding task only increases their energy consumption by about 5% over the baseline, which is known as the **default mode network (DMN)**. Apparently, our brain does not idle but is always working at an active level. This raises the obvious question: what is our brain working at so hard when we believe we are thinking about nothing?

One answer is that the DMN is activated by social cognition. Consensus is that humans are constantly evaluating their social environment even when we seem to be thinking about nothing at all. Consider for a moment what happens when you walk into a classroom for the first day of class. Should you sit in the front row or hide in the back? If you don't sit by a friend, will they take that as an insult? Should you sit by that person you find attractive, or will that be too obvious? Complex social decisions such as thinking about yourself, imagining what another person is thinking, considering a moral dilemma, or imagining an event that might happen in the future all activate specific centers in the brain that are part of the DMN. There are several regions thought to be a part of the DMN, but the posterior cingulate cortex and the precuneus as well as regions of the prefrontal cortex are thought to play essential roles in social cognition (**Figure 17.8**).

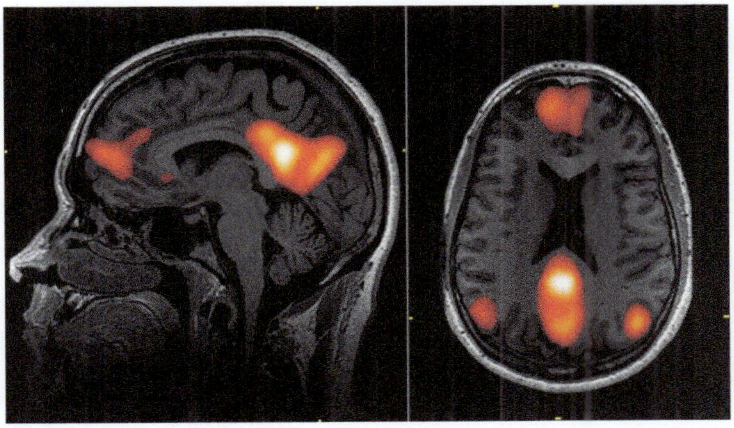

FIGURE 17.8 MRI of regions activated by the default mode network When the mind is not occupied by conscious tasks such as reading these words but is daydreaming, contemplating events, or lost in one's own thoughts, the DMN is actively but covertly processing information from the social and physical environment.

The DMN may be confusing at first, but a metaphor involving something you already know is helpful. Imagine for a moment that you are writing an assignment for a class on a computer, and you are thinking deeply before you begin to type. Although you are not moving a muscle, the computer is busy polling the keyboard to see if you have responded, managing your several open windows, and checking to see if an email has come in. These operations occur thousands of times each second without your awareness. We now believe that in the absence of mental tasks that entail significant cognitive demands (such as attending to your instructor), the DMN is actively but unconsciously processing information from your environment, and one important aspect of that environment is processing social information. Who is around you? What are their intentions? When you are in a group of people, regardless of whether they are family and close friends at home or complete strangers on a crowded sidewalk or train, your DMN is evaluating what their presence might mean to you. We now turn to a second system known as mirror neurons.

Mirror neurons

Many scientists believe that humans and their primate relatives have a neural system that enables them to imagine what another individual is doing or experiencing. In the 1980s and 1990s, Giacomo Rizzolatti and colleagues determined in an extensive series of studies that a set of neurons called **mirror neurons** in the premotor cortex of a macaque monkey activated when the monkey reached for an object such as a food treat and also when it merely observed a person or another monkey reaching for the same object. Even more interesting, these neurons did not fire when the object was simply moved. It was the purposeful action of reaching for the object that caused the neurons to fire, not the movement of the object itself.

The idea of mirror neurons is controversial, but studies measuring changes in blood flow or electrical activity in areas thought to be associated with social cognition have generally supported the idea that our brain activates to create an action and also as we observe another performing the same action.

Perhaps the strongest evidence in support of mirror neurons in humans occurred in a study of patients undergoing brain surgery. When an individual is suffering severe and unremitting epileptic episodes, a surgeon may recommend the removal of brain tissue that is the focal center of the seizures. The surgeon then places electrodes on the patient's brain to determine the borders of the seizure-producing tissue to avoid removing more than is essential. In one study, patients undergoing this procedure for medical reasons volunteered to perform a simple task and then observe a person performing that same task.

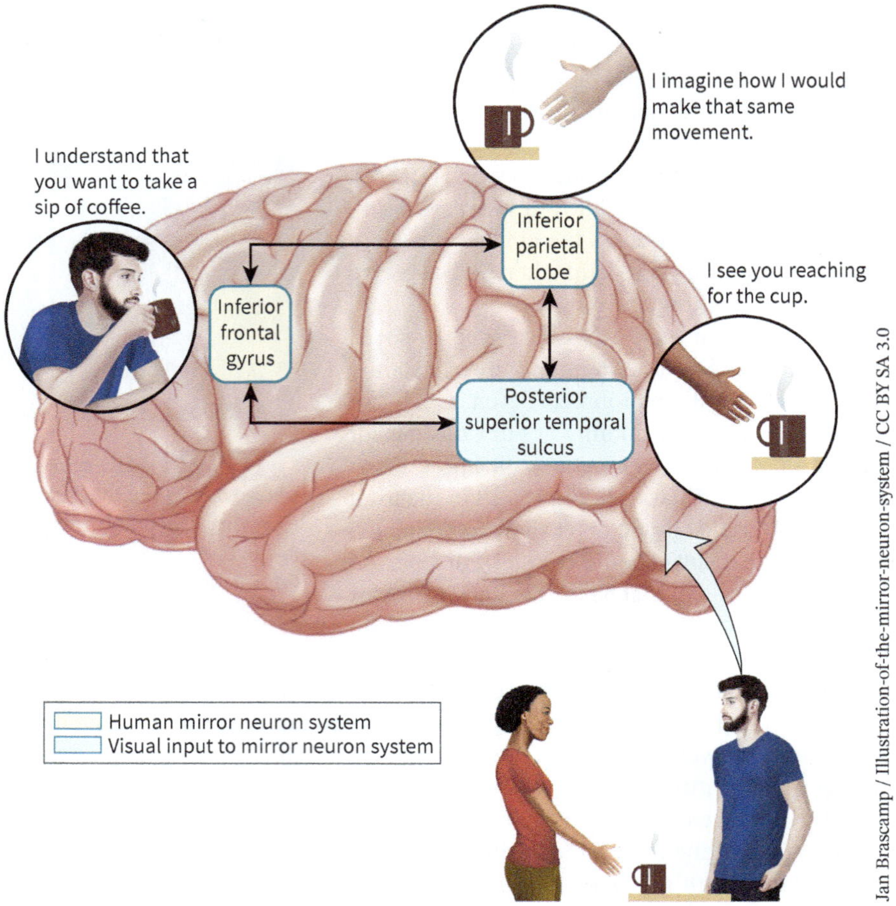

FIGURE 17.9 Mirror neuron system Mirror neurons activate when a person performs an action but also when watching another person perform that same action. The motor neuron system is thought to contribute to empathy.

In this study, some neurons increased their activity when the person both performed a task or observed someone performing the task. Other neurons increased activity only when the person performed the task. In other studies, a third set of neurons increased their activity only when observing another performing the task. These results argue for the existence of a system that engages different specialized neurons when we perform an action, when we watch an action being performed, and potentially when we imagine the action being performed (Figure 17.9).

Although the purpose and even existence of mirror neurons is hotly debated, a mirror neuron system could serve as not just a mechanism for vicarious or observational learning, but also a foundation of our ability to empathize.

17.2 Before You Go On

Section Summary

Facial expressions provide us with a way of communicating our internal state. Ekman and others propose that there are a small number of facial expressions common to all cultures, whereas others feel that cultures can adapt facial expressions to their unique needs. Smiles that convey sincere happiness differ in their neural origin and appearance from social smiles that signify agreement. Being able to infer the emotional state of another person through their facial expression is an adaptive skill, as is being able to trick others about one's intentions.

The brain is constantly analyzing, interpreting, and categorizing social information through the DMN. There is also evidence that the mirror neuron system plays a significant role in the development of empathy and TOM.

Comprehension Check

1. Paul Ekman would claim that when it comes to emotions _____.
   a. everyone has different facial expressions
   b. there are around 20 facial expressions common to us all

c. there are a small number of facial expressions common to us all

d. within a given culture there is a common set of expressions, but cultures themselves differ reliably

2. An important difference between the social and Duchenne smiles is that social smiles involve the _____, and Duchenne smiles involve the _____.

a. extrapyramidal system; pyramidal system

b. pyramidal system; pyramidal system and the basal ganglia

c. extrapyramidal system; extrapyramidal system and the basal ganglia

d. pyramidal system; extrapyramidal system and the limbic system

3. Evolutionary psychologists propose that humans may have evolved the desire to deceive because _____.

a. every mammal exhibits deceptive behavior, and it is part of our heritage

b. the limbic system is activated by deception and so provides a reward

c. one's membership in a group is not important and can easily be repaired

d. lying can provide an advantage to the liar, as long as the lie is not detected

4. You took a moment off from thinking about your schoolwork to just sit quietly and clear your mind. For those few minutes, you don't recall having any thoughts. Your understanding of the DMN would suggest that during that quiet time _____.

a. your blood pressure dropped

b. your brain was covertly analyzing your physical and social environment

c. you were smiling but were unaware that you were smiling

d. you might have nodded off for just a moment

5. The mirror neuron system is purported to activate when we _____.

a. are looking at ourselves in a mirror

b. see ourselves in a photograph or a video

c. perform an action or see another perform the same action

d. see something that is a mirror image of something else

17.3 Beauty and Attraction

LO 17.3 Summarize the ways that humans signal health and reproductive fitness.

Poets may tell us that beauty is in the eye of the beholder, but neuroscience would reply that evolution and the brain play a large part in our perception of beauty. In an earlier chapter, you learned highly symmetrical faces tend to be viewed more positively. We will now turn to a more detailed overview of the behavioral neuroscience of attraction. It is worth noting that until recently this research had almost always been carried out using heterosexual participants, so expanding this research into other populations is a fertile area of research.

Humans are a sexually dimorphic species; that is, males differ in size and appearance from their female counterparts. This divergence is a result of **sexual selection**, a special case of natural selection in which an organism chooses a mate based on physical and behavioral adaptations that increase its chances of reproductive success. Some birds grow large and luxurious tails, some fish build comfortable nests, and some animals engage in elaborate contests, all to prove to a potential mate that they are strong, healthy, and a good bet for passing on their DNA.

One theme woven throughout this text is that we live in a modern world, but we still have the brain of our ancient ancestors, which has not yet caught up with that world. Our earliest ancestors lived in an environment where starvation was a constant threat and predators abounded. What characteristics did humans evolve to signal to a prospective mate that we are healthy, potent, and fertile? You may be familiar with the concept of a poker tell, an unconscious behavior that gives away the person's hand. What are the tells for health or fertility that evolution has left for us?

External Tells for Health and Reproductive Fitness

From an evolutionary perspective, beauty is our brain's way of judging who among our potential partners is most likely to ensure that our DNA will be carried forward in an ancient world where survival was a daily struggle and life expectancy was short. Male handsomeness would therefore be a surrogate for strength and vitality as predictors of potency and the ability to provide resources. By contrast, female beauty would be a sign of health and fecundity, or reproductive ability. For both sexes these different traits are a sign of **reproductive fitness**, or the ability pass their genes on to offspring.

mimagephotos / Adobe Stock

Gage Skidmore / Wikimedia Commons /CC BY-SA 3.0

FIGURE 17.10 Hair and skin as signals for health A full head of thick, shiny hair and a clear complexion signal health. These traits are seen as beautiful because they suggest the health and fitness of a potential mate.

Skin and hair Consider for a moment that one characteristic of a beautiful person is that their skin is clear and their hair is full and lustrous (**Figure 17.10**). Our ancestors had to deal with all manner of parasites in their environment, and parasitic infestations affect one's skin and hair. When you look at a person with a clear complexion and flowing locks of shiny hair, your mind sees them as beautiful because your brain is assessing that they have a low parasite load and as such, are a good candidate for a mate. It is therefore not surprising that in 2021, U.S. consumers spent $86 billion on hair care and roughly $62.5 billion on cosmetics and skin care.

Face shape Facial dimensions also are sign for potency and fertility. A male would want to signal that he is virile, whereas a female would have an advantage if she could signal that she is fertile. How can humans signal their testosterone or estrogens levels? Testosterone during development leads to a broader chin, wider face, more prominent nose, bushier eyebrows, and a shorter forehead. In contrast, estrogen guides facial development toward a pointier chin, thinner face, smaller nose, thinner eyebrows, and a longer forehead (**Figure 17.11**). Those individuals seeking a more masculine or a more feminine partner would tend to be more attracted to a potential mate who signaled the corresponding trait.

Before we move on from the face to the body, consider for a moment the importance of the eyes in defining a face. It is often said that the eyes are the window to the soul, so it is not surprising that when people want to hide their emotions, they wear sunglasses. What can our eyes tell us about attraction? For students of behavioral neuroscience, the answer may lie in the pupil. When a person looks at someone they find appealing, this activates the amygdala, resulting in pupil dilation. Advertisers have long used this knowledge to their advantage; it is a common practice to enlarge the pupils of models in advertisements. The logic is that when a person views the image, their limbic system and DMN are unconsciously receiving the message that the person in the advertisement finds them attractive; as a result, presumably the advertiser's product would be more favorably received (**Figure 17.12**).

Body shape Beauty is more than skin deep; it includes our whole body. Here again we see evidence that our evolutionary past influences our judgment. Before you proceed, remind yourself that we are not talking about a body type that you and I should strive to reach. We are talking about a body type that millennia of evolutionary pressures have programmed our nervous system to seek out to enhance our chances of passing on our DNA. Such preferences are influenced by our evolutionary history but not dominated by it; culture does play a part. In our past when food was scarce, plumpness in a man or woman was considered an attractive

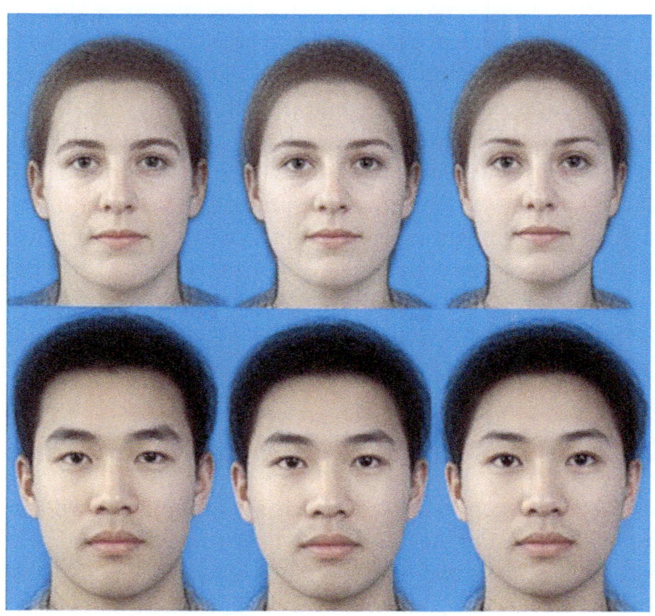

Scott, I., M et al (2014), National Academy of Science

FIGURE 17.11 Feminized and masculinized faces By manipulating characteristics such as facial width and length, chin, and eyebrows, faces can appear masculinized or feminized. The middle images show the actual person, the images to the left were altered to appear more masculine, and the images to the right were feminized.

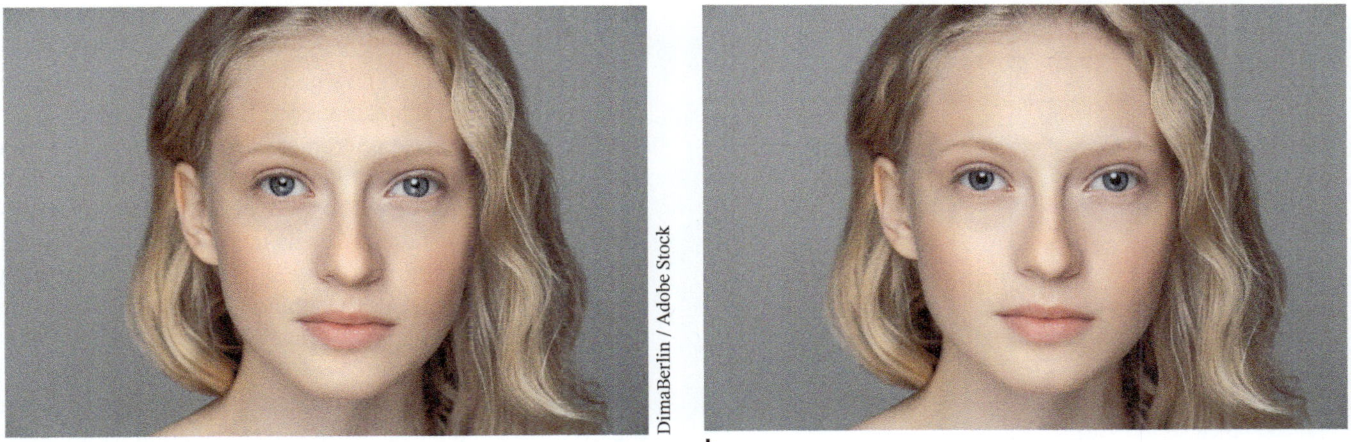

FIGURE 17.12 **Pupil dilation as a tell for attraction** These two images differ only in the size of the pupils. Which individual do you think is more attractive?

sign of wealth. Today, features such as skin tone, eye size, athletic or burly build, and facial tattoos may be considered beautiful in one culture but not in another.

In one study, heterosexual males and females were given software that allowed them to manipulate body characteristics to create what they perceived to be the most appealing bodies for the opposite sex. According to this study, the ideal body type for males is an inverted pyramid with broad shoulders and a small waist, a body shape that signals strength and vitality. The corresponding ideal body type for females has a waist-to-hip ratio of .70–.75, a sign of fertility and a higher likelihood among nulliparous women (that is, those who have never borne a child) of being able to conceive and carry a child to term. **Figure 17.13** presents a visual representation of the results.

That such body shapes are idealized does not mean they are healthy or even attainable; they may represent a state created by a combination of our evolutionary pressures and culture.

Ovulation and Attractiveness

From an evolutionary perspective, coitus other than at the time of ovulation is a wasted opportunity to pass on one's DNA. Many females in the animal kingdom signal ovulation through a distinct change in coloring or a conspicuous swelling of the genitals. In addition, there is a class of chemical messengers known as **pheromones** that use the olfactory system to signal physiological states, particularly sexual receptivity. Common household pets provide familiar examples of pheromonal communication; when the female comes into heat, she secretes a pheromone that

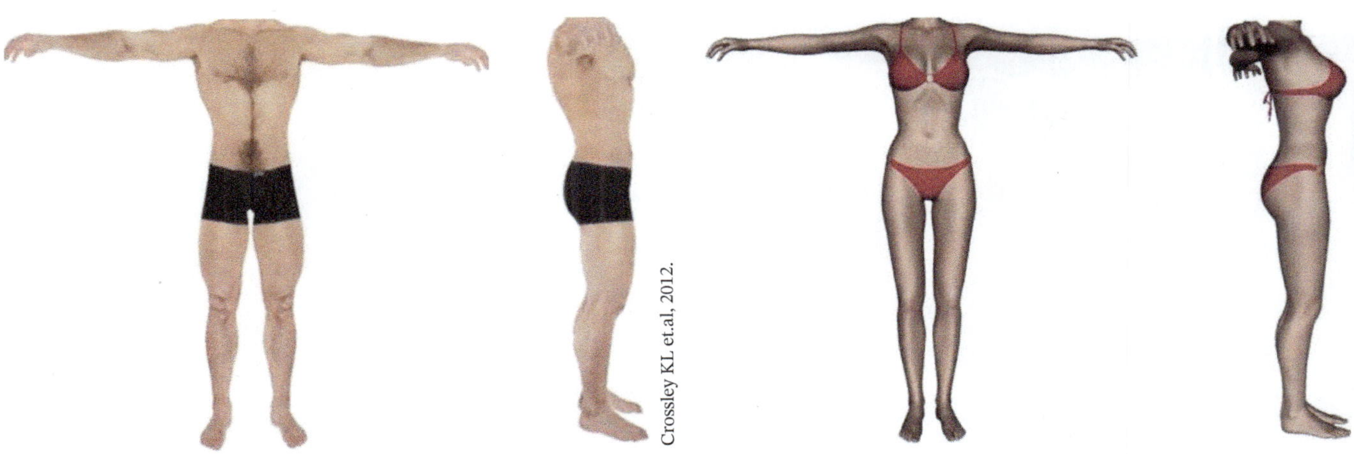

FIGURE 17.13 **Idealized male and female body shapes** **a.** This idealized male body shape, which signals strength and vitality, was created by heterosexual females. **b.** This idealized female body shape, which signals fertility, was created by heterosexual males.

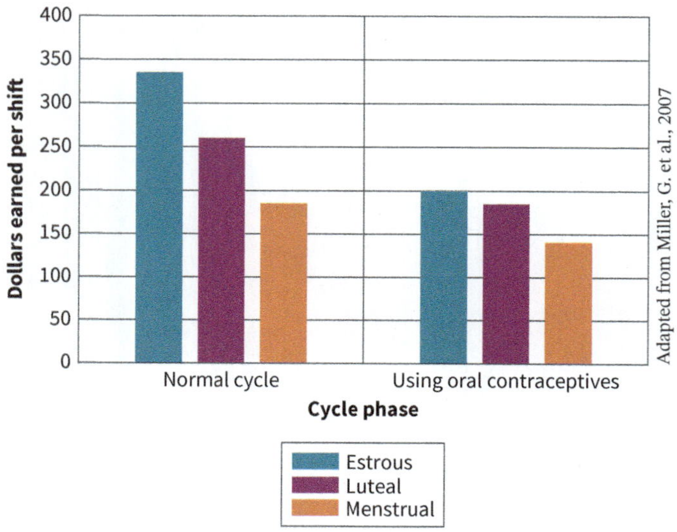

FIGURE 17.14 Tips as a function of the menstrual cycle Dancers not on oral contraceptives and experiencing a normal hormonal cycle received more tips around the point when ovulation occurred; tips decreased post-ovulation and reached their lowest point during menstruation.

males can sense at a distance. In another example, if you've ever walked a dog and wondered why it urinates every 100 paces instead of all at once, you now understand that they are laying down a pheromonal signal that says, "This is my territory."

Might humans use pheromones to signal fertility? In a study by researchers Saul Miller and Jon Maner, men smelled T-shirts worn by women who were either at or near ovulation or far from ovulation. The women were asked to avoid scented soaps or perfumes as well as aromatic foods such as garlic for the course of the study. Those T-shirts worn by women at or near ovulation were judged by the males as having a more attractive scent.

Perhaps the most remarkable experiment investigating the hypothetical male ability to detect ovulation used female exotic dancers working for tips. The dancers kept a record for 60 days of their daily tips and where they were in their ovulatory cycle; they were not aware of the purpose of the study. For those dancers who were not using oral contraceptives, their daily tips increased during estrus (ovulation), decreased during the post-ovulatory luteal phase, and were lowest during menstruation. Dancers who were taking oral contraceptives, which elevate progesterone levels to mimic pregnancy, did not show a similar effect (Figure 17.14).

Hormonal Influences on Pair Bonding

We began this discussion of the social neuroscience of attraction as the mechanism of sexual selection, but we have one more question to consider. Our evolutionary path led to a bigger brain at the expense perhaps of fearsome tooth and claw. As a result, a newborn human is quite defenseless from birth until several years of age. Given that reality, the survival of one's offspring is enhanced if the two individuals create a strong emotional bond that enables caregiving for an extended period (Figure 17.15).

You learned in a prior chapter that one of the effects of testosterone on both males and females is to stimulate libido. Researchers measured testosterone levels in males and females who reported whether they were in love at the time (Figure 17.16). Before you read more, take a moment to reflect on what mechanism nature might have created over time to strengthen the bond between a male and female partner. What behaviors might work in favor of building a longer-lasting relationship? In fact, males who reported that they were in love had lower testosterone levels, which might result in a lower interest in pursuing other relationships. Females in love showed the opposite pattern, suggesting that a strong emotional bond may be accompanied by an increased libido in the female.

FIGURE 17.15 Pair bonding as a human trait The creation of a stable bond between parents can enhance the survival of offspring.

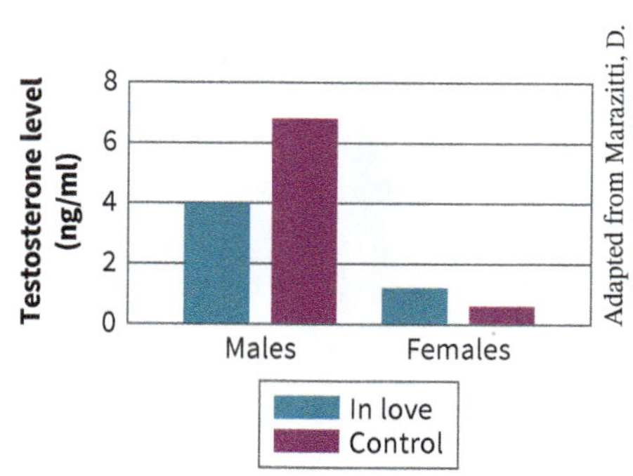

FIGURE 17.16 Testosterone levels by sex and relationship status Males reporting that they were in love had lower testosterone levels than uncommitted controls, whereas females who reported being in love showed an increase in testosterone.

Understanding the complex tapestry that is human behavior is a daunting task, and that is particularly true of topics such as how our brain creates the perception of beauty and feelings of attraction. Most of the currently available research used heterosexual individuals as participants, but this does not describe the entire human experience, so there is still much to learn.

17.3 Before You Go On

Section Summary

Understanding how our nervous system perceives some individuals as being more attractive is a question that behavioral neuroscience addresses. Sexual selection is a mechanism that accentuates characteristics of biological males and females to identify those mates most likely to be successful at passing on their genes. Differences in personal characteristics such as face, hair, skin, and body shape may be clues to a person's health and reproductive potential. There is some evidence that ovulating females are judged to be more attractive by heterosexual males. After a male–female pair bond is formed, hormone levels change to support that bond.

Comprehension Check

1. Which of these might be explained by sexual selection?

- **a.** Some male horses have a chestnut coat.
- **b.** Female deer do not have antlers, but males do.
- **c.** Male and female squirrels are the same size.
- **d.** The mating call of the common frog is loud.

2. A full head of hair and clear, unblemished skin are thought to be signs of what?

- **a.** youth
- **b.** high levels of testosterone in both males and females
- **c.** conspicuous wealth, which encourages attraction by others
- **d.** reproductive fitness

3. A small waist coupled with slightly larger hips in females is believed to be a sign of what?

- **a.** elevated testosterone, which enhances libido
- **b.** heightened estrogen levels, which stimulate maternal behavior
- **c.** increased fertility
- **d.** a greater interest in forming a romantic-pair bond

4. A study looking at tips given by males to female exotic dancers concluded that ______.

- **a.** males were more generous when the females were ovulating
- **b.** males were most generous immediately post-menses
- **c.** women on oral contraceptives were most appealing and received the most tips
- **d.** women on oral contraceptives perceived themselves to be more attractive

5. Males and females who self-disclose as being in love ______.

- **a.** show changes in testosterone levels that strengthen their relationship
- **b.** show an increase in testosterone that reflects frequent sexual activity
- **c.** show a decrease in testosterone due to frequent sexual activity
- **d.** soon return to post-relationship hormone levels

17.4 Human Social Networks

LO 17.4 Review the evidence for an evolutionary preference for ingroups and its negative consequences.

Although the scientific name for humans is *homo sapiens*, the Latin for thinking human, we could just as well be called *homo socialis*, or social human. One of our strengths as a species is our preference to join groups and signal our own group membership; the evidence suggests that this drive to belong to a group has an evolutionary basis. Being able to rapidly identify other group members confers an advantage, but hasty judgments about others in the absence of direct experience may lead to misclassification. Prejudice and stereotyping may also have an evolutionary basis, and our explicit attitudes may not line up with our implicit attitudes.

Evolutionary Advantages of Group Membership

Archeological evidence indicates that our ancestors lived in small groups that cooperated in hunting and gathering. If one hunter returned empty-handed, unable to feed his family, hunters who had better luck could help support them. Similarly, if an essential tool for scraping

FIGURE 17.17 Advantages of cooperative behavior A traditional barn raising is an example of how individuals benefit from group membership. Individuals cannot build their barns without help, so they help build the barns of others in the group knowing that eventually they will have their turn.

hides or butchering meat broke, they might borrow one from a neighbor until they could procure proper materials and had time to make another tool. Each individual family was responsible for themselves, but everyone's overall survivability improved if the group helped members who were experiencing temporary misfortune. That is, if you shared your food today with an unlucky hunter, then you would reasonably expect them to share with you in the future if you returned from the hunt empty-handed. This principle is known as **reciprocal obligation**, and it is an important advantage enjoyed by members of the group.

Under these circumstances, group loyalty and reciprocity were rewarded with tangible benefits. Individuals who shared precious resources indiscriminately (that is, without regard for payback) were less likely to survive in harsh times compared with individuals who only shared with those who reciprocated. Group membership clearly provided an evolutionary advantage, reinforcing cooperative behavior.

In our modern world, our day-to-day survival may not be in question, but we still benefit from cooperative behavior and group membership (Figure 17.17).

Signaling Group Membership MCAT

Humans feel a need to belong in groups, and they signal group membership in a variety of ways, including speech, clothing, and membership rituals. We join clubs, teams, and fraternities or sororities to be with people of similar interests. People desiring entrance to social groups are often required to submit to lengthy and even unpleasant initiation rituals whose purpose is to verify commitment to the ideals of the group. They may be required to wear clothes with signals that identify the person as a member of the group, or they may be expected to engage in silly behavior as proof of their dedication. Many employers, social groups, and even romantic relationships have specified trial periods such as a probationary period, initiation, or engagement during which the individual and the group interact while they determine mutual fitness. These rituals range from the serious to the absurd.

Signaling group membership to others is an important aspect of social behavior. We select clothing, attend specific events, choose certain foods or music, and use specific phrases to indicate to others whether we are like them or not like them. We cheer wildly at sports events for people we don't know and have little chance of ever meeting. We travel to events to spend time with others who enjoy the same music, political views, pets, or hobbies that delight us. Some of us paint our faces in the colors of our favorite teams or dress up as characters in books, television shows, and movies such as *Game of Thrones*, *The Hunger Games*, *Star Wars*, and *Lord of the Rings*.

We naturally identify ourselves as members of groups whose full membership we will never know: East-Coasters, Midwesterners, Jersey or Cali Girls, runners, nonsmokers, vapers, vintage film fans, fans of a given sports team. We wear proudly T-shirts and jackets of our favorite teams, colleges, or musical groups. Group identity is so important that we adopt fashion trends to signal our group membership to others; the daily garb of a Wall Street financier is distinctly different from that of a hipster from Brooklyn or a surfer from Malibu, yet all three would blend into a crowd of their peers. We are who we are with (Figure 17.18).

FIGURE 17.18 Clothing as a signal of group membership Individuals feeling a strong affiliation with a group may signal their membership through external signs such as clothing.

Ingroups and Outgroups MCAT

Take a moment to reflect on how you would describe yourself to others. You might define yourself by your major, year in school, individual residence hall or commuter, and as a fan (or not) of your college or university teams. We use social grouping labels to tell others who we are, and just as importantly, who we are not.

Individuals who share a common set of interests or beliefs are an **ingroup**, and those who are not in that group are in the **outgroup**. People simultaneously belong to many ingroups based on their interests, place of residence, and personal characteristics; this suggests that humans may have not only a natural affinity to categorize themselves and others via social groupings but a biological drive to do so. Defining yourself in contrast to an outgroup (who you are not) may help reinforce your identity and solidarity with your ingroup.

Image courtesy Stuttgart Regional Council, State Office for the Preservation of Monuments, Section 84, Photo E. Stauss

FIGURE 17.19 **Neolithic mass grave** The Talheim death pit in Germany dating from 7500 BCE contains the remains of 34 individuals who were the victims of violence. Unlike typical grave sites of the period in which the deceased were carefully arranged, sites such as these appear to be the result of a massacre.

Throughout history, much violence and strife has been based on such group dynamics. Archeologists have found several mass graves dating from 7500 BCE to 5000 BCE holding dozens and at times hundreds of slain individuals, all thrown into a common pit and all showing signs of blunt force trauma (Figure 17.19). These mass graves are interesting in terms of the demographic distribution of victims in that forensic analysis shows that adult women were conspicuously missing from the pits.

Based on this analysis, archeologists now believe that raids by competing tribes were conducted to steal women of child-bearing age and also children young enough to be assimilated as members of a new clan. If this theory is true, and the evidence supporting it is reasonably persuasive, then we might wonder if our ancestors viewed strangers emerging from the woods as a distinct threat.

There may be an evolutionary basis for viewing outgroups as threats. Those ancestors who met strangers with open arms may have had their hospitality rewarded with a fatal blow to the head, whereas individuals who were naturally suspicious of strangers were more likely to survive. In such a case, the genetic predisposition to welcome strangers would be less likely to be passed on to future generations.

If this were true, then people might be wary of individuals they have never met if (1) there is some difference such as skin color, language, or cultural difference that is readily apparent and so available as a signal of outgroup status and (2) the individual has had little or no direct contact with members of that group. Could humans have a genetically driven preference for the ingroup at the expense of the outgroup?

Prejudice and Stereotypes MCAT

One aspect of hostility toward outgroups is prejudice and stereotypes. **Prejudice** means to pre-judge; that is, to have an evaluative opinion about another that is not based on personal experience. Does prejudice have an evolutionary basis? Evolutionary psychologists have an interesting take on this question. Categorization to create **stereotypes**, or generalized beliefs about an object or person based on shared characteristics or group membership, might be hard-wired into our brains and therefore involuntary if not unconscious.

Consider for a moment the many minor decisions that you make in your daily life. Should we cross the street here or there? Have the vanilla or chocolate ice cream? Sit in this chair as opposed to that chair? If we stopped to ponder the implications of each alternative before acting, we would accomplish little in life. To facilitate such decision making, humans have evolved a powerful ability to categorize objects based on prior experience. As a result, many choices that we make are the result of unconscious mechanisms that require little or no mental

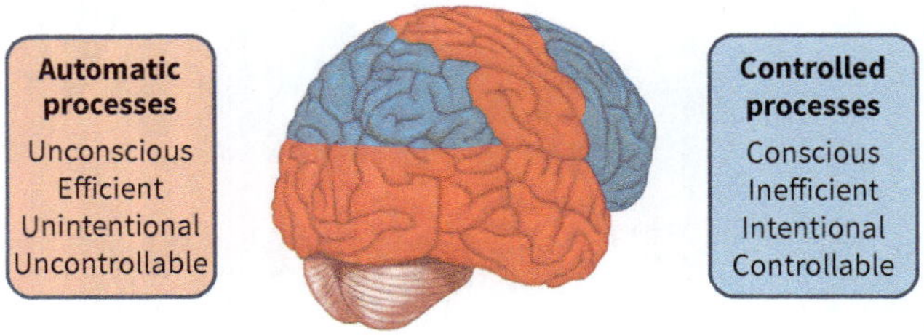

Reproduced with permission from Dr. Bob Spunt, Ph.D 2015.

FIGURE 17.20 Automatic versus conscious mental processes Controlled processes engage the frontal lobe and parietal lobe, whereas automatic processing involves greater representation in the motor strip and temporal lobe.

resources. In fact, automatic processing engages a different cerebral circuit than intentional processing, as illustrated in Figure 17.20.

The value of automatic categorization to create stereotypes is easily seen. Having once been bitten by a spider and experienced a painful reaction, one does not have to test the hypothesis by touching every other arachnid that is encountered. Stereotyping based on actual experience has some logic, but can categorizing occur as a result of repeated exposure to negative evaluations about a group? For example, if we encounter many images of and references to older people that include negative concepts such as illness, disability, and weakness, we may associate these traits with all members of that group, even though not all members exhibit those characteristics.

To take this idea one step further, if their culture regularly disparages the elderly through images and language, could a person develop negative attitudes toward older people that they are unaware of and that might even go against their consciously held beliefs? Social scientists refer to attitudes that we openly acknowledge as **explicit attitudes**. Attitudes we hold internally but are not yet aware of are referred to as **implicit attitudes**. A contradiction between explicit and implicit attitudes can express itself as **unconscious bias**, when we may perceive certain groups negatively and act on those perceptions without even being aware of it. These phenomena may be rooted in some of the same neural mechanisms that facilitate learning and cognition: long-term potentiation and priming.

Long-term potentiation (LTP) is when repeated activation of a neural connection leads to a physical change in the synapse. With repeated activation, the connection becomes stronger, more durable, and more easily activated. **Priming** is when a concept is activated in long-term memory (LTM) and that activation spreads to engrams (concepts) that are related. If your idea of a summer holiday always entails a cookout, seeing an advertisement for a cookout may cause you to spontaneously think of past holidays because the two concepts are linked in LTM. As a result of priming, activation of one concept happens more quickly when it is accompanied by a related concept and more slowly when accompanied by an incongruent concept. For example, having seen old age associated with infirmity many times, our brains find it easier to associate older people with infirmity than with athleticism (Figure 17.21).

ZUMA Press/ZUMA Press Inc/Alamy Stock Photo

FIGURE 17.21 Defying stereotypes Society often negatively categorizes the elderly, so the neural pathways that associate old age with infirmity are easily activated. By contrast, the brain must work harder to make a novel association between the category of elderly people and the athletic prowess that 103-year-old Julia "Hurricane" Hawkens and 100 year old Gerald Meyer demonstrate as competitive runners.

Psychologists Anthony Greenwald, Mahzarin Banaji, Brian Nosek, and their colleagues investigated how we might measure unconscious bias based on categories such as age, gender, ethnicity, and other defining characteristics of outgroups. To test this idea, they created the **implicit association test (IAT)** in an effort to explore purported unconscious bias against outgroups. With the two concepts of LTP and priming in mind, we can grasp the conceptual basis of the IAT. In the IAT, participants are asked to categorize images and words as positive or negative quickly and without reflection; reaction time to make the choice is recorded. The goal of the IAT is to measure the strength of positive and negative associations based on concepts such as race, age, and body type.

A simple example will show how the test works. A coffee–soda IAT would associate both coffee and soda with positive and negative concepts; a person who favors coffee over soda should take less time to evaluate a coffee-positive pair than a soda-positive pair (Figure 17.22). If the person holds implicit positive attitudes about coffee, then priming and LTP should speed up the time to evaluate the coffee-positive pair. By contrast, if a person has negative associations with soda, it will take longer for them to evaluate the soda-positive pair because they must work against priming and LTP. Even if

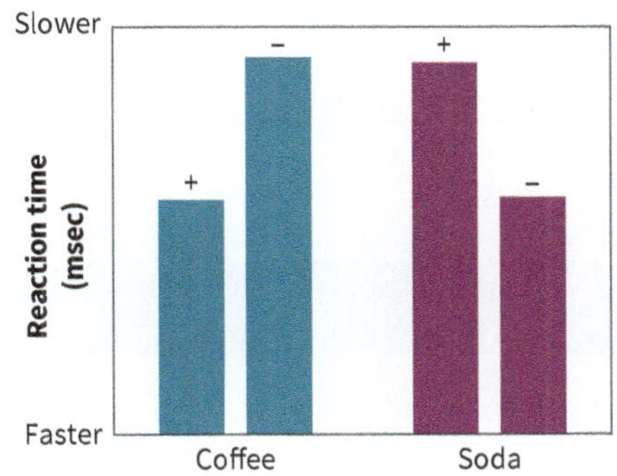

FIGURE 17.22 The implicit attitudes test **a.** In the coffee–soda IAT, participants evaluate images of coffee or soda in conjunction with positive or negative concepts. **b.** Shorter response times are thought to be indicative of an implicit association between the concepts and longer response times indicative of a conflict between explicit and implicit attitudes.

someone says they have positive feelings about soda (their explicit attitude), the speed of their reaction to the soda-positive pair will reveal their implicit anti-soda attitudes.

Such attitudes are relatively insignificant when it comes to beverages, but the IAT has been used to explore possible stereotyping and implicit bias in a wide range of applications such as gender, sexual orientation, religion, body type, political beliefs, age, and disability. It has attracted the greatest attention in applications exploring potential bias due to race. In this **Think About It: The Implicit Association Test**, you can consider the question of implicit attitudes for yourself.

Think About It | The Implicit Association Test

Jason Stitt / Adobe Stock

Research on implicit attitudes suggests that individuals can hold attitudes that they are unaware of and that even conflict with their stated beliefs. These implicit attitudes can affect behavior toward others. Completing the IAT results in a score where positive numbers are thought to indicate a positive attitude toward the group and negative numbers may represent a negative view. Banaji and others feel strongly that an IAT score that suggests implicit bias does not mean that person is biased; it only means that they may have unconscious attitudes not in line with their conscious beliefs.

The IAT is controversial. First, some believe that it provides a view into implicit attitudes, whereas others suggest that it reflects familiarity more than implicit attitudes. For example, many White individuals test as mildly discriminatory on the White–Black IAT but when the Black faces are of very familiar and highly regarded individuals such as Morgan Freeman, Beyoncé, Neil deGrasse Tyson, or Samuel L. Jackson, the race effect disappears, presumably because those faces are familiar. Furthermore, when the IAT was administered to White, Black, Asian, biracial White/Black, and biracial Asian/White, it was common for all groups to show a slight preference for their own ethnic identity.

Second, it seems reasonable to assume that a person's internal biases are relatively stable, so multiple tests over time should result in stable assessments; however, studies investigating test–retest reliability have been mixed in their conclusions.

Finally, some social neuroscientists point out that although people might have implicit biases, it is their actual behavior that matters. It is worth noting that since 1998, an online version of the IAT has collected data from more than 20 million individuals, and in general, bias scores have been decreasing over time.

You can experience the IAT for yourself by going to **ProjectImplicit.net** and reading about the IAT and the many versions of the IAT that have been developed. You can even take a test yourself if you are curious. Then answer these questions.

Questions

1. **The IAT is thought to reveal possible unconscious biases, but even assuming it does what it claims to do, is it the case that a person holding an unconscious bias will act in a biased manner? Briefly explain your answer.**
2. **If you thought that you might have an unconscious bias against some groups, how might that knowledge of a potential bias affect your daily behavior?**

17.4 Before You Go On

Section Summary

Humans are social beings who feel a strong need to affiliate with other humans, but the course of human history has demonstrated that despite our need to affiliate, we often consider some individuals that we have never met to be a threat. Mass graves from the neolithic era have been interpreted to indicate that violence was commonplace, so those who were naturally wary of strangers would have a greater chance to survive, which may explain prejudicial attitudes today. Finally, some social neuroscientists believe that humans can hold implicit attitudes that contradict their conscious beliefs.

Comprehension Check

1. What might be evidence of reciprocal obligation from our distant ancestors?
   - **a.** tools that were found at a grave site
   - **b.** a skeleton showing signs of several broken bones that were healed at time of death
   - **c.** evidence of the ability to grow crops
   - **d.** evidence of dwellings that held several people
2. Mass graves such as the Talheim Death Pit are thought to provide a clue about neolithic life because ______.
   - **a.** all the deceased died of natural causes, which shows the importance of social structures
   - **b.** the deceased were almost entirely children under the age of three
   - **c.** the remains of females of child-bearing age were conspicuously missing
   - **d.** all the deceased died violent deaths
3. During first-year orientation, Jamal meets another first-year student who is Asian, and Jamal thinks that he should form a study group for Calculus with that student. Jamal is demonstrating
   - **a.** prejudice
   - **b.** stereotyping
   - **c.** problem solving
   - **d.** explicit attitudes that differ from his implicit attitudes
4. The implicit association test purports to measure what?
   - **a.** whether implicit attitudes contradict explicit attitudes
   - **b.** amygdala activation
   - **c.** a person's conscious attitudes toward outgroups
   - **d.** attitude towards individuals that are well known to us all

17.5 Autism Spectrum Disorder and Emotional Blindness

LO 17.5 Discuss the neural and behavioral aspects of Autism Spectrum Disorder.

As with many topics in behavioral neuroscience, we can often learn something important from disorders where the usual neural mechanisms malfunction. In this chapter, we have discussed the neuroscience of emotions and the many ways in which people extract information about the inner state and intentions of others from their facial expressions. For any of us, correctly interpreting another's emotional state from their expression can be a challenge. What would it be like if the language of facial expressions was completely foreign to you? This was the challenge faced by Lindsay, whom you met in this chapter's *Stories of the Brain*. One of the most remarkable people I have met in my career, Lindsay is an intelligent and accomplished individual who has been diagnosed with autism spectrum disorder. We will end our discussion of the behavioral neuroscience of emotions by considering autism spectrum disorder and what we can learn from the social and emotional challenges these individuals face.

Autism Spectrum Disorder MCAT

The term *autism* was coined in 1910 by psychiatrist Eugen Bleuler, who also incorrectly categorized severely autistic behavior as a form of schizophrenia. The modern use of the word is ascribed to Leo Kanner as well as Hans Asperger in 1943. Asperger described four children who were intelligent, highly verbal, physically clumsy, and deficient in empathy and social skills. Today autism is recognized as including a range of neurodevelopmental disorders, which is reflected in its current name, **autism spectrum disorder (ASD)**. At one end of the spectrum, individuals who are profoundly autistic may lack language skills and struggle to coordinate motor movements. On the other end are individuals whose behavior is barely distinguishable from neurotypical individuals (**Figure 17.23**).

ASD affects perhaps 1% of the population and is more prevalent among males. The causes of autism spectrum disorder are not well understood. There is a clear genetic component, although at this time it is unclear if dysfunction of a few genes leads to major disruption or if the small but cumulative effects of many genes are responsible. Prenatal influences such as maternal diabetes and pregnancy after age 30 are among the conditions that have been thought to contribute to incidence of ASD. In the past, various causes of ASD have been proposed, including an emotionally distant maternal parent (a "refrigerator mom") and childhood vaccines, but both of these theories have been debunked.

The behavior of individuals with ASD often includes extreme devotion to a restricted set of interests and a compulsion for repetitive behaviors known as **stimming** that include hand flapping, rocking, or repeating sounds. The purpose of stimming may be to block out sensory input from the surroundings.

At the mildest level of the autism spectrum, we encounter individuals who are neurotypical in many respects with the notable exception of certain social and behavioral traits. Unlike those at the severe end of the autism spectrum, these individuals have an IQ that ranges from normal to above normal. Such individuals were formerly classified as having **Asperger's syndrome**; however, Asperger's is no longer recognized as a distinct diagnosis in the current DSM. For the purpose of clarity, we will use Asperger's syndrome to distinguish the experience of those individuals from others on the spectrum.

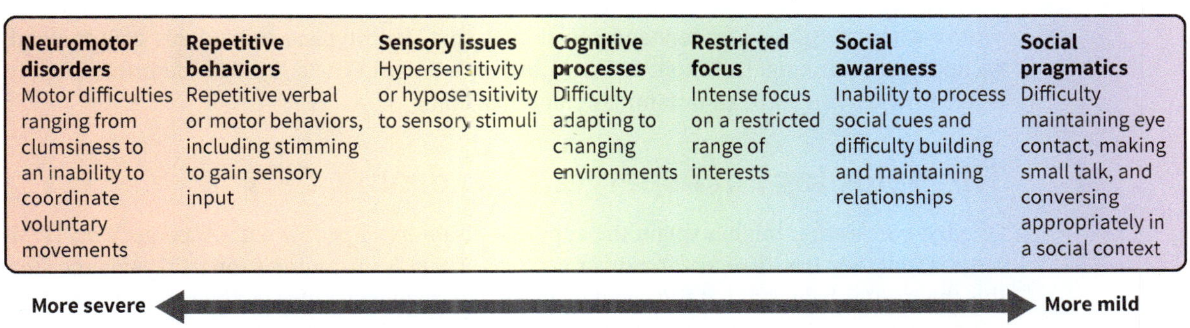

FIGURE 17.23 **The autism spectrum** Individuals on the autism spectrum range from those with extreme behavioral and cognitive disabilities on the far left to those on the far right who vary only slightly from neurotypical individuals.

FIGURE 17.24 **The Sally–Anne task** The Sally–Anne task requires an individual to switch their point of reference from their own perspective to that of another.

Difficulties in Processing Emotional Cues MCAT

Although people with Asperger's are highly functioning, they may experience social and personal difficulties in their relationships with others. Consider the following exchange between a male with Asperger's and his neurotypical spouse.

Husband: How was your day?

Wife: It was good.

Husband: Oh, by the way, your brother called. He is in the hospital.

Wife: What? How did this happen?

Husband: A car accident. So, are you ready to play tennis?

The wife was appalled at what she considered to be an extremely callous response on the part of her spouse. His response was that he understood that she would be upset, but there was nothing to be done about it, and playing tennis as they had originally intended might cheer her up. It wasn't that he did not care about her brother's accident, but that he was unable to mentally put himself into the mind of his spouse.

As is clear from this anecdote, individuals with Asperger's often struggle to see events from another's perspective. Earlier in the chapter you were reminded about theory of mind (TOM), or the ability to see events from the perspective of another person. The Sally–Anne task is used by developmental psychologists to determine a child's ability to take another's perspective. In this task, a child is shown a cartoon depicting two children, Sally and Anne (**Figure 17.24**). Sally puts a ball in a basket and departs; while she is gone, Anne moves the ball to an adjacent box. The child is then asked where Sally will look for the ball when she returns. By ages 4 to 5, most children recognize that Sally will look for the ball in the basket, where she believes it will be, rather than in the box, where it actually is; children with even mild Asperger's are delayed at successfully solving this puzzle.

It is important to recognize that individuals with Asperger's are not *lacking* in social emotions. In fact, they feel emotions strongly, but they struggle to comprehend the emotional status of others.

We began this chapter with a story about Lindsay, an individual with Asperger's syndrome who struggles to understand social cues that neurotypical individuals comprehend without even thinking. In **Check It Out: Lindsay's Social World**, Lindsay describes what it is like to be a stranger in an unfamiliar emotional land.

The Neurology of Asperger's Syndrome MCAT

One hypothesis that might explain the apparent disability of people with Asperger's on TOM tasks is that they have a defect in mirror neurons. There is intriguing supporting evidence for this theory. It is believed that EEG activity in the *mu* frequency range is a by-product of mirror neuron activity in humans. *Mu* suppression is customary while thinking about engaging in a simple motor task or watching a person perform that same action. Individuals on the autism spectrum do not show the same level of *mu* suppression, which means that their mirror neuron system is not being engaged by the task, thereby leading to a failure in TOM.

Check It Out | Lindsay's Social World

Feng Yu / Adobe Stock

What is it like to navigate the complex social cues that are a part of being human when you are an individual with Asperger's syndrome? In this video, you will hear for yourself how someone with Asperger's syndrome uses her intellectual talents to navigate through this challenging social world. Watch the video **Lindsay's Social World**, and then answer the questions.

1. In the video, Lindsay spoke about her difficulty interpreting social situations. Mention one example of a situation where Lindsay would likely struggle and a neurotypical individual would not.
2. Lindsay talks about how she has learned to compensate for her difficulties by relying on her strengths. Describe one such strength and how she uses it to deal successfully with social situations.
3. After watching this video, did your attitude toward individuals with Asperger's change in any way? Explain your answer.

This hypothesis was tested in a study conducted in 2006 by Ramachandran and Oberman that involved comparing the *mu* activity of autistic and neurotypical children on two tasks: in one they were asked to open and close one hand, and in the other they observed a video of a hand opening and closing. Both the autistic and the neurotypical controls showed clear *mu* suppression during the active task, but only the neurotypical controls showed *mu* suppression during the observational task. These results support the premise that a dysfunction of the mirror neuron system plays an important role in creating autistic behavior.

Another interesting explanation of the difficulty individuals with ASD have processing social cues is that there is a difference in the DMN, which ties multiple cortical regions together in a circuit that appears to be active in social cognition. In neurotypical individuals, there is a substantial level of specificity that defines the connections among regions of the DMN; this may not be true of individuals with Asperger's.

17.5 Before You Go On

Section Summary

Autism spectrum disorder is a developmental disorder characterized by cognitive and social interaction deficits. Individuals at the low end of the spectrum are profoundly disabled; in contrast, those with Asperger's syndrome on the high end of the spectrum are intellectually skilled but struggle to properly interpret social cues and often exhibit repetitive behaviors or restricted interests. The causes are not well understood, but there is clear evidence of neurological differences between ASD and neurotypical individuals.

Comprehension Check

1. Autism spectrum disorder is a developmental disorder characterized by ______.

- **a.** low self-esteem
- **b.** a wide range of intellectual and social skills
- **c.** low IQ and self-destructive behavior
- **d.** an emergence around age 20 or later

2. Individuals with Asperger's syndrome typically show ______.

- **a.** low IQ and obsessive behavior
- **b.** high IQ and depression
- **c.** average-to-low IQ and inability to focus
- **d.** average-to-high IQ and restricted interests

3. Why do children with ASD struggle to answer the Sally–Anne task correctly?

- **a.** They cannot remember the actions in the story.
- **b.** They are uncomfortable disagreeing with an adult.
- **c.** They focus on the cartoon and are unable to integrate the verbal description.
- **d.** They cannot see the event from another's perspective.

4. Which of these is a possible explanation for the social difficulties experienced by individuals with ASD?

- **a.** People with ASD cannot feel emotions.
- **b.** ASD may involve dysfunction in the mirror neuron system.
- **c.** People with ASD have low IQs.
- **d.** ASD may be caused by a hyperactive DMN.

Conclusion

We began this section with a provocative question: Are social behaviors learned through experience or culture, or is there a biological component that underlies and influences the expression of some social behaviors such as who we like and how we interpret the intentions of others? The goal of this chapter was to encourage you to keep an open mind about the role that your genes and your biology play in guiding social behavior. If our biology does influence social behavior, then knowing that positions us to understand our own behavior and that of others.

Humans share with other species the ability to communicate our internal state through facial expressions. There appears to be a basic set of these expressions, although there is evidence that culture can adapt expressions when necessary. Sometimes we use these expressions to signal happiness or agreement, but it is to our advantage to sometimes be less than truthful about our intentions. We signal more than internal states with our faces and bodies; we also signal vitality, fertility, and possibly even sexual interest.

Humans luxuriate in the warm feeling of belonging that comes from group identity. Group membership by definition means that some individuals are ingroup while others are outgroup. Evolutionary psychologists propose that this drive to categorize others as either in- or outgroup is a result of our ancestors' history, and recent neuroimaging and comparative research supports the view that we are innately prepared to differentiate others into in- and outgroup membership. Is it possible that in today's world, the automatic categorization of some into a threatening outgroup leads to unconscious negative attitudes? If so, awareness is the first step toward bringing our implicit attitudes in line with our explicit attitudes. Finally, there are some individuals who struggle to comprehend the social signals that come naturally to others, and understanding the neural basis of Asperger's syndrome can provide insights into the emotional processing of neurotypical individuals.

Critical Thinking Questions

1. You are likely aware of the fictional *Star Trek* character Mr. Spock, who is devoid of emotions. Do you think you would be better at making important choices if you could turn off your emotions? Why or why not?
2. Interpreting the inner emotional state of another is a complex task and culture plays a role. For example, the sense of appropriate personal space differs among cultures; some ethnic groups stand close while conversing, but others prefer more space. In some cultures, everyone may talk at once, whereas in others long silences are not a source of discomfort. Some cultures greet with a kiss or a hug, and others prefer a handshake or verbal acknowledgement. Have you ever been in a situation where you encountered a cultural norm that surprised you? Briefly describe the situation. If you have never had this experience, ask your friends or family if they have experienced such a situation.
3. We all swim in our culture, and it can be hard to think outside that box. In economically developed countries, people have more than enough to eat and sufficient resources to cover basic needs, but there are places in the world and times in our collective past where the search for resources is a daily struggle. Consider what an evolutionary psychologist would predict about the characteristics that would seem attractive in each of these situations.
4. Consider how signs or signals of group membership might lubricate the wheels of social interactions. Identify a behavior that you or your peers engage in that signifies group membership. Some examples might be appearance, clothing, ways of speaking, or rituals. How might such overt signals be adaptive in engaging with others?
5. Empathy for the feelings of others motivates people to help those in need. From an evolutionary perspective, why might a circuit that creates empathy be useful to a species whose survival is dependent on cooperation within the group?

Dig Deeper into Behavioral Neuroscience

17.1 Emotions

For anyone interested in emotions and the brain, these two books are a must read.

LeDoux, J. (1996). *The Emotional Brain: The Mysteries Underpinnings of Emotional Life*. Simon and Schuster, New York.

Damasio, A. (1999). *The Feeling of What Happens: Body and Emotion in the Making of Consciousness*. Harcourt-Brace New York.

Darwin's original work on the evolutionary basis of emotions is worth a read.

Darwin, C. (1872). *The Expression of the Emotions in Man and Animals*. John Murray. https://doi.org/10.1037/10001-000

The Lövheim cube is a novel perspective on the link between emotions and neurotransmitters, a topic that has not always been carefully considered outside the realm of psychiatric disorders.

Lövheim, H. (2012). A new three-dimensional model for emotions and monoamine neurotransmitters. *Medical Hypotheses 78*, 341–348. https://doi.org/10.1016/j.mehy.2011.11.016 PMID 22153577

17.2 Communicating Internal States

There is a wide range of opinions on whether there is a universal set of emotional expressions and if so, which are universal and which are cultural.

Ekman, P.; W. V. Friesen (1971). Constants across cultures in the face and emotion. *Journal of Personality and Social Psychology* (17): 124–129.

Ekman, P.; Friesen, W.V.; O'Sullivan, M.; et al (1987). Universals and cultural differences in the judgments of facial expressions of emotion. *Journal of Personality & Social Psychology* 53 (4): 712–717.

Elfenbein, H. A., & Ambady, N. (2002). On the universality and cultural specificity of emotion recognition: A meta-analysis. *Psychological Bulletin, 128*(2), 203–235.

Elfenbein, H. A., Mandal, M. K., Ambady, N., Harizuka, S., & Kumar, S. (2002). Cross-cultural patterns in emotion recognition: Highlighting design and analytical techniques. *Emotion, 2*(1), 75–84. https://doi.org/10.1037/1528-3542.2.1.75

17.3 Beauty and Attractiveness

Thinking about abstractions such as beauty and attractiveness from an evolutionary perspective is a fascinating task.

Barber, N. (1995) The evolutionary psychology of physical attractiveness: Sexual selection and human morphology. *Ethology and Sociobiology, 16*(5), 395–424. https://doi.org/10.1016/0162-3095(95)00068-2

Li, N., Tan, L., & Choy, B. (2021). Evolution, biology and attraction. *Oxford Research Encyclopedia of Psychology.* https://oxfordre.com/psychology/view/10.1093/acrefore/9780190236557.001.0001/acrefore-9780190236557-e-302

The hypothesis that a specific waist-to-hip ratio may be a tell for fertility and therefore innately attractive to heterosexual males has led to a considerable amount of research, not all of which supports the hypothesis. You can read a comprehensive review of that literature here.

Bovet, J. (2019). Evolutionary theories and men's preferences for women's waist-to-hip ratio: Which hypotheses remain? A systematic review. *Frontiers in Psychology, 10*, 1221. https://doi.org/10.3389/fpsyg.2019.01221

Miller et al.'s 2007 study on the relationship between tips given to exotic dancers and their ovulatory cycle is perhaps the oddest of all studies investigating implicit tells for fertility.

Miller, G., Tybur, J., & Jordan, B. (2007). Ovulatory cycle effects on tip earnings by lap dancers: Economic evidence for human estrus? *Evolution and Human Behavior, 28*(6), 375–381. http://dx.doi.org/10.1016/j.evolhumbehav.2007.06.002

In a study involving 325 men and women of different sexual orientations, it was found that pupil dilation was a reliable measure of attraction.

Rieger, G., & Savin-Williams, R. C. (2012). The eyes have it: Sex and sexual orientation differences in pupil dilation patterns. *PLoS ONE, 7*(8), e40256. https://doi.org/10.1371/journal.pone.0040256

17.4 Human Social Networks

Project Implicit is a clearinghouse for research using the Implicit Attitudes Test or IAT. You can learn more about the IAT by going to their site. https://www.projectimplicit.net/.

At this time, there is also a site at Harvard University.

Project Implicit (https://implicit.harvard.edu)

Some researchers have looked to see if the amygdala activates in response to viewing ethnic faces.

Phelps, E. A., O'Connor, K. J., Cunningham, W. A., Funayama, E. S., Gatenby, J. C., Gore, J. C., & Banaji, M. R. (2000). Performance on indirect measures of race evaluation predicts amygdala activation. *Journal of Cognitive Neuroscience. 12*, 729–738.

17.5 Autism Spectrum Disorder and Emotional Blindness

The National Institute of Mental Health, a part of the National Institutes of Health, provides a wide range of information about autism spectrum disorder.

https://www.nimh.nih.gov/health/topics/autism-spectrum-disorders-asd

The Asperger/Autism Network is a place where families can learn more about resources available to parents and families of autistic individuals.

www.aane.org

Mirror neurons are of great interest to behavioral neuroscientists, and this article provides an overview.

Rizzolatti, G., & Craighero, L.(2004) The mirror-neuron system. *Annual Review of Neuroscience 27*, 169–192. https://doi.org/10.1146/annurev.neuro.27.070203.144230. PMID: 15217330.

The default mode network or DMN is a topic of great interest to social and behavioral neuroscientists. These review articles provide a broad background.

Padmanabhan, A., Lynch, C. J., Schaer, M., & Meon, V. (2017, September). The default mode network in autism. *Biological Psychiatry and Cognitive Neuroscience Neuroimaging, 2*(6), 476–486. https://doi.org/10.1016%2Fj.bpsc.2017.04.004

Li, W., Mai, X., & Liu, C. (2014). The default mode network and social understanding of others: What do brain connectivity studies tell us? *Frontiers of Human Neuroscience*, https://doi.org/10.3389/fnhum.2014.00074.

Chapter 17 Before You Go On Answers

17.1 Before You Go On Answers

1. a. an internal mental state created by the nervous system, hormones, and neurotransmitters
2. c. physiology and emotion arise simultaneously
3. b claims that a person's intellectual or cognitive appraisal of the context for physiological responses is critical

4. d. cingulate, hippocampus, amygdala, hypothalamus, and thalamus
5. b. serotonin, dopamine, and norepinephrine

17.2 Before You Go On Answers

1. a. everyone has different facial expressions
2. a. extrapyramidal system; pyramidal system
3. d. lying can provide an advantage to the liar, as long as the lie is not detected
4. b. your brain was covertly analyzing your physical and social environment
5. c. perform an action or see another perform the same action

17.3 Before You Go On Answers

1. a. Some male horses have a chestnut coat.
2. d. reproductive fitness
3. c. increased fertility
4. a. males were more generous when the females were ovulating
5. d. soon return to post-relationship hormone levels

17.4 Before You Go On Answers

1. b. a skeleton showing signs of several broken bones that were healed at time of death
2. c. the remains of females of child-bearing age were conspicuously missing
3. b. stereotyping
4. a. whether implicit attitudes contradict explicit attitudes

17.5 Before You Go On Answers

1. b. a wide range of intellectual and social skills
2. d. average-to-high IQ and restricted interests
3. d. They cannot see the event from another's perspective.
4. b. ASD may involve dysfunction in the mirror neuron system.

Chapter 17 Glossary

Asperger's Syndrome A developmental disorder at the mild end of the autism spectrum characterized by difficulties in social interactions and restricted interests.

autism spectrum disorder (ASD) Any disorder that falls on the range of neurodevelopmental disorders that includes autism and Asperger's syndrome.

Cannon–Bard theory A theory of emotion stating that a physiological response occurs simultaneous to, and independent of, the emotional response and that the two together are what we experience as emotion.

default mode network (DMN) A hypothetical network of neural processing centers that work in the background to assess complex social relations.

Duchenne smile A smile that reflects genuine joy or mirth and goes through the limbic system.

emotion An internal or mental state created by the nervous system through the action of neural structures, neurotransmitters, and hormones.

empathy The ability to understand or to feel the emotional state of another.

explicit attitude An attitude about individuals and groups that affects our behavior that we consciously hold.

extrapyramidal motor system A name given to a circuit of cerebral centers and nerve tracts that are essential for the coordination of involuntary motor movements.

implicit association test (IAT) A controversial tool that is purported to assess implicit attitudes.

implicit attitude An unconscious attitude about people or groups that may affect our behavior.

ingroup A group to which a person identifies as being a member.

James–Lange theory A theory of emotion stating that an emotion follows a physiological response.

limbic system A collection of subcortical structures that are associated with emotions, motivation, and long-term memory.

long-term potentiation (LTP) A long-lasting or permanent increase in efficiency of a synaptic connection.

Lövheim cube A theory relating the relative effects of serotonin, dopamine, and noradrenaline to basic emotions.

mirror neuron A brain cell that responds when a person performs an action or witnesses another performing that action.

outgroup Those individuals who do not belong to the ingroup.

Papez circuit A neural circuit including the limbic system that regulates emotional response.

pheromone A chemical secreted by one individual and received by another that acts as a hormone to alter the behavior of the second individual.

prejudice A preconceived opinion that is not based upon experience.

priming The effect upon activation of a concept that prior activation of a related concept confers.

pyramidal motor system A name given to a circuit of cerebral centers and nerve tracts that are essential for the coordination of voluntary motor movements.

reciprocal obligation The exchange of goods or labor to members of a group needing assistance with the understanding that they incur an obligation in turn to others.

reproductive fitness The ability of an individual or a group to pass on their genes to subsequent generations.

Schachter–Singer theory A theory of emotion that includes the role of cognitively labeling physiological responses in the creation of the emotional experience.

sexual selection Natural selection through preference for individuals of the opposite sex who have certain characteristics.

social smile A smile that is used in social situations to signal agreement or accommodation.

stereotype A generalized belief about a category of people.

stimming Repetitive body movements such as rocking back and forth or hand flapping that are characteristic of autism spectrum disorder and thought to be self-stimulating.

Theory of mind (TOM) The ability to recognize that others have emotions, desires, intents, and wishes different from one's own, which is a sign of complex cognitive development.

unconscious bias Stereotypes about groups of people that are beneath the conscious level and lead to discrimination.

Index

A

A-delta (Aδ) fibers, 170
Abilify, 441
abnormalities, color vision
  anomalous trichromats, 136
  consequences of, 138
  incidence of, 137–38
  variation in number of cones, 136–37
abuse, 236
ACE. *See* adverse childhood event
acetylcholine (ACh), 67, 441
ACh. *See* acetylcholine
achromatopsia, 138
ACTH. *See* adrenocorticotrophic hormone
action potential, 42–44
activation-synthesis theory, 348
active transport, 40
acuity, visual, 114
AD. *See* Alzheimer's disease
adaptive strategy, 465–66
addiction, 73
  definition of, 236
  genetic basis for, 234–35
  risk factors of, 236–37
adenine–thymine (AT), 10
adenosine, 352
ADH. *See* antidiuretic hormone
ADHD. *See* attention-deficit-hyperactivity disorder
adipose tissue, 190
adrenal glands, 216–17
adrenocorticotrophic hormone (ACTH), 208
adverse childhood event (ACE), 303
afferent neurons. *See* sensory neurons
ageusia, 166
aging, sexuality and, 278
agonist, 74
agouti-related peptide (AgRP), 190
agraphia, 382–83
AgRP. *See* agouti-related peptide
AIS. *See* androgen insensitivity syndrome
alcohol, 73
  behavioral effects of, 245–47
  binge drinking, 244
  blood alcohol content, 245
  pharmacology, 244–45
  warning signs of alcohol poisoning, 245
alexia, 380–81
all-or-none law, 44–45
alleles, 7
allosomes, 10, 268
allostasis, 176
Alzheimer, Alois, 447
Alzheimer's disease (AD), 67, 306, 332, 447
American Sign Language (ASL), 373
amnesia, 397–98
AMPA receptor (AMPAR), 332–34
AMPAR. *See* AMPA receptor
amphetamines, 254–55
amplitude, 154–56
amygdala, 97
amyloid precursor protein (APP), 332
amyotrophic lateral sclerosis (ALS), 72
anatomical terms, 80–81
ancestry, speech in, 368–69
androgen insensitivity syndrome (AIS), 271
androgens, 217, 271
anencephaly, 308
aneurysm, 444
anhedonia, 428
animal kingdom, communication in, 366–67
anomic aphasia, 372
anorexia nervosa
  causes of, 195
  definition, 195
  treatment of, 196
anosmia, 162
antagonist, 74
anterior pituitary, hormones from, 208
anterograde amnesia, 397
antidepressants, 428–29
antidiuretic hormone (ADH), 208–9
anxiety disorders, 432–33
  general anxiety disorder (GAD), 433–34
  obsessive-compulsive disorder (OCD), 435–37
  panic disorder, 434–35
aphasia
  anomic, 372
  Broca's, 370
  conduction, 371
  types of, 373
  Wernicke's, 371
Aplysia (California sea slug), 329–30
apoptosis, 299
APP. *See* amyloid precursor protein
apperceptive agnosia, 407
arachnoid layer, 93
ARAS. *See* ascending reticular activating system
Archimedes, 392
arcuate fasciculus, 371
ascending reticular activating system (ARAS), 350–51
ASD. *See* autism spectrum disorder
asexual reproduction, 263
  Muller's ratchet, 263
  Red Queen dilemma, 263–64
  sexual dimorphism, 264
ASL. *See* American Sign Language
Asperger's syndrome, 479
  neurology of, 480–81
  processing emotional cues, 480
associative visual agnosia, 407
astigmatism, 114
attention
  aspects of, 399–400
  definition, 399
  hemineglect, 401–3
  neural basis of, 400–401
  Yerkes-Dodson curve, 400
attention-deficit-hyperactivity disorder (ADHD), 254, 309
attraction, biological basis for, 284–85
attractiveness, 471–72
audition. *See* hearing
auditory dorsal stream, 154
auditory processing, 153–54
autism spectrum disorder (ASD), 479
autism, term, 479
autonomic nervous system (ANS), 80, 102
autoreceptors, 63
autosomes, 10
axon, 33, 34

B

BAC. *See* blood alcohol content, 245
bacterial infection, 443
Banich, Marie, 412
barbiturates, 255
Bard, Philip, 460
Barrett, Syd, 439
Bartoshuck, Linda, 166
basal ganglia, 98
bed nucleus of the striata terminalis (BSTc), 4
behavioral tolerance, 242
behaviorism, 322
Belyaev, Dmitri, 21
benzodiazepines, 354
Berthold, Arnold, 202
beta amyloid plaques, 448
Big Bang, 2
biological sex
  allosomes, 268
  definition, 265–67
  genetic influences on, 268
biome, 2
bipedal, 4
biphasic effect, 243
biphasic sleep, 340
bipolar cells, 120
bipolar disorder
  definition, 430
  demographics, 431
  depressive episode, 431
  diagnostic criteria, 432
  environmental influences, 431
  genetic influences, 431
  manic episode, 431
  neurological basis, 431
  treatment for, 432
blindsight, 407
blood alcohol content (BAC), 245
BMI. *See* body mass index
body mass index (BMI), 192
body shape, external tell, 470–71
Bonham, John, 241
Borneo, 369
brain
  adolescent form of, 304–5
  age-related change of, 305–6
  anatomic differences between hemispheres of, 415–16
  auditory processing in, 153–54
  brainstem, 82, 84–85
  cerebral torque, 415
  cerebellum, 91–92
  cerebellar lobes, 92
  cerebellar neurons, 91–92
  cerebral cortex, 88–91
  cerebrovascular accident (CVA), 443–45
  cerebrum, 81
  dementia, 306–7
  developing brains at risk, 292
  development of nervous system, 80–82
  dorsal stream, 133–34
  frontal lobe, 82
  frontal operculum, 416
  gross anatomy of, 82–83
  gyrus, 82
  insula, 83
  learning and, 327–34
  medulla oblongata, 85
  medial forebrain bundle (MFB), 231
  mesocortical pathway, 231–32
  mesolimbic pathway, 231–32
  midbrain, 84
  music processing in, 385–87
  nucleus accumbens (NAcc), 230
  occipital lobe, 82
  parietal lobe, 82
  planes of, 83

brain (*Continued*)
pons, 85
prenatal development of, 290–94
reward centers in, 229–31
reward circuit, 232–33
and sexual orientation, 282–84
split-brain procedure, 416–17
sulcus, 82
temporal lobe, 82
traumatic brain injury (TBI), 445–47
ventral stream, 133–34
ventral tegmental area (VTA), 231
word in various languages, 377
brain, stories
addiction, 228
audition, 145
cognitive neuroscience, 392
disorders, 424
endocrine system, 201
evolution and genetics, 2
homeostasis, 175
language, music, and the brain, 366
learning and conditioning, 315
nervous system, 79
neurodevelopment across life span, 290
neurons, 28
sexual behavior, 262
sleep, dreaming, and circadian rhythms, 339
social neuroscience, 457
synapses and neurotransmitters, 52
vision of, 111
brainstem, 82, 351
Breiter, H. C., 230
Bremer, Frédéric, 349
Broca, Paul, 366
Broca's aphasia, 370
Broca's area, 370
Bronson, Charles, 447
BSTc. *See* bed nucleus of the striata terminalis

C
C fibers, 170
caffeine
abuse of, 254
behavioral effects of, 253–54
pharmacology of, 253
Cajal–Retzius cells, 298
Cajal, Santiago Ramón y, 30
California sea slug *(Aplysia)*, 329–30
Campbell, Glenn, 447
cannabidiol (CBD), 251
cannabinoids, behavioral effects of, 250–51
Cannabis sativa, 250
Cannon–Bard theory, 460
Cannon, Walter, 175, 460
canonical view, 404
Capgras delusion, 442
Caspi, Avshalom, 427
CBD. *See* cannabidiol
CBT. *See* cognitive behavioral therapy
CCK. *See* cholecystokinin
cells
bipolar cells, 120
Cajal–Retzius cells, 298
complex cells, 128–29
hair cells, 149
eukaryotic cells, 31
ganglion cells, 120
hypercomplex cells, 128–29
intermediate progenitor cell (IPC), 298
intrinsically photosensitive retinal ganglion cell (ipRGC), 119
koniocellular cell (K-cell), 123
magnocellular cells (M-cells), 123
parvocellular cells (P-cells), 123
progenitor cells, 296
radial glia cell (RGC), 298
simple cells, 128–29
stem cells, 296
visual cortex cells, 128–29
cellular sodium concentration, 184–85
central nervous system (CNS), 29, 80
basal ganglia, 98–100
cerebral cortex, 88–91
clearing metabolic waste from, 342–43
corpus callosum, 96
embryonic form of, 292
emotional response, 97–98
endocrine glands in, 206–11
glymphatic system, 94
hypothalamus, 96
limbic system, 96
major subcortical processing circuits, 95–100
memory formation/retrieval, 98
meninges, 93
support systems of, 92–95
thalamus, 96
ventricular system, 94
cerebral cortex, 88
cerebral torque, 415
cerebrospinal fluid (CSF), 94
cerebrovascular accident (CVA), 443–44
traumatic brain injury (TBI), 445–47
treating, 445
cerebrum, 81
cervix, 265
Charpentier, Emmanuelle, 14
chemical synapse, 56–59
chemosenses
gustation, 163–66
olfaction, 159–62
Cholinesterase, 448
choroid, 113
chromosomes, 9
chyme, 186
CIP. *See* congenital insensitivity to pain
circadian rhythm
definition, 358
melatonin, 358–59
mismatching with diurnal rhythms, 359–61
pineal gland, 358
sleep-wake cycle, 358–59
suprachiasmatic nucleus (SCN), 359
circumventricular organ (CVO), 184
classical conditioning
conditioned taste aversion, 320–21
definition, 317
discovery of, 316–18
innate behavior, 315–16
learned behavior, 315–16
Little Albert experiment, 319–20
clitoris, 265
Clostridium botulinum, 103
cognitive behavioral therapy (CBT), 429
cognitive neuroscience
attention, 399–403
executive function, 409–15
hemispheric specialization, 415–19
memory, 392–98
object recognition, 404–8
color vision, 16–17, 134–35
abnormalities in, 136–38
cones in, 135–36
constancy, 140–41
opponent process theory, 139–41
opponents, 140
sensitivity, 138–39
trichromatic theory, 135–36
commissurotomy, 416–17
complex cells, 128–29
compulsion. *See* obsessive-compulsive disorder (OCD)
concentration gradient, 39
conditioned response (CR), 317
conditioned stimulus (CS), 317
conditioned taste aversion, 320–21
conditioning, 316
classical, 315–21
definition, 316
operant, 322–27
conduction aphasia, 371
cones, 118
in color vision, 135–36
in eyes, 118–19
variation in number of, 136–37
congenital insensitivity to pain (CIP), 170
consciousness disorders, 437
schizophrenia, 438–41
continuous positive airway pressure (CPAP), 356
continuous reinforcement, 325
contrast sensitivity, 115
contrecoup, 446. *See also* traumatic brain injury (TBI)
Copeland, Aaron, 447
coprolalia, 451
cornea, 113
corpus callosum, 96
cortical plate (CP), 298
corticogenesis, 297–98
cortisol, 220–21
Cosmides, Leda, 459
COX. *See* cytochrome *c* oxidase
CP. *See* cortical plate
CPAP. *See* continuous positive airway pressure
CR. *See* conditioned response
cranial nerves, 103–6
Cretaceous–Tertiary (K–T) extinction, 3
cribriform plate, 160
Crick, Francis, 10
CRISPR-Cas9, 14
critical period, 303
Cruise, Tom, 382
CS. *See* conditioned stimulus
cutaneous senses, 167
A-delta (Aδ) fibers, 170
C fibers, 170
characteristics, 168
Meissner's corpuscles, 169
Merkel disks, 168
nociceptive pain, 17
Pacinian corpuscles, 169
pain, 169–170
pressure, 168–69
Ruffini's endings, 168
vibration, 168–69
CVA. *See* cerebrovascular accident
CVO. *See* circumventricular organ
cytochrome *c* oxidase (COX), 130
cytoplasm, 31

D
DA. *See* dopamine
Darwin, Charles, 458
Darwin's finches, 15
DBS. *See* deep brain simulation
deception, adaptive strategy, 465–66
decibel level, 145–46
deep brain simulation (DBS), 230, 429
default mode network (DMN), 466–67
deficits, hearing, 151–52
Deisseroth, Karl, 334
delta FosB gene, 234
delta-9 tetrahydrocannabinol. *See* THC
Dement, William, 346
dementia, 306–7
beta amyloid plaques, 448
definition, 447
diagnostic criteria, 447
environmental factors, 448
genetic factors, 448
neurofibrillary tangles, 448
neurological basis, 448
treatment, 448–49
dendrites, 33
dendritic spines, 34
dependence, 235
depolarization, 42

depressants, 252, 255
depression
anhedonia, 428
antidepressants for, 428–29
demographics, 425–26
diagnostic criteria, 425
environmental influences, 426
genetic influences, 426
major depressive disorder (MDD), 425
neural basis of, 426–28
nonpharmacological treatments for, 429
postpartum depression, 426
treatment of, 428–29
depressive episode, 431
developmental verbal dyspraxia, 368
DHT. *See* dihydrotestosterone
diabetes, 214–15
Diagnostic and Statistical Manual, 5th Edition (DSM-5-TR), 236, 425
Diamond, Marian, 302
dichotic listening test, 375
dichromats, 136
Dickinsonia, 3
diencephalon, 291
diet, evolution of, 19–20
diffusion tensor imaging (DTI), 46
dihydrotestosterone (DHT), 271
disease
Alzheimer's disease (AD), 447
Huntington's disease (HD), 2, 13
neurotransmission, 71–75
Parkinson's disease, 449–50
disorders
anxiety disorders, 432–37
attention-deficit-hyperactivity disorder (ADHD), 254, 309
autism spectrum disorder (ASD), 478–81
bipolar disorder, 430–32
consciousness disorders, 437–42
general anxiety disorder (GAD), 433–34
gustation, 166
hunger, 191–97
major depressive disorder (MDD), 425
mood disorders, 424–32
movement disorders, 449–51
neurodevelopment, 307–11
neurological disorders, 443–51
neurotransmission, 71–75
obsessive–compulsive disorder (OCD), 435–37
olfaction, 162
panic disorder, 434–35
post-traumatic stress disorder (PTSD), 221–23
seasonal affective disorder (SAD), 355
sex chromosomes, 268
sleep, 355–57
substance abuse, 235–36
thirst, 185
dissociative drugs, 256–57
diurnal rhythm, 358
diversity, genetic, 12
DMN. *See* default mode network
DNA (deoxyribonucleic acid), 9
replication of, 12
dogs, 21–22
dominant trait, 8
Doohan, James, 447
dopamine (DA), 194, 426
dopamine hypothesis, 440
dopaminergic system, 65–67
dorsal stream, 133–34
Doudna, Jennifer, 14
downregulation, 61
dreaming, 347–48
activation-synthesis theory, 348
definition, 347
lucid, 348
purpose of, 347–48
sleep paralysis, 347
drugs, 73
amphetamines, 254–55
barbiturates, 255
caffeine, 253–54
classic hallucinogens, 255–56
depressants, 255
dissociative drugs, 256–57
nicotine, 252–53
sleep and 353–55
DSM-5-TR. See *Diagnostic and Statistical Manual,* 5th Edition
Duchenne smile, 465
dura meter 93
dwarfism, 211
dyslexia, 381–82

E
ear, anatomy of, 146
basilar membrane, 149
cochlea, 148
external acoustic meatus, 147
hair cells, 149
helicotrema, 148
incus, 147
inner ear, 148–49
malleus, 147
middle ear, 147–48
organ of Corti, 149
ossicles, 147
outer ear, 147
oval window, 148
round window, 148
scala tympani, 148
scala vestibuli, 148
semicircular canals, 148
stapes, 147
tectorial membrane, 149
tympanic membrane, 147
Earth, history, 2–3
Eccles, John, 38
echoic memory, 393–94
Economo, Constantin von, 351
ectotherms, 178
ECT. *See* electroconvulsive therapy
ectoderm, 291
ED. *See* erectile dysfunction
EEG. *See* electroencephalography
efferent neurons. *See* motor neurons
Einstein, Albert, 382
electrical potential, 38
electrical synapse, 56–59
electroconvulsive therapy (ECT), 429
electroencephalography (EEG), 344
electromagnetic spectrum, 112
embryonic neurodevelopment, 291
emmetropia, 114
emotional cues, processing, 480
emotional response, 97–98
emotions
Cannon–Bard theory of, 460
definition, 457–58
empathy, 462
evolutionary basis of, 458–59
hormones and, 461–62
James–Lange theory of, 459–60
limbic system, 461
Lövheim cube, 462
neural basis, 460–62
neural circuits, 461
neurotransmitters and, 461–62
Schachter–Singer theory of, 460
theories of, 459–60
empathy, 462
encéphale isolé, 349
encephalitis, 443
encoding, 392
endocrine glands, 203
adrenal glands, 216–17
androgens, 217
in central nervous system, 206–11
estrogens, 217
glucoregulation, 213–15
gonads, 217
hyperthyroidism, 212–13
hypothalamus, 206–7
hypothyroidism, 212–13
pancreas, 213–15
parathyroid, 212
in periphery, 212–17
pineal gland, 207–8
pituitary gland, 207
pituitary hormones, 208–9
regulation of growth, 209–11
stress, 216–17
thymus, 212
thyroid, 212
endocrine system
CNS glands, 206–11
comparing to nervous system, 202–3
glands, 203–5
hormones, 203–5
hormones as messengers, 201–2
negative feedback system, 205
periphery glands, 212–17
stress response 218–24
endoderm, 291
endogenous opioids, 247–48
endoplasmic reticulum (ER), 32
endotherms, 178
energy produced through imbalances
concentration gradient, 39
electrical potential, 38–39
resting potential, 39
engram, 327
environment
adaptations to extreme environments, 17–18
color vision, 16–17
human diet, 19–20
natural selection, 15–16
episodic memory, 397
EPSP. *See* excitatory postsynaptic potential
equipotentiality, 328
erectile dysfunction (ED), 69, 278
estrogens, 217, 271
eukaryotic cells, 31
Everest, 18
evolution
color vision, 16–17
environment affecting, 15–20
genetic basis of, 7–15
humans, 2–6
mammals, 3–4
primates, 4
social behavior, 21–25
evolutionary psychology, 459
excitatory postsynaptic potential (EPSP), 330
executive function
areas of brain involved in, 411
aspects of, 409
definition, 409
evaluating, 412–13
neural basis of, 410
perseveration, 413
prefrontal cortex (PFC), 411
psychosurgery as treatment for, 413–15
Stroop task, 413
Tower of Hanoi, 413
Wisconsin Card Sorting Test (WCST), 412–13
exocrine gland, 203
exogenous, 73
exogenous opioids, 247–48
explicit attitude, 476
explicit memory, 397
external tells, 469–71
extinction, 324
extrapyramidal motor system, 65
extrastriate cortex, 131–33
exuberant synaptogenesis, 298
eye movements, reading, 379–80
eyes
anatomy of, 113–14
choroid, 113
color vision, 134–41
cones, 118–19
contrast sensitivity, 115
cornea, 113
focal plane, 114–15
fovea, 114
fundus, 114
ipRGC, 119
iris, 113

eyes (*Continued*)
macula, 114
optic nerve, 114
receptive fields, 120
retina, 114, 117–21
rods, 118–19
sclera, 113
subcortical centers, 122–26
visual acuity, 115
visual cortex, 126–34
visual processing stream, 116–17

F

face shape, external tell, 470
facial expressions, 463–64
Fallopian tubes, 265
FAS. *See* fetal alcohol syndrome
fat, 19
feedback systems, 177
female genitalia, 265
fetal alcohol syndrome (FAS), 308
fetus, neurodevelopment of, 293–94
FFA. *See* fusiform face area
FI. *See* fixed-interval reinforcement
fight-or-flight response, 219–20
figure-ground organization, 115
filiform, 164
filopodia, 300
fixation, 379
fixed-interval reinforcement (FI), 325–26
fixed-ratio reinforcement (FR), 326
focal plane, 114–15
foliate, 164
follicle-stimulating hormone (FSH), 208, 273
fornix, 98
FosB gene, 234
fovea, 114
fovea, distribution of cones across, 135–36
FOXP2 gene, 368
FR. *See* fixed-ratio reinforcement (FR)
Fregoli delusion, 442
frequency coding
place theory, 155
rate theory, 154
volley theory, 155–56
frontal lobe, 82
frontal operculum, 416
FSH. *See* follicle-stimulating hormone
functional classification, neurons, 34–35
fundus, 114
fungiform, 164
fusiform face area (FFA), 407

G

G protein, 59
GAD. *See* general anxiety disorder
Gage, Phineas, 446
Galvani, Luigi, 38
gamma-aminobutyric acid (GABA), 69–70
ganglia, 85
ganglion cells, 120
gap junction, 56
Garcia, John, 321
GAS. *See* general adaptational syndrome
gastrointestinal tract, 186–87
Gazzaniga, Michael, 417
Gehrig, Lou, 72
gender, 266–67
gene, 7
general adaptational syndrome (GAS), 221–22
general anxiety disorder (GAD)
definition, 433
demographics, 433
diagnostic criteria, 433
environmental influences, 433
genetic influences, 433
neurological basis, 434
treatment, 434
generalization, 325
genetic engineering, 11
genetically modified organisms (GMOs), 11
genetics, principles of
genetic variation, 12–14
heritability, 7–9
modern genetics, 9–12
geniculostriate pathway, 122–24
genotype, 8
geons, 404
Geschwind-Wernicke model of language, 372–73
GH. *See* growth hormone
GHIH. *See* growth hormone-inhibiting hormone
ghrelin, 189
GHRH. *See* growth hormone-releasing hormone
glioblasts, 296
gland, 203
endocrine gland, 203
exocrine gland, 203
hormones, 203–5
glia, 46
characteristics, 47
types of, 48
glucoregulation, 213–14
glutamate (Glu), 69
glymphatic system, 94
GnRH. *See* gonadotropin-releasing hormone
Gogh, Vincent van, 439
Golgi complex, 32
Golgi, Camillo, 30
gonadotropin-releasing hormone (GnRH), 273
gonads, 217
Goodall, Jane, 408
gray matter, 46
Green, Peter, 439
group membership
evolutionary advantages of, 473–74
signaling, 474
growth cones, 300
growth hormone (GH), 208–9
growth hormone-releasing hormone (GHRH), 209
growth hormone–inhibiting hormone (GHIH), 210
guanine–cytosine (GC), 10
guidepost neurons, 300
Guillain-Barré syndrome (GBS), 36
gustation. *See* taste
anatomy of taste, 163–64
disorders in, 166
physiology of taste, 163–64
Gymnema sylvestre, 165
gyrus, 82

H

habituation, 323
hair, external tell, 470
half-life, 240
hallucinogens, 252, 255–56
Haloperidol, 441
Hammond, Darrell, 439
Harnett, Mark, 33
Hatsopoulos, Nicholas, 87
Hawking, Stephen, 72
Hayworth, Rita, 447
health. *See* reproductive fitness
hearing
anatomy of ear, 146–49
auditory localization, 156–57
auditory processing, 153–58
deficits, 151–52
frequency coding, 154–56
physics of sound, 145–46
sound wave transmission, 149–51
timbre, 1345
vestibular system, 157–58
heat stroke, 180
Hebb, Donald, 330
hemianopsia, 125
hemifield, 122
hemineglect, 401–3
hemispheric specialization
anatomical differences between hemispheres, 415–16
cerebral torque, 415
commissurotomy, 416–17
left-handedness, 418–19
processing modes, 418
split-brain procedure, 416–17
Heston, Charlton, 447
heteroreceptors, 63
heterozygous, 7
Hickenlooper, John, 408
Hillenburg, Stephen, 72
hippocampus, 98
Hippocrates, 201
homeostasis
allostasis, 176
body balance/change, 175–76
definition of, 175
feedback systems, 177
hunger, 186–91
negative feedback system, 177
regulation of temperature, 178–82
satiety, 186–91
set point, 176
thirst, 183–86
homo sapiens. See humans
homo socialis (social human), 473. *See also* human social networks
homozygous, 7
Hooke, Robert, 29
hormonal influences
life-span sexual development, 273–78
prenatal sexual organization, 269–70
sexual differentiation, 272–72
hormone, 203, 204–5
hormones
anterior pituitary, 208
glands and, 203–5
as messengers, 201–2
posterior pituitary, 208–9
HP. *See* hypothalamic-pituitary (HP) axis
HPA. *See* hypothalamic-pituitary-adrenal (HPA) axis
HPG. *See* hypothalamic–pituitary–gonadal axis
human social networks
explicit attitude, 476
group membership, 473–74
implicit attitude, 476
implicit association test (IAT), 476
ingroups, 475
long-term potentiation (LTP), 476
outgroups, 475
prejudice, 475–77
priming, 476
reciprocal obligation, 474
stereotypes, 475–77
unconscious bias, 476
humans
evolution of, 4–6
evolution of diet of, 19–20
human orgasm, 281
human sleep patterns, 340–41
human social networks, 473–78
sexual dimorphism in, 264
humans, origins of
evolution of humans, 4–6
history of Earth, 2–3
mammals, 3–4
primates, 4
hunger
cholecystokinin (CCK), 187
chyme, 186
disorders of, 191–97
gastrointestinal tract (GI tract), 186–87
hormonal regulation of, 189
hyperphagia, 189
mechanisms regulating, 191
motivation to eat, 188
neural basis of, 188–89
treating disordered eating, 196–97
huntingtin (HTT), 14
Huntington's disease (HD), 2, 13
hyoid bone, 369

hypercolumns, 130
hypercomplex cells, 128–29
hyperglycemia, 213
hyperopia, 114
hyperphagia, 189
hyperpolarization, 42
hyperthermia, 180
hyperthyroidism, 212–13
hypnogram, 346
hypoglycemia, 214
hypothalamic–pituitary (HP) axis, 206
hypothalamic–pituitary–adrenal (HPA) axis, 218
hypothalamic–pituitary–gonadal (HPG), 273
hypothalamus, 96, 206–7
  sleep and, 351
hypothermia, 180–82
hypothyroidism, 212–13
hypoxia, 249

I

IAT. *See* implicit association test
iconic memory, 393–94
identity, delusions of, 442
IGF. *See* insulin-like growth factor
IM. *See* intramuscular (IM) injection
implicit association test (IAT), 476
implicit attitude, 476
implicit memory, 397
INAH-3. *See* third interstitial nucleus of the anterior hypothalamus
infarct, 444
ingroups, 475
inhalation, 239
innate behavior, 315–16
insomnia, 353
insula, 83
insulin-like growth factor (IGF), 210
intermediate progenitor cell (IPC), 298
intermediate zone (IZ), 298
intermittent reinforcement (IR), 325
internal stress
  deception as adaptive strategy, 465–66
  Duchenne smile, 465
  extrapyramidal motor system, 465
  facial expressions, 463–64
  pyramidal motor system, 464
  social smile, 464
  theory of mind (TOM), 466–68
interneuron communication, 55–56
interneurons, 35
intersex, 272
intoxicants
  alcohol, 244–47
  definition, 244
  marijuana, 250–51
  opioids, 247–49
intramuscular (IM) injection, 239
intranasal, 239
intravenous (IV) injection, 239
intrinsically photosensitive retinal ganglion cell (ipRGC), 119
Inuit, 17
ion exchange, 39–40
ionotropic receptors, 59
IPC. *See* intermediate progenitor cell
ipRGC. *See* intrinsically photosensitive retinal ganglion cell
IR. *See* intermittent reinforcement
iris, 113
ischemic stroke, 444
IV. *See* intravenous (IV) injection
IZ. *See* intermediate zone

J

James–Lange theory, 459–60
James, William, 459
Jenkins, Florence, 386

K

K-cell. *See* koniocellular cell
Kandel, Eric 329
karyogram, 9
Kennedy, Carrie Hill, 223
ketamine, 428
Knightley, Kiera, 382
koniocellular cell (K-cell), 123
Kullmann, Dimitri, 72

L

l-DOPA. *See* levodopa
labia, 265
lactation, 3
lactose intolerance, 19–20
Lamarck, 7
Lange, Carl 459
language
  animal kingdom communication, 366–67
  developmental verbal dyspraxia, 368
  evidence for speech in ancestry, 368–69
  evolution of, 366–69
  FOXP2 gene, 368
  hyoid bone, 369
  McGurk effect, 367
  neural basis for, 370–76
Lasker Public Service Award, 2
lateral geniculate nucleus (LGN), 122–24
  attributes of cells of, 124
lateral hypothalamic nucleus (LH), 188–89
learned behavior, 315–16
learning, 316
  AMPA receptor (AMPAR), 332–34
  definition, 316
  engram, 327
  equipotentiality, 328
  localization, 328
  long-term potentiation, 330–32
  mass action principle, 328
  memory storage, 327–29
  neuronal basis of, 329–30
  NMDA receptor (NMDAR), 332–34
Leborgne, Louis Victor, 366
LeDoux, Joseph, 457
left hemisphere, brain, 374–76
left-handedness, 418–19
leptin, 190
LeVay, Simon, 283
levodopa (l-DOPA), 450
LGN. *See* lateral geniculate nucleus
LH. *See* lateral hypothalamic nucleus
LH. *See* luteinizing hormone
life-span sexual development
  aging, 276
  menopause, 278
  menstrual cycle, 276–76
  parental behavior, 277–78
  puberty, 273–76
light
  describing, 112
  electromagnetic spectrum, 112
  physics of, 112
limbic system, 96, 461
Lincoln, Mary Todd, 438
Little Albert experiment, 319–20
lobotomy, 414
localization, 328
localization, auditory, 156–57
long-term depression (LTD), 331
long-term memory, 392, 396–97
long-term potentiation (LTP), 69, 476
long-term potentiation (LTP), 33
Lövheim cube, 462
LSD (lysergic acid diethylamide), 256
LTD. *See* long-term depression
lucid dreaming, 348
luteinizing hormone (LH), 208, 273
lysergic acid diethylamide. *See* LSD
lysosome, 32

M

M-cells. *See* magnocellular cells
Mackey, Allison, 303
macula, 114
magnocellular cells (M-cells), 123
major depressive disorder (MDD), 425
male genitalia, 265
mammals, 3
  comparison of brains of, 6
  evolution of, 3–4
mammillary bodies, 98
Maner, Jon, 471
manic episode, 431
MAO inhibitors, 428
Marie, Jean, 52, 61
marijuana, 250–51
mass action principle, 328
McGurk effect, 367
MDD. *See* major depressive disorder
MDMA, 256
medial forebrain bundle (MFB), 231
medial preoptic area (MPA), 282
medulla oblongata, 85
Meissner's corpuscles, 169
melatonin, 358–59
membrane potential, 39
memory, 343
  amnesia, 397–98
  echoic, 393–94
  encoding, 392
  episodic, 397
  explicit, 397
  iconic, 393–94
  implicit, 397
  long-term, 392, 396–97
  perceptual, 397
  procedural, 397
  semantic, 397
  sensory, 392–94
  storing, 327–29
three-store model of, 392–93
  working, 392, 395–96
menarche, 276
Mendel, Gregor, 7–9
meninges, 93
meningitis, 443
menopause, 278
MENSA, 418
menstrual cycle
  definition, 276
  menarche, 276
  ovarian cycle, 276
  uterine cycle, 277
Merkel disks, 168
mescaline, 256
mesencephalon, 291
mesocortical pathway, 231–32
mesoderm, 291
mesolimbic pathway, 231–32
messenger RNA (mRNA), 11
metabolic waste, clearance of, 342–43
metabotropic receptors, 59
MFB. *See* medial forebrain bundle
microsleep, 353
midbrain, 84
MIH. *See* Müllerian-inhibiting hormone
mild traumatic brain injury (mTBI), 445
Miller, Saul, 471
Milner, Peter, 229
mirror neurons, 467–68
mitochondrion, 32
modern behavior, biological basis for, 23
Molaison, Henry, 397–98
Moniz, Egas, 414
monochromats, 136
monophasic sleep, 340
mood disorders, 424
  bipolar disorder, 430–32
  depression, 425–29
Moon, 2
Morganucodon, 4

motor neurons, 35
movement disorders
coprolalia, 451
Parkinson's disease (PD), 449–50
tardive dyskinesia (TD), 450–51
Tourette's syndrome, 451
MPA. *See* medial preoptic area
mTBI. *See* mild traumatic brain injury
Muller's ratchet, 263
Müllerian ducts, 269–70
Müllerian-inhibiting hormone (MIH), 270
music
characteristics of, 384–85
evolutionary basis for, 383–84
model of musical processing, 386
processing, 385–87
Williams syndrome, 384
myelin, 35
myopia, 114

N

NAcc. *See* nucleus accumbens (NAcc)
narcolepsy, 355
Nash, John, 439
National Institute of Mental Health (NIMH), 439
National Institutes of Health, 151
National Transportation Safety Board (NTSB), 339
natural selection, 15–16
NDRI. *See* norepinephrine-dopamine reuptake inhibitor
NE. *See* norepinephrine
negative feedback system, 177
and endocrine system, 205
homeostasis, 177
neologisms, 371
nervous system, 79
brainstem, 84–85
comparing to endocrine system, 202–3
development of, 80–82
divisions of, 80–82
gross anatomy of brain, 82–83
spinal cord, 85–87
neural basis, language
Broca's area, 370–71
Geschwind-Wernicke model of language, 372–73
left-hemisphere contribution, 374–76
right-hemisphere contribution, 374–76
stuttering, 376
two-circuit model of language, 373–74
Wada test, 375
neural crest, 291
neural plasticity, 332
neural plate, 291
neural signal
action potential, 42–44
all-no-none law, 44–45
energy produced through imbalances, 38–39
ion exchange, 39–40
sodium-potassium pump, 40–42
neural tube, 291
neuroblasts, 296
neurodevelopment
apoptosis, 299
corticogenesis, 297–98
disorders related to, 307–11
embryonic neurodevelopment, 291
fetal neurodevelopment, 293–94
neurogenesis, 296
neuromigration, 296
as organizational process, 295–300
postnatal, 301–7
prenatal brain development, 290–94
synaptic reorganization, 299–300
synaptogenesis, 298
neurofibrillary tangles, 448
neurogenesis, 296
neurological disorders
bacterial infections, 443
cerebrovascular accident (CVA), 443–45
dementia, 447–49
viral infections, 443
neuromigration, 296
filopodia, 300
growth cones, 300
guidance for, 300
guidepost neurons, 300
pioneer neurons, 300
neuromuscular junction, 102–3
neuron
components of, 32–34
discovery of, 29–31
eukaryotic cell structure, 31–32
glia, 46–48
gray matter, 46
introduction to, 28–31
nervous system, 28–29
neural signal, 38–45
staining, 30
structure of, 31–37
types of, 34–37
white matter, 46
neuropeptide Y (NPY), 189
neuroplasticity, 301
neurotransmission
autoreceptors, 62–63
disorders related to, 71–75
downregulation, 61
heteroreceptors, 62–63
ionotropic receptors, 59–61
metabotropic receptors, 59–61
neurotransmitters, 64–71
treatments related to, 71–75
upregulation, 61
neurotransmitters, 53
dopaminergic system, 65–67
GABA, 69–70
glutamate, 69
nitric oxide, 69
noradrenergic system, 68
serotonergic system, 65
sleep and, 351–53
substance P, 70
neurotrophins, 299
neurulation, 291
defects in, 307–8
nicotine, 62
addictive properties of, 253
pharmacology of, 253
night terrors, 355
Nijinsky, Vaslav, 439
NIMH. *See* National Institute of Mental Health
nitric oxide (NO), 69, 330
nitrogen dioxide (NO_2), 69
nitrous oxide (N_2O), 69
NMDA receptor (NMDAR), 332–34
NMDAR. *See* NMDA receptor
NO. *See* nitric oxide
nociceptive pain, 170
nodes of Ranvier, 35
nonbenzodiazepines, 354
noradrenaline (NA), 68
noradrenergic system, 68
norepinephrine (NE), 68, 426
norepinephrine–dopamine reuptake inhibitor (NDRI), 428
Novichok, 72
NPY. *See* neuropeptide Y
NREM sleep, 345
NREM Stage 1, 346
NREM Stage 2, 346
NREM Stage 3, 346
NREM Stage 4, 346
NTSB. *See* National Transportation Safety Board
nucleotide, 10
nucleus, 31, 32
nucleus accumbens (NAcc), 230

O

obesity
body mass index (BMI), 192
causes of, 192–94
definition, 192
treatment of, 194–95
object recognition
blindsight, 407
canonical view, 404
captcha, 405
challenges to, 405
face-blind-ness, 408
fusiform face area (FFA), 407
geons, 404
primal sketch, 404
prosopagnosia, 407–8
visual, 404–5
visual agnosia, 406–7
obsession. *See* obsessive-compulsive disorder (OCD)
obsessive-compulsive disorder (OCD)
definition, 435
demographics, 436
diagnostic criteria, 435
environmental influences, 436
genetic influences, 436
neurological basis, 436
treatment, 436–37
occipital lobe, 82
OCD. *See* obsessive-compulsive disorder
Odiele, Hanne Gaby, 272
OFC. *See* orbitofrontal cortex
Olds, James, 229
olfaction
anatomy of, 159–61
anosmia, 162
circuit, 160
cribriform plate, 160
disorders of, 162
genetics and, 162
individual differences in, 161–62
olfactory bulb, 160
olfactory receptors, 161
physiology of, 159–61
olfactory bulb, 160
ontogenesis, 343
operant chamber, 322
operant conditioning
behaviorism, 322
conditioning phenomena, 323–25
discovery of, 322
operant chamber, 322
punishment, 322–23
reinforcement, 322–23
reinforcement schedule, 325–27
opioids
definition, 247
endogenous, 247–48
exogenous, 247–48
overdose deaths, 250
pharmacology of, 249
receptors, 248
relative potency of, 249
synthetic, 248–49
opponent process theory, 139–41
opsins, 118
optic chiasm, 122
optic nerve, 114
optogenetics, 334
oral administration, 239
orbitofrontal cortex (OFC), 231
orexin, 190, 352
organelles, 31
orgasm, evolutionary explanations for, 281
otoliths, 158
outgroups, 475
ovarian cycle, 276
ovaries, 265
ovulation, 471–72
oxytocin, 70

P

P-cells. *See* parvocellular cells
PAC. *See* primary auditory cortex
Pacinian corpuscles, 169
pain, 169–70
pair bonding, hormonal influences on, 472–73
pancreas, 213
panic disorder
definition, 434

demographics, 434–35
diagnostic criteria, 434
environmental influences, 435
genetic influences, 435
neurological basis, 435
treatment, 435
Papez circuits, 461
Papez, James, 96, 461
papillae, 163
paraphrastic errors, 371
parasympathetic nervous system (PNS), 102
parathyroid, 212
parental behavior, hormonal activation on, 277–78
parietal lobe, 82
Parkinson's disease (PD), 449–50
Parks, Rosa, 447
parvocellular cells (P-cells), 123
Pavlov, Ivan, 316–18
PD. *See* Parkinson's disease
penis, 265
penumbra, 444
perceptual memory, 397
Peretz, Isabelle, 385–86
peripheral nervous system (PNS), 29, 80
cranial nerves, 103–6
divisions of, 101–2
neuromuscular junction, 102–3
Permian–Triassic (P–T) extinction, 3
perseveration, 413
Pert, Candace, 247
PFC. *See* prefrontal cortex
pharmacodynamics
behavioral tolerance, 242
half-life, 240
potentiation, 240
sensitivity, 240–42
tolerance, 240–42
pharmacokinetics, 242–43
phenomena, conditioning, 323–25
phenotype, 8
pheromones, 471
phonemes, 377
phonological route, 380
photoreceptors, 117–19
pia meter, 94
pineal gland, 207–8, 358
pioneer neurons, 300
Pitt, Brad, 408
pituitary gland, 207
place theory, 155
plasma membrane, 32
Poe, Gina, 352
polarization, 39
polydipsia, 185
polysomnography, 344
pons, 85
post-traumatic stress disorder (PTSD), 221–23
posterior parietal cortex (PPC), 90
posterior pituitary, hormones from, 208–9
postnatal neurodevelopment
adolescent brain, 304–5
aging brain, 305–7
critical period, 303
environmental influences on, 301–4
neuroplasticity, 301
postpartum depression, 426
potentiation, 240
potentiation, long-term, 330–32
Prader-Willi syndrome (PWS), 175, 194–95
prefrontal cortex (PFC), 411
prejudice, 475
prenatal development
brain, 290–94
cellular level, 295–301
diencephalon, 291
ectoderm, 291
endoderm, 291
mesencephalon, 291
mesoderm, 291
neural crest, 291
neural plate, 291
neural tube, 291
neurulation, 291
prosencephalon, 291
rhombencephalon, 291
telencephalon, 291
teratogens, 291
prenatal sexual organization
genetic influences on biological sex, 268
homologous genital structures, 270
hormonal influences on, 270–72
Müllerian ducts, 269–70
sexual differentiation, 269–70
SRY gene, 269–70
Wolffian ducts, 269–70
presbyopia, 14
pressure, 168–69
primal sketch, 404
primary auditory cortex (PAC), 154
primary motor cortex (PMC), 89
primary sexual characteristics, 264
primary visual cortex (PVC), 122, 127–28
primates, 4
family tree, 5
primates, evolution of, 4
priming, 476
procedural memory, 397
Process C, 359
Process S, 359
progenitor cells, 296
prosencephalon, 291
prosody, 375
prosopagnosia, 407–8
proteins, 9
psilocybin, 256
psychiatric disorders, 424
psychology research, ethics in, 319–20
psychopharmacology
definition, 238
drugs delivery methods, 238–40
pharmacodynamics, 240–42
pharmacokinetics, 242–43
psychosurgery, 413–15
PTSD. *See* post-traumatic stress disorder
puberty
behavior changes, 275–76
definition, 273
environmental effects on, 274–75
HPG and, 273–75
secondary sex characteristics, 275
PVC. *See* primary visual cortex
PWS. *See* Prader-Willi syndrome
pyramidal motor system, 464

R

radial glia cell (RGC), 298
radial migration, 296
rapid eye movement sleep (REM sleep), 345
RAS. *See* renin-angiotensin system; reticular activating system
rate theory, 154
reading and writing, neural basis of
disorders, 380–83
eye movements, 379–80
fixation, 379
inventing reading and writing, 377
phonemes, 377
phonological route, 380
reading processes, 378–79
reading span, 379
saccade, 379
scanpath, 379
reading span, 379
reading, processes of, 378–79
Reagan, Ronald, 447
receptive fields, 120
recessive trait, 8
reciprocal obligation, 474
recovery from extinction, 325
Red Queen dilemma, 263–64
Reeve, Christopher, 79
refractory period, 42
refreshment, 342
reinforcement schedule, 325–27
REM sleep. *See* rapid eye movement sleep
renin-angiotensin system (RAS), 183–84
repetitive transcranial magnetic stimulation (rTMS), 429
repolarization, 42
reproduction
anatomy for, 264–65
asexual reproduction, 263
genitalia, 265
sexual reproduction, 263
reproductive anatomy, 264–65
reproductive fitness
attractiveness, 471–72
external tells for, 469–71
hormonal influences on pair bonding, 472–73
ovulation, 471–72
pheromones, 471
resting potential, 39
reticular activating system (RAS), 350–51
retina, 114
photoreceptors, 117–19
processing, 119–21
retinotopic mapping, 123
retrograde amnesia, 397
reuptake pumps, 56
reward
biological basis for, 228–29
centers in brain, 229–31
medial forebrain bundle (MFB), 231
mesocortical pathway, 231–32
mesolimbic pathway, 231–32
nucleus accumbens (NAcc), 230
reward circuit, 232–33
ventral tegmental area (VTA), 231
reward circuit
addiction as dysfunction of, 234–37
genetic basis for addiction, 234–35
and human behavior, 232–33
substance abuse disorders, 235–36
RGC. *See* radial glia cell
rhombencephalon, 291
ribosome, 11, 32
right hemisphere, brain, 374–76
risk factors, 236–37
Risperdal, 441
Rizzolatti, Giacomo, 467
rods, 118
Rozenzweig, Mark, 302
rTMS. *See* repetitive transcranial magnetic stimulation
Ruffini's endings, 168

S

saccade, 379
saccule, 158
SAD. *See* seasonal affective disorder
Sally–Anne task. *See* Asperger's syndrome
saltatory conduction, 35
SAMSA. *See* Substance Abuse and Mental Health Services Administration
satiety, 186
adipose tissue, 190
disorders of, 191–97
hormonal regulation of, 190
leptin, 190
mechanisms regulating, 191
SC. *See* subcutaneous (SC) injection
scanpath, 379
Schachter–Singer theory, 460
Schachter, Stanley, 460
schizophrenia, 310
anatomical basis for, 440
Capgras delusion, 442
cost of, 441
Cotard delusion, 442
definition, 438
delusion of identity, 442

schizophrenia (*Continued*)
- demographics of, 438–39
- diagnostic criteria, 438
- dopamine hypothesis, 440
- environmental influences, 439–40
- Fregoli delusion, 442
- genetic influences, 439–40
- neurological basis for, 440
- perceptual disturbances in, 438
- treatment of, 441

sclera, 113
SCN. *See* suprachiasmatic nucleus
scrotum, 265
SDN-POA. *See* sexually dimorphic nucleus of the medial preoptic area
SE. *See* serotonin
seasonal affective disorder (SAD), 355
second messengers, 60
secondary sexual characteristics, 264
selective serotonin reuptake inhibitor (SSRI), 428
Selye, Hans, 221
semantic memory, 397
sensitivity, 240–42
sensitization, 324
sensorimotor cortex, 89
sensorimotor homunculus, 90
sensory memory, 392–94
sensory neurons, 34
septal nuclei, 97
Seroquel, 441
serotonergic system, 65
serotonin (SE), 232, 426
serotonin–norepinephrine reuptake inhibitors, 428
set point, 176
sex chromosomes, disorders related to, 268
sexual behavior
- biological basis for attraction, 284–85
- cycles of sexual response, 281
- female sexual response, 279
- human orgasm, 281
- male sexual response, 279
- sexual orientation, 282–84

sexual differentiation, 269–70
- hormonal disruption of, 271–72

sexual dimorphism, 264
sexual orientation
- birth order effects, 284
- brain differences, 282–84
- definition, 282
- genetic influences, 284
- neural basis for, 283

sexual reproduction, 263
sexually dimorphic nucleus of the medial preoptic area (SDN-POA), 282
SFO. *See* subfornical area
shaping, 325
sheep, brain of, 93
Sherpa, 17–18
simple cells, 128–29
Singer, Jerome, 460
sinoatrial (SA) node, 58
skin, external tell, 470
skin. *See* cutaneous senses
Skinner, B. F., 322
sleep
- across species, 340
- biological basis of, 349–353
- biphasic, 340
- brainstem and, 351
- definition, 339
- disorders related to, 355–57
- disruptions, 353–57
- dreaming, 347–48
- drugs and, 353–55
- evolutionary patterns of, 339–41
- human sleep patterns, 340–41
- hypothalamus and, 351
- insufficiency, 356–57
- and memory, 343
- metabolic waste, 342–43
- microsleep, 353
- monophasic, 340
- neurotransmitters, 351–53
- ontogenesis, 343
- purpose of, 342–43
- refreshment, 342
- relating circadian rhythms to, 357–61
- reticular activating system and, 350–51
- stages of, 345–47
- study of, 344

sleep apnea, 356
sleep paralysis, 347
sleep-wake cycle, 358–59
Slovenia, 369
SMA. *See* supplementary motor area
Snyder, Solomon, 247
social behavior
- heritability of, 21–22
- modern behavior, 23
- survival benefits of, 22–23

social neuroscience
- attraction. *See* reproductive fitness
- autism spectrum disorder (ASD), 478–81
- beauty. *See* reproductive fitness
- emotions, 457–62
- human social networks, 473–78
- internal stress, 463–68

social smile, 464
sodium-potassium pump, 40–42
soma, 32
somatic nervous system (SNS), 80, 101–2
somatosensory cortex, 89
sound wave, transmission of, 149–51
sound, physics of, 145–46
speech, evidence for, 368–69
Sperry, Roger, 416–17
spina bifida, 307–8
spinal cord, 82, 85–87
split-brain procedure, 416–17
SRY gene, 269–70
SSRI. *See* selective serotonin reuptake inhibitor
stem cells, 296
stereocilia, 150
stereotype, 475
Sterling, Peter, 176
stimming, 479
stimulants, 252
stress, 216–17
- adrenal glands and, 216–17
- cortisol, 220–21
- fight-or-flight response, 219–20
- general adaptational syndrome, 221–22
- HPA axis, 218–19
- post-traumatic stress disorder, 221–23
- resistance to, 220

Stroop task, 413
structural classification, neurons, 34
stuttering, 376
subcortical centers
- damage to visual stream, 125
- geniculostriate pathway, 122–24
- tectopulvinar pathway, 124–25

subcutaneous (SC) injection, 239
subfornical area (SFO), 184
Substance Abuse and Mental Health Services Administration (SAMSA), 237
substance abuse disorder, 235–36
substantia nigra, 99
subventricular zone (SVZ), 298
sugar, 20
sulcus, 82
supplementary motor area (SMA), 385
suprachiasmatic nucleus (SCN), 359
substance P, 70
SVZ. *See* subventricular zone
sympathetic nervous system, 102
synapse, 30, 52–53
- chemical, 56–59
- electrical, 56–59
- interneuron communication, 55–56
- postsynaptic mechanism, 53–54
- presynaptic mechanism, 53

synaptic cleft, 53
synaptic plasticity, 69
synaptic reorganization, 299–300
synaptic vesicles, 53
synaptogenesis, 298
synaptopathy, 71–73
Synsepalum dulcificum, 165
synthetic growth hormone, 210
synthetic opioids, 248–49

T

tangential migration, 296
tardive dyskinesia (TD), 450–51
TAS2R38, 166
taste
- ageusia, 166
- anatomy of, 163–64
- cortical representation of, 164
- genetic influences on, 166
- gustation disorders, 166
- gustatory sense organs, 163–64
- individual differences in, 165–66
- papillae, 163
- physiology of, 163–64
- receptors, 165
- supertasters, 166–67
- taste bud, 163

taste bud, 163
TBI. *See* traumatic brain injury
TD. *See* tardive dyskinesia
tectopulvinar pathway, 124–25
Teitelbaum, Philip, 188
telencephalon, 291
telodendria, 33
temperature, regulation of. *See* thermoregulation
temporal lobe, 82
teratogens, 291
testes, 265
tetrachromats, 137
thalamus, 96
THC (delta-9 tetrahydrocannabinol), 250
theory of mind (TOM), 22
- default mode network, 466–67
- definition, 466
- mirror neurons, 467–68

thermoregulation
- ectotherms, 178
- endotherms, 178
- heat strokes, 180
- hyperthermia, 180
- hypothermia, 180–82
- neural basis, 178–79

third interstitial nucleus of the anterior hypothalamus (INAH-3), 282
thirst
- cellular sodium concentration, 184–85
- disorders related to, 185
- osmotic thirst, 184–85
- polydipsia, 185
- renin-angiotensin system (RAS), 183–84
- volumetric thirst, 183–84

Thorazine, 441
Thorndike, Edward, 322
threshold potential, 44
thymus, 212
thyroid, 212
thyroid-stimulating hormone (TSH), 208
tolerance, 240–42
TOM. *See* theory of mind
Tourette's syndrome, 451
Tower of Hanoi, 413
transcription errors, 12
transdermal administration, 239
transport proteins, 53
traumatic brain injury (TBI), 445–46
- symptoms, 446–47

treatments
- anorexia nervosa, 196
- bipolar disorder, 432

dementia, 448–49
depression, 428–29
general anxiety disorder (GAD), 434
neurotransmission, 71–75
obesity, 194–95
obsessive-compulsive disorder (OCD), 436–37
panic disorder, 435
schizophrenia, 441
trichromatic theory, 135–36
trichromats, 135–36
tricyclic antidepressants, 428
TSH. *See* thyroid-stimulating hormone
two-circuit model of language, 373–74

U
UCR. *See* unconditioned response
unconditioned response (UCR), 317
unconditioned stimulus (UCS), 317
unconscious bias, 476
United States, obesity in, 192
upregulation, 61
uterine cycle, 277
uterus, 265
utricle, 158

V
vagina, 265
vallate, 164
variable-interval reinforcement (VI), 326
variable-ratio reinforcement (VR), 326
variation, genetic, 12–14
vascular organ of lamina terminalis (VOLT), 184
ventral stream, 133–34, 154
ventral tegmental area (VTA), 67, 231
ventricles, 94
ventricular system, 94
ventricular zone (VZ), 298
ventromedial hypothalamic nucleus (VMH), 188–89
vesicle, 32
vestibular system, 157–58
VI. *See* variable-interval reinforcement
vibration, 168–69
Vinci, Leonardo Da, 382
viral infection, 443
vision
eyes, 113–14
physics of light, 112
visual agnosia, 406–7
visual cortex, 126–27
cells of, 128–29
columnar organization of, 130–31
dorsal stream, 133–34
extrastriate cortex, 131–33
primary visual cortex, 127–28
processing in, 128–31
regions of, 132
ventral stream, 133–34
visual object recognition, 404–5
visual processing stream, 116–17
visual stream, damage to, 125
vitamin D, 17–18
VMH. *See* ventromedial hypothalamic nucleus
volley theory, 155–56
VOLT. *See* vascular organ of lamina terminalis
volumetric thirst, 183–84, 184–85
VR. *See* variable-ratio reinforcement
VTA. *See* ventral tegmental area
vulva, 265
VZ. *See* ventricular zone

W
Wada test, 375
Wai, Louis, 438
Watson, John B., 10, 319–20
WCST. *See* Wisconsin Card Sorting Test
Wearing, Clive, 398, 443
Wernicke's aphasia, 371
Wernicke's area, 371
Wexler, Nancy, 2
white matter, 46
Williams syndrome, 310, 384
withdrawal, 235
Wolffian ducts, 269–70
Wong-Riley, Margaret, 130
working memory, 392, 395–96
Wozniak, Steve, 408
writing and reading. *See* reading and writing, neural basis of

Y
Yerkes-Dodson curve, 400

Z
Zyprexa, 441